Architectural Desktop 3.0/3.3

||||||||||||||||||||||||||||||||||||||||||||||||||||||||||||||||||||||||||||||||||||||||||

Basics Through Advanced

David A. Madsen

Faculty Emeritus
Former Chairman
Drafting Technology
Autodesk Premier Training Center

Clackamas Community College
Oregon City, OR

Former Board of Directors
American Design Drafting Association

Ron Palma

AEC Application Specialist
KETIV Technologies, Inc. DBA. IMAGINiT Technologies
Portland, OR

Prentice
Hall

Upper Saddle River, New Jersey
Columbus, Ohio

Editor in Chief: Stephen Helba
Executive Editor: Debbie Yarnell
Media Development Editor: Michelle Churma
Production Editor: Louise N. Sette
Production Supervision: Karen Fortgang, *bookworks*
Design Coordinator: Diane Ernsberger
Cover Designer: Jason Moore
Cover art: Jason Moore
Production Manager: Brian Fox
Marketing Manager: Jimmy Stephens

This book was set in Times Roman by STELLARViSIONs. It was printed and bound
by Courier Kendallville, Inc. The cover was printed by Phoenix Color Corp.

Architectural Desktop

Autodesk's Architectural Desktop is designed to aid the architect, designer, and drafter
through all phases of construction document creation. From the earliest stages of conceptual
design through design development and finally to the finished set of drawings, Architectural
Desktop provides the appropriate tools for getting the job done. Architectural Desktop
Release 3.0 is a set of architectural tools built into the AutoCAD 2000i platform, and
Architectural Desktop 3.3 is built into the AutoCAD 2002 platform. With generic AutoCAD
2000i, and 2002, you had tools such as lines, arcs, circles, and text to complete your design
needs. With the addition of object-oriented tools known as *AEC objects*, the design process
is moved to a new level. The AEC objects are divided into the three major phases of workflow: conceptual design, design development, and documented drawings. This text provides
you with the real-world knowledge needed to accomplish your architectural drawing needs
using Architectural Desktop.

Pearson Education Ltd., *London*
Pearson Education Australia Pty. Limited, *Sydney*
Pearson Education Singapore Pte. Ltd.
Pearson Education North Asia Ltd., *Hong Kong*
Pearson Education Canada, Ltd., *Toronto*
Pearson Educación de Mexico, S. A. de C.V.
Pearson Education—Japan, *Tokyo*
Pearson Education Malaysia Pte. Ltd.
Pearson Education, *Upper Saddle River, New Jersey*

10 9 8 7 6 5 4 3 2
ISBN: 0-13-093498-4

Preface

Architectural Desktop 3.0/3.3: Basics Through Advanced is a text and workbook that provides complete coverage of Architectural Desktop. This text is designed to introduce you to the AEC objects and commands used in Architectural Desktop to complete a set of construction documents. The chapters are arranged in an easy-to-understand format, beginning with basic topics and working toward advanced subjects. This text makes the assumption that you have a basic understanding of AutoCAD, although the most important AutoCAD commands relevant to Architectural Desktop are discussed for beginning AutoCAD users and as a review for trained AutoCAD users. This text provides a detailed explanation of the Architectural Desktop tools that aid in producing professional and accurate drawings. This text is intended for anyone who wants to learn how to use Architectural Desktop effectively to create real-world construction documents and drawings and can be used in secondary, postsecondary, and technical schools. This text provides the beginning student and the drafting or architectural professional with a complete understanding of every Architectural Desktop command using professional drafting methods and techniques. All software commands are presented in a manner that shows each prompt with examples of the exact input to be used. This text also contains the following variety of valuable features that help make learning Architectural Desktop easy:

- ✓ It is easy to read, use, and understand.

- ✓ Learning Goals are provided at the beginning of each chapter to identify what you will learn as a result of completing the chapter.

- ✓ Commands are presented in a manner that shows every prompt with the exact input you should use.

- ✓ Field Notes explain special applications, tricks, and professional tips for using Architectural Desktop.

- ✓ Exercises are provided throughout for you to practice as you learn Architectural Desktop.

- ✓ Residential and light commercial projects, included on the enclosed CD-ROM, allow you to learn how to use Architectural Desktop in your drafting, engineering, or architectural field.

- ✓ Chapter tests, included on the enclosed CD-ROM, provide you with the opportunity to comprehensively review each chapter by answering questions related to the chapter content.

- ✓ The Instructor's Resource Manual contains chapter test answers, exercise solutions, and project solutions.

Format of This Text

The format of this text helps you learn how to use Architectural Desktop both by example and through complete explanations of each feature. Your learning is also supported by exercises that allow you to practice while you learn specific content, and the resource CD-ROM, which includes chapter tests that provide you with an excellent way to review chapter content and projects that pull together everything you have learned in each chapter.

Accessing Architectural Desktop Commands

In addition to listing available command selection methods, all options are described within the text. For example, the PLINE command can be accessed by selecting **Polyline** from the **Draw** pull-down menu, by picking the **Polyline** button in the **Draw** toolbar, or by typing **PL** or **PLINE** at the Command: prompt. The keyboard letter used to access the menu and the command is shown underlined just as it is in the Architectural Desktop menu system.

When a command sequence is discussed, examples of the command are presented as if you are typing at the keyboard, or as it appears when you pick the command from the toolbar or pull-down menu. When the command is presented as if it is typed at the keyboard, any available keyboard shortcut is always given before the full command name. One goal of this book is to show you all the methods of using Architectural Desktop commands so that you can decide which methods work best for you.

Understanding the Command Format
Presented in This Text

When a command is issued at the keyboard, the command shortcut and full name are displayed in bold caps, and the return arrow symbol (↵) shows you when to press the [Enter] key. This is an example of how a command is displayed when entered at the keyboard:

Command: **PL** or **PLINE**↵

Although the text shows the typed command using bold caps for clear visual effect, you can type the command using lowercase letters if you wish.

When the command prompts are answered by typing a value, the typed information is also presented in bold. For example, the following PLINE command prompts are answered with coordinate values followed by pressing the [Enter] key:

Command: **PL** or **PLINE**↵
Specify start point: **40',50'**↵
Current line-width is 0'-0"
Specify next point or [Arc/Halfwidth/Length/Undo/Width]: **70',50'**↵
Specify next point or [Arc/Close/Halfwidth/Length/Undo/Width]: **70',62'**↵
Specify next point or [Arc/Close/Halfwidth/Length/Undo/Width]: **101'6",62'**↵
Specify next point or [Arc/Close/Halfwidth/Length/Undo/Width]: **101'6",86'**↵
Specify next point or [Arc/Close/Halfwidth/Length/Undo/Width]: **40',86'**↵
Specify next point or [Arc/Close/Halfwidth/Length/Undo/Width]: **C**↵
Command:

When you pick a point or use a technique other than typing, the prompt is followed by the suggested method presented in italics and angle brackets:

Command: **PL** or **PLINE**↵
Specify start point: *<pick a point>*
Current line-width is 0'-0"
Specify next point or [Arc/Halfwidth/Length/Undo/Width]: *<pick a point>*

After you have learned the different ways to respond to Architectural Desktop prompts, you can use the method that works best for you. For example, if the prompt reads

From point: *<pick a point>*

you can pick a point, as suggested, enter coordinate values, or use any other input method you prefer.

Presenting Command Options

When a prompt asks you to select a command option, the capitalized letter or letters within the option name must be entered at the keyboard. If a letter in more than one word is capitalized, you can enter either letter. The polyline command options are used in the following example:

```
Command: PL or PLINE↵
Specify start point: 40',50'↵
Current line-width is 0'-0"
Specify next point or [Arc/Halfwidth/Length/Undo/Width]: W↵
```

If you are given a prompt with a value in brackets ,<>, you can enter a new value or press [Enter] to accept the value in the brackets like this:

```
Command: PL or PLINE↵
Specify start point: <pick a point>
Current line-width is 0.0000
Specify next point or [Arc/Halfwidth/Length/Undo/Width]: W↵
Specify starting width <0.0000>: 1/2↵
Specify ending width <0.5000>: ↵
```

Identifying Key Words

Key words are set in ***bold italic*** for your ease in identification. These words relate to Architectural Desktop terminology and professional applications. These words are defined in the content of the text. Key words are an important part of your Architectural Desktop learning process, because they are part of the world of computer-aided drafting, drafting standards, and Architectural Desktop communication.

Special Features

Architectural Desktop 3.0/3.3: Basics Through Advanced contains several special features that help you learn and use Architectural Desktop, as explained in the following paragraphs. Most features are shown here as they appear in the text.

LEARNING GOALS

After completing this chapter, you will be able to:

◎ Use Architectural Desktop to prepare drawings for residential and/or light commercial construction.

◎ Answer questions related to Architectural Desktop.

◎ Use Architectural Desktop and drafting-related terminology.

◎ Do exercises as you learn Architectural Desktop.

◎ Use projects to learn Architectural Desktop and create construction documents.

Exercise P.1

1. An exercise is provided after each Architectural Desktop topic or command lesson.
2. The exercises allow you to practice what you have just learned. Practicing Architectural Desktop applications is one of the most important keys to learning the program effectively.
3. You should complete exercises at a computer while using Architectural Desktop to reinforce what you have just studied.
4. Exercises build on each other, allowing you to develop construction documents as you learn Architectural Desktop.
5. Exercises are saved to disk and can be used only as practice or as classroom assignments.

✏️ Field Notes

Field Notes are another special feature of *Architectural Desktop 3.0/3.3: Basics Through Advanced*. Field Notes are placed throughout the text to provide you with any one or more of the following advantages:

✓ Professional tips and applications

✓ Special features of Architectural Desktop

✓ Advanced applications

✓ Additional instruction about how certain features work

Resource CD-ROM

Included with this text is a resource CD-ROM to further enhance your learning and mastery of the Architectural Desktop skills presented in this text. It provides review questions and practice activities as well as informational resources. These valuable learning tools resources are:

Chapter Tests

There is a comprehensive test for each chapter. You can use the tests to check your understanding of the chapter content or to review the material. Answering the test questions is an excellent way to reinforce what you have learned.

Projects

Drafting projects are one of the most important ways to complete and solidify your learning and understanding of Architectural Desktop. It has been said many times, in terms of learning AutoCAD: "if you do not practice you will forget." The projects allow you to put to practice what you have just learned. The projects are different from the exercises, because they combine a variety of commands that are used in the current chapter and in past chapters. Exercises focus only on using the currently discussed command. The projects for every chapter are designed to provide Architectural Desktop practice for residential and light commercial construction. You or your instructor can select the projects that relate directly to your specific course objectives. The projects provided in this text are real-world architectural design projects. These projects allow you to finish one or more complete sets of construction documents for a residential or light commercial building as you progress through this text.

Appendices

The appendices for this text are designed to help you use Architectural Desktop in the architectural, engineering, drafting, and construction environments. The appendices contain the following valuable information:

A. AIA Layer Standards

B. Drawing Name Conventions

C. Typical Sheet Numbering Conventions

D. Introduction to CSI Format

E. File Management

F. Standard Tables

G. Wall Priorities

H. AEC Object Display Representations

I. Display Configurations and Their Uses

Basics Through Advanced

Architectural Desktop 3.0/3.3: Basics Through Advanced covers both basic and advanced topics in nearly every chapter. Chapter content begins with basic applications and progresses into advanced topics where appropriate. The basic material generally covers fundamental AutoCAD and Architectural Desktop applications, and the advanced coverage normally involves creating styles and other customization activities.

The content is well rounded for everyone at all levels of learning. The chapters are divided into the types of features used to create architectural drawings. In many cases an Architectural Desktop feature or command is placed in its own chapter, combining both basic and advanced topics. The chapter organization follows the order that is commonly used by an architect, designer, or drafter when creating construction documents in an architectural office.

Instructors like the idea of having both the basic and advanced material in the same text. This approach benefits the students who catch on quickly and want additional challenge. Even if the instructor chooses to cover only the basic material, the students who want to can explore on their own. This approach also leaves the idea of "extra credit" activities open for the instructor who wants to challenge advanced students.

The Next Step

Many instructors and students want to focus on the basic material, so they initially skip the advanced concepts. This text helps you with this objective by displaying The Next Step icon, shown in the margin, to identify advanced topics. The Next Step applications are optional topics that can be presented in advanced courses, or they can be used by the student who wants to go further into the Architectural Desktop software. This makes the entire textbook usable, because the content can be used in several basic through advanced courses or can be used by the student or professional who wants to explore every aspect of Architectural Desktop, now and in the future.

Instructor's Resource Manual

The Instructor's Resource Manual for this text includes the following items:

Introduction: How to Use This Text

Chapter Test Answers

Chapter Exercise Solutions

Chapter Project Solutions

Acknowledgments

We would like to express our appreciation of professional support from Jon Epley and Alan Mascord, Alan Mascord Design Associates, Inc., and SERA Architects. We would also like to express our appreciation to the reviewer of this text—Steven Dyer, Richland Community College (IL).

A message to students and teachers: please let us know how we can make this text better, and please send us drawing projects that you would like to have considered for the next edition of this text.

David Madsen
Ron Palma

Warning and Disclaimer

This book is designed to provide tutorial information about the Architectural Desktop computer program. Every effort has been made to make this book complete and as accurate as possible, but no warranty or fitness is implied.

The information is provided on an "as-is" basis. The authors and Pearson Education, Inc. shall have neither liability nor responsibility to any person or entity with respect to any loss or damage in connection with or arising from information contained in this book.

Contents

||||||||||||||||||||||||||||||||||||||||||||||||||||||||||||||||||||||||||||||||||||||||||||||||||

CHAPTER 1

Introduction to Architectural Desktop

LEARNING GOALS

After completing this chapter, you will be able to:

◎ Discuss the purpose and function of Architectural Desktop.

◎ Explain the similarities and difference between AutoCAD and Architectural Desktop.

◎ Describe conceptual design.

◎ Discuss design development.

◎ Describe documented drawings.

◎ Install Architectural Desktop.

◎ Start Architectural Desktop.

◎ Use the Today window.

◎ Use Active Assistance.

◎ Identify the features of the Architectural Desktop window.

◎ Dock and float toolbars.

◎ Use pull-down menus.

◎ Use shortcut menus.

◎ Access and use dialog boxes and modeless dialog boxes.

◎ Use the Command: window.

◎ Start a new drawing using a template.

◎ Explain the difference between using a template, starting from scratch, and using a wizard.

◎ Open an existing drawing.

◎ Save a drawing.

◎ Get help.

Architectural Desktop Release 3 is Autodesk's latest tool for the creation of architectural drawings and models. Architectural Desktop is a stand-alone software package that combines both the standard AutoCAD environment and drafting tools with special AEC (Architecture, Engineering, and Construction) objects. These objects understand how to relate intelligently with one another in the creation of a project. For example, a door object knows that it must reside inside a wall object and maintain properties about itself such as width, height, and type of material (see Figure 1.1).

The AEC objects maintain data and work cooperatively together throughout the drawing life cycle, and they can work consistently and interactively with standard AutoCAD commands, building on your existing AutoCAD knowledge. The AEC objects also allow you to continue with your existing 2D drawing practices in addition to providing you with 3D representations of the AEC objects for creation of 3D models (see Figure 1.2).

With the integration of AEC objects and AutoCAD commands, Architectural Desktop meets your architectural drafting needs, from conceptual design to the completion of the detailed construction documents, whether you are an architect, designer, or drafter. This text is designed to introduce you to the AEC objects and commands used in Architectural Desktop to complete a set of construction documents. The

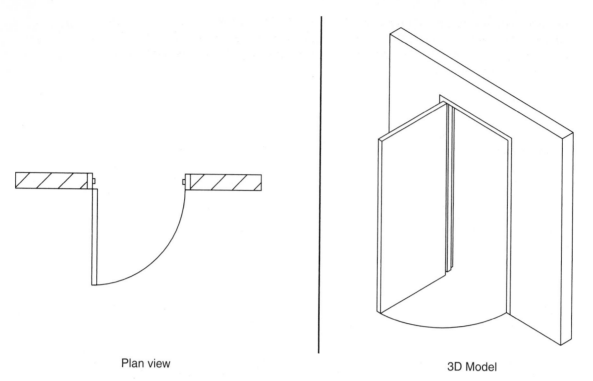

Plan view   3D Model

FIGURE 1.1   A door object understands that it must reside within a wall object.

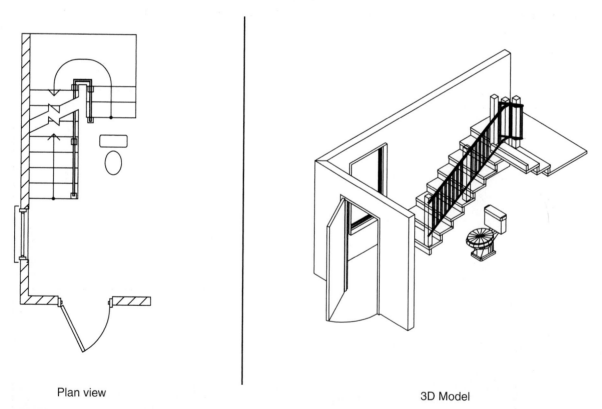

Plan view   3D Model

FIGURE 1.2   When you use AEC objects, 2D and 3D representations are displayed
based on your viewing direction.

chapters are arranged in an easy-to-understand format, beginning with basic topics and working their way to more advanced subjects. Although it is assumed that you have a basic understanding of AutoCAD, this text covers some advanced topics within AutoCAD and explains the Architectural Desktop tools that aid in producing professional and accurate drawings.

Architectural Desktop Overview

As indicated earlier, Architectural Desktop is a set of architectural tools built into the AutoCAD platform. Standard AutoCAD provides you with tools such as lines, arcs, circles, and text to complete your design needs. The addition of object-oriented tools known as AEC objects takes the design process to a new level. The AEC objects are divided into the three major phases of workflow: Conceptual Design, Design Development, and Documented Drawings.

The following sections summarize the three stages in developing an idea, refining the design, and documenting the drawings to complete a set of construction documents.

The Conceptual Design

Most architectural projects begin with a phase called *Conceptual Design*. In traditional design, concepts or ideas often consist of many hand-sketched drawings on paper. Once the concept has been finalized, the construction documents can be produced. Architectural Desktop advances the traditional design process by providing conceptual design tools. Objects known as *Mass Elements* are 3D shapes that can be brought together to form a massing study. These objects are object oriented, so the architect or designer can visually create an idea in three dimensions as well as plan in the two-dimensional environment (see Figure 1.3).

Once the design has been finalized, the 3D model can be divided into individual floors known as *Floor Slices,* as shown in Figure 1.4. The Design Development Phase can begin from the Floor Slices, by converting the slices into walls.

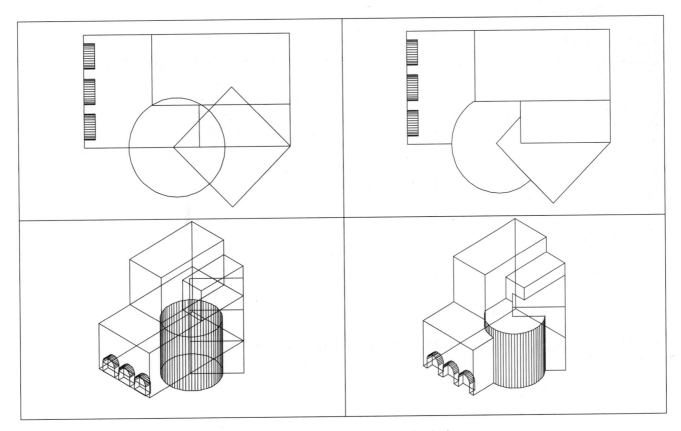

FIGURE 1.3   Mass Elements provide a means of creating a massing study for design.

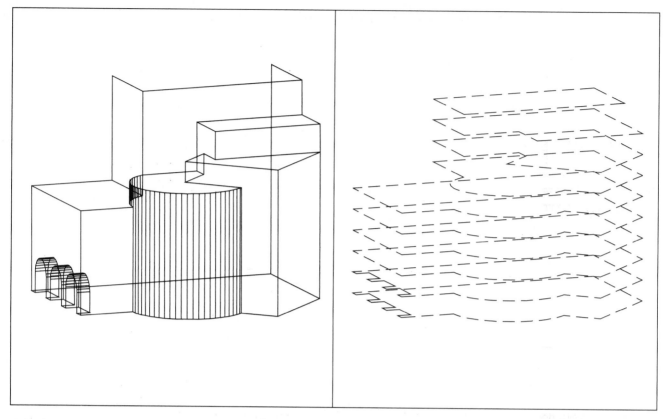

FIGURE 1.4　The massing study can be sliced into floors from which the walls can be derived.

The Design Development Process

The ***Design Development*** phase is the period during which the construction documents begin to take shape. Architectural Desktop provides tools and objects to aid in the creation of floor plans, elevations, and sections. Architectural elements such as walls, doors, windows, and stairs are drawn in this phase (see Figure 1.5).

Initially, the Design Development phase is still considered to be in the early design stage. As key elements such as walls and columns are created, the building design may go through several alterations before a final design is chosen. Once the final design is selected, additional details such as exact door and window sizes, proper stair locations, and correct wall thickness may be drawn in preparation for the next phase of work.

The Documented Drawings

After the building design has been fully developed, the construction documents can be finished with the addition of annotations such as dimensions, detail bubbles, and notes. Architectural Desktop includes many symbols that allow you to document your drawing with all the applicable information needed to build your idea (see Figure 1.6).

As you add AEC objects to your drawing, you can include information that can be extracted and sorted into schedules and inventories. This information can be updated automatically as any changes occur during the creation of the construction documents. A door and frame schedule is shown in Figure 1.7.

Before You Begin

Autodesk Architectural Desktop Release 3 includes two compact disks. One disk is the Architectural Desktop program, and the other is the Learning Assistance CD. The learning Assistance CD includes many tutorials, help files, and movie files that show you how to use AutoCAD commands. See the installation guide included with your software.

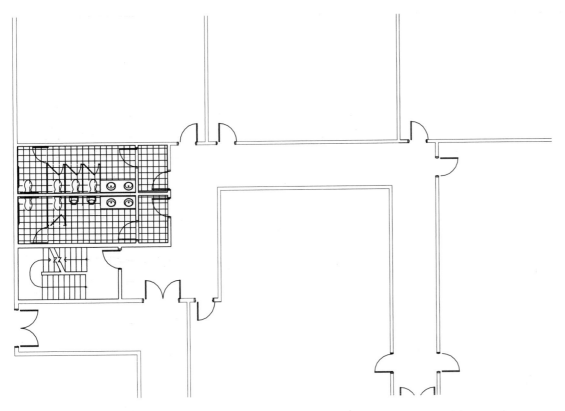

FIGURE 1.5 Walls, doors, and windows are drawn and arranged in the Design Development phase.

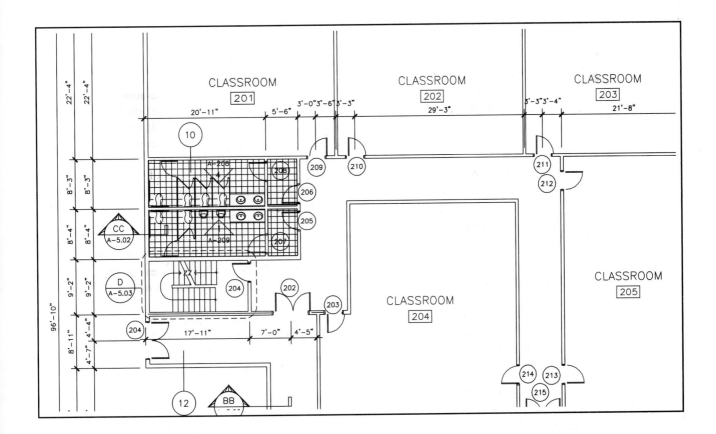

FIGURE 1.6 The Documentation Phase includes reference marks, labels, dimensions, and tags that are needed to construct your design.

FIGURE 1.7 Schedules can be extracted from the AEC objects and are updated as changes take place.

DOOR AND FRAME SCHEDULE

| MARK | DOOR SIZE | | | MATL | FIRE RATING LABEL | NOTES |
	WD	HGT	THK			
201	6'-0"	7'-0"	Right	METAL	2-HR	Double door
202	6'-0"	7'-0"	Left	METAL	2-HR	Double door
203	3'-0"	7'-0"	Left		1-HR	
204	3'-0"	7'-0"	Left		1-HR	
205	3'-0"	7'-0"	Left		1-HR	
206	3'-0"	7'-0"	Right		1-HR	
207	3'-0"	7'-0"	Right			
208	3'-0"	7'-0"	Left			
209	3'-0"	7'-0"	Left		1-HR	
210	3'-0"	7'-0"	Right		1-HR	
211	3'-0"	7'-0"	Right		1-HR	
212	3'-0"	7'-0"	Right		1-HR	
213	3'-0"	7'-0"	Left		1-HR	
214	3'-0"	7'-0"	Right		1-HR	
215	6'-0"	7'-0"	Left	METAL	2-HR	Double door

When you install Architectural Desktop, one of the initial windows prompts whether you want to install content as imperial or metric (see Figure 1.8). **Content** is defined as the blocks and symbols that are included with Architectural Desktop. When Metric is selected, the blocks and symbols are scaled metrically for any future drawings you create. The Imperial option installs both imperial and metrically scaled blocks and symbols.

Other options include how the content is to be arranged within Architectural Desktop, whether Autodesk Architectural Desktop or CSI MasterFormat. The Architectural Desktop option organizes the blocks and symbols into folders called Design, Documentation, and Schedules. Standard symbols such as

FIGURE 1.8 Select the features you want to configure for the Architectural Desktop installation.

Option to scale blocks and symbols with imperial or metric units

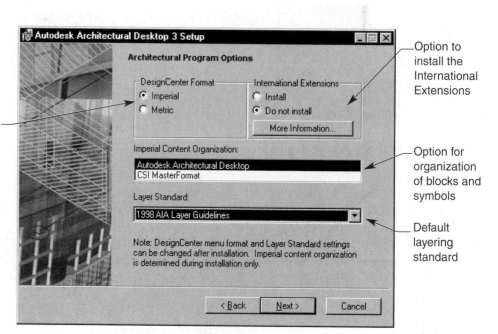

Option to install the International Extensions

Option for organization of blocks and symbols

Default layering standard

furniture and plumbing fixtures are organized under the Design folder. Annotation symbols are organized under the Documentation folder, and scheduling tags are organized under the Schedule folder.

The CSI (Construction Specification Institute) organization method categorizes the blocks and symbols into one of the standard CSI divisional folders. For example, Division 12 is where furniture blocks are organized, and Division 15 is where the plumbing fixtures can be found. See Appendix D for more information on the CSI organization.

Another option to consider is whether to install the *International Extensions*, which are additional tools that can be included only when you initially install Architectural Desktop. If you want to install the International Extensions at a later date, you must first uninstall Architectural Desktop, then reinstall it with International Extensions selected. This text includes both standard Architectural Desktop features as well as International Extension features.

The last option to consider is the default layering standard when starting new drawings. The options include the 1998 AIA Layering Guidelines and the British Standard. For more information regarding the AIA layering system see Appendix A. The British Standard is based on the BS1192 Layering Guidelines.

After you have selected your options, press the Next button to continue with the installation process.

Starting Architectural Desktop

After Architectural Desktop has been installed on your computer, a *desktop icon* is displayed on the Windows desktop. The term *desktop* refers to the background on your monitor when using a Windows operating system. The desktop is similar to an electronic desk where programs such as Architectural Desktop can be opened and work can be accomplished. To start Architectural Desktop, simply double-click the icon with your mouse (see Figure 1.9).

FIGURE 1.9   Double-click the Autodesk Architectural Desktop 3 icon to start Architectural Desktop.

Double-click means to tap (or "click") the left mouse button twice quickly.

An alternative method of starting Architectural Desktop is to select the Start button in the lower left corner of your Windows desktop. The Windows Start menu is displayed. Select Programs from the Start menu to display the Programs menu. Pick Autodesk Architectural Desktop 3 from the list, then select the Autodesk Architectural Desktop 3 icon to start the program (see Figure 1.10).

Using the Today Window

Once Architectural Desktop has been launched, the *Architectural Desktop 3.0 Today* window is displayed. The Today window is a start-up dialog box that allows you to open previously created drawings as well as to begin a new drawing (see Figure 1.11). The window is divided into three main areas: My Drawings, Bulletin Board, and Point A.

My Drawings
The **My Drawings** area contains the following three tabs:

- **Open Drawings**   This tab includes several methods for locating and opening existing drawing files. Initially, drawing files that were worked on earlier are listed in a history by date. Moving the cursor over one of the drawings displays a preview of the file (see Figure 1.12). Selecting a

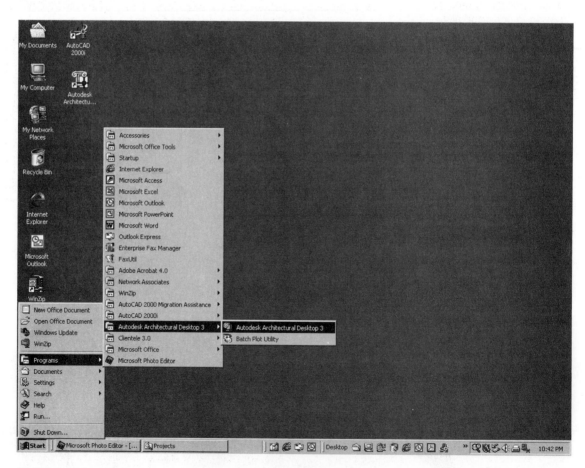

FIGURE 1.10   Pick Autodesk Architectural Desktop 3 from the Programs menu to begin working in Architectural Desktop.

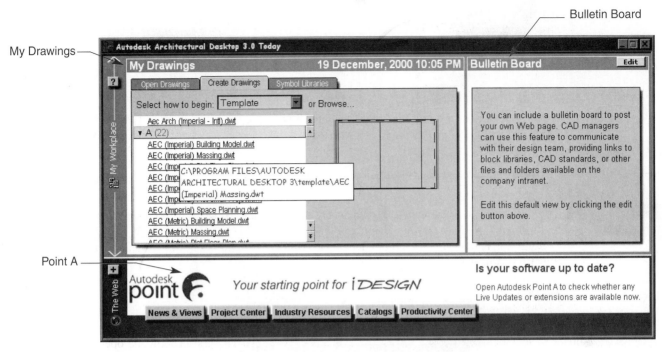

My Drawings

Bulletin Board

Point A

FIGURE 1.11 The Today window is displayed each time Architectural Desktop is started.

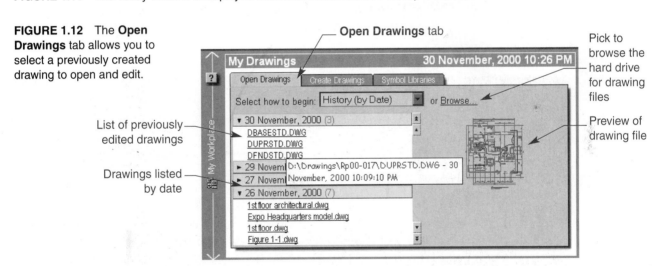

FIGURE 1.12 The **Open Drawings** tab allows you to select a previously created drawing to open and edit.

Open Drawings tab

Pick to browse the hard drive for drawing files

List of previously edited drawings

Drawings listed by date

Preview of drawing file

drawing in the history list by using the pick button of your mouse opens the drawing for additional editing or drawing. The **Browse...** link allows you to search hard drives or network locations to find and open a drawing file.

- **Create Drawings**   This tab includes a list of drawing templates from which a new drawing can be started. A *template* is a preconfigured drawing with settings that require little or no additional setup. The most recently used templates, from which a template can be selected, are listed first. Additional templates can be selected from the **A** button. Selecting a template from the list begins a new, clean drawing (see Figure 1.13).

- **Symbol Libraries**   This tab includes a list of sample drawing files that contain symbols or blocks you can use in your drawings. *Blocks* are named objects that are defined from one or more other objects. A door symbol is an example of a block (see Figure 1.14). This list can be customized to

FIGURE 1.13   The Create
Drawings tab allows you to start
a new drawing.

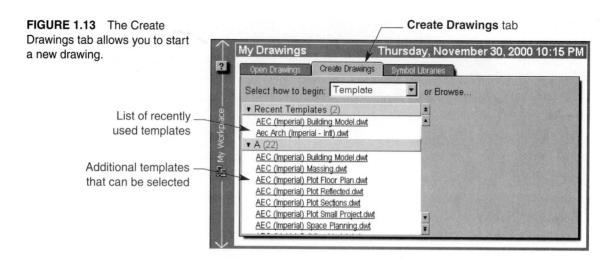

List of recently
used templates

Additional templates
that can be selected

Create Drawings tab

FIGURE 1.14   The Symbol
Libraries tab includes a list of
sample drawings that include
symbols you can add to your
drawing.

Symbol Libraries tab

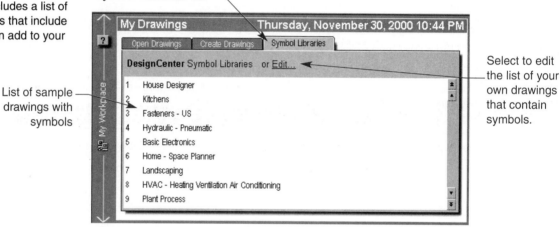

List of sample
drawings with
symbols

Select to edit
the list of your
own drawings
that contain
symbols.

display a list of drawing files you may access most often for your
own symbol libraries. Selecting a drawing from the list opens the
AutoCAD DesignCenter from which you can drag and drop symbols
into your drawing file.

The Bulletin Board

The **Bulletin Board** area allows you or CAD managers to include a link to a Web page or HTML document. You may want to include your personalized HTML file with links to other Web sites, a to-do list, or a
client list here. CAD managers may want to place the company Intranet home page here for their drafters
(see Figure 1.15).

Point A

The **Point A** area provides you with links to a personalized Web page at the Point A Web site hosted by
Autodesk. The buttons at the bottom of this area link directly to the Point A Web site (see Figure 1.16).
These links allow you to browse for information over the Internet, such as the latest codes or additional
block symbols.

   Access to the Point A Web site from within Architectural Desktop requires that you have a current
connection to the Internet. In order to obtain your free personalized Web page, you must also sign up at the
Point A Web site. Once you have registered a login name and a password, you can enter the site for the latest information required by your professional field. Figure 1.17 displays the authors' Point A Web page as
an example.

FIGURE 1.15 The Bulletin Board is used to display a custom HTML file or Web page.

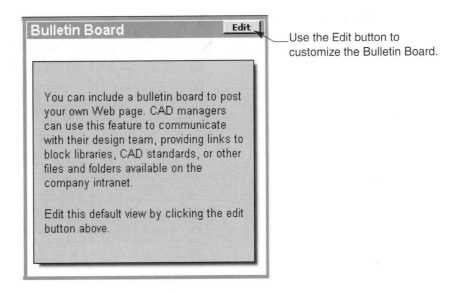

Use the Edit button to customize the Bulletin Board.

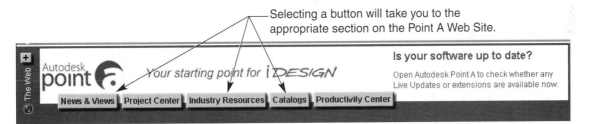

Selecting a button will take you to the appropriate section on the Point A Web Site.

FIGURE 1.16 Selecting a button will take you to the appropriate section on the Point A Web site.

Exercise 1.1

In this exercise you start Architectural Desktop and begin a new drawing based on a drawing template.

1. Start Architectural Desktop.
2. Select the Create Drawings tab to create a new drawing.
3. Select AEC arch (imperial).dwt from the A button.
4. Keep Architectural Desktop open for the next exercise. If you must quit, select Exit from the File pull-down menu, and pick No in the alert box.

Getting Assistance

By default, when Architectural Desktop is first started, the **Active Assistance** window is displayed (see Figure 1.18). This window provides you with helpful information about the type of command you are currently performing in the program. As you work, Active Assistance monitors the commands and routines that are being performed and updates you with information so that you can become proficient. Right-clicking in the window will display a shortcut menu from which you can set different options.

The shortcut menu allows you to browse forward or backward in the history of topics that have been displayed in your current session of Architectural Desktop. You can print topics, or you can return to the default Active Assistance topic or its "home."

Selecting the **Settings...** option from the shortcut menu displays the Active Assistance Settings dialog box. This dialog box allows you to specify whether Active Assistance will be turned on or off in future sessions of Architectural Desktop. Other options control how the Active Assistance is displayed. For example, Active Assistance may display only when you are looking at a dialog box, working with new commands not

Collaboration utilities for firms across the world

Codes and standards links

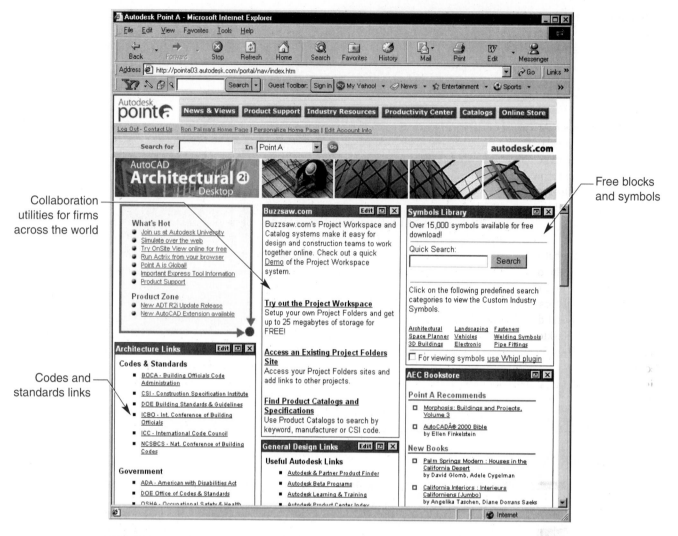

Free blocks and symbols

FIGURE 1.17   The Point A Web site provides the latest professional and CAD information for architects and designers.

FIGURE 1.18   The Active Assistance window provides helpful information as you work within Architectural Desktop.

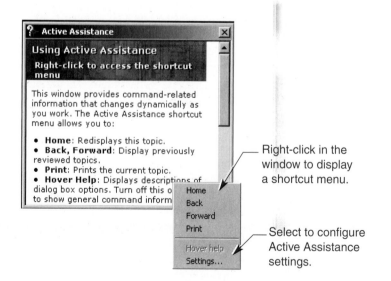

Right-click in the window to display a shortcut menu.

Select to configure Active Assistance settings.

found in Architectural Desktop, or for all commands. The Active Assistance window provides a quick and simple means of obtaining "how-to" information within Architectural Desktop. For more advanced help, see the help section later in this chapter.

The Architectural Desktop Window

As mentioned earlier, Architectural Desktop combines the standard AutoCAD drawing environment with architectural drafting tools. The standard AutoCAD tools are organized the same as they are in AutoCAD 2000i and 2002, the latest version of AutoCAD. The Architectural Desktop window is similar to other Windows-compliant programs in organization and handling. Picking the small control icon in the upper left corner displays a standard Windows control menu that allows you to minimize, maximize, and close Architectural Desktop.

The upper right corner of the Architectural Desktop window includes buttons for minimizing, maximizing, and closing the window. Each individual drawing file also includes these three buttons, allowing you to control each of the drawing windows parameters. As in other Windows programs, resizing of the windows requires that the mouse cursor be moved to the edge of the window until a double-arrow cursor appears indicating the direction the window can be stretched. Press and hold the left mouse button to stretch the window to the size desired.

Architectural Desktop uses the familiar Windows interface, which includes the use of toolbar buttons, pull-down menus, and dialog boxes. These items are discussed in detail in this chapter to familiarize you with the Architectural Desktop environment. Common commands such as NEW, OPEN, and PRINT are organized in the same locations found in other Windows programs, simplifying the learning process. Understanding and recognizing the layout, appearance, and proper use of the tools helps you master Architectural Desktop quickly. Figure 1.19 shows the Architectural Desktop window after a new drawing is started from the Today window.

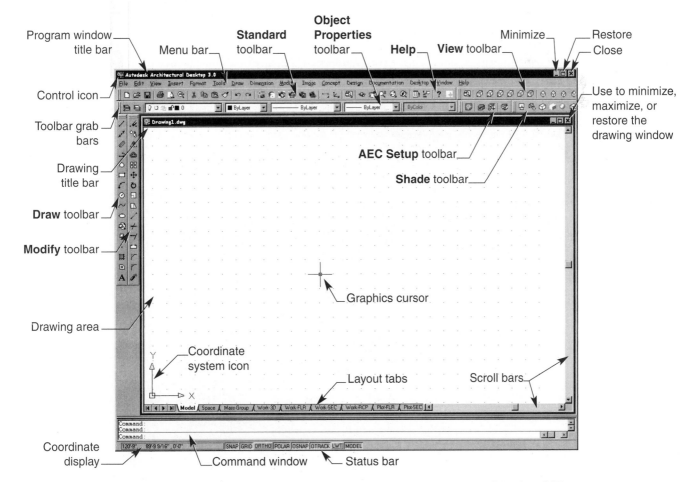

FIGURE 1.19   The Architectural Desktop window uses the organization of the AutoCAD drawing environment and adds in architectural tools.

The Architectural Desktop Window Layout

The Architectural Desktop window provides a large drawing area with tools arranged around the edges. By default the drawing window is bordered by *toolbars* at the left and top edges of the screen and by the *command window* at the bottom of the screen.

Field Notes

The proportions of the Architectural Desktop window vary depending on the current display resolution. The minimum recommended display resolution is 1024 × 768. A resolution of 1280 × 1024 allows you to see all the toolbars that are initially turned on in Architectural Desktop.

Many of the features within Architectural Desktop are not locked into a location but can be resized and moved around within the window. Thus, these features are considered to be *floating.* A standard Windows border at the top of a feature indicates that the toolbar or dialog box is floating above the drawing screen. The border along the top of the floating feature gives the drawing name, toolbar name, or dialog box name. When the border is highlighted, that feature is considered to be active. Active drawing windows are windows that can be drawn in. Active toolbars require that a toolbar button be selected in order to draw.

When Architectural Desktop is first started, the Architectural Desktop window is displayed floating on the Windows desktop. A smaller window within the Architectural Desktop window displays a drawing window floating in the program. Floating windows can be moved and resized as in other Windows programs. The drawing windows, however, can be adjusted only within the Architectural Desktop window (see Figure 1.20).

Some features, such as toolbars, can be *docked* in place. Docked toolbars do not display a border or title bar; instead, they display a *grab bar*. A grab bar consists of the two thin bars at the top or left edge of the toolbar. The *title bar* is the bar across the top of the toolbar that gives the name of the toolbar. A docked toolbar has no title bar and is embedded in the program window. To move a toolbar from a docked position to a floating position, double-click the grab bars. This places the toolbar in a floating position on your

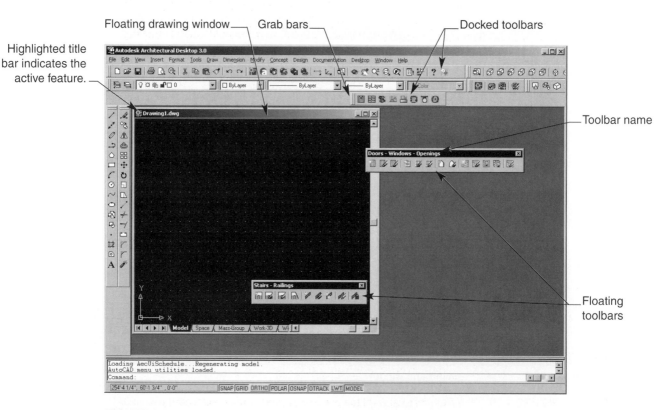

FIGURE 1.20   Floating and docked features within Architectural Desktop.

screen. A floating toolbar can be freely moved anywhere on the drawing window by picking and holding the left mouse button on the title bar. An alternative way of undocking a toolbar is to press and hold the left mouse button over the grab bars of the desired toolbar and drag the toolbar away from its docked position.

The procedure is similar for docking a toolbar from a floating position. Double-click the grab bars of a floating toolbar to dock it, or drag the toolbar by the grab bars to the edge of the screen where you want to dock the toolbar. When a toolbar is dragged, a thick outline of the toolbar is displayed attached to your pointer. Move the outline to an edge of the screen until it changes to a thin outline, then release the left mouse button. The thin outline lets you know that the toolbar is docked along the edge of the screen. Toolbars may be moved or docked in any position at any time as needed.

The following is a brief description of the components within the Architectural Desktop window. Refer to Figure 1.19 and your Architectural Desktop window as you review these features:

■ **Control icon**	Pressing this icon displays a menu for adjusting the Architectural Desktop window.
■ **Program window title bar**	This title bar indicates the that you are working in Architectural Desktop and displays the drawing title name if the drawing window is maximized within the Architectural Desktop window.
■ **Menu bar**	The menu bar displays a number of menu names. As with standard Windows menus, picking a menu name with your left mouse button causes a ***pull-down menu*** to appear. Each pull-down menu contains commands for drawing in Architectural Desktop.
■ **Crosshairs**	This is the primary means for drawing and editing within the drawing window. Similar to a Windows cursor, the crosshairs allow you to select specific points within your drawing.
■ **Standard toolbar**	The Standard toolbar contains a set of buttons, many of which are common to other Windows programs. When the mouse is used to move the crosshairs over a toolbar, the crosshairs change to the familiar Windows pointer. Once a button is picked, a command is executed in the currently active drawing window.
■ **Toolbars**	Toolbars are arranged around the Architectural Desktop window. Initially, only a few toolbars are displayed within the window. Additional toolbars are available and can be turned on or off as needed. See the toolbar section covered later in this chapter.
■ **Drawing area**	This is the area where all graphics are drawn.
■ **Scroll bars**	The scroll bars allow you to adjust your view of the drawing area by "sliding" the drawing from side to side.
■ **Coordinate system icon**	This icon indicates the current coordinate system drawing plane where you are drawing.
■ **Layout tabs**	There are two types of tabs within the drawing window. The **Model tab** is where your drawing is created and drawn. The other tabs, known as **Layout tabs**, allow you to set up how the drawing will be printed or plotted. Each Layout tab represents a sheet of plotted paper and can have a title block drawn on it.
■ **Command window**	The Command window defaults to a position at the bottom of the screen. This window displays user commands and Architectural Desktop prompts for input. This window serves as the primary means of communication between you and Architectural Desktop.

- **Status bar**   The status bar includes several buttons that indicate the current state of specific settings within Architectural Desktop. When a pull-down menu item is highlighted, a brief explanation for that item is displayed at the left end of the status bar.

- **Coordinate display**   This area, found on the status bar, displays the current X,Y, and Z location of the crosshairs within the drawing.

Exercise 1.2

1. Continue from Exercise 1.1, or start Architectural Desktop.
2. Select the Create Drawings tab to create a new drawing.
3. Select AEC arch (imperial).dwt from the A button.
4. Examine all the features discussed.
5. Keep Architectural Desktop open for the next exercise. If you must quit, select Exit from the File pull-down menu, and pick No in the alert box.

Using Pull-Down Menus

Pull-down menus are located on the menu bar across the top of the Architectural Desktop window. The menu bar includes several pull-down menu items: **File, Edit, View, Insert, Format, Tools, Draw, Dimension, Modify, Concept, Design, Documentation, Desktop, Window,** and **Help,** as shown in Figure 1.21. The first nine pull-down menus contain standard AutoCAD tools and commands, such as **Polyline, Erase,** and **Shade.** The next four pull-down menus, **Concept, Design, Documentation,** and **Desktop,** are specific to Architectural Desktop and contain tools specifically for creating architectural objects such as walls, windows, and doors. The last two pull-down menus are standard Windows menus for arranging drawing windows within Architectural Desktop, or for finding help on a particular topic.

Most of the available AutoCAD and Architectural Desktop commands can be found in both pull-down menus and toolbars; however, some commands may be found in only one of these locations. This is why it is very important to familiarize yourself with the layout of both pull-down menus and toolbars. Toolbars are discussed later in this chapter.

To access a pull-down menu, move your pointer to one of the pull-down menus and select it by clicking the left mouse button. A list of commands that can be selected by picking with your left mouse button is displayed (see Figure 1.22A). Some of the commands in a pull-down menu may have a small arrow beside them. This small arrow indicates that there is an additional menu available known as a *cascading menu* (see Figure 1.22B). Cascading menus contain additional commands specific to the command selected in the pull-down menu. To access a command from a cascading menu, pick the appropriate command with an arrow in the pull-down menu, then select the desired command from the cascading menu.

Some of the commands in the pull-down menus are followed by an ellipsis (…). Picking one of these commands displays a dialog box that includes options for performing the desired command. Dialog boxes are discussed later in this chapter. Once a command has been picked from a pull-down menu, the menu disappears, allowing you to continue with your selected command.

Field Notes

Commands also can be accessed by holding the left (pick) button down while moving the pointer to the desired command, then releasing the pick button.

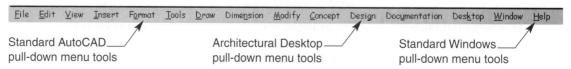

FIGURE 1.21   The Architectural Desktop pull-down menus include AutoCAD, Architectural Desktop, and Windows commands.

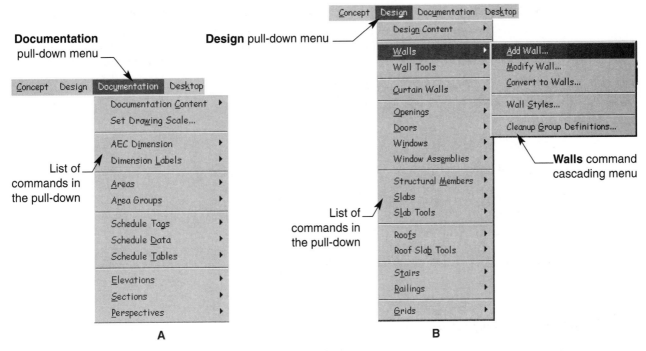

FIGURE 1.22 (A) A pull-down menu contains commands that can be selected in order to perform a function in Architectural Desktop. (B) Some commands have additional tools available in a cascading menu.

Accessing Pull-down Menus Using the Keyboard

Windows 95, 98, or Windows NT 4.0 users will notice that one character of each pull-down menu title is underlined. This indicates that the pull-down menu or command can be selected by pressing the [Alt] key plus the key of the letter that is underlined. For example, pressing the [Alt]+[G] key combination accesses the Design pull-down menu.

Field Notes

In Windows 2000 the underscore is not displayed by default. To display the underlined characters in a pull-down menu, first press the [Alt] key to display the underscores, then press the key of the underlined character.

Once a pull-down menu has been displayed using the [Alt]+[key] combination, a command can be selected simply by pressing the keyboard key indicated by the underscore in the desired command. These shortcut keystrokes are known as *menu accelerator keys.*

Field Notes

After a pull-down menu is displayed, commands and cascading menus can be selected by using the up, down, right, and left arrow keys on the keyboard. To access a cascading menu with the arrow keys, press the right arrow key when a command with a cascading arrow is highlighted. Use the arrow keys to scroll to the desired command. Press the [Enter] key to select the command once it has been highlighted.

Exercise 1.3

1. Continue from Exercise 1.2, or start Architectural Desktop as instructed in Exercise 1.2.
2. Using the mouse and pointer, open each of the pull-down menus and read the commands found in each menu without picking any of the commands.
3. Open each of the pull-down menus using the menu accelerator keys.
4. If you access a command or dialog box, press the [Esc] key on your keyboard to cancel the activity, and resume this exercise.
5. Keep Architectural Desktop open for the next exercise. If you must quit, select Exit from the File pulldown menu, and pick No in the alert box.

Using Shortcut Menus

Architectural Desktop makes extensive use of the AutoCAD shortcut menus introduced in AutoCAD 2000. Shortcut menus are used to accelerate and simplify accessing commands. Often referred to as *cursor menus* because they are displayed at the cursor location, these context-sensitive menus can be accessed by right-clicking. *Context-sensitive* means that the cursor menu displays commands that relate to an object or an area that you have picked.

Because the shortcut or cursor menus are context-sensitive, the options available in the menu vary based on the location of the pointer when you right-click. For example, right-clicking within the drawing window with no objects selected displays the shortcut menu shown in Figure 1.23A. If a wall object has been selected, then the cursor menu activated by right-clicking displays the menu in Figure 1.23B. Right-clicking in other areas of the Architectural Desktop window provides you with yet other options. Figure 1.23C displays the cursor menu displayed by right-clicking a Layout tab.

Shortcut menus act much like pull-down menus. To access a command or option, move the cursor to the command desired and pick with the left mouse button. Many of the options also include cascading shortcut menus as with the pull-down menus.

Field Notes

Shortcut menus (cursor menus) are discussed where they apply throughout this text. To familiarize yourself with this timesaving feature, try right-clicking at different times and in different areas of the Architectural Desktop window as you are learning how to use the Architectural Desktop commands.

Exercise 1.4

1. Continue from Exercise 1.3, or start Architectural Desktop as instructed in Exercise 1.2.
2. Access and view the commands available in the cursor menus by right-clicking in the following areas:
   - Drawing window
   - Command window
   - Layout tabs
   - Status bar
   - Toolbar buttons
3. Press the [Esc] key or pick in the drawing window away from the cursor menu to close the menu.
4. Keep Architectural Desktop open for the next exercise. If you must quit, select Exit from the File pull-down menu, and pick No in the alert box.

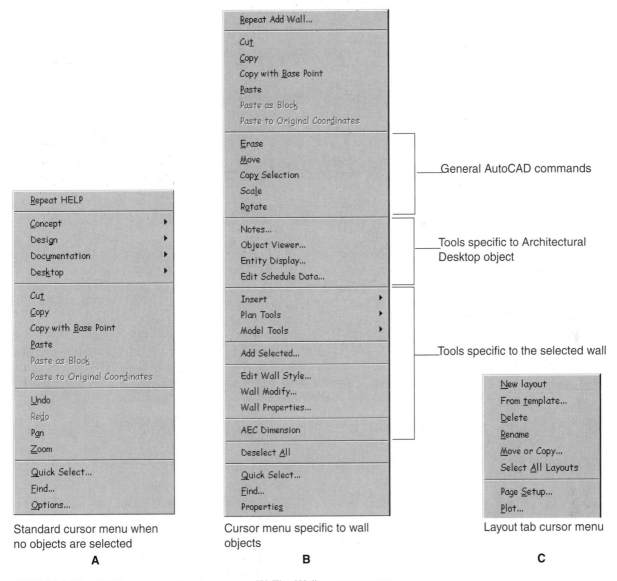

Standard cursor menu when no objects are selected

A

Cursor menu specific to wall objects

B

Layout tab cursor menu

C

FIGURE 1.23 (A) The standard cursor menu. (B) The Walls cursor menu. (C) The Layout tab cursor menu.

Using Toolbars

As mentioned in the preceding pull-down menu section, most of the AutoCAD and Architectural Desktop commands can be found in both toolbars and pull-down menus. When Architectural Desktop is started initially, only a few toolbars are displayed (see Figure 1.19). A mixture of AutoCAD and Architectural Desktop toolbars have been turned on for your use.

As the pointer is moved across toolbar buttons, a 3D button is displayed around a previously flat button. Holding the pointer motionless over a button for a moment displays a *tooltip*. The tooltip displays the name of the button in a small box next to the cursor. When a tooltip is displayed, a brief description of the button's function is displayed in the status bar at the bottom of the screen. Some buttons include a small black triangle in the lower right corner. These buttons are called *flyouts*. If the left mouse button is pressed and held over a flyout, an additional set of buttons is displayed.

Because Architectural Desktop is a combination of AutoCAD and Architectural Desktop tools, both AutoCAD and Architectural Desktop toolbars are provided. Different toolbars can be turned on or off as needed. Right-clicking over the top of a toolbar displays a list of toolbars available for either AutoCAD or Architectural Desktop (see Figure 1.24). Right-clicking over an AutoCAD toolbar produces a list of available AutoCAD toolbars; right-clicking over an Architectural Desktop toolbar provides a list of available Architectural Desktop toolbars.

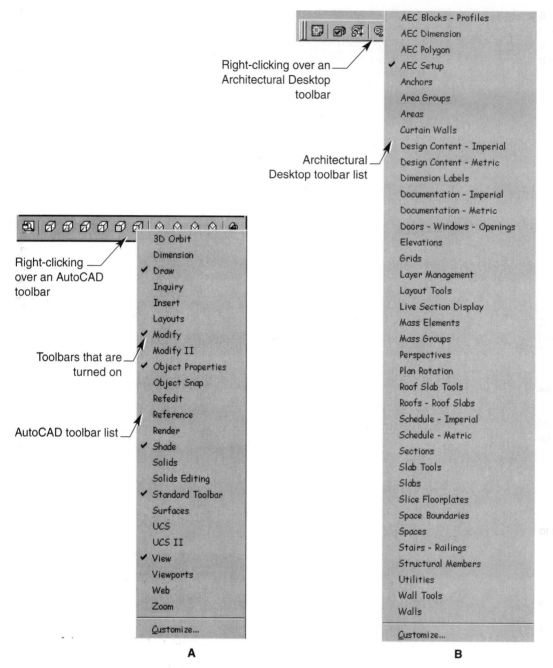

FIGURE 1.24   (A) Right-clicking over an AutoCAD toolbar displays a list of AutoCAD toolbars. (B) Right-clicking over an Architectural Desktop toolbar produces a list of Architectural Desktop toolbars.

A check mark next to a toolbar title in a displayed toolbar list indicates that the toolbar is already on and displayed in your Architectural Desktop window. Picking an unchecked toolbar title turns on the selected toolbar. As mentioned earlier, there are many cursor menus available. Right-clicking in a gray area of the screen where toolbars are docked displays the two sets of toolbar lists. Select the set that contains the toolbars you need to turn on additional toolbars (see Figure 1.25).

Toolbars can also be turned on or off from the **Customize** dialog box. To access this dialog box, pick the Toolbars… option from the Customize cascading menu in the Tools pull-down menu, or type **TOOLBAR** in the Command window. The Customize dialog box appears. In the Toolbars tab, select the appropriate Menu Group to display the toolbars you want to turn on or off. Pick the box next to the toolbar title in order to turn a toolbar on or off (see Figure 1.26).

FIGURE 1.25   Right-click in the gray area where toolbars are docked to display the two toolbar lists. Choose the list you need to access to turn on additional toolbars.

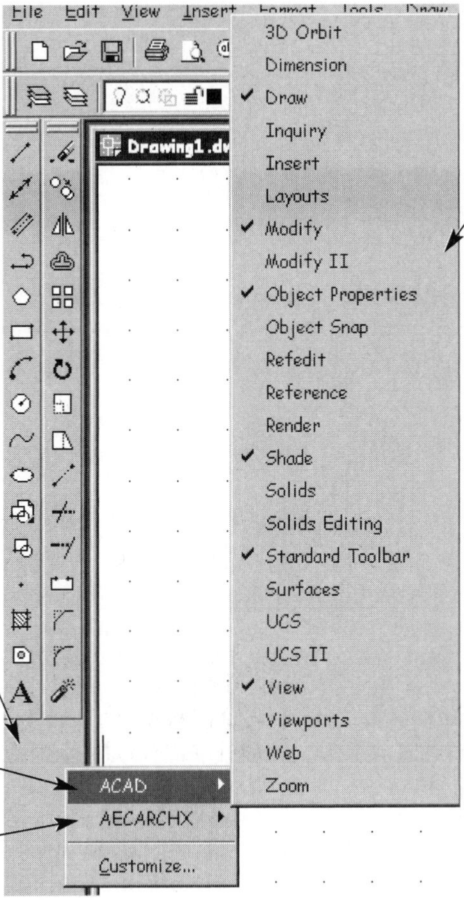

Right-click in the gray area to list both AutoCAD and Architectural Desktop toolbar lists.

AutoCAD toolbar list

Architectural Desktop toolbar list

AutoCAD toolbars

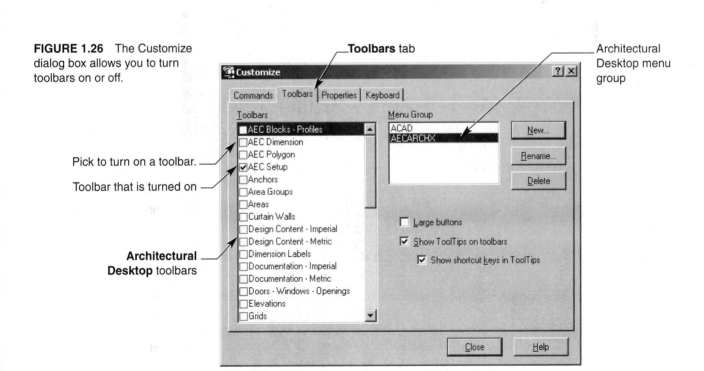

FIGURE 1.26   The Customize dialog box allows you to turn toolbars on or off.

Toolbars tab

Architectural Desktop menu group

Pick to turn on a toolbar.

Toolbar that is turned on

Architectural Desktop toolbars

Exercise 1.5

1. Continue from Exercise 1.4, or start Architectural Desktop as instructed in Exercise 1.2.
2. Right-click over the different toolbars initially displayed when Architectural Desktop is started. View the different toolbars available.
3. Find the Walls toolbar and pick it from the list.
4. Access the Customize dialog box.
5. Turn the Walls toolbar off.
6. Keep Architectural Desktop open for the next exercise. If you must quit, select E_xit from the F_ile pulldown menu, and pick No in the alert box.

Using Dialog Boxes

Dialog boxes are small windows that use a variety of methods to provide instructions or information to Architectural Desktop. Dialog boxes are used to control settings, to specify properties of objects, and in some cases to control many aspects of a particular object. By centralizing and controlling many settings in one location, dialog boxes help save time and increase productivity.

Dialog boxes contain several standard features for inputting information. Take a few minutes to review the following descriptions so you are familiar with the contents available in a dialog box:

- **Command buttons**   Command buttons are normally located at the bottom of the dialog box. The most common command buttons are **OK** and **Cancel,** as shown in Figure 1.27. Pressing the OK button saves the settings you made in the dialog box and closes the dialog box. Pressing the Cancel button closes the dialog box without saving your settings. Another common command button is the **Help** button. Pressing this button opens the Help window providing additional information specific to the dialog box. If a command button is *grayed-out,* that feature is not available.

    Buttons within a dialog box that have an ellipsis (…) display another dialog box when picked. The second dialog box is displayed on top of the original dialog box. You must make a selection or a change in this second dialog box before you can return to the first dialog box. You can close a dialog box at any time by picking the Cancel or OK buttons or by picking the close button **X** in the upper right corner of the dialog box.

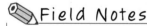 Field Notes

Pressing the **X** button in the upper right corner of a dialog box cancels any changes made in the dialog box. You should always pick the OK button in a dialog box so that Architectural Desktop will accept your changes. There may be times when an OK button is not available. In this case when you pick the **X** button the program accepts the changes you have made.

- **Tabs**   A dialog box tab is similar to an index tab used to separate the sections of a notebook or file drawer. Many of the dialog boxes contain two or more tabs or pages located at the top of the dialog box. Each tab contains a set of related options (see Figure 1.27). Picking a tab with the left mouse button makes it the current page from which options can be selected, set, or changed.

- **Radio buttons**   A radio button is one button within a group of buttons, only one of which can be selected at a time, much like a radio button in a car (see Figure 1.27).

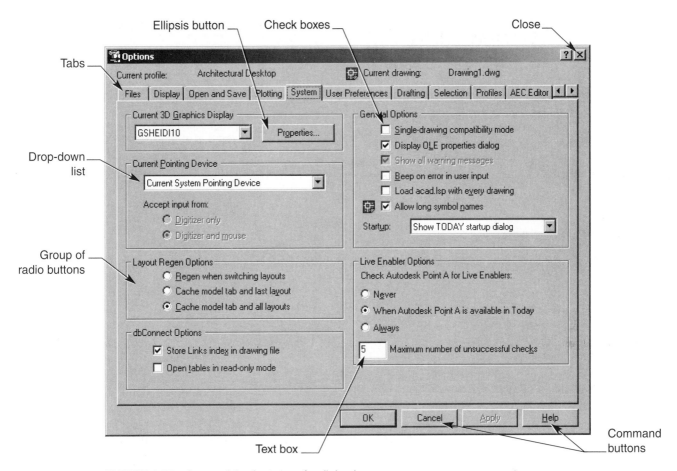

FIGURE 1.27   Some of the features of a dialog box.

- **Check boxes**

  A check box displays a ✓ when it has been selected or activated. Check boxes act like toggles that turn options on or off. Unlike radio buttons, multiple check boxes can be turned on or off.

- **Text boxes**

  Text boxes are areas where information can be entered from the keyboard. Text boxes are usually found when a value or a name is being specified (see Figure 1.27). To enter a value in a text box, place the pointer in the text box and press the pick button. When you pick inside a text box a blinking cursor is displayed, allowing you to enter a value. If there already is a value in a text box, picking in the text box highlights the entire value, allowing you to type a new value over it.

- **List boxes**

  A list box contains a list of items that can be selected. Items in list boxes can be selected by double-clicking the item or by first highlighting the item, then pressing the [Enter] key, as shown in Figure 1.28.

- **Drop-down lists**

  A drop-down list contains a list of options from which only one option can be selected. List boxes are displayed as a value in a text box with a drop-down arrow button beside the box. Pressing the arrow produces a list from which an option can be selected (see Figure 1.28).

- **Preview box or image tile**

  A preview box is an area of a dialog box that displays a "picture" of the item you select in a list box. An image tile is similar to a preview box except that the image tile is more

List box

Command
buttons within
a list box

Help button

Preview box

Command button

Command button

Text box and drop-down list

Drop-down list

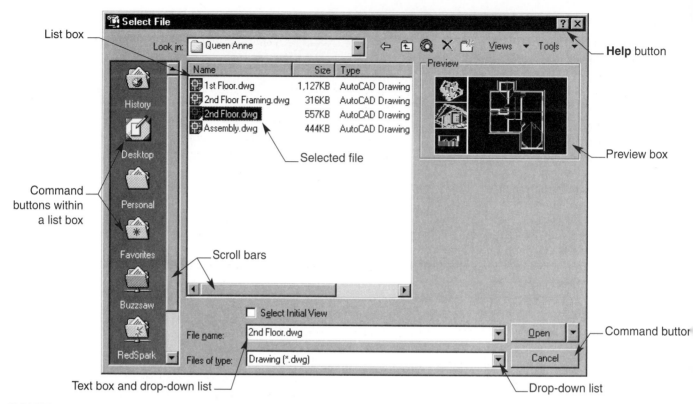

FIGURE 1.28   The Select File dialog box contains additional options to set.

interactive. In many of the image tiles, picking on the image changes a setting in the dialog box (see Figure 1.28).

- **Help**

Many of the dialog boxes include a Help button that displays a help section related to the dialog box you are currently accessing. Some dialog boxes include a ? button next to the close button in the upper right corner of the dialog box. If you are unsure of a feature in a dialog box, pick the ? button, then pick the feature you have a question about. A text box of help information is displayed next to the cursor, explaining the feature.

Exercise 1.6

1. Continue from Exercise 1.5, or start Architectural Desktop as instructed in Exercise 1.2.
2. Pick the Options… command in the Tools pull-down menu. Observe the features identified in the previous discussion.
3. You may make selections or change options in the dialog boxes if you desire, but this is not required.
4. You can close the dialog box at any time by picking the Cancel button or the **X** button in the upper right corner of the dialog box. If you activate a dialog box or feature and you wish to exit, press the [Esc] key on your keyboard to exit the command or activity.
5. Keep Architectural Desktop open for the next exercise. If you must quit, select Exit from the File pull-down menu, and pick No in the alert box.

Using Modeless Dialog Boxes

Architectural Desktop also includes another type of dialog box known as a *modeless dialog box* (see Figure 1.29A). Modeless dialog boxes are often displayed when an architectural object such as a wall, stair, or roof is being added or modified. Modeless dialog boxes contain many of the same features as a standard dialog box. The main difference between a standard dialog box and a modeless dialog box is that once you have selected the options, the modeless dialog box must remain on the screen until you are finished creating or modifying the object. When using a modeless dialog box you must pick once in the drawing window after setting the options to make the drawing window active before you can begin drawing or modifying.

Modeless dialog boxes include buttons that are not available in standard dialog boxes. Next to the close button in the upper right corner of the modeless dialog box is a *pushpin button*. When this button is depressed and the pushpin appears "pinned" to the screen, the modeless dialog box remains on the drawing screen as you create or modify the architectural objects. Deselecting, or "unpinning," the pushpin button minimizes the modeless dialog box while you create or modify the geometry (see Figure 1.29B). When you move the cursor over the minimized title bar the modeless dialog box reappears, allowing you to make any additional changes to the options.

The lower left corner of a modeless dialog box includes five buttons that are specific to the architectural object that is being created or modified. From left to right, the buttons are **Floating Viewer, Match, Properties, Undo,** and **Help**. A brief description of these buttons follows:

- **Floating viewer** The floating viewer displays a preview window of the type of object you are creating. This viewer displays objects in 2D or 3D, and in a wireframe or shaded mode (see Figure 1.30). The series of buttons across the top allows you to display the architectural object in a number of different shademodes as well as to adjust how the object is to be viewed.

  The drop-down list includes a list of preset views of the object such as the top or the front of the object, or an isometric viewing angle. The drop-down list above the preview allows you to select a preconfigured display for the objects that displays all the components or only certain components of the object. The display options are covered in greater detail in Chapter 5.

- **Match** This button allows you to select an existing architectural object in your drawing that changes the properties in the modeless dialog box to match the properties of the selected object.

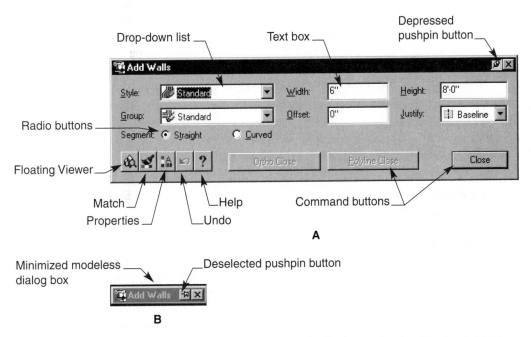

A

B

FIGURE 1.29 A modeless dialog box is displayed when architectural objects are created or modified.

Shademode buttons

Display options drop-down list

View buttons

Viewing angle options

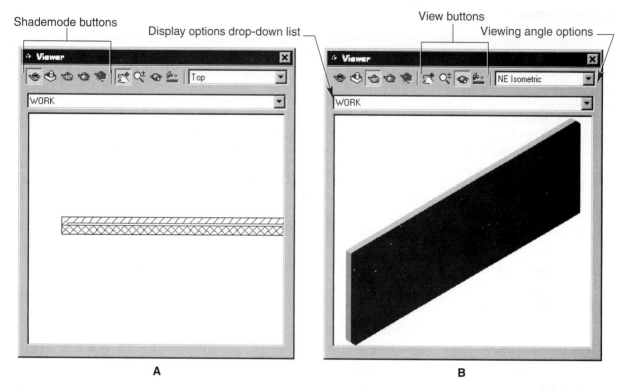

A   B

FIGURE 1.30   The floating viewer displays a preview of the architectural object being created or modified. (A) A wall being previewed from the top is displayed as a 2D wall. (B) A wall previewed from an isometric angle is displayed in 3D.

- **Properties**

  This button opens the Properties dialog box related to the particular architectural object you are creating. The Properties dialog boxes generally provide additional settings not available in a modeless dialog box.

- **Undo**

  This button will undo the operation you performed in the drawing window. For example, if you are drawing a wall, pressing the Undo button will undo the last point picked for the wall.

- **Help**

  This button opens the Help dialog box, which displays information related to the type of object being worked on.

Exercise 1.7

1. Continue from Exercise 1.6, or start Architectural Desktop as instructed in Exercise 1.2.
2. Pick the Add Wall... command from the Walls cascading menu in the Design pull-down menu. Observe the features identified in the preceding discussion.
3. Pick the Floating Viewer button. Use the Style drop-down list to choose different wall styles.
4. Close the dialog box at any time by picking the Close button. If you activate a feature and you wish to exit, press the [Esc] key on your keyboard to exit the command or activity.
5. Keep Architectural Desktop open for the next exercise. If you must quit, select Exit from the File pull-down menu, and pick No in the alert box.

The Command Window

The *Command window* is located at the bottom of the drawing screen by default and is the most important area of Architectural Desktop. This window is the line of communication between you and the program. When you activate a command from a toolbar or a pull-down menu the command is entered in the Command window at one of the Command: lines (see Figure 1.31). You then enter additional options in the Command window through the keyboard. You also may type the command at the Command: line, then press the [Enter] key. This tells the program that you want to execute the selected command. In return, the program prompts you for more information such as additional options to be set or where an object is to be created.

For example, in Figure 1.31, the command to add walls has been selected. Architectural Desktop then prompts you for the first point of the wall or to set additional parameters for the wall.

To set parameters at the Command window, type the uppercase character of the desired option. When you press the [Enter] key after typing the option character you are prompted for additional information.

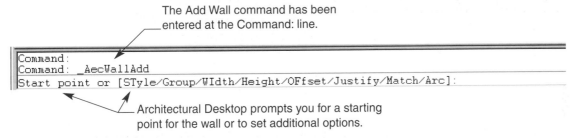

The Add Wall command has been entered at the Command: line.

```
Command:
Command: _AecWallAdd
Start point or [STyle/Group/WIdth/Height/OFfset/Justify/Match/Arc]:
```

Architectural Desktop prompts you for a starting point for the wall or to set additional options.

FIGURE 1.31 The Command window is used to tell the program to perform a command. In return, the computer prompts you for additional information.

Starting, Opening, and Saving Drawings

When you first start Architectural Desktop, the AutoCAD Today window is displayed, allowing you to begin work either by starting a new drawing or by opening an existing drawing. The following section discusses the different options for starting new drawings and opening existing drawings.

Starting A New Drawing

The My Drawings area of the AutoCAD Today window contains three tabs used to work with drawings. The **Create Drawings** tab within the My Drawings area is where different options for starting a new drawing can be found (see Figure 1.32). Below the tab is a drop-down list with three options for beginning a new drawing. These options include **Template, Start from Scratch,** and **Wizards**. Initially the Template option is displayed.

Using Template Drawings
A *template* is a drawing file that is preconfigured with standard settings, enabling you to begin work without having to set up the drawing. Architectural Desktop includes several preset templates. If the International Extensions are installed, additional templates are included. The templates are organized alphabetically under the button labeled A. Pressing the A button reveals the list of templates that can be selected.

All the template names begin with the acronym AEC, followed by the word Imperial or Metric in parentheses. Imperial templates begin a drawing in feet and inches, whereas metric templates begin a drawing in millimeters. The remaining parts of the template names briefly describe what the template may be used for. For example, "Arch (Imperial)" might be used mainly for construction documents using feet and inches, and "Massing" might be used for conceptual design. Choosing a template does not lock you into using the template only for the suggested purposes. Once a template is chosen from the list, Architectural Desktop begins a new drawing based on the selected template.

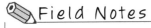 Field Notes

Each template includes a layout tab named **Template-Overview**. Selecting this tab gives the recommended uses of the included layout tabs in the template.

Starting options

List of recently used templates

Templates are organized alphabetically.

Preview window

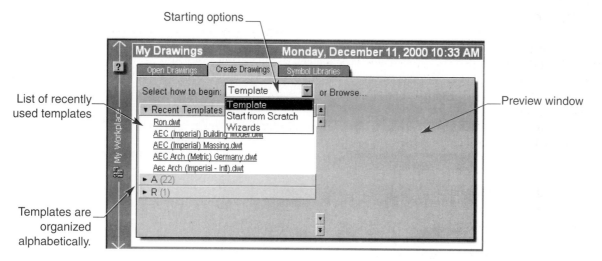

FIGURE 1.32   The My Drawings area of the AutoCAD Today window contains a drop-down list of different starting options.

If you need to begin another drawing, select the New button in the Standard toolbar, which again displays the AutoCAD Today window. Any templates that were previously used are rotated to a button named **Recent Templates,** making it easier to find the template for another new drawing.

✎ Field Notes

- As commands are discussed in this text the options for accessing the commands and the toolbar icons are provided in the margin if available. For example, to start a new drawing you can select the New button or the Today button from the Standard toolbar. You can also type either NEW or TODAY at the Command: prompt to start a new drawing. Another option is to select the NEW command by picking <u>N</u>ew… from the <u>F</u>ile pull-down menu, or by picking <u>T</u>oday from the <u>T</u>ools pull-down menu.

- It is recommended that you begin new drawings in Architectural Desktop by using a template. The Architectural Desktop templates include preconfigured settings and also many different styles of walls, doors, windows, and stairs to choose from when drawing these types of objects.

Exercise 1.8

1. Continue from Exercise 1.7, or start Architectural Desktop as instructed in Exercise 1.2.
2. Pick the New toolbar button in the Standard toolbar.
3. Select the Create Drawings tab.
4. From the A button, pick the AEC (Imperial) Building Model.dwt.
5. Find and select the Template-Overview tab.
6. The Template-Overview tab displays all the other layout tabs in the template and provides suggestions on their use. Note that zooming commands, which allow you to magnify and read the text on the Layout tab, are discussed later in this text.
7. Pick the Model tab.
8. Keep Architectural Desktop open for the next exercise. If you must quit, select E<u>x</u>it from the <u>F</u>ile pull-down menu, and pick No in the alert box.

Using Start from Scratch Drawings

When the **Start from Scratch** option is selected from the Create Drawings drop-down list two options are available: English and metric. These types of new drawings include only the basic settings for a new drawing from standard AutoCAD. The English option begins a new drawing that initially is not large enough for an architectural project until you change the size. The English option starts a drawing on a 12 × 9-in. drawing area working with decimal inches. The metric option starts the drawing on a 420 × 297-mm area.

When starting a drawing using the Start from Scratch option, it is recommended that you set options such as the size of the drawing, the type of drawing units (such as feet and inches versus decimal inches), and any styles needed before you begin drawing.

Using Wizard Drawings

The Wizards option also includes two options for beginning a new drawing: **Quick Setup** and **Advanced Setup**. When you select a Wizard, a dialog box is displayed from which you can choose different setting options for your new drawing. The settings that are included in the Wizard provide you with only a basic setup within Architectural Desktop such as the drawing area and the type of drawing units. You must import any architectural object styles into the drawing separately.

Opening an Existing Drawing

If you have drawings that were created earlier, you can use the **Open Drawings** tab in the My Drawings area of the AutoCAD Today window to locate the drawing file and open it (see Figure 1.33). Like the Create Drawings tab, the Open Drawings tab includes a drop-down list with several options for sorting recently opened drawings.

The initial option lists a history of recently opened drawings by date. The drawings below the drop-down list are listed by week. A button indicating the week the drawings were worked on is displayed, from which a drawing file can be chosen for opening. The other options in the drop-down list sort the recently used drawings by alphabetical filename, location, or most recently used. If a drawing is not listed in the Open Drawings tab, you can pick the Browse... hot link to display the **Select File** dialog box, shown in Figure 1.34.

The Select File dialog box displays the current directory or folder in which you are looking. You can browse through a list of folders and directories by picking the **Look in:** arrow. A list of files is displayed below the current folder drop-down list. An AutoCAD/Architectural Desktop drawing file is indicated in the list box by a red icon. Selecting a drawing file from the list provides a preview of the drawing file in the preview tile to the right of the dialog box. To open the highlighted drawing file for editing, pick the Open button in the lower left corner of the dialog box, or double-click the file name in the list box.

Next to the Open button is a drop-down arrow. Pressing the arrow before opening the drawing displays different opening options, such as opening the drawing as a read-only file or partially opening the file.

Along the left edge of the Select File dialog box are the following buttons:

■ **History**   This button displays a history list of drawing files that you have worked on in the past.

FIGURE 1.33 The Open Drawings tab can be used to sort through a list of recently used drawing files. Files not listed can be found by selecting the Browse... hot link.

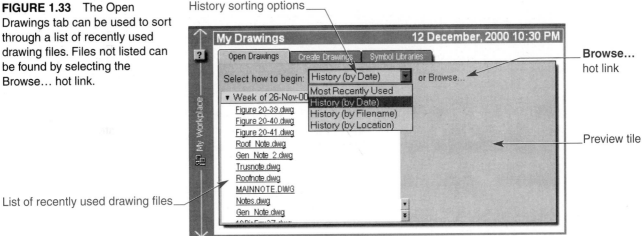

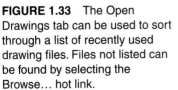

- **Desktop**   This button displays any files that are stored on your Windows desktop.

- **My Documents**   This button displays files that are stored in the My Documents folder of your computer.

- **Favorites**   This button lists any shortcuts present in your Favorites folder on your computer.

- **Buzzsaw**   This button takes you to the Buzzsaw.com Web site, where drawing files can be uploaded or downloaded to and from the Internet to collaborate with other architects, contractors, or clients. The Buzzsaw.com Web site is primarily aimed at the AEC industry.

- **Redspark**   This button takes you to the Redspark.com Web site, where drawing files can be uploaded or downloaded to and from the Internet to collaborate over the Internet. This Web site is primarily aimed at the mechanical engineering industry.

You can also open drawings by pressing the Open button in the Standard toolbar, typing **OPEN** at the Command: prompt, or by selecting Open... from the File pull-down menu. Any of these methods displays the Select File dialog box, shown in Figure 1.34.

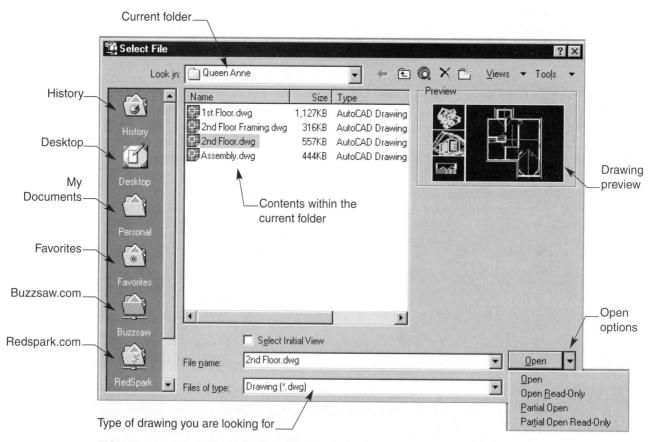

FIGURE 1.34   The Select File dialog box is used to browse through hard drives and network drives to find and open a drawing file.

Exercise 1.9

1. Continue from Exercise 1.8, or start Architectural Desktop as instructed in Exercise 1.2.
2. Use the Open button in the Standard toolbar to open a drawing.
3. Use the Look in: drop-down list to browse the \Program Files\Autodesk Architectural 3\Sample folder.
4. Select the 1st Floor.dwg file.
5. Enter **CLOSE** at the Command: prompt to close the drawing file.
6. Enter **TODAY** at the Command: prompt.
7. Select the Open Drawings tab, and use the Browse... hot link to browse to the Sample folder.
8. Select any of the drawing files to open.
9. Select Close from the Window pull-down menu to close the drawing.
10. Keep Architectural Desktop open for the next exercise. If you must quit, select Exit from the File pull-down menu, and pick No in the alert box.

Saving a Drawing

Whenever you work on a drawing file it is very important that you save the work you created to a file. As good practice, you should save a drawing approximately every 10 to 15 minutes. *This is very important!* By saving the drawing periodically, you have the latest version of the drawing available in case of a power failure, severe editing error, or other problems. If you save once every hour, a power failure could result in a loss of an hour's worth of work. Saving your work more regularly reduces the amount of lost work in case of a problem. It is also recommended that you back up your work at the end of the day to keep your work safe in another location if anything should happen to your hard drive. You can do this by saving your work twice in two different locations—a tape, Zip disk, CD, or at a minimum, a floppy disk.

Architectural Desktop includes two commands for saving your work: **SAVEAS** and **QSAVE**. In addition to these commands, Architectural Desktop displays a warning box if you attempt to close a drawing without saving. This dialog box gives you the option of saving or discarding the drawing before it is closed.

Using the SAVEAS Command

When you start a new drawing, the drawing does not have a file name. The title bar of the drawing shows the name Drawing#.dwg, where # represents a number, which is assigned consecutively. The SAVEAS command allows you to save the drawing with a unique name of your choosing. You can access this command by selecting Save As... from the File pull-down menu, by typing SAVEAS at the Command: prompt, or if this is the first time you are saving the drawing, by picking the Save button from the Standard toolbar. The **Save Drawing As** dialog box shown in Figure 1.35 appears.

This dialog box is similar to the Select File dialog box displayed in Figure 1.34. The difference is that the former is used for saving drawing files. Before you save the drawing, ensure that the folder you want to save into is displayed in the Save in: drop-down list at the top of the dialog box. By default, drawings are saved in an AutoCAD 2000 drawing format. If you need to save as another type of drawing, select the drawing type from the **Save as type:** drop-down list at the bottom of the dialog box. Enter a name for your drawing file in the **File name:** text box above the Save as type: drop-down list. It is important to note that you do not need to enter the .dwg file extension in the File name: text box, as AutoCAD automatically uses the .dwg extension. Finally, pick the Save button in the lower right corner of the dialog box.

Once a drawing has been saved, the name of the drawing file is displayed in the drawing title bar. The SAVEAS command is used in the following situations:

- The current drawing already has a name and needs to be saved with a different name.

- The drawing needs to be saved as a different type of drawing format file, such as an AutoCAD Release #14 or AutoCAD LT drawing file.

FIGURE 1.35   The Save
Drawing As dialog box is used to
save a drawing file to a location
on your hard drive or network
drive.

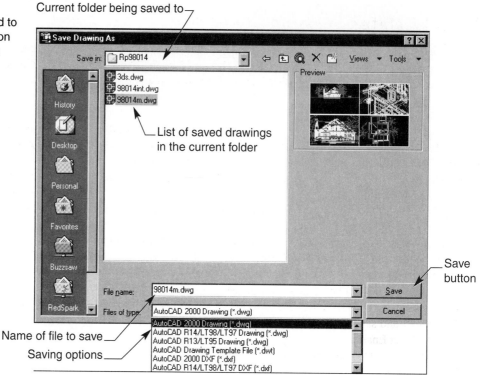

- The drawing needs to be saved to a different folder or network drive.

- A new drawing is started and needs to be saved with a name.

If you try to save a drawing using the same name and location as another drawing file, AutoCAD displays a warning box. You have the option of canceling the save operation, or replace the existing drawing with the one you are currently working on. If the drawing actually needs to be replaced, pick the Yes button to overwrite the existing file with the information in the current drawing. If you do not wish to overwrite the existing file, pick the No button to return to the Save Drawing As dialog box. Pick the Cancel button to cancel the save operation. **Be very careful**—picking the Yes button causes the existing drawing to be overwritten with the new drawing information. The existing drawing is gone forever.

Sharing Drawings with Other Users

Architectural Desktop 3 is based on the AutoCAD 2000i platform and Architectural Desktop 3.3 is based on the AutoCAD 2002 platform. When you save your Architectural Desktop drawing in the AutoCAD 2000.dwg file format, your drawing can be opened and worked on by other users with Architectural Desktop 3 or 3.3. If the drawing is going to be sent to a user with Architectural Desktop 2i, Architectural Desktop 2, or Architectural Desktop 1, you need to ensure that the AutoCAD variable **PROXYGRAPHICS** is turned on and that the drawing is saved as an AutoCAD R14.dwg before sending it to the user.

The PROXYGRAPHICS variable saves an image of proxy objects within the drawing file. A **proxy object** is an object that is a non-AutoCAD-created object such as all the Architectural Desktop objects. When PROXYGRAPHICS is set to 1, an image of the Architectural Desktop objects (proxies) is saved in the drawing so that users with an earlier version of Architectural Desktop or AutoCAD can view the objects. Proxy objects cannot be modified unless they are viewed in the original version of Architectural Desktop in which they were created. When PROXYGRAPHICS is set to 0, no proxy images are saved in the drawing. If the drawing is opened with an earlier version of Architectural Desktop, the Architectural Desktop Release 3 objects will not be present.

Field Notes

> If you do not intend on sharing your drawings with non–Architectural Desktop 3 users, you can set the PROXYGRAPHICS variable to 0. This decreases the file size of your drawing between 25% and 75%. The PROXYGRAPHICS variable is stored within the drawing file.

If your drawing will be shared with users of AutoCAD 2000i or 2002, you have two options for saving your drawing file. The first option is to set the PROXYGRAPHICS variable to 1 and save the drawing as an AutoCAD 2000.dwg file. Users can open and view the drawing but cannot make changes to the proxy objects. The other option is to have the AutoCAD 2000i or 2002 user download and install a free utility from Autodesk called the **Object Enabler**, which allows non–Architectural Desktop users to view and modify proxy objects. At the time of this writing, the current version of the Object Enabler for AutoCAD 2000i or 2002 users is Arch 3.3 OE.

If you are sharing your drawing with an AutoCAD 2000 user, you also have two options. The first option is to set the PROXYGRAPHICS variable to 1 and save the drawing as an AutoCAD 2000.dwg file. Users can open and view the drawing but cannot make changes to the proxy objects. The other option is to have the AutoCAD 2000 user download and install the OE 3.02 Object Enabler.

If your drawing is to be shared with an AutoCAD R14 user, the options are the same. Set PROXYGRAPHICS to 1 and save the drawing as an AutoCAD R14.dwg file, or have the user download and install the OE 1.22 Object Enabler.

If the drawing is to be saved as an AutoCAD R13.dwg file the only option is to set the PROXYGRAPHICS variable to 1.

Field Notes

> You can download the Object Enablers from the Autodesk Web site: www.autodesk.com. Use the Search option to locate and download the appropriate Object Enabler.

Using the QSAVE Command

QSAVE stands for *quick save*. The QSAVE command quickly saves a drawing after it has initially been saved with a name. When the QSAVE command is used to save a drawing, Architectural Desktop saves the drawing without first displaying the Save Drawing As dialog box because the program already knows the drawing has a name and is saved to a particular directory. This command can be accessed by picking the Save button from the Standard toolbar, typing QSAVE at the Command: prompt, or by selecting Save from the File pull-down menu.

If this is a new drawing that does not have a name, the preceding options display the Save Drawing As dialog box. Select a directory in which to save the drawing, enter a name for the drawing, and specify a drawing format type. After the drawing is named, these options save the drawing without displaying the Save Drawing As dialog box.

Exercise 1.10

1. Start Architectural Desktop as instructed in Exercise 1.2.
2. From the Create Drawing tab, select the AEC (Imperial) Building model.dwt.
3. Use the SAVEAS command to save the drawing file.
4. Save the drawing with the name EX1-10.dwg and an AutoCAD 2000.dwg format.
5. Pick the Save button to save the file.
6. The drawing name should be displayed in the title bar.
7. Pick the Save button in the Standard toolbar.
8. Select Close from the Window pull-down menu to close the drawing.

Finding Help

If you have trouble understanding how a command works, or if you need additional information about a feature within Architectural Desktop, help is one click away. There are several methods of getting help with AutoCAD and Architectural Desktop subjects. The simplest way is to press the [F1] function key on your keyboard. The **AutoCAD Architectural Desktop Help** window appears. This window is a powerful on-line help system (see Figure 1.36).

In addition, you can use the following options to access the AutoCAD Architectural Desktop Help:

✓ Select **Architectural Desktop <u>H</u>elp** from the **<u>H</u>elp** pull-down menu.

✓ Type **HELP** at the Command: prompt.

✓ Pick the **Help** button in the **Standard** toolbar.

The AutoCAD Architectural Desktop Help window includes the following five tabs:

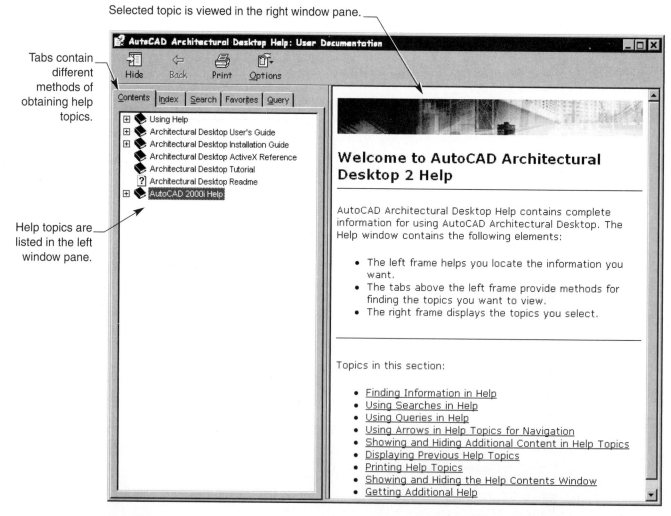

FIGURE 1.36   The AutoCAD Architectural Desktop Help window provides many avenues to different help topics.

- **Contents**   This tab presents a list of topics and subtopics in a treelike configuration in the left window pane. You can browse through topics by selecting the book icons and subtopics listed. When a topic is selected from the list, its topic contents are displayed in the right window pane.

- **Index**   This tab displays an alphabetic listing of keywords contained in the Contents tab. AutoCAD and Architectural Desktop topics can be found in this list. When you type the letters of a keyword in the text box at the top of the tab, the help system begins sorting the topics in alphabetical order.

- **Search**   This tab allows you to do a search for a help topic based on a keyword. Enter the keyword topic you are looking for, such as WALLS, in the text box at the top of the tab. Select the List Topics button, and the help system lists a series of related topics in the left pane. The topic you select to view is shown in the right pane.

- **Favorites**   This tab provides you with a list of bookmarked entries to the help system. After you have found a topic that you want to bookmark, select the Favorites tab. The bottom of the tab includes an Add button that is used to add the help topic page to a favorites or bookmarked list in the left pane. Select a bookmark from the list, then pick the Display button to display the topic in the right window pane.

- **Query**   This tab is used to search for information in a question format. This system allows you to enter a question in the text box based on everyday language. The drop-down list below the text box allows you to specify which help documentation manual to search when searching for your question.

There are also buttons along the top of the window that allow you to hide the tab pane, move backward through the list of topics you looked up, print your topic, and use a few additional options for viewing the help topics. Right-clicking in the right topic pane also provides you with a help shortcut list. You may want to visit the **Tutorials** section in the Contents tab for additional information on the use of Architectural Desktop.

 Field Notes

> If you are unfamiliar with the Windows help system, you may want to browse through this help window. Much AutoCAD and Architectural Desktop information can be found that is outside the scope of this textbook.

Exercise 1.11

1. Start Architectural Desktop as instructed in Exercise 1.2.
2. From the Create Drawing tab, select the AEC (Imperial) Building model.dwt.
3. Access the AutoCAD Architectural Desktop Help window.
4. In the Contents tab, browse through the list of topics found in the Using Help section.
5. Pick on Finding Information in Help topic.
6. Review the topic in the right window pane.
7. Look through the list of help topics in the Architectural Desktop User's Guide.
8. Pick the Close button in the upper right corner of the dialog box.
9. Close your drawing.

|||||||||||| CHAPTER REVIEW

 Use the CD-ROM to test your knowledge and skills.

Chapter Test

To check your understanding of the content provided in this chapter, access the Test file in the CH01 folder of the CD-ROM that accompanies this text.

Chapter Project

To practice the Architectural Desktop skills presented in this chapter, access the Project files in the CH01 folder of the CD-ROM that accompanies this text. The project files are in pdf format and include sample drawings and instructions for completing each project.

CHAPTER 2

Basic Drafting Techniques

LEARNING GOALS

After completing this chapter, you will be able to:

◎ Use model and layout space.

◎ Switch between model and layout space.

◎ Set up a drawing.

◎ Establish units.

◎ Set plotting scales.

◎ Establish architectural layers.

◎ Control how default AEC objects appear in a drawing.

◎ Draw polylines.

◎ Use the Cartesian coordinate point-entry systems.

◎ Draw with Ortho on.

◎ Use direct-distance entry.

◎ Use tracking and polar tracking.

◎ Create drawings using object snaps and running object snaps.

◎ Print or plot a drawing.

As with any job, a mastery of the tools and techniques required by the profession will advance you in your career. Such is the case with Architectural Desktop. There are a number of commands, both AutoCAD and Architectural Desktop, that once mastered will save you time and simplify the drafting process. One time-saving device is to use templates when you begin a new project.

As mentioned earlier, Architectural Desktop includes a number of templates that contain preestablished settings that allow you to quickly begin drawing. Two of the templates that you may use most often are the AEC Arch (Imperial) and the AEC (Imperial) Building Model templates. These two templates include many configured settings such as Layout tabs for plotting, Object styles for walls, doors, and windows, and layering styles.

No matter how you begin a drawing in Architectural Desktop, there are a few commands and features that you should understand before drawing.

An Introduction to Model Space and Layout Space

Every drawing in Architectural Desktop includes a **Model tab** found at the bottom of the drawing screen. The Model tab is the drafting area where your model or drawing is created. Any drawings such as floor plans, elevations, sections, and details are drawn in the Model tab area. Everything that is drawn in *model space* is drawn at full scale, 1:1 (see Figure 2.1).

Depending upon the template or drawing file, at least one **Layout tab** will also be present. A Layout tab, also known as *paper space,* is where the drawing is laid out for plotting. This is the area where your drawing/model is given a scale, the title block is added, and possible annotation notes are created so that

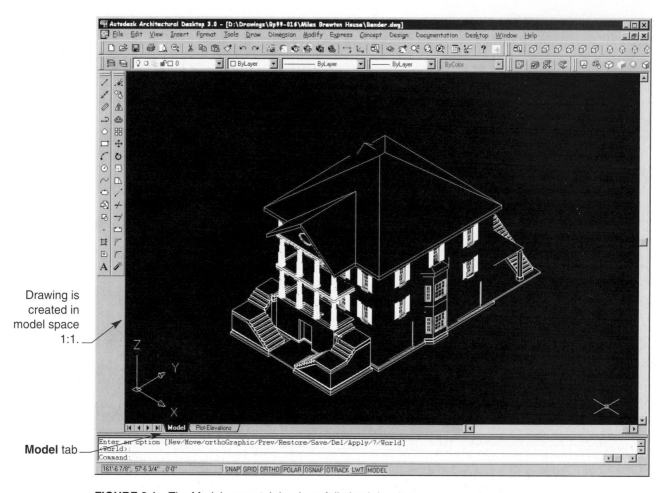

Drawing is
created in
model space
1:1.

Model tab

FIGURE 2.1   The Model space tab is where full-sized drawings are created.

the drawing can be plotted. One Layout space tab can contain many views in your drawing at different scales. For example, a detail sheet may have many views of different parts of the drawing at different scales. A drawing can have only one Model tab, but a drawing can include multiple Layout tabs, each representing a different sheet in a series of construction documents (see Figure 2.2).

Switching between tabs is possible throughout the development of the construction documents by picking the desired tab from the bottom of the drawing screen. Right-clicking over a tab gives you a menu of options that include creating new Layout tabs, page setup, and plotting the contents of the tab. Layouts and plotting are discussed in greater detail later in the text.

Exercise 2.1

1. Start Architectural Desktop.
2. Select the AEC Arch (Imperial).dwt.
3. Access the different Layout tabs. Notice that some templates include a title block, and some do not. Use the arrows to the left of the Model tab to view all the Layout tabs.
4. Select the Template Overview tab. Quickly review this tab.
5. Select the Model tab. Notice that the Model tab does not display a paper representation.
6. Save the drawing as EX2-1.dwg.

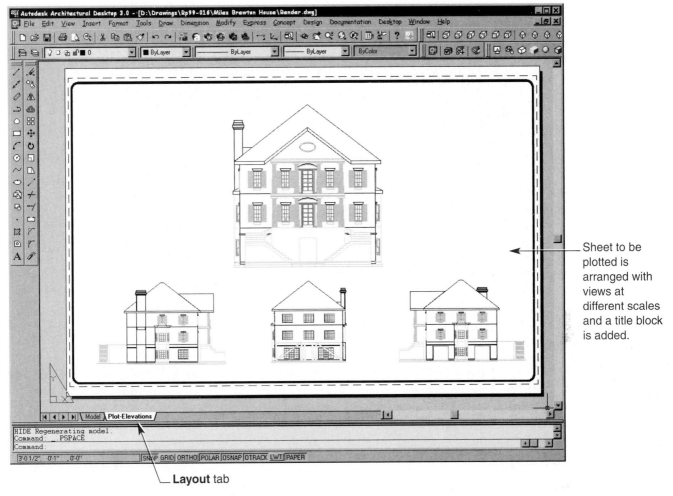

Sheet to be plotted is arranged with views at different scales and a title block is added.

Layout tab

FIGURE 2.2 Layout space is where the drawing from model space is given a scale, a title block is added, and views are arranged on the "paper" for plotting.

Setting Up the Drawing

One of the first steps you should take after starting a new drawing is to set up the drawing. When setting up a drawing in Architectural Desktop, things to consider are the desired drawing units, the scale at which you intend to plot the drawing, how Architectural Desktop objects are placed on layers, and how the objects appear when you draw. The **Drawing Setup** dialog box can be used to set these variables.

Use one of the following options to access the Drawing Setup dialog box:

✓ Pick the **Drawing Setup** button on the **AEC Setup** toolbar.

✓ Select **Drawing Setup...** from the **Desktop** pull-down menu.

✓ Enter **AECDWGSETUP** at the Command: prompt.

The Drawing Setup dialog box shown in Figure 2.3 appears. The dialog box is divided into four tabs or areas: **Units, Scale, Layering**, and **Display**. The following sections describe each tab.

The Units Tab

Every object that is drawn in Architectural Desktop is measured in drawing units. For example, one unit may represent 1 inch or 1 foot. Refer to Figure 2.3 when reviewing this section. You can choose the type of

Set the layering key style.

Set the drawing scale.

Set the drawing units.

Control the display of objects.

Specify how objects from other drawings will be scaled when inserted.

Specify drawing units.

Choose the type and accuracy of drawing units.

Choose the type and accuracy of angled drawing units.

Specify the type, precision, and suffix for areas.

Specify the type, precision, and suffix for volumes.

Check to apply settings to new drawings.

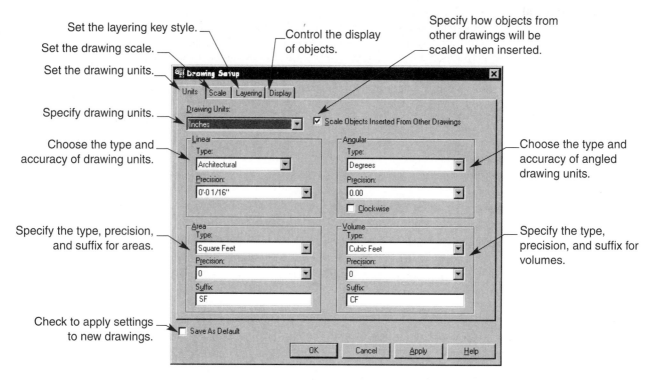

FIGURE 2.3 The Drawing Setup dialog box is used to set up a new drawing.

units that you intend to use for drawing from the Drawing Units: drop-down list. Once the drawing units have been selected, a warning box appears indicating that the drawing units have changed and prompts whether you want to scale the existing objects in your drawing to the new settings (see Figure 2.4).

The **Inches** option allows you to use architectural units (feet and inches), in which 1 unit is equal to 1″, and 12 units is equal to 12″ or 1′-0″. The other drawing units are feet, millimeters, centimeters, decimeters, and meters. With these options, one drawing unit is equal to 1′-0″, 1mm, 1cm, 1dm, or 1m, respectively. When imperial units (feet and inches) are being used, it is most common to choose inches as the drawing units. When metric units are being used, it is most common to choose millimeters as the drawing units.

There may be times when you want to copy objects from an existing drawing into your current drawing. The **Scale Objects Inserted From Other Drawings** check box allows you to scale objects from another drawing to the scale of your current drawing. If this option is unchecked, objects from another drawing are inserted at their original drawing scale.

The **Linear** area of the Units tab includes drop-down lists for the type and precision of length values you enter when creating geometry. Descriptions of the Type: options Architectural, Decimal, Engineering, Fractional, and Scientific follow:

FIGURE 2.4 After you change drawing units a warning is displayed allowing you to scale existing objects.

This will scale existing objects in your model space drawing to reflect the new drawing units.

This will scale existing objects in your layout/paper space drawing to reflect the new drawing units.

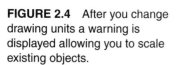

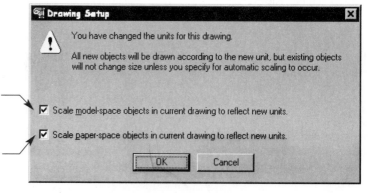

- **Architectural**   Architectural units are most commonly used when creating drawings in Imperial (feet and inches) units. Length values for geometry are entered into Architectural Desktop using feet and inches. For example, if a wall is being drawn that is twelve feet six inches in length, a value of 12'-6" is entered into the Command: line. Feet are specified by placing the symbol for feet (') after the numerical value, and inches are specified by placing the symbol for inches (") after the value. The Precision drop-down list includes options for rounding off measured length values. Other methods of entering architectural unit values include the following: 72", 72, 6'0", 6', 12'6, 24'6-1/2" and 18'3-3/4. Note that the inch symbol (") is not required when specifying inch values. Also notice that a hyphen (-) must be used to separate inch and fractional values. Architectural units are available only when inches are the drawing units. Architectural is the most common type used when creating imperial (feet and inches) drawings.

- **Decimal**   The Decimal type is available for all the drawing units. This type of unit is most commonly used when specifying inch or metric length values. Methods of entering length values include whole numbers or decimal numbers: 144, 24.5, 256.75, 58.015, and 96.50, for example. Notice that the inch symbol (") is not used when using the Decimal type.

- **Engineering**   The Engineering type is available when inches are the drawing units. This type of unit is commonly used for drawing civil engineering projects such as site plans, dam and bridge construction, and geographical maps. Length values are entered as feet and decimal inches. Formats used for length input include 68.5', 72'-6.5, 54'8.75, and .25', for example. Notice that a hyphen (-) can be used to separate the feet from the decimal inch value.

- **Fractional**   The Fractional type is available when inches are the drawing units. This type of unit is used for drawings that use fractional parts of a common unit of measure. Length measurements are entered as whole numbers and fractions: 36-1/2, 24-3/8.

- **Scientific**   As with the Decimal type, Scientific units are available for any of the drawing units. The Scientific type is typically used in chemical engineering and astronomy. Length values are entered as a base number followed by E+xx, where the *xx* represents the power of 10 by which the base number is multiplied. For example, 10.5E+06, 25.4E+12.

The **Precision**: drop-down list includes values of precision that can be chosen. The precision values change based on the type of units selected. The value that you specify determines how length values will be rounded off when displayed in the coordinates area at the lower left of the Architectural Desktop window.

The **Angular** area in the Units tab specifies how angles are to be entered when angled shapes are drawn or angular units are specified when geometry is created. As with the Linear area, a number of types with associated precisions are available. The following types are described next: Degrees, Degrees/Minutes/Seconds, Grads, Radians, and Surveyor.

- **Degrees**   The Degrees type is often used in architectural or mechanical drawing. This system is based on decimal degrees, where 360° equals a complete circle. Architectural Desktop begins measuring angles from the right beginning with angle 0° and measures them in a counterclockwise direction, as shown in Figure 2.5. Angular values are entered into Architectural Desktop as decimal values such as 0, 45, 90, 135, and 180.

- **Degrees/Minutes/Seconds**   This type of angular unit is occasionally used in architectural and structural drafting when more precision is required

FIGURE 2.5 Architectural Desktop measures all angles in a counterclockwise direction beginning with angle 0 to the right.

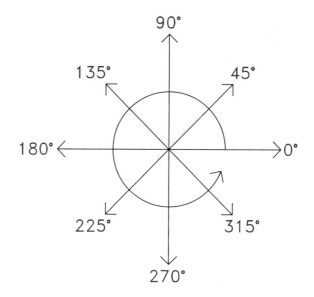

between degree angles. This type is commonly used in civil drafting. There are 60 minutes (′) in one degree, and 60 seconds (″) in one minute. Angular values are entered as xxdxx′xx″, where xx equals a number, and *d* stands for the degree symbol. Some examples of angular values are the following: 25d39′54″, 274d01′59″, and 359d59′59″.

■ **Grads**

Grads is an abbreviation of *gradient*. A full circle measured with gradients includes 400 grads. A 90° angle is equal to 100 gradients. Angular values are entered by typing the number angle followed by a *g*. Examples of gradient values include 153g, 372g, 50g.

■ **Radian**

The Radian type is based on the relations p=180°, 2p=360°, and p/2=90°. Angular values are calculated by multiplying or dividing a number by π, and an *r* is placed after the calculated value. Examples of angular values using this type of unit include 3.925r, 1.57r, and 6.28r.

■ **Surveyor**

This type of angle is commonly used when measuring bearings for plot plans and maps. A ***bearing*** is a direction based on one of the quadrants of a compass. Bearings are measured beginning from north or south, with units in degrees (d), minutes (′), and seconds (″), to the east or west (see Figure 2.6). An angle that was measured from 32°24′33″ from the north toward the east is entered as N32d24′33″E. Other values may be entered as S49d15′39″W, N58d59′13″W, and S39dE, for example.

As with the linear units, each angular type has a different precision value related to the type of angle selected. Selecting a precision determines how the angle is rounded off in the coordinates display in the status bar. Angles can also be measured in a clockwise direction instead of counterclockwise by placing a ✓ in the **Clockwise** check box.

 Field Notes

It is advisable always to measure your angles in a counterclockwise direction, because most AutoCAD users are trained to draw this way. Changing the direction to clockwise moves the 45° angle to the SE quadrant from the NE quadrant.

FIGURE 2.6 Surveyor units are measured in degrees in the appropriate compass quadrant.

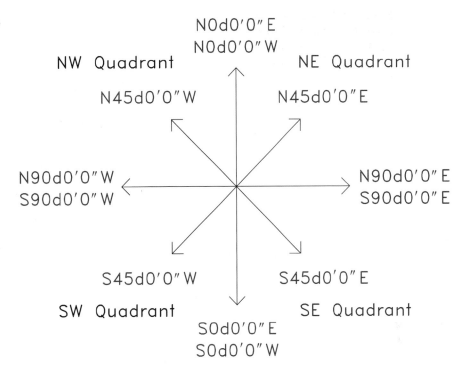

The **Area** section in the Units tab controls how areas are measured within Architectural Desktop. The Type: drop-down list reflects a squared value based on the drawing units selected. When inches or feet are used, the available area types are **Square Inches, Square Feet**, and **Square Yards**. If a metric drawing unit is selected, the available area types include **Square Millimeters, Square Centimeters, Square Decimeters,** and **Square Meters**. The Precision: drop-down list specifies how areas are rounded off when calculated. The Suffix text box also changes based on the Drawing Units selected. This allows you to enter a suffix that is displayed after an area is calculated. A *suffix* is a special note or application that is placed behind the value.

The **Volume** area of the Units tab is similar to the Area section, except that the former controls how volumes of objects are calculated. As with the Area section, the volumes change based on the type of drawing units selected. After selecting the settings, press the Apply button to apply the settings to your drawing.

The **Save As Default** check box is used to save your settings as the default for new drawings you begin by starting from scratch or with templates that do not have setup information configured.

Exercise 2.2

1. Open EX2-1.dwg.
2. Access the Drawing Setup dialog box.
3. Select Millimeters from the Drawing Units drop-down list. What happens?
4. Select Inches from the Drawing Units drop-down list. What happens?
5. Select the Engineering Linear type, and review the precision values available.
6. Select the Architectural Linear type, and review the precision values available. Set the precision to 1/32″.
7. Select the Surveyors Angular type, and review the precision values available.
8. Select the Degrees Angular type, and review the precision values available.
9. Apply the changes to the drawing.
10. Keep the file open or exit the dialog box by pressing the OK button and saving the drawing as EX2-2.

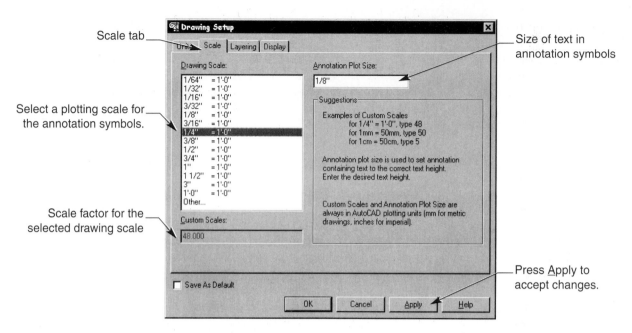

FIGURE 2.7   The Scale tab is used to determine a plotting scale
for the annotation blocks in Architectural Desktop.

Using the Scale Tab

The **Scale** tab of the Drawing Setup dialog box is used to set a plotting scale for annotation blocks that are
included with Architectural Desktop (see Figure 2.7). **Annotation blocks** are symbols that are drawn at full plotted scale. An example is a detail bubble that has a diameter of 1-1/4″. When the detail bubble is inserted into
model space, it is inserted by a scale factor. When a drawing scale is selected from the **Drawing Scale:** list, the
scale factor is entered in the **Custom Scales:** text box. For example, a drawing scale of 1/8″ = 1′-0″ has a scale
factor of 96 (1′-0″ ÷ 1/8″ = 96). A drawing scale of 1/4″ = 1′-0″ has a scale factor of 48 (1′-0″ ÷ 1/4″ = 48).

By this process, the detail bubble is scaled by the scale factor, (1-1/4″ bubble × 48 scale factor = 2′-6″
scaled bubble). This method allows a standard-sized detail bubble to be used in multiple drawings that have different scales. When the drawing is subsequently plotted, the full-scale drawing and the scaled annotation blocks
are scaled down by the scale factor, fitting them on a real size sheet of paper on a layout tab (see Figure 2.8).

FIGURE 2.8   The annotation
block is scaled up in the Model
tab and scaled down to the
proper scale in the paper space
layout.

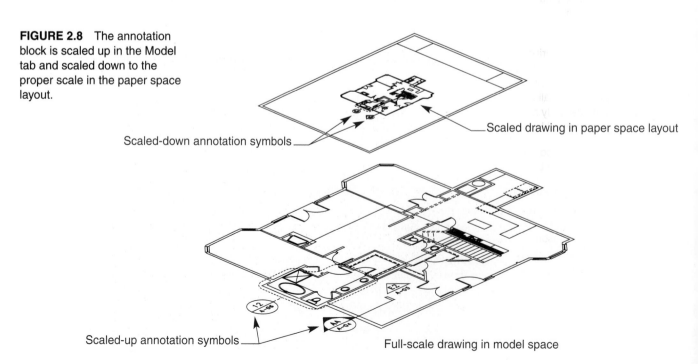

Depending on the type of drawing units selected in the Units tab, a number of scales are available in the <u>D</u>rawing Scale list. For example, if metric units are selected, the scales listed are common metric scales. If imperial units are selected, the scales listed are common imperial scales. Both lists also provide you with an Other… option. This option allows you to enter a custom scale factor in the <u>C</u>ustom Scales: text box.

The **<u>A</u>nnotation Plot Size:** text box is used to enter a text height for annotation symbols. The default is 1/8″. This makes the text in Architectural Desktop annotation blocks 1/8″ high when plotting from a Layout tab. Any value you enter here will modify the height of text in an annotation block when plotted.

Exercise 2.3

1. Continue from the previous exercise, or open EX2-2.
2. Access the Scale tab in the Drawing Setup dialog box.
3. Select the 1/8″=1′-0″ scale. Note that the scale factor displayed in the <u>C</u>ustom Scales: text box is 96.
4. Select the 1/4″=1′-0″ scale. Note that the scale factor displayed in the <u>C</u>ustom Scales: text box is 48.
5. Enter 3/32″ in the <u>A</u>nnotation Plot Size: text box.
6. Keep the drawing open for the next exercise, or save the exercise as EX2-3.

Establishing Architectural Layers

The **Layering** tab of the Drawing Setup dialog box is used to control how Architectural Desktop assigns layers to AEC objects (see Figure 2.9) There are two parts to layering: **layering standards** and **layer keys**. *Layering standards* are a set of rules that determine how a layer name is created. For example, the **AIA Long Format** included with Architectural Desktop has rules that create a layer name in the format Discipline field-Major field-Minor field-Status field: for example, A-WALL-DIMS-NEWW. The *layer keys* tie the layer standards to AEC objects. For example, the WALL key is assigned to the Wall object. The WALL key uses the layer standard to determine that Wall objects are automatically drawn on a layer named A-Wall. Each AEC object is assigned a layer key that in turn places the object on a layer. Layering standards and layer keys are described in greater detail later in this text.

The Layering tab is used to assign a **layer key style** to AEC objects. The **Layer Standards/Key File to Auto-Import:** text box is used to specify a drawing that contains layer standards and layer key styles that are automatically loaded when a new drawing is started. Architectural Desktop is installed with a default

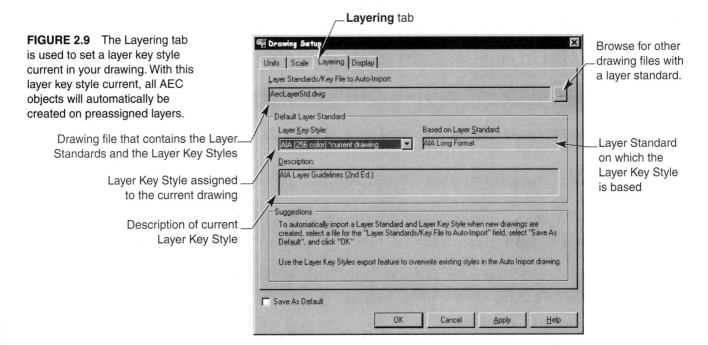

FIGURE 2.9 The Layering tab is used to set a layer key style current in your drawing. With this layer key style current, all AEC objects will automatically be created on preassigned layers.

Drawing file that contains the Layer Standards and the Layer Key Styles

Layer Key Style assigned to the current drawing

Description of current Layer Key Style

Layering tab

Browse for other drawing files with a layer standard.

Layer Standard on which the Layer Key Style is based

layer standard drawing file, named AecLayerStd.dwg. This layer standard includes the AIA layering standards, two types of British layering standards, and associated layer key styles. This file can be found in the \Autodesk Architectural Desktop 3\Content\Layers folder. The ellipsis (...) button can be used to browse other folders for customized layer standards and layer key styles.

 Field Notes

> If the International Extensions are installed, the AecLayerStd.dwg is supplemented with additional German layering standards and layer key styles.

The Layer <u>K</u>ey Style: drop-down list includes a number of layer key styles: AIA (256 color), BS1192 AUG Version 2 (256 color), BS1192 Descriptive (256 color), and Standard. The AIA (256 color) layer key style is based on the AIA Long Format layering standards. This key style assigns AIA layer names to AEC objects. For example, the WIND key assigns a window object to the A-Glaz layer. The two types of BS1192 layer key styles are British layering standards that assign layer names to AEC objects. The Standard key style assigns layer names based on the key name. For example, a layer key named WALL places a wall object on the Wall layer.

The **Based on Layer Standard:** text box displays the layer standard that the selected layer key style is using to create layer-naming conventions. The **<u>D</u>escription:** text box provides a description for the selected layer key style. Both of these text boxes are grayed out and provide additional information only as you select a layer key style to use. As with the previous two tabs, the Save As Default check box is used to assign the selected key style as the default for new drawings that do not have a key style already assigned.

Exercise 2.4

1. Continue from the previous exercise, or open EX2-3.
2. Access the Layering tab in the Drawing Setup dialog box.
3. Ensure that the Layer Standards/Key File to Auto-Import: text box contains the AecLayerStd.dwg file. If the text box has another file it is referencing, use the ellipsis (...) button to browse to the \Autodesk Architectural Desktop 3\Content\Layers folder and select the AecLayerStd.dwg file.
4. Select the AIA (256 color) key style from the Layer Key Style: drop-down list.
5. Press the Apply button to apply your settings to your drawing file.
6. Keep the drawing file open for the next exercise, or save the drawing as EX2-4.

Using the Display Tab

The **Display** tab of the Drawing Setup dialog box is used to control how default AEC objects appear in a drawing when they are first created (see Figure 2.10). As discussed in the previous section, every AEC object is assigned to a layer. When an object is placed on a layer, the layer governs properties of the object such as color, linetype, and lineweight. With Architectural Desktop, each AEC object is then placed on an appropriate layer through the use of layer key styles. The Display tab allows you to further control how that AEC object displays when the object is viewed from different angles or from different display configurations.

A *display configuration* allows you to view your drawing in a number of different ways. When a template is used, the default display configuration is set to Work. The Work display configuration displays a wall as a 2D flat wall when viewed from the top view, with the colors and linetypes of the layer where it is placed. When the same wall is viewed from an elevation or isometric direction, the wall is displayed as a 3D wall with colors that reflect the type of building material the wall is representing. The Work_Reflected display configuration displays walls with header lines over the doors and windows. A number of display configurations are included with the templates, from which a different configuration can be selected at any time as you are drawing. The Architectural Desktop Display System is covered in greater detail in Chapter 5.

The information included in the Display tab is the first step to controlling how AEC objects appear in your drawings. This first step is called ***Display Representations***. Each AEC object in Architectural Desktop

Display tab

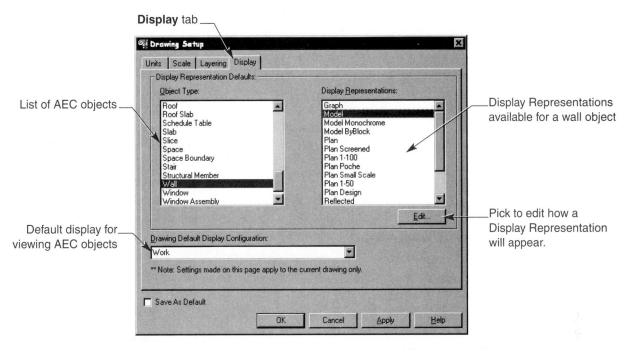

FIGURE 2.10 The Display tab is used to control additional properties of an AEC object, furthering the display control provided by layers.

has one or more display representations. Most of the AEC objects have a **Plan** (2D) representation and a **Model** (3D) representation. Many of the symbol blocks have a **General** representation.

The **Object Type:** list along the left edge of the tab is a list of all the AEC objects included with Architectural Desktop. The scroll bar allows you to scroll through the list of objects in order to view their representations. When an object type is highlighted on the left, the object's representations are listed in the Display Representations: list on the right side of the tab, as shown in Figure 2.11. A door object has multiple display representations. For example, a Plan representation door is displayed when the top of the door is looked down on from a top view when using the Work display configuration. The Model representation door is displayed when the door is looked at from an angle other than the top view when using the Work display configuration.

FIGURE 2.11 Each AEC object may have one or more representations.

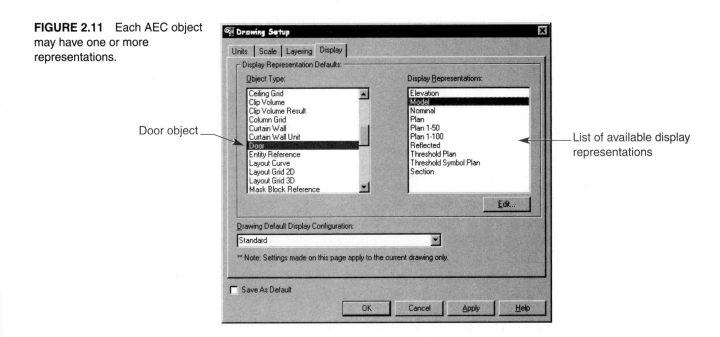

Field Notes

If the International Extensions are installed, objects will have more display representations than a standard installation. Custom representations can also be created, increasing the number of default display representations.

Using the door object as an example, assume that you want to change the way a Model (3D) door is supposed to display in your drawing. First, select the door object from the Object Type: list, then select the Model display representation from the list at the right (see Figure 2.11). Next select the Edit… button to edit the way the Model door is to appear in the drawing. The **Entity Properties** dialog box shown in Figure 2.12 appears. Model view doors include five individual components: Door Panel, Frame, Stop, Swing, and Glass. Each of these components is a separate piece of the door that can be turned on or off, the color or linetype can be changed, and other properties can be modified. Although these individual properties can be changed, they are all tied together into one object, similar to a block in AutoCAD.

FIGURE 2.12   The Entity Properties dialog box lists the components of an object that are displayed based on the display representation selected.

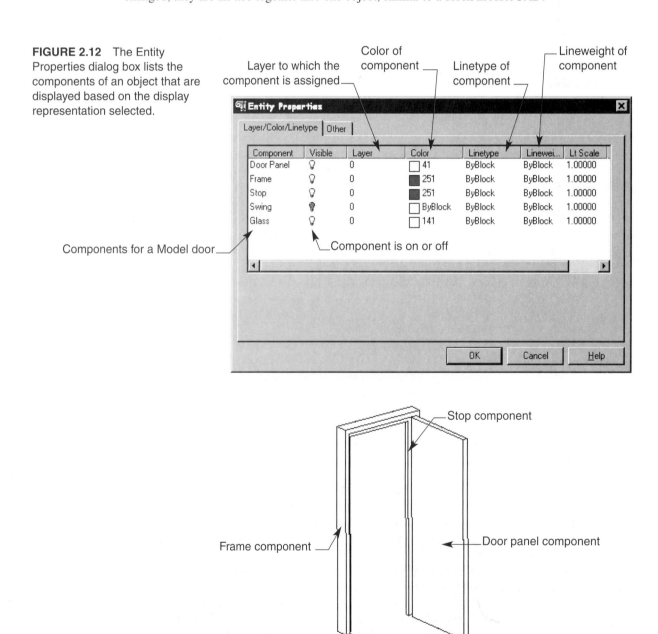

Components for a Model door

Component is on or off

Layer to which the component is assigned

Color of component

Linetype of component

Lineweight of component

Component	Visible	Layer	Color	Linetype	Linewei...	Lt Scale
Door Panel	♀	0	41	ByBlock	ByBlock	1.00000
Frame	♀	0	251	ByBlock	ByBlock	1.00000
Stop	♀	0	251	ByBlock	ByBlock	1.00000
Swing	♀	0	ByBlock	ByBlock	ByBlock	1.00000
Glass	♀	0	141	ByBlock	ByBlock	1.00000

Stop component

Frame component

Door panel component

Swing component

Object components can be turned on or off, which causes them to display or not to display. In Figure 2.12, for example, the swing component for a door will be turned off. This indicates that if the model door representation is viewed, the door swing component is turned off. Yet, if you look at the plan door representation, the door swing is turned on. When a component's layer is set to 0, that component is assigned to the layer where the overall object was created. Layers can be assigned to components, but the component belongs to both the assigned layer and the layer assigned to the overall object. Colors, linetypes, lineweights, and linetype scales can be assigned to the individual components by picking the appropriate icon or word in the Entity Properties dialog box. When you are finished assigning properties to components, press the OK button, which returns you to the Display tab, where you can configure other default display representations.

Field Notes

When first learning Architectural Desktop, you may decide to use the default settings for display representations of AEC objects. As you begin to become more familiar with how the display system works, you may begin to establish color, plotting, and rendering standards for the AEC objects.

Once you are finished setting up the AEC objects default display settings, you can select the default display configuration to be used in the drawing from the **Drawing Default Display Configuration:** drop-down list at the bottom of the Display tab. This sets the default configuration in the drawing and determines what object representations are viewed.

Field Notes

It is strongly recommended that you use a preconfigured template such as the AEC Arch (Imperial).dwt template when first beginning a new drawing. If you use a template instead of starting from scratch or using a wizard, the display representations and the default display configuration have already been established at the most desirable settings. If you start from scratch or use a wizard there are no display configurations available, and you have to either import configurations into your drawing or create new ones.

Exercise 2.5

1. Continue from the previous exercise, or open EX2-4.
2. Access the Display tab in the Drawing Setup dialog box.
3. Locate and pick the Door object. Pick the Model representation, then press the Edit… button. In the Entity Properties dialog box, ensure that the door swing component is turned off. Change the color of the Door Panel, Frame, and Stop to color 31. Change the Glass component to color 151. Press OK when finished.
4. Locate and pick the Window Object. Pick the Model representation, then press the Edit… button. In the Entity Properties dialog box change the color of the Frame and Sash to color 31. Change the Glass component to color 151. Press OK when finished.
5. Pick the Plan representation for the windows and press the Edit… button. Turn off the glass component. Press OK when finished.
6. Select the Work display configuration from the drop-down list at the bottom of the Display tab.
7. Apply the changes and exit the Drawing Setup dialog box.
8. Use SAVEAS to save the drawing as EX2-5. You can also save the drawing as your own template by using the SAVEAS command and selecting the AutoCAD Drawing File (*.dwt) extension in the Files of type: drop-down list at the bottom of the Save Drawing As dialog box.

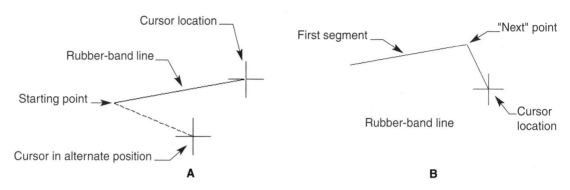

FIGURE 2.13   The PLINE command is used to create line segments. (A) Pick a starting point with the pick button. (B) Pick a second point to draw a line between the two points.

Drawing Basics

Earlier versions of AutoCAD required the use of simple shapes in order to produce a set of construction documents. Architectural plans were created with lines, arcs, and circles. Among these shapes an object known as a ***polyline*** was developed to help increase productivity. Architectural Desktop continues to use the polyline, and the polyline can be an important component in developing drawings with Architectural Desktop objects. Walls, roofs, floor slabs, and railings can be created from polylines. Custom shapes for doors, windows, and endcaps at the end of walls are generated from polylines. The polyline remains as one of the most important tools from the pre–Architectural Desktop era.

Field Notes

> Current AutoCAD users should note that the Architectural Desktop shapes have been developed to enhance current drawing philosophy. You can mix lines, arcs, circles, and polylines with AEC objects to create your construction documents. The polyline enhances the use of many AEC objects.

If you are a new or an experienced user of AutoCAD, this section reviews some basic drafting techniques used in AutoCAD and Architectural Desktop.

   The word *polyline* indicates that multiple (*poly*) lines (*line*) can be drawn. The AutoCAD polyline thus creates multiple line segments that are joined together at their ends to create one total entity. Use one of the following options to access the Polyline command:

✓ Picking the Polyline button in the Draw toolbar.

✓ Selecting Polyline from the Draw pull-down menu.

✓ Type **PLINE** or **PL** at the Command: prompt.

AutoCAD then prompts you to specify a starting point. This is where the polyline must begin. Moving the crosshairs around changes the coordinate display in the status bar, which indicates the position of your crosshairs in the drawing. To begin drawing the polyline, move the mouse to a location in the drawing and pick with the left mouse button to establish the first point of the polyline (see Figure 2.13A). Notice that a "rubber-band" line is started from the first point and is attached to your crosshairs. As you move the crosshairs around the drawing screen, the rubber-band line changes size as it follows the crosshair movement. Also notice that the Command: prompt now asks you to specify the next point. Pick a point in the drawing to draw a line segment (see Figure 2.13B). Picking more points with the pick button creates additional line segments. When you are finished drawing line segments, you can right-click to get a cursor menu from which you can use the Enter or Cancel commands to end the PLINE command. The [Enter], [Esc], or space-bar keys also end the PLINE command.

The following command sequence was used to create the shape in Figure 2.14:

FIGURE 2.14 Two line
segments drawn with the PLINE
command. Both lines are joined
forming one entity.

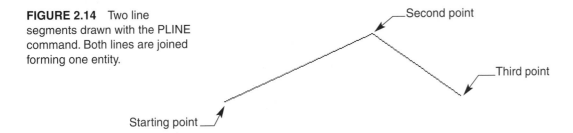

Command: **PLINE** or **PL** ↵
Specify start point: *(pick a point)*
Current line-width is 0'-0"
Specify next point or [Arc/Halfwidth/Length/Undo/Width]: *(pick a point)*
Specify next point or [Arc/Close/Halfwidth/Length/Undo/Width]: *(pick a point)*
Specify next point or [Arc/Close/Halfwidth/Length/Undo/Width]: ↵
Command:

After you have picked the initial point, you are prompted to pick a next point or [Arc/Close/Halfwidth/Length/Undo/Width]. Words that are displayed at the Command: prompt within brackets ([ ]) are called *options*. You can select an option at any time during the drawing process when one is provided. For example, to close a polyline with at least two line segments, type the uppercase letter of the Close option at the Specify next point or [Arc/Halfwidth/Length/Undo/Width]: prompt. This adds a line segment between the starting point and ending point of the last line segment and ends the polyline command. The command sequence for using the Close option is as follows:

Command: **PL** or **PLINE** ↵
Specify start point: *(pick a point)*
Current line-width is 0'-0"
Specify next point or [Arc/Halfwidth/Length/Undo/Width]: *(pick a point)*
Specify next point or [Arc/Close/Halfwidth/Length/Undo/Width]: *(pick a point)*
Specify next point or [Arc/Close/Halfwidth/Length/Undo/Width]: **C** ↵
Command:

Right-clicking also provides you with a cursor menu from which you can select the Close option. The other Polyline options are discussed in detail later in this section. For now, the starting points, next points, and Close option are used to describe how to draw accurately.

Exercise 2.6

1. Open EX2-5, or begin a new drawing from a template.
2. Use the PLINE command to draw a few line segments.
3. Press the [Enter] key to end the command
4. Use the PLINE command to draw more line segments.
5. Right-click and select Enter from the cursor menu.
6. Save the drawing as EX2-6.dwg.

Using the Point-Entry Methods

Point entry refers to the method of picking or obtaining locations for entities or objects such as polylines and walls. Architectural Desktop is built on top of AutoCAD, so the point-entry techniques can be directly applied to AEC objects. Becoming familiar with these techniques is very important, as they provide the basis for creating and modifying geometry. Points in Architectural Desktop/AutoCAD are drawn within a system known as the *Cartesian coordinate system*.

The Cartesian coordinate system is based on three intersecting axes: X = horizontal axis, Y = vertical axis, and Z = height axis. These three axes intersect one another at a point known as the *origin* (see Figure

FIGURE 2.15 The Cartesian coordinate system is based on three axes representing horizontal, vertical, and height point locations (X,Y,Z).

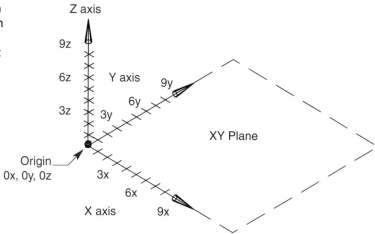

2.15). In standard 2D drafting practices such as floor plans, objects are drawn in the XY plane. The Z axis gives objects drawn in the XY plane a height.

When drawing in a top or plan view (the default when a new drawing is started), you are looking down the Z axis perpendicular to the XY drawing plane. This view provides a flat drawing plane in which geometry can be created. Points in a drawing are located in relation to an origin, where X=0, Y=0, and Z=0 (0,0,0) (see Figure 2.15). When you are drawing on the XY plane, the origin (0x,0y) divides the coordinate system into four quadrants (see Figure 2.16). When a new drawing is started in Architectural Desktop, the origin (0x,0y) is located in the lower left corner of the drawing. This places point locations in the upper right quadrant, where the X and Y values are positive. There are three methods of selecting points using the Cartesian coordinate system: absolute coordinates, relative coordinates, and polar coordinates.

Using Absolute Coordinates

You locate points in your drawing with the **_absolute coordinates_** method by measuring distances from the origin (0,0) in the XY plane. For example, assume that a polyline is to be started 20'-0" to right of the 0 X axis, and 15'-0" up from the 0 Y axis. The absolute coordinate you enter at the Specify start point: prompt is 20',15' (see Figure 2.17). This begins the polyline 20'-0" to the right and 15'-0" up from the origin. If the next polyline point is to be 18'-0" long to the right, the absolute coordinate is 38',15' (X=20'+18',Y=15') for this point. When Cartesian coordinate methods are used, the axis values are always entered with the X value first, followed by the Y value, with a comma separating the two values (X,Y), and measurements are always made from the origin.

FIGURE 2.16 The point where the 0x axis, 0y axis, and the 0z axis cross one another is known as the origin. It is used as a reference for establishing new points in the XY plane.

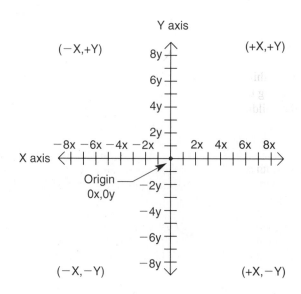

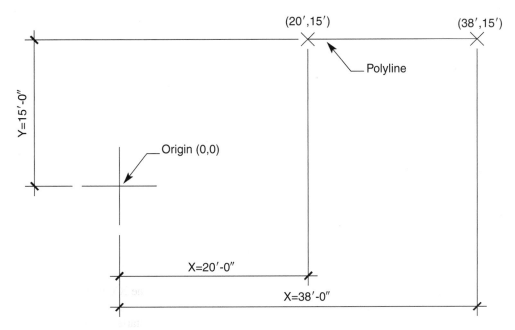

FIGURE 2.17 Absolute coordinate points are always measured from the origin. An X,Y value is required.

FIGURE 2.18 The building outline created using absolute coordinates.

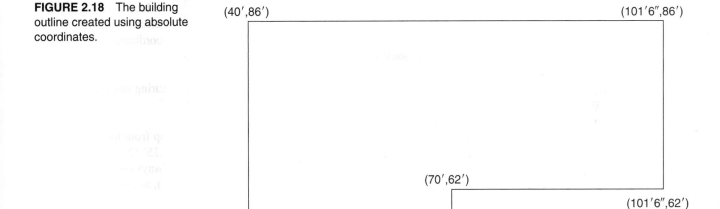

The important thing to remember when using absolute coordinates is that any point locations are always measured along the X and Y axis from the origin. Refer to the following command and point placements to create the building outline shown in Figure 2.18:

```
Command: PL or PLINE ↵
Specify start point: 40',50' ↵
Current line-width is 0'-0"
Specify next point or [Arc/Halfwidth/Length/Undo/Width]: 70',50'
Specify next point or [Arc/Close/Halfwidth/Length/Undo/Width]: 70',62' ↵
Specify next point or [Arc/Close/Halfwidth/Length/Undo/Width]: 101'6",62' ↵
Specify next point or [Arc/Close/Halfwidth/Length/Undo/Width]: 101'6",86' ↵
Specify next point or [Arc/Close/Halfwidth/Length/Undo/Width]: 40',86' ↵
Specify next point or [Arc/Close/Halfwidth/Length/Undo/Width]: C↵
Command:
```

Exercise 2.7

1. Begin a new drawing using a template.
2. Use the PLINE command to enter the following absolute coordinates to create a building outline.

Point	Coordinates	Point	Coordinates
1	90′,78′	8	110′,50′
2	100′,78′	9	108′,48′
3	100′,86′	10	100′,48′
4	133′6″,86′	11	98′,50′
5	133′6″,62′-6″	12	90′,50′
6	118′,62′-6″	13	Close
7	118′,50′		

3. Watch where the next vertex point goes when entering a new coordinate.
4. Save the drawing as EX2-7.dwg.

Using Relative Coordinates

The *relative coordinates* method is similar to the absolute coordinates method because points are measured along the X and Y axes. The difference is that relative coordinate points are measured from the previous location rather than from the origin. For example, when drawing the outline of a new structure with a PLINE, you start the polyline by picking a point in the drawing with the pick button. To draw the front of a building that is 23′-6″ long using relative coordinates, you measure the next point 23′-6″ along the X axis and 0″ along the Y axis from the previous point. You must use the @ symbol in front of the X,Y values to indicate that you are drawing X=23′-6″ and Y=0″ from the previous point: for example, @23′6,0, as shown in Figure 2.19. Select the @ symbol by holding the [Shift] key and pressing the [2] key on your keyboard.

In the Cartesian coordinate system points are either positive or negative along the different axes from the origin (see Figure 2.16). When relative coordinates are used, the previous point is defined as the origin for the new point. For example, a line that is drawn horizontally to the left has a negative value for the X axis: (@-X,0). A line drawn vertically up has a value of X=0 and a positive Y value: (@0,Y). A line drawn vertically down has a negative Y value: (@0,-Y). The following sequence is used to draw an octagon for the gazebo shown in Figure 2.20:

```
Command: PL or PLINE ↵
Specify start point: (pick a point)
Current line-width is 0′-0″
Specify next point or [Arc/Halfwidth/Length/Undo/Width]: @8′,0 ↵
Specify next point or [Arc/Close/Halfwidth/Length/Undo/Width]: @4′,4′ ↵
Specify next point or [Arc/Close/Halfwidth/Length/Undo/Width]: @0,8′ ↵
Specify next point or [Arc/Close/Halfwidth/Length/Undo/Width]: @-4′,4′ ↵
```

FIGURE 2.19 Relative coordinates measure X and Y distances from the previous point rather than from the origin. Use @X,Y values to enter the new points.

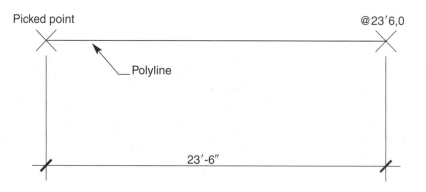

FIGURE 2.20 A PLINE is used to lay out a gazebo outline using relative coordinates.

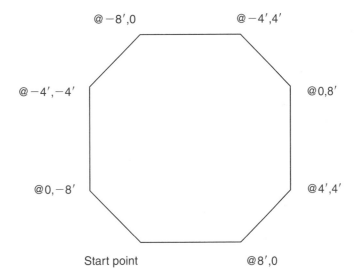

Specify next point or [Arc/Close/Halfwidth/Length/Undo/Width]: **@-8',0** ↲
Specify next point or [Arc/Close/Halfwidth/Length/Undo/Width]: **@-4',-4'** ↲
Specify next point or [Arc/Close/Halfwidth/Length/Undo/Width]: **@0,-8'** ↲
Specify next point or [Arc/Close/Halfwidth/Length/Undo/Width]: **C** ↲

Exercise 2.8

1. Begin a new drawing using a template.
2. Use the PLINE command to enter the following relative coordinates to create a building outline.

Point	Coordinates	Point	Coordinates
1	Pick a point	8	@2',2'
2	@24',0	9	@4',0
3	@0,16'	10	@0,24'
4	@3'6,3'6	11	@-38',0
5	@3'6,0	12	@0,-12'
6	@2',-2'	13	@-7',0
7	@6',0	14	Close

3. Watch where the next vertex point goes when entering a new coordinate.
4. Save the drawing as EX2-8.dwg.

Using Polar Coordinates

The **polar coordinates** method is based on a distance and a compass angle for the direction the line is drawn. As with relative coordinates, the @ symbol is used in front of polar coordinate values to specify the next point in relation to the current point. In order to specify a point using polar coordinates, the distance of the line and the angle at which the line is drawn is required: (@distance<angle). The < symbol is used between the distance and the angle value. Consistent with AutoCAD conventions, angles are measured in a counterclockwise direction with the 0° angle to the right (east). The following sequence is used to create the diagram shown in Figure 2.21:

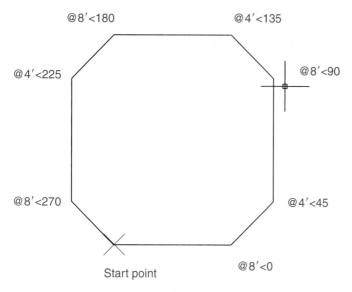

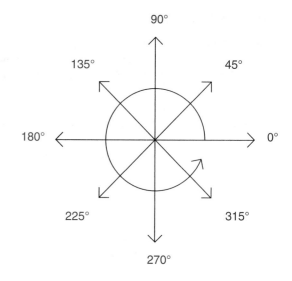

FIGURE 2.21   Polar coordinates utilize a distance and an angle to specify new point locations.

Command: **PL** or **PLINE_**
Specify start point: (*pick a point*)
Current line-width is 0'-0"
Specify next point or [Arc/Halfwidth/Length/Undo/Width]: **@8'<0** ↵
Specify next point or [Arc/Close/Halfwidth/Length/Undo/Width]: **@4'<45** ↵
Specify next point or [Arc/Close/Halfwidth/Length/Undo/Width]: **@8'<90** ↵
Specify next point or [Arc/Close/Halfwidth/Length/Undo/Width]: **@4'<135** ↵
Specify next point or [Arc/Close/Halfwidth/Length/Undo/Width]: **@8'<180** ↵
Specify next point or [Arc/Close/Halfwidth/Length/Undo/Width]: **@4'<225** ↵
Specify next point or [Arc/Close/Halfwidth/Length/Undo/Width]: **@8'<270** ↵
Specify next point or [Arc/Close/Halfwidth/Length/Undo/Width]: **C** ↵

Exercise 2.9

1. Begin a new drawing using a template.
2. Use the PLINE command to enter the following polar coordinates to create a building outline.

Point	Coordinates	Point	Coordinates
1	Pick a point	8	@24'<180
2	@24'<0	9	@16'<225
3	@16'<45	10	@8'<270
4	@8'<90	11	@8'<180
5	@8'<0	12	@16'<225
6	@16'<45	13	Close
7	@24'<90		

3. Watch where the next vertex point goes when entering a new coordinate.
4. Save the drawing as EX2-9.dwg.

Using the Ortho Mode

Ortho mode forces the polyline line segments to be drawn horizontally or vertically, which is advantageous when you are creating rectangular shapes, because the corners are always square. Ortho mode can be turned on or off by picking the **ORTHO** button in the status bar, by pressing the [F8] key, or by using the [Ctrl] +[L] key combination.

After Ortho mode has been turned on, any new straight polyline segments are limited to horizontal or vertical movement. If an angled line other than horizontal or vertical is desired, relative or polar coordinates can be used to specify the next point. Ortho mode is not limited to the PLINE command but can be used when creating walls or when using some editing commands.

Exercise 2.10

1. Begin a new drawing using a template.
2. Use the PLINE command to pick new vertex points for the polyline.
3. Try turning Ortho mode on and off as you draw new polyline segments.
4. Save the drawing as EX2-10.dwg.

Using Direct Distance Entry

The previous sections explained the use of the three Cartesian coordinate methods of specifying point locations for your lines. *Direct Distance Entry* is another method of entering point locations. Although not considered a Cartesian coordinate method, direct distance entry may be used to quickly lay out geometry in your drawing. Direct distance entry can be used after an initial point has been selected, and a rubber-band line is displayed. To use this method, drag the mouse (rubber-band line) in the direction you want the line to be drawn, then enter the distance (length) for the line at the Specify next point prompt. This process creates a line segment the correct length and at the correct angle that was specified, by dragging the mouse.

Direct distance entry works best when used in combination with the Ortho mode or Polar Tracking, which is explained in the next section. The following sequence is used to draw the building outline shown in Figure 2.22, using a combination of direct-distance entry and Ortho mode:

Command: **PL** or **PLINE** ↵
Specify start point: *(pick a point)*
Current line-width is 0′-0″

FIGURE 2.22 Using Direct Distance Entry and Ortho mode to quickly create an outline for a building.

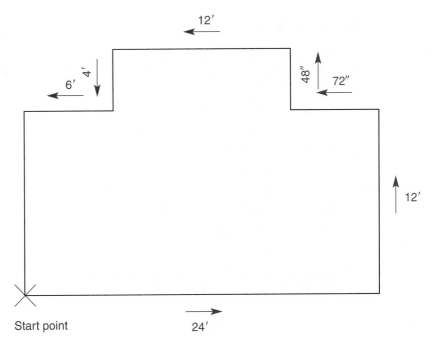

Start point   24′

Specify next point or [Arc/Halfwidth/Length/Undo/Width]: *(drag crosshairs to right)* **24′** ↵
Specify next point or [Arc/Close/Halfwidth/Length/Undo/Width]: *(drag crosshairs up)* **12′** ↵
Specify next point or [Arc/Close/Halfwidth/Length/Undo/Width]: *(drag crosshairs to left)* **72** ↵
Specify next point or [Arc/Close/Halfwidth/Length/Undo/Width]: *(drag crosshairs up)* **48** ↵
Specify next point or [Arc/Close/Halfwidth/Length/Undo/Width]: *(drag crosshairs to left)* **12′** ↵
Specify next point or [Arc/Close/Halfwidth/Length/Undo/Width]: *(drag crosshairs down)* **4′** ↵
Specify next point or [Arc/Close/Halfwidth/Length/Undo/Width]: *(drag crosshairs to left)* **6′** ↵
Specify next point or [Arc/Close/Halfwidth/Length/Undo/Width]: **C** ↵

 Field Notes

> Use direct-distance entry in combination with Ortho mode or polar tracking to quickly lay out geometry. This method can be used whenever Architectural Desktop expects a point coordinate value when you are using drawing or editing commands.

Using Polar Tracking

Polar Tracking is similar to drawing straight segments with the Ortho mode, except you are not limited to horizontal or vertical directions. The use of polar tracking causes the crosshairs to "snap" to a predefined angle increment. Polar tracking can be turned on by picking the **POLAR** button on the status bar or by pressing the [F10] key. Notice that either the Ortho mode or Polar Tracking can be used, but both cannot be used at the same time. If one of the buttons is picked, the other is turned off.

Polar tracking provides a dotted-line visual aid at a predefined angle when the crosshairs are moved into the angle. Once you have started a command and picked a point, you can move the crosshairs around to indicate a new angle for the line segment. If a dotted line appears, your line segment is drawn along the angle of this line (see Figure 2.23). The dotted line is known as the *alignment path*. The alignment path indicates the predefined tracking angle for drawing geometry. Initially, polar tracking is set to track along 0°, 90°, 180°, and 270°

Other alignment angles can be specified by choosing an increment angle in the Drafting Settings dialog box (Figure 2.24). Use one of the following options to access the dialog box:

✓ Right-click on the **POLAR** button on the status bar, and choose **Settings...** from the cursor menu.

✓ Pick **Drafting Settings** in the **Tools** pull-down menu.

✓ Type **DSETTINGS** at the Command: prompt, then pick the **Polar Tracking** tab.

New alignment paths can be specified by changing the angle in the Increment angle drop-down list. Every time the cursor is moved toward the specified increment angle, an alignment path appears. Figure 2.24 indicates that every 45° angle becomes aligned when the cursor is moved around the starting point. Additional angles can be specified by entering a new angle in the Additional angles list box. The use of direct distance entry in conjunction with polar tracking can greatly enhance drafting speed.

FIGURE 2.23 Polar tracking tracks along a predefined angle, so that new points may be entered along an alignment path or angle.

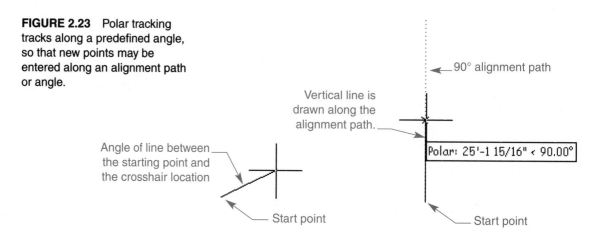

Polar Tracking tab

Turn polar tracking on/off.

Set an angle to align to incrementally.

Align only to this angle in addition to the increment angle.

Specify an additional alignment angle.

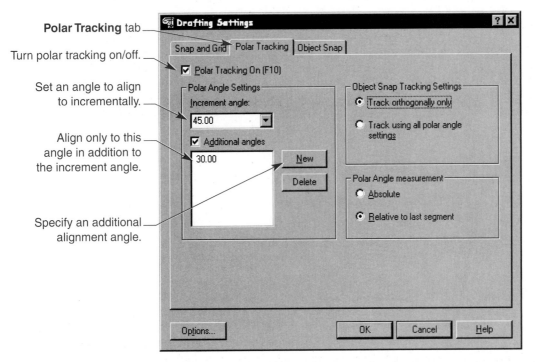

FIGURE 2.24 The Drafting Settings dialog box allows you to specify a new increment angle for the alignment paths when using polar tracking.

Exercise 2.11

1. Begin a new drawing using a template.
2. Set the increment angle for Polar Tracking to 45.
3. Draw the following diagram using a mixture of polar tracking and Ortho mode.
4. Save the drawing as EX2-11.dwg.

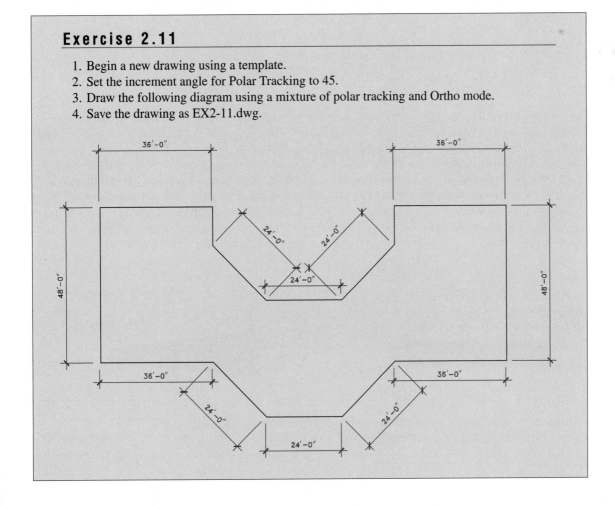

🖊️Field Notes

> Try using different point-entry techniques as you draw. Some methods may be better in certain situations than others. Keep in mind that you can use any of the techniques when creating geometry. For example, you may start a polyline with an absolute coordinate, then draw it with a mixture of relative and direct-distance entry methods.

Using Polyline Options

As discussed earlier in this chapter, polylines are important in Architectural Desktop because they can be used to generate other architectural objects such as walls, roofs, and spaces. The following discussion focuses on the options included with the Polyline command to enhance the use of polylines.

Creating Wide Polylines

When using the PLINE command, you are always first prompted to specify a starting point. After the starting point has been selected, a few options become available at the prompts. One of the choices available is the **Width** option, which allows you to draw wide polylines.

To enter an option, enter the uppercase letter(s) of the desired option at the Command: prompt. The Width option requires you to specify a starting and ending width. Specifying the same width for the starting and ending prompts creates a polyline equal to that width along its length. Specifying two different widths causes the polyline to taper. The following command sequence is used to draw the polyline shown in Figure 2.25A:

```
Command: PL or PLINE ↵
Specify start point: (pick a point)
Current line-width is 0.0000
Specify next point or [Arc/Halfwidth/Length/Undo/Width]: W ↵
Specify starting width <0.0000>: 1/2 ↵
Specify ending width <0.5000>: 1/2 ↵
Specify next point or [Arc/Halfwidth/Length/Undo/Width]: (pick an ending point)
Specify next point or [Arc/Close/Halfwidth/Length/Undo/Width]: (pick another point or [Enter] to
    exit the command)
```

Figure 2.25B displays a polyline with a different starting and ending width. Once the polyline width has been set, any new polylines drawn are created using the last ending width. If you need to draw polylines without a width, set the values for the starting and ending width to 0.

The **Halfwidth** option of the PLINE command allows you to specify the width of the polyline measured from the center of polyline to one side. As with the Width option, a starting half-width and ending half-width must be entered. Figure 2.26 displays a polyline created with the Halfwidth option. Notice that the same values are entered for both polylines but the half-width polyline is twice as wide as the polyline created with the Width option.

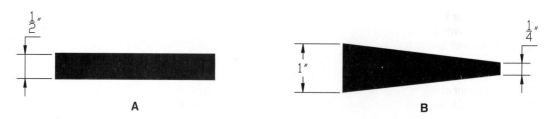

A   B

FIGURE 2.25   (A) A wide polyline with a starting and ending width of 1/2″. (B) A wide, tapered polyline with a starting width of 1″ and an ending width of 1/4″.

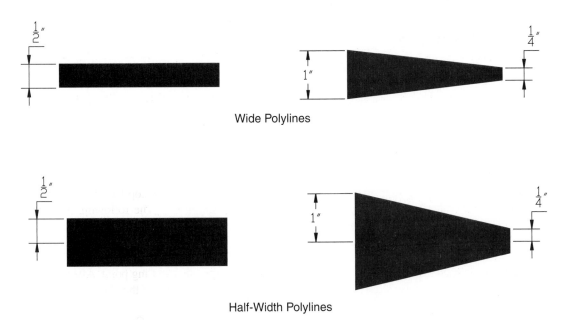

FIGURE 2.26   The Halfwidth option specifies what half of the total width of the polyline will be.

Using the Length Option

The polyline **Length** option creates a polyline with a specified length, using the same angle as the previously drawn polyline segment. After initially creating a polyline, reenter the PLINE command and pick a starting point. Enter the Length option and specify a length:

Command: **PL** or **PLINE** ↵
Specify start point: *(pick a point)*
Current line-width is 0′-0″
Specify next point or [Arc/Halfwidth/Length/Undo/Width]: L ↵
Specify length of line: **24′6** ↵
Specify next point or [Arc/Close/Halfwidth/Length/Undo/Width]: ↵

The new polyline is drawn at the same angle as the previous polyline, but with the length that you specified. If there are not any previously drawn polylines, then the polyline is drawn at the 0° angle.

Creating Polylines with Arcs

Up to this point the PLINE command has been used to create straight-line segments. The PLINE command also includes an option that allows you to create arced polylines. As with the other PLINE options, once a starting point has been selected, the **Arc** option becomes available. After the Arc option has been accessed, additional options specifically for polyline arcs become available. The following sequence draws the line-arc-arc-line polyline segment displayed in Figure 2.27:

Command: **PL** or **PLINE** ↵
Specify start point: *(pick a point)*
Current line-width is 0′-0″
Specify next point or [Arc/Halfwidth/Length/Undo/Width]: **@24′<0** ↵
Specify next point or [Arc/Close/Halfwidth/Length/Undo/Width]: **A** ↵
Specify endpoint of arc or
[Angle/CEnter/CLose/Direction/Halfwidth/Line/Radius/Second pt/Undo/Width]:
@12′,-12′ ↵
Specify endpoint of arc or
[Angle/CEnter/CLose/Direction/Halfwidth/Line/Radius/Second pt/Undo/Width]:

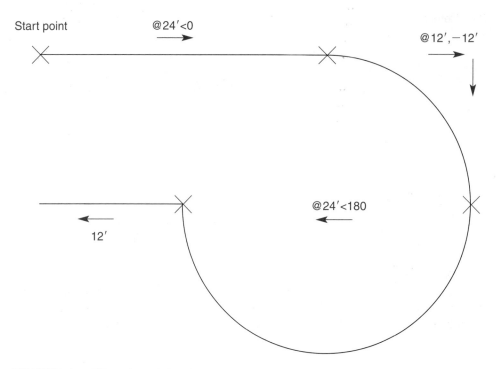

FIGURE 2.27   When the polyline Arc option is used, the arcs are connected to the straight-line segments like other straight-line polylines.

@24'<180 ↵
Specify endpoint of arc or
[Angle/CEnter/CLose/Direction/Halfwidth/Line/Radius/Second pt/Undo/Width]: **L** ↵
Specify next point or [Arc/Close/Halfwidth/Length/Undo/Width]: *(pull cursor in direction of travel)*
    12' ↵
Specify next point or [Arc/Close/Halfwidth/Length/Undo/Width]: ↵

Once the Arc option has been initiated, the standard polyline options are replaced with options that are related to arcs. The following is a brief description of each of the Arc options:

- **Angle**   This option specifies the included angle for the arc measured from the start point (see Figure 2.28A).

- **Center**   This option specifies the center point for the arc segment (see Figure 2.28B).

- **Close**   This option is similar to the standard Close option, except that the polyline closes using a polyline arc (see Figure 2.28C).

- **Direction**   This option pecifies the starting direction for the polyline arc. The second point indicates the tangent direction; the third point is the end of the arc (see Figure 2.28D).

- **Halfwidth**   This option specifies the half-width for the polyline arc.

- **Line**   This option returns the standard polyline options and resumes drawing straight-line segments (see Figure 2.28E).

- **Radius**   This option specifies the radius for the polyline arc (see Figure 2.28F).

- **Second pt**   This option requires that a second and third point be picked in order to draw the polyline arc (see Figure 2.28G).

- **Undo**   This option undoes the previous arc segment.

- **Width**   This option specifies the width for the polyline arc segment.

FIGURE 2.28   The different
polyline arc options.

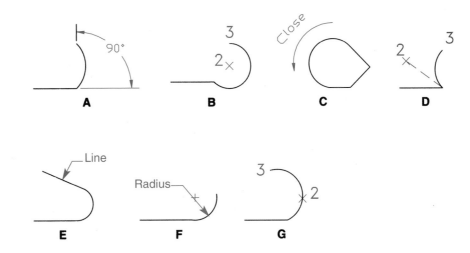

Undoing Previously Drawn Segments

When using the PLINE command, you can use the **Undo** option to remove a previously drawn line or arc segment. The Undo option allows you to "undo" your polyline segments as many times as there are arc or line segments in the polyline. Each time the Undo option is used, the last segment is erased.

Exercise 2.12

1. Begin a new drawing using a template.
2. Draw the following figure using one polyline.
3. Use a mixture of Cartesian coordinate methods and direct distance entry when drawing.
4. Save the drawing as EX2-12.dwg.

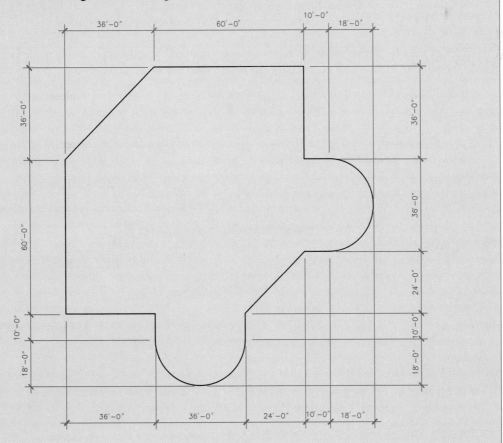

Drawing Accurately

Object snaps, also called *osnaps*, are one of the most useful tools in Architectural Desktop. ***Object snaps*** are drafting aids that "snap" the crosshairs to an exact point or place on an object as you draw. Osnaps increase accuracy, productivity, and drafting ability by allowing you to specify exact points when drafting.

    AutoSnap™ is a feature of AutoCAD that works with object snaps to display a marker at the object snap location when you move the cursor to indicate the desired object or snap point. T he AutoS nap feature is turned on when Architectural Desktop is installed. When AutoS nap OSNAP is active, yellow markers are displayed along existing geometry, pointing out specific points that can be snapped to as new objects are being drawn. These visual cues identify the current object snap mode that is being found.

    Two examples of the object snaps provided by AutoSnap are shown in Figure 2.29. In Figure 2.29A the endpoint of a polyline segment has been found, and if it is picked, the new polyline is "snapped" to the end of the existing polyline. Another object snap is the intersection osnap. Figure 2.29B shows the marker for an intersection of two polylines. If the intersection is picked, the new polyline is snapped to the position.

Using the Object Snaps Modes

Only a few object snap modes are available by default, but additional modes can be selected. Object snap modes can be accessed by one of the following methods:

- **Prompt line**   You can enter object snap modes at the Command: prompt when you are prompted for a point location. Enter the first three letters of the desired osnap to activate it.

- **Shortcut menu**   You can select object snap modes from a shortcut menu when you are prompted for a point location (see Figure 2.30). Activate the menu by holding down the [Shift] key and then right-clicking.

- **Toolbar**   You also can select object snap modes from one of the osnap buttons on the **Object Snap** toolbar shown in Figure 2.31. Access this toolbar by right-clicking on an existing AutoCAD toolbar button, then select Object Snap.

    Each of the object snap modes includes a marker that indicates the type of osnap that is being referenced. The table in Figure 2.32 summarizes the object snap modes in Architectural Desktop. Some of the

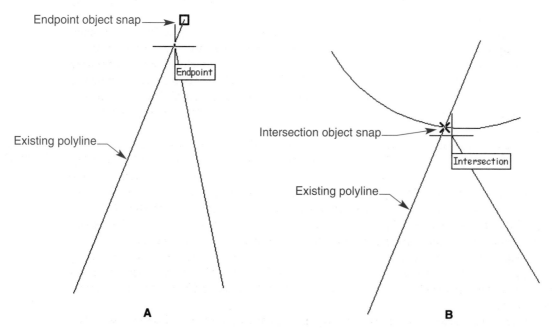

FIGURE 2.29   (A) The endpoint osnap is used to snap geometry to the end of an object.
(B) The intersection osnap is used to snap geometry to the intersection of two objects.

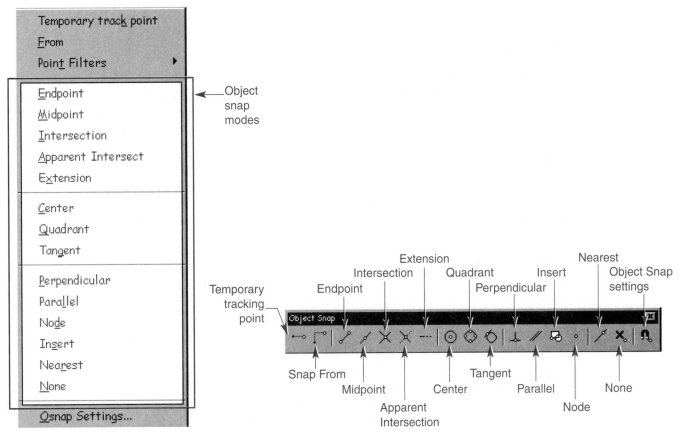

FIGURE 2.30 The osnap shortcut menu can be accessed by pressing the [Shift] key and right-clicking.

FIGURE 2.31 Object snap modes can be accessed from the Object Snap toolbar.

object snap modes do not work on all objects. Practice drawing polylines and using different osnaps to snap to specific locations on existing geometry.

As previously indicated, some object snaps are available by default for use as you draw new geometry. When new geometry is created, and the crosshairs cross over existing geometry, Architectural Desktop finds and displays the osnap points for the object snaps that are turned on. A function known as ***running osnap*** indicates that object snaps are on and automatically displays symbols as the crosshairs are moved across existing geometry. This function can be turned on and off by pressing the OSNAP button in the status bar. When running osnaps are turned off, Architectural Desktop does not look for specific points to snap your new geometry.

To view the available running osnaps right-click over the OSNAP button, and select <u>S</u>ettings… from the cursor menu. The Object Snap tab is displayed in the Drafting Settings dialog box (see Figure 2.33). Checking an osnap mode turns on that object snap, so that as new geometry is drawn Architectural Desktop also considers searching for the new point. You can use different combinations of osnaps as you draw. For example, the most common osnap is Endpoint. This osnap can be combined with the Midpoint, Perpendicular, and Extension osnaps when linear geometry is being referenced when drafting. The Center, Quadrant, and Endpoint osnaps might be used in combination when referencing curved geometry.

If running osnaps are on, you are not limited to using these osnaps. You can override the running osnap for the next point by selecting the desired nonrunning osnap from the Object Snap toolbar, by entering the desired osnap at the Command: prompt, or by holding the [Shift] key and right-clicking to access the Osnap shortcut menu.

Each of the object snap modes is described next. Try each mode as you draw to determine how they work. You may find some that are more useful than others. Remember that osnaps are used to help create accurate geometry by snapping new geometry to exact locations on existing objects.

Object Snap Modes

Mode	Marker	Button	Description
Endpoint	 □	 /	Finds the nearest endpoint of a line, arc, elliptical arc, spline, ray, solid, wall, door, window, stair, railing, or roof
Midpoint	 △	 /	Finds the middle point of any object having two endpoints, such as a line, arc, elliptical arc, spline, ray, solid, wall, door, or window
Center	 ○	 ⊙	Locates the center point of a curved object such as a circle, arc, polyline arc, ellipse, elliptical arc, or curved wall
Quadrant	 ◇	 ◈	Picks the closest of the four quadrant points that can be found on circles, arcs, ellipses, elliptical arcs, and curved walls (Not all these objects may have all four quadrants.)
Intersection	 ✕	 ✕	Picks the intersection of two objects
Apparent Intersection	 ⊠	 ✕	Selects a visual intersection between two objects that appear to intersect on screen in the current view but may not actually intersect each other in 3D space
Extension	 ▪▫▫	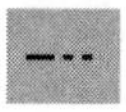 ▫▫▫	Finds a point along the imaginary extension of an existing line or arc (This mode is used in conjunction with another object snap such as Endpoint or Midpoint.)
Insertion	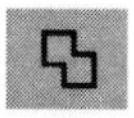 ⌐	 ⌐	Finds the insertion point of text, blocks, doors, windows, openings, and window assemblies
Perpendicular	 ⌐	 ⊥	Finds a point that is perpendicular to an object from the previously picked point
Parallel	 //	 //	Used to find any point along an imaginary line parallel to an existing line or polyline
Tangent	 ⊙	 ⊙	Finds points of tangency between radial and linear objects
Nearest	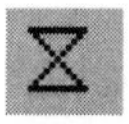 ⊠	 /	Locates a point on an object closest to the crosshairs
Node	 ⊗	 ∘	Picks a point object on a dimension, point object, or most of the Architectural Desktop objects
None		 ✕∘	Turns the running object snap off for the current point to be picked

FIGURE 2.32   This table displays the marker and the toolbar button as well as a description for each osnap mode.

Using Endpoint Object Snap

In many cases polyline, wall, floor, or roof points need to be connected to the endpoints of existing objects. In this situation use the **Endpoint** object snap by moving the crosshairs close to the end of the existing line or arced segment to be picked. A small square symbol marks the endpoint that will be snapped to when you press the pick button.

The following command sequence is used to connect a polyline to the end of the existing polyline arc shown in Figure 2.34:

    Command: **PL** or **PLINE** ↵
    Specify start point: (pick a starting point)
    Current line-width is 0'-0"

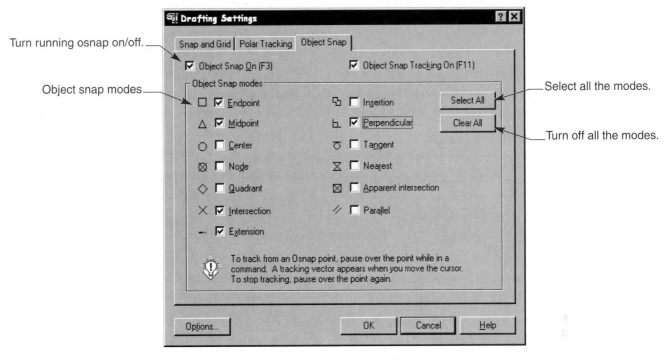

FIGURE 2.33   The Object Snap tab in the Drafting Settings dialog box specifies which osnaps are to be used as running (always on) osnaps.

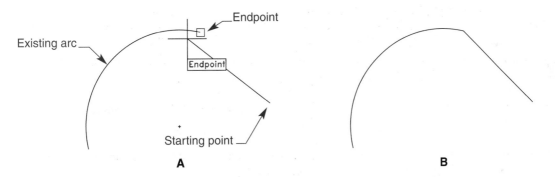

FIGURE 2.34   Snapping a new polyline to the endpoint of a polyline arc.

Specify next point or [Arc/Halfwidth/Length/Undo/Width]: *(pick the Endpoint button, type **END**, or pick **Endpoint** from the **Object Snap** cursor menu, move the crosshairs near the end of the existing object until the marker is displayed, and pick)*
Specify next point or [Arc/Close/Halfwidth/Length/Undo/Width]: ↵

Endpoint osnap is used to quickly snap to the endpoint of many types of objects. It is often chosen as a running osnap.

Using Midpoint Object Snap
Midpoint object snap is used to snap to the middle of a line or arc segment or to the middle of a wall, door, window, or other AEC object in much the same way that Endpoint osnap is used.

The following sequence is used to snap a polyline segment from the middle of a line to the middle of an arc as shown in Figure 2.35:

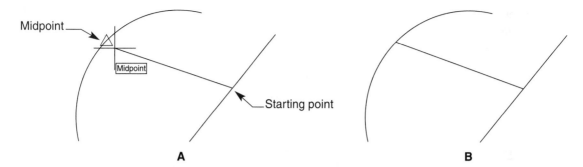

FIGURE 2.35   Midpoint osnap snaps new points to the middle of existing objects.

Command: **PL** or **PLINE** ⏎
Specify start point: *(pick the **Midpoint** button, type **MID**, or pick **Midpoint** from the Object Snap cursor menu, move the crosshairs toward the middle of the line segment, and pick)*
Current line-width is 0'-0"
Specify next point or [Arc/Halfwidth/Length/Undo/Width]: *(pick the **Midpoint** button, type **MID**, or pick **Midpoint** from the **Object Snap** cursor menu, move the crosshairs toward the middle of the arc segment, and pick)*
Specify next point or [Arc/Close/Halfwidth/Length/Undo/Width]: ⏎

Exercise 2.13

1. Open EX2-12.dwg.
2. Use Endpoint osnap to draw the horizontal and vertical lines shown in the following diagram.
3. Use Midpoint osnap to draw the angled line shown in the following diagram.
4. Save the drawing as EX2-13.dwg.

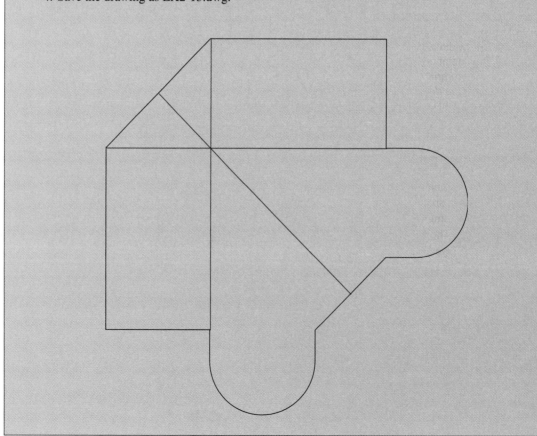

FIGURE 2.36   Center osnap is used to snap to the center of curved objects.

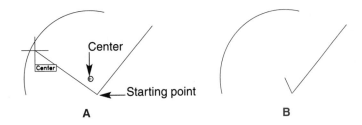

Using Center Object Snap

Center osnap is used to snap new geometry to the center of a circle, arc, ellipse, elliptical arc, curved wall, or spiral staircase. After selecting Center osnap, move the crosshairs onto the existing curved object. Moving the crosshairs to the center of the curved object may not always access the centerpoint for the curved object, which is why it is important to move the crosshairs on top of the curved object that has the centerpoint you are trying to access. A small circle marker is displayed when the centerpoint is found.

The following sequence is used to draw the diagram in Figure 2.36:

Command: **PL** or **PLINE** ↵
Specify start point: *(pick the endpoint of the polyline)*
Current line-width is 0'-0"
Specify next point or [Arc/Halfwidth/Length/Undo/Width]: *(pick the Center button, type **CEN**, or pick **Center** from the **Object Snap** cursor menu, move the crosshairs onto of the arc segment, and pick when the center marker appears)*
Specify next point or [Arc/Close/Halfwidth/Length/Undo/Width]: ↵

Using Quadrant Object Snap

Quadrant osnap is used to snap new geometry to one of the quadrants of a circle, arc, ellipse, elliptical arc, curved wall, or spiral staircase. Quadrants are located at 0°, 90°, 180°, and 270°. To select a quadrant, move the crosshairs toward the desired quadrant and pick when the marker is displayed.

The following sequence is used to snap a polyline to a quadrant as shown in Figure 2.37:

Command: **PL** or **PLINE** ↵
Specify start point: *(pick the endpoint of the polyline)*
Current line-width is 0'-0"
Specify next point or [Arc/Halfwidth/Length/Undo/Width]: *(pick the Quadrant button, type QUA, or pick Quadrant from the Object Snap cursor menu, move the crosshairs onto of the arc segment near the desired quadrant, and pick when the quadrant marker appears)*
Specify next point or [Arc/Close/Halfwidth/Length/Undo/Width]: ↵

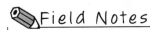Field Notes

If you use the Center and the Quadrant object snaps together as running osnaps, the new geometry will always snap to a quadrant. If you press the [Tab] key while the Quadrant marker is displayed, Architectural Desktop cycles through the other available osnaps in the general area of the crosshairs. The Center marker is displayed if you cycle the markers. Pick when the Center marker is displayed.

FIGURE 2.37   Quadrant osnap is used to snap to one of four quadrants on a curved object.

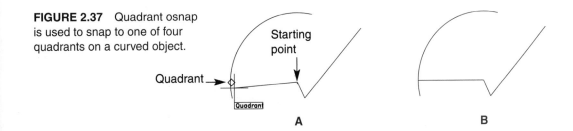

Exercise 2.14

1. Open EX2-13.dwg.
2. Use Center osnap to draw the angled polyline from the center of one arc to the center of the other arc as shown in the following diagram.
3. Use Quadrant osnap to draw the horizontal and vertical lines from a quadrant to the center of the arcs.
4. Save the drawing as EX2-14.dwg.

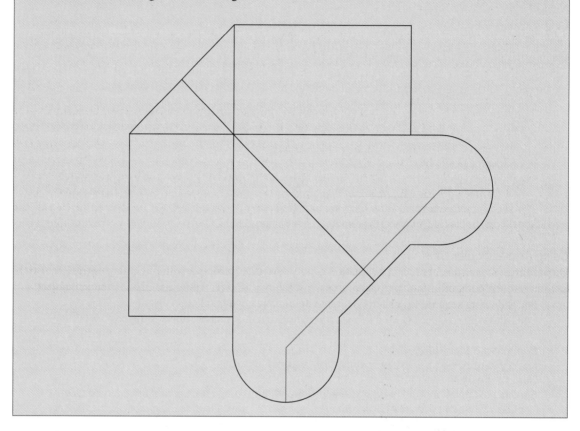

Using the Intersection Object Snap

Intersection osnap is used to snap to the intersection of two or more objects. The marker for an Intersection osnap is a small **X.** This marker appears when the crosshairs are near the intersection of objects. If the crosshairs are not near an intersection but are close to an object, the Intersection marker is followed by an ellipsis (**X...**), which indicates that Architectural Desktop is looking for the intersection of this object and another object. If you pick the crosshairs when the **X...** marker is displayed, you must pick another object that intersects with this object. This allows you to snap the new geometry to the intersection of both objects.

The following sequence is used to draw the polyline segment in Figure 2.38:

Command: **PL** or **PLINE** ↵
Specify start point: *(pick the center of the arc)*
Current line-width is 0'-0"
Specify next point or [Arc/Halfwidth/Length/Undo/Width]: *(pick the Intersection button, type **INT**,*
    *or pick **Intersection** from the **Object Snap** cursor menu, move the crosshairs to the intersec-*
    tion of the arc and the line, and pick when the intersection marker appears)
Specify next point or [Arc/Close/Halfwidth/Length/Undo/Width]: ↵

Using Apparent Intersection Object Snap

The *apparent intersection* is a point where two objects created in two different planes appear to cross each other based on the currently displayed viewing angle. **Apparent Intersection osnap** is used to select the point

FIGURE 2.38 Intersection osnap is used to snap to the intersection of two or more objects.

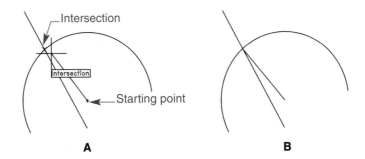

at which the two objects appear to cross each other. When you select Apparent Intersection, the first object you select establishes the plane where the new geometry will be drawn. Selecting the object that appears to cross the first selected object's path starts the new geometry at the apparent intersection of the two objects.

Using Extension Object Snap

Extension object snap is used to locate a point along an imaginary extension of an existing object. This osnap is activated by picking the Extension button, by entering EXT at the Command: prompt, or by selecting Extension from the Object Snap cursor menu. This osnap differs from the previously discussed osnaps in that it is not picked like the other osnaps. An initial point, known as the *acquired point*, is found by moving the crosshairs to the object that is to be used for the extension point. When the object is found, a + symbol marks the extension location. Moving the crosshairs causes a dashed path to appear along which the crosshairs can move. The actual snapped point is then selected along the ***extension path***. Picking a point along the path allows the new geometry to be snapped to that location from the extension of an existing object (see Figure 2.39).

Using Perpendicular Object Snap

Perpendicular osnap is used to draw a new object perpendicular to an existing object. Perpendicularity is measured to the point of intersection, so it is possible to snap an object perpendicular to a curved object.

The following sequence is used to draw the diagram in Figure 2.40:

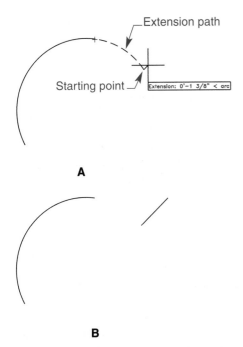

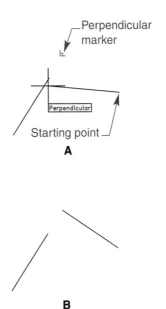

FIGURE 2.39 Extension osnap is used to find an imaginary path along which the crosshairs may move to establish a point for the new geometry.

FIGURE 2.40 Perpendicular osnap is used to snap to a point perpendicular to an existing object.

Command: **PL** or **PLINE** ↵
Specify start point: *(pick a point)*
Current line-width is 0'-0"
Specify next point or [Arc/Halfwidth/Length/Undo/Width]: *(pick the Perpendicular button, type*
   ***PER**, or pick **Perpendicular** from the **Object Snap** cursor menu, move the crosshairs on top*
   of the object, and pick when the perpendicular marker appears)
Specify next point or [Arc/Close/Halfwidth/Length/Undo/Width]: ↵

Exercise 2.15

1. Open EX2-14.dwg.
2. Erase the interior polylines drawn previously.
3. Use Endpoint, Midpoint, Intersection, Perpendicular, and Extension osnaps to finish the following diagram.
4. Save the drawing as EX2-15.dwg.

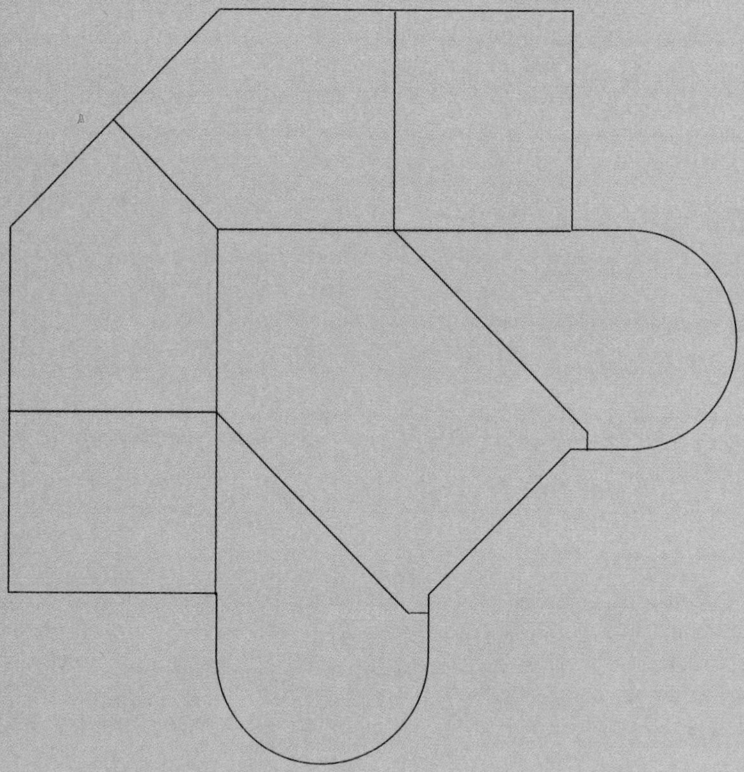

Using Tangent Object Snap

The word ***tangent*** refers to a linear or curved object that touches an arced entity at only one point. **Tangent osnap** aligns objects tangentially. Usually when you are aligning an object with the tangent location of a curved entity, the starting point of the object has already been specified. Tangent osnap is then used to select the entity that designates the tangency point. In situations where a linear object's starting point and ending point are to be tangent to two curved objects, Tangent osnap is used to establish the starting point's tangency location. In this case Tangent osnap is followed by an ellipsis (…), indicating that Architectural Desktop needs the ending point's location in order to locate the tangency point for the starting point. This type of tangent is called a *deferred tangent*. Once both points are known, the tangency points are calculated, and the linear object is drawn correctly.

The following sequence is used to drawn a polyline tangent to the polyline arc in Figure 2.41:

FIGURE 2.41 Tangent osnap is used to obtain the tangency point between a new object and an arced entity.

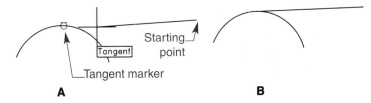

Command: **PL** or **PLINE** ↵
Specify start point: *(pick a point)*
Current line-width is 0′-0″
Specify next point or [Arc/Halfwidth/Length/Undo/Width]: *(pick the **Tangent** button, type **TAN**, or pick **Tangent** from the **Object Snap** cursor menu, move the crosshairs on top of the curved object, and pick when the tangent marker appears)*
Specify next point or [Arc/Close/Halfwidth/Length/Undo/Width]: ↵

Using Parallel Object Snap

Parallel osnap is used to locate a point along an imaginary path that is parallel to an existing linear object. Parallel osnap is similar to Extension osnap because it requires you to acquire the object that is to be referenced as the parallel edge. When Parallel osnap is used, the starting point for the linear object must first be specified. Once you have identified the start point, enter the Parallel osnap option, and move the crosshairs over the object that you are referencing as the parallel edge. When the Parallel marker is displayed, the line has been acquired. Move the crosshairs away from the referenced line. As you do so an imaginary parallel path similar to the extension path is displayed. The ending point for the line can now be selected along the parallel path.

The following sequence is used to make one polyline parallel to the existing polyline in Figure 2.42:

Command: **PL** or **PLINE** ↵
Specify start point: *(pick a point)*
Current line-width is 0′-0″
Specify next point or [Arc/Halfwidth/Length/Undo/Width]: *(pick the Parallel button, type PAR, or pick Parallel from the Object Snap cursor menu, move the crosshairs on top of the linear object. When the Parallel marker appears, move the crosshairs until the parallel path is displayed. Move the crosshairs along the path, then pick the desired ending point.)*
Specify next point or [Arc/Close/Halfwidth/Length/Undo/Width]: ↵

Using Node Object Snap

A *node* is a specific point on an object in Architectural Desktop. Similar to Endpoint osnap, **Node osnap** snaps to objects with defined nodes. AutoCAD objects such as points and dimensions have nodes that can be snapped to. Most of the Architectural Desktop objects have nodes along many of the common locations of the object. For example, a wall object has nodes at its starting-, ending-, and midpoints; a door or win-

FIGURE 2.42 Parallel osnap is used to draw new linear objects parallel to an existing linear object.

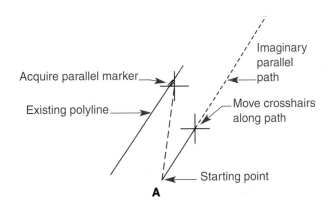

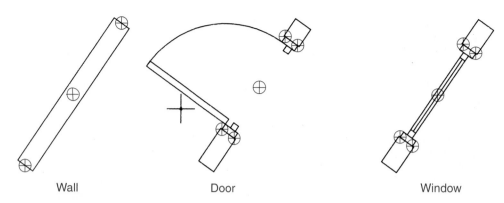

Wall   Door   Window

FIGURE 2.43   Node osnap is used to snap to a specific defined point on an object.

dow has node points along the frame and at the center of the object. Figure 2.43 displays the Node marker at the node locations for a wall, door, and window.

Using Nearest Object Snap

Nearest osnap is used to snap geometry to a point along another object that is closest to the crosshairs. This osnap is used the least, because it does not find an accurate location for snapping. This object snap should be used when geometry that is being drawn must be closed for a reason, such as to close an area so it can be hatched. This osnap should not be used as a running osnap, because it locates a point closest to the crosshairs and not an accurate point.

Tracking a New Location from Known Points

Tracking is a system that is used to locate new points from existing object points. Tracking can be used by acquiring a point, then moving the crosshairs along either the X or Y coordinate axis of the acquired point to pick a new point along this axis. This mode of tracking is known as *object snap tracking* or *Otrack*. Object snap tracking can be turned on or off by using the OTRACK button in the status bar or by pressing the [F11] key. Object snap tracking must always be used in conjunction with running osnaps, because the acquired points are referenced from an osnap.

To track along the X or Y axis of an endpoint, you must turn on Endpoint osnap in the Drafting Settings dialog box, then acquire the endpoint by holding the crosshairs over the desired "tracking" endpoint and wait for a + to display on the endpoint. This indicates that the point has been acquired. Move the crosshairs along the tracking path and pick the new point. The new point and the existing point should now be in line as shown in Figure 2.44.

FIGURE 2.44   To track along the axis of an existing point, first acquire the endpoint, then move the crosshairs along the tracking path.

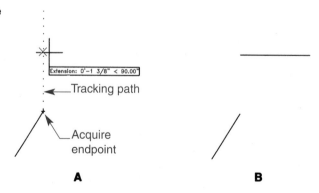

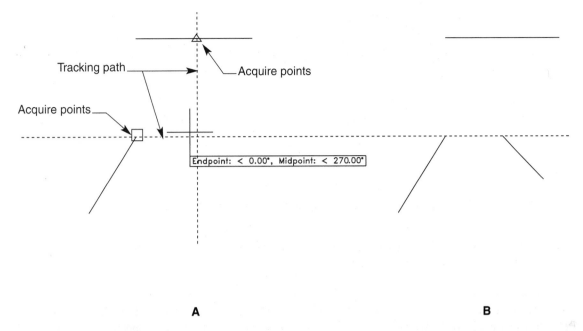

A   **B**

FIGURE 2.45   Multiple points can be used for Object Snap tracking. First, acquire all the desired points, then move the crosshairs until multiple tracking paths display. When multiple paths are displayed, pick to locate the new point.

Multiple points can also be acquired and tracked along in order to locate a point where the existing points' tracking paths cross. To do this, acquire the points as you would a single point, then move the crosshairs until several tracking paths are displayed. When multiple paths are displayed use the pick button to pick the point where the paths cross. Figure 2.45 shows how the midpoint and endpoint of two different objects are acquired in order to locate their intersection along tracking paths.

Field Notes

Object Snap tracking is similar to Extension osnap. The difference is that Extension osnap tracks along the same orientation of the acquired object as if it were continuing to be drawn. The OTRACK feature tracks along the X or Y axis of the acquired point. Use both options to enhance your drafting abilities and to save time while drawing.

Exercise 2.16

1. Start a new drawing using AEC Arch (Imperial).dwt.
2. Draw the layout sketch for a new building as shown in the following figure.
3. Use the osnaps discussed in the previous section.
4. Try using Object Snap tracking in conjunction with direct distance entry to locate new starting points. [*Hint*: Acquire a point, pull the mouse in the tracking direction, and enter how far away the new point needs to be away from the existing point.]
5. Do not dimension.
6. Save the drawing as EX2-16.dwg.

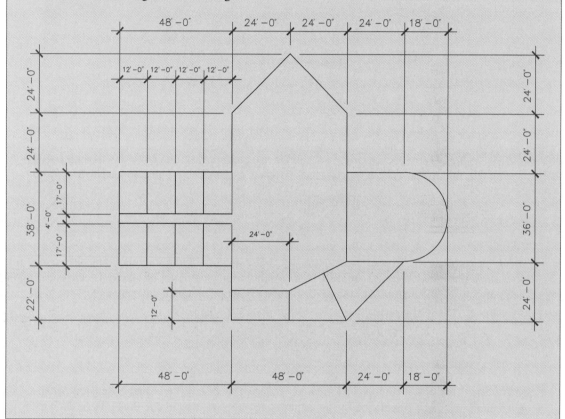

Printing Basics

As indicated at the beginning of this chapter, model space is the area where your full-scale drawing is created. Layout space is where the title block, border, and the scaled drawing are placed. When you have finished creating a drawing it often needs to be plotted or printed as a ***hardcopy***, which the printed version of a computer file. Plotting can be accomplished from the Model tab or from a Layout tab. This section introduces you to plotting basics from the Model tab. Use of layout space is explained in detail later in this text.

      To plot a drawing from the Model tab, make sure that model space is the active drawing area. Select the Plot button, type PLOT or press the [Ctrl]+[P] key combination at the Command: prompt, select Plot… from the File pull-down menu, or right-click the Model tab and select Plot… from the cursor menu. The Plot dialog box shown in Figure 2.46 appears. This dialog box includes two tabs: **Plot Device** and **Plot Settings**.

The **Plot Device** tab is used to specify the plotter and pen tables for the plot. The **Plot Settings** tab is used to specify the size of paper and the scale for the plot.

The first thing to do is to assign the plotter to be used for plotting the drawing. In the **Plotter configuration** area, select from a list of plotters/printers. All printers or plotters configured to your Windows operating system or in the Architectural Desktop system are available. Selecting a plotter assigns the plotters and also designates the paper sizes available for the specified plotter.

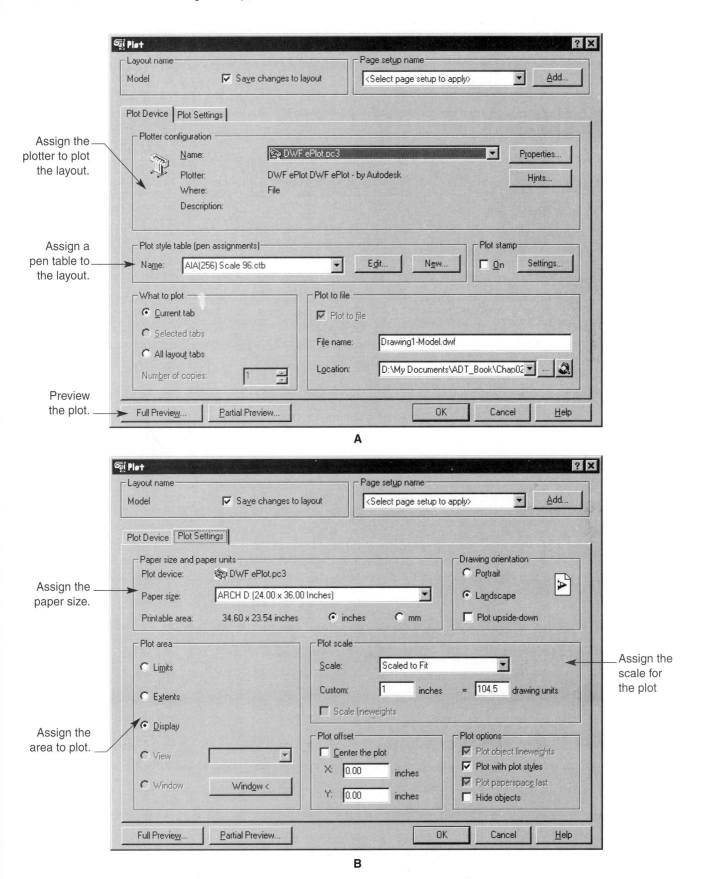

Assign the plotter to plot the layout.

Assign a pen table to the layout.

Preview the plot.

A

Assign the paper size.

Assign the area to plot.

Assign the scale for the plot

B

FIGURE 2.46 The Plot dialog box is used to assign a plotter, paper size, and a scale for the drawing and to plot the drawing.

After the plotter has been assigned, a *plot style table* (pen assignments) can be assigned to the Model or Layout tab. A ***plot style table*** is used to assign pen weights and printing colors to geometry colors within your drawing. For example, if Architectural Desktop "sees" red-colored geometry, it may plot that geometry as a red line if using a color table or as a black line if using a monochrome table. Some tables interpret geometry colors and plot them with grayscale colors. Line weights are preset for many of the colors in Architectural Desktop. These tables can be modified to suit your own needs. This topic is fully discussed later in the text.

Select the Plot Settings tab to assign the paper size, the area to be plotted, and the scale of the drawing. The **Paper size and paper units area** includes a list of paper sizes supported by the chosen plotter. Select a paper size from the drop-down list.

The Plot area is used to define what will be plotted. The options in this area are described next:

- **Limits** Initially, when a drawing is started, an area known as Limits defines a scaled piece of paper in model space. If a drawing is started using one of the AEC templates, the limits are defined by the grid dots displaying in the drawing screen. If you want everything within the limits plotted, then select this option.

- **Extents** This option plots to the farthest outside edge of the geometry in the drawing.

- **Display** This option plots what is currently being displayed on screen.

- **View** If any views were defined in the drawing, a particular view can be selected from the list, to be plotted.

- **Window** This area is grayed out until the Window< button is selected. This option returns you to the current space (model or layout) so that a windowed area to plot can be specified. Anything within the windowed area is then plotted.

The **Plot scale** area is used to assign a scale to the plotted drawing. A drop-down list with different scales is available, including Scaled to Fit. This option scales the drawing so that it fits within the area of the selected paper.

After you have selected these settings, you can pick the Full Preview… button to display a preview of the plotted drawing before plotting. It is advisable to preview the plot before plotting because what you see in the preview is what you get when the drawing is plotted. To exit the plot preview, right-click and select Exit from the cursor menu. Make any changes to your settings if needed, and check the plot again with the Full Preview… before plotting. Once the plot is configured as desired, pick the OK button to send the plot to the plotter. You also can send the file directly to the plotter from the plot preview by right-clicking and picking the Plot option.

Exercise 2.17

1. Open EX2-15.dwg.
2. Plot the drawing.

CHAPTER REVIEW

Use the CD-ROM to test your knowledge and skills.

Chapter Test

To check your understanding of the content provided in this chapter, access the Test file in the CH02 folder of the CD-ROM that accompanies this text.

Chapter Project

To practice the Architectural Desktop skills presented in this chapter, access the Project files in the CH02 folder of the CD-ROM that accompanies this text. The project files are in pdf format and include sample drawings and instructions for completing each project.

| |

CHAPTER 3

Drawing Walls, Doors, and Windows

LEARNING GOALS

After completing this chapter, you will be able to:

◎ Create walls.

◎ Convert existing geometry into walls.

◎ Clean up walls.

◎ Work with AutoCAD editing tools.

◎ Use Grips to modify objects.

◎ Create objects with the OFFSET command.

◎ Use the TRIM and EXTEND commands.

◎ Modify walls.

◎ Create wall cleanup groups.

◎ Add doors to your drawing.

◎ Modify doors.

◎ Add windows to your drawing.

◎ Modify windows.

◎ Create and modify openings.

◎ Draw window assemblies.

◎ Modify window assemblies.

A typical architectural project is divided into three major phases: conceptual design, design development, and construction documentation (or documented drawings). As indicated in Chapter 1, the design development phase is the period during which the construction documents begin to take shape. Architectural Desktop includes tools and objects that aid in the creation of these documents.

AEC objects such as the wall object are representations of real-world building components. The wall object understands how an architect perceives and draws the wall for a floor plan in a 2D or top view while also providing a realistic representation of the wall in 3D for an elevation or perspective view (see Figure 3.1). Walls in conjunction with other objects, such as doors and windows, are used together to provide a comprehensive set of documents that can be used to construct the actual building.

This chapter is devoted to the basic objects used in the design development phase. The creation and modification of the wall, door, window, opening, and window assembly objects are discussed.

The Wall Object

The *wall object* is one of the most important features in Architectural Desktop, because it provides the basis of the construction documents. Walls are similar to the line or polyline objects created in standard AutoCAD; however, one benefit of wall objects over lines or polylines is that a wall is a single object rather than two parallel lines representing a wall. Wall objects also can be quickly modified to reflect changes in

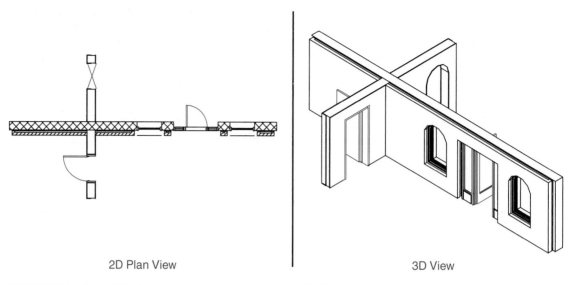

FIGURE 3.1   The AEC objects can display one way in a 2D plan view, yet as a "real" entity in 3D view.

the width or height of the wall after it has been drawn. Walls can be converted from layout geometry such as lines, polylines, arcs, and circles or drawn by selecting a starting and ending point, much like drawing polylines or lines.

Adding Walls to Your Drawing

The **Add Walls** modeless dialog box is used to create wall objects. Use one of the following options to display the Add Walls modeless dialog box, shown in Figure 3.2:

✓ From the **Design** pull-down menu, select **Walls,** then pick **Add Wall….**

✓ Select the **Add Wall** button in the **Walls** toolbar.

✓ Right-click in the drawing area, and select **Design, Walls,** then pick **Add Wall….**

✓ Type **AECWALLADD** at the Command: prompt.

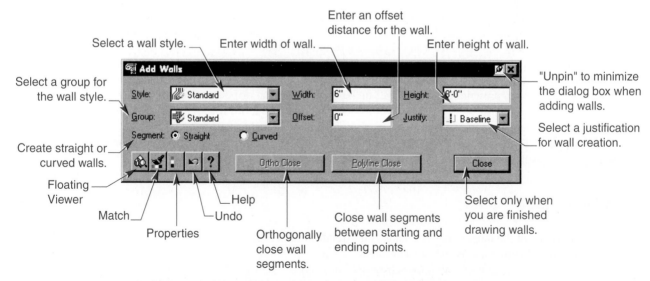

FIGURE 3.2   The Add Walls dialog box is used to select the wall style (type), width, height, and justification for new walls.

To draw a wall, first select the parameters for the wall from the Add Wall dialog box. After selecting the parameters, pick once in the drawing to make the drawing active, then pick a point to establish the starting point for the wall, and pick an ending point for the new wall. Repeat this process to draw the desired number of walls. When you are finished drawing walls, right-click and select Enter, or press the Close button in the Add Walls dialog box. The following is a description of the settings in the Add Walls dialog box:

- **Style**   This drop-down list includes several different styles that represent different types of walls. Some styles are available for use if you start a new drawing using a template. Starting from scratch or opening a legacy drawing includes only the Standard style. A *legacy drawing* is a previously created drawing. Other styles can be imported and created through the Style Manager, which is discussed in Chapter 12.

- **Width**   Here you enter a width for the wall. Some wall styles may have this option grayed out because the style designates a set wall width, as in the CMU-8 wall, which represents an 8″-wide concrete masonry unit (CMU) block.

- **Height**   Here you enter a height for the wall. Walls in a 2D plan view display the wall in two dimensions regardless of variations in height. A 3D view displays the specified height.

- **Group**   This drop-down list allows you to select a group in which to assemble a series of walls. Walls that are in the same group clean up intersections and corners between the walls automatically. Initially, the Standard cleanup group is the only group available; other cleanup groups can be created. This topic is discussed later in this chapter.

- **Offset**   This option draws the wall segment offset the distance specified from the points picked for the starting and ending points of the wall.

- **Justify**   The starting and ending points picked in the drawing screen draw the wall based on one of the following four wall justifications (see Figure 3.3):

- **Left**   This justification draws the wall along the left edge as the wall is drawn from left to right.

- **Center**   This justification draws the wall along the center.

- **Right**   This justification draws the wall along the right edge as the wall is drawn from left to right.

- **Baseline**   All wall styles are created in relationship to a baseline. The *baseline* usually represents a logical point from which to measure a wall. For example, an 8″-wide CMU wall with a 4″ brick veneer style may be built so the baseline is located at the outside edge of the CMU, with the brick veneer on the other side of the baseline. This aids in accurately drawing the CMU wall to actual dimensions, because the brick veneer is not considered a structural component. In many situations this may be the most desired justification when drawing walls.

FIGURE 3.3 Walls are created in relation to a justification.

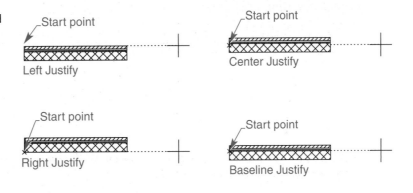

FIGURE 3-4 The Ortho Close
feature can aid in quickly closing
a series of walls.

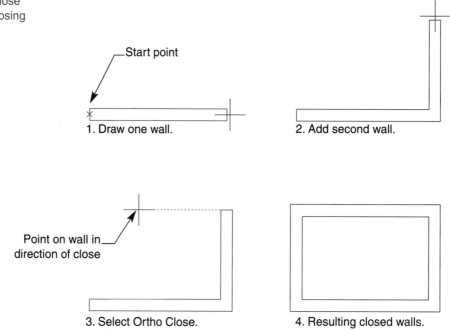

Start point

1. Draw one wall.   2. Add second wall.

Point on wall in
direction of close

3. Select Ortho Close.   4. Resulting closed walls.

- **Segment**   This area contains two radio buttons that designate the type of wall segment to be drawn. The **Straight** option draws a straight wall, which requires two points (starting and ending) to be picked. The **Curved** option draws a curved wall, which requires a start point, a midpoint, and an end point to be picked.

- **Ortho Close** This button is grayed out until two or more wall segments have been drawn. This option adds two walls between the first and last wall segments created. You are prompted to select a point in the direction in which to close the wall segments. The point you select extends a wall segment until it is perpendicular to the edge of the first wall segment, then adds the last wall segment (see Figure 3.4).

- **Polyline Close** This button is also grayed out until two or more wall segments have been drawn. This option, similar to the polyline Close option, adds one wall segment between the starting point of the first wall segment and the ending point of the last wall segment.

- **Close**   This button closes the dialog box. If no walls have been drawn, pressing this button ends the Add Walls command. This button should be pressed only when you are finished drawing walls.

- **Pushpin Button** This button, described in Chapter 1, is similar for all modeless dialog boxes. When button is depressed the dialog box remains "pinned" to the screen as you draw walls (see Figure 3.5A). Deselecting the button with the

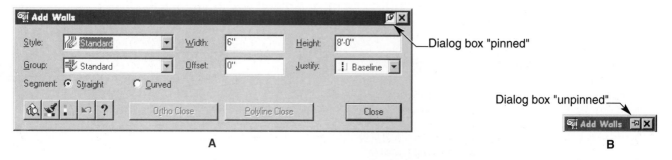

Dialog box "pinned"

Dialog box "unpinned"

A   B

FIGURE 3.5 The pushpin button is a new feature to Architectural Desktop Release #3. "Unpinning" the pin minimizes the dialog box, providing you with more room in which to draw AEC objects.

left mouse button "unpins" and minimizes the dialog box when crosshairs are away from the dialog box (see Figure 3.5B). This gives you more drawing space. To maximize the dialog box, move the crosshairs over the title bar that remains on screen, and the dialog box reappears.

The following group of buttons in the lower left corner of the Add Walls modeless dialog box are similar to all Add or Modify modeless dialog boxes for other AEC objects.

- **Floating Viewer** Use this button to display the floating viewer shown in Figure 3.6. The floating viewer is used to display the selected style (wall, door, or window). Selected styles can be viewed from many directions. The drop-down list in the upper right corner contains a list of preset views. The top view displays the selected style as it appears in a plan (2D) view. Selecting the appropriate buttons at the top of the dialog box shades the selected object style. The drop-down list above the viewer window allows you to see how the object appears using a different display configuration. Right-clicking in the viewer provides you with a cursor menu with additional options.

- **Match** This button prompts you to select an object to match its properties for the new object you are creating. For example, if you are in the Add Walls dialog box, and you pick the Match button, you are prompted to Select a wall to match:. Select an existing wall, then choose an option to be matched from the prompt *Match [Style/Group/Width/Height/Justify] <All>:*. "All" replace all the values in the Add Walls dialog box with those from the selected wall.

- **Properties** This button takes you into the properties for the new objects you are creating. Objects can be given a description such as 2-hr wall, Fire door, or any other desired description. Additional properties are available for the type of object you are adding.

- **Undo** This button will undo the new object that is placed in the drawing, similar to using Undo in AutoCAD.

FIGURE 3.6 The floating viewer is used to preview how the selected object style will appear in different views.

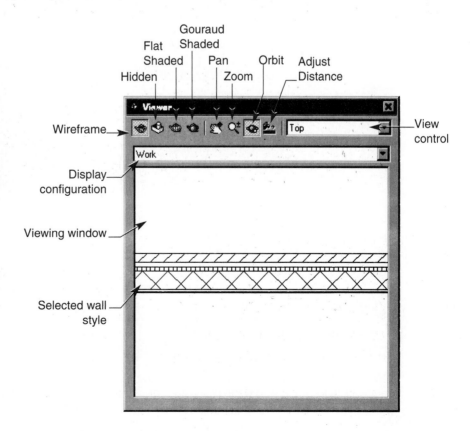

- **Help**　　　　　　　This button displays the help section related to the dialog box that is currently open. Use Help when you have a question on how to do something in Architectural Desktop.

Field Notes

The group of buttons just discussed is similar for other AEC object dialog boxes. The floating viewer is especially helpful in determining how a selected style will appear in 2D and 3D views.

While you are selecting and changing options in the Add Walls dialog box notice that the modeless dialog box is active. The highlighted title bar indicates an active dialog box or window. As you make changes in the dialog box the drawing window remains inactive. To add a wall into the drawing, pick once in the drawing window to make the drawing active. The title bar for the drawing is highlighted. Now, pick a starting point for the wall. Select points by entering a coordinate, picking a location with the mouse, or snapping the crosshairs to an existing object. To finish drawing the wall, select an ending point (see Figure 3.7).

Each time you select an ending point, that point becomes the starting point for the next segment. As you draw walls you can adjust the options for the next segment by picking the Add Walls dialog box to make it active, changing the value of an option, then picking once in the drawing window to make it active again, and selecting the next point for the wall. As walls are drawn an arrow is displayed next to the wall segment. This

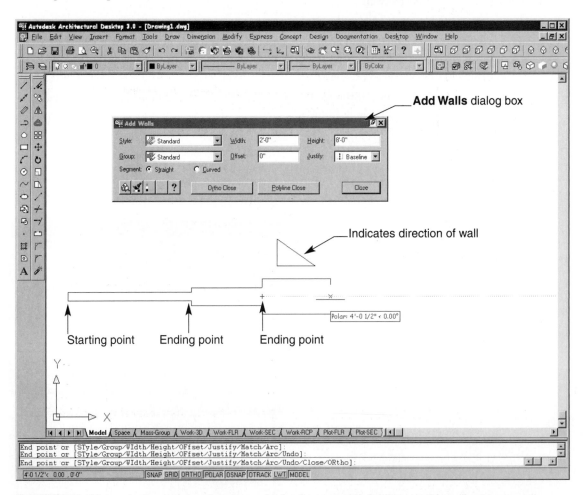

FIGURE 3.7　Pick a starting point and ending point for the new wall. Additional picks will add new wall segments. Access the **Add Walls** dialog box before selecting another endpoint to change any of the parameters for the next wall segment.

arrow indicates the direction in which the wall is being drawn. The wider end of the arrow indicates the side of the wall that is the starting point, and the point of the arrow indicates the side that is the ending point.

When you have finished drawing walls, select the Close button in the Add Walls dialog box, right-click and select <u>E</u>nter, press the [Enter] key, or press the [Esc] key. Any one of these methods ends the process of adding walls. If you draw a wall with the wrong style, width, justification or height, you can modify the wall later. Continue drawing new walls until you are finished. Modifying walls is discussed later in this chapter.

 Field Notes

> If you are trying to snap a wall to a specific justification point, use the Node object snap. A wall has three node points located along the justification line used to create the wall: one at the starting point, one at the midpoint, and one at the ending point for the wall.

Exercise 3.1

1. Start a new drawing using the AEC Arch (Imperial).dwt.
2. Using the CMU-X wall style, draw four 6″-wide walls with the four different justifications similar to part A in the figure.
3. Using the Stud-X wall style draw part B of the figure. Start with a 6″-wide wall, change it to a 12″-wide wall, then back to a 6″-wide wall. You can vary the height if desired.
4. Using the Concrete-8 Concrete-16x8–footing wall style draw part C of the figure. Draw two of the wall segments, then use Ortho Close to add the final two wall segments.
5. Access the Add Walls dialog box again, and use the floating viewer to preview the wall style. Use the Match button to select the CMU-X wall created using the baseline justification. Choose ALL to match all the wall properties in the dialog box. Draw the walls in part D of the figure. Use Polyline Close to close the wall segments.
6. Save the drawing as EX3-1.dwg.

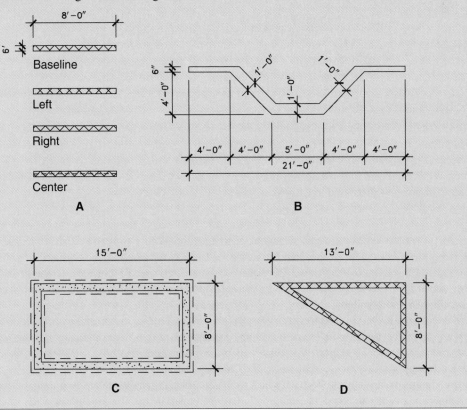

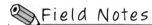Field Notes

> You can add a new wall to the drawing that is based on the same style, width, and height of an existing wall in your drawing. To do this, select the wall that contains the settings you want to repeat, then right-click and select Add Selected… from the cursor menu. The Add Walls dialog box appears with the settings from the selected wall.

Converting Existing Geometry to Walls

Walls can also be created by converting existing geometry. Lines, arcs, circles, and polylines can all be converted to walls. Use one of the following options to convert existing geometry into walls

   ✓ Pick the **Design** pull-down menu, select **Walls,** then **Convert to Walls….**

   ✓ Select the **Convert to Walls** button in the **Walls** toolbar.

   ✓ Right-click in the drawing area, and select **Design, Walls,** then **Convert to Walls….**

   ✓ Type **AECWALLCONVERT** at the Command: prompt.

After the command is entered, the following prompt sequence is displayed:

    Select lines, arcs, circles, or polylines to convert into walls: *(select geometry to be converted)*
    Select lines, arcs, circles, or polylines to convert into walls: *(select additional geometry to be converted or press [Enter] to stop selecting geometry)* ↵
    Erase layout geometry? [Yes/No] <N>: *(select Y to erase the geometry or [Enter] to keep the geometry)* ↵

Select the AutoCAD objects that you want to convert into walls. Only lines, arcs, circles, and polylines can be converted. The starting and ending points for the AutoCAD geometry are used as the starting and ending points for the walls. The existing geometry is used as the location edge for the wall justification. When you are finished selecting geometry to convert, press the [Enter] key. The next prompt asks whether to erase the layout geometry. *Layout geometry* is the geometry that is being converted. Entering a Y erases the geometry from the drawing. Entering N keeps the original objects. Make a selection and press the [Enter] key.

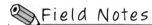Field Notes

> It is usually a good idea to place layout geometry on a "construction" layer that can be turned on/off, or frozen/thawed. Then, when you convert the geometry into walls, keep the geometry, because it indicates how the justification places the new wall in relation to the existing geometry. When you have finished converting the geometry to walls you can freeze the layer. The creation of layers is discussed in Chapter 6.

After you have kept or erased the layout geometry, the **Wall Properties** dialog box is displayed as shown in Figure 3.8. The tabs of the Wall Properties dialog box control how the new walls are to be created. The **Style** tab allows you to select the type of wall to be created (see Figure 3.8). If you are not sure what the style looks like, use the Floating Viewer button to preview the wall styles. Once you have selected a style, pick the Dimensions tab.

The **Dimensions** tab is used to specify the width, height, length, radius and justification for new walls (see Figure 3.9). The **Length**: text box is ignored, because the existing geometry is being used to specify the length of the new walls. If there are no curved entities, the radius text box is grayed out. Select the parameters for the new wall segments, then select the Cleanups tab.

The **Cleanups** tab is used to specify how walls clean up. *Cleanup* refers to automatic fixing of wall intersections so lines do not cross (see Figure 3.10). The check box at the top allows walls to cleanup with each other when they cross paths or meet at corners. If there are cleanup groups available, the new walls

Style tab

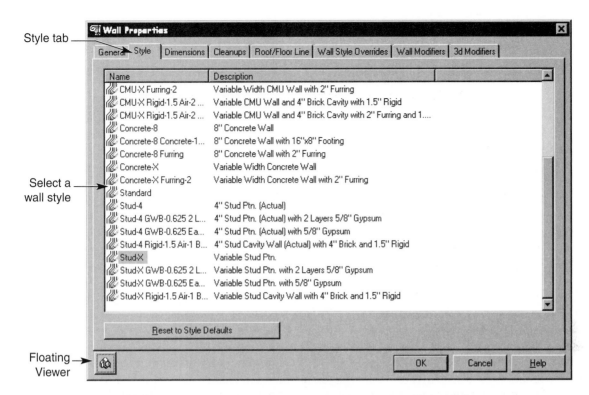

Select a wall style

Floating Viewer

FIGURE 3.8 After the layout geometry has been selected and kept or erased, the Wall Properties dialog box is displayed. The Style tab is active initially. Select a style for the new walls.

Dimensions tab

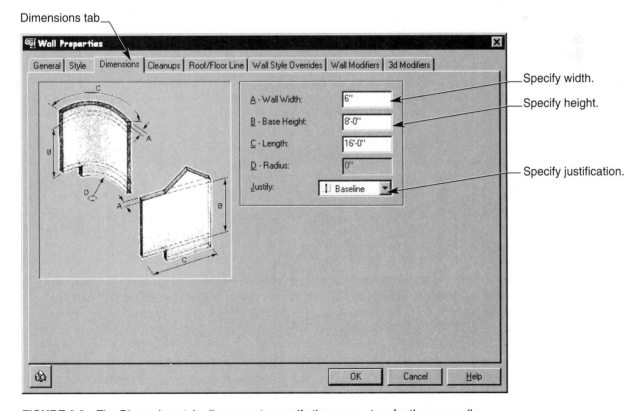

Specify width.

Specify height.

Specify justification.

FIGURE 3.9 The Dimensions tab allows you to specify the parameters for the new walls.

Cleanups tab

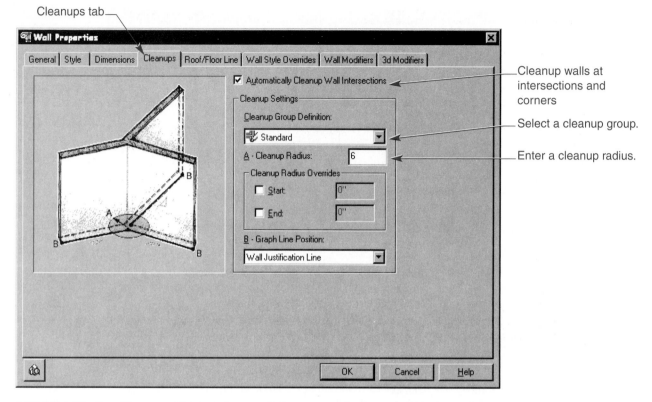

Cleanup walls at intersections and corners

Select a cleanup group.

Enter a cleanup radius.

FIGURE 3.10   The Cleanups tab is used to specify how new walls cleanup with other walls at intersections and corners.

can be added to a cleanup group. Refer to the Group description in the previous Add Walls topic. Creating Cleanup groups is discussed later in this chapter.

The cleanup radius is very important to walls. Anytime a wall intersects with another wall, the two walls try to cleanup with each other. A *cleanup radius* is placed at both endpoints of the justification line and at the point where two or more wall justification lines cross each other (see Figure 3.11). In most situations a cleanup radius of 0″ cleans up walls. This requires that the wall justification lines touch or cross each other. There may be instances when walls do not cleanup properly when using a left, right, or center justification is used. When walls do not cleanup, a red "defect" symbol is displayed. In these situations the cleanup radius can be modified to a different value. Usually a cleanup radius that is equal to the width of the wall cleans up the walls. Cleaning up wall intersections is discussed in greater detail later in this chapter.

FIGURE 3.11   (A) Three walls with a mixture of left and right justifications and a cleanup radius equal to 0″ return a defect symbol, and the walls do not cleanup correctly. (B) The same three walls with the same justifications and a cleanup radius equal to the width of the largest wall removes the defect symbol and cleans up the walls.

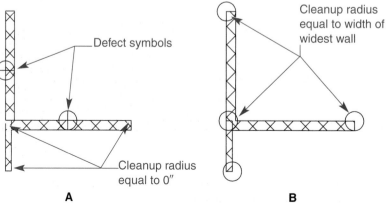

Defect symbols

Cleanup radius equal to width of widest wall

Cleanup radius equal to 0″

A

B

Field Notes

> It is highly recommended that the cleanup radius be set equal to your most common wall width. This helps eliminate defect symbols and helps cleanup walls. Individual walls that do not cleanup can be drawn with a larger or smaller cleanup radius to get the walls to cleanup.

The remaining tabs contain advanced parameters for controlling how the wall is to appear in both a plan and a model view. These topics are discussed later in this text. When you are finished setting parameters in the Style, Dimensions, and Cleanups tabs, press the OK button to finish the process of converting the lines into walls. If you determine that a different wall style or wall width needs to be used, you can modify the walls to reflect your desired output. Modification of walls is discussed later in this chapter. If the walls do not get placed correctly into the drawing relative to the existing geometry, reverse the starting and ending points for the walls. This "flips" the edge of the walls to the other side of the geometry.

Reversing Wall Endpoints

Occasionally when you add walls or convert geometry to walls, the edge of a wall component may be on the wrong side of the justification line. In this situation the endpoints of the wall can be reversed, causing the edge of the wall to flip to the other side of the justification line. The easiest way to reverse a wall or several walls is to first select the walls to be reversed, then right-click. The cursor menu for walls is displayed as shown in Figure 3.12. Select Reverse from the Plan Tools cascade menu. This process reverses the endpoints of the walls.

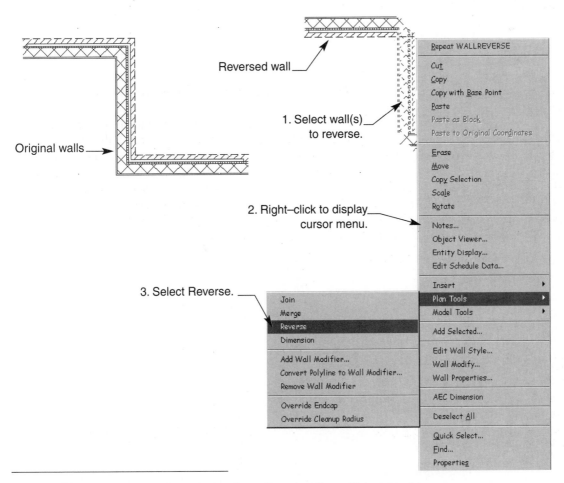

FIGURE 3.12   To reverse the endpoints for walls, select the wall(s), right-click, and select Reverse from the Plan Tools cascade menu.

Use one of the following methods to reverse walls.

✓ Pick the **Reverse Wall Start/End** button in the **Wall Tools** toolbar.

✓ Pick the **Design** pull-down menu, select **Wall Tools,** and then select the **Reverse Wall Start/End**.

✓ Right-click, and select **Design, Wall Tools,** and then **Reverse Wall Start/End**.

✓ Type **AECWALLREVERSE** at the Command: prompt.

Exercise 3.2

1. Start a new drawing using the AEC Arch (Imperial).dwt or AEC Arch (Imperial-Int).dwt.
2. Set the drawing scale to 1/4″=1′-0″
3. Draw the building outline shown in the figure with polylines.
4. Convert the polylines into walls.
5. Use the Stud-X wall style with a 6″ width, 8′-0″ height, and a baseline justification for the exterior walls. Use a 6″ cleanup radius. [*Note*: The polyline should be located on the "outside" of the walls. If the wall segments happen to be on the wrong side of the polyline, use the AecWallReverse command to reverse the endpoints of the wall(s)].
6. Use the Stud-X wall style with a 4″ width, 8′-0″ height, and a center justification for the interior walls. Use a cleanup radius of 4″.
7. Save the drawing as EX3-2.dwg.

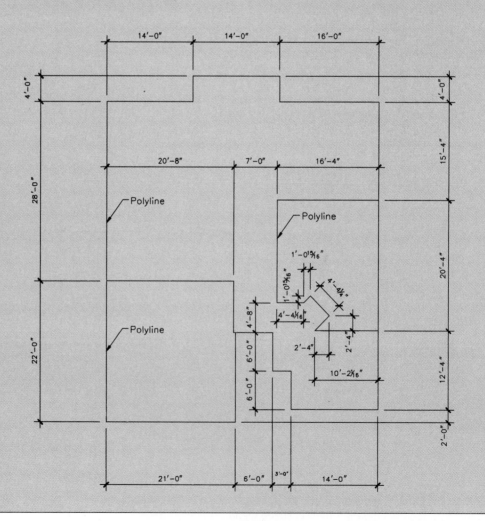

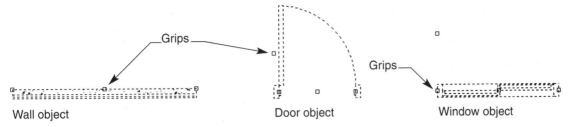

FIGURE 3.13 Grips are located around different points on AutoCAD geometry or AEC objects.

Using AutoCAD Editing Tools

In many situations you will draw a new wall only to realize later that the wall really needed to be a little longer, was in the wrong location, or simply did not need to be added. Standard AutoCAD commands can be used to modify AEC objects. The simplest way to modify AEC objects is with grip modes. Selecting an object when there is not a command active highlights the object and places boxes on the object indicating the grip points (see Figure 3.13).

Using Grips to Modify Objects

Grips appear on geometry only if the object has been picked without a command active. The objects become dashed (highlighted) and grips are displayed, indicating that the object has been selected and is ready to be modified. Grips are typically present at strategic points on an object. For example, grips are located at the starting, ending and midpoints of a wall object (see Figure 3.13).

The grips at this stage are referred to as *warm* grips, which means they are ready to be modified. Selecting several objects places these objects in a ***selection set***. All the objects in a selection set are affected by the modifying command applied.

You can remove from the selection set selected objects that you do not want to be modified by holding down the [Shift] key and repicking the object. This removes the highlighting on the object but keep the grips active. This state is known as a *cold* grip (see Figure 3.14). Reselecting a cold gripped object adds it back into the warm selection set so that it also can be modified.

FIGURE 3.14 Objects can be removed from a "warm" state and placed in a "cold" state so that they will not be modified.

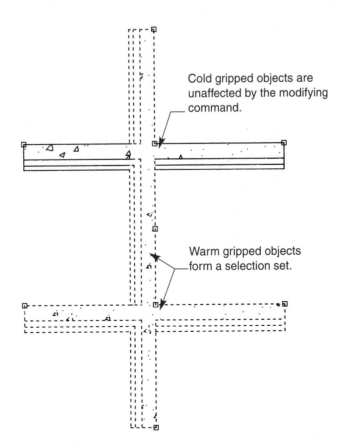

Cold gripped objects are unaffected by the modifying command.

Warm gripped objects form a selection set.

FIGURE 3.15 "Hot" grips indicate the object or point at the grip box is currently being modified.

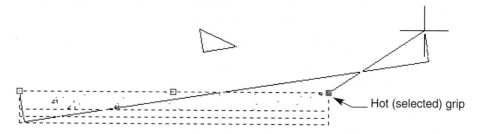

Hot (selected) grip

Picking a grip box with the left mouse button results in a color-filled grip (see Figure 3.15). These grips are called *hot* grips because they have been selected and are currently being modified. Depending on the type of grip mode or modifying command you apply, the whole object or just the grip point will be modified. Objects in a selection set are all affected by picking one grip.

Selecting a grip enables the grip-editing feature of AutoCAD. There are five commands that can be executed through the use of grips: **Stretch, Move, Rotate, Scale**, and **Mirror**. Each of these commands also includes a **Copy** option in which objects can be copied when one of the five commands is performed. When a grip is first picked, the Command: prompt displays the following:

** STRETCH **
Specify stretch point or [Base point/Copy/Undo/eXit]: (*pick a new location for the grip point*)

This sequence activates the Stretch grip mode. To stretch the selected object, move the cursor around and pick a new location. This relocates the hot grip, which in turn stretches the object as shown in Figure 3.16. If a midpoint grip is selected on a wall or line, the object is moved instead of stretched.

Field Notes

> Some objects, such as block symbols, doors, and windows, cannot be stretched. The stretch command typically works on linear and arced objects such as walls, polylines, arcs, and circles.

The Stretch grip mode includes some options that can be used to enhance the use of the command. The following options apply to all grip modes:

- **Base point**   When a hot grip is selected, that point becomes the base point. The *base point* is the point used as a reference location with the Move, Rotate, Scale, and Mirror grip modes. When you use the Stretch grip mode, the base point is the point that is being relocated. Select a new base point by entering B at the Command: prompt, pressing [Enter], and picking a new location with the pick button.

- **Copy**   Using the Copy option copies the objects as you are using one of the grip modes. Enter C at the prompt to activate this feature. When the Copy option is selected, the grip mode is displayed at the prompt with the word (Multiple)

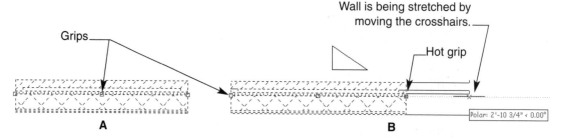

FIGURE 3.16   Grip editing can be used on AEC objects to quickly modify geometry. (A) Select an object to stretch. (B) Selecting a grip turns on the grip-editing mode and defaults to the STRETCH command. Move the crosshairs to relocate the grip and stretch the object.

beside it. This indicates that multiple copies are being made as you are performing the grip mode command.

- **Undo**   To undo the previous operation, enter U at the prompt and press the [Enter] key.

- **eXit**   Type X and press [Enter] at the prompt to exit the grip mode command and return to a blank Command: prompt.

If you need to stretch several ends of different walls at the same time, you can select multiple grip points. To do this, hold the [Shift] key while picking the grip boxes to be stretched. When you are finished selecting points, let go of the [Shift] key and pick on top of one of your hot grips. This allows you to stretch all the endpoints at the same time (see Figure 3.17).

Right-clicking while in a grip mode or with a hot grip active displays a menu from which different editing commands and/or options can be selected (see Figure 3.18). With the cursor menu displayed, select a grip mode (Stretch, Move, Rotate, Scale, Mirror), or select an option for the current grip mode. Select the Exit option to close the cursor menu and return to the drawing window. Clear grips from the objects by pressing the [Esc] key once.

Using the Move Grip Mode

If you need to move an object with grips, first select the object, then pick a grip to use as the base point, and right-click to display the grip mode menu. Select the Move grip mode from the menu. The following prompt appears:

```
** MOVE **
Specify move point or [Base point/Copy/Undo/eXit]: (pick a new location for the "warm" object(s))
```

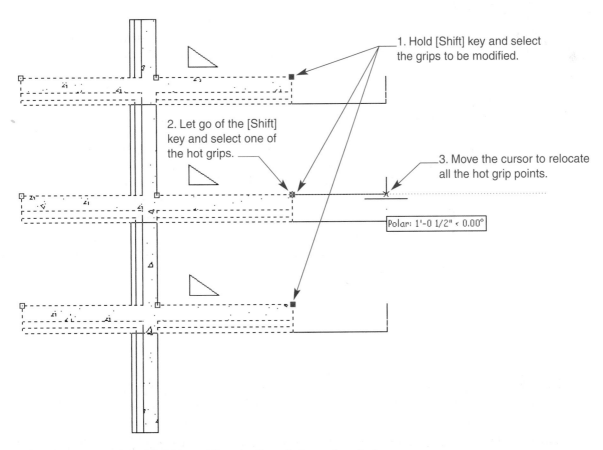

FIGURE 3.17 Hold the [Shift] key while selecting grip boxes to select more than one hot grip to be stretched.

FIGURE 3.18   After selecting a grip for editing, right-click to display a menu of other grip-editing commands.

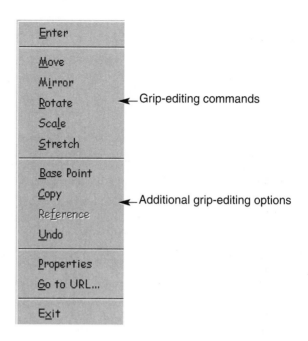

◄— Grip-editing commands

◄— Additional grip-editing options

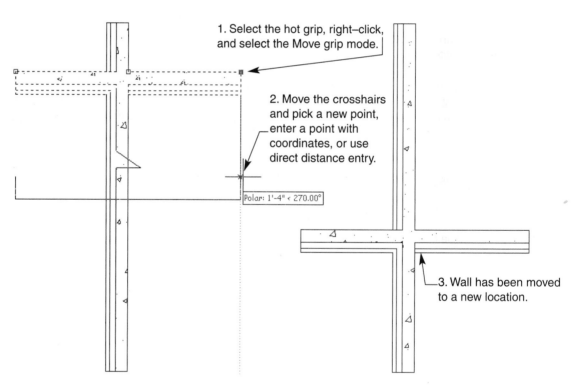

1. Select the hot grip, right–click, and select the Move grip mode.

2. Move the crosshairs and pick a new point, enter a point with coordinates, or use direct distance entry.

3. Wall has been moved to a new location.

FIGURE 3.19   The Move grip mode is used to move the selected object.

The hot grip becomes the base point or the point from which the object is being moved. You can use an object snap to snap the object to a new location or enter coordinates to move the object (see Figure 3.19). Using direct distance entry, pull the crosshairs in the direction you would like to move the object, enter the distance to move the object, and press the [Enter] key. If you picked the wrong grip point to use as a base point, or if you need to use a point away from the object as the base point, use the Base point option to select a new base point.

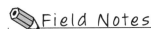
Field Notes

With the exception of the Stretch grip mode, any "warm" objects in a selection set are affected by the selected grip mode. In the preceding example, any highlighted objects would be moved with the Move grip mode.

Using the Grip Mode Copy Option

The Copy option is found in each of the grip modes. After entering a grip mode, right-click to select the Copy option. This allows you to make new copies of objects as you are stretching or moving them. Once the Copy option has been selected, the grip mode prompt changes. The following sequence shows how to use the Copy option with the Move grip mode:

```
** MOVE **
Specify move point or [Base point/Copy/Undo/eXit]: (right-click and select Copy or) C ↵
** MOVE (multiple) **
Specify move point or [Base point/Copy/Undo/eXit]: (pick a new location for the copied objects)
```

Using the Rotate Grip Mode

Objects can also be rotated with grips. When a grip has been selected, it becomes the axis point from which the selected objects are rotated. To access the Rotate grip mode, select an object or objects, pick a grip to make it hot, right-click, then select Rotate from the list. The following is displayed in the command window:

```
** ROTATE **
Specify rotation angle or [Base point/Copy/Undo/Reference/eXit]: (enter a rotation angle or pick
    a point designating the rotation angle)
```

The Rotate grip mode includes an option called **Reference**. This option is used when an object is already rotated at an unknown angle and it needs to be rotated to a new angle. To use this option, right-click while in the Rotate grip mode, and select Reference from the menu. The following prompt sequence is used to rotate an object at an unknown angle to a 45° angle (see Figure 3.20):

```
** ROTATE **
Specify rotation angle or [Base point/Copy/Undo/Reference/eXit]: R ↵
Specify reference angle <0.00>: (pick an endpoint)
Specify second point: (pick the other endpoint)
** ROTATE **
Specify new angle or [Base point/Copy/Undo/Reference/eXit]: 45 ↵
```

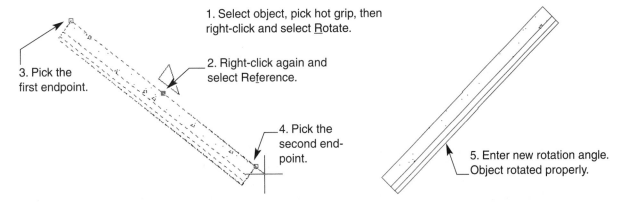

1. Select object, pick hot grip, then right-click and select Rotate.

3. Pick the first endpoint.

2. Right-click again and select Reference.

4. Pick the second end-point.

5. Enter new rotation angle. Object rotated properly.

FIGURE 3.20   The Reference option in the Rotate grip mode can be used to rotate an object from an unknown angle to a known rotation angle.

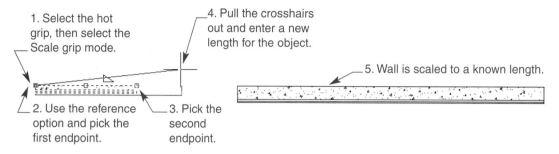

1. Select the hot grip, then select the Scale grip mode.

2. Use the reference option and pick the first endpoint.

3. Pick the second endpoint.

4. Pull the crosshairs out and enter a new length for the object.

5. Wall is scaled to a known length.

FIGURE 3.21   The Scale grip mode can be used to resize an object.

Using the Scale Grip Mode

An object can also be scaled with the Scale grip mode. The hot grip becomes the base point from which the object is scaled. After selecting the grip, right-click, and select Scale from the cursor menu. Move the crosshairs to increase or decrease the size of the object. You can also enter the desired scale factor at the prompt.

Like the Rotate grip mode, the Scale grip mode includes a Reference option. If an object is drawn to a size that is not known, and it needs to be drawn to a specific length, the Reference option can be used to scale the object. The following prompt sequence is used to scale the wall in Figure 3.21 to a known length:

```
** SCALE **
Specify scale factor or [Base point/Copy/Undo/Reference/eXit]: R ↵
Specify reference length <0'-1">: (pick an endpoint)
Specify second point: (pick the other endpoint)
** SCALE **
Specify scale factor or [Base point/Copy/Undo/Reference/eXit]: 12' ↵
```

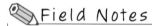 Field Notes

The Scale grip mode is useful for scaling the unknown length of an object to a specified length. If you need to adjust the length of a wall, it may be better to use the Properties button in the Modify Walls dialog box to set the length of the wall. This option scales the length of the wall without scaling the width too. See the Modifying Walls section.

Using the Mirror Grip Mode

The Mirror grip mode is used to reverse or create a mirror image of an object or set of objects. To use this grip mode, select the objects to be mirrored, then select a grip to be used as the first point of the mirror line. The *mirror line* is the line from which the objects are mirrored (see Figure 3.22). After selecting the grip point, right-click and select the Mirror option.

The hot grip becomes the first point of the mirror line. Moving the crosshairs creates a mirror line between the hot grip and the crosshairs. The selected objects are mirrored around the mirror line. Picking a location for the second point of the mirror line causes the objects to be reversed. Pulling the crosshairs vertically and picking a point causes the objects to be mirrored along the vertical mirror line. The following prompt sequence is used to mirror the objects in Figure 3.22:

```
** MIRROR **
Specify second point or [Base point/Copy/Undo/eXit]: (pick a second point for the mirror line)
```

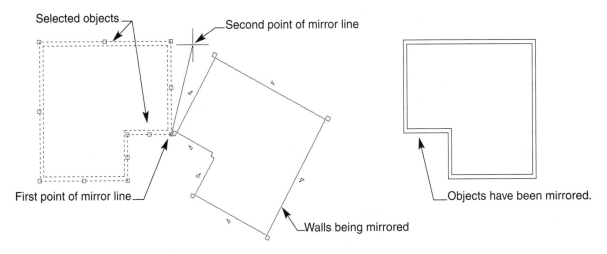

FIGURE 3.22   The Mirror grip mode is used to mirror or reverse the way objects are drawn.

Exercise 3.3

1. Start a new drawing using the AEC Arch (Imperial).dwt.
2. Draw the walls in the figure.
3. Use the Stretch grip mode to stretch the wall in part A of the figure 2'-0" to the right.
4. Use the Move grip mode with the Copy option to make a copy of the wall in part B of the figure.
5. Use the Rotate grip mode to rotate the wall 30° in part C of the figure.
6. Use the Scale option to scale the wall in part D of the figure so that its length is 12'-0".
7. Use the Mirror grip mode to mirror the walls in part E of the figure vertically to the right.
8. Save the drawing as EX3-3.dwg.

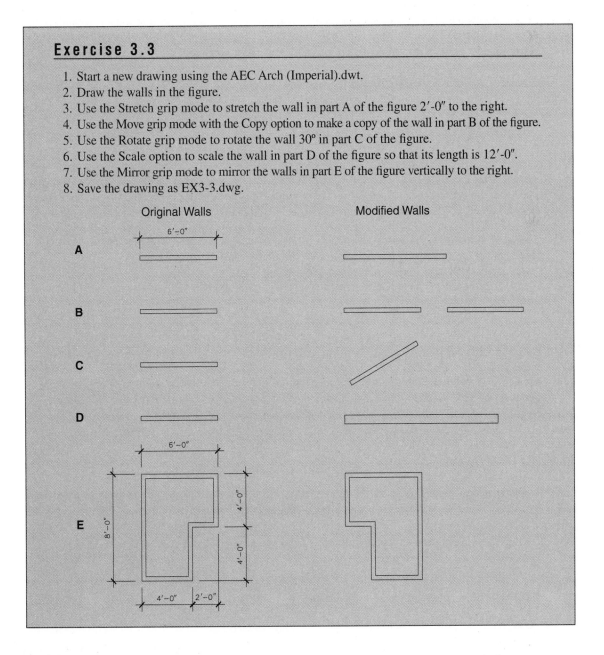

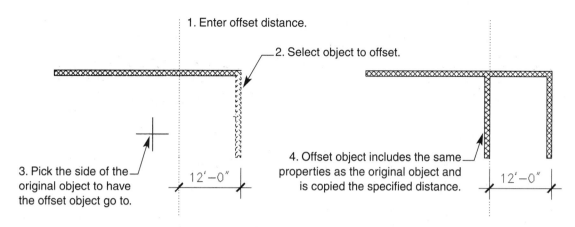

FIGURE 3.23   The Offset command can be used to quickly offset walls in order to lay out a floor plan.

Using the OFFSET Command

The OFFSET command creates a new object that is a copy of an object set at a specified distance away from the original object. This command can be used to quickly lay out a set of walls or polylines for rooms within a building. Use one of the following options to access this command:

✓ Select **Offset** from the **Modify** pull-down menu.

✓ Select the **Offset** button in the **Modify** toolbar.

✓ Type **OFFSET** or **O** at the Command: prompt.

After entering the OFFSET command, you need to specify a distance from the original object to the new copied object. After specifying a distance, you are prompted to select an object. Select the object that is to be copied. Once the object has been selected, you are prompted for the side of the original object that the new object is to be copied to. The following prompt sequence is used to offset a wall 12′-0″ away in Figure 3.23:

Command: **O** or **OFFSET** ↵
Specify offset distance or [Through] <Through>: (*enter a distance*) **12′** ↵
Select object to offset or <exit>: (*pick the original object*)
Specify point on side to offset: (*pick either side of the original object*)
Select object to offset or <exit>: (*pick another object to offset or [Enter] to end the command*)

✎ Field Notes

The OFFSET command is very useful for laying out floor plans. First create the desired walls, then use OFFSET to offset the original walls to lay out room sizes. Measure the offset distance from the justification line of the wall to the justification line of the new wall.

Using the TRIM Command

Occasionally, walls may need to be trimmed away from another object. The TRIM command is similar to a paper cutter. The paper cutter has a cutting plane (edge) that is used to trim paper. With the TRIM command, an edge is first selected as a cutting edge. Then, anything that crosses through the edge can be trimmed as shown in Figure 3.24.

Use the following options to access this command:

✓ Pick **Trim** from the **Modify** pull-down menu.

✓ Pick the **Trim** button on the **Modify** toolbar.

✓ Type **TRIM** or **TR** at the Command: prompt.

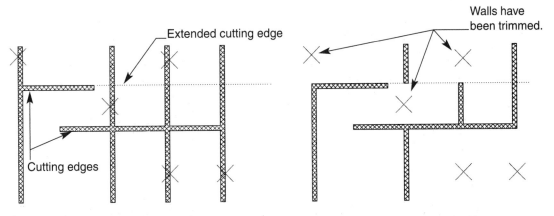

FIGURE 3.24 The TRIM command is used to trim objects away from a cutting plane (edge).

After entering the TRIM command, you need to select a cutting edge or edges. Several edges can be picked at once to use as cutting edges. If walls are selected, the justification line becomes the edge of the cutting plane. The TRIM command works only on objects that cross through the cutting plane; however, the command includes an **Edge** option. Entering this option allows you to specify an imaginary extended cutting edge (see Figure 3.24). The **Extend** option is used to create the imaginary extended cutting edge. Any object crossing through the extended cutting edge can also be trimmed. The following prompt sequence is used to trim the walls in Figure 3.24 from a cutting edge and an extended cutting edge:

Command: **TR** or **TRIM** ↵
Current settings: Projection=View, Edge=None
Select cutting edges ...
Select objects: *(select a cutting edge)*
1 found
Select objects: *(select another cutting edge or [Enter] to stop picking cutting edges)*
1 found, 2 total
Select objects: ↵
Select object to trim or shift-select to extend or [Project/Edge/Undo]: **E** ↵
Enter an implied edge extension mode [Extend/No extend] <No extend>: **E** ↵
Select object to trim or shift-select to extend or [Project/Edge/Undo]: *(pick object to trim)*
Select object to trim or shift-select to extend or [Project/Edge/Undo]: *(pick object to trim)*
Select object to trim or shift-select to extend or [Project/Edge/Undo]: *(pick object to trim)*
Select object to trim or shift-select to extend or [Project/Edge/Undo]: *(pick object to trim)*
Select object to trim or shift-select to extend or [Project/Edge/Undo]: *(pick object to trim)*
Select object to trim or shift-select to extend or [Project/Edge/Undo]: ↵

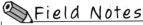Field Notes

> The TRIM command in AutoCAD 2000i and Architectural Desktop 3 and 3.3 allow you to use the cutting edges as a boundary edge to extend to when you hold the [Shift] key and select an object to extend to the boundary edge. The Extend command is discussed next.

Using the Extend Command
Whereas the TRIM command allows you to trim walls and other AutoCAD geometry away from a cutting plane, the **EXTEND** command extends the end of a wall, polyline, arc, or line to a boundary edge. With the **EXTEND** command, an edge is first selected as a boundary edge. Other objects can then be selected to extend their length to the boundary (see Figure 3.25). In most situations the boundary edge must be an edge that will be intercepted by the extending object. As with the TRIM command, the Edge option can be used to extend an imaginary line to which other objects can be extended.

Use one of the following options to acess the EXTEND command:

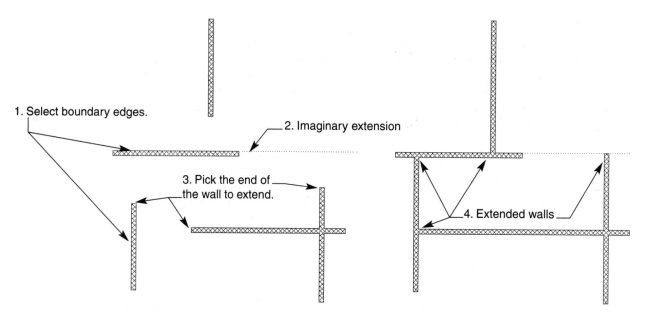

FIGURE 3.25   The EXTEND command is used to extend the length of an object to the boundary edge of another object. If the boundary edge does not cross the path of the object being extended, the Edge option can be used to create an extended edge.

✓ Select **Extend** from the **Modify** pull-down menu.

✓ Pick the **Extend** button in the **Modify** toolbar.

✓ Type **EXTEND** or **EX** at the Command: prompt.

The following sequence is used to extend the objects in Figure 3.25:

Command: **EX** or **EXTEND** ↵
Current settings: Projection=View, Edge=Extend
Select boundary edges ...
Select objects: *(select boundary edge)*
1 found
Select objects: *(select boundary edge)*
1 found, 2 total
Select objects:
Select object to extend or shift-select to trim or [Project/Edge/Undo]: **E** ↵
Enter an implied edge extension mode [Extend/No extend] <Extend>: **E** ↵
Select object to extend or shift-select to trim or [Project/Edge/Undo]: *(end of wall to extend)*
Select object to extend or shift-select to trim or [Project/Edge/Undo]: *(end of wall to extend)*
Select object to extend or shift-select to trim or [Project/Edge/Undo]: *(end of wall to extend)*
Select object to extend or shift-select to trim or [Project/Edge/Undo]: *(end of wall to extend)*
Select object to extend or shift-select to trim or [Project/Edge/Undo]: ↵

✎ Field Notes

When used on wall objects, the TRIM and EXTEND commands use the justification line of the wall as either the cutting edge or the boundary edge. If the edges are extended through the Edge option, the imaginary extension is from the justification line.

Additional standard AutoCAD commands such as Erase, Move, Copy, Rotate, Scale, Mirror, Stretch, Break, Array, and Properties work on wall objects. Most of these commands also work on other AEC objects. For additional help with standard AutoCAD commands, look up ***Editing Drawings*** in the Architectural Desktop Help, or refer to the textbook recommended in the preface to this text.

Modifying Walls

Standard AutoCAD commands are not the only commands that can be used to modify walls. Occasionally in the design process, you may find that a wall or several walls need to have their width, height, style, or cleanup group changed. Rather than erasing the faulted walls, you can use the **Modify Wall** command to change existing walls to reflect your design needs. To quickly modify walls, select the walls you wish to change, then right-click. A menu specific to wall objects appears (see Figure 3.26). Select the Wall Modify… command.

In addition, you can use one of the following to modify existing walls:

✓ Pick the **Design** pull-down menu, select **Walls,** then the **Modify Wall…** command.

✓ Select the **Modify Wall** button from the **Walls** toolbar.

✓ Right-click in the drawing area, and select **Design, Walls,** then the **Modify Wall…** command.

✓ Type **AECWALLMODIFY** at the Command: prompt.

The Modify Walls dialog box appears, allowing you to change the wall style, the width and height of the wall, wall justification, and the group (see Figure 3.27). Make any desired changes to the selected walls, then press the Apply button to see if the changes are truly what you want before closing the dialog box. Modifying the justification for a wall can be particularly helpful, especially after converting walls from existing geometry.

FIGURE 3.26 Right-clicking after selecting one or more walls displays a cursor menu that is specific to wall objects. To modify the walls, select the Wall Modify… command.

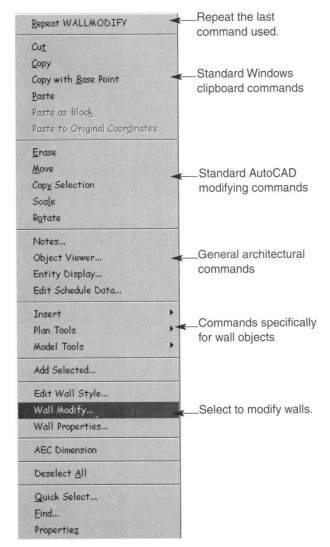

Repeat the last command used.

Standard Windows clipboard commands

Standard AutoCAD modifying commands

General architectural commands

Commands specifically for wall objects

Select to modify walls.

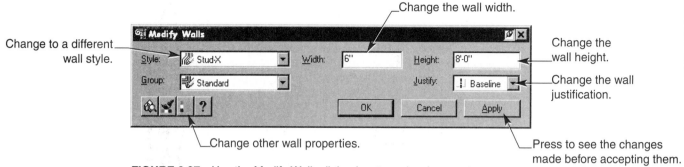

FIGURE 3.27 Use the Modify Walls dialog box to make changes to your walls instead of re-creating new walls.

Exercise 3.4

1. Open EX3-2.dwg.
2. Add in the remaining walls displayed in the figure.
3. Use AutoCAD modifying commands like TRIM, EXTEND, and OFFSET to create new walls. Interior walls should use the Stud-X wall style, be 4″ wide, 8′-0″ high, and use a center justification.
4. Try using object tracking, discussed in Chapter 2, to draw in new walls.
5. Use the Modify Wall command when necessary to make the walls appear as in the figure. Use the Trim and Extend command when necessary.
6. Save the drawing as EX 3-4.dwg.

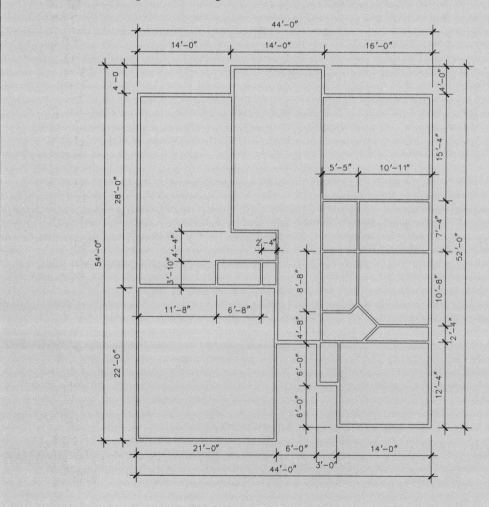

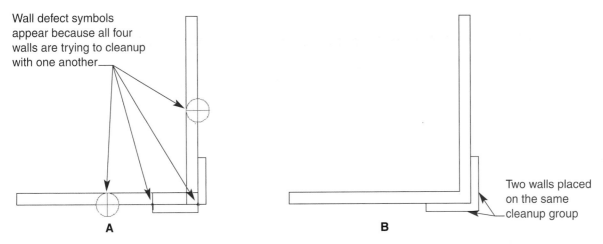

Wall defect symbols appear because all four walls are trying to cleanup with one another

Two walls placed on the same cleanup group

A   **B**

FIGURE 3.28   (A) The Standard wall style is used to create the four walls, which display defect symbols when placed beside one another. (B) The smaller outside walls have been assigned to a cleanup group definition that cleans only the outermost two walls. The inner walls still belong to the Standard cleanup group definition.

Creating Wall Cleanup Group Definitions

Occasionally, you may have the desire to create two walls of the same style side by side. Usually, if walls are created side by side, Architectural Desktop tries to cleanup both walls, which may cause a wall defect symbol to be displayed (see Figure 3.28A). Using *cleanup group definitions* is a method of grouping common types of walls so that only walls within the same group cleanup with one another (see Figure 3.28B).

Wall cleanup group definitions can be created by providing a new definition name. Use one of the following options to access the Style Manager dialog box:

✓ Pick the **Design** pull-down menu, select **Walls,** then the **Cleanup Group Definitions...** command.

✓ Select the **Cleanup Group Definitions** button on the **Walls** toolbar.

✓ Right-click in the drawing area, and select **Design**, **Walls**, then the **Cleanup Group Definitions...** command.

See Figure 3.29.

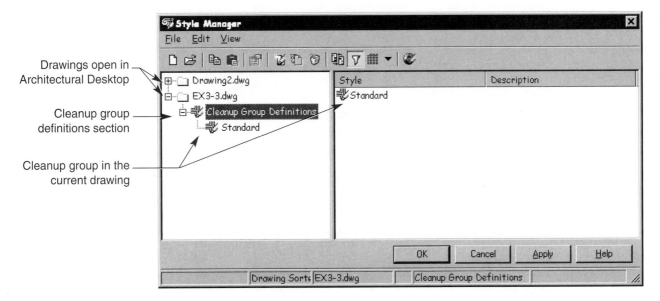

Drawings open in Architectural Desktop

Cleanup group definitions section

Cleanup group in the current drawing

FIGURE 3.29   The Style Manager dialog box can be used to create a new Cleanup Group Definition.

New Style button

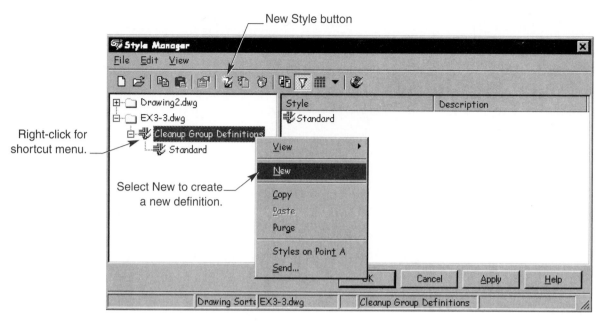

Right-click for
shortcut menu.

Select New to create
a new definition.

FIGURE 3.30   Create a new definition name that can be assigned to new walls or existing walls.

To create the new definition, select the New Style button, or right-click over cleanup group definitions in the tree view on the left and select the New option (see Figure 3.30). This creates a new definition in the list box on the right. Enter a new name for the definition. The next time a wall is drawn or modified, the new group definition will be available in the Group: drop-down list.

Field Notes

Architectural Desktop includes other wall styles that represent countertops and cabinets. Create a cleanup group definition for these types of wall styles so they do not cleanup with structural walls. Corner trim and quoins can be created with a wall style, too. Create cleanup group definitions so that these items do not cleanup with structural walls. You do not want countertops and cabinets to cleanup with walls, because you want them to appear independent from the wall symbol. (*Quoins* are the corner masonry of a building, typically used decoratively on the exterior.) Wall style creation is covered in Chapter 12.

Working with Wall Cleanups

As indicated earlier in this chapter, the wall cleanup radius is very important in how walls cleanup at intersections and corners. By default, Architectural Desktop sets the wall cleanup radius to 0″. In most situations this is sufficient to cleanup walls. This radius requires that the justification points you pick when drawing walls touch or cross each other in order for wall cleanup to work properly. In some situations where three or more walls meet, the 0″ wall cleanup radius has difficulty cleaning up the walls (see Figure 3.31). In these situations try changing to a larger cleanup radius.

The default cleanup radius can be changed to any other value in the Options dialog box, shown in Figure 3.32. Use one of the following methods to access the Options dialog box:

✓ Select **Options...** from the **Tools** pull-down menu.

✓ Right-click in the drawing area and select **Options** from the cursor menu.

✓ Type **OPTIONS** at the Command: prompt.

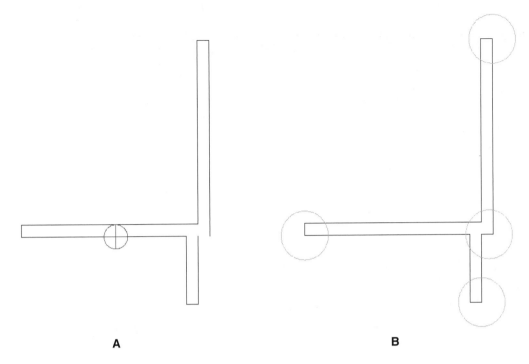

A   **B**

FIGURE 3.31   (A) Three walls meeting close together with a 0″ cleanup radius result in a defect symbol. (B) The same three walls with a 12″ cleanup radius.

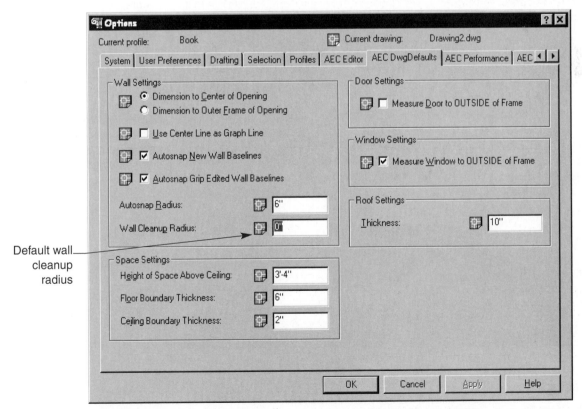

FIGURE 3.32   The default wall cleanup radius can be specified in the AECDwgDefaults tab within the Options dialog box.

Use the arrow buttons in the upper right of the Options dialog box to see all the Architectural Desktop settings tabs. Select the AECDwgDefaults tab. Change the value in the Wall Cleanup Radius: text box to reflect the new default wall cleanup radius.

 Field Notes

> It is recommended that you use a cleanup radius that is equal to your most common wall width. This will help reduce wall defect problems. If you still receive the defect symbols when you create walls, try increasing or decreasing the cleanup radius for the defect walls and any walls that touch or cross the defect walls.

If you find that walls continue to display defect symbols, you can override the cleanup radius for individual walls. To override the default cleanup radius, select the defect wall, right-click, and select Wall Modify… from the cursor menu. In the Modify Walls dialog box, select the Properties button to display the Wall Properties dialog box (see Figure 3.33). Select the Cleanups tab and make adjustments to the cleanup radius for the selected wall or walls.

You can specify the cleanup radius for the entire wall in the A - Cleanup Radius text box. You also can assign a cleanup radius to the starting and ending points of walls by checking on the Start: or End: check boxes, then entering a radius in the appropriate text box. When you have finished entering a new radius, press OK to exit the Wall Properties dialog box, which returns you to the Modify Walls dialog box. Press Apply to see the walls cleanup. If they do not cleanup, select the Properties button again and increase or decrease the cleanup radius. You may need to try this a few times to get a cleanup radius that works properly.

Use one of the following options to override the starting and ending point radii for walls:

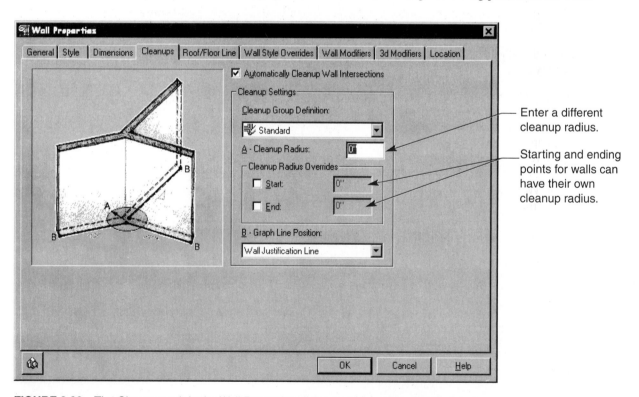

FIGURE 3.33   The Cleanups tab in the Wall Properties dialog box is used to change the cleanup radius for defect walls.

✓ Select the **Override Cleanup Radius** button from the **Wall Tools** toolbar.

✓ Pick the **Design** pull-down menu, select **Wall Tools,** and then pick the **Override Cleanup Radius** command.

✓ Right-click, and select **Design, Wall Tools,** and then the **Override Cleanup Radius** command.

✓ Select the wall(s), then right-click. **Select Override Cleanup Radius** from the **Plan Tools** cascade menu.

✓ Type **AECWALLAPPLYCLEANUPRADIUSOVERRIDE** at the Command: prompt.

The following prompt sequence is used to override the cleanup radius for the starting or ending point for a wall:

Command: **AECWALLAPPLYCLEANUPRADIUSOVERRIDE** ↵
Select a wall: *(select the defect wall(s))*
Select a point: *(pick the starting or ending point of the wall to attach the override)*
Enter new radius*<default>*: *(enter a new radius)*

Viewing the Cleanup Radius

Occasionally you may want to view the cleanup radii for different walls in your drawing to help determine where wall cleanups are occurring. Use one of the following options to view the radii and the justification lines for a wall:

✓ Select the **Toggle Wall Graph Display** button in the **Wall Tools** toolbar.

✓ Pick the **Design** pull-down menu, select **Wall Tools,** then pick the **Toggle Wall Graph Display** command.

✓ Right-click in the drawing area, and select **Design, Wall Tools,** and then the **Toggle Wall Graph Display** command.

✓ Type **AECWALLGRAPHDISPLAYTOGGLE** at the Command: prompt.

This displays any cleanup radii and the justification lines for the walls that you have overridden and may aid in the proper cleanup for your walls. To switch back to a normal display mode, select the Toggle Wall Graph Display again to reverse the display.

Tips for Working with Walls

The following are some tips to aid in the proper cleanup of walls:

- Walls must be created in the same XY drawing plane.

- Walls must be in the same cleanup group.

- When two walls are being cleaned up, the justification points must be within the cleanup radius.

- When three or more walls that are close together are being cleaned up, the cleanup radius must encompass the justification lines and each side of the walls.

- A cleanup radius that is equal to the wall width is useful.

Exercise 3.5

1. Open EX3-4.dwg.
2. Add the fireplace using the Stud-X wall style for the outside walls in the figure. Make the width 4″ and the height 8′-0″. Use the TRIM command to trim out the rear wall from the fireplace.
3. Create a cleanup group definition named Firebox.
4. Add the firebox for the fireplace using the Brick-X wall style and the Firebox cleanup group. Make the walls 2″ wide and 4′-0″ high.
5. If you have any wall defect symbols, adjust the cleanup radii.
6. Save the drawing as EX3-5.dwg.

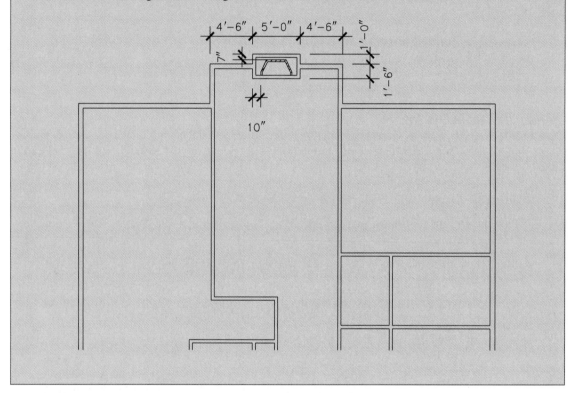

Creating Door Objects

Door objects in Architectural Desktop are AEC objects that interact with wall objects. A door understands that it belongs in a wall, and after it is placed in a wall becomes constrained to only wall objects. When doors are placed inside walls, the opening for the door is cut out of the wall. This is convenient for creating floor plans, because the desired effect is to display the door without the header over the door opening.

Adding Doors to Your Drawing

You can add doors into a drawing by first accessing the **Add Doors** dialog box. Use one of the following options to begin adding doors to your drawing:

✓ Select the **Design** pull-down menu, select **Doors**, then the **Add Door** command.

✓ Pick the **Add Door** button from the **Doors-Windows-Openings** toolbar.

✓ Right-click, and select **Design, Doors,** then the **Add Door** command.

✓ Type **AECDOORADD** at the Command: prompt.

The Add Doors dialog box shown in Figure 3.34 is displayed.

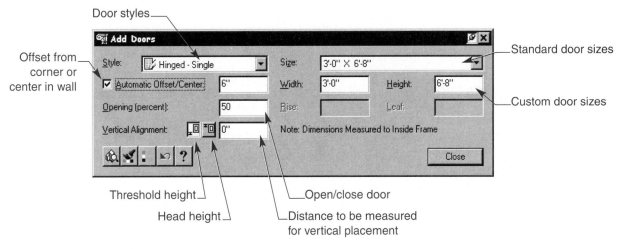

FIGURE 3.34 The Add Doors dialog box is used to insert doors into the drawing.

Another way to access the Add Doors dialog box is to first select a wall, right-click, then select Door… from the Insert cascade menu. The Add Doors dialog box is similar to the Add Walls dialog box. A description of the components within the dialog box follows.

- **Style**   As with wall objects, Architectural Desktop includes different door styles in the AEC Arch (Imperial).dwt. Choose a style from the list to add to your walls.

- **Size**   Some door styles have preconfigured sizes that can be chosen from this drop-down list. These sizes are considered to be the most commonly used in industry. Select a size from the drop-down list or enter your own size in the area below.

- **Automatic Offset/Center**   When this checkbox is unchecked, the doors that are added to the walls can be placed randomly along the wall. Once this checkbox is selected, the text box to the right becomes available. Entering a distance in this text box inserts doors that distance away from the corner of two walls to the edge of the door or from other openings in the wall. If the crosshairs are moved toward the center of a wall for the door placement, the Automatic Offset/Center centers the door within that wall segment.

- **Opening (percent)**   This text box allows you to open or close the door. An entry of 0 closes the door, and an entry of 100 opens the door as far as that particular door can be opened. The default is a door halfway open.

- **Vertical Alignment**   This area is used to measure where the door is placed vertically within a wall. The two buttons indicate how the door is measured vertically for placement in the wall.

- **Threshold Height**   This button takes the value in the text box and measures the door placement from the wall bottom to the threshold (bottom) of the door.

- **Head Height**   This button uses the value in the text box and measures the door placement from the wall bottom to the header (top) of the door.

- **Width**   This text box allows you to enter a custom width for the door. The width from the standard size is overwritten.

■ **Height**	This text box allows you to enter a custom height for the door. The height from the standard size is overwritten.
■ **Rise**	This text box allows you to enter a custom rise for nonrectangular doors. The rise from the standard size is overwritten.
■ **Leaf**	This text box allows you to enter a custom leaf width for uneven doors. The *leaf* controls the width of one of the doors in this style of door.

As in the Add Walls dialog box, select any configurations desired for the type of door to be added to the drawing. When the changes have been made, pick once in the drawing to make the drawing active, then select a wall to place the door. Once a wall has been selected, the door can be freely moved around but remains only within walls. To place the door, pick with the pick button. The door is placed into the drawing within the wall, and an opening is cut into the wall.

You can accurately place the swing of the door with slight movements of the crosshairs before picking the door location (see Figure 3.35). This works best when Automatic Offset/Center is used to place the door. As the crosshairs are moved to a location for the door, the door "locks" into place. Move the crosshairs to each side of the wall to see the swing change sides. Moving the crosshairs slightly also mirrors the hinge location for the door. When you have found the swing and hinge location, pick the left mouse button to insert the door into the wall. The door can be modified later if the hinge of the door is placed wrong.

You can insert as many doors into the drawing as needed. Once one door has been inserted into a wall, another door is attached to the crosshairs for insertion. You can change the style, size, and opening before inserting another door. Remember that each time you pick the Add Doors dialog box to make a change for new doors, the dialog box becomes active. Pick once in the drawing to make the drawing active before you continue adding doors. When you are finished inserting doors into the walls, press the [Enter] key, or pick the Close button in the Add Doors dialog box.

You can also add doors into the drawing without inserting them into walls. To do this, access the Add Doors dialog box, and configure any settings for the new door. When you pick once in the drawing, making the drawing active, you are prompted to *Select wall, space boundary or RETURN:.* Pressing [Enter] before selecting a wall allows you to insert the door into the drawing without attaching it to a wall. After you insert the door you are prompted for a rotation angle. Specify the rotation angle for the new door. This places a door into the drawing independent of a wall.

Field Notes

Inserting doors into the drawing and not into a wall can be useful when working on legacy drawings created with standard lines. This allows you to use Architectural Desktop doors without having to convert your old drawings into Architectural Desktop walls.

FIGURE 3.35 Slight movements of the crosshairs allow you to place where the hinge and swing of the door are to be located.

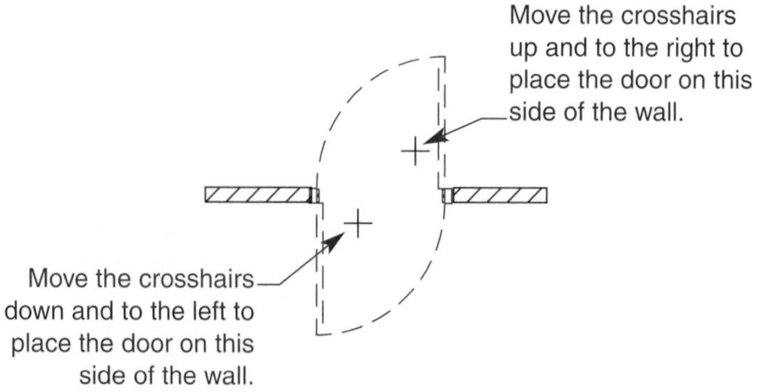

Move the crosshairs up and to the right to place the door on this side of the wall.

Move the crosshairs down and to the left to place the door on this side of the wall.

Field Notes

You can insert a new door into the drawing based on the parameters of an existing door. First, select the door that contains the style, width, and height that you want to use for the new door, then right-click and select Add Selected from the cursor menu. The Add Doors dialog box is displayed with the appropriate settings from the selected door.

Exercise 3.6

1. Open EX3-5.dwg.
2. Place doors using the sizes and styles indicated in the figure.
3. Use the Automatic Offset/Center option to place the doors accurately.
4. Do not try and change the swings if they come out on the wrong side. You can modify them in the next section.
5. Save the drawing as EX3-6.dwg.

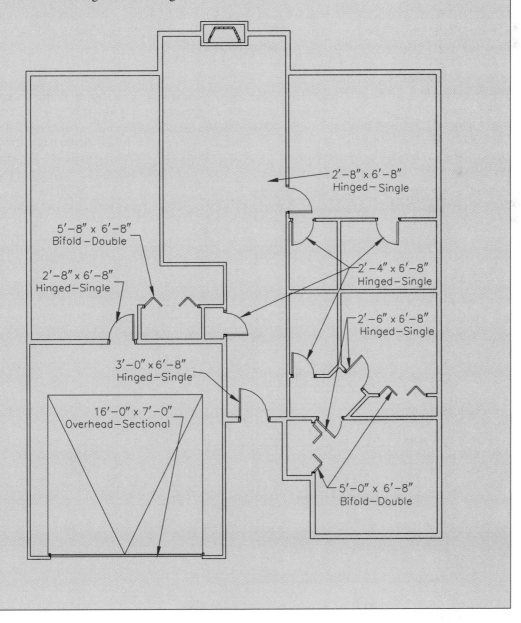

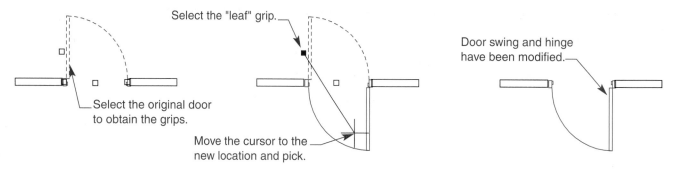

FIGURE 3.36   Use grips to modify the swing and the hinge of the door leaf after the door has been placed into the drawing.

Using AutoCAD Commands on Doors

Like wall objects, doors can be modified with most of the AutoCAD commands. The simplest change that can be made to a door is to reverse the side of the wall where the door is swinging, or the hinge side for the door leaf. To do this, first select the door that needs to be changed, then use the door leaf grip and the Stretch grip mode to modify its location (see Figure 3.36). Selecting the grip allows you to move the crosshairs and pick the new location of the door leaf. In Architectural Desktop, the side of the wall where the door leaf is located is considered the "inside" of the room.

Using the Properties Command

The **Properties** command can be helpful in modifying Architectural Desktop objects. Properties such as the layer and insertion position of the object can be modified as well as properties such as width, height, and style. The way to enter the Properties command is to double-click the object that needs to be modified. The Properties dialog box shown in Figure 3.37 is displayed. The object that you double-clicked becomes a

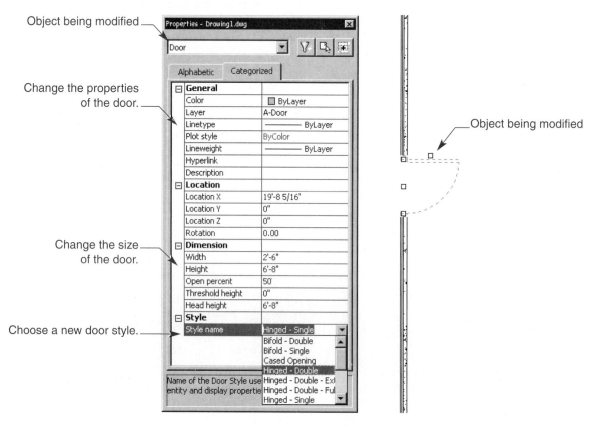

FIGURE 3.37   The Properties dialog box is used to change the properties of AEC objects and AutoCAD geometry.

gripped object with the grip boxes displayed. At the top of the Properties dialog box, the type of object selected is displayed in the drop-down list.

Use one of the following options to open the Properties dialog box:

✓ Select the object to be modified, right-click, and select **Properties** from the cursor menu.

✓ Select the **Modify** pull-down menu, and select **Properties** from the list.

✓ Pick the **Properties** button from the standard toolbar.

✓ Press the [Ctrl]+[1] key combination.

✓ Type **PROPERTIES** at the Command: prompt.

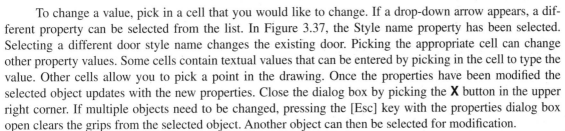

To change a value, pick in a cell that you would like to change. If a drop-down arrow appears, a different property can be selected from the list. In Figure 3.37, the Style name property has been selected. Selecting a different door style name changes the existing door. Picking the appropriate cell can change other property values. Some cells contain textual values that can be entered by picking in the cell to type the value. Other cells allow you to pick a point in the drawing. Once the properties have been modified the selected object updates with the new properties. Close the dialog box by picking the **X** button in the upper right corner. If multiple objects need to be changed, pressing the [Esc] key with the properties dialog box open clears the grips from the selected object. Another object can then be selected for modification.

For additional information on the Properties dialog box, look up ***Editing Object Properties*** in Architectural Desktop Help.

Modifying Doors

As the design process develops, you may decide that the door you originally placed in your drawing needs to be a different size or style. Like wall objects, door objects can be modified through the **Modify Doors** dialog box (see Figure 3.38). The Modify Doors dialog box can be used to change the style, size, opening percentage, and the vertical alignment. The easiest way to display this dialog box is to first select the door to be modified, right-click, then select Door Modify… from the cursor menu. Make any changes to the door as needed. When finished making changes to the door press the Apply button to see the changes take effect, then press the OK button.

Use one of the following options to access the Modify Doors dialog box:

✓ Select the **Design** pull-down menu, select **Doors,** then the **Modify Doors…** command.

✓ Pick the **Modify Door** button in the **Doors-Windows-Openings** toolbar.

✓ Right-click in the drawing area, and select **Design, Doors,** then the **Modify Doors…** command.

✓ Type **AECDOORMODIFY** at the Command: prompt.

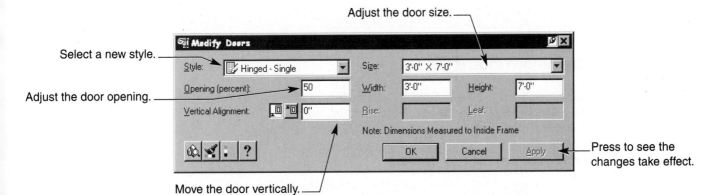

FIGURE 3.38   The Modify Doors dialog box is used to modify a door after it has been placed in the drawing.

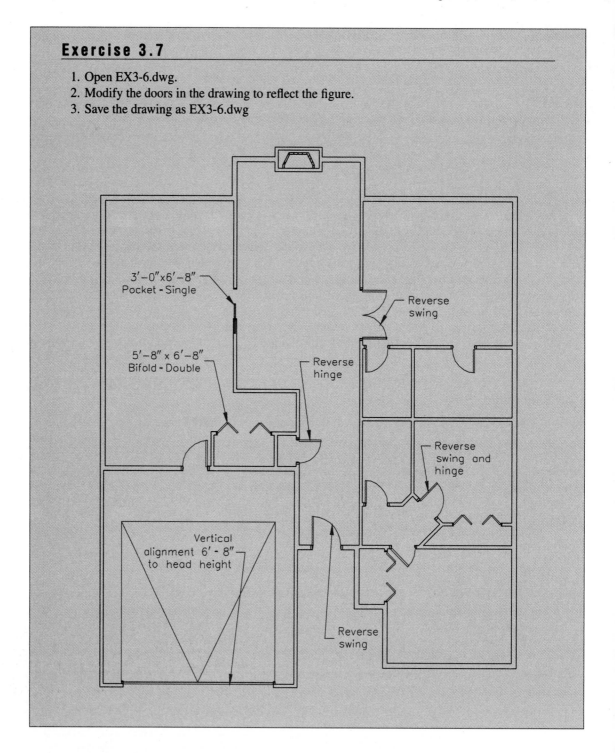

Exercise 3.7

1. Open EX3-6.dwg.
2. Modify the doors in the drawing to reflect the figure.
3. Save the drawing as EX3-6.dwg

Creating Window Objects

Like door objects, window objects understand how to interact with wall objects. Once a window has been inserted into a wall, it becomes constrained to the wall. Like doors, window objects cut a hole out of the wall, so the header over the door does not display in a plan view.

Adding Windows to Your Drawing

Use one of the following options to display the **Add Windows** dialog box, shown in Figure 3.39:

Window styles

Horizontal placement of
window from a corner or
centered on the wall

Open or close window.

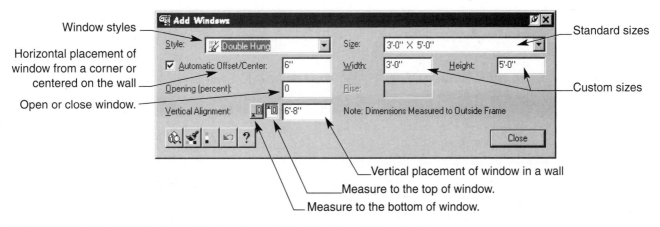

Standard sizes

Custom sizes

Vertical placement of window in a wall
Measure to the top of window.
Measure to the bottom of window.

FIGURE 3.39   The Add Windows dialog box is used to add windows into wall objects.

✓ Select the **Design** pull-down menu, select the **Windows** cascade menu, then pick **Add Window….**

✓ Pick the **Add Window** button from the **Doors-Windows-Openings** toolbar.

✓ Right-click in the drawing area, and select **Design, Windows, Add Window...** from the cursor menu.

✓ Select a wall, right-click, then select **Window...** from the **Insert** cascade menu.

✓ Type **AECWINDOWADD** at the Command: prompt.

The Add Windows dialog box looks similar to the Add Doors dialog box. Set any of the parameters as you did for the door objects.

To place the window in the drawing, pick once in the drawing window to make the drawing active, then pick a wall to add the window. The window is constrained to the wall, so any movement of the crosshairs moves the window along wall objects. When the window is in the appropriate location, click the left mouse button to insert the window. Place as many windows in the drawing as needed. Changes can be made to the style, width, or height at any time. Just remember to pick in the drawing first to make the drawing active before trying to place the windows. Use the Automatic Offset/Center option to offset the new windows a distance from any wall corner or from other windows or doors. To place the window in the center of a wall segment, use the Automatic Offset/Center option and move the crosshairs to the center of the desired wall. Pick to place the window.

You can also add windows into the drawing without inserting them into walls. To do this, access the Add Windows dialog box. Configure any settings for the new window. When you pick once in the drawing, making the drawing active, you are prompted to *Select wall, space boundary or RETURN:*. Pressing [Enter] before selecting a wall allows you to insert the window into the drawing without attaching it to a wall. After you insert the window you are prompted for a rotation angle. Specify the rotation angle for the new window. This places a window in the drawing independent of a wall.

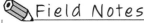Field Notes

Inserting windows into the drawing and not into a wall can be useful when working on legacy drawings created with standards lines. This allows you to use Architectural Desktop windows without having to convert your old drawings into Architectural Desktop walls.

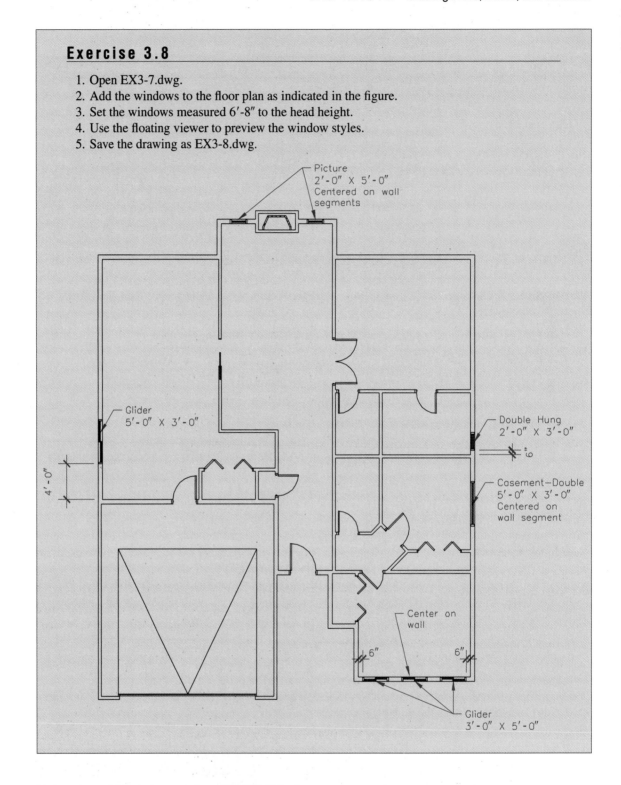

Exercise 3.8

1. Open EX3-7.dwg.
2. Add the windows to the floor plan as indicated in the figure.
3. Set the windows measured 6'-8" to the head height.
4. Use the floating viewer to preview the window styles.
5. Save the drawing as EX3-8.dwg.

Picture
2'-0" X 5'-0"
Centered on wall
segments

Glider
5'-0" X 3'-0"

Double Hung
2'-0" X 3'-0"

Casement—Double
5'-0" X 3'-0"
Centered on
wall segment

4'-0"

Center on
wall

6"   6"

Glider
3'-0" X 5'-0"

Using AutoCAD Commands to Modify Windows

As with wall and door objects, AutoCAD modifying commands can be used to modify window objects. Grip editing can be used to ensure that the window is on the right side of the wall (see Figure 3.40). The grip box that is away from the window indicates the inside of the window. This becomes more apparent when casement windows are open or when double-hung or single-hung windows are viewed in an elevation view, and the top window pane is on the wrong side of the wall. Other standard commands such as the grip modes, Erase, Copy, Move, Rotate, Scale, Mirror, and Properties can also be used to modify window objects.

Windows can be relocated in a wall through the use of grips. To do this, pick a grip to make it hot, and stretch the end of the window to the new location. Door and window frames can be overlapped to

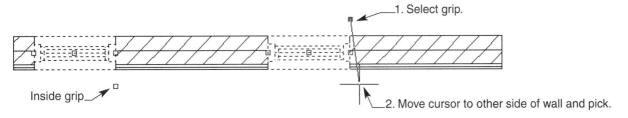

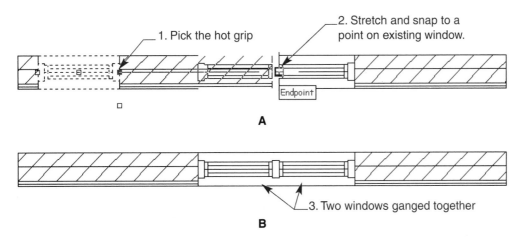

FIGURE 3.40 The grip away from the window represents the inside of the window. Use grip editing to modify its location.

FIGURE 3.41 Using grips to move one window next to another window.

"gang" them together (see Figure 3.41). After selecting the hot grip, use object snaps to snap the grip point to a new location next to another window.

Modifying Windows

During the design phase of a project, specifications may call for different types of doors or windows after they have been drawn, or sizes of windows may change to meet the client's wishes. Just like walls or doors, windows can be modified after they have been placed in the drawing. The **Modify Windows** dialog box is used to make these changes (see Figure 3.42). To access this dialog box, select the window to modify, right-click, and select Window Modify... from the cursor menu.

In addition, you can use one of the following methods to access the Modify Windows dialog box:

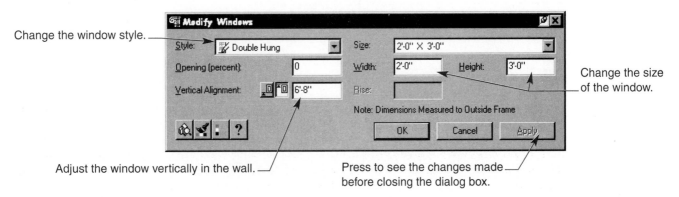

FIGURE 3.42 The Modify Windows dialog box is used to make changes to an existing window. Properties such as style, width, height, and vertical alignment can be changed.

✓ Open the **Design** pull-down menu, and select **Windows, Modify Window….**

✓ Pick the **Modify Window** button from the **Doors-Windows-Openings** toolbar.

✓ Right-click, and select **Design, Windows, Modify Window…** from the cursor menu.

✓ Type **AECWINDOWMODIFY** at the Command: prompt.

Make any adjustments to the properties, then press the Apply button to see the changes applied before exiting the dialog box.

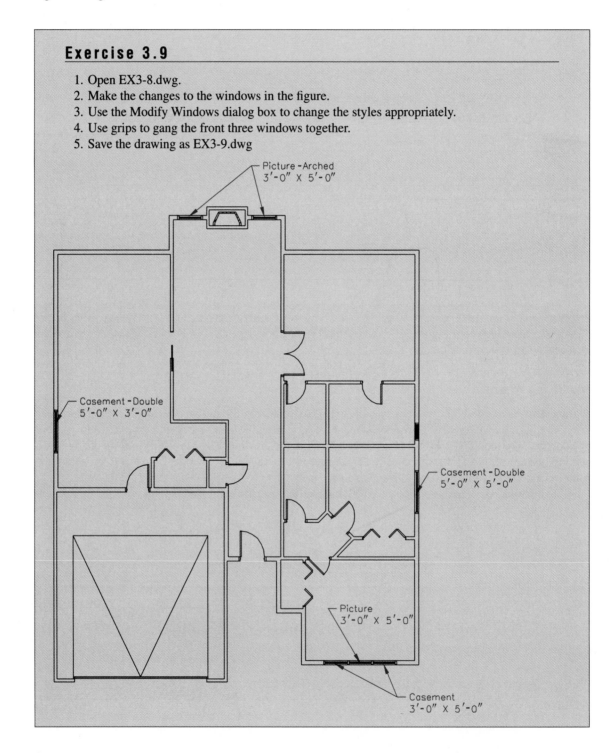

Exercise 3.9

1. Open EX3-8.dwg.
2. Make the changes to the windows in the figure.
3. Use the Modify Windows dialog box to change the styles appropriately.
4. Use grips to gang the front three windows together.
5. Save the drawing as EX3-9.dwg

Picture - Arched
3'-0" X 5'-0"

Casement - Double
5'-0" X 3'-0"

Casement - Double
5'-0" X 5'-0"

Picture
3'-0" X 5'-0"

Casement
3'-0" X 5'-0"

Creating an Opening Object

An *opening object* is an object that is used to cut a hole out of a wall similar to a door or window. An opening, however, does not include a frame, leaf, or glazing. Openings can be of many shapes and sizes and are represented in a floor plan view as a cross in a wall where the opening is placed.

Adding Openings

Openings can be added in a wall by selecting a wall, right-clicking, and picking the Insert option then the Opening... command. The **Add Openings** dialog box, shown in Figure 3.43, is displayed. In addition, you can use one of the following methods to access the Add Openings dialog box:

- ✓ Select the **Design** pull-down menu, and pick **Openings,** then **Add Opening....**

- ✓ Pick the **Add Opening** button from the **Doors-Windows-Openings** toolbar.

- ✓ Right-click and select **Design, Openings, Add Opening...** from the cursor menu.

- ✓ Type **AECOPENINGADD** at the Command: prompt.

The Add Opening dialog box is similar to the Add Doors or Add Windows dialog boxes. Instead of a style of opening, however, you can select a predefined shape by picking the **Predefined Shape:** radio button and selecting a shape from the Predefined Shape: drop-down list. You can also use custom shapes in the opening by selecting the Custom Shape: radio button and selecting a shape from the drop-down list. Use the floating viewer to preview the shapes. The custom shapes use an object known as an **AECProfile**. You can create your own custom shapes and use them as AECProfiles in many Architectural Desktop objects. Creating custom shapes is discussed in Chapter 7.

After selecting the shape, determine the size for the opening, the offset if desired, and the vertical alignment. When you are ready to add the opening into the drawing, pick once in the drawing to make the drawing active, then select a wall. Move the opening to the desired location and pick. When you are finished adding openings to the drawing, press [Enter] or the Close button on the Add Openings dialog box to end the command.

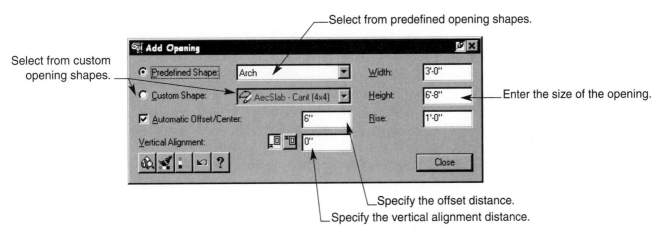

FIGURE 3.43   The Add Opening dialog box is used to add openings into a wall that do not include a frame, door leaf, of glazing.

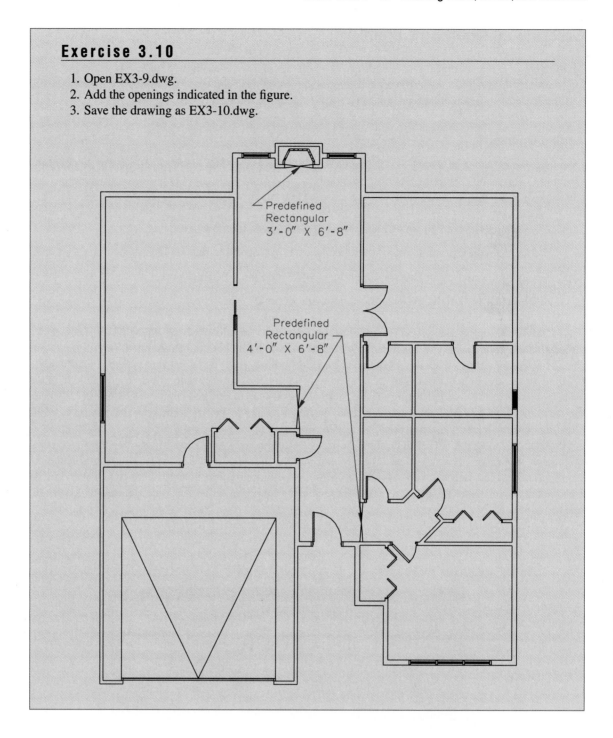

Exercise 3.10

1. Open EX3-9.dwg.
2. Add the openings indicated in the figure.
3. Save the drawing as EX3-10.dwg.

Predefined
Rectangular
3'-0" X 6'-8"

Predefined
Rectangular
4'-0" X 6'-8"

Modify Openings

Openings can also be changed through the Modify Openings dialog box (see Figure 3.44). The easiest way to access this dialog box is to select the opening(s) first, then right-click and select Opening Modify... from the cursor menu. In addition, you can use one of the following options to access the Modify Openings dialog box:

✓ Pick the **Design** pull-down menu, select **Openings,** then **Modify Openings....**

✓ Pick the **Modify Opening** button on the **Doors-Windows-Openings** toolbar.

✓ Right-click, and select **Design, Openings,** then **Modify Opening...** from the cursor menu.

✓ Type **AECOPENINGMODIFY** at the Command: prompt.

FIGURE 3.44 The Modify
Opening dialog box is used to
modify the shape and size of an
existing opening.

Make any changes desired, such as a custom shape instead of a predefined shape, or change the opening size or the vertical alignment. When finished modifying the opening, press the Apply button to apply the changes before pressing the OK button.

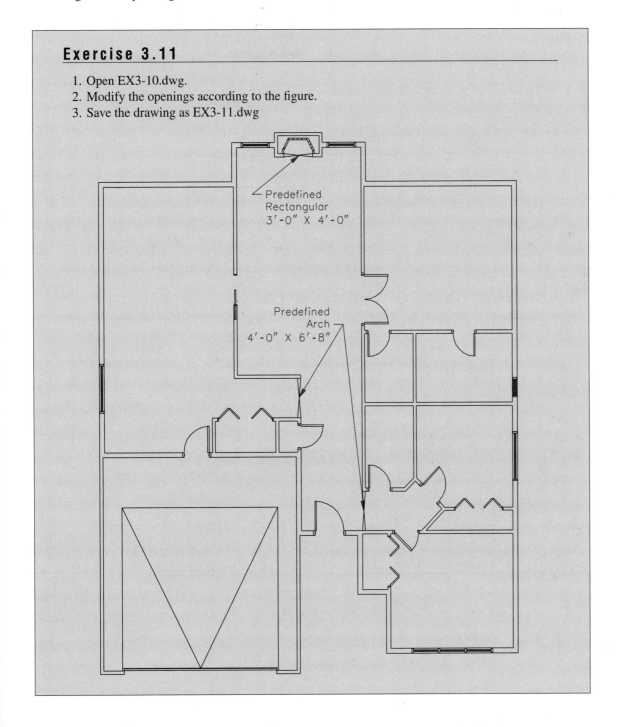

Exercise 3.11

1. Open EX3-10.dwg.
2. Modify the openings according to the figure.
3. Save the drawing as EX3-11.dwg

Creating Window Assembly Objects

Window assembly objects create several windows or a combination of windows and doors that are grouped together to form one object. Window assemblies can be used in the creation of a commercial building where several windows can be found side by side. In residential applications the window assembly can group together a door and sidelight windows, or other designs as desired. The AEC Arch (Imperial).dwt contains some preconfigured window assemblies that can be used immediately. Architectural Desktop includes additional window assembly styles that can be loaded into your drawing.

Adding Window Assemblies

Like door, window, and opening objects, a window assembly can be added to a drawing by right-clicking. Select Insert, then Window Assembly…. The **Add Window Assemblies** dialog box shown in Figure 3.45 is displayed.

In addition, you can use one of the following options to insert Window Assemblies into a wall:

✓ Open the **Design** pull-down menu, select **Window Assemblies,** and select **Add Window Assembly….**

✓ Pick the **Add Window Assembly** button in the **Doors-Windows-Openings** toolbar.

✓ Right-click and select **Design, Window Assemblies, Add Window Assembly….**

✓ Type **AECWINASSEMBLYADD** at the Command: prompt.

In the Style: list box, select the style to be used. Use the floating viewer to preview the selected styles. The floating viewer is designed to view the window assembly makeup. Some of the styles include a door. These styles do not display the door until the entire window assembly is attached to the wall.

The width and height values represent the overall size of the assembly. Each style has a default size. If the size is changed, the width or height text box changes, and the text is displayed in a red color in the dialog box, indicating that a different size has been entered. If the window assembly is to be located a distance from a corner, specify the Automatic Offset/Center distance. The window assembly can also be placed in the wall vertically measured a distance to the sill height or to the head height.

After you have specified the values, pick inside the drawing once to activate the drawing, then pick a wall in which to insert the window assembly. Move the cursor around your walls until you have found the proper location, then pick to insert the window assembly. The wall is cut out and the assembly added to the wall. Press [Enter] or the Close button to stop placing window assemblies. If a window assembly with a door has been used, the swing of the door can be changed using grips, as previously discussed. Doors within a window assembly can also be changed by selecting the door, right-clicking, and selecting Door Modify… from the cursor menu.

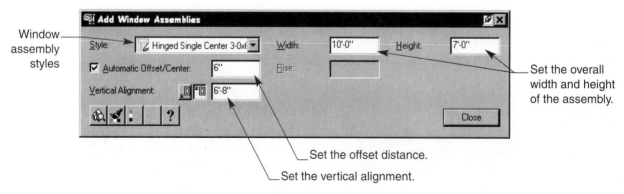

FIGURE 3.45   The Add Window Assembly dialog box is used to add window assemblies into a wall.

Exercise 3.12

1. Open EX3-11.dwg.
2. Add the window assemblies indicated in the figure.
3. Set the vertical alignment measured 6'-8" to the head height.
4. Center the window assemblies on the walls.
5. Save the drawing as EX3-12.dwg.

Hinged single
center 3-0x6-8
+sidelights
12'-0" X 6'-8"

Hinged single
center 3-0x6-8
+Sidelights
12'-0" X 6'-8"

Modifying Window Assemblies

Window assemblies can also be changed in ways similar to the way doors or windows are modified. To modify a window assembly, first select the assembly, then right-click and select Window Assembly Modify…. It is important to select the assembly and not the door. If the window assembly includes a door, selecting the door object accesses the Door Modify… command in the cursor menu. The **Modify Window Assembly** dialog box is displayed, allowing you to make changes to the assembly (see Figure 3.46).

In additon, you can use one of the following options to access the Modify Window Assembly dialog box:

✓ Pick the **Design** pull-down menu, and select **Window Assemblies,** then **Modify Window Assembly….**

✓ Pick the **Modify Window Assembly** button in the **Doors-Windows-Openings** toolbar.

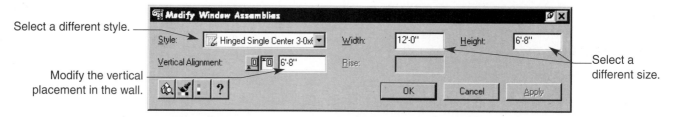

Select a different style. —

Modify the vertical
placement in the wall. —

Select a
different size.

FIGURE 3.46   The Modify Window Assembly dialog box is used to modify the
style, size, and vertical alignment of the assembly.

✓ Right-click, and select **Design, Window Assemblies, Modify Window Assembly….**

✓ Type **AECWINASSEMBLYMODIFY** at the Command: prompt.

Make any changes necessary, and pick the Apply button to apply the changes before you close the dialog
box. Press [Enter] or the Close button to close the dialog box when you have finished making modifications.

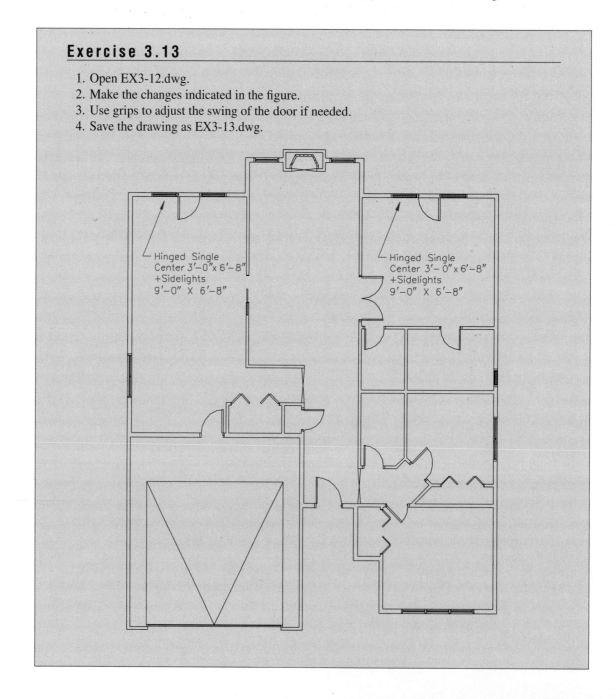

Exercise 3.13

1. Open EX3-12.dwg.
2. Make the changes indicated in the figure.
3. Use grips to adjust the swing of the door if needed.
4. Save the drawing as EX3-13.dwg.

Hinged Single
Center 3'−0"x 6'−8"
+Sidelights
9'−0" X 6'−8"

Hinged Single
Center 3'−0"x 6'−8"
+Sidelights
9'−0" X 6'−8"

CHAPTER REVIEW

Use the CD-ROM to test your knowledge and skills.

Chapter Test

To check your understanding of the content provided in this chapter, access the Test file in the CH03 folder of the CD-ROM that accompanies this text.

Chapter Project

To practice the Architectural Desktop skills presented in this chapter, access the Project files in the CH03 folder of the CD-ROM that accompanies this text. The project files are in pdf format and include sample drawings and instructions for completing each project.

|||||||||||||||||||||||||||||||||||||||||||||||||||||||||||||||||||||||||||||||||||||||||

CHAPTER 4

Drawing Display Commands

LEARNING GOALS

After completing this chapter, you will be able to:

◎ Use the ZOOM command and its options.

◎ Perform real-time zooms.

◎ Use the PAN command.

◎ Apply 3D Orbit to change your views.

◎ Use projection methods, shading, and visual aid tools.

◎ Employ display commands transparently.

◎ Shade your 3D model.

◎ Regenerate your drawing.

◎ Use views in your drawing.

◎ Establish preset views.

◎ Create model space viewports.

◎ Use the Object Viewer.

◎ Create perspectives and movie files.

When you create drawings in Architectural Desktop, the drawings are drawn at full scale. This enables you to draw with precision and to catch design flaws at a scale that is in real units. As you draw, situations can arise that require you to look more closely at your work, to work on a small detail, or to move farther away from your drawing to see the total drawing. When developing a design for a new building or completing construction documents you may need to view different parts of the drawing in many different ways. Architectural Desktop includes many tools that let you view your work from different angles, display your objects as 2D drawings or 3D models, and use different types of visualization methods. This chapter is devoted to the visualization tools in Architectural Desktop Release 3 that make you productive as you create construction documents.

Using the Zoom Commands

The ability to "get closer to your work" in Architectural Desktop is related to the functions of a camera. When taking pictures with a camera, you typically *zoom in* closer to the subject or *zoom out* to encompass the entire scene. The same process and terminology is used in Architectural Desktop. Zooming allows you to move around the drawing and display various areas of the design. The **ZOOM** command is a very helpful tool that can be used often as you design your building and complete the construction documents. The different zoom options are discussed in this section.

FIGURE 4.1 The Zoom options can be found on the Standard toolbar and in the Zoom flyout, in the Zoom toolbar, from the pull-down menus, or by entering **ZOOM** at the Command: prompt.

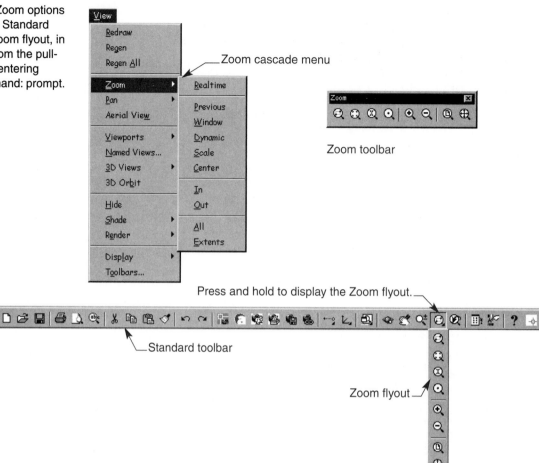

Using the ZOOM Options

There are eight separate zooming options within the ZOOM command. All these options can be found on the Standard toolbar, within the ZOOM command, or by picking the View pull-down menu, then selecting the Zoom cascade menu. The Zoom toolbar can also be used to select most of these options (see Figure 4.1).

Entering ZOOM at the Command: prompt displays the following prompt with all the options:

Command: **Z** or **ZOOM** ↵
Specify corner of window, enter a scale factor (nX or nXP), or
    [All/Center/Dynamic/Extents/Previous/Scale/Window] <real time>:

The zooming options are described next.

   ■ **All**   This option zooms to the edge of the drawing limits. When a drawing is started from a template a series of dots known as the grid indicates the drawing limits. Drawing limits are used to set up the available drawing area within a drawing file. The available area can be modified with the LIMITS command. If objects are drawn beyond the drawing limits, the All option zooms to the edge of your geometry.

   ■ **Center**   This option is used to center a selected point on the center of your screen. When accessing this option you are prompted to select a point that will be centered on the screen. After you have selected the point, you can specify how close or far the display will be away from your line of sight. The following prompt sequence is displayed when this option is used:

Command: **Z** or **ZOOM** ↵
Specify corner of window, enter a scale factor (nX or nXP), or
   [All/Center/Dynamic/Extents/Previous/Scale/Window] <real time>: **C** ↵
Specify center point: *(pick a point to be centered)*
Enter magnification or height <current>: *(enter a zooming distance or magnification)*

You can enter a magnification by specifying a number followed by an *x,* such as 2x. Entering a number without an *x* specifies a height above the drawing. A default height value is displayed in brackets and can be selected by pressing the [Enter] key.

- **Dynamic**   This option is used to resize what you are currently looking at and to specify a new area of the drawing in which you would like to view. When this option is selected, the drawing screen displays your entire drawing with a few rectangles (see Figure 4.2). The large blue dashed rectangle represents the drawing limits and shows where your drawing fits in relationship to the limits. A green dashed rectangle represents the current display you were viewing before entering the Dynamic option. Your cursor also changes to a rectangle with an **X** in the middle, which represents the current size of your drawing screen. Moving the cursor around the drawing and pressing [Enter] moves the drawing screen display to that area.

You can resize the cursor box to make the drawing screen larger or smaller by picking the left mouse button once. This changes the cursor to a rectangle with an arrow. Moving the mouse resizes the represented drawing screen. Picking the left mouse button again locks the size of the represented drawing screen and allows you to move to another part of the drawing. Press [Enter] to display the selected portion of the drawing.

- **Extents**   This option zooms the drawing screen to the edge of the outermost geometry in the drawing. This is the portion of the drawing that contains objects.

- **Previous**   This option is not included in the Zoom toolbar but can be found on the Standard toolbar. Using this option returns the previous display or zoom. You can go back to the last 10 displays, one at a time.

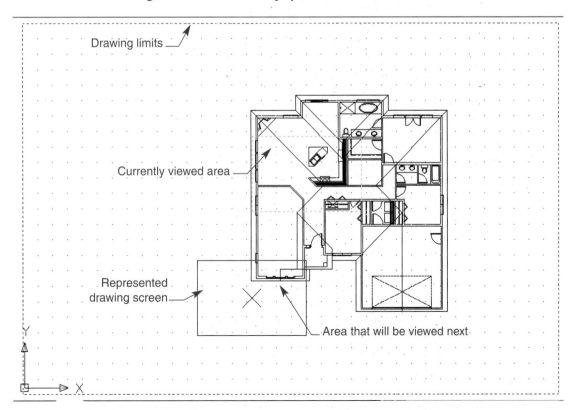

FIGURE 4.2   The Dynamic option allows you to resize and change what is displayed in the drawing screen

■ **Scale**   When you use this option you are prompted to enter a scale factor. This option serves two purposes. The first is to scale the current view in the drawing by a factor. For example, if you want to zoom into the drawing "three times" the current view size, then enter 3X at the prompt.

     The other purpose is to set the drawing up for plotting from Layout space. This option scales the view in model space relative to the view in layout space. To change the scale of a drawing in layout space, you enter a number at the prompt followed by XP. For example, a drawing that is to be plotted at 1/4″=1′-0″ is scaled in layout space to 1/48XP. Setting a drawing up in layout space is discussed later in this text.

■ **Window**   This option is one of the more commonly used zooming options. This option displays portions of your drawing within a window or box that you establish. When you use this option, you are prompted to pick the first corner of a box. After you have picked the first corner, you are prompted for the opposite corner of the box. Picking the opposite corner of a box creates the window. Anything within the window is what is displayed on the drawing screen. The Window option is also one of the defaults when the ZOOM command is first entered. Immediately selecting two points of a box instead of entering W at the Zoom prompt allows you to automatically use the Window zoom option.

■ **<realtime>**   The default option allows zooming into and out of the drawing in real time. This button can be found on the Standard toolbar. This option is discussed in the next section.

■ **In**   This option is available only from the Zoom toolbar or the <u>V</u>iew pull-down menu. This option zooms into the drawing at a scale factor of 2X.

■ **Out**   This option also is available only in the Zoom toolbar or the <u>V</u>iew pull-down menu. This option zooms out of the drawing at a scale factor of .5X.

Using Realtime Zoom

As indicated earlier, the <realtime> option zooms in or out of the drawing in realtime or dynamically. This means that you can see your drawing change size as you are performing this zooming feature. Use one of the following options to activate realtime zoom:

✓ Press the Zoom Realtime button on the Standard toolbar.

✓ Select the <u>V</u>iew pull-down menu.

✓ Pick the Zoom cascade menu, then <u>R</u>ealtime.

✓ Right-click and select Zoom from the cursor menu.

The crosshairs are changed into a magnifying glass with a + and a − sign (see Figure 4.3).

     By pressing and holding down the left mouse button (pick button), you can zoom the display in or out by moving the mouse up or down. By releasing the pick button, you can reposition the cursor for additional zooming when you press and hold the pick button again. Moving the mouse up zooms into the drawing, and moving the mouse down zooms out of the drawing. When you have found the appropriate display, exit the command by pressing the [Esc] button or by right-clicking and selecting Exit from the cursor menu.

     Right-clicking with the Zoom Realtime cursor active displays a cursor menu with some different zooming options (see Figure 4.4). A check mark beside an option indicates the current mode. These options are described next.

FIGURE 4.3 Accessing the
Zoom Realtime command
changes the crosshairs into a
magnifying glass

FIGURE 4.4 The Realtime cursor menu.

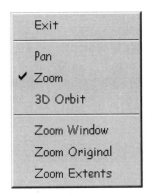

- **Exit**   This option exits you from any of the realtime commands.

- **Pan**   This option activates the Pan Realtime command, which adjusts the view of the drawing by sliding it from side to side. When this option is selected, the Zoom Realtime cursor changes to a hand icon. This option is discussed in the next section.

- **Zoom**   This is the Zoom Realtime option. This option can be selected from the Pan Realtime or the 3D Orbit icons.

- **3D Orbit**   This option changes the point of view around the drawing. With the Pan or Zoom options, the display is changed based on the current viewing direction from your eye to the drawing. With the 3D Orbit option, the viewing direction from your eye to the drawing or model can be adjusted. This option is discussed later in this chapter.

- **Zoom Window**   This option changes the cursor into a rectangle with an arrow. Unlike with the standard Window option, to create a zooming window, move the mouse to the first point of the desired zooming window. Press and hold the pick button and drag the cursor to create a window. When you release the pick button, the windowed area becomes the new displayed area.

- **Zoom Original** This option restores the original view of the drawing before any realtime zooming was performed.

- **Zoom Extents** This option zooms to the edges of your geometry.

Exercise 4.1

1. Open EX3-13.dwg.
2. Zoom into the drawing using Zoom Realtime.
3. Use Zoom Extents to zoom to the edges of your drawing.
4. Use Zoom Window to zoom into the southeastern corner of the building.
5. Use Zoom Previous to display the previous zoom.
6. Use Zoom All to see the building and the entire drawing area.
7. Save the drawing as EX4-1.dwg.

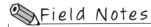Field Notes

If you have a wheel mouse, scrolling the wheel allows you to zoom in and out of your drawing. Double-clicking the wheel button performs a Zoom Extents of the drawing.

Using the Pan Command

The **PAN** command is used to adjust the display of the drawing without changing the magnification of the drawing. This command allows you to slide the drawing around on the screen, for example, after you have zoomed into the drawing and a part of the drawing lies outside of the display screen. Use one of the following options to access the PAN command:

✓ Pick the **Pan Realtime** button from the **Standard** toolbar.

✓ Right-click in the drawing screen, and select **Pan** from the cursor menu.

✓ Access the **View** pull-down menu, select the **Pan** cascade menu, and select **Real Time.**

✓ Type **PAN** or **P** at the Command: prompt.

Once the command has been activated, the crosshairs change to a hand icon.

As with the Zoom Realtime command, press and hold the pick button to slide the drawing up or down, or from side to side. Release the pick button to reposition the cursor for additional panning of the drawing. Right-clicking displays the Realtime cursor menu shown in Figure 4.4. Selecting an option from the cursor menu changes the cursor icon appropriately. To exit the command, press the [Esc] key or right-click and select Exit from the cursor menu.

If you pan to the edge of the drawing, a bar is displayed on one side of the hand cursor. If this happens, additional panning in that direction cannot be continued until the drawing has been regenerated. Using the **REGEN** command allows you to continue panning in that direction. REGEN is discussed later in this chapter.

 Field Notes

If you have a wheel mouse, pressing and holding down the wheel while moving the mouse allows you to slide the view around the drawing.

The scroll bars in the lower left corner also allow you to pan the drawing.

Exercise 4.2

1. Open EX4-1.dwg.
2. Use the PAN command to slide the drawing from right to left.
3. Use the right-click shortcut menu to switch between the realtime PAN and ZOOM commands.
4. Use the scroll bars to center the drawing on the drawing screen.
5. Save the drawing as EX4-2.dwg.

Using 3D Orbit to Change Your View

The **3D Orbit** command is used to dynamically get a different view or perspective of your drawing. It is very useful when working with 3D models.

Use the following options to access this command:

✓ Select the **3D Orbit** button in the **Standard** or **3D Orbit** toolbar.

✓ Pick the **View** pull-down menu, then select **3D Orbit.**

✓ Type **3DORBIT** or **3DO** at the Command: prompt.

A green circle with four smaller circles at the quadrants is displayed in the center of your drawing screen. This is known as the *Arcball* (see Figure 4.5). The Arcball is used to twist and rotate your view around the drawing. The crosshairs also change to one of four icons, depending on where the cursor is placed in the drawing.

To rotate the drawing, place the cursor in one of the four locations described in Figure 4.5, then press and hold the pick button. Your drawing rotates around and is displayed from a new angle. After you have obtained the new view, press [Esc] or right-click and select Exit to end the command. Notice that once the view is changed from a plan (top) view of the floor plan, the Architectural Desktop objects change into 3D elements with the values you specified when you added them to the drawing (see Figure 4.6).

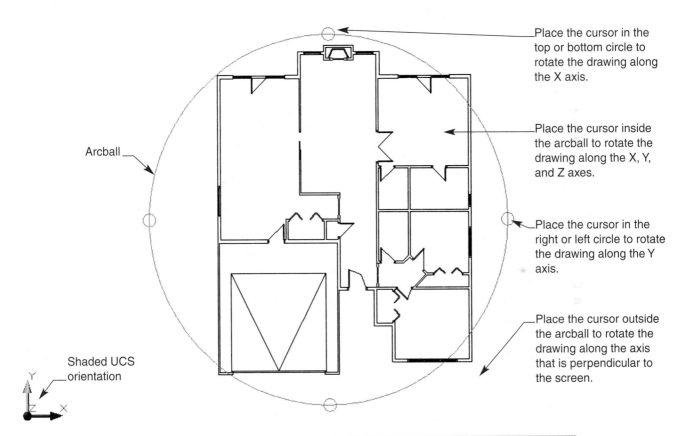

Place the cursor in the top or bottom circle to rotate the drawing along the X axis.

Place the cursor inside the arcball to rotate the drawing along the X, Y, and Z axes.

Place the cursor in the right or left circle to rotate the drawing along the Y axis.

Place the cursor outside the arcball to rotate the drawing along the axis that is perpendicular to the screen.

Arcball

Shaded UCS orientation

Cursor	Appearance	Description
Two ellipses		This cursor icon is displayed when the cursor is inside the arcball. Pressing and holding the pick button while moving the mouse rotates the drawing along the X, Y, and Z axes.
Circular		This cursor icon is displayed when the cursor is outside the arcball. Pressing and holding the pick button while moving the mouse rotates the drawing along the axis that is perpendicular to the screen.
Horizontal ellipse		This cursor icon is displayed when the cursor is placed in the left or right circles on the arcball. Pressing and holding the pick button while moving the mouse rotates the drawing along the Y axis.
Vertical ellipse		This cursor icon is displayed when the cursor is placed in the top or bottom circles on the arcball. Pressing and holding the pick button while moving the mouse rotates the drawing along the X axis.

FIGURE 4.5   The arcball is displayed when the 3D Orbit command is accessed. The crosshairs also take on one of four shapes depending on the location of the cursor in the drawing.

FIGURE 4.6 Architectural Desktop objects display as 3D elements when the view is changed from a plan view.

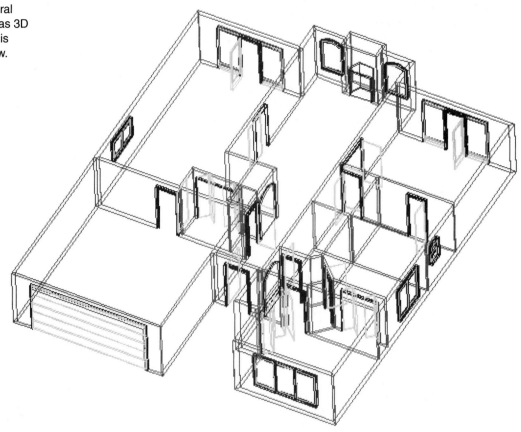

Projection Methods, Shading, and Visual Aid Tools

The 3D Orbit command includes many options for controlling how your model appears in the display screen. These options include projection methods such as parallel or perspective projections, shading and hiding objects, and tools such as a compass and grid. These options can be accessed when the 3D Orbit command is active by right-clicking and selecting the options from the cursor menu shown in Figure 4.7.

The 3D Orbit options are described next.

- **Projection**   A 3D model projection can be parallel or perspective. In the parallel mode, sides of the building are parallel to each other. In perspective mode, the sides of objects point or project to a vanishing point. When using the perspective mode, you can pan and zoom while you are in the 3D Orbit mode only. If you exit the 3D Orbit mode, you cannot edit, zoom, or pan in the drawing until you set the mode back to parallel. Figure 4.8 displays the difference between the parallel and perspective modes on your model.

- **Shading Modes** There are six shading options available in this shortcut menu: four types of shading that can be applied to your 3D models, as well as a hide mode and the default wireframe. These shading modes can also be accessed in the Shade toolbar. These options are discussed in detail later in this chapter.

- **Visual Aids**   The Visual Aids shortcut menu includes three options that assist you in relating the chosen view to the UCS (User Coordinate System) and aid in determining which is the positive and negative Z axis. These visual aids are described next.

FIGURE 4.7 The 3D Orbit cursor menu contains many options for controlling how the model is viewed in the display screen.

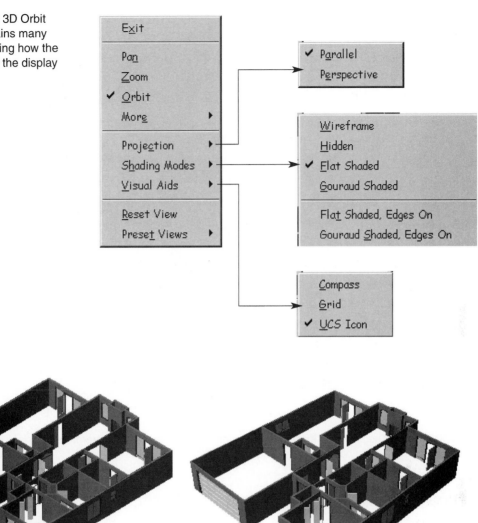

Parallel Projection

Perspective Projection

FIGURE 4.8 With parallel projection the sides of the building are parallel to one another. With perspective projection, the sides of the building project to a vanishing point.

- **Compass**   When this option is selected, a spherical 3D compass is displayed within the arcball. Tick marks and labels indicate the positive X, Y, and Z axes (see Figure 4.9A).

- **Grid**   This option displays a grid that is the same size as the limits. It is displayed at the current XY drawing plane, which by default is located at the 0″ Z axis. This provides a good "groundline" for visualization but does not print (see Figure 4.9B).

- **UCS Icon**   This option turns on or off the shaded UCS icon. This icon is used to help understand where the positive X, Y, and Z axes are pointing. Remember that the X and Y axes determine the drawing plane.

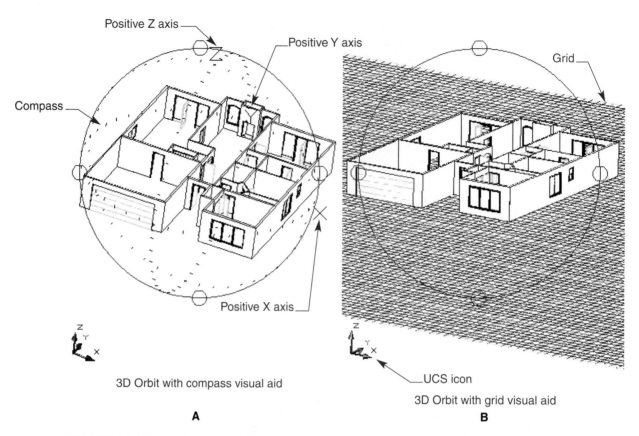

FIGURE 4.9   The compass and grid visual aids can be used to help determine where the positive axes are located as the model is turned around.

Exercise 4.3

1. Open EX4-2.dwg.
2. Use the 3D Orbit command to rotate the drawing.
3. Right-click to exit the 3D Orbit command.
4. Use the 3D Orbit command to try the different shading options.
5. While in the 3D Orbit command, try the different visual aids.
6. Change the projection to perspective.
7. Exit the 3D Orbit command. Try to use a Zoom option. Can you zoom in the drawing?
8. Save the drawing as EX4-3.dwg.

3D Orbit Preset View Options

In addition to allowing you to rotate your model around to get your own view in the drawing, the 3D Orbit command has some preset viewing options that can be used to create different views of the drawing. Right-clicking after initiating the 3D Orbit command gives you the 3D Orbit shortcut menu shown in Figure 4.10. The following options are available:

- **Reset View**   This option resets the view to the view displayed before the 3D Orbit command was entered.

FIGURE 4.10 The Preset view options can be found at the bottom of the 3D Orbit shortcut menu.

Return to the initial view before using 3D Orbit.

Select a user-defined view; available only if named views are present in the drawing.

E_xit	
Pa_n	
_Zoom	
✓ _Orbit	
Mor_e	▶
Proje_ction	▶
S_hading Modes	▶
_Visual Aids	▶
_Reset View	
Prese_t Views	▶
_Saved Views	▶

_Top
_Bottom
_Front
Bac_k
_Left
_Right
S_W Isometric
S_E Isometric
_NE Isometric
NW _Isometric

Preset orthographic views

Preset isometric views

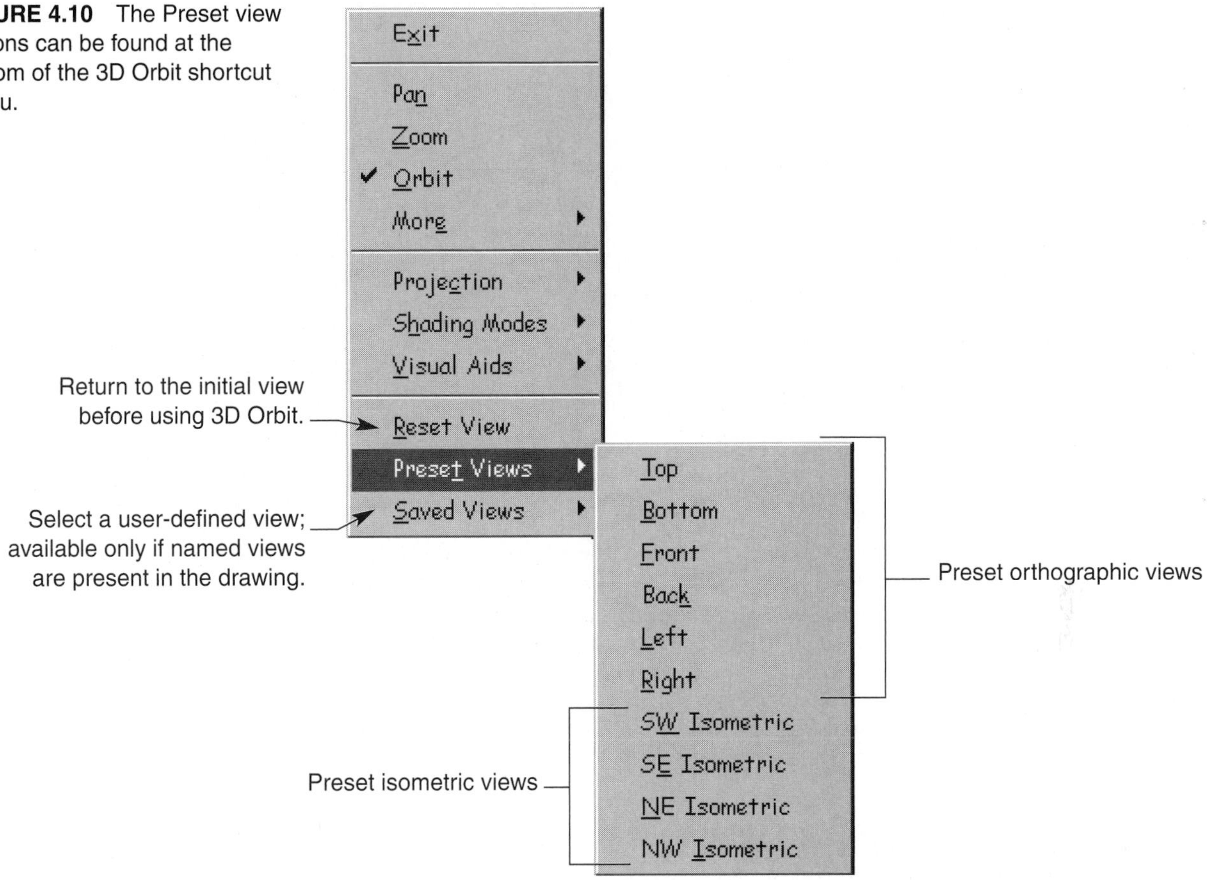

- **Preset Views**   This cascade menu contains a list of preset viewing directions. There are six orthographic views. Four of these views display the front, right, back, and left sides of a building. The top view is used to see the floor plan view. When the top view is selected in Architectural Desktop, the AEC objects revert to 2D objects. There are also four preset isometric views that can be used to display 3D views. Preset views are discussed later in this chapter.

- **Saved Views**   This option is available only if there are named views present in the drawing. Named views are views that are defined by you. Named views are discussed in detail later in this chapter.

There are additional options that can be used to adjust the view of your model in the 3D Orbit shortcut menu. These options can be found in the **Mor_e** cascade menu. Some of these options include **Adjust Distance** and **Swivel Camera**, which adjust the view in perspective projection. The **Continuous Orbit** is used to "spin" the model continuously by holding down the pick button, moving the mouse in the direction the model is intended to spin, and releasing the pick button. Other options found here include the clipping plane options. These options create an imaginary clipping plane that allows you to remove a portion of the model from the front or back of a clipping plane to create a section. These tools can be used to create sections of a building, but Architectural Desktop includes specific tools that accomplish sectioning better. Sections will be covered in detail later in this text.

Exercise 4.4

1. Open EX4-3.dwg.
2. Use the 3D Orbit command to pick some of the preset views.
3. Use the Reset View option.
4. Change the projection back to Parallel.
5. Select the Top preset view.
6. Change the Shading Mode to Wireframe.
7. Enter **SHADEMODE** at the Command: prompt then type 2D↵. This sets the 2D wireframe option current in the drawing. See the following Field Notes.
8. Save the drawing as EX4-4.dwg.

 Field Notes

After the 3D Orbit command is used, the UCS icon maintains a colored 3D appearance even if the shading mode is changed back to a wireframe mode. Any arced objects in the drawing appear segmented to help AutoCAD's display performance. To change the UCS back to the standard color and symbol and to "round out" arced segments, enter **SHADEMODE** at the Command: prompt and type the **2D wireframe** option. Shading modes are discussed later in this chapter.

Using Transparent Display Commands

In AutoCAD/Architectural Desktop, starting a new command usually requires that the current command either be finished or canceled. Selecting a command from a button or a pull-down menu usually cancels the current command and begins the selected command. Transparent commands function without canceling the currently active command.

A *transparent command* temporarily halts the active command so another command can function inside the active command. When the transparent command is completed, the interrupted command is resumed. Many of the display commands such as ZOOM, PAN, and 3DORBIT can be used transparently.

Suppose a wall needs to be drawn that extends somewhere off the screen. One option would be to cancel drawing the wall, zoom out to see more of the drawing, and then start the wall command again. A more efficient method is to start the wall command, then to use PAN or ZOOM transparently while in the wall command. The following gives an example prompt sequence of this situation:

Command: **AECWALLADD**↵
Start point or [STyle/Group/WIdth/Height/OFfset/Justify/Match/Arc]: *(pick a starting point for the wall)*
End point or [STyle/Group/WIdth/Height/OFfset/Justify/Match/Arc]: *(pick the Pan Realtime button or type 'PAN)*
>>Press ESC or ENTER to exit, or right-click to display shortcut menu: *(pan to the new location, then right-click and select Exit)*
Resuming AECWALLADD command.
End point or [STyle/Group/WIdth/Height/OFfset/Justify/Match/Arc]: *(pick an ending point for the wall)*
End point or [STyle/Group/WIdth/Height/OFfset/Justify/Match/Arc/Undo]: ↵

In the preceding prompt sequence the AECWALLADD command was interrupted to accommodate the PAN command. The double prompt (>>) indicates that a command has been interrupted while a transparent command is being used. The transparent command must be completed before the original command is resumed. The ZOOM, PAN, and 3DORBIT commands can be entered transparently by selecting the appropriate toolbar button. A typed transparent command must be preceded by an apostrophe ('), such as **'PAN.** This indicates to AutoCAD that the command being entered is a transparent command. If the command is typed without the apostrophe, AutoCAD returns a "2D point or option keyword required" prompt, indicating that it did not understand the command you entered.

Shading the 3D Model

As indicated in the 3D Orbit section discussed earlier, the model can be shaded in a number of different ways. There are a total of seven different shading options, compared with the six options in the 3D Orbit command. Use one of the following methods to access the different shading options (see Figure 4.11):

✓ Select an option from the Shade toolbar.

✓ Pick the View pull-down menu, then select the Shade cascade menu.

✓ Type **SHADEMODE** at the Command: prompt.

The shading options are described next.

- **2D Wireframe**   This option displays 3D objects as a normal wireframe view. Lines and curves are used to define the edges of the 3D objects. Curved objects appear as curves, not segmented straight lines such as the door swings on door objects. The UCS is also displayed as a 2D icon (see Figure 4.12A).

- **3D Wireframe**   This option displays the same type of wireframe view as the 2D Wireframe; however, curved objects become "straight-

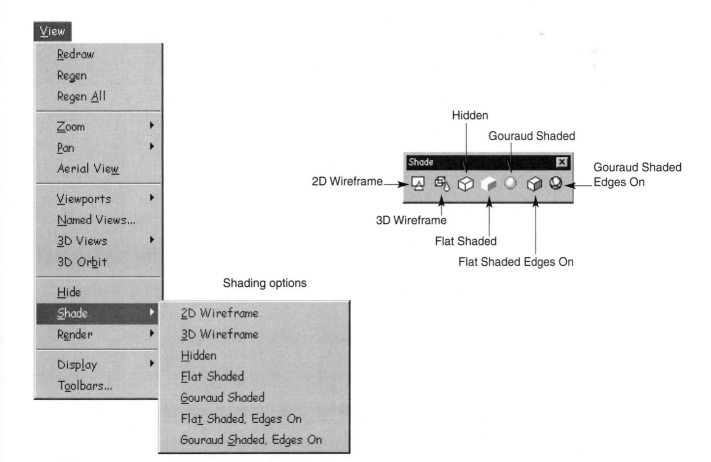

FIGURE 4.11 The shading options can be selected from the Shade cascade menu or from the Shade toolbar.

FIGURE 4.12 The different shading modes applied to the same objects.

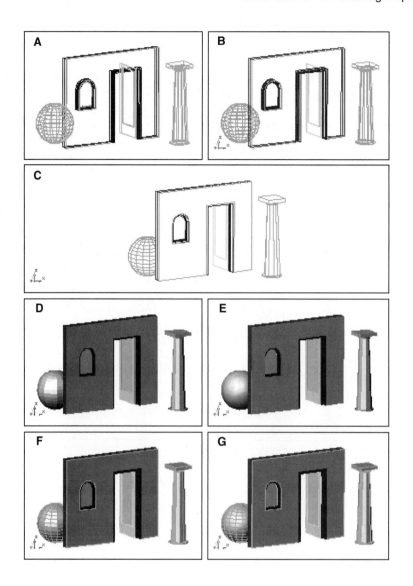

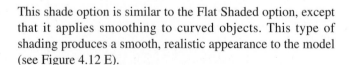

ened" to simplify the display. The UCS icon is also displayed as a colored 3D icon (see Figure 4.12B).

	■ **Hidden**	This option hides lines and arcs that are behind a solid object such as a wall object (see Figure 4.12C).
	■ **Flat Shaded**	This option applies shading to the surface of 3D objects. The shade is displayed in the object's color. The edges of the geometry are not displayed. Smoothing is also not applied to objects, causing curved objects to appear faceted (see Figure 4.12D).
	■ **Gouraud Shaded**	This shade option is similar to the Flat Shaded option, except that it applies smoothing to curved objects. This type of shading produces a smooth, realistic appearance to the model (see Figure 4.12 E).
	■ **Flat Shaded, Edges On**	This option is the Flat Shaded option with wireframe edges displayed (see Figure 4.12F).
	■ **Gouraud Shaded, Edges On**	This option is the same as the Gouraud Shaded option with the edges of the geometry displayed (see Figure 4.12G).

Field Notes

> When working in a top view so that the 2D objects are displayed in the floor plan, it is best to use a 2D wireframe shade. This ensures that the curved objects appear curved in the floor plan.
>
> Shading is used only as an aid to the development of a design. Shaded views cannot be plotted with a shade mode applied. In this situation a rendering needs to be created. A hidden view can be plotted but must be set up correctly first. Plotting hidden views is discussed later in this text.

Regenerating the Display

Occasionally when you are zooming in and out of your drawing, arced objects appear to be segmented. Architectural Desktop "segments" curved objects in the drawing because it can redisplay straight-line segments faster than it can calculate curved objects for display. As mentioned earlier, this is also the situation when a shading mode other than 2D wireframe is used. It is important to note that the objects are not straight-line objects but true curved objects that do not plot as curved objects. Architectural Desktop simply displays curved objects this way to increase your productivity instead of having you wait for the objects to be calculated correctly.

If the straight segments are distracting to you while you are working, you can regenerate the drawing to update the display. The **REGEN** command is used to regenerate the drawing. Use one of the following options to access this command:

✓ Select the <u>V</u>iew pull-down menu, then pick Re<u>g</u>en.

✓ Type **REGEN** or **RE** at the Command: prompt.

The arced objects are regenerated in the drawing, displayed as curved objects.

Field Notes

> The REGEN command does not affect curved objects when a shading mode other than 2D wireframe is used.

The REGEN command is also used when you can no longer pan to the side or when you have zoomed out so far that zooming no longer has any effect on what you see. The REGEN command recalculates all drawing objects and their placement in the drawing. It also regenerates the display based on the current zoom magnification. If you are working on the drawing while in a shaded mode, you may also notice some discrepancies when editing objects. Use REGEN to fix any display problems.

Using Views in the Drawing

As discussed earlier in this chapter, the 3D ORBIT command is used to get a different view of the drawing, and the ZOOM command allows you to view different areas of your drawing. Architectural Desktop also includes a tool that can be used to select a preset view such as an elevation, top view, or an isometric view. The tool that controls the view currently being displayed is known as the **VIEW** command. The VIEW command can also be used to create custom views of your drawing.

Views are similar to a camera picture. With a camera, a picture or "view" of a subject is taken. This picture is then "displayed" on photographic paper so it can be viewed at any time after the picture was taken. With the VIEW command, "pictures" are taken of different areas of the drawing so they can be viewed later in the development of the drawing. This saves time when working on a project, because a view can be "restored" without having to zoom into the area. Use one of the following options to access this command:

✓ Go to the <u>V</u>iew pull-down menu and select Named Views….

✓ Pick the Named Views button from the Standard or View toolbars.

✓ Type **VIEW** or **V** at the Command: prompt.

The View dialog box shown in Figure 4.13 is displayed.

The View dialog box includes two tabs: **Named Views** and **Orthographic & Isometric Views**. The Named Views tab is used to create and view custom named views on the drawing screen. As new views are created the views are displayed in the list area. The Orthographic & Isometric Views tab is used to select a preconfigured view and set it current in the drawing screen. The following is a description of the options within each tab:

- **Named Views**   This tab lists custom views in the drawing. A *custom view* can be a view of a part of the drawing, such as a view zoomed into the kitchen

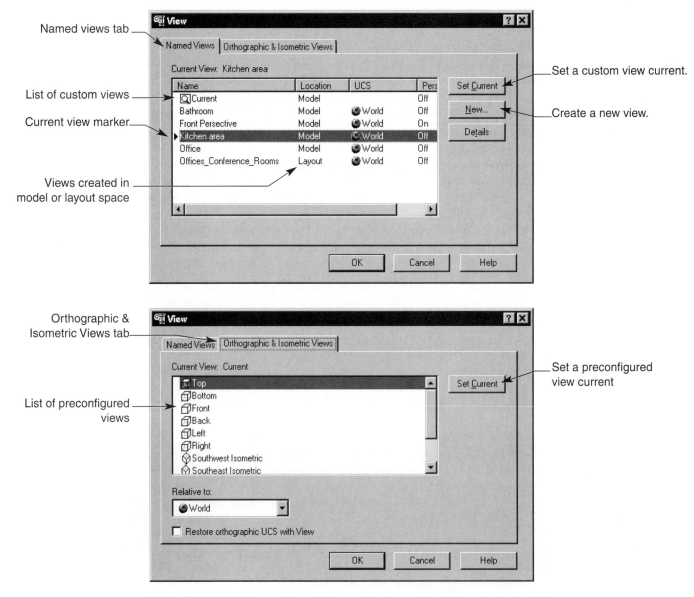

FIGURE 4.13   The View dialog box is used to create and restore views in the drawing.

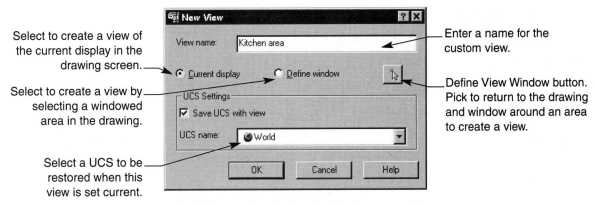

Select to create a view of the current display in the drawing screen.

Select to create a view by selecting a windowed area in the drawing.

Select a UCS to be restored when this view is set current.

Enter a name for the custom view.

Define View Window button. Pick to return to the drawing and window around an area to create a view.

FIGURE 4.14 The New View dialog box is used to create a new view in the drawing.

area of the floor plan, or a view of the entire floor plan. The following options are available in the Named Views tab:

- **Set Current** This button is used to restore a view after it has been created in the list box. Drawings can have as many views created as needed. You can have a view of each room inside your floor plan, or a particular angle that was created through the 3D Orbit command. Selecting a view from the list, then selecting the Set Current button restores the view in the drawing screen.

- **New...** This button allows you to create a named view in the drawing. Pressing this button displays the **New View** dialog box shown in Figure 4.14. To create the view, enter a name in the View name: text box. View names can be up to 255 characters in length. Next, select the Current display or the Define window radio buttons. The **Current display** radio button creates a view of what is currently being displayed on the drawing screen. The **Define window** radio button highlights the Define View Window button. Picking this button allows you to window around an area of the drawing that you wish to be created as a view.

  User coordinate settings can also be saved with a view. The User Coordinate System sets up the current drawing plane (XY plane) in the drawing. If you have named user coordinate systems in the drawing, these UCS names are available in the drop-down list. The World Coordinate System is always available and is usually used when creating views. When you have finished creating the new view, press the OK button. The new view name is displayed in the Named Views list box. To set the view current in the drawing screen, highlight the appropriate view and press the Set Current button.

- **Details** When this button is selected, a dialog box containing the current views information is displayed. Information such as the area, direction, and lens length are provided.

After a view name has been created, the named view appears in the Named Views list. Select the appropriate view and press the Set Current button to restore the view in the drawing screen. Named views can be renamed and deleted by highlighting the view name in the list box, right-clicking, and selecting the appropriate option from a shortcut menu. Selecting Rename allows you to rename the named view. Selecting Delete deletes the view name from the list.

- **Orthographic & Isometric Views** This tab includes preset views that can be selected to change the viewing direction of the model or drawing.

These are the same views that can be found in the 3D Orbit shortcut menu. Views include the top view (floor plan view), bottom view (worm's eye view of the floor plan), four elevation views, and four isometric views.

- **Set Current**   Similar to the Named Views tab. To set a preset view current, first select the appropriate view, then press the Set Current button. The preconfigured view is restored in the drawing.

- **Relative to**   This drop-down list allows you to select a named user coordinate system to restore with your view. The World Coordinate System is always available, and is used in most situations.

Anytime you need to restore a view, access the View dialog box, select the appropriate named view or one of the preset views, and set them current. The drawing screen is changed to reflect the area of the drawing that is defined by the view.

Field Notes

Views can be of great value when working on architectural projects. Often a drawing can be so large that you find yourself zoomed into a particular area. When you need to work on a different area of the drawing, you can restore a view quickly, saving the time of zooming out of the drawing, panning to the correct location, and zooming into the desired area. In architecture, views are commonly created of different parts of the floor plan, a specific area in an elevation, or of a particular angle or perspective that is important.

Exercise 4.5

1. Open EX4-4.dwg.
2. Create a view named Floor Plan.
3. Zoom into the kitchen area and create a view named Kitchen.
4. Set the Kitchen view current. Press OK.
5. Set the Floor Plan view current.
6. Create a view named Master Suite of the master bedroom and the master bathroom using the Define Window radio button. Window around the master bedroom and the master bathroom. Set the view current.
7. Set the Floor Plan view current and create a view of the other rooms in the drawing.
8. Set the Southeast Isometric view current from the Orthographic & Isometric Views tab.
9. Save the drawing as EX4-5.dwg.

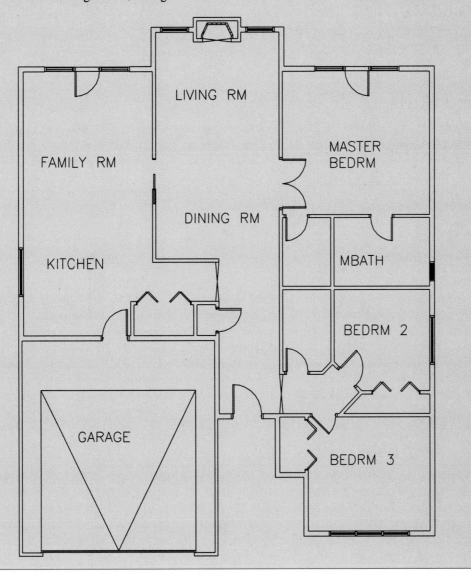

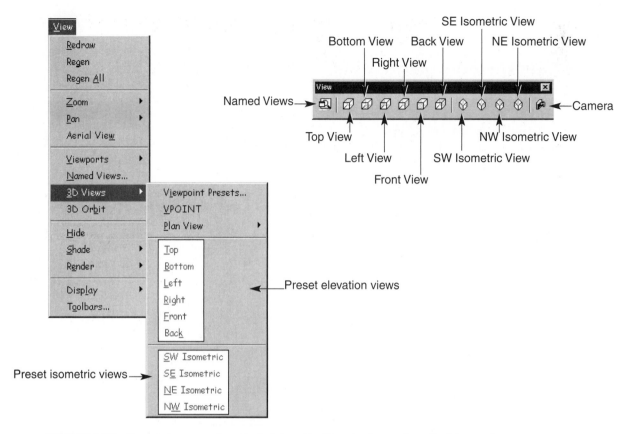

FIGURE 4.15   Preset views can be selected from the View toolbar or from the View pull-down menu.

Using Preset Views

The View dialog box is used to manage views and set different views current on the drawing screen. During the development of your design, you may need to view the elevations for the building to make sure the building is being created correctly, or you may want to see an isometric view of the model.

     The View toolbar includes all the preset views within Architectural Desktop (see Figure 4.15). You can also select preset views by accessing the View pull-down menu, selecting the 3D Views cascade menu, then an appropriate preset view. Selecting a view button displays the appropriate view on the drawing screen. When an elevation view is selected, the User Coordinate System adjusts so the drawing plane is perpendicular to your line of sight. This allows you to draw in the appropriate drawing plane without having to manually adjust the UCS.

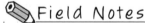✎Field Notes

> When you set an elevation view current, the UCS is rotated so the positive Z axis is pointing straight at you, and the XY plane is perpendicular to your line of sight. Changing to an isometric view from an elevation view maintains this UCS configuration. If you need to work on the floor plan in the isometric view, you may be limited to the drawing plane that was established in the elevation view. This causes difficulty in moving, copying, or drawing objects in the isometric view. To adjust this, set the top view current before setting an isometric view current. This places the UCS back into the World Coordinate System where you first created the walls.

Using Model Space Viewports

As you have been drawing in Architectural Desktop, objects have been created and displayed on the drawing screen. The drawing screen is considered to be a viewport in AutoCAD. A *viewport* is a window in the drawing. Architectural Desktop can be configured to display several viewports in the drawing. Each viewport can then be assigned to a different view or zoomed area of the same drawing. This aids in the development of a drawing, as a 2D construction document and as a 3D model.

There are two types of viewports available in Architectural Desktop: *tiled viewports* (model space), and *floating viewports* (layout space). Viewports in model space are considered to be tiled, because any new viewports created are tiled against one another, much like ceramic tiles. Viewports in layout space are considered to be floating because the viewports do not have to be tiled against one another and can be moved around the layout paper to arrange the different views (see Figure 4.16). This section focuses on the tiled viewports in model space. Floating viewports are discussed in the plotting chapter.

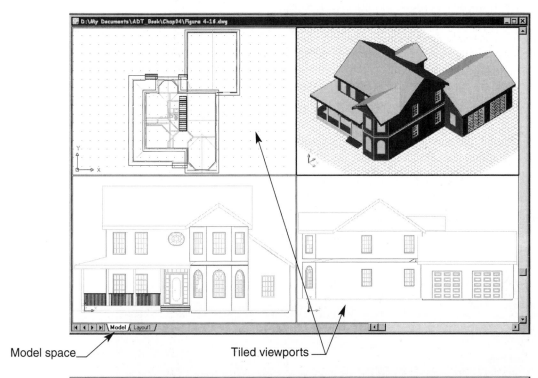

Model space ⌐   Tiled viewports ⌐

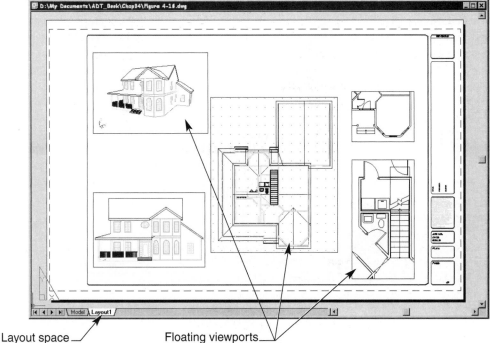

Layout space ⌐   Floating viewports ⌐

FIGURE 4.16   Viewports in model space are tiled against one another. Viewports in layout space can be moved around the paper in order to arrange the different views for plotting.

By default, a single viewport is displayed when a new drawing is started. Use one of the following options to create viewports in the drawing:

✓ Access the **View** pull-down menu, pick the **Viewports** cascade menu, then pick **New viewports....**

 ✓ Pick the **Display Viewports Dialog** button in the **Viewports** toolbar.

✓ Type **VPORTS** at the Command: prompt.

The Viewports dialog box shown in Figure 4.17 is displayed.

The Viewports dialog box includes two tabs: **New Viewports** and **Named Viewports**. The New Viewports tab is used to create a new viewport configuration. The Named Viewports tab is used to restore a named viewport configuration in the drawing screen. The options for each tab are described next.

■ **New viewports**   This tab is used to configure how the drawing screen is split into different viewports. Enter a name for the viewport configuration in the New name: text box. Names can contain up to 255 characters and are used to describe what the viewport configuration looks like or is used for.

■ **Standard viewports**   This list box includes preset viewport configurations. The **Single** option displays one viewport in the drawing screen. Other viewport configurations divide the drawing screen into two, three, or four viewports. When a viewport configuration is selected, a preview of how the viewport configuration appears is displayed to the right.

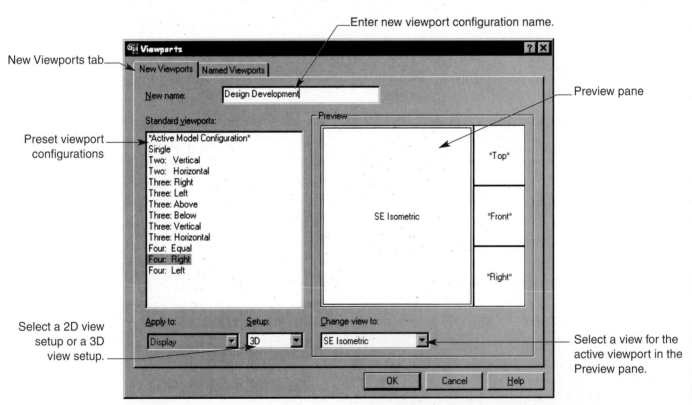

FIGURE 4.17   The Viewports dialog box is used to divide model space into tiled viewports.

- **Apply to**   This drop-down list specifies where the viewport configuration will be applied. Initially, this drop-down list is grayed out as the viewport configuration is applied to the drawing screen. If you already have more than one viewport on the drawing screen, accessing the Viewports dialog box again allows the Apply to: drop-down list to be available. The options include **Display** or **Current Viewport**. When Display is selected, the new viewport configuration is applied to the overall drawing screen. When Current View is selected, the viewport configuration is applied to the active viewport in the drawing, further dividing that viewport into smaller viewports. Viewports can be divided into smaller viewports for a total of 48 viewports in model space.

- **Setup**   This drop-down list also has two options: **2D** and **3D**. Selecting 2D from the list applies a top view to all the viewports. Selecting 3D adjusts the views in the viewports, allowing you to have a top view in one viewport and an isometric and/or elevation view in another viewport. When one of these options is selected, the preview pane displays the view that is applied to that viewport (see Figure 4.17).

- **Preview pane**   The preview pane displays how the drawing window is divided into different viewports. Each viewport is then assigned a view in the drawing. Picking inside a viewport in the Preview pane makes that viewport active so a view can be applied to that viewport. Views are assigned to viewports from the **Change view to:** drop-down list.

- **Change view to:**   This drop-down list includes the preset views mentioned earlier. If there are named views in the drawing, these views are also available in this list. Pick a viewport in the Preview pane to which you want to assign a view, then select the appropriate view. The view name is displayed within the viewport in the Preview pane.

Press the **OK** button when you have finished configuring the viewport assignments. Architectural Desktop divides the drawing screen into viewports based on the configuration you selected in the Viewports dialog box and assigns the appropriate views to each viewport.

- **Named Viewports**   This tab contains a list of the named viewport configurations that you create. Selecting a named viewport configuration from the list displays the viewport configuration in the preview pane. Press OK to have the drawing screen divided into the selected viewport configuration.

After you have divided the drawing screen into viewports, one of the viewports becomes "active." The active viewport is considered current and is designated by a dark outline around its edge. The crosshairs are available only in the active viewport. Object creation, zooming, and panning can be performed in the current viewport. To change to a different viewport, move the mouse to an "inactive" viewport and pick to make it active. Drawing commands can be initiated in one viewport and finished in a different viewport by picking in a different viewport while the command is active.

You also can create viewports by selecting an option from the Viewports cascade menu under the View pull-down menu. You can access the Viewports dialog box at any time to further divide the active viewport into smaller viewports, or you can change the drawing screen back into a single viewport by selecting the Single viewport configuration.

Exercise 4.6

1. Open EX4-5.dwg.
2. Use the Viewports dialog box to divide the drawing screen into two vertical viewports, using a 3D setup.
3. Pick in the Top viewport in the preview pane to make the viewport active. Set the Kitchen view current from the Change view to: drop-down list. Press OK when finished.
4. Access the Viewport dialog box and select the Three: Right viewport configuration using a 3D setup, and apply the viewport configuration to the current viewport. Name the viewport configuration Four Viewports. Press OK.
5. Access the Viewport dialog box and set the drawing screen back to a single viewport using a top view.
6. Access the Viewport dialog box and select the Four: Right viewport configuration. Use the 3D setup. Change the large viewport to the Floor Planview and the top view viewport to a SE Isometric view.
7. Save the drawing as EX4-6.dwg.

Using the Object Viewer

The *Object Viewer* is a viewing window that is a combination of the floating viewer window introduced in Chapter 3 and the 3D ORBIT command. The Object Viewer is a dialog box that allows you to view portions of your drawing in different views and shading modes. It can also be used to set a chosen view current in the active viewport or drawing screen.

To access the Object Viewer, first select objects in the drawing that you would like to view at different angles. After the objects have been selected, right-click and select **Object Viewer...** from the cursor menu. The Object Viewer window, shown in Figure 4.18, is displayed. In addition, you can use the following options to access the Object Viewer:

✓ Pick the **Desktop** pull-down menu, select **Utilities**, then **Object Viewer....**

✓ Pick the **Object Viewer** button in the **Utilities** toolbar.

✓ Right-click the drawing screen and select **Desktop, Utilities, Object Viewer...** from the cursor menu.

✓ Type **AECOBJECTVIEWER** at the Command: prompt.

All these methods prompt you to select the objects you want to see in the Object Viewer.

Figure 4.18 shows the buttons available for adjusting the view, and the shortcut menu that is displayed when you right-click within the viewer. The buttons are similar to the buttons found in the 3D Orbit, Zoom, and Shade toolbars. In addition to these buttons is a button called **Set View**. Selecting this button sets the view in the current viewport to match the viewing angle in the Object Viewer. This can be helpful when creating a perspective view of your drawing.

In addition to the viewing and shading buttons are two drop-down lists. The first drop-down list in the upper right corner of the viewer includes the preset viewing directions Top, Front, and NW Isometric. Selecting a preset view updates the Object Viewer window to reflect the viewing direction for the objects that are currently being viewed.

The other drop-down list is the **Display Configurations** list. This list contains the display configurations available in the drawing. For example, the Work display configuration displays AEC objects as 2D objects in the top view, and as 3D objects in an elevation or isometric view. The Work_Reflected display configuration displays the AEC objects in a top view as they appear in a reflected ceiling plan. In the Work_Reflected display configuration, doors and windows are not displayed, and the header over the wall openings is displayed over the door and window locations. Other display configurations include plotting configurations, conceptual design configurations, and a render configuration. Each of the display configurations is designed to show the AEC objects only as they should appear based on the type of display configuration chosen. Display configurations are discussed in greater detail later in this chapter.

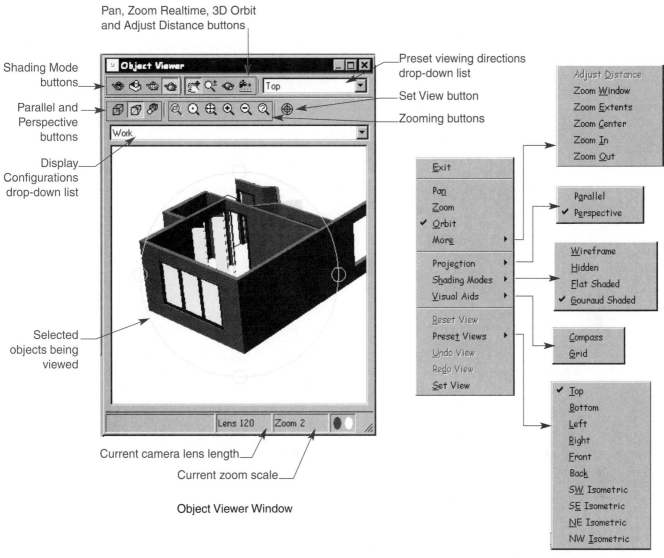

FIGURE 4.18 The Object Viewer window is similar to the Floating Viewer window. Additional buttons and a right-click menu allow you to view the selected objects in a number of different angles and shade modes.

When you are finished viewing your design or a portion of your design, press the X button in the upper right corner to close the Object Viewer window.

Exercise 4.7

1. Open EX4-6.dwg.
2. In the upper right viewport, select a few AEC objects. Right-click and select Object Viewer....
3. Use the 3D Orbit and Perspective buttons to create a perspective view.
4. Set the view current with the Set View button. Exit the Object Viewer when done.
5. Move to the bottom right viewport. Access the Object Viewer window.
6. Create any desired view and set it current in the viewport.
7. Save the drawing as EX4-7.dwg.

Creating Perspectives and Movie Files

So far in this chapter you have learned several methods of visualizing your drawing. The 3D Orbit and the Object Viewer utilities have been used to obtain standard views such as elevation, and isometric views. These commands have also been used to establish perspective views into your drawing. Architectural Desktop includes additional utilities that make perspective creation even easier. The use of cameras in Architectural Desktop allows you to specifically identify where the viewer "is standing" in the drawing, and where the viewer is looking. After a camera is placed in the drawing, the camera location and target view can be adjusted to get "just the right view."

Use one of the following options to place a camera in the drawing:

✓ Select the **Documentation** pull-down menu, pick **Perspectives**, then **Add Camera**....

 ✓ Pick the **Add Camera** button in the **Perspectives** toolbar.

✓ Type **AECCAMERAADD** at the Command: prompt.

✓ Right-click in the drawing screen, select **Documentation, Perspectives, Add Camera...** from the cursor menu.

The Add Camera dialog box shown in Figure 4.19 is displayed.

After the Add Camera dialog box is displayed, enter a name for the camera. If a name is not specified for a camera, the default camera is named Camera01, additional cameras are named Camera02, Camera03,.... Once you have designated the camera, enter a zoom length. The zoom length can be any value between 8 and 500. The smaller the zoom length value, the farther away the perspective appears from the camera. A larger zoom length brings the perspective closer to the model.

The final step before placing the camera is to set the eye level. The eye level is the height the camera is placed off the current X,Y drawing plane. The average human eye level is 5'-6" off of the ground. If you would like to create the perspective view once the camera is placed in the drawing, check on the **Generate View After Add** check box. If you are working in a single viewport, the view is automatically created after the camera is placed in the drawing. If you are using several different viewports, you are prompted to pick in the viewport that creates the perspective.

Once you have specified all the parameters for the camera, pick once in the drawing to make the drawing current, then pick a spot in the drawing where you would like to place the camera. After setting the camera location, you are required to select a target point. The *target point* is the point in your drawing on which you would like to focus the camera. After the camera has been situated, a camera object is placed in the drawing (see Figure 4.20A).

The camera can be adjusted with grips. Four grips are placed on the camera object. One grip is on the camera, which controls the camera placement, and three are at the end of the target cone. The middle grip on the target cone adjusts the target location; the two flanking grips adjust the field of view (FOV) (see Figure 4.20B).

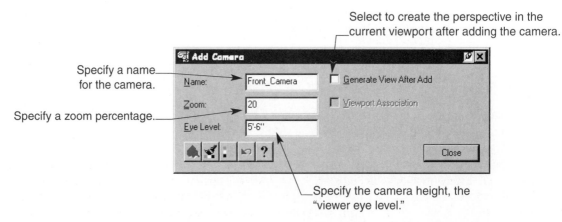

FIGURE 4.19   The Add Camera dialog box is used to set the camera name, zoom length, and height.

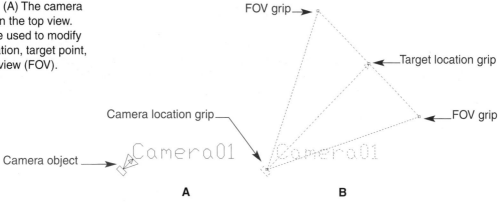

FIGURE 4.20. (A) The camera object as seen in the top view. (B) Grips can be used to modify the camera location, target point, and the field of view (FOV).

You can place a camera in the drawing without creating a perspective if you wish. If you would like to generate the perspective after the camera is created, select the camera, right-click, and select **Create View**. If you are working with a single viewport, the perspective is created immediately. If you are working with multiple viewports, you are prompted to pick the viewport you would like the perspective generated in. In addition, you can use one of the following options to create camera view:

✓ Select the **Documentation** pull-down menu, pick **Perspectives**, then **Create Camera View**.

✓ Pick the **Create Camera View** button in the **Perspectives** toolbar.

✓ Right-click in the drawing screen and select **Documentation, Perspectives, Create Camera View** from the cursor menu.

✓ Type **AECCAMERAVIEW** at the Command: prompt.

Field Notes

After a camera is created, a view is created in the View dialog box. This view can be restored in any viewport in the drawing.

Exercise 4.8

1. Open EX4-7.dwg
2. Add a camera to the front of the building. Target the front of the building.
3. Name the camera Front_Camera. Set the zoom to 25 and the eye level to 5'-6".
4. Create the camera view in the upper right viewport.
5. Add a camera in the foyer, targeting the living room.
6. Name the camera Foyer_Camera. Set the zoom to 35 and the eye level to 5'-0".
7. Create the camera view in the lower right viewport.
8. Save the drawing as EX4-8.dwg.

Modifying Camera Objects

Select the camera object, then right-click and select **Camera Modify** from the cursor menu to display the **Modify Camera** dialog box shown in Figure 4.21. In the Modify Camera dialog box, the camera object can be renamed and the zoom length adjusted. When you are finished making the adjustments, press the OK button to adjust the camera. After a camera has been adjusted, you may want to re-create the camera view in another viewport by selecting the camera, right-clicking, and selecting Create View from the cursor menu. Pick the viewport that you want to have contain the camera view.

FIGURE 4.21 The Modify
Camera dialog box is used to
rename a camera and adjust the
zoom percentage.

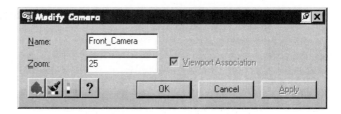

The camera name, zoom percentage, and location can also be adjusted in the **Camera Properties** dialog box shown in Figure 4.22. Use one of the following options to access this dialog box:

> ✓ Pick on the **Documentation** pull-down menu, move to **Perspectives**, and select **Camera Properties…**.
>
> ✓ Pick the **Camera Properties** button in the **Perspectives** toolbar.
>
> ✓ Select the camera object, right-click, and select **Camera Properties…** from the cursor menu.
>
> ✓ Right-click and select **Documentation, Perspectives,** then **Camera Properties…**.
>
> ✓ Type **AECCAMERAPROPS** at the Command: prompt.

The following settings are found in the Camera Properties dialog box:

- **General** This tab is used to give the camera a description, such as "Front perspective facing north." Notes and property sets can also be assigned to the camera. Selecting the **Notes…** button allows you to enter any additional notes to the camera object. A *Property set* is a collection of values that can be added to AEC objects. Property sets are similar to AutoCAD attributes. Property sets are discussed later in this text.

- **Dimensions** This tab is similar to the Modify Camera dialog box. The name of the camera object can be modified, and the zoom length can be changed here.

- **Location** This tab is used to control where the camera object is situated within the drawing. The areas in this tab are described next.

  - **Relative To** If you have user-defined User Coordinate Systems in the drawing, both radio buttons are available. The first radio button, **World Coordinate System,** indicates that the camera location is placed in relation to the default World UCS XY plane. The second radio button, **Current Coordinate System,** indicates that the camera has been placed in relation to the current UCS XY plane.

  - **Insertion Point** This area includes the three axes, X, Y, and Z, The value in each of these text boxes indicates the camera position relative to the UCS selected in the Relative To: area. Generally, the values display the World Coordinate System values. The height of the camera can be adjusted by changing the Z axis value.

  - **Normal** This area controls the plane in which the camera is created. To place the camera in the XY plane, enter a 0 for the X and Y axes and a 1 for the Z axis. To place the camera in the XZ plane, enter a 0 for the X and Z axes and a 1 for the Y axis. To place the camera in the YZ plane, enter a 0 for the Y and Z axes and a 1 in the X axis.

  - **Rotation** This area indicates the cameras rotation in the current XY plane.

Once you have made any changes in the Camera Properties dialog box, press the OK button to apply the changes to the camera object. To update the view in a viewport, select the camera, right-click, and select Camera View from the cursor menu. Pick a viewport for the view to be displayed.

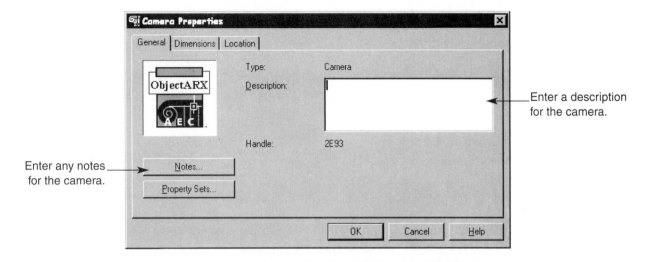

Enter a description
for the camera.

Enter any notes
for the camera.

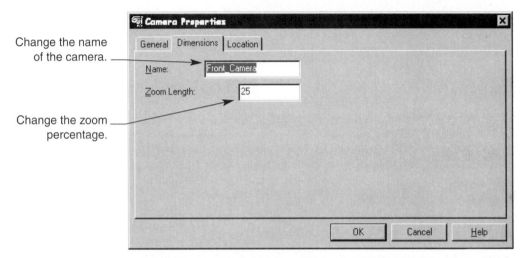

Change the name
of the camera.

Change the zoom
percentage.

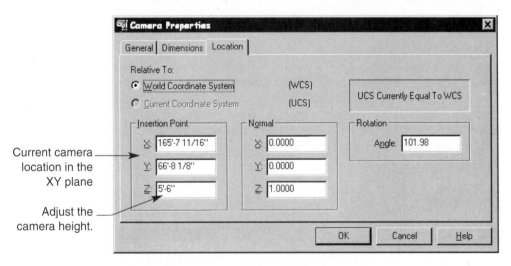

Current camera
location in the
XY plane

Adjust the
camera height.

FIGURE 4.22   The Camera Properties dialog box.

Adjusting the Camera View

Placing a camera in the drawing is as simple as selecting a zoom length, height, and a camera position and target point. Adding cameras in the drawing may not always yield the desired perspective. Cameras and their view can be adjusted through the use of the **Adjust Camera Position** dialog box. Use one of the following options to access this dialog box:

    ✓ Select the **Documentation** pull-down menu, pick **Perspectives**, then select **Adjust Camera View….**

    ✓ Select the camera to adjust, right-click, and pick the **Adjust View...** option from the cursor menu.

    ✓ Pick the **Adjust Camera View** button from the **Perspectives** toolbar.

    ✓ Right-click then pick **Documentation, Perspectives,** then **Adjust Camera View** from the cursor menu.

    ✓ Type **AECCAMERAADJUST** at the Command: prompt.

The Adjust Camera Position dialog box shown in Figure 4.23 is displayed.

 Field Notes

> If you select a camera in a nonperspective viewport and then access the Adjust Camera Position dialog box, the camera view is displayed in the current viewport. When you access the Adjust Camera View command, you are prompted to select a camera or press the [Enter] key to select a camera from a list. This can be helpful if you currently are in a perspective viewport where a camera cannot be selected. Press the [Enter] key to display a list of cameras in the drawing and select a camera, and the view is generated in the current viewport.

The Adjust Camera Position dialog box contains a number of buttons that can be used to adjust the camera position, angle, and zoom length. At the bottom of the dialog box are three text boxes that are used in conjunction with these buttons (see Figure 4.23). The **Forward, Back, Left, Right, Up** and **Down** buttons control the position of the camera. The **Step:** text box is used to control how large a step is taken when the camera position is adjusted. For example, 1'-0" moves the camera forward 1'-0" or up 1'-0".

    The **Turn Left, Turn Right, Look Up,** and **Look Down** buttons control the angle of the camera. The **Angle:** text box is used to control how large an angle is used when the angle of the camera is adjusted. The **Zoom In** and **Zoom Out** buttons control the zoom length of the camera. The **Zoom:** text box controls how much zoom length is applied when zooming into or out of the drawing.

    The **Hide** button is used to perform a hide in the current viewport. Any objects behind lines, doors, or other AEC object are not displayed in the viewport when a hide operation is performed. When the **Auto**

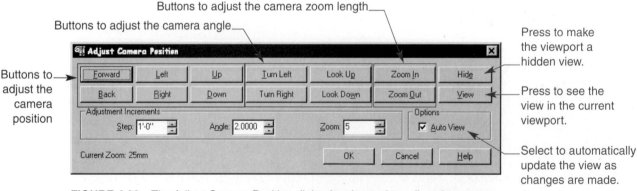

FIGURE 4.23   The Adjust Camera Position dialog box is used to adjust the camera position and the target, which will automatically update in the selected viewport.

View check box is selected, any changes to the camera position, angle, or zoom are automatically updated in the current viewport to present you with the latest viewing angle. If the Auto View check box is unchecked, the **View** button can be used to update the current viewport with the latest changes made to the camera.

As the camera is adjusted, the actual camera is repositioned to reflect your changes. When you are finished adjusting the camera view press the OK button to close the dialog box. As changes are made, a new named view is created in the View dialog box to reflect the camera view.

Exercise 4.9
_____

1. Open EX4-8.dwg.
2. Use the Camera Properties dialog box to adjust the Foyer_Camera so that it is 6'-0" off the XY plane.
3. Make the upper right viewport active.
4. Access the Adjust Camera View command. Press [Enter] to select the Front_Camera from the camera list.
5. Use the Adjust Camera position dialog box to adjust the Front_Camera to create a front perspective of your building.
6. Exit the dialog box when finished.
7. Save the drawing as EX4-9.dwg.

Creating Movie Files

As indicated in the last section, the camera objects can be used to create a perspective view of your model. Architectural Desktop also includes a tool that allows you to create a simple walk-through of the model using camera objects and a walk path.

Before a walk-through can be created, a path needs to be established in the drawing for the camera object to follow. Paths can be created with a polyline that has been splined to make the walk path smooth. To create the path, use the PLINE command to select points that establish the camera's path, as shown in Figure 4.24. After the path has been created with a polyline, the PEDIT command can be used to spline the polyline. The following sequence is used to turn the polyline in Figure 4.24 to a splined polyline:

```
Command: PE or PEDIT ↵
Select polyline or [Multiple]: (select the polyline)
Enter an option [Close/Join/Width/Edit vertex/Fit/Spline/Decurve/Ltype gen/Undo]: S ↵
Enter an option [Close/Join/Width/Edit vertex/Fit/Spline/Decurve/Ltype gen/Undo]: ↵
```

After entering the PEDIT command, select the polyline to spline, then enter S at the prompt. This changes the straight-line segmented polyline into a polyline with curves. Creating the splined polyline allows the camera to move smoothly along the path.

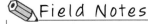 **Field Notes**

> A walk path can also be created with any AutoCAD object such as an arc, line, or spline. If a spline is used, the points designated for a spline line can be adjusted in the Z axis to allow the camera to move up and down. If a polyline is used, the camera maintains a constant level pass along the Z axis of the polyline.

In addition to the walk path, the camera can be focused on a target path. When you are creating a walk-through movie, you can focus the camera on one point or target it along a path. This process creates the illusion of walking through the model and focusing on a different area of the model as you are walking. You can create the target path in the same manner as the walk path, using polylines, lines, arcs, or splines, or you can even offset the walk path.

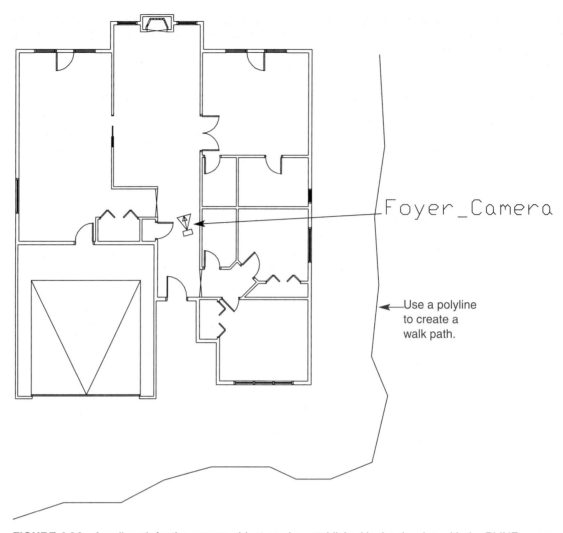

FIGURE 4.24   A walk path for the camera object can be established in the drawing with the PLINE command.

    If the walk path and target path are drawn so they are located on the XY plane, the camera will be focused on the ground and not on the building. To have the camera focus on the path and to be able to view the drawing at the same time you need to move the target path up along the Z axis. To move the camera up, use the MOVE command. The following sequence can be used to move the target path up on the Z axis to 5'-0'' above the 0Z axis:

> Command: **M** or **MOVE** ↵
> Select objects: *(select the walk and target path entities)*
> 1 found
> Select objects: ↵
> Specify base point or displacement: *(pick a basepoint anywhere in the drawing)*
> Specify second point of displacement or <use first point as displacement>: **@0,0,5'** ↵ *(this will move the target 0'' along the X axis, 0'' along the Y axis, and up 5'-0'' on the Z axis)*

After the walk path and the target path have been established, the camera can be added to the drawing. Use the Add Camera command to add the camera to the desired end of the walk path. When selecting the target location for the camera, select the desired end of the targeting path.

    Use one of the following options to create the movie file:

✓ Access the **Documentation** pull-down menu, go to **Perspectives**, and select **Create Video....**

✓ Pick the **Create Video** button from the Perspectives toolbar.

✓ Right-click and select **Documentation, Perspectives, Create Video...** from the cursor menu.

✓ Type **AECCAMERAVIDEO** at the Command: prompt.

If you have several cameras in the drawing, you are prompted to select a camera for the video or to press [Enter] for a list of cameras. After you have selected the camera, the **Create Video** dialog box is displayed as shown in Figure 4.25.

The Create Video dialog box contains the following settings:

- **Camera Path**   This area is used to set the parameters for the camera. The **Pick Path<** button is used to return to the drawing screen and pick the camera path object. If a path is selected, you are prompted for a path name. Entering a path name places the named path in the drop down list. The **Pick Point<** button is used if the camera will be stationary. Selecting this button displays the coordinate location in the drop-down list.

- **Target Path**   This area is used to set the parameters for the target. The **Pick Path<** button allows you to select the object to be used as the target path. Once the path has been selected, you are prompted to enter a name for the target path. The Pick Point< button is used if the camera will be focused on one location. Selecting this option places the coordinates of the target point in the drop-down list.

- **Regen**   This drop-down list contains five different options that control how the model appears in the movie. The None option displays the model in wireframe mode. The Hide option hides lines behind any surfaces. Three different shading modes are also available: Shade - 256 Color, Shade - 256 Edge, and Shade Filled.

- **Frames**   This area controls the number of frames and rates to be used in the movie file. Frames are the number of steps to be taken along the path. The total number of frames to be used can be entered in the **Number:** text box. The more frames that are entered, the smoother the walk-through. The **Rate:** text box controls how long in milliseconds the movie is to focus on one frame. A maximum rate of 25 milliseconds per frame is allowed.

- **Options**   This area includes two options: **Dry Run** and **Corner Deceleration**. The Dry Run check box is used to preview the camera as it moves along the path. If this option is selected, press the OK button to preview the camera moving and focusing along the paths. After previewing the dry run enter the Create Video command again to create the video. The Corner Deceleration check box slows the camera down while moving around corners.

When you are finished selecting the options for the video, pick the OK button to create the video. The **Camera View File** dialog box is displayed, where a location on the hard drive and name for the video file must be entered. After a location and a name for the .avi file have been specified, the **Video Compression**

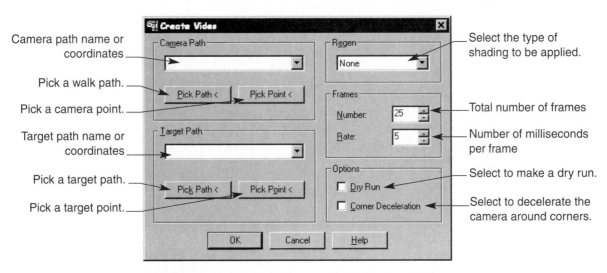

FIGURE 4.25   The Create Video dialog box is used to create a movie file of the drawing.

dialog box is displayed. Select a compressor utility to compress the frames to create a video. Architectural Desktop begins compiling each frame in the drawing and compressing it to the .avi file. When Architectural Desktop is finished compressing the file you are prompted to preview the file.

Exercise 4.10

1. Open EX4-9.dwg.
2. Add a polyline to be used as a walk path. Use the PEDIT command to spline the polyline. See the figure.
3. Add a polyline for the target path. Spline the polyline. See the figure.
4. Use the MOVE command to move the target path and the walk path up 5′-0″ from the XY plane. (*Hint:* pick the target path, select a basepoint, and enter @0,0,5′ for the displacement point.)
5. Add a camera named Video, using a zoom of 30, and a height of 5′-0″ to an endpoint of the walk path. Select an endpoint on the target path for the camera target point.
6. Enter the Create Video command to create a video following the walk path and the target path.
7. Use the Shade Filled Regen mode. Make the video 100 frames long with a rate of 25. Select Corner Deceleration.
8. Enter a name and location for the video.
9. Select the Cinepak Codec by Radius compressor.
10. Preview the .avi file.
11. Save the drawing as EX4-10.dwg.

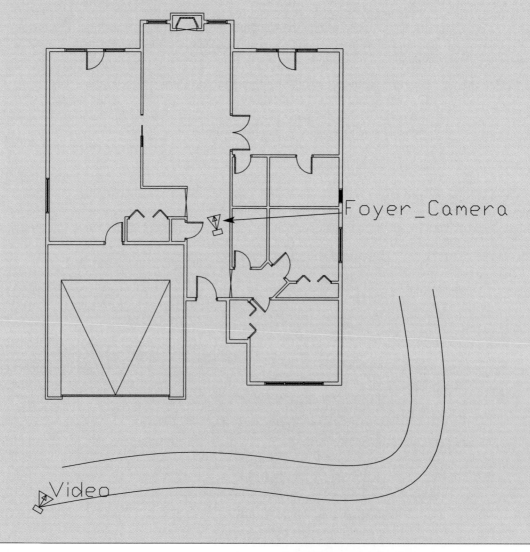

CHAPTER REVIEW

Use the CD-ROM to test your knowledge and skills.

Chapter Test

To check your understanding of the content provided in this chapter, access the Test file in the CH04 folder of the CD-ROM that accompanies this text.

Chapter Project

To practice the Architectural Desktop skills presented in this chapter, access the Project files in the CH04 folder of the CD-ROM that accompanies this text. The project files are in pdf format and include sample drawings and instructions for completing each project.

|||||||||||||||||||||||||||||||||||||||||||||||||||||||||||||||||||||||||||||||||||||||

CHAPTER 5

The Architectural Desktop Display System

LEARNING GOALS

After completing this chapter, you will be able to:

◎ Control how individual AEC objects are displayed.

◎ Group together individual display representations.

◎ Assign a display set to a particular viewing direction.

◎ Edit display representations.

◎ View display representations.

◎ Use display configurations.

◎ Use custom display settings in other drawings.

◎ Apply the display system.

◎ Control AEC display by object.

As you have created walls, doors, windows, and other AEC objects in the drawing the objects have "understood" how to display themselves based on your viewing direction. For example, a wall or a door in a top view has been displayed as a 2D object. When the view is changed to an elevation or isometric view, the same object is displayed as a 3D object (see Figure 5.1). The method of displaying AEC objects is known as the *display system*.

The display system controls how AEC objects are displayed in all viewports in model space or in individual viewports in layout space. With this system, different displays of AEC objects can be created, such as architectural displays that show floor plans, reflected ceiling plans, elevations, 3D models, or schematic plans.

The display system is divided into three separate elements, which work together to form a particular architectural display. The three elements are *display representations,* which control how individual AEC objects are displayed; *display sets,* which group together individual display representations; and *display configurations,* which assign a display set to a particular viewing direction. The display configuration is then applied to model space and individual layout space viewports. These three pieces are hierarchical in nature. The end result is that display configurations may include several display sets assigned to different viewing directions, which in turn, may include several groups of display representations.

Display Representations

The first level of the display system is the display representation, which controls how the individual AEC object parts are displayed. For example, a wall object includes a plan representation, which displays the 2D wall, and a model representation, which displays the 3D wall. Each AEC object includes a varying number of display representations. The display representations for AEC objects can be controlled in the **Drawing Setup** dialog box, shown in Figure 5.2. Chapter 2 briefly explained the components within the **Display** tab in the Drawing Setup dialog box.

The Display tab includes different areas for display control. The left side of the tab includes the **Object Type:** list, which lists all the AEC objects within Architectural Desktop. Selecting an object from

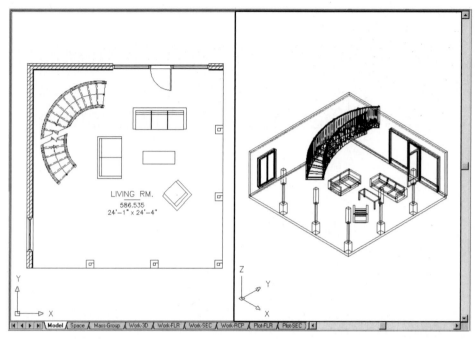

2D Objects in a Top View   3D Objects in an Isometric View

FIGURE 5.1 AEC objects can be displayed in a number of different ways. Top views display objects as 2D representations. Elevations and isometric views display objects as 3D representations.

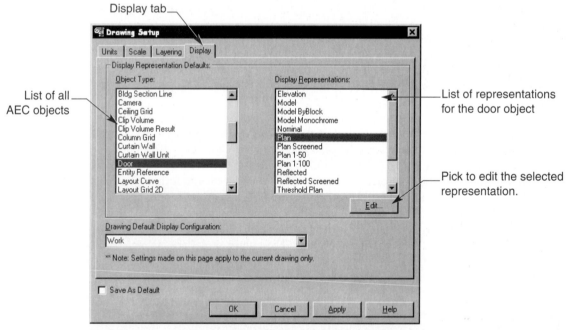

FIGURE 5.2 The Display tab in the Drawing Setup dialog box controls the display of the individual representations for each AEC object.

the Object Type: list provides a list of the different representations for the object in the **Display Representations:** list on the right. Figure 5.2 displays the different representations for the door object. Each representation controls a different aspect of the doors. For example, the **Plan** representation includes the different components of the door in 2D. The **Model** representation controls how the 3D components for the door are displayed in a model view. Appendix I lists the representations for each object, and their proposed usage.

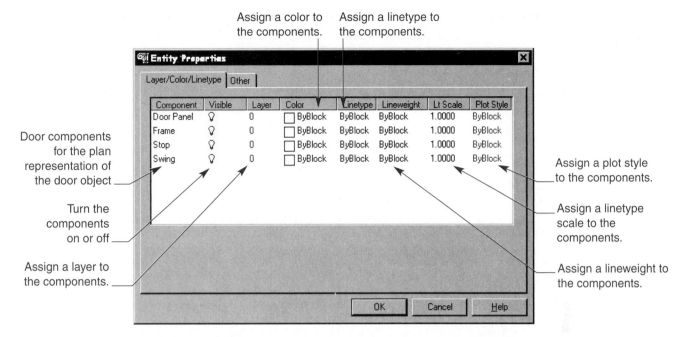

FIGURE 5.3 The Entity Properties dialog box controls the components of an object in the appropriate display representation.

To edit the representations, select a representation from the Display Representations: list, then pick the Edit... button. The **Entity Properties** dialog box, shown in Figure 5.3, is displayed. Depending on the object and the representation selected, the available tabs and list of components vary. Figure 5.3 shows the list of components for the door object using the Plan representation.

The Entity Properties dialog box is used to control how the default objects appear when they are created in the drawing. The **Layer/Color/Linetype** tab is used to control how the object components appear in the selected representation. In Figure 5.3, the Door-Plan representation components are displayed for modification. The following columns are included in this tab:

- **Components** This column includes a list of the components available for the selected representation. This list may vary between plan, model, and other representations depending on the AEC object.

- **Visible** This column allows you to turn a component on or off for the selected representation. There may be situations as you draw in which you do not want a particular component to be displayed in a view. For example, if you do not want the door stop to be displayed in a plan representation you can turn off the Stop component in the plan representation for the door object. To turn a component on or off, pick on the lightbulb symbol.

- **Layer** By default most of the components in a representation are set up with layer 0 in this column. This allows the components of an object to be placed on the layer onto which the object is being inserted. For example, when a door is inserted into the drawing, it is placed on the A-Door layer. Because the plan representation for the door has assigned layer 0 to the components, all the components for a plan door take on the A-Door layer. To assign a layer to a component, pick the 0 symbol for that component. A list of layers currently in the drawing is displayed from which you can select the layer to assign to a component. Layers are discussed in Chapter 6.

- **Color** This column controls the color that is displayed for the component. Select the color swatch to access the **Select Color** dialog box, from which you can select a color and assign it to a component.

- **Linetype**   This column assigns linetypes to components. To change the linetype assigned to a component, pick on the word ByBlock to access the **Select Linetype** dialog box. This dialog box lists any linetypes loaded in the current drawing. Selecting a linetype assigns the linetype to the component.

- **Lineweight**   This column assigns a lineweight to a component. Pick the word ByBlock to assign a different lineweight to the component.

- **Lt Scale**   Components can be assigned their own linetype scale. This scale multiplies the value of the current LTSCALE in the drawing by the value you enter in this column.

- **Plot Style**   By default this column is grayed out. If you are using plot style tables instead of color tables to plot the drawing, then a plot style can be assigned to a component. Plotting, plot style tables, and color tables are covered in Chapter 21.

When you are finished setting the values for the selected representation, press the OK button in the Entity Properties dialog box. You are returned to the Drawing Setup dialog box open to the Display tab. You may need to walk through each of the display representations of one type of AEC object in order to set up how the default display settings for the objects should appear.

 Field Notes

Many of the AEC objects include multiple display representations. These representations are displayed only if you select the appropriate display configuration as you draw or set the drawing up for plotting. The most common representations that are modified for AEC objects include plan, model, and reflected. Door and window objects include an elevation representation that is used when viewing the drawing from an elevation view. Some objects do not include a plan representation; in these cases the general representation is used.

Exercise 5.1

1. Start a new drawing using the AEC Arch (Imperial).dwt or the AEC Arch (Imperial-Int).dwt.
2. Access the Drawing Setup dialog box. Pick the Display tab.
3. Select the door object and the plan representation. Pick the Edit... button.
4. Change the Swing component color to Color 4 (cyan). Press OK.
5. Select the door object and the model representation. Pick the Edit... button.
6. Change the Frame, Stop, and Door Panel components to Color 33. Change the Glass component to Color 151. Turn off the Swing component. Pick the Other tab.
7. In the Other tab, place a check in the Override Open Percentage check box. Enter a value of 0 in the text box below the check box. This setting will close the doors when they are viewed in a model view. Press the OK button.
8. Exit the Drawing Setup dialog box.
9. Draw a wall 12'-0" long using the standard style. Insert a door using the standard style. The door should appear with a cyan-colored swing. Select an isometric view. The door should appear closed, with the frame, stop, and door panel using Color 33.
10. Save the drawing as EX5-1.dwg.

Using Display Manager to Edit Display Representations

Architectural Desktop includes a utility known as the **Display Manager** that can be used to organize and configure all three pieces of the display system. Use one of the following options to access the Display Manager dialog box, shown in Figure 5.4:

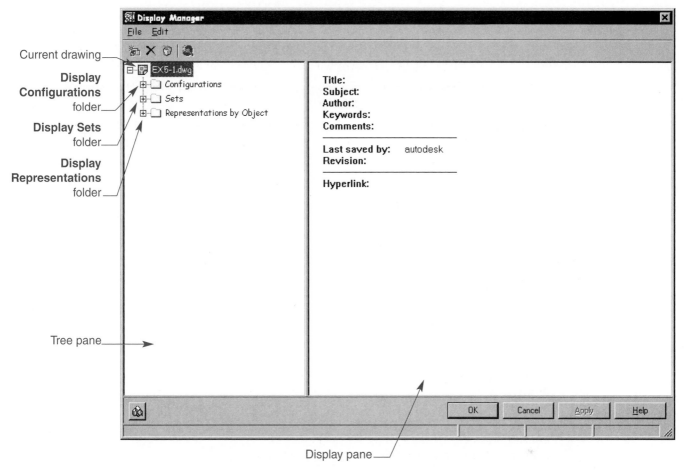

Current drawing

Display Configurations folder

Display Sets folder

Display Representations folder

Tree pane

Display pane

FIGURE 5.4   The Display Manager dialog box is used to organize and configure the three parts of the display system.

✓ Select the **Desktop** pull-down menu, and select **Display Manager....**

✓ Pick the **Display Manager** button from the **AEC Setup** toolbar.

✓ Right-click and select **Desktop, Display Manager....**

✓ Type **AECDISPLAYMANAGER** at the Command: prompt.

The Display Manager is divided into two panes: the **tree-view pane** and the **display pane**. The tree-view pane displays the three parts of the display system in three different folders. The **Configurations** folder includes display configurations that get applied to model space or to individual layout space viewports. The **Sets** folder includes displays sets that are assigned to viewing directions within a single display configuration. The last folder, **Representations by Object,** includes a list of objects with their associated display representations.

To view the individual object display representations, press the + symbol to the left of the Representations by Object folder. A tree of all the AEC objects within Architectural Desktop is displayed. If you pick the Representations by Object folder, a table is displayed in the display pane on the right (see Figure 5.5). The table view includes a list of the AEC objects along the left edge, and a list of the display representations available within Architectural Desktop. A properties symbol at the intersection of an AEC object and a display representation indicates that the representation is available for the object. In Figure 5.5, the intersection of the Elevation display representation column and the Door object row results in a properties symbol, indicating that the Elevation representation is assigned to the door object.

Moving the cursor over a properties symbol and right-clicking displays a cursor menu with two menu items, **Delete** and **Edit...**. The Delete option allows you to delete a user-defined display representation.

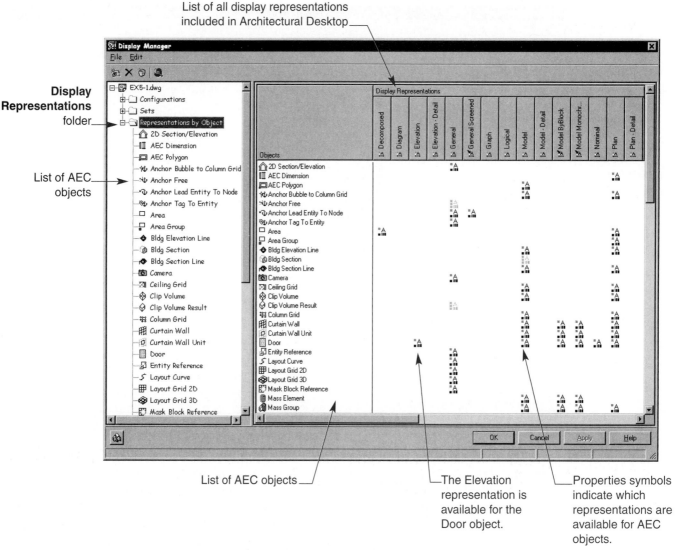

FIGURE 5.5   The Display Manager displays a list of AEC objects in the left pane, and a table view of the representations available for each AEC object.

Creating a user-defined display representation is discussed later in this chapter. Selecting the Edit… option displays the **Entity Properties** dialog box for the representation on which you right-clicked. This is the same Entity Properties dialog box displayed when display representations are edited from the Drawing Setup dialog box (see Figure 5.3). By right-clicking over the different properties symbols, you can configure the default display representations. Modifying the representations from the Drawing Setup dialog box or from the Display Manager results in the same configuration.

Exercise 5.2

1. Open EX5-1.dwg.
2. Access the Display Manager. Select the Representations by Object folder.
3. Find the properties symbol at the intersection of the Wall object row and the Model Display Representations column. Right-click and select Edit....
4. Change the color of Boundary 1 to Color 3 (green).
5. Find the properties symbol at the intersection of the Window object row and the Model display representations column. Right-click and select Edit.... Change the frame and sash color to Color 6 (magenta). Exit the Entity Properties dialog box, then the Display Manager.
6. Create two viewports using the VPORTS command. Set one viewport to a top view and the other to an isometric view.
7. Add a window using the standard window style to your wall. Look at the drawing in an isometric view. What colors are the wall and the window? Look at the drawing in a top view. How do the wall and window appear in the top view?
8. Save the drawing as EX5-2.dwg.

Viewing the Display Representations by Object

If you pick on top of an AEC object in the left pane (tree view) of the Display Manager a different table is displayed in the display pane (see Figure 5.6). This table displays the list of representations available for the selected object along the left edge. A list of the display sets can be found at the top of the table. This table

List of display representations for the window object ⎯⎯⎯⎯⎯   ⎯⎯⎯List of display sets in the drawing

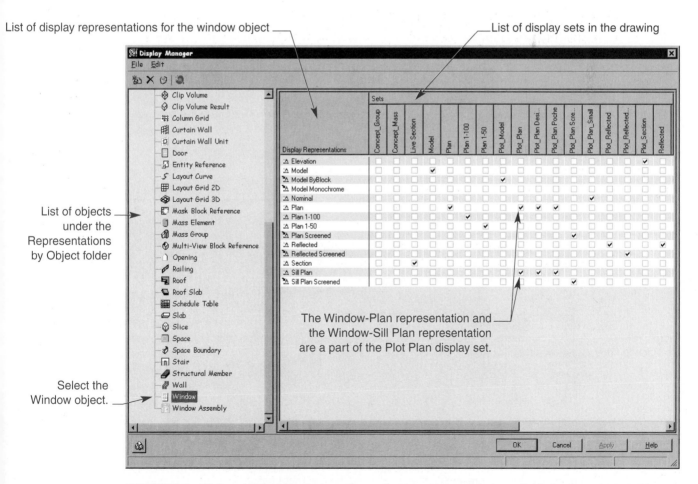

FIGURE 5.6   Pick an object from the tree view to list the representations available for the selected object and display the Display Set to which the representations belong.

indicates which individual display representations belong to a particular display set. Figure 5.6 shows the representations for a window. A check in a box indicates that the display representation is part of the listed display set.

The intersection of a display representation row and a display set column reveals a check box. Picking the check box causes the display representation to be added to the display set. Display sets group together several individual display representations for use in a display configuration.

Architectural Desktop includes a good selection of display representations for each AEC object; however, sometimes you may want to create your own display representation. To create a display representation, first select an existing representation to copy, then right-click over the representation. The cursor menu with several options shown in Figure 5.7 is displayed. Select the **Duplicate** option at the top of the list. This creates a new representation for the object that is listed in the tree view. You can enter a custom name for the representation and edit it if you desire.

The shortcut menu in Figure 5.7 also includes the following options for the selected display representation:

- **Delete**   Allows you to delete a custom representation. Default representations cannot be deleted.

- **Rename**   This option allows you to rename a custom representation.

- **Select All** Selecting this option places a check in all the check boxes for the selected representation. The representation is then added to all the display sets.

- **Clear All**   This clears all the check boxes for the highlighted representation.

- **Edit...**   This option displays the Entity Properties dialog box, where the representation can be configured.

 Field Notes

When you use either the AEC ARCH (Imperial) or AEC ARCH (Imperial-Int) template, the display representations are already configured to display your objects desirably. You may decide that the default representations meet your needs for drafting and plotting; however, this system does allow you to customize what and how objects are displayed in your drawings.

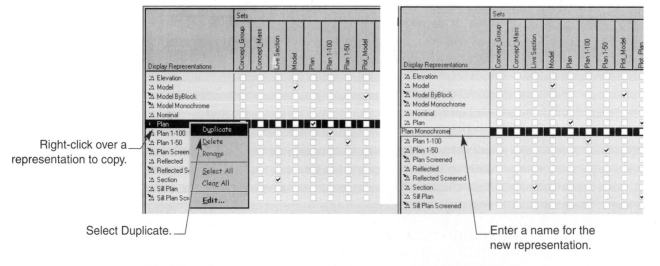

FIGURE 5.7 New display representations can be created by right-clicking over an existing representation and selecting the Duplicate option.

Exercise 5.3

1. Open EX5-2.dwg.
2. Access the Display Manager.
3. Select the Representations by Object folder, then pick on the Door object from the tree-view list.
4. Right-click on the plan representation for the door object in the display pane and select Duplicate. Name the customized representation Plan Monochrome. Edit the representation so that all the door components are using Color 7 (white). Do not assign it to a display set.
5. Select the Wall object from the list below the Representations by Object folder.
6. Right-click on the plan representation for the wall object in the display pane, and select Duplicate. Name the customized representation Plan Monochrome. Edit the representation so that all the Boundary components are using Color 7 (white). Do not assign it to a display set.
7. Select the Window object from the list below the Representations by Object folder.
8. Right-click on the plan representation for the window object in the display pane, and select Duplicate. Name the customized representation Plan Monochrome. Edit the representation so that all the window components are using Color 7 (white). Do not assign it to a display set.
9. Exit the Display Manager. Save the drawing as EX5-3.dwg.

Display Sets

The next level of the display system is the display set. A ***display set*** assembles individual display representations into a group or set. Generally, a display set combines similar representations. For example, all the plan representations for AEC objects can be grouped together into a display set named Plan Set, and all the model representations can be grouped together into a display set named Model Set. Display sets are then used in a display configuration, where the viewing direction is assigned to a display set.

The **Display Manager** is used to create and configure display sets. The **Sets** folder includes a listing of all the display sets in your drawing. Selecting the Sets folder displays a list of the display sets in the display pane on the right (see Figure 5.8). Right-clicking on a display set in the right pane of the Display Manager displays a shortcut menu that can be used to delete or rename a display set.

The toolbar at the top of the Display Manager dialog box includes the following four buttons:

- **New**   This button allows you to create a new display set or display configuration.

- **Delete**   This button deletes custom display sets or display configurations.

- **Purge**   This button removes unused display sets or display configurations from the drawing.

- **Display Systems on Point A** If you have a live Internet connection, this button connects you to the Point A web site, from which display sets and display configurations can be imported into your drawing.

Pressing the ⊞ beside the Sets folder reveals a tree listing of all the display sets. When a display set is selected from the tree view, a table is displayed in the display pane, as shown in Figure 5.9. The display pane includes two tabs: **General** and **Display Control**. The General tab is used to rename the display set and provide a description of what the display set is to be used for. The Display Control tab displays the table view, in which display representations can be selected to be a part of the display set.

The right side of the table view lists the AEC objects in Architectural Desktop. Along the top of the table, the display representations in the drawing are listed. The intersection of an AEC object and a display

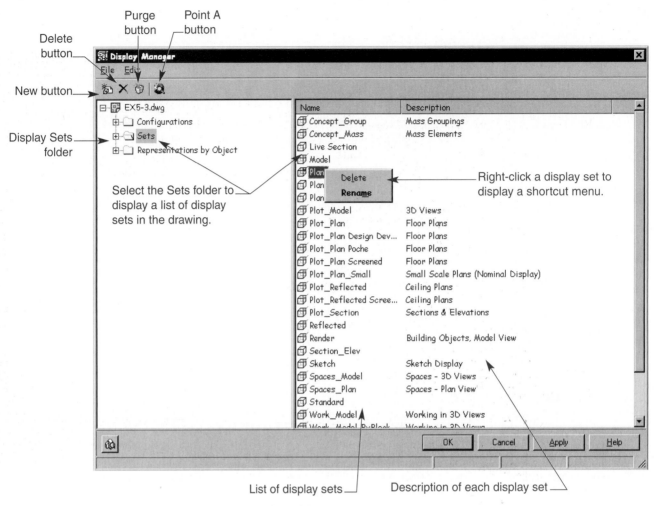

FIGURE 5.8 Select the Sets folder to display a list of display sets in the current drawing. Right-click a named set to delete or rename the display set.

representation produces a check box. The check box indicates that the display representation is available for the AEC object. Selecting the check box adds a check mark in the box. The check mark indicates that the display representation is grouped into the selected display set.

Typically, common representations are grouped together. For example, the display set named Work_Plan, places a check mark in all the Plan representations, which groups all the AEC object–Plan representations. All the Model representations are grouped into the Work_Model display set. Some AEC objects do not include a Plan or Model representation. In these cases the General representation is added to the Work_Plan and Work_Model display sets.

Display sets can be configured to display a mixture of display representations. For example, the Sketch display set groups together the Door-Nominal, Window-Nominal and Wall-Plan sketch display representations. Custom display sets can be created by picking the Sets folder then picking the New button, or by right-clicking over the Sets folder and selecting New from the cursor menu. A new display set is created in the tree view. Enter a new name, highlight the new display set, then pick the desired representations to group into the new display set. The display set can be renamed by selecting the set from the tree view and right-clicking to display a cursor menu from which the Rename option can be selected. The display set can also be renamed and given a description in the General tab.

After the display set is created and has been assigned display representations, it is ready to be assigned to a viewing direction in a display configuration.

Description view tab   Table view tab   Display representations list

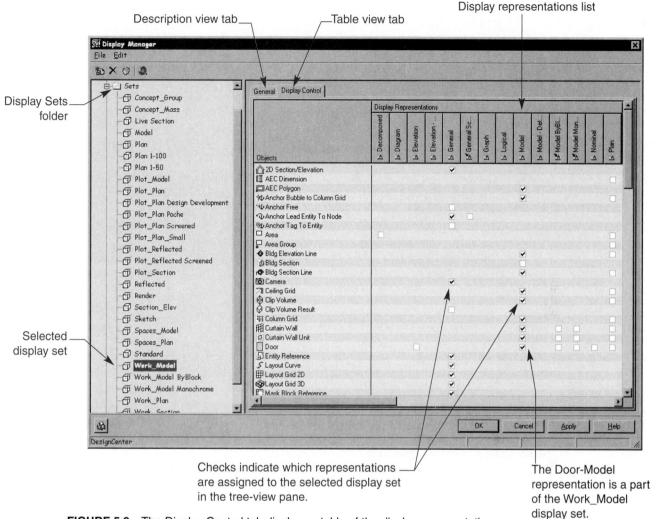

Display Sets folder

Selected display set

Checks indicate which representations
are assigned to the selected display set
in the tree-view pane.

The Door-Model
representation is a part
of the Work_Model
display set.

FIGURE 5.9   The Display Control tab displays a table of the display representations
assigned to the display set.

Field Notes

The display sets included in the AEC ARCH (Imperial) and AEC ARCH (Imperial-Int) templates have been configured into typical types of groups. These groups are then assigned to a display configuration. Custom display sets are created in order to view only specific types of objects in a particular viewing direction.

Exercise 5.4

1. Open EX5-3.dwg.
2. Access the Display Manager.
3. Select the Sets folder, then pick on the + symbol to display the tree-view list.
4. Right-click on the Sets folder and pick <u>N</u>ew from the cursor menu. Name the display set Always Plan. In the General tab, enter the following description: *Displays the wall-plan, door-plan, and window-plan representations.*
5. In the Display Control tab, check on the door-plan, the wall-plan, and the window-plan representations.
6. Right-click on the Sets folder and pick <u>N</u>ew from the cursor menu. Name the display set Work_Plan Monochrome. In the General tab, enter the following description: *Displays plan representations in monochrome.*
7. In the Display Control tab, check on the door-plan monochrome, the wall-plan monochrome, and the window-plan monochrome representations.
8. Exit the Display Manager and save the drawing as EX5-4.dwg.

Using Display Configurations

The last piece of the display system is the display configuration. A ***display configuration*** assigns a display set to a particular viewing direction. As seen in Figure 5.1, different display representations are displayed based on the viewing directions. In a top view, objects are displayed as 2D objects, and in an isometric view, the same objects are viewed as 3D representations. It is important to note that display representations and display sets are not dependent on a viewing direction. The display representations control how the individual representations appear in drawing. The display sets then group the display representations together for use in the display configurations.

As with display representations and display sets, the **Display Manager** is used to create and configure display configurations. The **Configurations** folder is used to organize the display configurations. Selecting the Configurations folder displays a table view in the display pane (see Figure 5.10). The table view shows the display sets in the current drawing and which display configurations they are assigned to.

Cube-shaped viewing direction symbols with different sides colored indicate the viewing directions to which the display sets are assigned. A cube with its top shaded indicates that the display set is assigned to the top view. A cube with any of its four sides colored indicates that the display set is assigned to the appropriate left, front, right, or back viewing directions. A cube with all its sides colored indicates that the display set is used in all elevation viewing directions. A cube without color indicates that the display set is used in any view other than the standard top, bottom, or elevation views.

Right-clicking on top of a display configuration displays the cursor menu shown in Figure 5.11. By selecting a display configuration and right-clicking, you can assign the configuration to the current viewport in layout space, assign it as the default configuration in model space, or delete or rename it. Display configurations included with Architectural Desktop cannot be deleted.

Pressing the + symbol to the left of the Configurations folder lists all the display configurations in the tree view. Selecting a display configuration from the tree view displays the properties for the configuration in the Configuration tab in the display pane (see Figure 5.12).

Selecting the display configuration from the tree view displays the properties for the selected display configuration in the right pane. The configuration properties include two tabs: **General** and **Configuration**. The General tab is used to rename the configuration and provide a description for the display configuration. The Configuration tab is used to assign display sets to viewing directions.

In Figure 5.12, the Work display configuration has been selected. The Configuration tab displays a list of standard viewing directions. These include top, bottom, left, right, front, and back. The default view direction indicates any view other than the standard viewing directions. The default viewing direction could be an isometric view, a perspective view obtained with a camera, or a view created from the 3D Orbit command.

Each viewing direction is then assigned a display set. In Figure 5.12, the top view is assigned to the Work_Plan display set. This display set displays plan representations when the drawing is viewed from the top view. The elevation views are assigned the Work_Section display set when the drawing is viewed with

List of display configurations
in the current drawing

List of display sets in
the current drawing

Select the
Configurations
folder to display
a table in the
display pane.

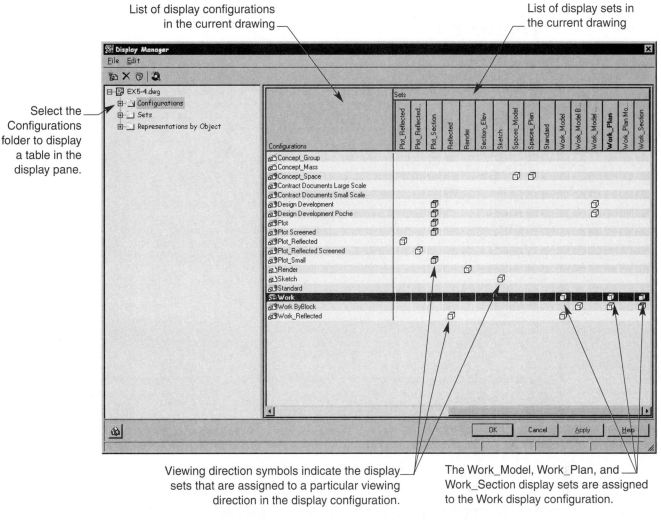

Viewing direction symbols indicate the display
sets that are assigned to a particular viewing
direction in the display configuration.

The Work_Model, Work_Plan, and
Work_Section display sets are assigned
to the Work display configuration.

FIGURE 5.10   Select the Configurations folder to display a table view indicating
which display sets are assigned to a particular display configuration.

FIGURE 5.11   Right-click over a
display configuration in the table
view to display a cursor menu.

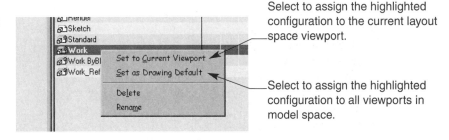

Select to assign the highlighted
configuration to the current layout
space viewport.

Select to assign the highlighted
configuration to all viewports in
model space.

an elevation view. The default viewing direction is assigned to the Work_Model display set, which displays model representations in any view other than the standard views. If a viewing direction is not assigned a display set, then the display set in the default viewing direction is used.

To assign a display set to a viewing direction, pick the area under the **Display Representation Set** column for the appropriate view direction. A drop-down list appears with a list of available display sets (see Figure 5.13). Select the desired display set to assign it the viewing direction.

Many of the display configurations included with Architectural Desktop have only one or two display sets assigned to viewing directions. The most common settings assign a display set to the top view, and another to the default view. The elevation views then use the default display set assigned to the default view.

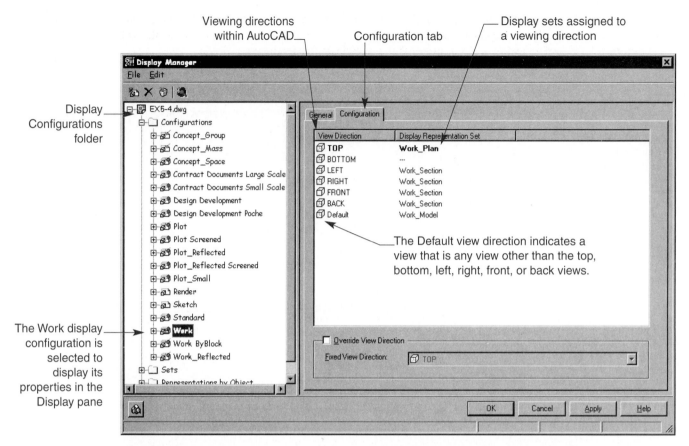

FIGURE 5.12   Select a display configuration from the tree view to display the configurations properties.

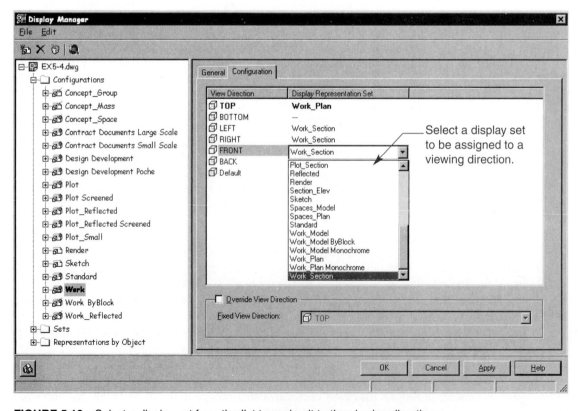

FIGURE 5.13   Select a display set from the list to assign it to the viewing direction.

At the bottom of the Configuration tab is a check box labeled **Override View Direction**. Picking this check box overrides the display sets assigned to the standard viewing directions and uses the display set assigned to the default viewing direction for all viewing directions in the drawing.

To create a custom display configuration, highlight the **Configurations** folder, pick the **New** button or right-click, and select **New** from the cursor menu. The properties for the new display configuration are displayed. In the General tab, enter a description for the display configuration. In the Configuration tab assign display sets to the appropriate viewing directions. When you are finished creating display configurations, press the Apply button to apply the changes to the drawing, then press the OK button.

Exercise 5.5

1. Open EX5-4.dwg.
2. Access the Display Manager.
3. Select the Configurations folder. Press the + symbol to list the display configurations in the tree view.
4. Right-click over the Configurations folder and select New. Name the display configuration Always 2D.
5. Select the "Always 2D" display configuration in the tree view. In the General tab, enter the following description: *Always displays 2D walls, doors, and windows in all viewing directions.*
6. Assign the Always Plan display set to the default viewing direction. Check the Override View Direction check box at the bottom of the Configuration tab.
7. Create another display configuration named Monochrome Colors. In the General tab enter the following description: *Displays plan and model representations as a monochrome color.*
8. Assign the Work_Plan Monochrome display set to the top view. Assign the Work_Model Monochrome display set to the default viewing direction.
9. Exit the Display Manager dialog box and save the drawing as EX 5-5.dwg.

Using Custom Display Settings in Other Drawings

The preceding sections described how to create custom display representations, display sets, and display configurations. There may be situations in which you need to import a display configuration that does not exist in your current drawing. This most often occurs when starting a drawing from scratch or opening a drawing created without Architectural Desktop. You can import display configurations and display sets from other Architectural Desktop drawings. Custom display representations are imported automatically when you import a display set or display configuration.

To import display configurations, right-click on the Configurations folder and select the **Import/Export...** option. The Import/Export dialog box shown in Figure 5.14 is displayed. The left side of the dialog box displays the Current Drawing: list, which contains the display configurations in the current drawing. Use the Open... button to browse and open a drawing file that contains the display configurations you would like to import. Drawing files or drawing template files can be opened.

After a file is opened, a list of display configurations is displayed in the External File list to the right. Select the display configurations you would like to import to the current drawing, and pick the <<<Import button to import the display configurations. When display configurations are imported into a drawing, their associated display sets are also imported into the current drawing.

Display configurations can also be exported to an external file. The New... button can be used to create a new external file. After selecting New... you must specify a location and name for the new drawing file. To export display configurations, highlight the display configurations from the Current Drawing: list, then select the Export>>> button. The display configurations are exported to the external file. When you are finished importing or exporting display configurations press the OK button to return to the Display Manager.

Display sets can be imported and exported in the same manner. To access the Import/Export dialog box for display sets, right-click over the Sets folder in the Display Manager, and select the Import/Export...

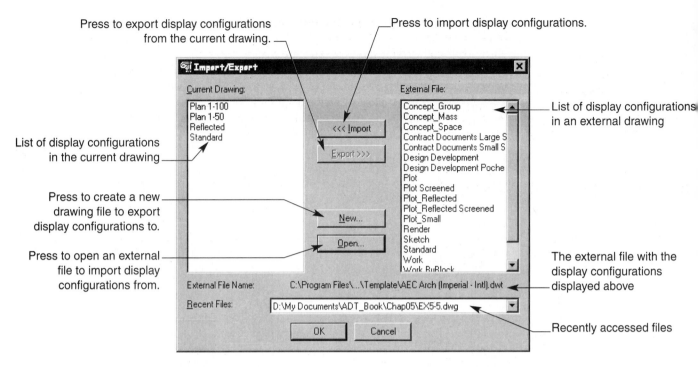

Press to export display configurations from the current drawing.

Press to import display configurations.

List of display configurations in an external drawing

List of display configurations in the current drawing

Press to create a new drawing file to export display configurations to.

Press to open an external file to import display configurations from.

The external file with the display configurations displayed above

Recently accessed files

FIGURE 5.14   The Import/Export dialog box is used to import and export display configurations and display sets into and out of the current drawing.

option. The Import/Export dialog box for display sets appears. The dialog box is the same as the dialog box for the display configurations except that display sets are listed instead. When display sets are imported or exported the associated display representations are also imported or exported.

Exercise 5.6

1. Open EX5-5.dwg.
2. Access the Display Manager.
3. Right-click over the Configurations folder and select Import/Export... from the cursor menu.
4. Press the Open...button to browse and open EX4-10 created in Chapter 4.
5. Export the Always 2D and the Monochrome Colors Display configurations.
6. Close the Import/Export dialog box.
7. When prompted if you want to save EX4-10.dwg press the Yes button.
8. Close the Display Manager and save the drawing as EX5-6.

Applying the Display System

So far in this chapter you have learned the three parts of the display system. The fundamental element that is used by the drafter when creating drawings is the display configurations. Display configurations can be assigned globally to all viewports in model space or to individual viewports in layout space. Use one of the following options to assign a display configuration to model space:

✓ Access the **Des<u>k</u>top** pull-down menu and pick the **<u>S</u>elect Display...** command.

✓ Right-click and select **Des<u>k</u>top,** then **<u>S</u>elect Display....**

✓ Type **AECSELECTDISPLAY** at the Command: Prompt.

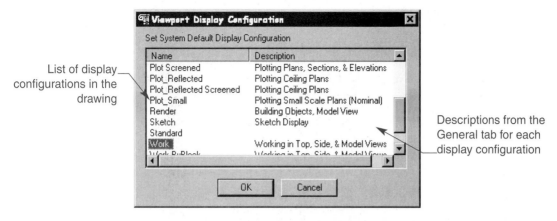

FIGURE 5.15 The Select Display dialog box includes a list of all the display configurations in the drawing. Selecting a display configuration and pressing the OK button applies the display configuration to all model space viewports.

The Select Display dialog box shown in Figure 5.15 appears.

Select a display configuration from the dialog box, then press the OK button. The display configuration is applied to all the viewports in model space. Changing the viewing directions in the drawing displays the AEC objects differently based on the display configuration selected.

An alternative way of setting a display configuration current is to pick the Select Current Display Configuration button in the AEC Setup toolbar. A list of display configurations in the current drawing including any customized display configurations appears (see Figure 5.16). Pick one of the display configurations from the list to set it current in model space.

Setting display configurations in layout space viewports is discussed in the plotting chapter later in this text.

FIGURE 5.16 Pick the Select Current Display Configuration button to display a list from which a display configuration can be made current.

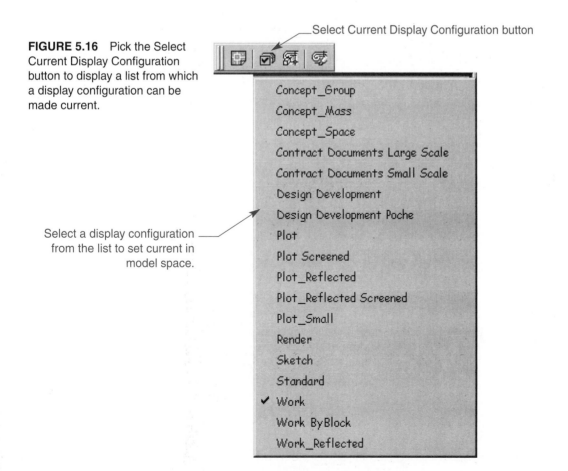

Exercise 5.7

1. Open EX5-6.dwg.
2. From the Select Display dialog box, select the Monochrome Colors display configuration. What happens to wall, door, and window objects? Change the viewing directions. How are the objects displayed?
3. Pick the Select Current Display Configuration button, and select the Always 2D display configuration. What happens to the wall, door, and window objects? Does changing the view display the objects as 3D objects?
4. Pick the Select Current Display Configuration button, and select the Work display configuration. How are the objects displayed in a top view versus an isometric view?
5. Save the drawing as EX5-7.dwg.

Field Notes

Appendix I, *Display Configurations and Their Uses,* lists the default display configurations included in the templates and provides a description for using each.

Controlling AEC Object Display by Object

By using display representations you have been able to configure how default AEC objects appear when viewed with display configurations. The display representations control only how the default representation for each object appears. AEC object display can additionally be controlled by the AEC objects style or by the individual AEC object in the drawing. This system of overriding the default display representation is known as **Entity Display**.

Every AEC object in Architectural Desktop can have its default representation overridden through Entity Display. Anytime you select one AEC object and right-click in the drawing, the Entity Display… command becomes available (see Figure 5.17). Picking Entity Display… from the cursor menu displays the Entity Display dialog box shown in Figure 5.18.

Field Notes

Selecting more than one AEC object, right-clicking, and picking Entity Display… displays only the AutoCAD Props tab. When overriding the display representations display, select one AEC object, right-click, and select Entity Display…..

The Entity Display dialog box includes two tabs: AutoCAD Props and Display Props. The **AutoCAD Props** tab is used to control AutoCAD properties for the selected object. The **Display Props** tab is used to attach overrides to the display representation and configure how the overrides display an objects representation.

The following is a description of the components within the Display Props tab:

- **Display representation** This drop-down list contains all the display representations for the selected AEC object. When an object in the drawing is selected the representation for that view appears in the display representation list with an asterisk (*) beside the representation name. Selecting any representation from the list allows you to modify the properties for that representation.

- **Property Source** This column lists the different levels of display control. The system default is the lowest level of display control and is controlled by the default display representations. Object Style is the second level

FIGURE 5.17 Selecting an AEC object then right-clicking displays a similar menu with the Entity Display... command available. This cursor menu is specific to door objects.

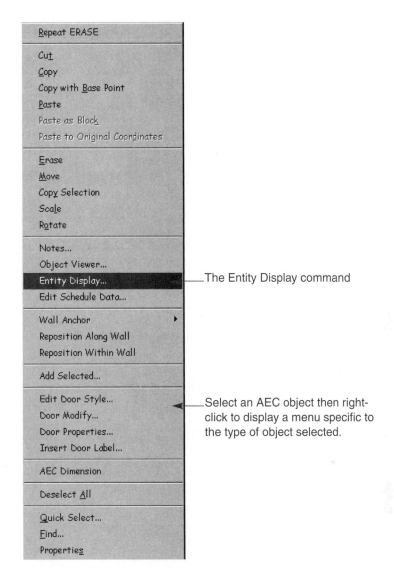

Repeat ERASE

Cut
Copy
Copy with Base Point
Paste
Paste as Block
Paste to Original Coordinates

Erase
Move
Copy Selection
Scale
Rotate

Notes...
Object Viewer...
Entity Display...   ——— The Entity Display command
Edit Schedule Data...

Wall Anchor   ▶
Reposition Along Wall
Reposition Within Wall

Add Selected...

Edit Door Style...   ◀——— Select an AEC object then right-
Door Modify...   click to display a menu specific to
Door Properties...   the type of object selected.
Insert Door Label...

AEC Dimension

Deselect All

Quick Select...
Find...
Properties

of display control. Selecting this property source allows you to control how the selected object's style is displayed for all objects using the same style. The last property source is the individual Object. Selecting this property source allows you to control how the one selected object is supposed to appear in the drawing.

- **Display Contribution**   This column indicates which level of control is being applied to the selected object's display. A hollow arrow with the word *Empty* beside it indicates that there is no override attached to the property source. As overrides are attached the arrow becomes a filled arrow with the word *Control* beside it indicating the property source that is controlling the display representation. If an override is attached, the arrows below the controlling arrow are displayed with a red **X** and the word Overridden indicating that the property source is being overridden by a higher property source.

- **Attached**   This column displays a check mark in the box if an override has been attached by the Property Source.

- **Attach Override**   This button attaches overrides to a property source. To attach an override, highlight the appropriate property source (System Default, Style, or Object) that is going to set the override, then

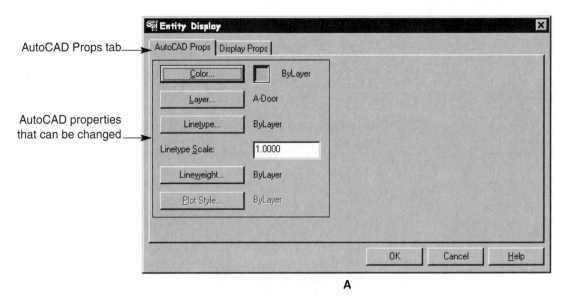

AutoCAD Props tab

AutoCAD properties
that can be changed

A

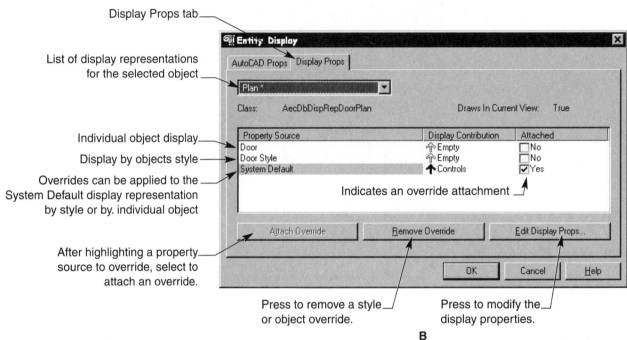

Display Props tab

List of display representations
for the selected object

Individual object display

Display by objects style

Overrides can be applied to the
System Default display representation
by style or by. individual object

After highlighting a property
source to override, select to
attach an override.

Indicates an override attachment

Press to remove a style
or object override.

Press to modify the
display properties.

B

FIGURE 5.18 The Entity Display dialog box includes two tabs. (A) The AutoCAD Props tab controls general display settings such as layer, color, and linetype. (B) The Display Props tab sets overrides to the selected AEC object's default display representation by overriding the display by style or by the individual object.

press the Attach Override button. A check mark is placed in the appropriate check box under the Attached column.

- **Remove Override** This button removes an override from a property source.

- **Edit Display Props…** Selecting this button with an override attached displays the Entity Properties dialog box for the selected representation (see Figure 5.19). Any changes made to the components here result in modification of the representation display in the drawing. The default dis-

Adjust any additional display settings.

Make changes to the components.

Door components

Door display properties

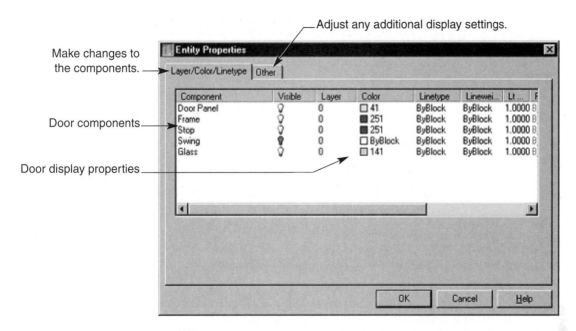

FIGURE 5.19 The Entity Properties dialog box for the door object. Changes made here override the system default display representation.

play representation is not changed for the default AEC objects but is applied only to the selected object's level override (by style or by object). The default representation can be changed by not applying an override to the system default property source, then selecting the Edit Display Props… button.

Field Notes

> Many of the AEC styles in Architectural Desktop override the system default display representations by style. You can apply an override by object to these style overrides or remove the style override.

Selecting the Edit Display Props… button displays the Entity Properties dialog box. This dialog box is the same dialog box used to set up the default display representations. The difference is that the values you change here affect the level of display you selected in the Display Props tab. Components can be turned on or off or assigned colors, linetypes, lineweights, linetype scale, and plot styles. When you are finished changing the display for the current display level or override, press the OK button to return to the Display Props tab. You can assign additional overrides to other display representations by selecting the appropriate representation from the display representation list, assigning an override, and selecting the Edit Display Props… button.

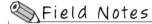

Field Notes

- When the plan representation for wall objects is edited, a wall style can have up to 20 different boundaries. The **boundaries** refer to each individual component within a wall. For example, a wall that included brick, CMU, and stud components would have a total of three boundaries. Each boundary also includes a **hatch component**. Hatch components can be turned on or off by selecting the appropriate hatch component. For example, Boundary 1 includes Hatch 1 component. If you want to turn on the hatch pattern for Boundary 1, turn on the lightbulb symbol next to Hatch 1. The hatch can also be editing by selecting the Hatch tab and picking the hatch swatch next to the hatch component you want to modify. Modifying the hatch of a wall style is discussed in greater detail later in this text.

- Often, when the display of objects is changed the drawing does not reflect the display changes. This usually means that the drawing needs to be regenerated. Enter **REGEN** at the Command: prompt to regenerate the display. The REGEN command works only in the current viewport. Use the **REGENALL** command to regenerate in all the viewports.

- With AEC objects the REGEN command may not be enough to regenerate the updated display of the objects. If you are unsuccessful using the REGEN command, use the **OBJRELUPDATE** command. This command prompts you to select entities to update or to press [Enter] to update everything. Selecting entities then pressing [Enter] updates only the selected object's display. Pressing the [Enter] key without selecting anything updates everything in the drawing in all viewports. This is similar to the REGEN command specifically for AEC objects.

Exercise 5.8

1. Open EX4-10.dwg.
2. Select the front door, right-click, and select Entity Display from the cursor menu.
3. Access the Display Props tab. Select the Threshold Plan display representation from the drop-down list.
4. Select the door property source. Attach an override, and pick the Edit Display Props… Button.
5. In the Layer/Color/Linetype tab turn on the Threshold B component. Select the Other tab. Change the C-Extension to a value of 2″. Change the D-Extension to a value of 2″. Press OK to return to the Entity Display dialog box. Exit the Entity Display dialog box.
6. Select a window object, right-click, and select Entity Display… from the cursor menu.
7. Access the Display Props tab. Select the Plan display representation from the drop-down list. Pick the System Default property source, then select the Edit Display Props… button.
8. Turn off the glass component. Press OK to return to the Entity Display dialog box. Exit the Entity Display dialog box.
9. Select a wall object, right-click, and select Entity Display… from the cursor menu.
10. Access the Display Props tab. Select the Model display representation from the drop-down list. Select the Wall Style property source, and attach an override if none is attached. Select the Edit Display Props… button.
11. Change the Boundary 1 (Stud) component to Color 7 (white). Press OK to return to the Entity Display dialog box. Exit the Entity Display dialog box.
12. Save the drawing as EX5-8.dwg

CHAPTER REVIEW

Use the CD-ROM to test your knowledge and skills.

Chapter Test

To check your understanding of the content provided in this chapter, access the Test file in the CH05 folder of the CD-ROM that accompanies this text.

Chapter Project

To practice the Architectural Desktop skills presented in this chapter, access the Project files in the CH05 folder of the CD-ROM that accompanies this text. The project files are in pdf format and include sample drawings and instructions for completing each project.

CHAPTER 6

Layer Management

LEARNING GOALS

After completing this chapter, you will be able to:

◎   Use Architectural Desktop layers in your drawing.

◎   Access and use the Architectural Desktop Layer Manager.

◎   Create and use layer standards.

◎   Establish new layers.

◎   Set a layer current.

◎   Create layer filter groups.

◎   Add user groups.

◎   Use snapshots.

◎   Create and use layer keys.

◎   Use layer key overrides.

◎   Import and export layer standards.

Using layers in Architectural Desktop is similar to establishing a group of transparent overlays. Objects that share common properties such as color linetype and lineweight may be grouped together onto one layer. In architectural drawing, AEC objects that represent a real world object are grouped together onto the same layer. For example, windows can have the same color and linetype and represent a real-world window, so they can be organized under a layer named Windows or Glazing.

Managing multiple objects in the drawing becomes easier by grouping common items together. The most common property of layers is the color. Colors help differentiate the types of information in the drawing. All walls in the drawing may be assigned a specific color on the Wall layer that is different from the color of the objects on the Door layer. Linetypes can define different types of drafting elements. For example, hidden lines can represent demolition walls, or a gas linetype can represent the gas utilities in the building.

The use of layers also allows specification of how objects are to be displayed in the drawing at any giving moment. Entire groups of objects on different layers can be turned on or off to create many different types of drawings. For example, a drawing that is given to a client may display only the walls, doors, and windows for a preliminary plan sheet. The same drawing may also be given to the contractor, who requires more information such as display of the dimensions, legends, and detail flags, so the building can be constructed (see Figure 6.1). Turning layers on or off and combining sets of layers allows these two different types of plans to be created from the same drawing file.

In standard AutoCAD, layer names and their properties must be manually created by the drafter. As objects such as polylines, arcs, text, and dimensions are drawn, a layer needs to be created and the object assigned to the appropriate layer. As AEC objects are drawn in Architectural Desktop, the layer name, color, and linetype are automatically created, and objects are automatically assigned to an appropriate layer.

The Architectural Desktop layering system is built on two main elements: a *layer standard* and *layer keys*. A **layer standard** is a set of rules that govern how a layer name is to be created. In Architectural Desktop, layer names are created in accordance with the AIA layering guidelines. See Appendix A for more

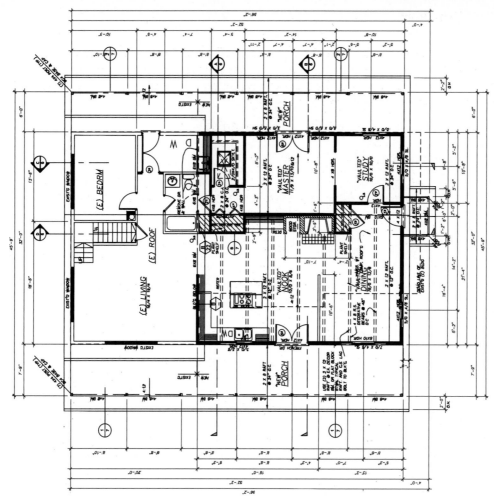

Contractor drawing with dimension, notes, and detail layers turned on

Client drawing with dimension, notes, and detail layers turned off

FIGURE 6.1 A drawing of a building may have different sets of layers displayed in order to create two separate plans from one drawing file.

information on the AIA layering standard. *Layer keys* interpret the layer standard and assign a layer name to AEC objects. When an AEC object is added to the drawing, the layer name and properties assigned to the object are then created and added to the drawing, and the object is automatically placed on the layer. This saves layer management time, because layers are created as you draw.

Introduction to the Layer Manager

Architectural Desktop includes a utility known as the **Layer Manger**, where layers can be created and their properties assigned. Within Layer Manager, additional utilities are provided for the daily management of layers. Use one of the following options to access the Layer Manager dialog box, shown in Figure 6.2:

✓ Pick the **Desktop** pull-down menu, move to **Layer Management,** and select **Layer Manager...** from the list.

✓ Pick the **Layer Manager** button on the Layer Management toolbar.

✓ Right-click and select **Desktop, Layer Management, Layer Manager....**

✓ Type **AECLAYERMANAGER** at the Command: prompt.

✎ Field Notes

> The Architectural Desktop Layer Manager dialog box is an advanced version of the standard AutoCAD Layer Properties Manager. Although they are similar, the Architectural Desktop Layer Manager provides additional tools not found in the AutoCAD Layer Properties Manager.

Once activated, the Layer Manager dialog box provides a list of the layers currently in the drawing. The dialog box is divided into two separate panes. The left pane displays lists of **filter groups** and **user**

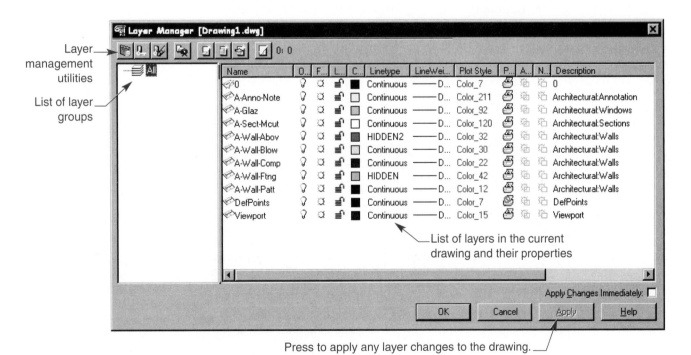

FIGURE 6.2 The Layer Manager dialog box is used to organize and manage layers in the drawing. Additional utilities include creating and sorting layers, as well as establishing a layer standard.

groups. Filter groups and user groups combine sets of layers together for easy management. Initially the only group available is the **All** layer group. This group combines all the layers in the drawing, which are displayed in the right pane. Creating filter groups and user groups is discussed later in this chapter.

The right pane lists the individual layer names and groups and displays their current properties, such as color, linetype, and current status. As AEC objects are created in the drawing, new layers are added to the drawing, and their properties are automatically assembled in the Layer Manager dialog box. A description of each column in the right pane follows (see Figure 6.3):

- **Name**

  This column provides a list of layers in the drawing by name. The layer names are sorted alphabetically. Layer names can be created with up to 255 characters. To the left of the layer name is an icon. A different icon is displayed depending on the status of the layer. These icons and their descriptions are explained in the table in Figure 6.4.

- **O... (On)**

  This column, represented by a lightbulb, controls the visibility of a layer. Picking the lightbulb symbol beside the desired layer turns the layer on or off. Layers that are turned on are visible in the drawing screen. Layers that are turned off are not visible in the drawing screen. Objects on layers that are turned off are regenerated with the rest of the drawing.

- **F... (Freeze in all VP)**

  This column, represented by a sunshine symbol, also controls the visibility of layers. Picking the sunshine symbol beside the desired layer thaws or freezes the layer. Layers that are thawed are visible in the drawing screen. Layers that are frozen are not visible in the drawing screen. Freezing layers helps speed up the ZOOM, PAN, and 3D ORBIT commands. Layers that are frozen are also not regenerated with the rest of the drawing. If there are layers that will not be shown for long periods, the Freeze option should be used instead of the Off option, to speed up the performance of Architectural Desktop.

- **L... (Lock)**

  This column, represented by an unlocked or locked padlock, controls the accessibility of the layer. Picking the padlock symbol beside the desired layer locks or unlocks the layer. Layers that are unlocked allow editing of objects on the layer. Layers that are locked do not

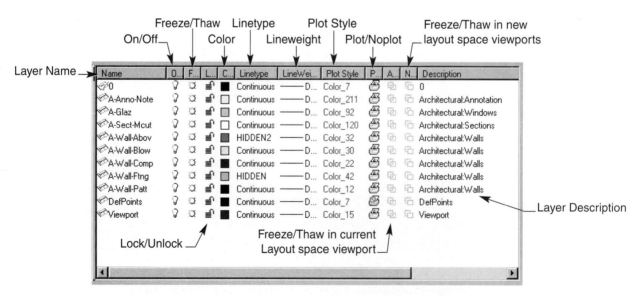

FIGURE 6.3   The right pane includes a list of columns that control a different property or status for a layer.

Icon	Represents	Description
	Unused layer	Layers can be created in a drawing yet not be assigned to any geometry. This icon identifies the layer as not being assigned to any geometry.
	Used layer	Once an object(s) have been assigned to a layer, this icon is displayed beside the layer name to indicate the layer has been assigned to some geometry.
	Current layer	A used layer icon with a green check mark indicates that the layer is the current layer. Any AutoCAD entities that are drawn will be placed on the current layer.
	Layer group	Once a layer group has been created, this icon will appear in both the left and right pane of the Layer Manager. This icon represents any layers that have been grouped together in the Layer Manager.
	Layer filter	This icon represents a list of layers that meet the criteria of a set of layer sorting rules. For example a layer filter if sorting for all the layers using the color red may be added to a layer filter.
	Deleted layer	This icon is displayed next to a layer if the layer has been chosen to be deleted. A layer cannot be deleted if it has geometry assigned to it.

FIGURE 6.4

allow objects on the layer to be selected for editing. Locked layers "lock" the objects in place with their current properties.

- **C... (Color)** This column is used to assign a color to a layer. The color of the layer is represented in the swatch box with the color name or number beside the swatch. Picking the swatch beside the desired layer accesses the **Select Color** dialog box, shown in Figure 6.5. Pick a color from the dialog box to be used for the layer. When a color is selected, the color name or number is entered in the **Color:** text box at the bottom of the dialog box. Pressing OK returns the Layer Manager dialog box displaying the new color in the color swatch.

- **Linetype** This column is used to assign a linetype to a layer. The linetype assigned to the layer is represented by the name of the linetype. Picking the linetype name beside the desired layer accesses the **Select Linetype** dialog box (see Figure 6.6). Selecting a linetype from the list assigns the selected linetype to the layer. Linetypes can be listed by all the linetypes currently loaded in the drawing or by all the linetypes in the acad.lin file.

- **LineWei...** This column is used to assign a lineweight to a layer. Lineweights add width to the objects drawn on the layer. Picking the name or number of the lineweight beside the desired layer accesses the **Lineweight** dialog box, shown in Figure 6.7. Selecting a lineweight from the list assigns the lineweight to the layer.

- **Plot Style** This column is used to assign a plot style to a layer. By default, plot styles are grayed out. When a drawing is plotted, pen tables are used to interpret the layer color and translate it into a finished hard-copy plot. Before you begin a new drawing you can choose to use **color pen tables** or **plot style tables**. If plot styles are chosen, then a plot style can be assigned to a layer from this column. Pen tables and plotting are discussed in Chapter 21.

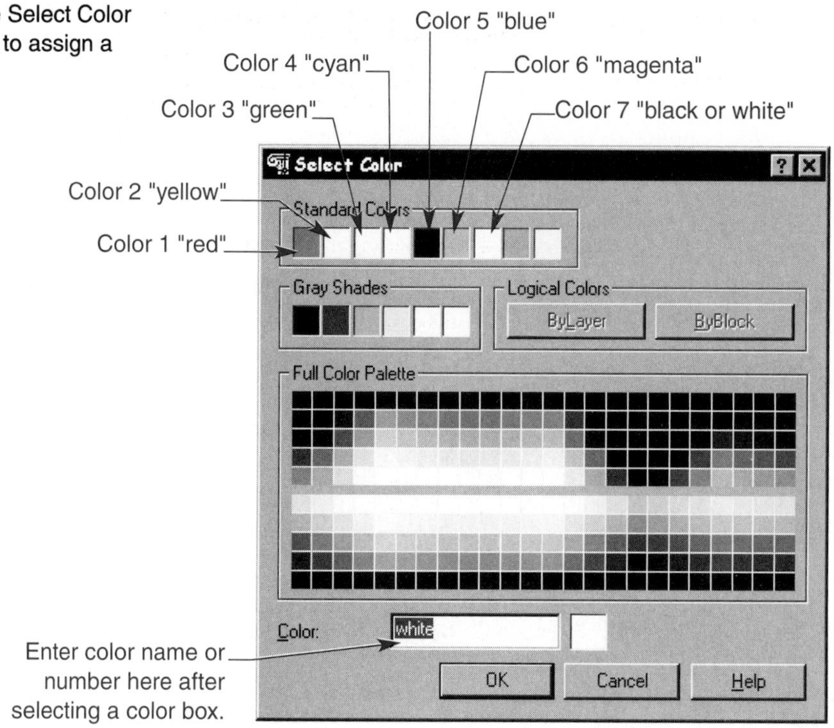

FIGURE 6.5 The Select Color dialog box is used to assign a color to a layer.

Color 5 "blue"

Color 4 "cyan"

Color 6 "magenta"

Color 3 "green"

Color 7 "black or white"

Color 2 "yellow"

Color 1 "red"

Enter color name or number here after selecting a color box.

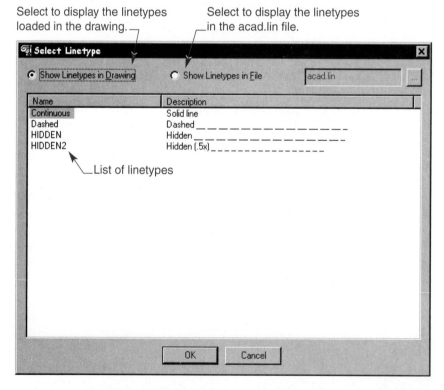

FIGURE 6.6 The Select Linetype dialog box is used to select a linetype that is assigned to a layer.

Select to display the linetypes loaded in the drawing.

Select to display the linetypes in the acad.lin file.

List of linetypes

FIGURE 6.7 The Lineweight dialog box is used to select a lineweight to be assigned to a layer.

List of lineweights

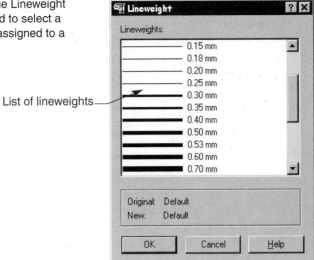

- **P... (Plot)** This column, represented by a printer symbol, is used to determine whether a layer will plot onto a piece of paper. Picking the printer symbol beside the layer causes the layer to be a plotting or non-plotting layer. When the plot symbol is picked, objects on the layer plot unless they are turned off or frozen. When the no-plot symbol is turned on, objects on the layer are not plotted, regardless of being turned on or thawed.

- **A... (Active VP Freeze)** This column, represented by a sunshine with a rectangle, controls the visibility of layers in the active layout space viewport. This column is available only when a layout space viewport is active. Picking this symbol beside the desired layer causes the layer to be thawed or frozen in the active viewport. Freeze and thaw in the active viewport option is different from the standard freeze and thaw option. Freezing in the active viewport freezes the layer in the active layout space viewport only, not the entire drawing.

- **N... (New VP Freeze)** This column, represented by a sunshine with a filled rectangle, controls the visibility of layers in newly created layout space viewports. This column is available only when a layout space viewport is active. Picking this symbol beside the desired layer causes the layer to be thawed or frozen in new layout space viewports. When the sunshine symbol is displayed, the layer is visible when a new viewport is created in layout space. When the snowflake symbol is displayed the layer does not appear in the new viewport created in layout space.

- **Description** This column provides a description of the layer. The description is automatically generated when a new layer is created. Depending on the layer standard being used the description can be quite detailed.

The **Apply Changes Immediately:** check box immediately adjusts the layer properties and the display of the layers as soon as they are changed. If this box is unchecked no changes are made to the layer until the Layer Manager dialog box is closed. After making any changes to the layer(s) press the Apply button so the changes are displayed appropriately in the drawing screen.

FIGURE 6.8 Right-click in the right pane to display this cursor menu.

Select All
Clear All

Rename Layer
Make Current
Delete
Remove From Group
Change Layer Standard ▶
Change Description

Field Notes

When you make a change to a layer, it may not show up immediately. You may need to use the REGEN, REGENALL, or OBJRELUPDATE commands to update the layer changes in the drawing.

Right-clicking in the right pane of the Layer Manager dialog box produces a cursor menu with the following options that aid in controlling layers (see Figure 6.8):

- **Select All**   This option highlights all the layers in the right pane. This option can be used to freeze all the layers at the same time or to make all the layers one color or linetype.

Field Notes

If you need to make several changes to multiple layers, highlight all the layers by holding down the [Ctrl] key as you pick on top of the layer names. The selected layers are highlighted, allowing you to make changes to these layers at the same time.

- **Clear All**   This option clears the highlight from any selected layers.
- **Rename Layer**   This option allows you to rename the selected layer.
- **Make Current**   Highlighting a layer, right-clicking, and selecting this option sets the layer as the current layer. Any non–Architectural Desktop object is drawn on the current layer.
- **Delete**   This option deletes the layer if it currently does not have any objects assigned to it. If you try to delete a "used" layer, a warning is displayed letting you know that the layer cannot be deleted.
- **Remove From Group**   If this option is available, the highlighted layer can be removed from the group. Creating and managing filter groups and user groups is discussed later in this chapter.
- **Changing Layer Standard**   This option displays a list of layer standards in the drawing. The current layer standard is listed with a check mark beside it. When a layer standard is current, newly created layers adhere to the naming rules of the layer standard.
- **Change Description:**   Selecting this option allows you to change the description in the Description column for the selected layer.

Exercise 6.1

1. Open EX5-8.dwg.
2. Access the Layer Manager dialog box. Turn the A-Door layer off. Press the Apply button, then the OK button. What happens to the doors?
3. Access the Layer Manager. Turn the A-Door layer back on. Press the Apply button, then the OK button. You may have to use REGEN to redisplay the door objects.
4. Lock the A-Anno-Nplt layer. Try to move the Foyer_Camera object. Can the camera object be moved?
5. Unlock the A-Anno-Nplt layer, then freeze the layer. Press the Apply button, then the OK button.
6. Save the drawing as EX6-1.dwg

Using and Creating Layer Standards

Layer standards are used to establish individual, project, or office layer naming conventions. A layer standard contains a set of rules that determines how a new layer name is to be created. For example, the *AIA Long Format* layer standard included with Architectural Desktop contains rules that divide the layer name into four parts or fields separated by hyphens: Discipline code-Major code-Minor code-Status code.

The *discipline code* designates the discipline that created the geometry, such as A=Architectural and S=Structural. The *major code* designates the type of assembly or construction material being used, such as WALL or DOOR. The *minor code* designates a further description of the major code, such as FULL or PART. The *status code* is used to indicate the state of the geometry being drawn, such as DEMO for demolition work, or NEWW for new work. Together these fields create a layer name such as A-WALL-FULL-DEMO. Appendix A explains the AIA layering standard and provides examples of its usage.

New layer names can be created that adhere to these rules or to a set of rules that create a custom layering standard. As AEC objects are added to the drawing, the layer where the objects belong is created, and the objects are assigned to the layer name. AutoCAD objects such as polylines, text, and dimensions must be assigned to a layer as they are drawn. To do this, first set the desired layer current, then draw the AutoCAD object. Newly drawn AutoCAD objects are assigned to the current layer.

Many architectural offices have adopted the AIA layer guidelines for their CAD drawings. Some offices have their own standards, and some do not use a layer standard at all. The use of a layer standard can be very helpful in organizing your CAD geometry and can make it easier to change the display of objects in the drawing.

Architectural desktop includes three layer standards: AIA Long Format, BS1192-Aug Version2 Format, and BS1192 Descriptive. The AIA Long Format is the default layer standard when an Architectural Desktop drawing is started from a template and is based on the *CAD Layer Guidelines, Second Edition,* from the American Institute of Architects (AIA). The two BS1192 layer standards are British standards. If the International Extensions are installed (see Chapter 1 for additional information), five additional layer standards are installed: Generic Architectural Desktop Format, DIN 276 Format, ISYBAU Short Format, ISBAU Long Format, and STLB Format. The Generic Architectural Desktop Format is based on European layering guidelines. The other four layer standards are based on German layering guidelines.

In addition to these layer standards, custom layer standards may be created or modified from the existing layer standards. Two popular layer standards in residential work replace the discipline code in the AIA layer standard with a floor field, which indicates the floor level of the building. This field may by an alphabetic or numeric field. For example, M-Window-Demo may indicate the main-floor window that is being demolished, or the layer name 3-Door-Neww may indicate the third-floor door that is new construction. No matter which standard you use, newly created layer names follow the rules that are set up in the layer standard.

Creating a New Layer Standard

To create a new layer standard, display the Layer Manager dialog box. Across the top of the dialog box are a set of buttons that are used to manage layers. The first button is the Layer Standards button. Select this button to display the **Layer Standards** dialog box, shown in Figure 6.9.

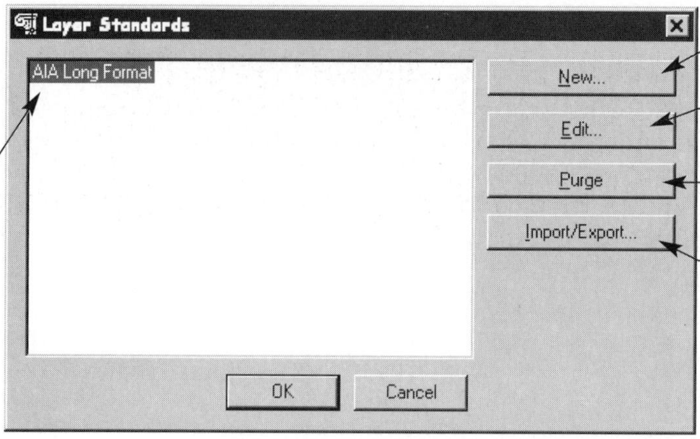

FIGURE 6.9   The Layer
Standards dialog box is used to
create or modify a layering
standard.

List of layering
standards in the drawing

Create a new
layer standard.

Edit an existing
layer standard.

Purge existing
layer standards.

Import a layering
standard into the
drawing or export a
layering standard
to another drawing.

FIGURE 6.10   The Create
Layer Standard dialog box.

Enter a name for
the new standard.

Select to base the new standard
on an existing standard.

The Layer Standards dialog box includes a list of layering standards in the current drawing. Along the right side of the dialog box are buttons to create, edit, purge and import or export layer standards. To create a new layer standard pick the New... button. The Create Layer Standard dialog box, shown in Figure 6.10, is displayed.

Enter a new name for the layer standard. If the layer standard is to be based on an existing standard, select the **Based On:** check box, then choose a standard from the list. If the new layer standard is not based on an existing standard, the layer standard will have to be created from scratch. If the standard is based on an existing standard, you can edit the standard to reflect your rules. After you enter a name and press the OK button, the Layer Standards dialog box is displayed with the new standard in the list.

Select the new layer standard from the list and pick the Edit... button. The **Layer Standard Properties** dialog box is displayed. This dialog box is used to set up the layer naming rules for the layer standard. The dialog box includes four tabs: *Component Fields, Edit Descriptive Fields, Edit Descriptions,* and *Description Specification*. The four tabs and their uses are described in the following sections.

Using the Component Fields Tab

The Component Fields tab is shown by default and allows you to set up how a layer name will be created (see Figure 6.11). The top of the tab displays the current layer standard being modified. The drop-down list includes a list of standards in the current drawing. The main part of the tab is used to list the component fields in a layer name, that is, the AIA standard four component fields. The rules governing how the field name is created can be found in this area. At the bottom of the tab are three buttons. The **Delete** button deletes a layer field name, the **Add above selected item** and **Add below selected item** buttons control where a new field name is to be created in the list above.

The following creates a layer standard for residential use that includes four fields: a Floor field, an Object Type field, Sub-Object Type field, and a Status field. An example of a layer name using this layer standard is M (main floor), Walls (wall objects), Dims (dimensions), New (new construction) = M-Walls-Dims-New.

When a new layer standard is created, the **Default value** component field is already entered in the layer fields list area. To create a new field, highlight the Default value component field, and press the Add above selected item button. A new component field is created at the top of the list. Enter the name for the new component field and press [Enter]. In this case, Floor is entered as the component field name in Figure 6.12.

Once the component field name has been created, the rules for the component field can be set up by selecting a value for the field name under each one of the columns. Picking on a value displays a cursor menu with options that can be selected. The Default column is the only column that requires you to enter a value from the keyboard. The following describes each column and how it affects the component field name:

Component Fields tab

Layer standard being edited

Layer field name rules

Add above the selected item.

Add below the selected item.

Delete

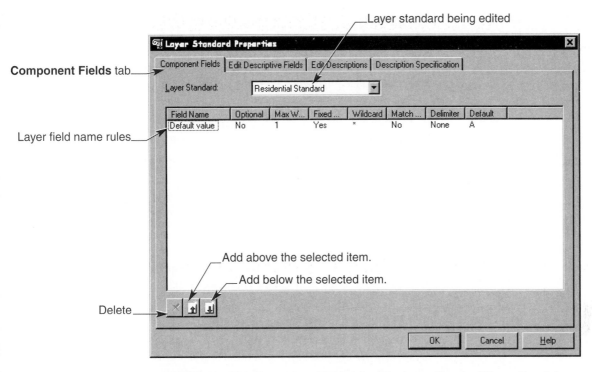

FIGURE 6.11   The Component Fields tab of the Layer Standard Properties dialog box is used to create the rules for the layer names.

Is the field name a fixed width?

Does the layer description match the layer name?

Layer Field name

Floor layer field

Is the field optional?

Maximum number of characters

What types of characters are used for the field name?

What is the default for the field name?

Is there a delimiter in front of the field name?

Add above the selected item.

Add below the selected item.

Delete button

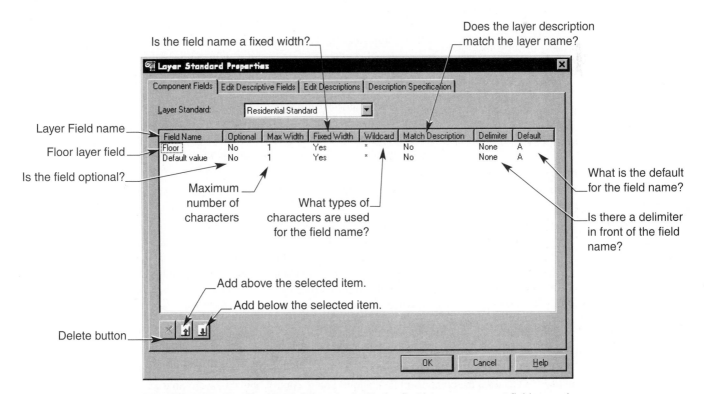

FIGURE 6.12   The Residential Standard with the first layer component field named Floor is ready to have rules applied.

- **Field Name**

This column is used to identify the name of a layer component field. Component field names are entered after selecting one of the Add buttons from the bottom of the tab. To rename a component field, first highlight the field name, then pick the field name again. This places a box around the component field name so that you can rename the field.

- **Optional**

This column indicates whether the component field is an optional field. Picking on the value displays a menu containing a Yes and a No. Selecting Yes makes the field an optional field that is not required when creating a new layer name.

- **Max Width**

This column identifies the maximum number of characters that can be used for the component field. Selecting a value displays a menu of several numbers representing the total number of characters to be used that can be selected.

- **Fixed Width**

This column specifies whether the component field must use the total number of characters from the Max Width column or if the number of characters used in the field can be shortened. Picking the value displays a menu with a Yes and a No. Selecting Yes requires that the component field name use the maximum number of characters.

- **Wildcard**

This column specifies the type of characters that can be used in the component field. Selecting the value displays a menu list of options. These options include Alpha (alphabetic characters), Numeric (number characters), Any (letters, numbers, or special characters), and Other…. The Other wildcard displays a dialog box in which certain characters can be entered separated by commas.

- **Match Description**

This column recognizes the layer component field only if it exactly matches a description. For example, the AIA layering standard dictates that component field names are created in the order, discipline, major, minor, and status. The minor and status component fields are optional fields, so it is possible to create a layer that includes the discipline, major, and minor component fields, or the discipline, major, and status component fields. If the value for Match Descriptions is set to Yes for the status component field, and a layer is to be created with the discipline, major, and status component fields, the status component field descriptions are checked against the descriptions of the minor component field to determine if there are any matches. If there are any matches, the layer name is invalid. This column is most commonly set to No to avoid this complication.

- **Delimiter**

This column specifies a character to be placed in front of the layer component field for clarity between the layer field names. Picking on the value displays a list of valid delimiters for a layer field name. These options include $, - (hyphen), _(underscore), and none. None is most commonly used in front of the first layer field, because a delimiter is not necessary.

- **Default**

This column allows you to enter a default field name that appears when layer names are created. These values can be left blank.

Remember, each layer component field in the layer standard is a part of the overall layer name. Picking on top of a value for the component field allows you to control how the layer name can be created. Component field names can be added above or below a selected component field to create the layer component field name order. The completed Component Fields tab for the sample residential standard is shown in Figure 6.13.

FIGURE 6.13 The Residential Standard has been assigned four layer fields, two of which are optional, all of which have varying rules applied.

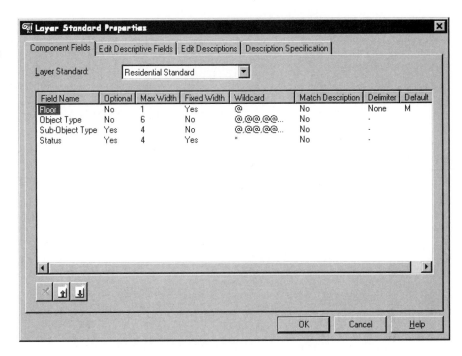

Exercise 6.2

1. Start a new drawing using the AEC Arch (Imperial).dwt or AEC Arch (Imperial-Int).dwt.
2. Access the Architectural Desktop Layer Manager. Select the Layer Standards button.
3. Create a new layer standard named Residential Standard.
4. Select and edit the Residential Standard.
5. Use the following table to create four layer component fields with the following properties:

Field Name	Optional	Max Width	Fixed Width	Wildcard	Match Description	Delimiter	Default
Floor	No	1	Yes	Alpha	No	None	M
Object Type	No	6	No	Alpha	No	—	Walls
Sub-Object Type	Yes	6	No	Alpha	No	—	Dims
Status	Yes	4	Yes	Any	No	—	Demo

6. Delete the Default Value layer field.
7. Press the OK button when you have finished creating the layer component fields and adjusting their properties. Press the OK button again in the Layer Standards dialog box to return to the Layer Manager dialog box.
8. Exit the Layer Manager dialog box and save the drawing as EX6-2.dwg.

FIGURE 6.14   The Edit
Descriptive Fields tab is used to
create descriptive field names for
use in the Edit Descriptions tab.

Edit Descriptive Fields tab   **Edit Descriptions** tab

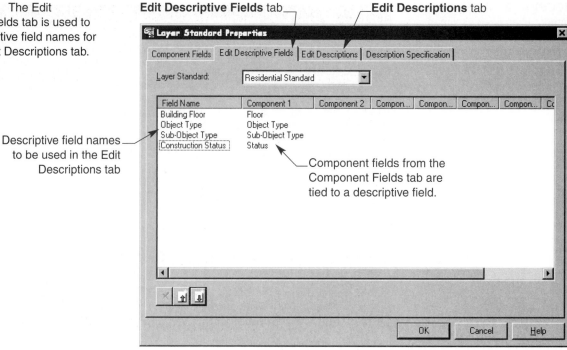

Descriptive field names
to be used in the Edit
Descriptions tab

Component fields from the
Component Fields tab are
tied to a descriptive field.

Using the Edit Descriptive Fields Tab

This tab is used to set up descriptive field names in the layer standard. *Descriptive field names* are different from the component field names created in the Component Fields tab. The descriptive fields are used in the Edit Descriptions tab, where layer names and descriptions are added to the layer standard. When a descriptive field name is created, a component field from the Component Fields tab is assigned to it. This ties the component field to the descriptive field used in the Edit Descriptions tab (see Figure 6.14).

As indicated in Figure 6.14, a descriptive field is added to the Field Name column. The descriptive field can be a brief description of what the component field means in a layer name. For example, the Building Floor descriptive field in Figure 6.14 indicates that the first part of the layer standard is using the Floor component field and that it designates the floor level in a layer name. Descriptive fields can also match the component field name that is assigned, because the Object Type descriptive field is assigned the Object Type component field.

Creating the descriptive field names is similar to creating the component field names. Highlight the Default value descriptive field, then press the Add above selected item button to add a new descriptive field. Type a name and press the [Enter] key. After a descriptive field name is created, it needs to be assigned one of the component fields created in the Component Fields tab. Under the column Component 1, pick the space that coincides with the descriptive field row you are editing (see Figure 6.15). A list of component field names in the layer standard that can be assigned to the descriptive field is displayed. Select a component field from the list and move to the next descriptive field.

Additional component fields can be added to the same descriptive field by selecting the appropriate space below the Component 2 column. This is most often done in layering systems where the layer names are hierarchical. Normally, one descriptive field is made for each component field, as shown in Figure 6.14.

FIGURE 6.15 Picking in the space where a Component column and a Descriptive Field row cross each other results in a cursor menu that allows you to assign a component field to a descriptive field.

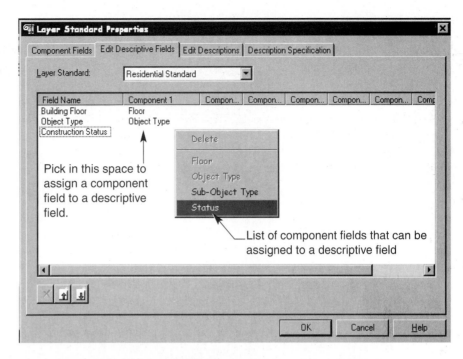

Exercise 6.3

1. Open EX6-2.dwg
2. Access the Architectural Desktop Layer Manager. Select the Layer Standards button.
3. Select and edit the Residential Standard. Pick the Edit Descriptive Fields tab.
4. Use the following table to create four descriptive fields. Assign a component field to the appropriate descriptive field.

Field Name (descriptive fields)	Component 1 (component fields)
Building Floor	Floor
Object Type	Object Type
Sub-Object Type	Sub-Object Type
Construction Status	Status

5. Delete the Default value descriptive field.
6. Press the OK button when you have finished creating the layer descriptive fields and assigning the appropriate component fields. Press the OK button again in the Layer Standards dialog box to return to the Layer Manager dialog box.
7. Exit the Layer Manager dialog box and save the drawing as EX6-3.dwg.

Using the Edit Descriptions Tab

The Edit Descriptions tab is used to assign layer descriptions to each descriptive field. The *layer description* is the actual part of the layer that is used for the layer name. Each part (description) of the layer is assigned to a component field by means of the descriptive field. For example, the Construction Status descriptive field may have the descriptions Neww, Demo, Extg, and Phs1 that become a part of the Status component field. When a layer is created, the layer name is made up of descriptions for each component field, such as M-Walls-Demo.

To edit the descriptions, first access the Edit Descriptions tab (see Figure 6.16). At the top of the tab is the **Layer Standard:** drop-down list. The layer standard currently being modified is listed here. To the right is the **Field to Edit:** drop-down list. This list includes a list of the descriptive fields created in the Edit Descriptive Fields tab. Selecting a descriptive field from the list displays the layer component names and their descriptions, as shown in Figure 6.16.

Initially, the only description that is displayed is the values you entered in the Default column in the Component Fields tab. A layer component can have as many parts or descriptions as needed for the creation

FIGURE 6.16 The Edit
Descriptions tab allows you to
create and manage the parts
(components) of a layer name
and provide a description for
each part.

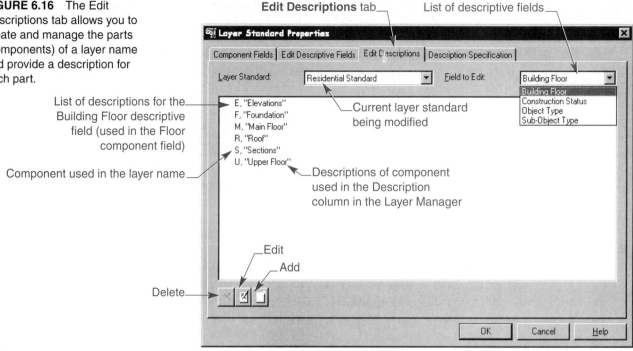

of new layer names. For example, in Figure 6.16, the Floor layer component field (Building Floor descriptive field), has six descriptions representing the six floor levels of a residential building. This allows the creation of M- layers, U- layers, and F- layers, for example.

To add your own descriptions, select a descriptive field from the Field to Edit: drop-down list. Select the Add button at the bottom of the tab. The **Add Description** dialog box, shown in Figure 6.17 is displayed. Enter a value in the **Value for Object Type:** text box. This value is available and used in the creation of the layer name. In the **Description:** text box enter a description for the layer part name. Pick the Apply button to enter the layer component name and the description beside it (see Figure 6.17). Enter as many names as needed when assembling layer names for your standard. When you are finished with one

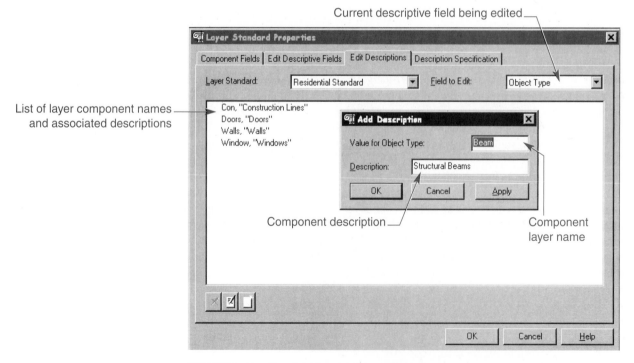

FIGURE 6.17 The Add Description dialog box is used to enter component names
and descriptions for each descriptive field.

descriptive field, move to the next descriptive field in the Field to Edit: drop-down list, and create the layer component names and descriptions for that descriptive field.

When entering layer component field names, you are limited to the total number of characters specified in the Max Width column in the Component Fields tab. As you refine the layer standard, you can use the Edit button to edit the description for a layer component name. Use the Delete button to delete layer component names and descriptions.

Field Notes

The layer standard is a ***living document***. This means that it can continually be edited and updated to reflect your individual, industry, project, or office layering standards. Additional descriptions can be added or deleted as required.

Exercise 6.4

1. Open EX6-3.dwg.
2. Access the Layer Manager dialog box. Select the Layer Standards button and edit the Residential Standard.
3. Select the Edit Descriptions tab and create the following descriptions for the Building Floor descriptive field (Floor component field):

Layer Component Name (Building Floor descriptive field)	Description	Layer Component Name (Building Floor descriptive field)	Description
A	Attic	P	Paperspace
B	Basement	R	Roof
E	Elevation	S	Sections
F	Foundation	U	Upper Floor
M	Main Floor		

4. Create the following descriptions for the Object descriptive field (Object component field):

Layer Component Name (Object descriptive field)	Description
Anno	Annotations
Area	Areas
Beam	Beams
Brace	Braces
Ceilng	Ceiling Objects
Colmns	Columns
Commo	Communications
Constr	Construction Lines
Ctrl	Control Systems
Detail	Details
Door	Doors
Elev	Elevation Objects
Eqpmnt	Equipment
Fire	Fire Protection

Exercise continues on next page

Exercise continued

Floor	Floor Information
Furn	Furniture
Grid	Layout Grid
Hvac	HVAC objects
Light	Lights
Mass	Massing Objects
Plant	Planting
Roof	Roofs
Sect	Section Objects
Walls	Walls
Window	Windows

5. Create the following descriptions for the Sub-Object descriptive field (Sub-Object component field):

Layer Component Name (Sub-Object descriptive field)	Description
Accs	Access
Assmby	Assembly
Case	Casework
Ceilng	Ceiling Objects
Dims	Dimensions
Fixt	Misc Fixtures
Full	Full Height
Group	Groups
Hrail	Hand Rails
Iden	Identification
Legn	Legends, Schedules
Move	Movable
Note	Notes
Nplt	Non-plotting
Patt	Patterns
Pfix	Plumbing Fixtures
Power	Power Designations
Prht	Partial Height
Revs	Revisions
Strs	Stairs
Swtch	Switches
Symb	Symbols
Ttbl	Titleblocks

6. Create the following descriptions for the Construction Status descriptive field (Status component field):

Layer Component Name (Construction Status descriptive field)	Description	
Demo	Demolished Items	
Exst	Existing Items	
Futr	Future Items	
Neww	New Items	*Exercise continues on next page*

Exercise continued

Phs1	Phase 1 Work
Phs2	Phase 2 Work
Phs3	Phase 3 Work

7. Press the OK button when you have finished editing the descriptions. Press the OK button again in the Layer Standards dialog box to return to the Layer Manager dialog box.
8. Exit the Layer Manager dialog box and save the drawing as EX6-4.dwg.

Using the Description Specification Tab

The Description Specification tab is used to configure how descriptions of layer names appear in the Description column of the Layer Manager (see Figure 6.18). As component names are put together to form a layer name the corresponding layer descriptions are also assembled and displayed in the Description column of the Layer Manager. The Description Specifications tab controls how the descriptions are assembled.

The Description Specification tab is divided into two columns. The first column, **Prior Text,** indicates a symbol that is used to separate the components of a description. The **Field** column is used to organize the order in which the descriptions are created. To add a description field, select the Add above (or Add below) the selected item button. The first descriptive field is added to the list. Selecting the descriptive field allows you to select a different descriptive field. In the **Prior Text** column, the word New is entered. When a layer is created, the description is preceded by the word New. To change this, pick the word New and press the [Backspace] key.

Additional descriptive fields can be entered, and a symbol or other text can precede the description. In Figure 6.18, the Building Floor descriptive field does not have a character for the prior text; the Object Type descriptive field has a colon placed in the Prior Text column. These settings create a description that displays the Building Floor description, a colon, then the Object Type description, such as Main Floor:Walls.

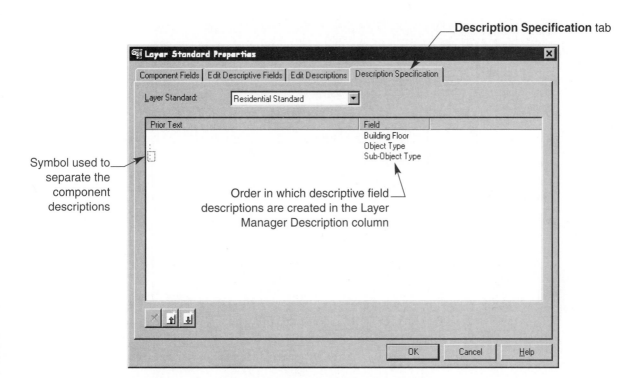

FIGURE 6.18   The Description Specification tab is used to organize how the descriptions appear in the Description tab in the Layer Manager dialog box.

Exercise 6.5

1. Open EX6-4.dwg.
2. Access the Layer Manager dialog box. Select the Layer Standards button and edit the Residential Standard.
3. Select the Description Specification tab and organize the descriptions as follows:

Prior Text	Field
	Building Floor
:	Object Type
:	Sub-Object Type
:	Construction Status

4. Press the OK button when you have finished editing the descriptions. Press the OK button again in the Layer Standards dialog box to return to the Layer Manager dialog box.
5. Exit the Layer Manager dialog box and save the drawing as EX6-5.dwg.

When you are finished creating the layer standard, new layers can be created with the Layer Manager. Layer names can also be assigned to AEC objects through the layer key styles. Layer key styles are discussed later in this chapter. Keep in mind that the layer standard is a standard that is created in the current drawing; It is not a system setting. Layer standards can be imported into other drawings or into a template. Importing and exporting layer standards is also discussed later in this chapter.

Creating New Layers

The preceding discussion walked you through the process of creating a layer standard. Now that a layer standard has been created, new layers can be established through the Layer Manager dialog box by selecting the New Layer button. The **New Layer** dialog box is displayed, as shown in Figure 6.19.

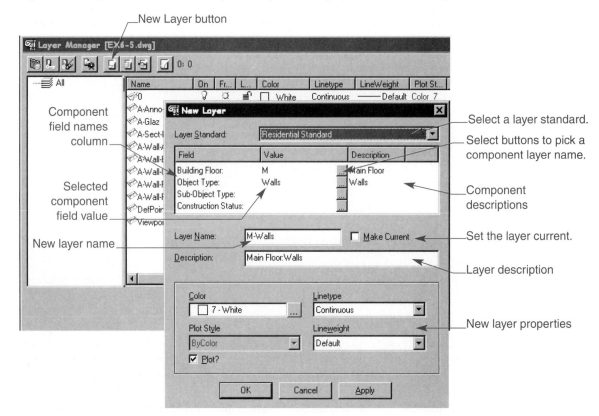

FIGURE 6.19   The New Layer dialog box is used to create a new layer in the drawing.

FIGURE 6.20 The Choose a
Pre-Specified Value dialog box
displays a list of layer field
names and their descriptions for
the ellipsis button selected.

List of component names and
descriptions for the Building
Floor descriptive field

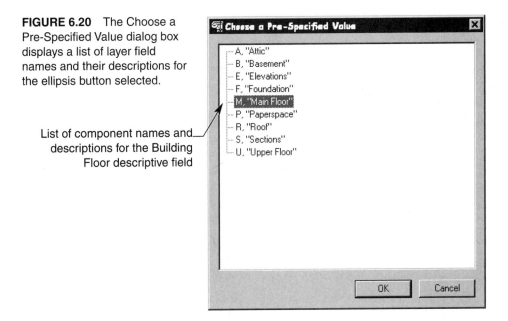

The current layer standard is displayed at the top of the New Layer dialog box. Depending on the layer standard selected, component fields are listed under the **Field** column. The **Value** column displays the default value that was set up in the Layer Standard Properties Default column. Also included in the Value column is a set of ellipsis (…) buttons. There is one ellipsis button for each component field. Selecting a button displays the **Choose a Pre-Specified Value** dialog box (see Figure 6.20).

A list of layer component names and their descriptions from the Edit Descriptions tab in the Layer Standard Properties dialog box is displayed. Select a description from the list to assemble the new layer name, then press the OK button, which returns you to the New Layer dialog box, where the selected value is displayed in the Value column. As values are assembled, the layer name begins to take shape in the **Layer Name:** text box. Below the layer name, the description is also assembled in the **Description:** text box, based on the values selected from the ellipsis buttons. The **Make Current** check box is used to set the newly created layer as the current layer in the drawing. When a layer is set current, new AutoCAD geometry is placed on the layer. This is not true for AEC Objects. AEC Objects are automatically placed on layers when created.

Field Notes

Generally new layers are created for geometry that is not an AEC object. As AEC objects are drawn, a predefined layer name is created, and the object is assigned to the layer through layer keys. Layer keys are discussed later in this chapter. In some cases, a new layer needs to be created for an AEC object that is placed on a different layer than desired.

At the bottom of the New Layer dialog box are drop-down lists to control the properties of the new layer. The new layer can be assigned a color, linetype, lineweight, plot style (if style tables are being used), and can be a plotting or nonplotting layer. A check mark in the **Plot?** check box indicates that items drawn on the layer will plot. Once the layer name and the properties have been specified, press the Apply button to create the layer in the Layer Manager dialog box.

Setting a Layer Current

By default, whenever you start a new drawing, Layer 0 is set as the current layer. Any non-AEC object that is drawn is placed on this layer. All objects that are drawn should be placed on a layer in your drawing, because this aids in controlling the visibility of objects in the drawing. For example, suppose a few layout polylines are to be drawn. A layer named M-Constr could be used for the layout polylines. To do this, first create the layer if it is not available. After the layer has been created, highlight the layer, right-click, and select the Make Current option.

The layer current is set current in Architectural Desktop so that any AutoCAD geometry is now drawn with this layer's properties, such as color and linetype. If an AEC object is drawn, it is automatically drawn and placed on a preset layer regardless of the current layer.

Another way of setting a layer current in the Layer Manager dialog box is to highlight the layer and pick the **Make selected layer current** button. Picking this button places a green check mark on the layer name symbol and sets the layer current.

Exercise 6.6

1. Open EX6-5.dwg.
2. Access the Layer Manager dialog box, and select the New Layer button.
3. Create the following layers:

Layer Name	Color	Linetype	Lineweight	Plot/No-Plot
M-Anno-Note	160	Continuous	Default	Plot
M-Anno-Revs	1 (Red)	Continuous	Default	No-Plot
M-Constr	71	Continuous	Default	No-Plot
U-Anno-Legend	160	Continuous	Default	Plot
U-Fire-Iden	2 (Yellow)	Continuous	Default	Plot
U-Walls-Demo	30	Hidden	Default	Plot

4. Set the M-Constr layer current.
5. Press the Apply button to apply the layers to the drawing, then press the OK button to exit the Layer Manager dialog box.
6. Save the drawing as EX6-6.dwg.

Creating Layer Filter Groups

A *layer filter group* is a routine that filters layers into groups that meet specific criteria. Filter groups can combine layers according to the state of the layer (on/off, frozen/thawed, locked/unlocked), the layer properties, or by layer names. For example, a filter group can be created that includes all layers in the drawing using the yellow color.

Two types of filter groups can be created: dynamic and static. *Dynamic filter groups* automatically update when a change to the properties of a layer is made. As new layers are created they are added to the dynamic filter group based on their properties. A *static filter group* does not automatically update when changes are made to layers. Only the layers that were filtered at the time the static filter group was created are listed in the group. New layers are not added to the group automatically.

To create a filter group, press the **New filter group (off All)** button. The **Layer Filter Properties** dialog box, shown in Figure 6.21, is displayed. At the top of this dialog box are two radio buttons, **Dynamic** and **Static**. Choose the type of filter group you would like to create. After selecting the type of filter group, enter the name for the filter group. Figure 6.21 shows the creation of a dynamic filter group with the name Yellow Layers. The dialog box also includes the following five tabs that allow you to control how layers are filtered into groups:

- ▪ **States**   Unchecking the **Ignore this Filter** check box makes this tab available. Layers can be filtered by their current state. For example, a filter that grouped together all the layers that were turned off would have the **Off** radio button selected, as shown in Figure 6.22.

- ▪ **Colors**   To filter layers by color, remove the check mark in the Ignore this Filter check box. Colors in the current drawing are listed under **Available Color(s):**. Select the color or colors to be filtered for, then pick the Add> button to add the colors to the **Selected Color(s):** list (see Figure 6.23). If a color that is to be fil-

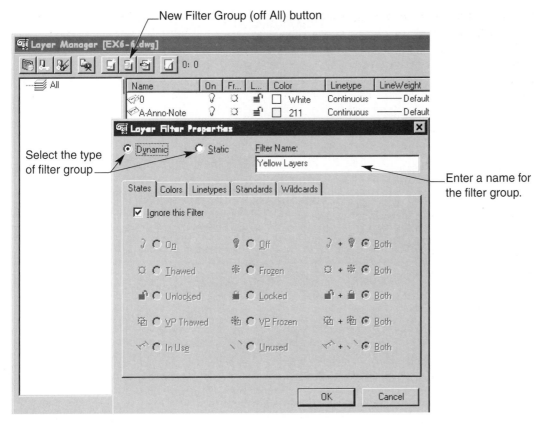

FIGURE 6.21 The Layer Filter Properties dialog box is used to set up a layering filter.

FIGURE 6.22 The States tab filters layers by their current state.

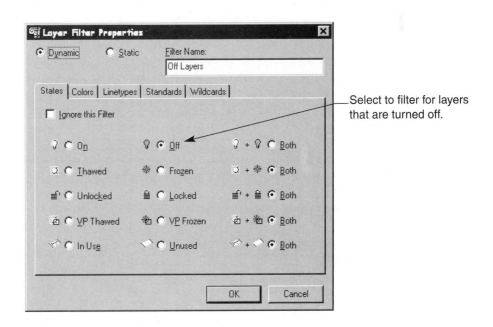

FIGURE 6.23   The Colors tab filters layers by color.

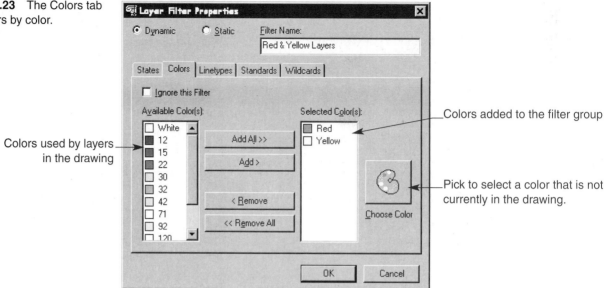

Colors added to the filter group

Colors used by layers in the drawing

Pick to select a color that is not currently in the drawing.

tered is not in the A̲vailable C̲olor(s): list, pick the C̲hoose Color button to select a color for filtering. Colors can be added or removed from the Selected Color(s): list.

- **Linetypes**   To filter layers by specific linetypes remove the check mark in the I̲gnore this Filter check box. Linetypes currently being used by layers are listed in the **A̲vailable for Filtering** list box. Select the linetypes to be filtered, and press the A̲dd button. The filtered linetypes are listed in the **Already i̲n Filter** list box, as shown in Figure 6.24.

- **Standards**   This tab filters for layers of a certain layer standard and using specific fields. Figure 6.25 shows filtering for layers using the Residential Standard and the Building Floor and Object Type descriptive fields.

- **Wildcards**   This tab is used to filter for layers with a particular name or part of a name. To filter for parts of a layer name, enter the part of the layer name being filtered for in the drop-down list on the right, followed by an asterisk (*). An asterisk is

FIGURE 6.24   The Linetypes tab is used to filter for layers using specific linetypes.

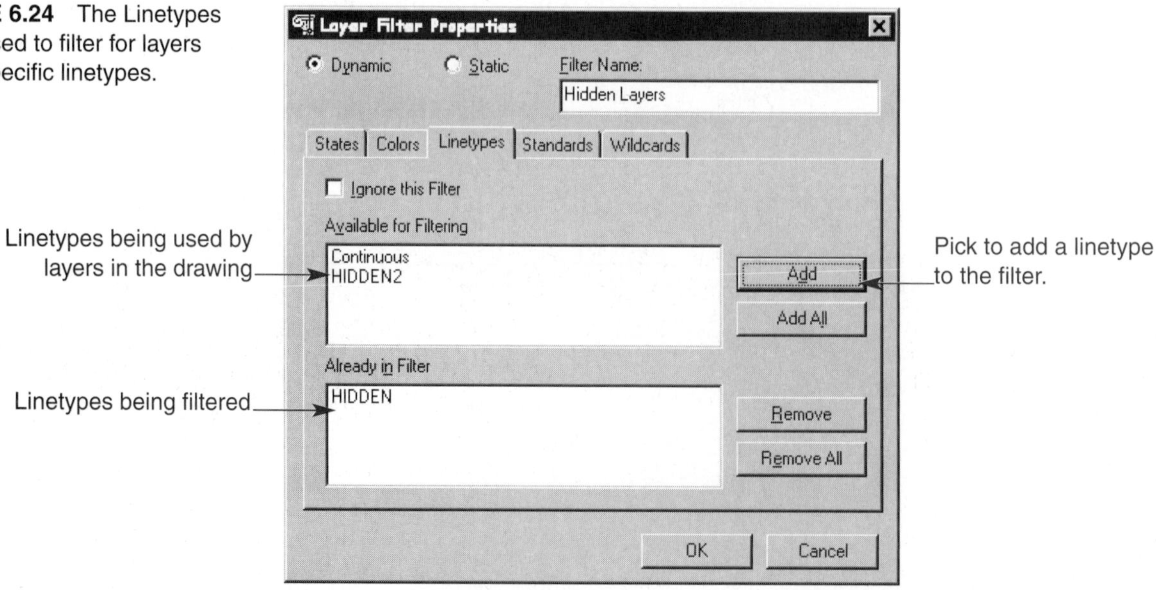

Linetypes being used by layers in the drawing

Pick to add a linetype to the filter.

Linetypes being filtered

FIGURE 6.25 The Standards tab is used to filter for layers using specific descriptive fields from a particular layer standard.

Select layer standard to filter through.

Descriptive fields in the drawing

Descriptive fields being filtered for

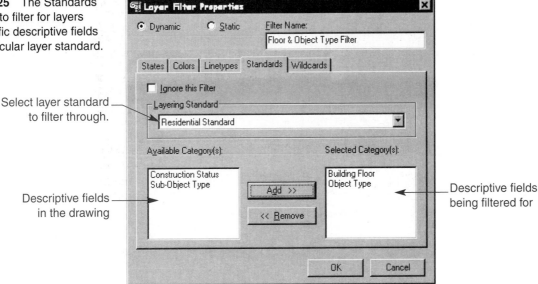

FIGURE 6.26 The Wildcards tab is used to filter for layer names using all or part of the layer name.

Layer names must match the filter in the right drop-down list.

M-* filters for all layers beginning with the letter M followed by a hyphen.

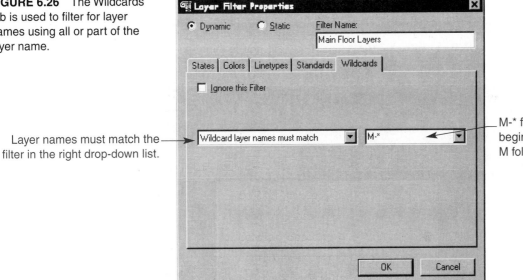

also known as a ***wildcard character***. The wildcard indicates any additional layer name characters. For example, Figure 6.26 shows filtering for Main Floor layers for all layer names that begin with M-. The wildcard after the hyphen indicates any additional characters after the M-. The wildcard can be used at the beginning, middle, or end of the filtered name.

After establishing a filter, press the OK button to create the filter group. The filter group name and a symbol are created in the left pane of the Layer Manager dialog box (see Figure 6.27A). If the All group is selected, the list of filters also appears in the right pane. If a filter is selected from the list in the left pane, the filtered layers are displayed in the right pane (see Figure 6.27B).

When the filter groups are displayed in the right pane, the entire group can be turned on/off, frozen/thawed, locked/unlocked by picking the appropriate symbol in the right pane. Right-clicking a filter group in the left pane also displays a cursor menu for the filter group. Selecting an option from the menu also allows you to turn on/off, freeze/thaw, or lock/unlock the layers in the filter group.

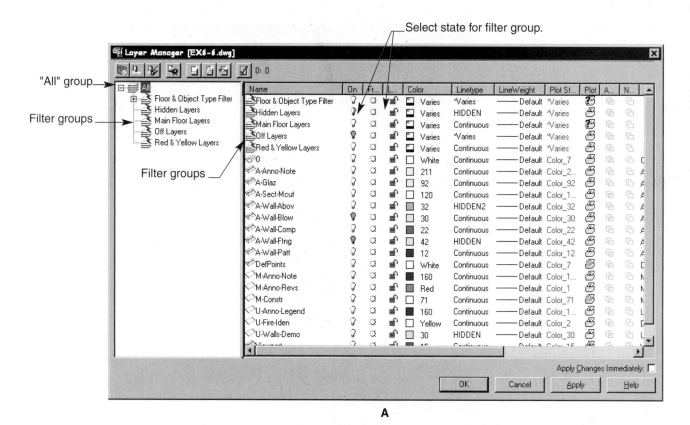

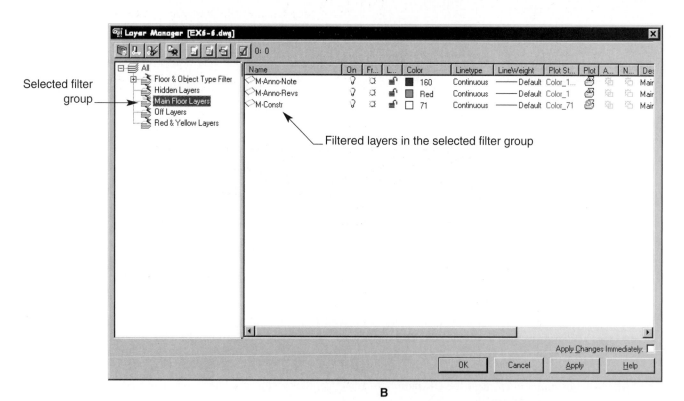

FIGURE 6.27   (A) After the filters are created, the filter group name is displayed in the left pane.
(B) Selecting a filter from the left pane displays the list of layers that meet the filter criteria.

Field Notes

A filter group does not have to include all filter information in each of the tabs. Complex filters can be created by selecting multiple filters from each tab. For example, a filter can be created that filters for all Upper floor layers that are red, by using the Wildcards and Colors tabs.

Exercise 6.7

1. Open EX6-6.dwg.
2. Access the Layer Manager dialog box.
3. Create a filter group named Main Floor that filters for layers that begin with the letter M.
4. Create a filter group named Upper Floor that filters for layers that begin with the letter U.
5. Select the Upper Floor filter group in the left pane. Which layers are displayed in the right pane? Select the Main Floor filter group in the left pane. Which layers are displayed in the right pane?
6. Select the filter named All. In the right pane select the thaw icon for the Upper Floor filter.
7. Save the drawing as EX6-7.dwg.

Adding User Groups

In addition to filter groups, which automatically filter for specific layer information, ***user groups*** can be created, which group together layers of your choice. To create a user group, select the **New user group (off All)** button. A user group is added in the left pane, where you enter a name for the group.

Initially, the new user group does not contain any layers. Select the All group to see a list of all the layers in the drawing. Highlight the layers that will be added to the new user group. With the layers highlighted, drag and drop them from the right pane onto the user group. The layer is added to the user group (see Figure 6.28).

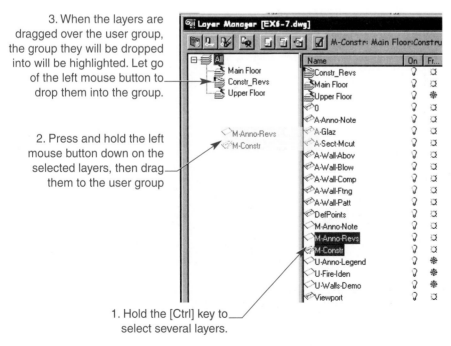

FIGURE 6.28 After creating the new User Group, highlight the layers to be grouped together, then drag and drop them onto the user group in the left pane.

Using Snapshots

Occasionally when creating a drawing, you may need to reference other objects on different layers in relation to the items you are currently drawing. For example, suppose you are drawing an upper floor plan of a building to fit over the top of the first floor. In this situation you may need to reference the first-floor wall layers in order to know where to draw the upper-floor walls. Typically, you will freeze all the first-floor layers except the first-floor wall layer, change the color so you can focus on the upper floor wall layer, and lock the layer so any editing of the upper-floor walls will not affect the first-floor walls. When you have completed the upper floor, you may need to thaw the layers, change the first-floor wall color back to its original color, and unlock it.

If the design changes, you may need to freeze the layers again, change the colors, and lock the first-floor wall layer so that you can make changes to the upper-floor walls. By using *snapshots* the process of freezing/thawing, and changing the colors becomes very simple and saves a lot of time.

A *snapshot* is similar to a photograph of a scene in time. The snapshot remembers the state of your layers at the time the snapshot was created. Using the previous example, you could create a layer state in your drawing with all the layers thawed, unlocked, and on and the colors, linetypes, and lineweights set. After creating the snapshot, you could change the layers in the drawing, changing all the layers to a single color, and freezing and thawing other layers. Then, you could create another snapshot of the current state of the layers. At any time you could restore the original state of the layers through the **Snapshots** dialog box.

To create a snapshot, select the Snapshots button in the Layer Manager dialog box. The Snapshots dialog box, shown in Figure 6.29, is displayed. The Snapshot dialog box contains the following options:

- **New...**   This button displays the Snapshot dialog box. Enter a name for the new snapshot, then press the OK button. This creates a new snapshot based on the current state of your layers.

- **Restore**   This button restores the state of the layers in a selected snapshot. First, select the snapshot from the list, then pick the Restore button to restore the state.

- **Edit...**   Snapshots can be edited after the snapshot has been taken. To do this, first select a snapshot you wish to edit, then pick the Edit... button. The **Snapshot Edit:** dialog box, shown in Figure 6.30, is displayed. Any property for a layer in this dialog box can be selected to change its state. Modifying any of the properties updates the snapshot. Layers can be added to or deleted from the snapshot by

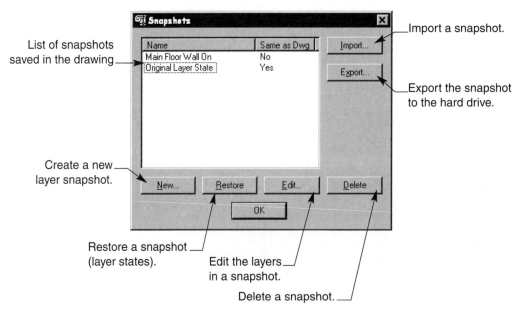

FIGURE 6.29   The Snapshots dialog box is used to create snapshots of the states of the layers at the time the snapshot is created.

FIGURE 6.30 The Snapshot Edit: dialog box is used to modify the properties of layers within a snapshot.

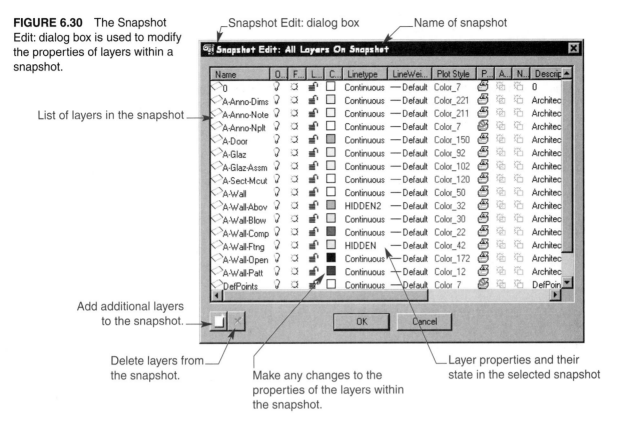

Snapshot Edit: dialog box

Name of snapshot

List of layers in the snapshot

Add additional layers to the snapshot.

Delete layers from the snapshot.

Make any changes to the properties of the layers within the snapshot.

Layer properties and their state in the selected snapshot

using the Add or Delete buttons in the lower left of the dialog box. This does not add new layers to or delete layers from the drawing; it affects only the snapshot.

- **Delete**   This button deletes the highlighted snapshot from the drawing.

- **Export...**   This button opens the Save As dialog box for snapshots. A snapshot and its layer property settings can be saved as an .ssl file for use in another drawing file.

- **Import...**   Selecting this button allows you to import a saved .ssl file created with the Export... option. When a snapshot .ssl file is imported, the layers and their properties are imported into the drawing. If the current drawing contains the same layers with different properties, then the imported snapshot becomes the current layering state.

Exercise 6.8

1. Open EX6-1.dwg.
2. Access the Layer Manager dialog box.
3. Select the Snapshots button, and create a new snapshot named Original Colors.
4. Press OK to return to the Layer Manager.
5. Change the colors of all the layers to 7 (White).
6. Select the Snapshots button, and create a new snapshot named White Colors.
7. Press OK to return to the Layer Manager. Press OK to exit the Layer Manager.
8. Use the REGENALL to regenerate all the viewports. Are all the colors in the top view white? The colors in the elevation and isometric views will probably still be colored according to settings in the Entity Display for model representation objects.
9. Access the Layer Manager, and select the Snapshots button again. Restore the Original Colors snapshot.
10. Exit the Layer Manager and save the drawing as EX6-8.dwg.

Creating and Using Layer Keys

As AEC objects are added to a drawing a predefined layer is created, and the objects are placed on this layer. This saves you the time of setting different layers current before you draw objects. Every AEC object and symbol that is included with Architectural Desktop has this automatic layering capability. In order for this process to work, a feature called *layer keys* is used.

A *layer key* is the map between the layer standard and an AEC object. The layer key interprets the layer standard and assigns a layer to the object. For example, by default the AIA layer standard and layer key style are used in new drawings from the AEC Arch (Imperial) or AEC Arch (Imperial-Int) templates. A layer key named Wall is assigned to the wall object. Whenever a wall object is drawn, the layer key looks to the layer standard, creates a layer named A-Wall, and assigns the wall object to this layer.

Layer key styles can be created from scratch or by copying an existing layer key style and modifying it to reflect your needs. Use one of the following options to start creating a layer key style:

✓ Select the **Desktop** pull-down menu, move to the **Layer Management** cascade menu, and select **Layer Key Styles…**.

✓ Pick the **Layer Key Styles** button on the **Layer Management** toolbar.

✓ Right-click and select **Desktop, Layer Management, Layer Key Styles…**.

✓ Pick the **Layer Key Styles** button in the **Layer Manager**.

✓ Type **AECLAYERKEYSTYLE** at the Command: prompt.

The **Style Manager** dialog box with the layer key styles displayed appears as shown in Figure 6.31.

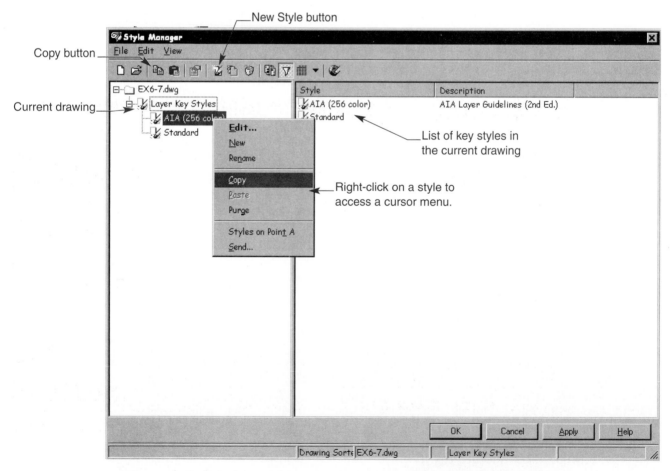

FIGURE 6.31   The Style Manager dialog box is used to organize all AEC styles in the drawing. When the layer key styles are accessed, the Style Manager displays the ones in the current drawing.

New layer key styles can be created by selecting the **New Style** button in the Style Manager dialog box. A new layer key style that can be edited is created in the Layer key styles list. Another way of creating a new layer key style is to copy an existing style and modify it. This can save a lot of time, because the original layer key style has already been configured, and the copy can be modified very quickly.

To create a new layer key style based on an existing style, highlight the key style in the left or right pane, and select the **Copy** button in the Style Manager, or right-click and select **Copy**. After you have copied the existing style, right-click and select **Paste** or pick the **Paste** button in the Style Manager. A new layer key style with the same name and the number 2 in parenthesis is created. Right-clicking on the new style and selecting **Rename** from the cursor menu allows you to rename the copied key style.

Double-click the new style in the right pane to edit the style or select the style, right-click, and select **Edit...** from the cursor menu. The **Layer Key Style Properties** dialog box, shown in Figure 6.32, is displayed. The dialog box includes two tabs, a **General** tab and the **Keys** tab. The General tab can be used to rename the key style and provide a description, such as which layer standard is being referenced. The Keys tab includes a list of layer keys that Architectural Desktop uses for all the AEC objects and symbols, as well as the properties for the automatic layers.

The Layer Key Style Properties dialog box includes a list of all the layer keys Architectural Desktop has assigned to all the AEC objects and symbols. The **Description** column indicates the types of objects to which the layer keys are assigned. The **Layer** column includes an ellipsis (...) button that can be selected to assign a layer name to a layer key in much the same way that a new layer is created in the Layer Manager. The properties for each automatic layer can be configure next to the appropriate layer name.

At the bottom of the Layer Key Style Properties dialog box is a drop-down list with the layer standards in the current drawing. The layer standard that is being used by the layer key style is the default standard in the list. Selecting a different layer standard to use in the key style generates a warning dialog box (see Figure 6.33). This dialog box indicates that the layer descriptions currently in the layer key style are

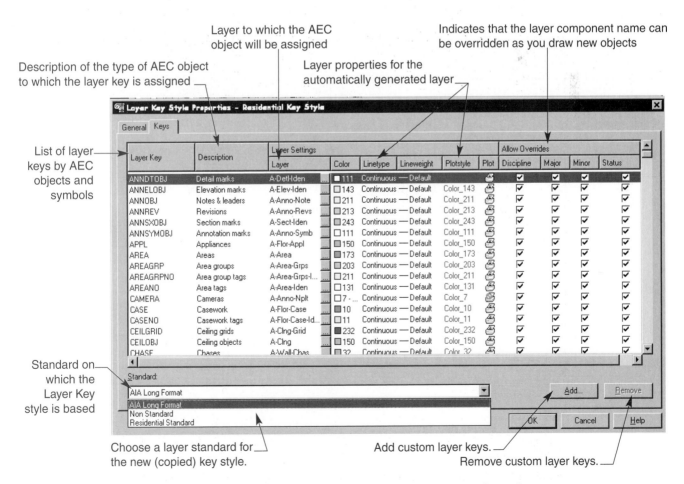

FIGURE 6.32 The Layer Key Style Properties dialog box is used to configure the layer, and its properties, that is created automatically for AEC objects.

FIGURE 6.33 Changing the layer standard that a layer key style is referencing generates a warning box that indicates that the exiting layer descriptions may not match the new layer standard you have selected.

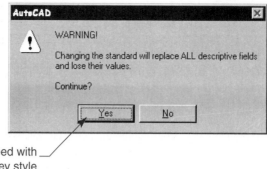

Press Yes to proceed with modifying the layer key style

based on an existing layer standard and that by changing the standard the layer descriptions may not match the selected layer standard. Press the <u>Y</u>es button to proceed. This resets the layer standard that the layer key style is referencing for the layer names it assigns to AEC objects.

The layer key style now references a different layer standard for the creation of the layer names. Notice that the layer names have not changed; however, notice that in Figure 6.34 the Residential layer standard is now referenced, and the A-Wall layer no longer stands for Architectural Walls but for Attic Walls. To change the layer names for all the AEC objects, select the ellipsis (…) button beside the layer name to access the Layer Name dialog box. In the Layer Name dialog box, select the appropriate ellipsis buttons beside the layer component name to assign a new value.

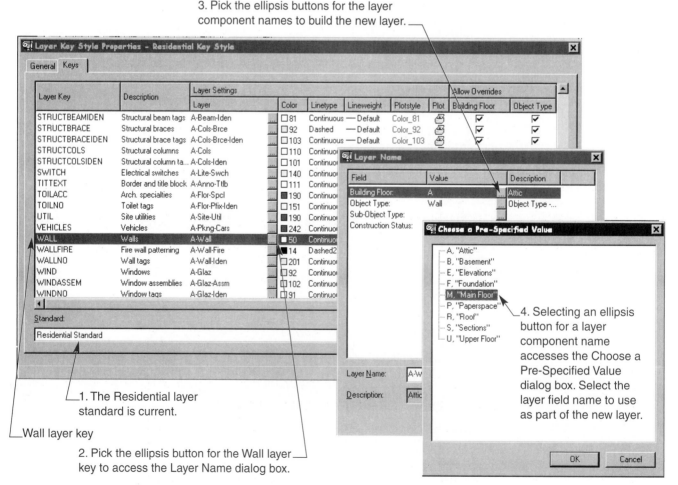

3. Pick the ellipsis buttons for the layer component names to build the new layer.

4. Selecting an ellipsis button for a layer component name accesses the Choose a Pre-Specified Value dialog box. Select the layer field name to use as part of the new layer.

1. The Residential layer standard is current.

Wall layer key

2. Pick the ellipsis button for the Wall layer key to access the Layer Name dialog box.

FIGURE 6.34 With the Residential layer standard current, begin reassigning the layer names to the layer keys. Use the ellipsis button for each layer to build the layer names.

When you have finished assigning the new layer name to a layer key press the OK button to return to the Layer Key Style Properties dialog box to assign the next layer to the next layer key. After remapping the new layer names to the layer keys, press the OK button in the Layer Key Style Properties dialog box to return to the Styles Manager. Press the OK button to exit the dialog box.

 Field Notes

> The <u>A</u>dd... button allows you to create additional layer keys in the key style. You can assign new layer keys to custom symbols you create. When you use a custom symbol such as a lavatory symbol the symbol can be inserted automatically on a layer using the new layer key. Creating symbols is covered in the next chapter. Use the <u>R</u>emove button to remove custom layer keys you may have created. This button does not remove Architectural Desktop layer keys.

Before you can draw new AEC objects, you need to set the layer key style current in the drawing. You can do this in the **Drawing Setup** dialog box under the **Layering** tab (see Figure 6.35). Use one of the following options to access the Drawing Setup dialog box Layering tab:

✓ Pick the **Des<u>k</u>top** pull-down menu, move to the **<u>L</u>ayer Management** cascade menu, and pick **<u>S</u>elect Layer Standard....**

✓ Pick the **Select Layer Standard** button on the **Layer Management** toolbar

✓ Right-click, and select **Des<u>k</u>top, <u>L</u>ayer Management, <u>S</u>elect Layer Standard....**

✓ Type **AECDWGLAYERSETUP** at the Command: prompt.

Now, select the appropriate layer key style from the Layer Key Style: drop-down list.

Any new AEC object that is created should now be placed on the remapped layer name. If you are drawing items such as polylines, create a new layer for the polyline in the Layer Manager dialog box, set it current, and draw the polyline. If a layer is set current when you draw AEC objects, the layer is ignored, and the AEC object is placed on the layer designated by the layer key style.

FIGURE 6.35   Select the layer key style to be used for new AEC objects.

Select a layer key style to be used for new AEC objects.

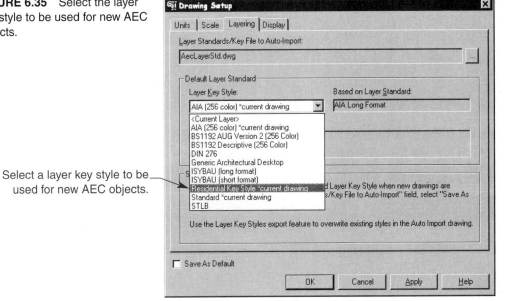

Field Notes

When creating a new layer key style using your layer standard, it is recommended that you copy an existing layer key style and modify the copy. This saves time, because colors and linetypes have already been assigned, and all that you need to do is update the layers assigned to the layer keys to match your layer standards.

Exercise 6.9

1. Open EX6-7.dwg.
2. Access the Layer Key Styles section in the Style Manager.
3. Right-click over the AIA (256) color key style. Select the Copy option from the cursor menu. Right-click in the right pane to paste the copy into the drawing.
4. Double-click the copied key style to access the Layer Key Style Properties dialog box.
5. In the General tab, rename the key style Residential Key Style. Enter the following description: "Base on the Residential Layer Standard."
6. In the Keys tab, select the Residential Standard from the Standard: drop-down list. Pick the Yes button in the warning dialog box.
7. Rename each layer for all the keys using the ellipsis buttons. For all the layer names use the M- floor designation, that is, M-Walls, M-Doors, M-Windows, and so on.
8. When you have finished renaming all the layers, exit the dialog boxes.
9. Set the Residential key style current in the Drawing Setup dialog box.
10. Draw a wall, and insert a door and a window into the wall. Did the wall get inserted on the M-Walls layer? The door on the M-Doors layer? The Window on the M-Window layer?
11. Save the drawing as EX6-9.dwg.

Using Layer Key Overrides

Once a layer standard and layer key style have been created, any new AEC objects that are drawn are placed on the layers designated by the layer key style. Through the use of *layer key overrides*, the layer where an AEC object is drawn can be overridden.

The preceding section created a layer key style for the Residential layer standard. With this layer key style current, new AEC objects are placed on layers that begin with the letter M, which indicates the Main Floor. If you want to create objects on the Upper Floor layers, you will need to override the current layer keys, so that the M- layer component is replaced with the U- layer component.

Use one of the following options to set the layer overrides:

✓ Select the **Desktop** pull-down menu, **Layer Management,** then **Layer Key Overrides….**

✓ Pick the **Layer Key Overrides** button on the **Layer Management** toolbar.

✓ Right-click, and select **Desktop, Layer Management, Layer Key Overrides….**

✓ Pick the **Layer Key Overrides** button in the **Layer Manager**.

✓ Type **AECLAYERKEYOVERRIDE** at the Command: prompt.

The Layer Key Overrides dialog box, shown in Figure 6.36, is displayed.

The top of the Layer Key Overrides dialog box includes a drop-down list of the layer key styles in the drawing. The current layer key style is the default in this list. The list box in the center of the dialog box includes a list of the component/descriptive field names from the layer standard. The **Override** column includes an ellipsis (…) button for each field. Selecting an ellipsis button for a component field accesses the Choose a Pre-Specified Value dialog box containing the descriptions for the selected component (see Figure

FIGURE 6.36   The Layer Key
Overrides dialog box allows you
to set overrides to layer name
components.

Current layer key style

List of component /
descriptive fields

Enable the use of overrides

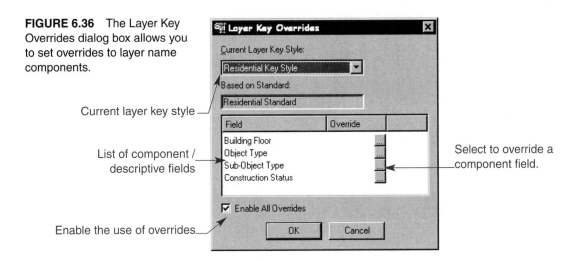

Select to override a
component field.

FIGURE 6.37   Select a value to
use as an override to the layer
key style.

List of values for the
Building Floor field

Select an override.

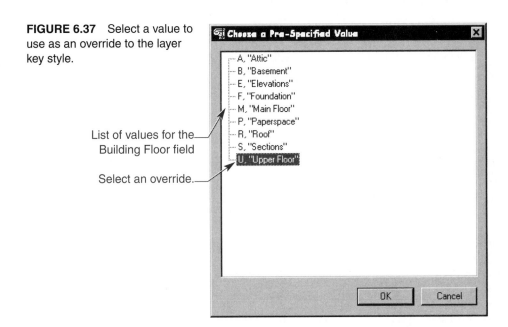

6.37). Selecting a value from the list and picking the OK button sets the selected value as the override to the layer key style layer names.

Once an override has been selected, any new objects that are drawn are created with the layer names from the layer key style, with the override set substituted for any layer field names. For example, in Figure 6.37, the Upper Floor value is selected. Any new objects created now substitute U- for the M- in the layer name. If a layer override has been set, it remains as the overriding value until the value is changed or the layer key overrides are turned off.

✎Field Notes

If there is a component field in the layer name that should never be overridden, such as the Object Type field in the Residential standard, the layer key style can be adjusted to never use an override for a component. The **Allow Overrides** column in the Layer Key Style Properties dialog box includes a check mark next to each field in a layer name (see Figure 6.32). Unchecking a component field does not allow overrides to be set for the component.

After a layer key override has been set, the override can be turned off so that AEC objects generate layers based on the layer key style again. Use one of the following options to turn the overrides on or off:

✓ Select the **De<u>s</u>ktop** pull-down menu, **<u>L</u>ayer Management,** then **O<u>v</u>errides On/Off.**

✓ Pick the **Overrides On/Off** button on the **Layer Management** toolbar.

✓ Right-click, and select **De<u>s</u>ktop, <u>L</u>ayer Management, O<u>v</u>errides On/Off.**

The override can be turned back on again by selecting the button or pull-down menu again.

If objects have accidentally been drawn with an override, or if you would like to change the layer key used by an object, the object can be remapped to a different layer key. When an object is remapped to a different layer key, the object takes on the properties such as color and linetype that the layer key is assigned in the layer key style. Use one of the following options to remap layers to a different layer key:

✓ Select the **De<u>s</u>ktop** pull-down menu, **<u>L</u>ayer Management,** then **<u>R</u>emap Object Layers**.

✓ Pick the **Remap Object Layers** button on the **Layer Management** toolbar.

✓ Right-click, and select **De<u>s</u>ktop, <u>L</u>ayer Management, <u>R</u>emap Object Layers.**

✓ Type **AECREMAPLAYERS** at the Command: prompt.

Once you have elected to remap an object's layer key you are prompted to enter a layer key name or to use the default layer key assigned to the object. The following prompt is issued after the command is entered:

Command: **AECREMAPLAYERS** ↵
Select objects to re-layer: *(select objects to be remapped)*
Select objects to re-layer: ↵
Enter a Layer Key or [?/byObject]: *(enter a layer key name, list the layer keys in the drawing, or use the default layer key)*

The ? option lists the layer keys available in the drawing. Press the [F2] key to review the list of layer keys in the AutoCAD Text window. Press [F2] again to return to the drawing screen. The **byObject** option can be used to remap an object that was drawn with an override and needs to be placed on the default layer key layer. This command can also be used on standard AutoCAD geometry to assign the geometry to an Architectural Desktop layer key.

Exercise 6.10

1. Open EX6-9.dwg.
2. Set an override to the Building Floor component field.
3. Select the U, "Upper Floor" value. Press the OK button. This will set an override to the first field name in the layer.
4. Add a wall, door, and window. Are the objects now on the U-Walls, U-Door, U-Window layers?
5. Turn the override off. Add another wall. Which layer is the wall on?
6. Use the Remap Object Layers command to remap the upper-floor objects to reflect that they are on the main floor.
7. Save the drawing as EX6-10.dwg.

Using Layer Standards in Other Drawings

The preceding discussions have described how to create a layer standard, layer key styles, and new layers. After a standard and a key style have been created you may want to use them in other drawings or in future drawings.

225

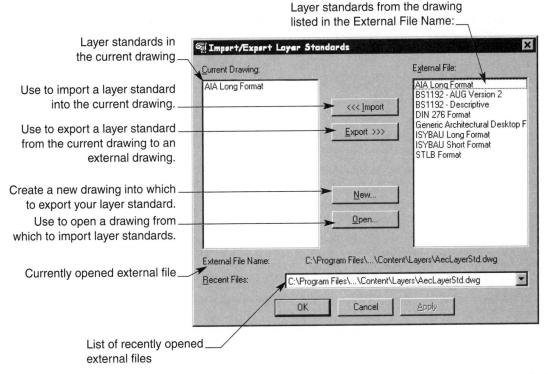

Layer standards from the drawing
listed in the External File Name:

Layer standards in
the current drawing

Use to import a layer standard
into the current drawing.

Use to export a layer standard
from the current drawing to an
external drawing.

Create a new drawing into which
to export your layer standard.
Use to open a drawing from
which to import layer standards.

Currently opened external file

List of recently opened
external files

FIGURE 6.38 The Import/Export Layer Standards dialog box is used to import or to export layer standards.

Importing and Exporting Layer Standards

A layer standard can be imported into the current drawing by accessing the Layer Standards dialog box from the Layer Manager (see Figure 6.9). In the **Layer Standards** dialog box, select the **Import/Export...** button to display the **Import/Export Layer Standards** dialog box, shown in Figure 6.38.

The left side of the Import/Export Layer Standards dialog box includes a list of layer standards in the current drawing. Use the Open... button to browse for a drawing that includes the layering standard you would like to import. Once an external drawing has been opened, a list of layer standards included in the external file is listed on the right. Select the layer standard or standards that you would like to import into the current drawing. Pick the **<<<Import** button to import the standards into the current drawing.

A layer standard can also be exported to another drawing file. You can create a new drawing that includes the exported layer standard or open an existing drawing to which the standard can be exported. Select the layer standard or standards to export from the current drawing and pick the **Export>>>** button to export the layer standard.

When you have finished importing or exporting layer standards press the OK button to return to the Layer Standards dialog box. Layer standards that are not in use can be purged from the drawing by selecting the layer standard from the list and selecting the **Purge** button.

Importing and Exporting Layer Key Styles

Importing a layer standard is important for use in your drawing, but it is not very useful without the layer key style. Layer key styles can also be imported into the current drawing or exported to another drawing. In order to import or export layer key styles, access the Style Manager for layer keys (see Figure 6.39). Use the Open button to browse for the drawing that contains the desired layer key style.

The newly opened drawing and the current drawing are listed in the left pane of the Style Manager. Expand the layer key styles in both drawings by picking the + symbols. To import or export a layer key style from one drawing to another, select the desired layer key style drag and drop it to the desired file. In Figure 6.39, the Residential key style is selected from EX6-10.dwg, then dragged over the EX6-8.dwg with the left mouse button and dropped. The Residential key style is then added to the current drawing. When

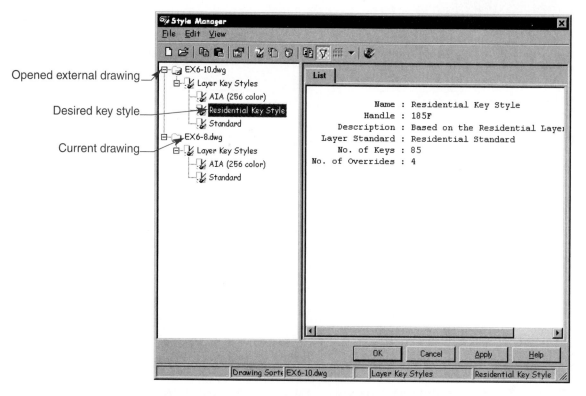

FIGURE 6.39   Open a drawing that contains the desired layer key style, then drag and drop it onto the current drawing.

you have finished importing a layer key style, close the drawing from which you imported the layer key style by right-clicking over the drawing name and selecting the Close option.

The only thing left to do is set the layer key style current before adding any new AEC objects. Access the Drawing Setup dialog box to do this, to ensure that the new objects are properly placed on the correct layers.

Exercise 6.11

1. Open EX6-8.dwg.
2. Import the Residential Standard layer standard from EX6-10.dwg.
3. Drag and drop the Residential key style from EX6-10 into the drawing.
4. Set the Residential key style as the current key style.
5. Thaw the A-Anno-Nplt layer.
6. Use the Remap Object Layers command to map the cameras, camera path and target path, and the layout lines to the CAMERA layer key.
7. Remap all the walls to the WALL layer key, all the doors to the DOOR layer key, all the openings to the OPENING layer key, and all the windows to the WIND layer key.
8. Create a filter group for the M-* layers. Freeze the M-Anno_Nplt layer.
9. Delete any unused layers in the Layer Manager dialog box.
10. Save the drawing as EX6-11.dwg.

CHAPTER REVIEW

Use the CD-ROM to test your knowledge and skills.

Chapter Test

To check your understanding of the content provided in this chapter, access the Test file in the CH06 folder of the CD-ROM that accompanies this text.

Chapter Project

To practice the Architectural Desktop skills presented in this chapter, access the Project files in the CH06 folder of the CD-ROM that accompanies this text. The project files are in pdf format and include sample drawings and instructions for completing each project.

CHAPTER 7

Architectural Symbols and Structural Elements

LEARNING GOALS

After completing this chapter, you will be able to:

◎ Use the AutoCAD DesignCenter.

◎ Organize design content.

◎ Insert documentation blocks.

◎ Create standard AutoCAD blocks and Wblocks.

◎ Use the INSERT command.

◎ Create, use, and modify multi-view blocks.

◎ Produce mask blocks.

◎ Add custom symbols to the DesignCenter.

◎ Create AEC profiles.

◎ Generate and use structural members.

◎ Insert structural member styles.

As construction documents are created, many types of architectural symbols are used to convey the drafter's ideas and to enhance the CAD drawing. An ***architectural symbol*** is a group of individual lines, arcs, and circles that are constructed together to form a single object. In standard AutoCAD these symbols are called **blocks**. Blocks can represent any type of item or feature—for example, a symbol or block that looks like a sink or shower that is used repeatedly in the drawing.

Architectural Desktop builds on the standard AutoCAD block principle and uses blocks to create what is known as a **multi-view block** (see Figure 7.1). A ***multi-view block*** is similar to other AEC objects in that it is displayed as a 2D block in a top view, a 2D elevation block in an elevation view, and a 3D block in a model or isometric view. Architectural Desktop includes over 750 imperial-scaled multi-view blocks, 500 metrically scaled multi-view blocks, and if the international extensions have been installed, an additional 650 metric blocks. Many of these blocks can be used in either imperial or metric drawings. The AutoCAD DesignCenter is used to access the list of blocks.

The AutoCAD DesignCenter

The **AutoCAD DesignCenter**, also known as **DesignCenter**, is a powerful drawing information manager that provides a simple tool for accessing Architectural Desktop multi-view blocks, standard AutoCAD blocks, and other content from any drawing (see Figure 7.2). Content from the DesignCenter can be added to the drawing by dragging and dropping.

Organizing Design Content

The multi-view block content found in the DesignCenter can be organized in one of two ways: Autodesk Architectural Desktop or CSI MasterFormat (see Figure 7.3). When Architectural Desktop is first installed one of the setup pages includes these options. You have the choice of organizing the multi-view blocks

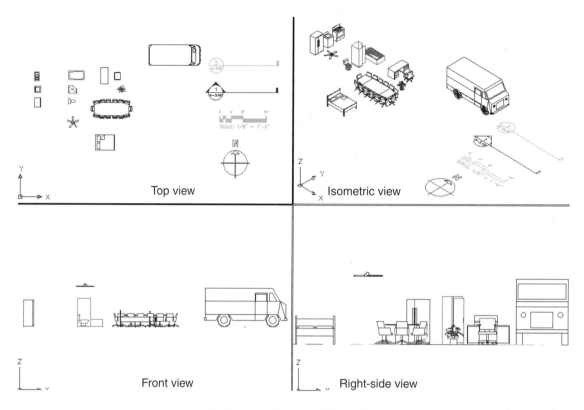

Architectural Desktop Multi-View Blocks

FIGURE 7.1   Multi-View Blocks have different representations depending on the viewing direction.

under standard folders such as Appliances, Furniture, and Plumbing by selecting the Autodesk Architectural Desktop option or organizing them under specific division folders such as Division 11 - Equipment, Division 12 - Furnishings, and Division 15 - Mechanical by choosing the CSI MasterFormat option. See Appendix D for additional information regarding the CSI MasterFormat.

After you choose an option at installation, the multi-view blocks are organized appropriately. The DesignCenter can be accessed by picking the Design pull-down menu, selecting the Design Content cascade menu, and selecting a section (see Figure 7.4). The pull-down menu differs for the two installation organizations. The CSI MasterFormat organzation is used in this text.

Using the DesignCenter

There are two types of multi-view blocks: *design blocks* and *documentation blocks*. **Design blocks** are symbols of real-world architectural objects, such as plumbing fixtures, furniture, and casework. Design blocks are originally drawn at full scale, so a bathtub symbol inserted into the drawing is displayed at the actual size of the bathtub. **Documentation blocks** are symbols used to document the drawings, such as detail bubbles, revision clouds, and drawing tags. Documentation blocks are originally drawn full scale to a plotted sheet of paper. When documentation blocks are inserted into model space, they are scaled by the scale of the drawing. If the documentation blocks are inserted into layout space, they are inserted full scale.

Design blocks can be accessed by selecting the **Design** pull-down menu and the **Design Content** cascade menu, then picking one of the sections. The DesignCenter is opened to the appropriate design content folder. Documentation blocks are accessed by selecting the **Documentation** pull-down menu and **Documentation Content**, then picking one of the sections. The DesignCenter is opened to the appropriate documentation folder. In addition, you can use one of the following options to open the AutoCAD DesignCenter:

✓ Select the **Tools** pull-down menu and select the **AutoCAD DesignCenter Ctrl+2** command.

✓ Select the **AutoCAD DesignCenter (Ctrl+2)** button on the **Standard Toolbar.**

✓ Type **ADCENTER, ADC** at the Command: prompt.

✓ Press **[Ctrl]+[2]** at the Command: prompt.

The **Custom** button lists the folders containing the multi-view blocks.

Folders containing multi-view blocks

Opened folder

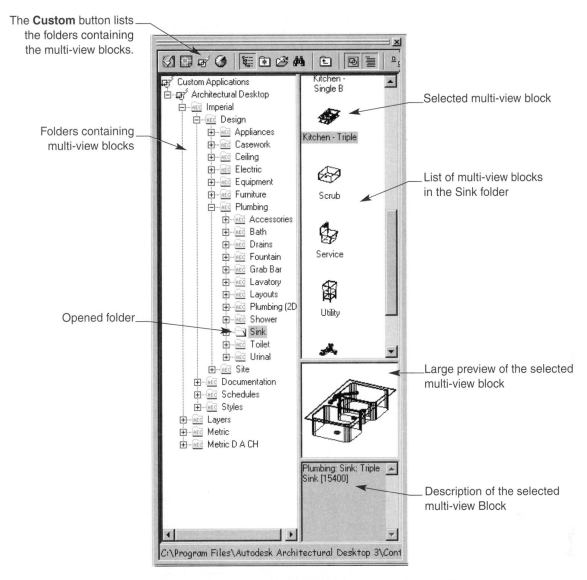

Selected multi-view block

List of multi-view blocks in the Sink folder

Large preview of the selected multi-view block

Description of the selected multi-view Block

FIGURE 7.2 The AutoCAD DesignCenter is used to browse for content such as multi-view blocks and to add items to the drawing by dragging and dropping.

FIGURE 7.3 One of the setup pages displayed during the installation of Architectural Desktop gives you the option of organizing the multi-view blocks under standard Architectural Desktop folders or under the CSI MasterFormat division folders.

Specify the type of formatting for the blocks.

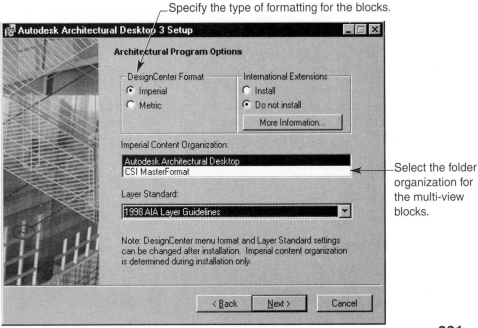

Select the folder organization for the multi-view blocks.

FIGURE 7.4 (A) Multi-view blocks organized under the Autodesk Architectural Desktop organization. (B) Multi-view blocks organized under the CSI MasterFormat organization.

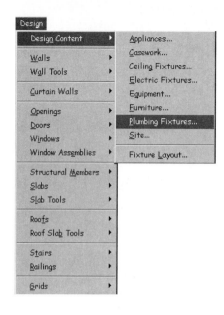

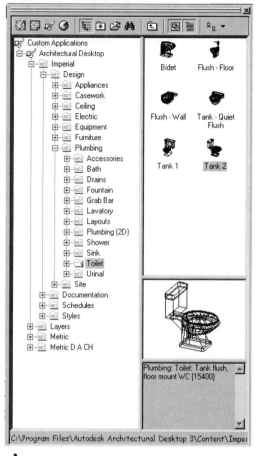

A

Autodesk Architectural Desktop Organization

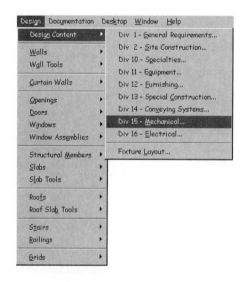

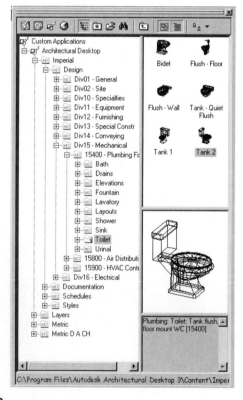

B

CSI MasterFormat Organization

If the DesignCenter is not accessed from the Design or Documentation pull-down menus, the DesignCenter opens looking at the last accessed folder.

Inserting Design Blocks

In addition to accessing DesignCenter from the Design pull-down menu, you can access it by selecting a button from the **Design Content-Imperial** or **Design Content-Metric** toolbars, or by right-clicking and selecting Design, then Design Content, and picking a section. Figure 7.5 displays the different folder sections in the two toolbars.

Once a block section has been selected, the DesignCenter window is displayed on the drawing screen, looking into the selected folder. Initially this window is docked on the left side of the AutoCAD drawing screen, as shown in Figure 7.6A. The window is similar to a standard Windows screen. Moving the cursor over any border in the DesignCenter window changes the cursor to a double-arrow cursor, which indicates that the window or windowpane can be stretched. To stretch the size of the window or to change the border edge location to make the panes larger, press the left mouse button while dragging the mouse. At the top of the DesignCenter window is a double grab bar. Double clicking on this bar undocks the window and turns the DesignCenter into a floating window, as shown in Figure 7.6B. The floating DesignCenter can be repositioned anywhere on the drawing screen.

The tree-view pane along the left edge of the DesignCenter is used to browse through the different multi-view block folders. If the DesignCenter has been accessed from the Design or Documentation pull-down menus or from the Design Content toolbars, the DesignCenter is displayed looking in the folder that was selected. Folders can be accessed by picking on top of a folder in the tree view, which displays a list of blocks in the preview pane to the right.

The upper right side of the window displays a small preview of the blocks within the selected folder. Picking a preview icon from the small preview pane displays a larger preview of the block in the lower left side of the window. This larger preview pane allows you to spin the block around by moving the mouse into the preview area, holding down the left mouse button, and moving the cursor. Right-clicking in the larger preview pane displays a cursor menu from which different shading modes and views can be selected. Below the large preview pane is a description pane. This pane displays a description of the selected block.

When you have found a block that you would like to use in the drawing, drag and drop it into the drawing. In order to drag and drop a block from the DesignCenter, select the block's small icon from the small preview pane. Press and hold the left mouse button and drag the cursor into the drawing screen. Let go of the mouse button to insert the block into the drawing. Some of the blocks require you to enter a value or a rotation angle. Be sure to read the Command: prompt to see what type of information Architectural Desktop is asking for. If you are prompted for a rotation angle, enter the angle, or move the mouse and pick the rotation angle. Once the block has been inserted into the drawing, a layer is automatically added to the drawing, and the block is placed on the layer. Each multi-view block is assigned to a layer key. The layer that is assigned to the layer key is the layer on which the block is inserted. If you drag and drop blocks into the drawing, you may have to use the MOVE command to place the block in the correct location.

An alternative way of inserting a block into the drawing is by double-clicking the small preview icon. Either the **Add Multi-View Block** dialog box or the **Insert** dialog box is displayed depending on the type of block selected (see Figure 7.7). The dialog boxes display the name of the multi-view block or blocks that are being inserted in a drop-down list at the top of the screen. The scale of the block is displayed for the X axis, Y axis, and Z axis (for design blocks this is typically X=1, Y=1, and Z=1). You can enter the scale for the block, or you can enter the scale after the block is inserted. A rotation angle can be preset, or you can

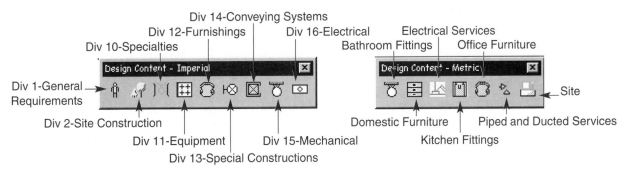

FIGURE 7.5 The Design Content-Imperial and Design Content-Metric toolbars can be used to access the different folders containing multi-view blocks.

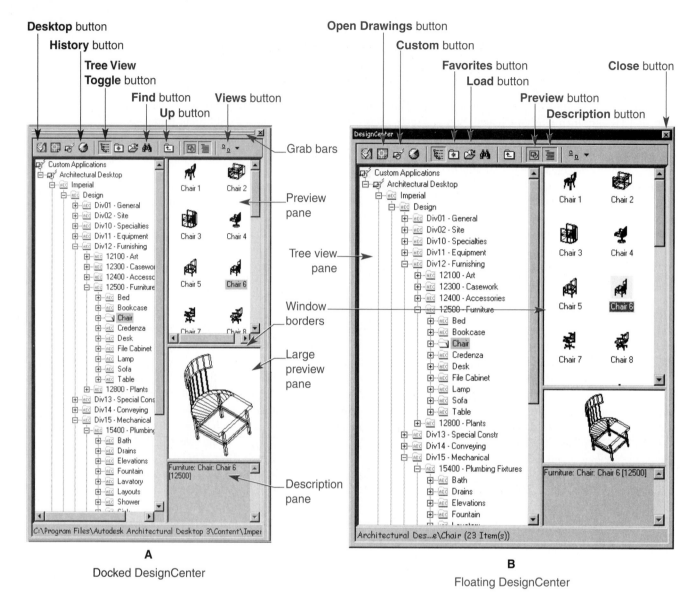

FIGURE 7.6   (A) Initially, the DesignCenter is displayed docked to the left of the drawing screen. (B) Double-clicking the grab bars changes the DesignCenter to a floating window.

choose to specify the rotation angle after the block has been inserted. After selecting the information in the Add Multi-View Blocks dialog box, pick once in the drawing to make the drawing active, then pick a location for the block. Enter any values needed to complete the insertion of the block. If the block you are inserting uses the Insert dialog box, press the OK button to insert the block into the drawing.

Multi-view block being inserted⎯⎯⎯⎯⎯⎯⎯⎯⎯⎯⎯⎯⎯Preset rotation angle

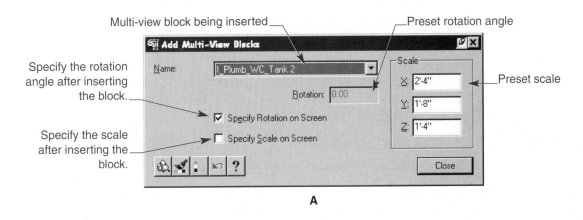

Specify the rotation
angle after inserting
the block.

Specify the scale
after inserting the
block.

⎯Preset scale

Standard block being inserted⎯⎯⎯⎯

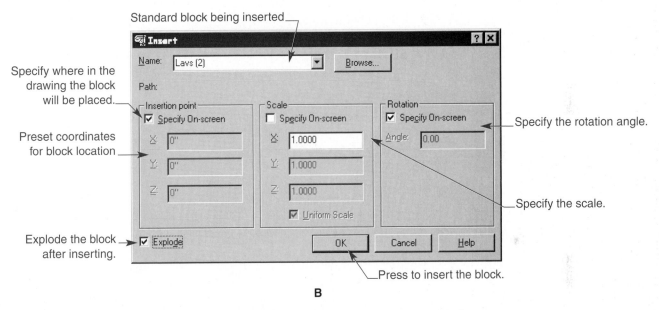

Specify where in the
drawing the block
will be placed.

Preset coordinates
for block location

Explode the block
after inserting.

Specify the rotation angle.

Specify the scale.

Press to insert the block.

FIGURE 7.7   (A) The Add Multi-View Blocks dialog box is used to insert multi-view blocks.
(B) The Insert dialog box is used to insert standard blocks into the drawing.

Field Notes

- Some of the multi-view blocks in the DesignCenter are 2D blocks that do not appear as 3D blocks in a model view.
- The fixture layout blocks are blocks of typical layouts for bathrooms. Using the EXPLODE command on these items allows you to stretch the countertops or bathroom stall partitions to meet the sizes required by your design. Use caution when exploding blocks. If a multi-view block is exploded, it loses its 2D and 3D display capabilities. Typically, a standard block is composed of multi-view blocks and other AEC objects. When exploded, these pieces are no longer grouped together but retain their 2D and 3D display capabilities.

Exercise 7.1

1. Open EX6-11.dwg.
2. Add a 30″ × 60″ bathtub to the main bathroom in the figure by dragging and dropping. Add a 36″ × 48″ shower to the master bathroom by dragging and dropping. Use the MOVE command to move the tub and shower into the proper locations. Drag and drop a toilet into the bathrooms.
3. Find the double sinks in the Layouts folder. Double-click to add this block to the master bathroom. Select the Explode option before inserting the block. Use grips to stretch the length of the counter to fit inside the bathroom.
4. Add a washer and dryer to the utility closet by double-clicking to insert.
5. Add a refrigerator, stove, sink, and dishwasher to the kitchen area. Do not insert casework into the drawing. You will draw counters and cabinets later in the text.
6. Add a water heater to the garage.
7. Add any additional blocks such as furniture if desired.
8. Save the drawing as EX7-1.dwg.

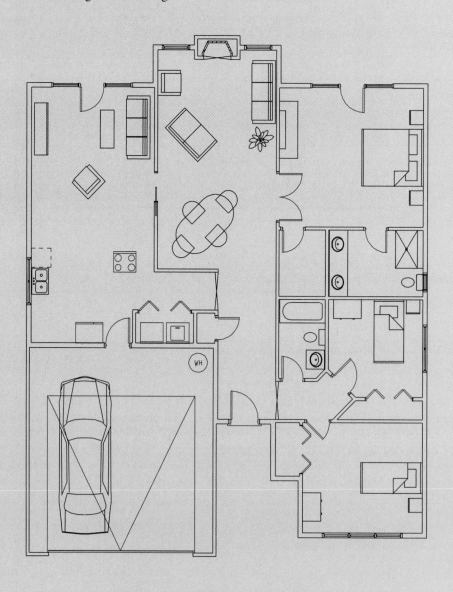

FIGURE 7.8 The Scale tab in the Drawing Setup dialog box is used to select a drawing scale for the scaling of documentation blocks.

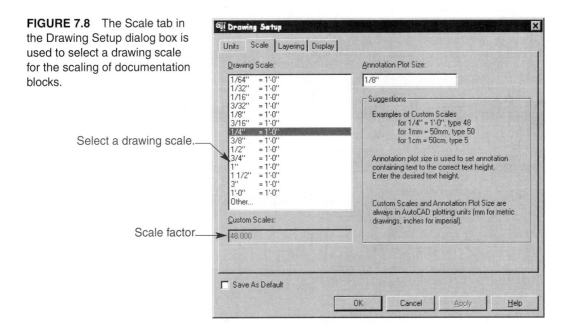

Select a drawing scale.

Scale factor

Inserting Documentation Blocks

As mentioned previously, documentation blocks are inserted into the drawing by the drawing scale. To set the drawing scale, access the **Drawing Setup** dialog box, discussed in Chapter 2. The **Scale** tab is used to specify the drawing scale. The scale that is selected calculates a scale factor that is used to scale up the documentation blocks (see Figure 7.8). For example, the drawing scale of 1/4″=1′-0″ is calculated to a scale factor of 48. This means that a documentation block inserted into model space is scaled up 48 times. When the drawing is plotted it will be shrunk down to 1/48th of its real size in model space, producing a drawing that is scaled to 1/4″=1′-0″. This in turn shrinks the scaled-up documentation block so it appears as a real-sized block on a plotted drawing.

If the desired plotting scale is not available in the Scale tab, selecting the Other... scale makes the **Custom Scales**: text box available. Enter a scale factor for your drawing. This affects the scale of the documentation blocks being inserted. Scale factors are the reciprocal of the drawing scale. For example, the scale factors for a 1/8″=1′-0″ and a 1/4″=1′-0″ scaled drawing are calculated using the following method:

1/8″ = 1′-0″	1/4″ = 1′-0″
.125″ = 12″	.25″ = 12″
12″ ÷ .125″ = scale factor of 96	12″ ÷ .25″ = scale factor of 48

Appendix F provides a list of common scales and their scale factors for use in the Scale tab. After the scale factor has been determined and applied to the drawing, any new documentation blocks that are inserted are scaled by the resulting scale factor.

Documentation blocks can be inserted by selecting the Documentation pull-down menu and picking the Documentation Content cascade menu, then selecting a section. Documentation blocks can also be accessed by selecting a button from the Documentation-Imperial or Documentation-Metric toolbars, or by right-clicking and selecting Documentation and Documentation Content, and picking a section. Figure 7.9 shows the different folder sections in the two toolbars.

Like the design blocks, documentation blocks can be added to the drawing by the drag-and-drop technique or by double-clicking on the block preview. Many of the documentation blocks include a routine for the block. For example, a Title Mark block prompts you to fill out the symbol bubble values and select points for the line under the title of the drawing. A Revision Cloud block prompts you to select a starting point and an ending point for the revision cloud. It is very important to read the Command: prompt line to see what Architectural Desktop is asking you to do. When a block is finished being inserted it is automatically placed on a layer designated by the layer key style.

FIGURE 7.9   The Documentation-Imperial and Documentation-Metric toolbars can be used to access the different folders containing documentation blocks.

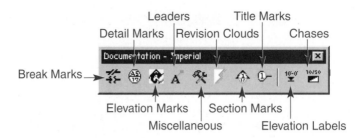

FIGURE 7.10   Double-click on a documentation block with attributes to change the attribute values in the Edit Attributes dialog box.

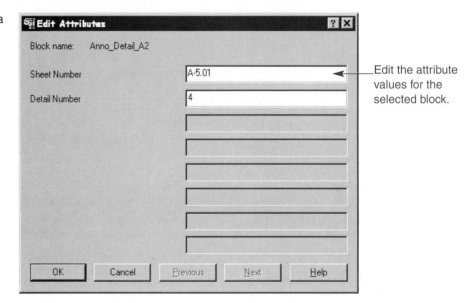

Many of the blocks have values that are variable. When inserting a documentation block that prompts you for a value, type the desired value. These values are known as ***attributes***. To change the attribute value for a block after it has been inserted, double-click on the block to access the **Edit Attributes** dialog box, shown in Figure 7.10. The values can be edited for the selected block. Press the OK button when you are finished to accept your changes and update the block.

Field Notes

- Documentation blocks that include text are inserted using the current text style. If you would like to use a specific type of font for documentation blocks, create a text style and set it current before adding documentation blocks. See the AutoCAD help system for more information on creating a text style.
- Most of the documentation blocks are 2D only and are displayed only as flat blocks in an elevation or model view.

Exercise 7.2

1. Open EX7-1.dwg.
2. Set the drawing scale to 1/4″=1′-0″.
3. Add a title mark to the bottom of the drawing named Floor Plan.
4. Add some detail marks around the drawing, as shown in the figure.
5. Add a north arrow to the drawing. North arrows can be found under the Miscellaneous folder.
6. Save the drawing as EX7-2.dwg.

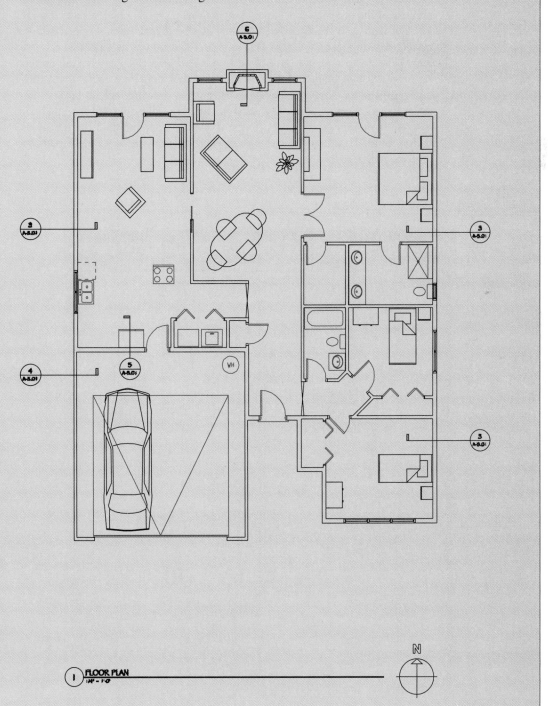

Using Additional DesignCenter Features

In addition to being utilized to insert Architectural Desktop blocks, the DesignCenter can be used to access other parts of existing drawings to be added to the current drawing. Items such as layers, layouts, dimension styles, text styles, and linetypes can be dragged from existing drawings in other drives, such as a hard drive, network drive, floppy disk, CD, or Zip drives, and dropped into the current drawing. Figure 7.6 identifies the following buttons used in the DesignCenter window:

- **Desktop**   This button is used to browse for drawings in other drives. The tree-view pane (left pane) is used to browse through folders similar to the way Windows Explorer is used. Selecting a folder that contains drawing files displays the full preview of the drawing. Double-clicking a drawing in the preview pane displays a list of icons representing elements that can be dragged and dropped into the current drawing (see Figure 7.11). The folder can also be expanded in the tree-view pane and a drawing selected from the tree view. Double-clicking an icon displays a list of items that can be used in other drawings. Items that can be dragged from an existing drawing to the current drawing include blocks, dim-styles, layers, layouts, linetypes, textstyles and xrefs. To drag and drop the item, first select the item to be added to the drawing, press and hold the left mouse button to drag it to the drawing screen, then let go of the left mouse button to drop the item into the drawing.

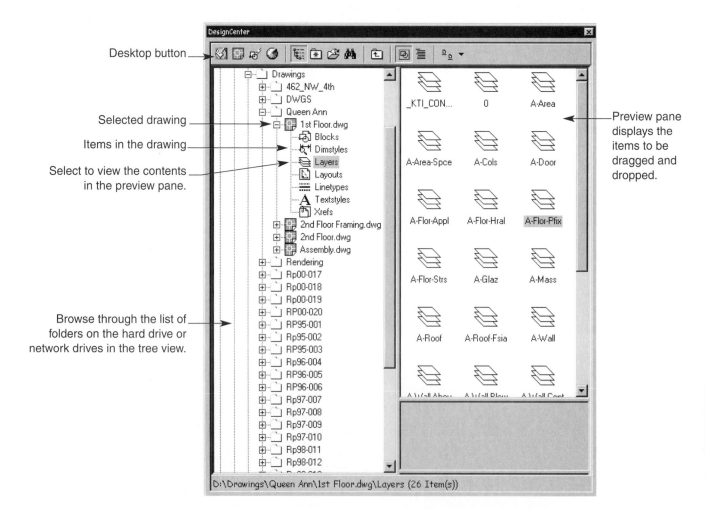

FIGURE 7.11   The Desktop button allows you to browse through folders to look for a drawing that includes parts of a drawing that you would like to use in the current drawing.

Field Notes

If blocks are dragged and dropped into the drawing, the blocks are inserted on the current layer. If you are dragging and dropping a block from the Architectural Desktop Content folder while viewing the content from the Desktop icon, the AEC block loses its "smartness" and is not inserted on the correct layer, nor scaled by the drawing scale. If you are going to use the Architectural Desktop blocks, then use the Custom button in the DesignCenter to view the Architectural Desktop content.

Field Notes

AutoCAD Hatches and Images files can be viewed in the DesignCenter and dragged and dropped into the drawing. Review the Command: line to properly insert these types of objects. To view a sample icon of a hatch locate a *.pat file that contains hatches. A preview of the hatches is displayed, allowing you to drag them into the drawing. Images can also be viewed and dragged and dropped into the drawing. Image files that can be used include files with extensions such as .bmp, .tga, .jpg, .pcx. .gif, and .tif files.

▪ **Open Drawings**   This button displays a list of drawings currently open in Architectural Desktop. Similar to the information displayed with the Desktop button, blocks, dimstyles, layers, layouts, linetypes, textstyles, and xrefs can be dragged from an existing drawing and dropped into the current drawing (see Figure 7.12).

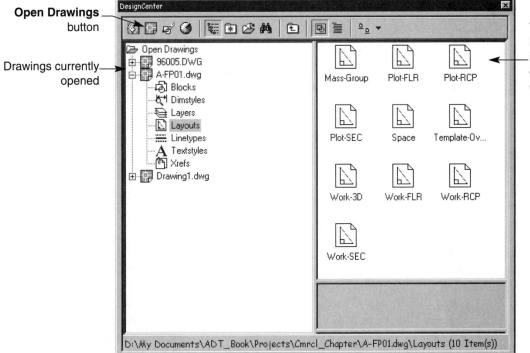

FIGURE 7.12   Select the Open Drawings button to display a list of the drawings open in Architectural Desktop. Information can be dragged from one drawing and dropped into the current drawing.

Custom button

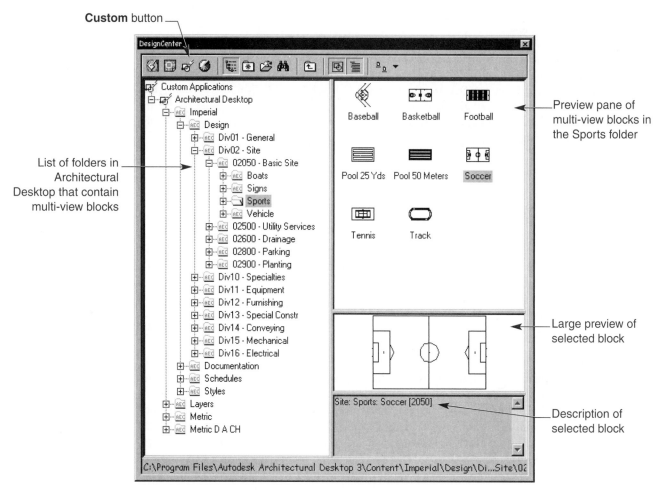

List of folders in
Architectural
Desktop that contain
multi-view blocks

Preview pane of
multi-view blocks in
the Sports folder

Large preview of
selected block

Description of
selected block

FIGURE 7.13   The Custom button displays the default view when blocks are
accessed from the Design or Documentation pull-down menus.

	■ **Custom**	This button displays a list of the Architectural Desktop multi-view blocks and their organized folders. This view is displayed when a type of block to add is selected from the Design or Documentation pull-down menus or toolbars. Dragging and dropping a block from the preview pane inserts the block on the appropriate layer. Each block in this list has been assigned a layer key and looks to the layer key style for the layer on which it is supposed to be inserted (see Figure 7.13).
	■ **History**	This button lists a history of drawings that have been accessed in Architectural Desktop. Double-clicking a drawing from the list on the right displays the Desktop view with the drawing file open (see Figure 7.14). Select items from the drawing that you would like to use.
	■ **Tree View Toggle**	This button minimizes the left tree-view pane, providing more room for the display of the previews.
	■ **Favorites**	This button displays the Autodesk folder under the Favorites folder within Windows. Shortcuts to other folders or files can be placed into the Favorites folders for quick access. The Autodesk folder also includes shortcuts to the acad.pat and acadiso.pat files that include hatch patterns that can be dragged and dropped.
	■ **Load**	This button displays the **Load DesignCenter Palette,** which is similar to the Open dialog box. You can browse for drawings on the hard drive, on network drives, or over the Internet for drawing files with content.

FIGURE 7.14 The History button lists drawings previously used to obtain different types of content for your drawings.

History button

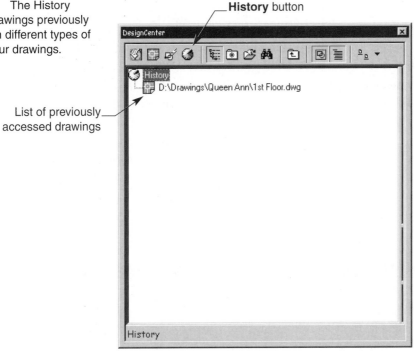

List of previously accessed drawings

■ **Find**

This button displays the **Find** dialog box, as shown in Figure 7.15. This dialog box is used to search for drawings, blocks, dimstyles, textstyles, hatches, linetypes, layouts, and xrefs in drawing files. Specify the directory to search and the name of the item you are searching for. A list of matches is displayed and can be dragged and dropped into the drawing.

Specify the directory to search through.

Select the type of content to search for.

Enter the name of the content you are searching for.

Content that matches the search criteria

Press to start the search.

Press to begin a new search.

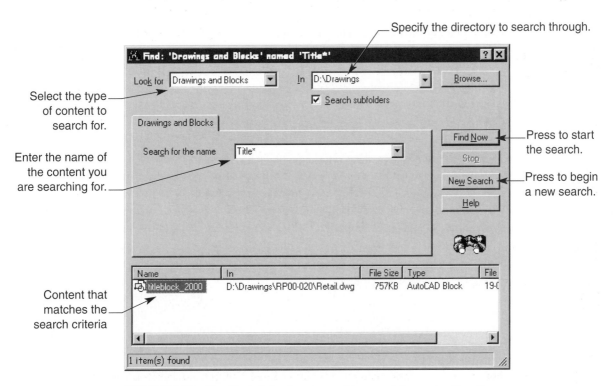

FIGURE 7.15 The Find dialog box is used to search for content in other drawings that can be added to the current drawing.

- **Up**   Pressing this button while in the Desktop view shows the contents of
                         folders up to the previous level.

- **Preview**   Initially, this button is depressed. This button turns the large preview
                         pane on or off in the DesignCenter.

- **Description**   This button, when depressed, displays a description pane below the large
                         preview pane. All content found under the Custom button includes a
                         description. Any custom blocks that are created can have a description
                         display in this pane.

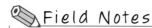

- **Views**   This button includes a drop-down menu with a list that allows you to
                         choose how the previews will appear in the preview pane. Content can
                         be viewed as large icons, small icons, as a list of the content, or as the
                         details of the content.

✎ Field Notes

Remember that blocks inserted from the Custom button are inserted with an appropriate layer.
Documentation blocks inserted from the Custom view are also scaled by the drawing scale.
Design and documentation blocks inserted from the Desktop view lose all the automatic for-
matting such as automatic layer and scale.

Creating Standard AutoCAD Blocks

Recall that a *block* is a group of individual objects that are created together to represent a type of symbol.
The advantage of blocks is that you can draw one group of objects once and use it many times in the drawing without having to re-create the initial geometry. Once the block is created it can be inserted into a drawing through the **INSERT** command or dragged and dropped from the DesignCenter. There are a few
differences between standard AutoCAD blocks and Architectural Desktop blocks. Standard AutoCAD
blocks are inserted on the current layer and reflect either 2D or 3D objects, depending on how the original
geometry was constructed. Architectural Desktop blocks, are inserted onto predefined layers and are displayed as 2D in a top view and as 3D in a model view.

Block Creation Criteria

Before drawing the block, you need to determine the following criteria:

1. Which layer should the geometry within the block be created on?

2. What color should the initial geometry be?

3. What linetype should the initial geometry be?

4. What lineweight should the initial geometry be?

When standard blocks are inserted into a drawing, they are placed on the layer that is current at the time the
block is inserted; the exception is if a block is inserted from the Custom view in the DesignCenter. If the
initial block geometry was drawn on layer 0, the block is placed on the current layer when inserted and is
displayed with the color and linetype of the current layer. If the geometry for the block was created on a
layer *other* than 0, then the geometry in the block belongs to the current layer *and* to the layer(s) on which
the initial geometry was created. To avoid having the geometry attached to two layers, the initial geometry
should be created on layer 0.

After determining the layer where the initial geometry will be created, you need to decide how the
block will obtain its color, linetype, and lineweight. If the block geometry is drawn with a color, linetype, or
lineweight other than **ByLayer** or **ByBlock**, those properties are displayed in the block, overriding the

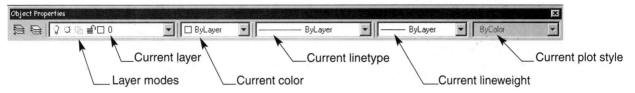

FIGURE 7.16 The Object Properties toolbar includes several drop-down menus to control the properties of newly created geometry.

color, linetype, and lineweight of the layer where the block is inserted. Drawing the initial geometry with a color, linetype, and lineweight different from the layer 0 properties is least desirable, as most situations rely on the layer to designate the properties for the block.

When geometry for a block is created with a ByLayer color, linetype or lineweight, it is displayed with the color, linetype, and lineweight of the layer where the block is inserted. If you try to change the color, linetype, or lineweight of the block, the properties do not change, because the geometry is looking for the color, linetype, and lineweight of the layer where it is inserted. If you are planning to use the ByLayer properties, set the properties current in the Object Properties toolbar prior to creating any geometry (see Figure 7.16).

Another option for block geometry properties is ByBlock. When objects are drawn with a ByBlock color, linetype, or lineweight, the block is inserted on the current layer and assumes the properties of the layer, as with the ByLayer option, but the color, linetype, and lineweight of the block can be changed through Entity Display or the Properties window. This is the preferred method of creating blocks for use in Architectural Desktop. The ByBlock colors, linetypes, and lineweights in a block can be overridden when tying a block into an Architectural Desktop multi-view block. The creation of a multi-view block is covered later in this chapter.

Creating the Block Components

Prior to drawing the geometry that will be used in the block, set layer 0 as the current layer, and the ByBlock color, linetype, and lineweight drop-down lists current. This ensures that the geometry within the block can be used in standard blocks and multi-view blocks. Draw the geometry that will become the block symbol using polylines, arcs, circles, or AEC objects such as walls, doors, or windows.

Blocks are inserted into the drawing at the ***insertion point***. When blocks are dragged and dropped or inserted, the block is attached to the crosshairs at the insertion point on the block, which is a logical point on the block. The insertion point should be determined before grouping the geometry together. The insertion point is specified at the time the block is created. Figure 7.17 displays a few blocks with their insertion points identified. Notice that the insertion point does not need to be drawn and is usually specified using osnaps.

FIGURE 7.17 Some symbols that have been created as blocks. Note the insertion point has been located.

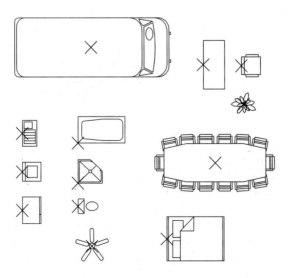

X Denotes the insertion point

Creating the Block

Once the geometry for the block has been drawn, a block symbol can be created. Use one of the following options to turn the geometry into a block:

✓ Open the **Draw** pull-down menu, select the **Block** cascading menu, then pick **Make....**

✓ Pick the **Make Block** button in the **Draw** toolbar.

✓ Type **B**, **BLOCK**, or **BMAKE** at the Command: prompt.

The **Block Definition** dialog box, shown in Figure 7.18, is displayed. Use the following steps to create a block:

1. In the **Name:** text box, type a name for the block, such as SPA-6040. The name cannot exceed 255 characters and can include numbers, letters, and spaces as well as the dollar sign ($), hyphen (-), and underscore (_). The block name can be a description of the block or a series of numbers and letters indicating the type of symbol the block represents.

2. The **Base point** area is used to specify the insertion point for the block. Type the absolute coordinates for the insertion base point in the text boxes, or select the **Pick point** button to use the crosshairs to select an insertion point. Remember, this is the point at which the block will be attached to your crosshairs when it is inserted in the drawing.

3. The **Objects** area is used to select the geometry that will be grouped together in the block definition. Pick the **Select objects** button to return to the drawing screen, and use the crosshairs to select the geometry for the block definition. Press [Enter] when you are done. The Block Definition dialog box reappears, and the number of objects selected is shown in the Objects area. The **Quick Select** button can be used to filter for a number of objects that meet specific criteria.

4. The Objects area also includes three options for controlling the original geometry selected. You can specify whether to retain, convert, or delete the selected objects. The **Retain** radio button retains the original objects on the drawing screen. The **Convert to block** radio button

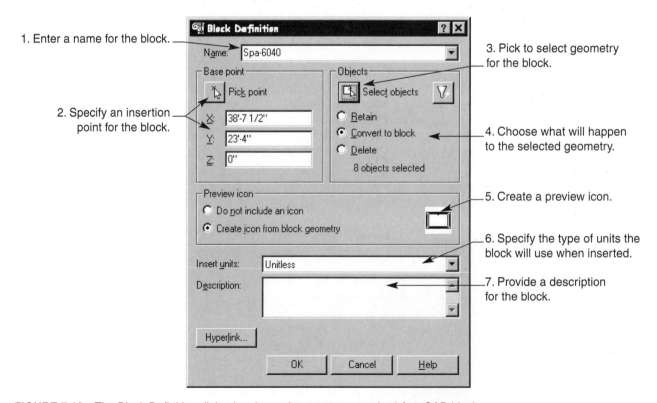

FIGURE 7.18   The Block Definition dialog box is used to create a standard AutoCAD block.

is used to convert the selected geometry into a block at the same location. The **Delete** radio button removes the original geometry from the drawing screen.

5. The **Preview icon** area is used to either create a preview icon from the block geometry or not. The preview icon is used when viewing the DesignCenter from the Desktop view. If you choose to create a preview icon, an image of the block is displayed to the right.

6. The **Insert units:** drop-down list specifies the type of units the AutoCAD Design Center uses when inserting the block. This option affects standard AutoCAD blocks only and has no effect on blocks inserted from the Custom view in the DesignCenter.

7. In the **Description:** text box, type a description to help identify the block for easy reference, such as *This is a 6'-0" × 4'-0" Spa Tub.*

8. After you have finished defining the block, pick the OK button.

The block is now ready for use through the DesignCenter. Locate the block in the Open Drawings view or the Desktop view, and drag and drop it into the current drawing. When a block is created, a *block definition* is generated in the drawing. The block definition resides in the drawing and can be used in other drawings by dragging and dropping the block from the DesignCenter.

Exercise 7.3

1. Start a new drawing using the AEC Arch (Imperial) or AEC Arch (Imperial-Int) template.
2. Set layer 0 current. Set the color, linetype, and lineweight drop-down lists to ByBlock.
3. Create the blocks displayed in the figure.

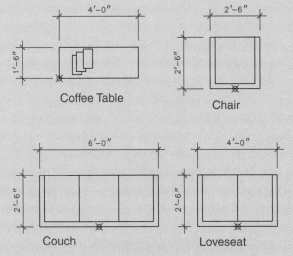

4. Create a layer named A-FURN and set current.
5. Drag and drop the new blocks into the drawing through the DesignCenter.
6. Save the drawing as EX7-3.dwg.

Creating a Wblock

When a block is created it resides within the drawing where it was generated. The **WBLOCK** *(write block)* command allows you to create a drawing (.dwg) file from a block definition, from a group of individual objects in a drawing, or from the entire drawing and place it into a folder on your hard drive or network drives. The resulting drawing file can then be inserted as a block into any drawing by dragging and dropping the drawing file from the DesignCenter or by using the **INSERT** command.

To create a Wblock type **WBLOCK** at the Command: prompt. The **Write Block** dialog box, shown in Figure 7.19, is displayed. Using the Write Block dialog box is similar to using the Block Definition dialog box. An insertion point for the drawing file needs to be specified, and the objects that will be included in the new drawing file need to be selected. The following are the parts of the Write Block dialog box:

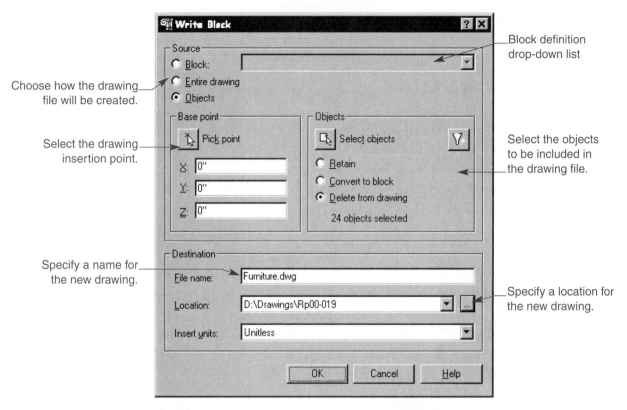

Choose how the drawing file will be created.

Select the drawing insertion point.

Block definition drop-down list

Select the objects to be included in the drawing file.

Specify a name for the new drawing.

Specify a location for the new drawing.

FIGURE 7.19   The Write Block dialog box is used to create a new drawing file from a block definition, from a group of objects in the drawing, or from the whole drawing.

- **Source area**   This area at the top of the dialog box includes the following three methods of creating a new drawing file:

  - **Block**   When this radio button is selected, the drop-down list to the right becomes available. A block definition from the current drawing can be selected from the drop-down list and turned into a new drawing file.

  - **Entire drawing**   Selecting this radio button creates a new drawing file from everything that is thawed in the current drawing. This is an effective way to purge the drawing of excess items such as layers, text styles, or block definitions that are not being used in the drawing. This is a good option to use at the end of a project when the unused drawing information needs to be purged.

  - **Objects**   This radio button allows you to select the items that need to be "blocked out" into a separate drawing file. This is a good option to use when you have drawn something that could be used again in a future drawing. For example, suppose you have created a standard bathroom layout for a commercial building, and it is typical of layouts that you draw for commercial buildings. By creating a drawing file from these objects, you do not have to create bathroom layouts from scratch again; just insert the bathroom drawing file into the drawing.

  - **Base point**   This area is similar to the Block Definition dialog box. Like blocks, drawing files that are inserted into a drawing have an insertion point. The insertion point for the drawing can be selected by pressing the **Pick point** button or by typing an absolute coordinate for the X, Y, and Z axes. This option is available only when the **Objects** radio button is selected. The block option uses the block definition's insertion point, and the **Entire drawing** radio button assumes the insertion point is located at 0,0,0.

- **Objects**   Again similar to the Block Definition dialog box, this area allows you to select the items to write to the hard drive as a drawing file. This option is available only when the Objects radio button is selected.

- **Destination**   This area controls the following options:

  - **File name**   This text box is used to specify the new drawing file's name. If you are creating a drawing file from a block definition, *do not* name the drawing the same as the block name. This causes conflicts when you try to insert the drawing into another drawing.

  - **Location**   This text box is used to specify the location for the new drawing file. The entire path and folder name must be included in the text box. If you are unsure of the path, use the ellipsis (...) button to browse the hard drive for the folder location.

  - **Insert units**   This text box is used to select the type of units that will be used when inserting the drawing file.

Exercise 7.4

1. Open EX7-2.dwg.
2. Access the Wblock command.
3. Select the Objects radio button.
4. Set the insertion point at X = 0, Y = 0, and Z = 0.
5. Select only the exterior walls; do not include any interior walls. Do include the interior garage walls. Select the Retain option.
6. Name the new drawing EX7-4_Foundation. Save the drawing to the same folder that contains the EX7-2 drawing file.
7. Specify the units as inches. Press the OK button when finished. This will create a drawing file of the exterior walls that can be changed into foundation walls.
8. Wblock out the furniture blocks. Do not include bathroom or kitchen blocks. Use the insertion point of 0,0,0, and select the Delete from drawing option. Name the file EX7-4_Furniture, and save it in the same folder as the EX7-4_Foundation and EX7-2 drawing files.
9. Save the drawing as EX7-4flr-pln.dwg.

Using the INSERT Command

In addition to using the DesignCenter to insert blocks or drawing files into the current drawing, you can also use the INSERT command. Use one of the following options to access this command:

✓ Pick the **Insert** pull-down menu, and select **Block....**

✓ Pick the **Insert Block** button from the **Draw** toolbar.

✓ Type **INSERT, DDINSERT,** or **I** at the Command: prompt.

The Insert dialog box, shown in Figure 7.20, is displayed. The following are the parts of the Insert dialog box:

- **Name**   This drop-down list includes a list of all the block definitions in the current drawing. Select the block definition to insert by selecting the definition name.

- **Browse...**   Select this button to browse through the hard drive or network drives for a drawing file to insert.

- **Insertion point**   This area is used to specify the insertion location for the block or drawing file:

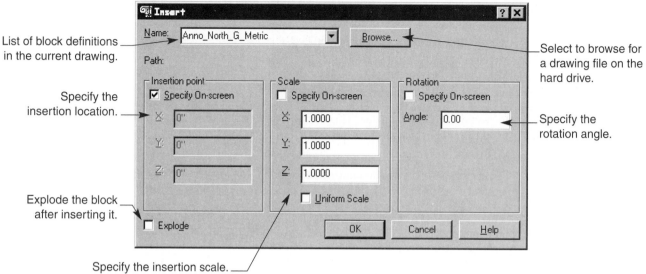

List of block definitions in the current drawing.

Specify the insertion location.

Explode the block after inserting it.

Select to browse for a drawing file on the hard drive.

Specify the rotation angle.

Specify the insertion scale.

FIGURE 7.20 The Insert dialog box is used to insert block definitions from the current drawing or drawing files from a hard drive.

- **Specify On-screen** Selecting this check box allows you to "pick" the location of the block or drawing file with your crosshairs, after you press the OK button.

- **X** When the Specify On-screen check box is unchecked, this option allows you to specify the absolute X axis location for the block.

- **Y** When the Specify On-screen check box is unchecked, this option allows you to specify the absolute Y axis location for the block.

- **Z** When the Specify On-screen check box is unchecked, this option allows you to specify the absolute Z axis location for the block.

- **Scale** This area controls the scale of the block along the X, Y, and Z axes:

  - **Specify On-screen** Selecting this check box allows you to enter the scale of the block or drawing file at the Command: prompt. The scale along each axis can be controlled individually.

  - **X** When the Specify On-screen check box is unchecked, this option scales the block along the X axis.

  - **Y** When the Specify On-screen check box is unchecked, this option scales the block along the Y axis.

  - **Z** When the Specify On-screen check box is unchecked, this option scales the block along the Z axis.

  - **Uniform Scale** This check box scales the Y and Z axis of the block uniformly with the scale specified for the X axis.

- **Rotation** This area is used to control the insertion rotation angle of the block:

  - **Specify On-screen** Selecting this check box allows you to enter the rotation angle for the block or drawing file at the Command: prompt

  - **Angle** When the Specify On-screen check box is unchecked, the rotation angle of the block can be specified.

- **Explode** When this check box is selected, the block or drawing file is exploded into the original geometry that was used to create the block or drawing file at the time the block is inserted. This may be helpful when the block needs to be modified to work within the drawing where it is being inserted.

After specifying all the parameters for the insertion of the block, press the OK button. If any of the Specify On-screen check boxes were selected, you must read the Command: prompt for the information that Architectural Desktop needs. For example, if the Insertion point Specify On-screen check box was selected, you are prompted to *Specify an insertion point*. Either enter in absolute coordinates, or select a point with the crosshairs.

Field Notes

Before inserting a block, always set the layer current where the block will be inserted. The block then adopts the properties of the layer it is inserted on.

Exercise 7.5

1. Start a new drawing using the AEC Arch (Imperial) or AEC Arch (Imperial-Int) template.
2. Use the INSERT command to browse for and insert the EX7-4_Foundation drawing.
3. Insert the drawing at 0,0,0; set the scale to 1 on the X axis; select the Uniform Scale check box; and set the rotation angle to 0. This will insert the foundation to the same coordinates as the main floor plan so that the walls line up with one another.
4. Use the Explode check box to explode the drawing at the time of insertion.
5. Modify the walls to use the Concrete-8 Concrete-16x8-Footing wall style. Modify the height of the walls to 30″ high.
6. Change to an isometric view. Use Entity Display to change the plan display representations cut plane to 24″ high. Change back to a top view.
7. Perform any additional modifications to the walls to reflect the figure.
8. Save the drawing as EX7-5fnd.dwg.

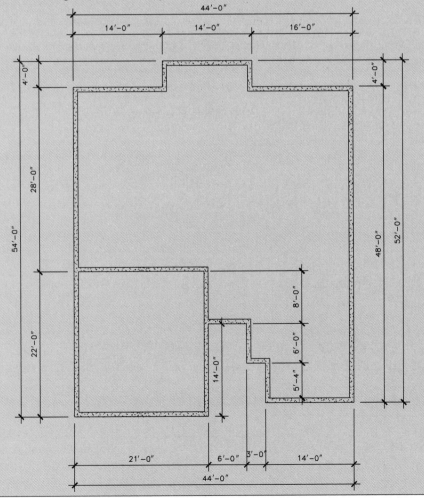

Creating Multi-View Blocks

As indicated earlier in this chapter, a multi-view block is an AEC object that is displayed as a 2D block in a top view, as a 2D block in an elevation view, and as a 3D block in a model view. The creation of a multi-view block builds on the standard AutoCAD block principle. The multi-view block definition uses a standard AutoCAD block for each viewing direction. The standard blocks are created first then tied together in the multi-view block definition. For example, a toilet multi-view block may use five separate blocks, each assigned to a different viewing direction, as shown in Figure 7.21. When the blocks are combined, the resulting multi-view block displays one of the blocks when viewed from a particular direction.

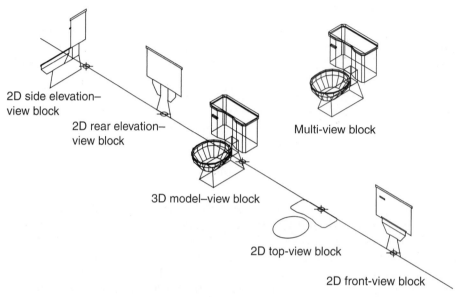

FIGURE 7.21   A multi-view block uses several standard AutoCAD blocks that are displayed based on the viewing direction.

FIGURE 7.22   Each block is drawn separately with its insertion point picked such that when the blocks are assembled into the multi-view block definition, the insertion points all line up.

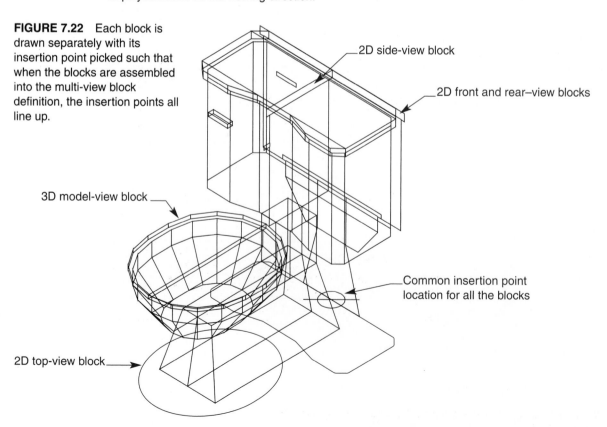

Notice in Figure 7.21 that each side of the toilet is drawn, then turned into a standard AutoCAD block. Also, the insertion point of each block is located such that if all the blocks are assembled together, the insertion points will line up (see Figure 7.22).

Creating a Multi-View Block

The following discussion takes you through the steps of creating the blocks necessary for a custom multi-view block, then discusses the process of adding the blocks into a multi-view block definition. The example multi-view block represents a wood framing member that as a center line is displayed in a top view, a cut piece of framing in a front view, a piece of framing in a side view, and a 3D wood framing member in a model view (see Figure 7.23).

The goal for creating this block is that it must be able to be scaled differently along the X, Y, and Z axes to provide different framing sizes. A new drawing is started from scratch, because it is a smaller scaled drawing, and the blocks need to fit within a 1″ cube so that the block can be scaled to 2× its size in the X axis to make the framing member 2″ wide, 8× in the Z axis to make the member 8″ tall, and 72× in the Y axis to make the framing member 6′-0″ long. This scaling creates a 2 × 8 framing member that is 6′-0″ long.

After you start a new drawing, the first step in creating a multi-view block is to draw the geometry for the individual standard blocks. Each of the blocks needs to be oriented appropriately in 3D space so the **VPORTS** command can be used to separate the drawing screen into four equal viewports each looking at a different viewing direction. After the viewing direction of the viewports has been selected, the drawing plane is adjusted so the XY plane is perpendicular to your view, to aid you in the process of creating the elevation-view blocks (see Figure 7.24).

1. The rules for creating blocks are the same as mentioned in the previously discussed block section. Draw the geometry on layer 0, use ByBlock colors, linetypes, and lineweights unless a part of the block needs to use a specific color, linetype, or lineweight. The top-view block appears as a single center line. Before drawing the line, load the Center linetype. After the linetype has been loaded, set layer 0 as the current layer, and set the ByBlock color, linetype, and lineweight current in the drop-down lists. In the top view viewport, draw a vertical line 1″ in length. To assign the Center linetype to the line, select the line, right-click, and select Entity Display…. In the Entity Display dialog box change the linetype from ByBlock to Center. Change the linetype scale to .5. The first piece for the multi-view block has been created.

2. Next, pick in the front-view viewport to make it active. Draw a 1″ × 1″ rectangle with an **X** extending from corner to corner of the rectangle. This creates the front-view geometry for the multi-view block. After the front view geometry has been drawn, pick in the right-view viewport. Draw a 1″ × 1″ rectangle in this viewport. Notice that as you draw the front- and right-view geometry, it appears oriented correctly in the isometric viewport.

3. The final piece that needs to be drawn is the 3D cube, which represents a 3D wood framing member. Pick in the isometric viewport to make the viewport active. Use the **3D** command to create a 3D box. The command sequence is as follows:

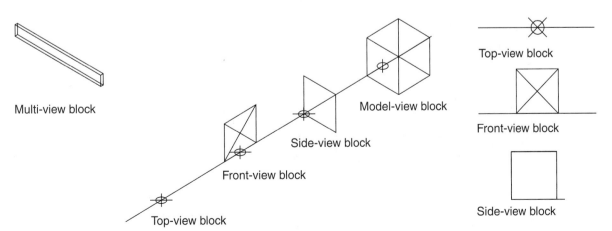

Multi-view block

Model-view block

Side-view block

Front-view block

Top-view block

Top-view block

Front-view block

Side-view block

FIGURE 7.23   The desired multi-view block for the wood framing member.

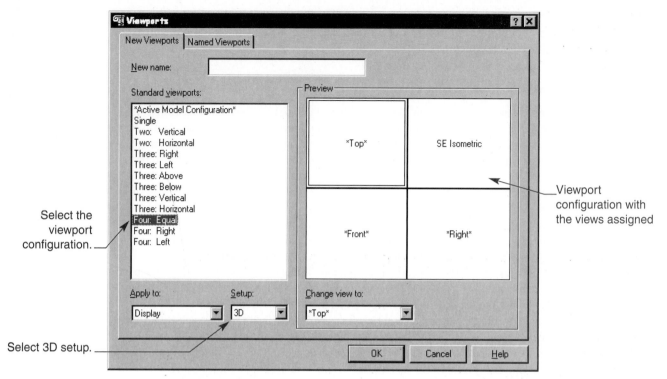

FIGURE 7.24   Use the Viewports dialog box to split the drawing screen into viewports with a preset view assigned to each.

Command: **3D** ↵
Initializing... 3D Objects loaded.
Enter an option
[Box/Cone/DIsh/DOme/Mesh/Pyramid/Sphere/Torus/Wedge]: **B** ↵
Specify corner point of box: *(pick a location in the isometric view)*
Specify length of box: **1** ↵
Specify width of box or [Cube]: **C** ↵
Specify rotation angle of box about the Z axis or [Reference]: **0** ↵
Command:

A three-dimensional cube is created in the drawing.

4. The final step before creating the multi-view block is to turn the four sets of geometry drawn into separate blocks. While remaining in the isometric view, ensure that the drawing plane is in the World Coordinate System by using the following command sequence:

Command: **UCS** ↵
Current ucs name: *WORLD*
Enter an option [New/Move/orthoGraphic/Prev/Restore/Save/Del/Apply/?/World]
<World>: **W** ↵

Now you can make the blocks. When naming the blocks for a multi-view block, it is a good idea to create names that reflect the viewing direction of the blocks. For example, the front-view block may be named Frmg_Front. Insertion points need to be selected for each block so the geometry lines up with each other. Use the BLOCK command to create a block for each set of geometry. Figure 7.25 shows the names and insertion points for each block.

5. Once the blocks have been created, the multi-view block definition can be assembled. Use one of the following options to open the **Style Manager** dialog box, with the **Multi-View Block Definitions** section open (see Figure 7.26):

FIGURE 7.25 The name and insertion point for each block.

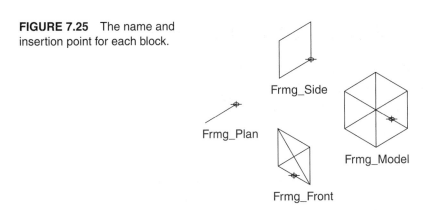

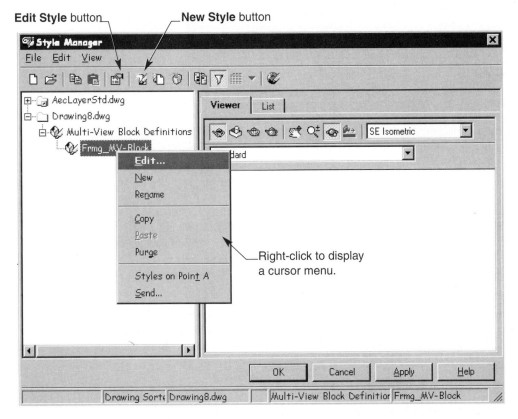

FIGURE 7.26 The Style Manager dialog box is used to create new styles and multi-view block definitions.

✓ Access the **Desktop** pull-down menu, select the **Multi-View Blocks** cascade menu, and pick **Multi-View Block Definitions….**

✓ Pick the **Multi-View Block Definitions** button in the **AEC Blocks - Profiles** toolbar, right-click, and select **Desktop, Multi-View Blocks, Multi-View Block Definitions….**

✓ Type **AECMVBLOCKDEFINE** at the Command: prompt.

Pick the New Style button, or right-click in the right pane, and select New to create a new multi-view block definition. Enter a name for the multi-view block.

6. To edit the definition, select the multi-view block name in the left viewport, right click, and select Edit…, or pick the Edit Style button. The Multi-View Block Definition Properties dialog box, shown in Figure 7.27, is displayed. This dialog box is used to assemble the standard blocks under a display configuration for each viewing direction.

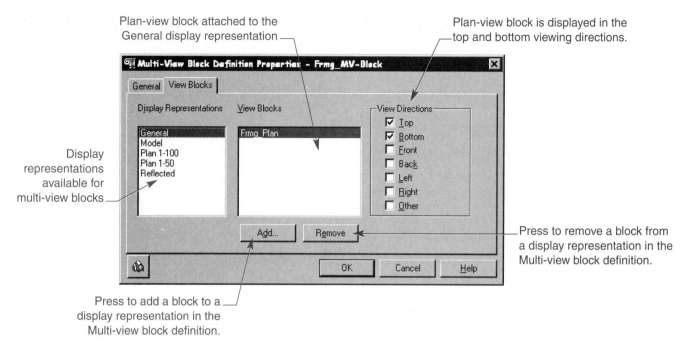

FIGURE 7.27   The Multi-View Block Definition Properties dialog box is used to define the multi-view block.

The left side of the dialog box lists the display representations used by the multi-view block definition. These include General (used in top views and some elevation-view configurations), Model (used for model views and some elevation-view configurations), Plan 1-100 and Plan 1-50 (used in top views and some elevation-view configurations), and Reflected (used in reflected ceiling-plan views). The blocks that were created need to be assigned to one or more of the display Representations in order to be displayed properly.

The Frmg_Plan block needs to be displayed in a top view. The General display representation is used for the top view, so highlight the General display representation on the left, then pick the Add... button. The **Select A Block** dialog box is displayed. Pick the Frmg_Plan block from the list. Press the OK button after selecting the block. The block will be listed in the View Blocks area, as shown in Figure 7.27. The final part is to assign a viewing direction to the block. The block will be viewed from the top view, so select the Top check box in the View Directions area. You would also like to see the block as a line if you are looking from the bottom, so also select the Bottom check box. Uncheck all the other check boxes, because you do not want to see the block in any of the other viewing directions.

7. The final step is to finish assigning the remaining blocks to an appropriate display representation and viewing direction. The Frmg_Side block should be seen from the Left and Right viewing directions. The Frmg_Front block should be seen in the Front and Back viewing directions. The Frmg_Model block should be seen in the Other viewing direction. Figure 7.28 displays the blocks that are assigned to the different display representations.

Adding the Multi-View Block to a Drawing

When you are finished assigning the blocks to display representations and viewing directions, press the OK button to exit the dialog box to return to the Style Manager dialog box. Exit the Style Manager dialog box. To use the multi-view block in other drawings, first start another drawing. Access the Multi-View Block Definitions... command. In the Style Manager, the drawings that are currently open are displayed. Multi-view block definitions can be dragged from one drawing to another. To do this, first list the multi-view block in the right pane by selecting the appropriate drawing, then selecting the Multi-View Block Definitions icon. Drag the icon to the other drawing to add the multi-view block definition to the new drawing. Exit the Style Manager when you are finished.

Use one of the following options to add the block to the drawing:

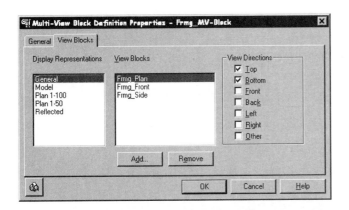
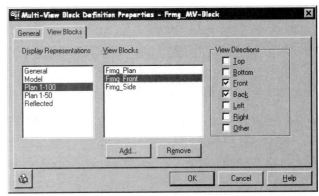
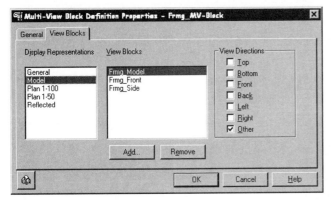
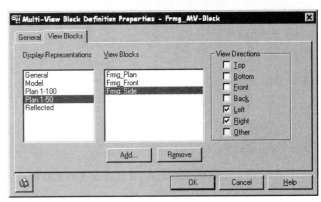

FIGURE 7.28 The different blocks assigned to each display configuration with their viewing directions assigned.

✓ Pick the **Desktop** pull-down menu, select the **Multi-View Blocks** cascade menu, and pick **Add Multi-View Block….**

✓ Pick the **Add Multi-View Block** button in the **AEC Blocks - Profiles** toolbar.

✓ Right-click, and select **Desktop, Multi-View Blocks, Add Multi-View Block….**

✓ Type **AECMVBLOCKADD** at the Command: prompt.

The **Add Multi-View Blocks** dialog box, shown in Figure 7.29, is displayed. Select the desired block from the list. The scale can be specified after the block is inserted or it can be preset in the dialog box. The rotation angle can also be specified after the block is inserted or preset in the dialog box. After specifying all the parameters, pick once in the drawing to make the drawing active, then pick a location for the block.

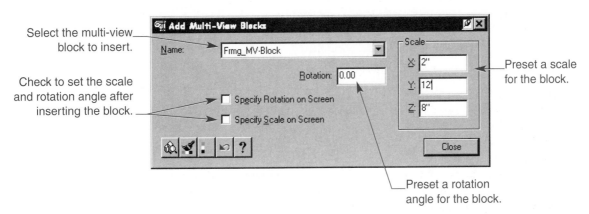

Select the multi-view block to insert.

Check to set the scale and rotation angle after inserting the block.

Preset a scale for the block.

Preset a rotation angle for the block.

FIGURE 7.29 The Add Multi-View Blocks dialog box is used to insert multi-view blocks.

✎ Field Notes

> If blocks are created so they originally can fit within a 1″ cube, as in the previous example, a scale can be specified along the X, Y, and Z axes to differently size symbols from the same definition. Blocks are originally drawn at full scale, so they are probably inserted at a scale of X=1, Y=1, and Z=1, unless these values are changed during insertion.

Multi-view blocks can also be assembled to be inserted from the DesignCenter. Adding a multi-view block to the DesignCenter is discussed later in this chapter.

Modifying a Multi-View Block

After a multi-view block has been inserted into the drawing, the scale, rotation angle, and the multi-view block being used can be modified. Use one of the following options to modify a multi-view block inserted into the drawing:

- ✓ Select the **Desktop** pull-down menu, pick the **Multi-View Blocks** cascade menu, and select **Modify Multi-View Block…**.

- ✓ Pick the **Modify Multi-View Block** button in the **AEC Blocks - Profiles** toolbar.

- ✓ Right-click, and select **Desktop, Multi-View Blocks, Modify Multi-View Block…**.

- ✓ Select the multi-view block first, right-click, and select **Multi-View Block Modify…**.

- ✓ Type **MVBLOCKMODIFY** at the Command: prompt.

The Modify Multi-View Blocks dialog box is displayed.

The scale and rotation angle for the multi-view block can be changed by entering the appropriate values in the text boxes. A different multi-view block can also be used instead, by selecting a different multi-view block from the Name: drop-down list.

Exercise 7.6

1. Start a new drawing from scratch using English units.
2. Create the framing member multi-view block described in this section.
3. Save the drawing as EX7-6A.dwg.
4. Start a new drawing using the AEC Arch (Imperial) or AEC Arch (Imperial-Int) template.
5. Drag the multi-view block definition from the EX7-6A.dwg into the current drawing.
6. Insert the multi-view block with a scale of X=2, Y=8′-0″, Z=6.
7. Insert the multi-view block with a scale of X=2, Y=4′-0″, Z=4.
8. Preview the block in different viewing directions.
9. Modify the two blocks so the scale for the X axis equals 4.
10. Save the drawing as EX7-6B.dwg.

Creating Mask Blocks

A *mask block* is an AEC object that is used to cover or "mask" AEC objects to which it is attached. Mask blocks are the electronic CAD equivalent of correction fluid. Mask blocks can be used to clean up parts of your drawing that cannot be adjusted by other means. Architectural Desktop uses many light fixture blocks as mask blocks to mask out ceiling grid lines in a reflected ceiling grid plan.

Another use of mask blocks is with old CAD drawings that use two lines to represent a wall. The two lines do not cleanup with an AEC wall object when they are intersected. Use a mask block to mask out the AEC wall to make the two different types of objects cleanup with each other, as shown in Figure 7.30.

FIGURE 7.30   Use a mask block to cleanup intersections between AEC wall objects and AutoCAD lines.

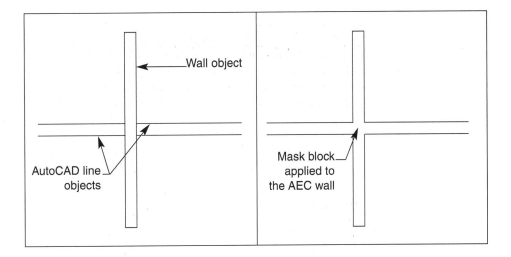

Mask blocks mask out only AEC objects and have no effect on standard AutoCAD geometry. They are inserted on the current layer, which can be frozen or set to a non-plotting layer, so the mask blocks do not appear on the finished drawing.

Creating a Mask Block

A mask block is made from a closed polyline that has been turned into a mask block definition. To create a mask block, first use a polyline to draw the shape that will be used to mask an AEC object. When you have finished drawing the polyline shape, use the Close option to close the polyline (see Figure 7.31).

After the polyline shape is created, the mask block can be defined. This is done in the **Mask Blocks** section of the Style Manager. Use one of the following options to display the Style Manager open to the Mask Blocks section, as shown in Figure 7.32:

- ✓ Pick the **Desktop** pull-down menu, select the **Mask Blocks** cascade menu, then pick **Mask Block Definitions…**.
- ✓ Pick the **Mask Block Definitions** button on the **AEC Blocks - Profiles** toolbar.
- ✓ Right-click, and select **Desktop, Mask Blocks, Mask Block Definitions…**.
- ✓ Type **AECMASKDEFINE** at the Command: prompt.

Pick the New Style button or right-click in the right pane, and select <u>N</u>ew to create a new mask block definition. After you have named the new definition, select the definition name in either pane, right-click, and select Set <u>F</u>rom… in the cursor menu, or press the Set From button. This takes you to the drawing screen, where you are prompted to select a closed polyline. The following prompt sequence is used to define a mask block after selecting the Set From option:

FIGURE 7.31   To create a mask block, first draw the shape of the mask with a closed polyline.

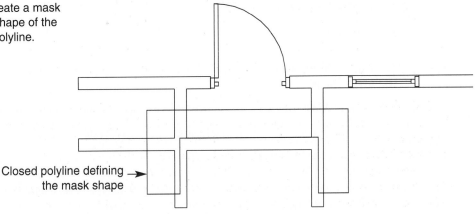

Closed polyline defining → the mask shape

New Style button ___   ___ **Set From** button

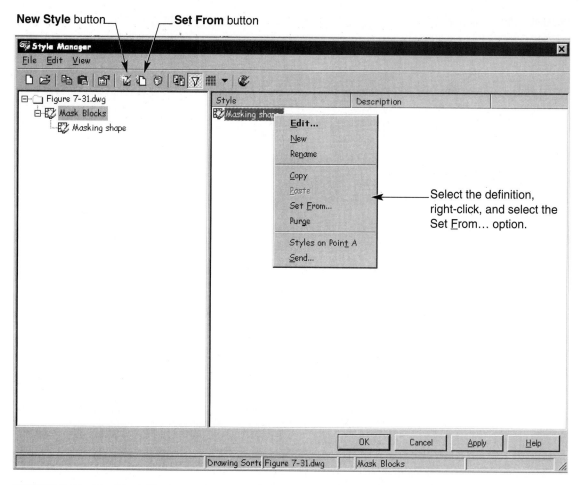

FIGURE 7.32   The Mask Blocks section of the Style Manager. Press the New Style button to create a new definition. Right-click and select Set From… to set the definition from the closed polyline.

Select a closed polyline: *(select the polyline shape)*
Add another ring? [Yes/No] <N>: **N** ⏎
Insertion base point: *(select an insertion point)*
Select additional graphics: *(pick any additional objects to use on the block)*

The first prompt requires you to select the polyline that is to be used as the masking object. After selecting the polyline, you are prompted to add another ring. This allows you to select an inner closed polyline shape that does not mask AEC objects. The insertion point is similar to the insertion point for a standard block; this is, the point at which the block is attached to the crosshairs at the time of insertion. Additional graphics can be selected to enhance the look of the block (see Figure 7.33).

Once the definition has been created, it can be inserted into the drawing and attached to the AEC objects it is supposed to mask. Use one of the following options to insert the mask block:

✓ Select the **Desktop** pull-down menu, pick the **Masking Blocks** cascade menu, then pick **Add Mask Block….**

✓ Pick the **Add Mask Block** button on the **AEC Blocks - Profiles** toolbar.

✓ Right-click, and select **Desktop, Masking Blocks, Add Mask Block….**

✓ Type **AECMASKADD** at the Command: prompt.

The Add Mask Blocks dialog box, shown in Figure 7.34, is displayed. This dialog box is similar to the Add Multi-View Blocks dialog box.

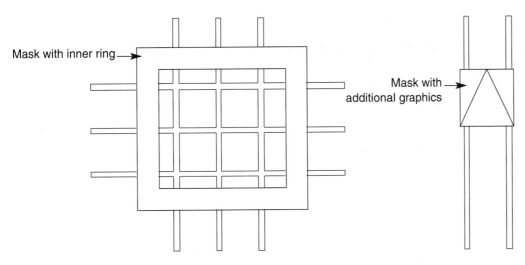

FIGURE 7.33   Some examples of mask blocks using "another ring" and "additional graphics."

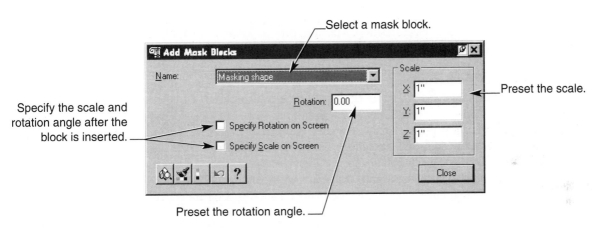

FIGURE 7.34   The Add Mask Blocks dialog box.

After the block has been inserted, it must be attached to the AEC objects that it will mask. One way of attaching the mask block to the AEC objects is to first select the mask block, right-click, and select the Attach option. You are then prompted to select an AEC object. Select the AEC object to be masked. After the object is selected the Select Views dialog box appears with a list display representations for the type of object you selected. Select the desired display representation, then press the OK button to mask the AEC object. In addition, you can use one of the following options to attach a mask block:

✓ Select the **Des̲ktop** pull-down menu, pick the **Masking Blocks** cascade menu, then select **Attach Mask to O̲bjects....**

✓ Pick the **Attach Mask to Objects** button on the **AEC Blocks - Profiles** toolbar.

✓ Right-click, and select **Des̲ktop, Masking B̲locks, Attach Mask to O̲bjects....**

✓ Type **AECMASKATTACH** at the Command: prompt.

The finished mask block after being attached to AEC walls is displayed in Figure 7.35.

The mask block can also be modified after being inserted and attached to objects. The process is similar to modifying a multi-view block. To modify a mask block, select the mask block, right-click, and pick Mask Modify from the cursor menu. The **Modify Mask Blocks** dialog box is displayed. The block used, and the scale and rotation angle, can be modified in the same manner as multi-view blocks. In addition, you can use one of the following options to access the Modify Mask Blocks dialog box:

FIGURE 7.35 The mask block after being attached to AEC objects.

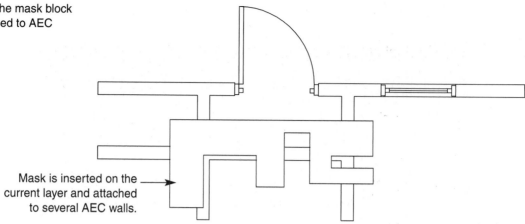

Mask is inserted on the current layer and attached to several AEC walls.

✓ Select the **Desktop** pull-down menu, pick the **Masking Blocks** cascade menu, then select **Modify Mask Block….**

✓ Pick the **Modify Mask Block** button on the **AEC Blocks - Profiles** toolbar.

✓ Right-click, and select **Desktop, Masking Blocks, Modify Mask Block….**

✓ Type **MASKMODIFY** at the Command: prompt.

Adding Custom Symbols to the DesignCenter

Earlier in this chapter you were introduced to the AutoCAD DesignCenter. The Custom view option of the DesignCenter allows you to add AEC blocks into your drawing that are automatically placed onto a layer, and some of which are scaled depending on the drawing scale. Custom blocks, multi-view blocks, and mask blocks can be added to the DesignCenter through a wizard known as **Create AEC Content Wizard**.

Use one of the following options to access the Create AEC Content Wizard utility:

✓ Pick the **Desktop** pull-down menu, and select **Create AEC Content….**

✓ Pick the **Create AEC Content** button in the **AEC Blocks - Profiles** toolbar.

✓ Right-click, and select **Desktop, Create AEC Content….**

✓ Type **AECCREATECONTENT** at the Command: prompt.

The first page of the Create AEC Content Wizard, shown in Figure 7.36, is displayed.

The first page contains the following components:

- **Content Type**   This area contains five separate radio buttons: Block, Drawing, Multi-View Block, Masking Block, and Custom Command. Select the type of content that you want to add to the DesignCenter. The Custom Command option is discussed later in this text.

- **Current Drawing**   After a type of content is selected from the Content Type area, a list of blocks in the current drawing appear in this list box. Select the block name that is to be added into the DesignCenter.

- **Add>>>**   After selecting a block from the Current Drawing list box, use the Add>>> button to add the block to the Content File list box.

- **<<<Remove**   This button is used to remove a block from the Content File list box.

- **Content File**   This list box lists the block that will be added to the DesignCenter.

- **Command String**   This box is used to enter a command macro for a block to perform when it is inserted into the drawing. This section is discussed later in this text.

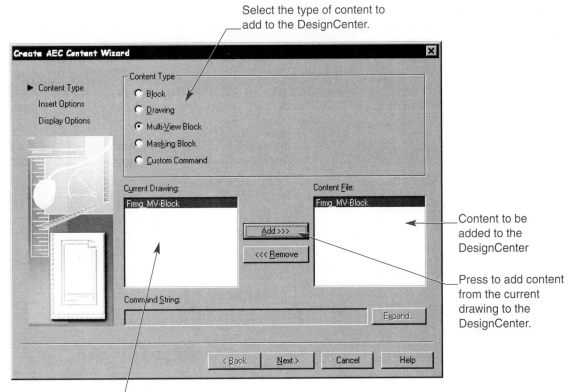

List of blocks, multi-view blocks, and
mask blocks in the current drawing

FIGURE 7.36   The first page of the Create AEC Content Wizard.

When you have finished selecting the options for the first page, press the Next> button. The second page of the Create AEC Content Wizard is displayed (see Figure 7.37). The second page contains the following components:

- **Insert Options**   This area controls how the block will be inserted. The following options are available:

  - **Explode On Insert** When this check box is selected, the block is inserted and exploded. This is generally not a good option to select, as it defeats the purpose of the block symbol. Do not select this option for multi-view blocks or mask blocks, as the blocks will no longer be "smart."

  - **Preset Elevation** An elevation height for the Z axis can be specified in the text box. This may be helpful if a block needs to be placed at a specific height. For example, suppose a block of a potted plant is created, and it needs to sit on top of a countertop that is 36″ off the ground. To do this, enter 36″ for the preset height, and the plant block will be inserted 36″ off the ground.

  - **Anchor Type**   This drop-down list includes a number of AEC objects to which the block can **anchor**. If an object is selected, the block is inserted and tied to the AEC object. If the AEC object is moved, the block remains with the object.

  - **Scale**   This area sets the scale for the inserted block. The following options are available:

    - **X, Y, and Z**   These text boxes allow you to preset a scale along each of the axes as the block is inserted.

Block insertion options

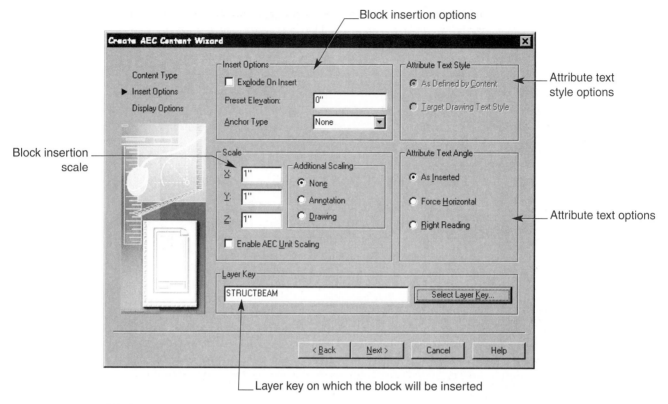

Block insertion scale

Attribute text style options

Attribute text options

Layer key on which the block will be inserted

FIGURE 7.37   The second page of the Create AEC Content Wizard.

- **Additional Scaling** This area contains the following three radio buttons and controls any additional scaling factors:

  - **None** This radio button scales the block based on the values entered in the $\underline{X}$, $\underline{Y}$, and $\underline{Z}$ text boxes.

  - **Annotation** This option scales the block relative to the Annotation Plot Size specified in the Drawing Setup dialog box. This option scales one unit in the block to reflect the Annotation Plot Size units. Use this option when creating blocks of notes.

  - **Drawing** This option scales one unit in the block by the drawing scale factor. The drawing scale is selected in the Drawing Setup dialog box.

  - **Enable AEC Unit Scaling** This check box scales a block drawn with a different set of units, such as millimeters, up to the current Drawing Units setting under the Units tab in the Drawing Setup dialog box. If the block units equal the Drawing Units setting, the block is inserted at the actual scale at which it was drawn.

- **Attribute Text Style** This area is available if the block contains attributes. An **attribute** is a textual piece of information that can be applied to a block to display a note, name, or other value. For more information regarding attributes see the on help in the AutoCAD 2000i user guide. If the block includes attributes, the text can be inserted using the original text style or displayed with the current text style in the current drawing.

- **Attribute Text Angle** This area controls how attributes in a block are displayed after being inserted into the drawing.

■ **Layer Key**   This check box allows you to select a layer key that is assigned to the block content. When the block is inserted from the custom area of the DesignCenter, the block is placed on the layer that is assigned to the layer key in the Layer Key Styles dialog box. Use the Select Layer Key... button to select a layer key in the current drawing.

When you have finished setting the parameters for the block in the second page of the Create AEC Content Wizard, press the Next> button to move to the final page of the wizard, shown in Figure 7.38. This page contains the following options:

■ **File Name**   Use this area to specify the saved location and the name of the block. In order for the custom content to be accessed from the Custom view in the DesignCenter, and to be automatically placed on a layer, the block needs to be saved under the \\Program Files\Autodesk Architectural Desktop 3\Content folder. You may wish to create your own folder named Custom Content under the \\Content folder to store all your customized blocks. Use the Browse... button to browse through the folders on the hard drive to specify the path to the block.

■ **Icon**   This are specifies the preview icon used in the DesignCenter. The **New Icon...** button is used to select a .bmp image to use as a preview image. The **Default Icon...** button uses the current view in the drawing as the preview image for the block. The preview image is displayed to the left of the buttons.

■ **Detailed Description**   This area is used to enter a description for the block that is displayed in the DesignCenter.

■ **Save Preview Graphics** This check box saves the preview image with the block.

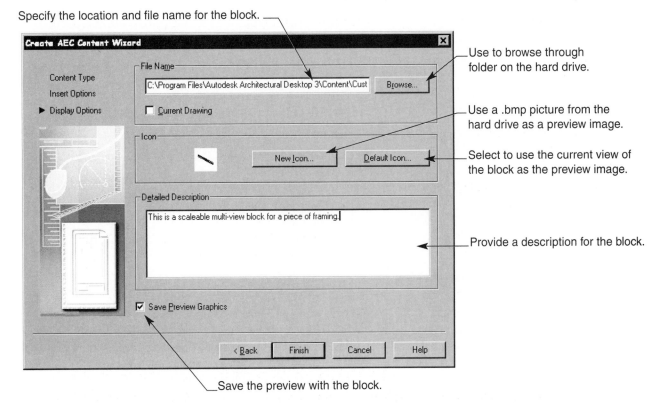

FIGURE 7.38   The third and final page of the Create AEC Content Wizard.

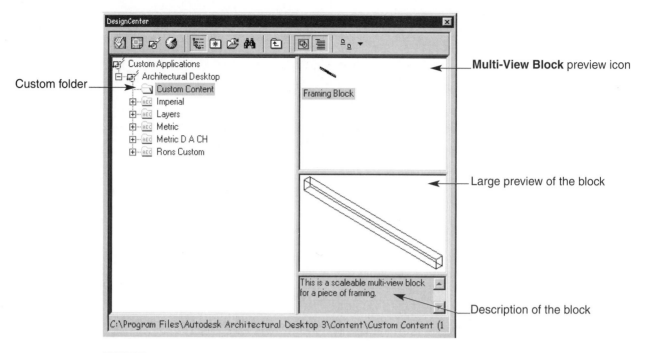

Custom folder

Multi-View Block preview icon

Large preview of the block

Description of the block

FIGURE 7.39   The finished block has been assembled and placed under a Custom Content folder in the DesignCenter.

When you are finished with all the settings for the block, pick the Finish button to have Architectural Desktop process the block and add it to the DesignCenter. Once the block has been processed, the block can be accessed from the Custom view in the DesignCenter (see Figure 7.39). Select the block, and drag and drop it into the drawing, or double-click the preview icon to specify the insertion criteria.

Exercise 7.7

1. Open EX7-6B.
2. Erase one of the blocks in the drawing. Set one of the isometric views current.
3. Access the Create AEC Content Wizard command.
4. Select the Multi-View Block option.
5. Add the framing multi-view block to the content file.
6. Set the scale to X=1, Y=1, Z=1, and the additional scaling to None.
7. Select the Structbeam layer key.
8. Place the custom block in a folder under the \\Program Files\Autodesk Architectural Desktop 3\Content folder, and give the multi-view block a name.
9. Select the Default icon button and enter a description for the block.
10. Process the block by pressing the Finish button.
11. Save the drawing as EX7-7.

Creating AEC Profiles

An *AEC profile* is a predefined shape that can be used in the construction of custom walls, doors, windows, stair railings, fascias and soffits, and mass elements. All these AEC objects use standard shapes but can quickly be modified through the use of a profile. A *profile* is a closed polyline shape that is turned into an AEC profile. Figure 7.40 provides some examples of profiles used to create custom AEC objects.

To create an AEC profile, draw the desired shape or design with a polyline using the Close option to close the polyline. After the shape has been drawn the AEC profile can be created. To create the AEC profile, use one of the following options to display the Style Manager looking at the Profiles section, as shown in Figure 7.41:

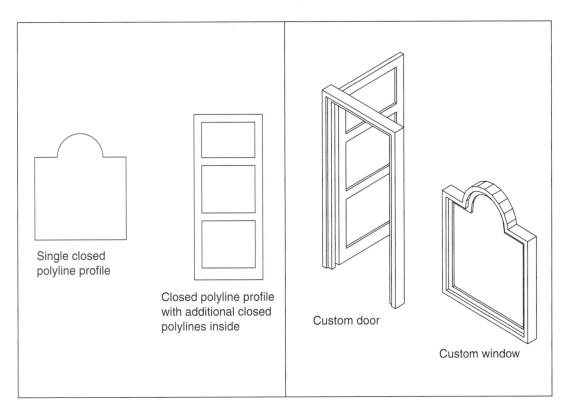

Single closed
polyline profile

Closed polyline profile
with additional closed
polylines inside

Custom door

Custom window

FIGURE 7.40 Some uses for custom AEC profile shapes.

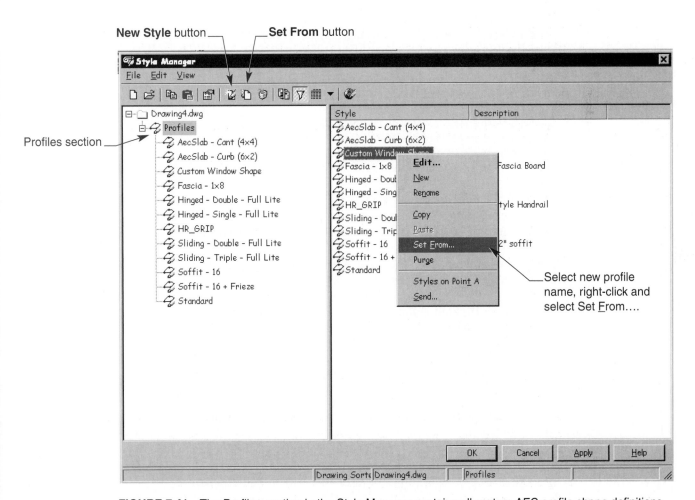

FIGURE 7.41 The Profiles section in the Style Manager contains all custom AEC profile shape definitions.

✓ Select the **Desktop** pull-down menu, pick the **Profiles** cascade menu, and select **Profile Definitions…**.

✓ Pick the **Profile Definitions** button in the **AEC Blocks - Profiles** toolbar.

✓ Right-click, and select **Desktop, Profiles, Profile Definitions…** from the cursor menu.

✓ Type **AECPROFILEDEFINE** at the Command: prompt.

Use the New Style button to create a profile name, or right-click in the right pane and select New. Enter a name for the new shape. After the new profile name has been created, the profile needs to be defined from the polyline shape. Select the profile name, right-click, and select Set From… in the cursor menu, or pick the Set From button. This returns you to the drawing screen, where you are prompted to select a polyline object. The following command sequence is used to define the window shape in Figure 7.40:

```
Select a closed polyline: (pick a polyline shape)
Add another ring? [Yes/No] <N>: N ↵
Insertion Point or <Centroid> (pick an insertion point or press Enter for the center of the
     shape) ↵
Command:
```

This defines the polyline shape as an AEC profile for use in the creation of other objects. Complex profiles such as the door profile can be created as well. The outermost polyline defines the overall shape of the custom object, and the inner polylines cut holes into the shape. The command sequence for the custom door profile is as follows:

```
Select a closed polyline: (pick the outermost polyline shape)
Add another ring? [Yes/No] <N>: Y ↵
Select a closed polyline: (pick an internal polyline shape)
Ring is a void area? [Yes/No] <Y>: Y ↵
Add another ring? [Yes/No] <N>: Y ↵
Select a closed polyline: (pick an internal polyline shape)
Ring is a void area? [Yes/No] <Y>: Y ↵
Add another ring? [Yes/No] <N>: Y ↵
Select a closed polyline: (pick an internal polyline shape)
Ring is a void area? [Yes/No] <Y>: Y ↵
Add another ring? [Yes/No] <N>: N ↵
Insertion Point or <Centroid>: ↵
```

Some custom objects such as the Edge Styles require that the profile shape have a specific insertion point. In most cases the Centroid option can be used when defining custom AEC objects. Creating custom objects from a profile is discussed later in this text. For now, it is important that you understand how to create the profile.

Profiles can be inserted into the drawing as standard polylines so they can be modified and redefined as new AEC profiles. Use one of the following options to insert a profile as a polyline:

✓ Select the **Desktop** pull-down menu, pick the **Profiles** cascade menu, and select **Insert Profile as Polyline…**.

✓ Pick the **Insert Profile as Polyline** button in the **AEC Blocks - Profiles** toolbar.

✓ Right-click, and select **Desktop, Profiles, Insert Profile as Polyline…** from the cursor menu.

✓ Type **AECPROFILEASPOLYLINE** at the Command: prompt.

The Profile Definitions dialog box, where a profile definition can be selected to be inserted, is displayed.

Exercise 7.8

1. Open EX7-4Flr-Pln.
2. Draw the diagram in the figure. It will be used later to custom create a door style.

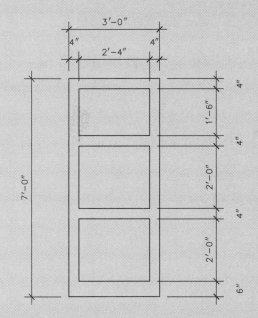

3. Define the polylines as an AEC profile, making the inner polylines voids.
4. Save the drawing as EX7-8Flr-Pln.dwg.

Creating Structural Members

Structural members are objects in Architectural Desktop that allow you to create beams, columns, and braces. The shapes are similar to AEC profiles, representing a cross section of the structural member but are defined through a structural catalog. The *Structural Member Catalog* is organized similar to industry standard structural catalogs. Structural member shapes include concrete, steel, and timber (see Figure 7.42).

Use one of the following options to access the Structural Member Catalog, shown in Figure 7.42:

✓ Pick the **Design** pull-down menu, select the **Structural Members** cascade menu, and select **Structural Member Catalog....**

✓ Pick the **Structural Member Catalog** button from the **Structural Members** toolbar.

✓ Right-click, and select **Design, Structural Members, Structural Member Catalog...** from the cursor menu.

✓ Type **AECSMEMBERCATALOG** at the Command: prompt.

The Structural Member Catalog is divided into three different panes. The left pane includes the structural catalogs for both imperial-unit members and metric members. Pressing the + beside the catalog name expands the tree view of each catalog. The lower right pane provides a list of the structural shapes available

Generate Member Style button   Preview pane

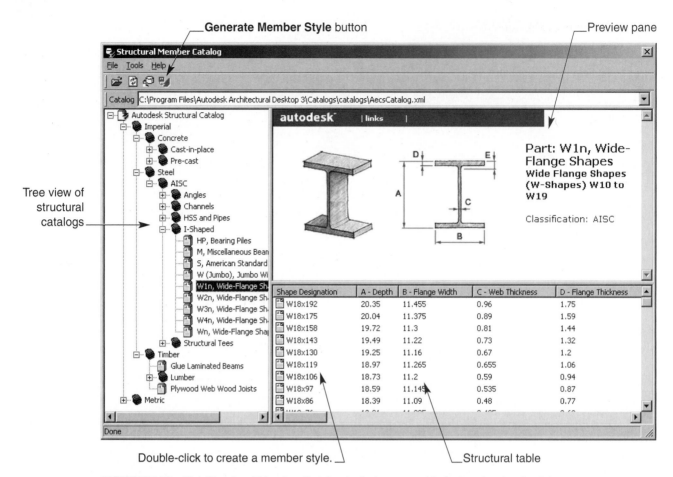

Tree view of structural catalogs

Double-click to create a member style.   Structural table

FIGURE 7.42 The Structural Member Catalog includes several industry structural catalogs.

for each catalog, along with structural information related to the structural member in a table. The upper right pane includes a preview of the type of structural member selected in the lower right pane.

In order to use a structural member in the drawing, you must first create a style for the shape. To create the structural shape style, browse through the catalog for the desired structural member. Once you have found the member, you can create the style by double-clicking on the member in the lower right pane. Highlighting the structural element in the table and picking the Generate Member Style button also creates the member style. The Structural Member Style dialog box is displayed. Enter a name for the structural member style. (Special characters cannot be used in the member style name.) You can add as many structural member styles into your drawing as needed. You can also combine structural members to create a member style composed of more than one structural element.

Exercise 7.9

1. Open EX7-5Fnd.
2. Access the Structural Member Catalog.
3. Select the Imperial\Concrete\Cast-in-place\Circular Columns structural catalog. Create a style of the 20″-diameter column. Name the style 20in Conc Pad.
4. Select the Imperial\Timber\Rough Cut Lumber structural catalog. Create a style of the 4x4R. Name the style 4x4 Post.
5. Select the Imperial\Timber\Rough Cut Lumber structural catalog. Create a style of the 4x8R. Name the style 4x8 Girder.
6. Exit the catalog and save the drawing as EX7-9Fnd.dwg.

Using Structural Members

After some structural member styles have been created, they can be used to design columns, beams, and braces. Structural members can be drawn by specifying a starting and ending point for a structural shape. The distance between the two points is filled with the shape of the structural member style. Structural members can also be created by converting straight, curved, or multisegmented polylines into a structural member. The process for adding a beam, brace, or column is similar for the three objects.

Adding a Beam

A *beam* is a horizontal structural member. Beam objects in Architectural Desktop require a starting and ending point. Use one of the following options to add a beam into the drawing:

✓ Select the **Design** pull-down menu, pick the **Structural Members** cascade menu, and select **Add Beam….**

✓ Pick the **Add Beam** button from the **Structural Members** toolbar.

✓ Right-click, and select **Design, Structural Members, Add Beam...** from the cursor menu.

✓ Type **AECSBEAMADD** at the Command: prompt.

The Add Beam modeless dialog box, shown in Figure 7.43, is displayed.
The following properties need to be set before a beam is drawn:

- **Style:** Select a structural member style to use for the beam. The list displays the styles created from the Structural Member catalog.

- **Start Offset:** This value specifies an offset point from the start of the beam. This is where the starting point of the beam is attached to the crosshairs when the starting location for the beam is specified (see Figure 7.44).

- **End Offset:** This value specifies an offset from the end of the beam. This is where the ending point of the beam is attached to the crosshairs when the ending location for the beam is specified (see Figure 7.44).

- **Justify:** Select a justification for the beam. The blue square indicates the side of the beam that is attached to the crosshairs as the beam is drawn.

- **Roll:** The *roll value* rotates the beam along its length. The *length* is the distance between the starting and ending points of the beam. This is also known as the X axis for the beam (see Figure 7.44).

After specifying the properties for the beam, pick once in the drawing screen to make the drawing active. Select a starting point and ending point for the beam. You can continue to add beams by selecting new starting and ending points. When you are finished drawing beams, press the Close button on the Add Beam modeless dialog box, or press the [Enter] key to end the command.

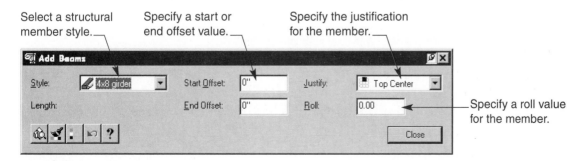

FIGURE 7.43 The Add Beam modeless dialog box is used to create a horizontal beam object.

FIGURE 7.44   The properties of
a beam object.

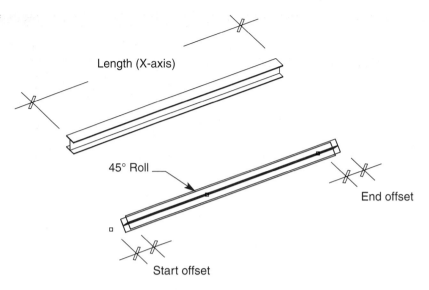

Length (X-axis)

45° Roll

End offset

Start offset

Converting a Polyline to a Beam

Like the wall objects, polylines, lines, and arcs can be converted into structural members. To do this, first draw the polyline shape desired—straight, curved, multisegmented, or a combination of straight and curved, then use one of the following options to convert a polyline into a beam:

- ✓ Select the **Design** pull-down menu, pick the **Structural Members** cascade menu, and select **Convert to Beam....**

- ✓ Pick the **Convert to Beam** button from the **Structural Members** toolbar.

- ✓ Right-click, and select **Design, Structural Members, Convert to Beam...** from the cursor menu.

- ✓ Type **AECSBEAMCONVERT** at the Command: prompt.

After entering the command, select the polylines, lines, or arcs to be converted. When you are finished selecting polylines press [Enter]. You are then asked if you want to erase the layout geometry. Once you have specified whether to erase the layout geometry, the Structural Member properties dialog box is displayed. Select the structural member style to use and press the OK button.

Adjusting the Member Properties

The properties of an individual or a number of structural members can be adjusted by using the Member Properties command. Use one of the following options to access this command:

- ✓ Pick the **Design** pull-down menu, select the **Structural Members** cascade menu, and pick **Member Properties....**

- ✓ Pick the **Member Properties** button from the **Structural Members** toolbar.

- ✓ Right-click, and select **Design, Structural Members, Member Properties...** from the cursor menu.

- ✓ Type **AECSMEMBERPROPS** at the Command: prompt.

FIGURE 7.45 The Structural Member Properties dialog box is used to adjust the properties of one or more structural members.

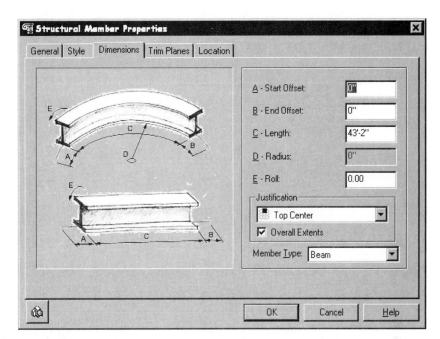

Select the structural members whose properties need to be adjusted and press [Enter]. The **Structural Member Properties** dialog box, shown in Figure 7.45, is displayed. The Structural Member Properties dialog box can also be accessed by first selecting the members, right-clicking, and picking Member Properties… from the cursor menu. The dialog box contains the following tabs:

- **General**   This tab is used to specify a description for the structural member.

- **Style**   This tab includes a list of structural styles in the drawing. Select a different style to change the structural members shape.

- **Dimensions**   This tab controls the dimensional properties for the structural member.

- **Trim Planes**   This tab is used to trim the ends of a structural member. Different angles can be specified to miter the ends of structural members so they fit together accurately.

- **Location**   This tab allows you to specify the location of the structural member in the X, Y, and Z coordinates. When beams, braces, and columns are created, the starting points are placed on the current XY plane. If a beam is to be a specified distance from the ground, modify the Z insertion point to reflect its correct position.

Standard AutoCAD commands also can be used to modify structural members. The TRIM and EXTEND commands can be used to lengthen or shorten the members. Grip editing works to stretch, rotate, and copy the structural members, and the ARRAY or OFFSET commands can be used to create members that need to be parallel, such as in a floor joist layout plan.

Exercise 7.10

1. Open EX7-9Fnd.
2. Add foundation beams to the plan as shown in the figure.
3. Use the 4x8 girder structural member style.
4. Set the justification to Top Center.
5. Use TRIM, EXTEND, and OFFSET as required. The ends of the beams should extend into the foundation walls 3″. Use the grip stretch command to modify the ends of the members.
6. Use the Member Properties command to change the insertion Z axis to 30″. This should make the top of the girders flush with the top of the foundation wall.
7. Use Entity Display to change the plan display representation for the structural style. Set the linetype for the axis component to the Center linetype. Turn off the Visible Comp 1 component.
8. Save the drawing as EX7-10Fnd.dwg.

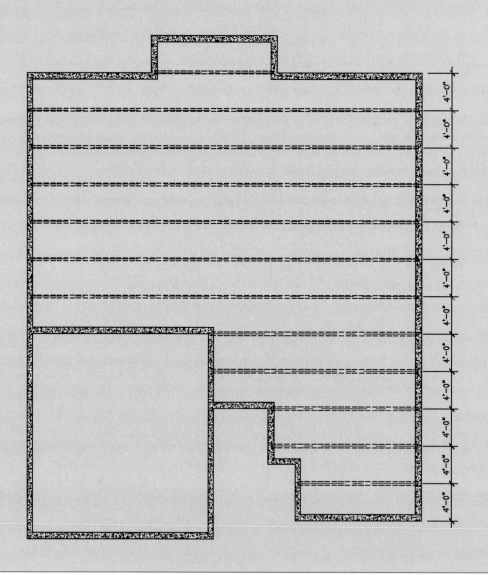

Adding a Brace

A *brace* is a structural element placed at an angle, such as a rafter in a roof. Use one of the following options to add a brace object, which is the same process as adding a beam:

✓ Select the **Design** pull-down menu, pick the **Structural Members** cascade menu, and select **Add Brace....**

✓ Pick the **Add Brace** button on the **Structural Members** toolbar.

✓ Right-click, and select **Design, Structural Members, Add Brace...** from the cursor menu.

✓ Type **AECSBRACEADD** at the Command: prompt.

The **Add Braces** modeless dialog box, shown in Figure 7.46, is displayed.

The **Add Braces** modeless dialog box is similar to the Add Beams modeless dialog box. The following two properties are exclusive to brace structural members:

- ▪ **Rise** This value specifies the total vertical distance between the brace starting point and the brace ending point.

- ▪ **Specify on Screen** This check box allows you to enter the rise for the brace by specifying a 3D point in vertical space, either by entering a 3D coordinate or by snapping to an object at a height different from the start point.

Like beam structural members, braces can be converted from polylines, lines, or arcs. This technique can be effective for creating a curved vaulted ceiling (see Figure 7.47). Use the **Convert to Brace** command to convert lines, arcs, or polylines into a brace with one of the following options:

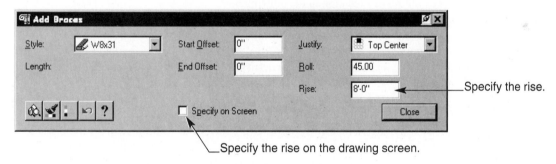

FIGURE 7.46   The Add Braces modeless dialog box is used to specify the properties for brace structural members.

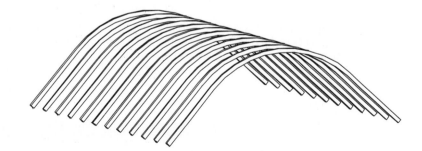

FIGURE 7.47   Example of a brace converted from a curved polyline.

✓ Select the **Design** pull-down menu, pick the **Structural Members** cascade menu, and select **Convert to Brace….**

✓ Pick the **Convert to Brace** button on the **Structural Members** toolbar.

✓ Right-click, and select **Design, Structural Members, Convert to Brace…** from the cursor menu.

✓ Type **AECSBRACECONVERT** at the Command: prompt.

Adding a Column

A **column** is a vertical structural member often used as a post to support a beam or brace. Columns can also be used as vertical stud members in a wall. Use one of the following options to create a column:

✓ Pick the **Design** pull-down menu, select the **Structural Members** cascade menu, and pick **Add Column….**

✓ Pick the **Add Column** button on the **Structural Members** toolbar.

✓ Right-click, and select **Design, Structural Members, Add Column…** from the cursor menu.

✓ Type **AECSCOLUMNADD** at the Command: prompt.

The **Add Columns** modeless dialog box, shown in Figure 7.48, is displayed. The properties for controlling the structural member are the same as in the beams and braces modeless dialog boxes. The Length property is a property that is exclusive to columns. It controls the height of the column object. Keep in mind that the length property is the X axis for the structural member. If the column is rotated horizontally, the height becomes the length for a beam. When adding the column you are prompted to *Select a grid or Return.* Press the [Enter] key to insert the column at a location in the drawing.

Like the beams and braces, columns can be converted from lines, arcs, and polylines using one of the following options:

✓ Select the **Design** pull-down menu, pick the **Structural Members** cascade menu, and select **Convert to Column….**

✓ Pick the **Convert to Column** button on the **Structural Members** toolbar.

✓ Right-click, and select **Design, Structural Members, Convert to Column…** from the cursor menu.

✓ Type **AECSCOLUMNCONVERT** at the Command: prompt.

Complex Structural Member Styles

As indicated earlier in this section, structural member styles are created from the Structural Member Catalog. Several individual members can be combined to form complex structural member styles that are referred to as **structural assemblies**. Architectural Desktop also includes some complex structural assemblies that can be used, and structural assemblies can be downloaded from the Point A Web site.

FIGURE 7.48 The Add Columns modeless dialog box controls the properties for column objects.

Specify the height for the column.

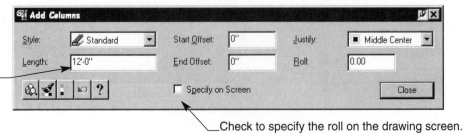

Check to specify the roll on the drawing screen.

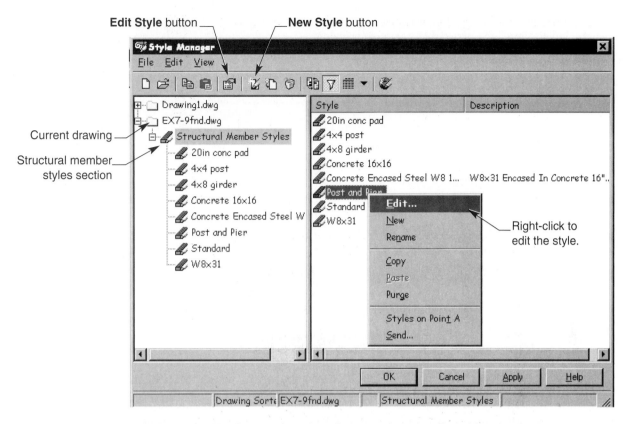

Edit Style button ____ ____**New Style** button

Current drawing

Structural member
styles section

Right-click to
edit the style.

FIGURE 7.49 The Structural Member Styles section in the Style Manager is used to create
and edit structural member styles.

Creating a Custom Structural Assembly

To create a new structural assembly style use one of the following options to display the Style Manager
with the Structural Member Styles section open, as shown in Figure 7.49:

✓ Pick the **Design** pull-down menu, select the **Structural Members** cascade menu, and select
Member Styles….

✓ Pick the **Member Styles** button from the **Structural Members** toolbar.

✓ Right-click, and select **Design, Structural Members, Member Styles…**.

✓ Type **AECSMEMBERSTYLE** at the Command: prompt.

Select the New Style button to create a new structural style. After creating the new style, highlight it,
and right-click to access the cursor menu. Pick the Edit… option. The **Structural Member Style
Properties** dialog box is opened for the style being edited (see Figure 7.50).

The following components can be found in the Structural Member Style Properties dialog box table-
view headers and are used to configure individual structural members in relation to one another in the
assembly:

- **Component**   This is the name of the component within the assembly. The
name does not have to match the actual individual structural
style name being used. Use this column to describe the structural member in the assembly.

- **Start Shape and End Shape**   The table is divided into two main categories, Start Shape
and End Shape. The columns under each area control the
parameters for the start point and end point of each shape in
the assembly. The starting and ending shapes can be the
same, or they can be different.

FIGURE 7.50 The Structural
Member Style Properties dialog
box is used to create structural
assemblies from individual
member styles.

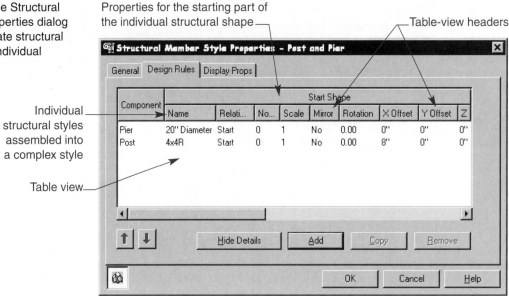

Properties for the starting part of
the individual structural shape

Table-view headers

Individual
structural styles
assembled into
a complex style

Table view

- **Name**   This column can be found in both the Start Shape and the End Shape cate-
                  gories. The name refers to the name of the individual structural style being
                  used. Selecting the name displays a drop-down list with all the individual
                  structural styles available in the drawing. Select the desired structural shape
                  to be added in the assembly.

- **Relative to** This column can be found in both the Start Shape and the End Shape cate-
                  gories. The column specifies the position of the structural shape relative to the
                  starting or ending point of the assembly.

- **Node**   Found in both the Start Shape and the End Shape categories, the node column
                  places structural shapes along different points of the assembly known as
                  nodes. A single segment within an assembly has two nodes, one at the begin-
                  ning of the shape and one at the end of the shape. Thus, an assembly with
                  three shapes has six nodes, one at each end of the shape, that can be assigned
                  an additional shape. The starting point of the first shape in the assembly is
                  known as Node 0. The next point along the assembly is considered Node 2,
                  and so on.

- **Scale**   This column is also found in both the Start Shape and End Shape categories.
                  The column controls the scale of the individual structural member at either
                  end of the shape.

- **Mirror**   This column mirrors the structural shape. This column is also found in both
                  categories.

- **Rotation**   This column rotates the individual shape within the assembly. The column is
                  also found in both categories.

- **X**   This column is found in both categories and offsets the individual structural
                  shape along the X axis, which is the length of the assembly.

- **Y**   This column also is found in the Start Shape and End Shape categories. The
                  values specified offset the structural shape in the Y direction perpendicular to
                  the X axis.

- **Z**   Found in both categories, this column offsets the structural shape in the Z
                  direction, perpendicular to the X axis.

- **Priority**   This column can be found at the end of the table. Each individual structural
                  shape can have a different priority assigned. A high priority is a low number,
                  and a low priority is a high number. Structural shapes that start and end at the

same point are mitered based on their assigned priority. A structural component with a high priority overrides a structural component with a low priority where they cross each other.

- **Add**　　　　This button adds another component or individual structural style to the assembly.

- **Copy**　　　This button allows you to copy the values of an existing component in the assembly for use as another component in the assembly.

- **Remove**　　This button removes individual components from the assembly.

- **Hide Details**　This button hides the table values for the components within the assembly.

Exercise 7.11

1. Open EX7-10Fnd.
2. Create a new Structural Style named Post and Pier. Highlight the style, right-click, and select Edit....
3. Create the style with the following values:

Components	Start Shape								
	Name	**Relative to**	**Node**	**Scale**	**Mirror**	**Rotation**	**X Offset**	**Y Offset**	**Z Offset**
Pier	20″ Diameter	Start	0	1	No	0.00	0″	0″	0″
Post	4x4R	Start	0	1	No	0.00	8″	0″	0″
	End Shape								
	Name	**Relative to**	**Node**	**Scale**	**Mirror**	**Rotation**	**X Offset**	**Y Offset**	**Z Offset**
Pier	20″ Diameter	Start	0	1	No	0.00	8″	0″	0″
Post	4x4R	Start	0	1	No	0.00	1′-10″	0″	0″
	Priority								
Pier	300								
Post	500								

continues on next page

Exercise 7.11 continued

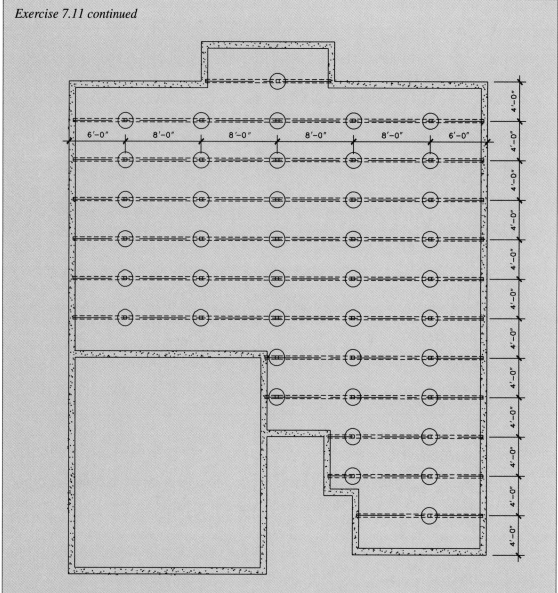

4. Use the Add Column command to add the Post and Pier structural style to the foundation drawing. Make the columns 22″ high. See the figure for placement.
5. Save the drawing as EX7-11.dwg.

Importing Styles

Architectural Desktop includes several preconfigured structural member styles. Some are complex assemblies such as steel bar joists and precast concrete planks, and others are simpler structural shapes such as round columns with flared tops, and concrete encased steel members (see Figure 7.51).

In order to import these styles into the drawing, access the Member Styles... command to open the Styles Manager dialog box. Then use one of the opening options in the Style Manager to open a drawing with additional structural member styles:

✓ Select the File pull-down menu and pick the **Open Drawing...** command.

✓ Pick the **Open Drawing** button.

✓ Select the drawing file folder, right-click, and select **Open....**

FIGURE 7.51 Some additional complex assembly structural shapes included with Architectural Desktop.

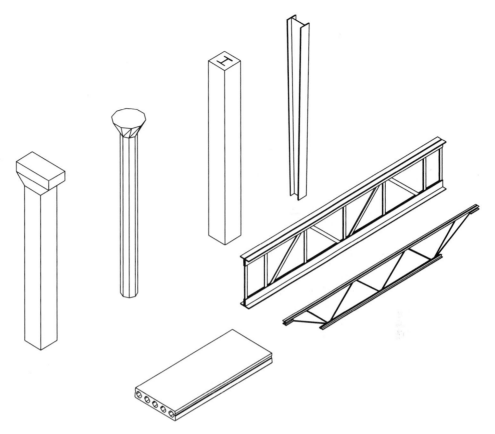

Now, browse through the hard drive to the following path and folder: \\Program Files\Architectural Desktop 3\Content\Imperial\Styles. This folder contains many drawing files full of different types of styles.

There are two drawings full of additional structural members. These drawings are the Bar Joist Styles (Imperial).dwg and the Member Styles (Imperial).dwg. Select one of the drawings, then press the Open button to return to the Styles Manager with the new drawing file opened (see Figure 7.52). To import a style from the newly opened source drawing into your current drawing, expand the Structural Member Styles list in the source drawing, find the desired style, and drag and drop it into your current drawing folder. When you are finished dragging styles from a source drawing, close the drawing by highlighting the folder in the left pane, right-clicking, and selecting the Close option.

Additional styles that are scaled metrically can be found in the \\Program Files\Architectural Desktop 3\Content\Metric\Styles.dwg. If you have installed the international extensions, additional structural styles can be found in the \\Program Files\Architectural Desktop 3\Content\Metric DACH\Styles.dwg.

Another source for additional structural styles is the Point A Web site. Within the Styles Manager dialog box, select the **Styles on Point A** button. The styles on the Point A page are displayed in the right pane, as shown in Figure 7.53. If there are styles available, a preview image of the style is available. Drag and drop the file from the preview image into the current drawing folder to download the structural style. You must have an active Internet connection for this to work. If you do not have an Internet connection, then this page is unavailable.

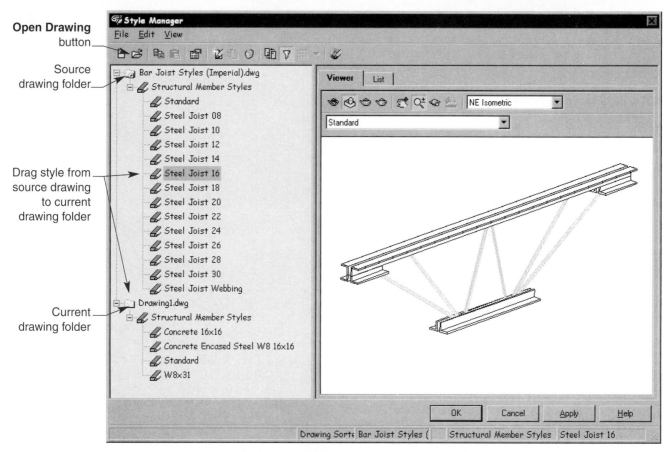

FIGURE 7.52 The Styles Manager displaying the current drawing and a source drawing. Drag and drop the desired style into the current drawing folder.

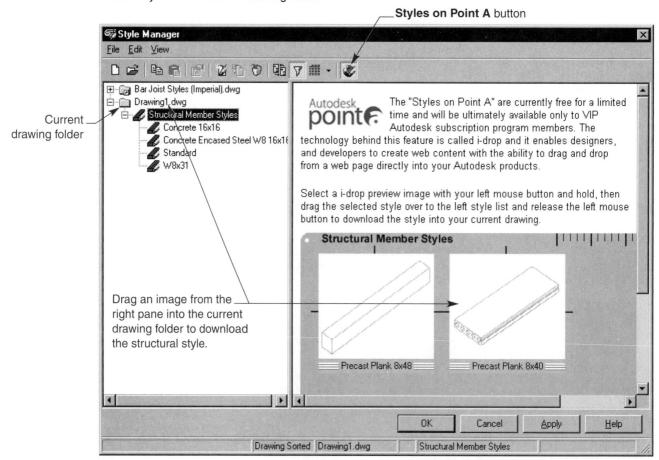

FIGURE 7.53 Styles can be downloaded from the Point A Web site by selecting the Styles on Point A button.

CHAPTER REVIEW

Use the CD-ROM to test your knowledge and skills.

Chapter Test

To check your understanding of the content provided in this chapter, access the Test file in the CH07 folder of the CD-ROM that accompanies this text.

Chapter Project

To practice the Architectural Desktop skills presented in this chapter, access the Project files in the CH07 folder of the CD-ROM that accompanies this text. The project files are in pdf format and include sample drawings and instructions for completing each project.

CHAPTER 8

Dimensioning Drawings

LEARNING GOALS

After completing this chapter, you will be able to:

- ◎ Create AutoCAD dimension styles.
- ◎ Modify dimension styles.
- ◎ Establish dimension overrides.
- ◎ Set a dimension style current.
- ◎ Import dimension styles from other drawings.
- ◎ Use the AutoCAD dimensioning commands.
- ◎ Grip-edit dimensions.
- ◎ Use the Properties command to edit dimensions.
- ◎ Create Architectural Desktop dimensions.
- ◎ Draw AEC dimensions.
- ◎ Create AEC dimension styles.

Dimensions in a set of construction drawings are provided to describe the size, shape, and location of features within a building or structure. Dimensioning is one step in the process of creating construction documents. Dimension objects include numerical sizes with extension lines and arrowheads, leader lines with notes, and tags with a size note.

Architectural Desktop is built on the AutoCAD platform, so the standard dimensioning routines are accessible for the dimensioning of AEC objects. Available commands allow you to dimension linear distances, angles, and arcs. Notes with a leader line and arrow can be added to the drawing to point to a specific feature or area in the drawing. Dimensioning styles are also provided to control the height, width, text style, and spacing of individual components within a dimension. Building on this platform, Architectural Desktop includes automatic dimensioning routines that use dimensioning styles to dimension the AEC objects in the drawing and automatically place them on the appropriate layers.

As geometry in the drawing is dimensioned the dimensions reflect the actual size of the feature being measured. This is important to the drafter, because the accuracy of the objects in the drawing is reflected in the dimensions being created.

Dimension Styles

Dimension styles control the individual elements used with a single dimension object. A dimension style controls through the use of more than 70 settings, which include controlling the height of the text, type of units to be displayed, and the type of arrowheads used. A drawing may have multiple dimension styles each with its own specific settings and uses. A *dimension style* is the saved configurations of these settings.

The use of different dimension styles is required when dimension settings need to reflect a specific appearance for the dimensions in the drawings. For example, a dimension style used on site plans may use the Romans text font placed between dimension lines and capped with arrowheads, as shown in Figure 8.1A. Another dimension style for floor plans may use the Stylus BT text font placed above the dimension

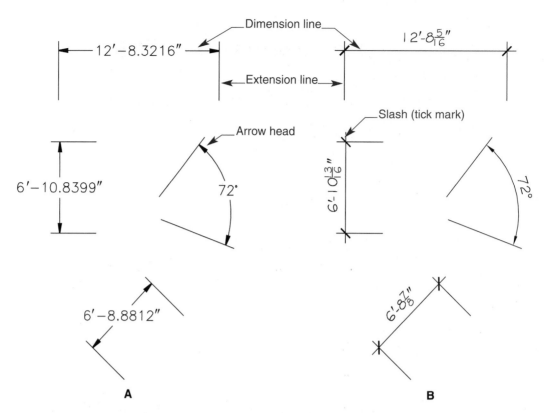

FIGURE 8.1   Different types of dimension styles can be created to produce a unique appearance.

line, with slashes (tickmarks) placed at the dimensioning locations rather than arrowheads (see Figure 8.1B).

Dimension styles are included in the templates provided by Architectural Desktop. The AEC Arch (Imperial) template includes a dimensioning standard named Aec_Arch_I. This dimension style is configured to meet most architectural office standards. Another dimension standard named AEC_Stair is included and is used by the AEC Stair objects. These objects and the use of this style are covered later in this text. If the International Extensions have been installed, use of the AEC Arch (Imperial-Int) template includes the same styles mentioned previously plus two additional dimension styles: AEC Dimension (Large) and AEC Dimension (Small). These two additional styles provide dimensions that are scaled differently than the standard Aec_Arch_I dimension style.

Creating Dimension Styles

Whether you use the dimension styles included in the templates or create your own dimensioning style, it is important to understand the parts of the dimension style and how they affect the outcome of dimensions in the drawing. As the style is created or modified, think about how the style will be implemented in the drawing and the types of objects that will be dimensioned. Once the dimension style has been configured, it should be used throughout the drawing project to maintain a consistent appearance in the construction documents.

Dimension styles are created, modified, and managed using the **Dimension Style Manager** dialog box (see Figure 8.2). Use one of the following options to access this dialog box:

✓ Select the **Dimension** pull-down menu and pick **Style….**

✓ Pick the **Format** pull-down menu and select **Dimension Style….**

✓ Select the **Dimension Style** button in the **Dimension** tool bar.

✓ Type **D, DST, DDIM, DIMSTY,** or **DIMSTYLE** at the Command: prompt.

The Dimension Style Manager dialog box includes the following properties:

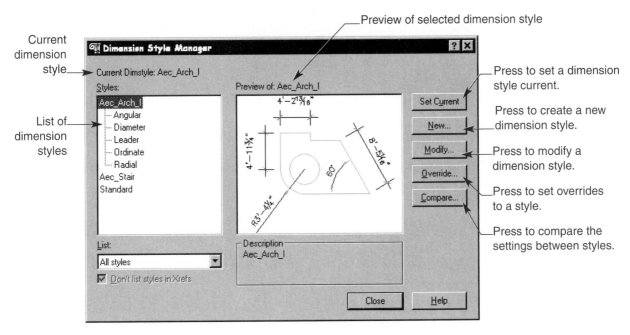

FIGURE 8.2   The Dimension Style Manager dialog box is used to manage and create dimension styles.

- **Styles:**   This list box includes a list of all the dimension styles in the current drawing.

- **List:**   This drop-down list includes options that control the styles listed in the Styles: list box.

- **Don't list styles in Xrefs:** This check box becomes available when Xreferenced drawings are used. If this check box is checked, dimension styles from referenced drawing files will not be displayed. Xrefs are discussed later in this text.

- **Preview of:**   This area provides a preview of the dimensions in the currently selected dimension style in the Styles list box.

- **Description:**   This area provides a description of the selected dimension style.

- **Set Current:**   This button sets the selected dimension style as the current or active dimension style in the drawing. The current style is used for any new dimensions created in the drawing or by dimensions that are updated.

- **New…:**   Select this button to create a new dimension style.

- **Modify…:**   When this button is picked, the selected dimension style from the Styles area can be modified.

- **Override:**   This button allows the selected dimension style to have overrides attached to one or more of its properties.

- **Compare:**   This button compares the property values of two dimension styles.

Creating a New Dimension Style

To create a new dimension style, select the New… button in the Dimension Style Manager dialog box. The Create New Dimension Style dialog box, shown in Figure 8.3, is displayed. This dialog box includes the following options:

FIGURE 8.3 The Create New Dimension Style dialog box is used to create a dimension style.

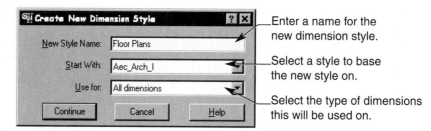

Enter a name for the new dimension style.

Select a style to base the new style on.

Select the type of dimensions this will be used on.

- **New Style Name:**   This text box is used to enter a new name for the dimension style.

- **Start With:**   This drop-down list is used to select an existing dimension style for use as a basis for the new dimension style. The properties from the selected dimension style become the default properties for the new dimension style. If this is a drawing that has been started from scratch, the Standard dimension style is used.

- **Use for:**   This drop-down list allows you to select the type of dimensions this style will be used for. Initially, you must configure a dimension style for *All dimensions* before you can create a style for individual dimension types. These are known as ***parent dimensions***. Parent dimensions set up the rules for all dimensions. After the parent dimensions have been set up, overrides can be applied to dimension types.

After you have finished specifying the parameters in this dialog box, press the **Continue** button to create and configure the new dimension style.

Configuring the New Dimension Style

Once a dimension style has been created and the Continue button pressed, the New Dimension Style dialog box is displayed open to the **Lines and Arrows** tab (see Figure 8.4). The tab is divided into four setup areas plus a preview pane that displays how the dimension style appears on dimensions. The four areas in the Lines and Arrows tab include Dimension Lines, Extension Lines, Arrowheads, and Center Marks for Circles.

Some parameters include text boxes for numerical values, check boxes, radio buttons, and drop-down lists. It is important to note that any numerical value entered should reflect the actual size that can be measured on a finished plotted drawing. For example, if your office standards say that all arrowheads need to be $^1/_8''$ in length on the plotted drawing, then the arrowhead size should indicate $^1/_8''$. The following describes each area of the Lines and Arrows tab:

- **Dimension Lines**   This area is used to control the look and behavior of the dimension line. ***Dimension lines*** are the lines that are drawn between extension lines and have an arrowhead or other object placed at the ends.

  - **Color:**   By default, dimensions are placed on the current layer, which designates the color of the dimension. The dimension line color can be specified here to override the layers color.

  - **Lineweight:**   By default, dimensions use the lineweight of the layer where they are drawn. The lineweight for the dimension line can be overridden in this drop-down list.

  - **Extend beyond ticks:**   This value specifies the distance a dimension line can extend past an arrowhead or tick mark. Place a 0 here for no extension beyond the extension line. Some offices prefer to have the dimension line extend past the extension lines, with $^1/_8''$ commonly used.

  - **Baseline spacing:**   This value specifies the distance between two dimension strings when the baseline dimensioning command is used.

FIGURE 8.4 The Lines and Arrows tab is used to configure the dimension lines, extension lines, arrowheads, and centerline marks.

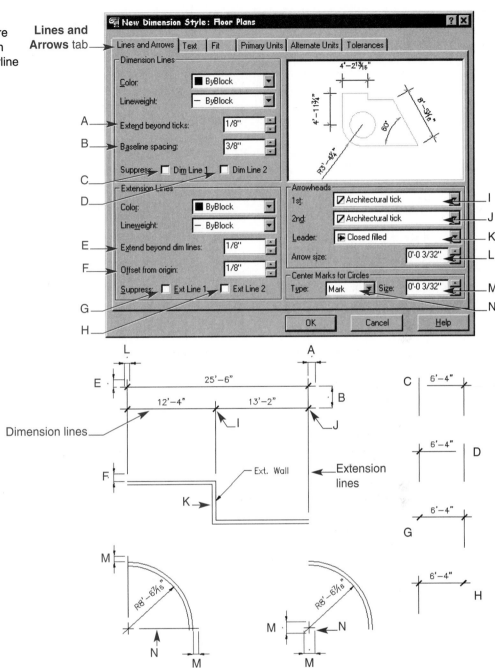

- **Suppress Di̲m Line 1**   A check in this box removes the dimension line for the first dimension line. The first and second dimension lines are indicated by the first and second points picked to be dimensioned.

- **Suppress Di̲m Line 2**   A check in this box suppresses the second dimension line.

- **Extension Lines**   This area is used to control how extension lines appear and behave. *Extension lines* are the lines that are drawn to either side of a dimension line and project from the feature being dimensioned.

- **Colo̲r:**   Similar to the dimension line color, a color can be assigned to extension lines that will override the layer color that they are drawn on.

- **Linewe̲ight:**   The extension line lineweight can be controlled by selecting a value from the drop-down list.

- **E<u>x</u>tend beyond dim lines:** The value entered here controls how far past the dimension line the extension lines will be drawn.

- **O<u>f</u>fset from origin:** This is the distance from the actual point being dimensioned to the beginning of the extension line. A $1/16''$ extension is common.

- **<u>S</u>uppress <u>E</u>xt Line 1** Checking this box removes the extension line for the first point selected.

- **Suppress Ext Line 2** Checking this box removes the second extension line for the second point picked.

- **Arrowheads** This area is reserved for the types of arrowheads to be used for the first and second points of the dimension line, and for leader lines. The size for the arrowheads is also found here:

  - **1s<u>t</u>:** Select an arrowhead to be used for the first point of the dimension line.

  - **2n<u>d</u>:** Select an arrowhead for the second point of the dimension line.

  - **<u>L</u>eader:** Pick the arrowhead to be used at the end of leader lines.

  - **Arrow s<u>i</u>ze:** Enter a value for the length of the arrowhead selected.

- **Center Marks for Circles** This area is used to configure how the center marks at the center of circles or arcs should appear.

  - **T<u>y</u>pe:** This drop-down list includes three options. The **None** option does not create a center mark at the center of a curved feature being dimensioned. The **Mark** option creates a small cross at the center of a curved feature. The **Line** option extends the centerlines past the edge of the curved feature.

  - **Si<u>z</u>e:** This value controls the length of the center mark. If a mark is being created, the size reflects the size of the overall center mark. If a line is being used, the size reflects the distance of the centerline past the edge of the curved feature.

As changes are made the preview pane updates and displays a preview of how the value affects the display of the dimensions. When you are finished, move to the next tab or press the OK button to accept the parameters and return to the Dimension Style Manager dialog box.

Exercise 8.1

1. Start a new drawing using the AEC Arch (Imperial) or the AEC Arch (Imperial-Int) template.
2. Access the Dimension Style Manager dialog box.
3. Create a new style named Floor Plans based on the AEC_Arch_I dimension style. Use it for all dimensions.
4. In the Lines and Arrows tab, make the following adjustments:
   - <u>B</u>aseline spacing: $1/2''$.
   - <u>L</u>eader: Closed filled.
   - Arrow S<u>i</u>ze: $1/8''$.
   - Type: Line.
   - Si<u>z</u>e: $1/8''$.
5. Press the OK button to accept the values.
6. Set the Floor Plans dimension style current. Press Close to return to the drawing.
7. Save the drawing as EX8-1.dwg.

Modifying a Dimension Style

Once a dimension style has been created and configured, it can be modified at any given time. You may find that the values you specified for the dimensions are not reflected as desired on the finished plot, or office standards may have changed to using dots instead of closed filled arrowheads. If you need to modify a dimension style, first access the Dimension Style Manager dialog box, select the dimension style to be modified, then pick the <u>M</u>odify... button. This opens the **Modify Dimension Style** dialog box, shown in Figure 8.5. This dialog box is exactly the same as the New Dimension Style dialog box. Select the tab that needs to be modified and make the desired changes.

The **Text** tab is used to control how the dimension text appears in a dimension. The following values can be adjusted to configure the appearance of the dimension text:

- **Text Appearance**   This area is used to control the appearance of the text within the dimension object.

- **Text style:**   This drop-down list provides a listing of text styles currently available in the drawing. Select a text style for use in the dimension style. If there

FIGURE 8.5   The Modify Dimension Style dialog box is exactly the same as the New Dimension Style dialog box. Make any changes needed to modify the dimension style. This figure displays the Text tab.

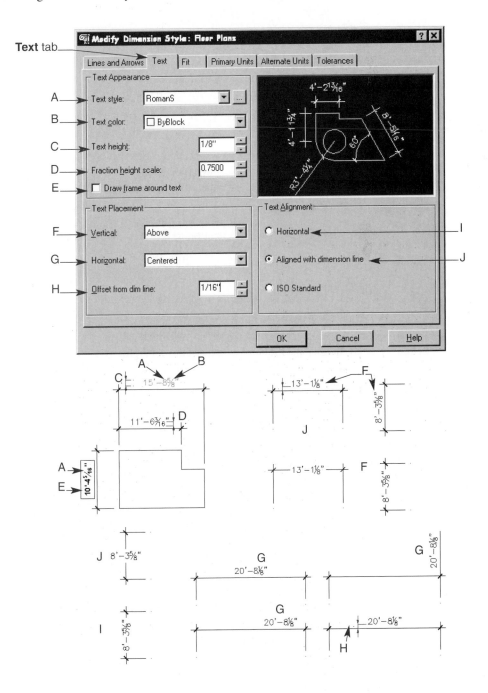

Create a new text style

Rename a text style

Delete a text style

List of text styles in the drawing

Font type for the text style

Apply different effects to control the look of the text style.

Pick to set the text style current in the drawing.

Pick to Close the Text Style dialog box.

Set the text style height (0″ recommended).

Additional styles for the selected font

Preview of the text style

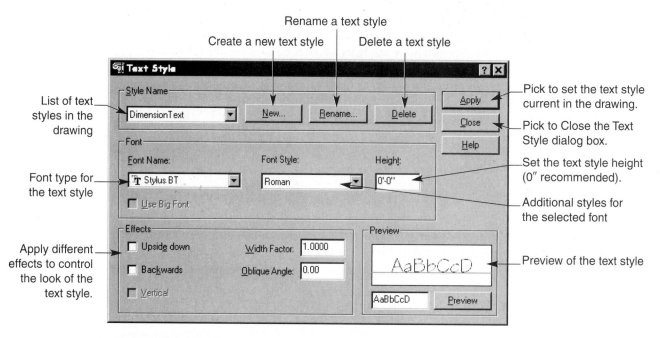

FIGURE 8.6   The Text Style dialog box is used to create a new text style in the drawing for use in notes and dimension styles.

are no text styles available, or if a text style needs to be created for the dimensions, use the ellipsis (…) button to create a new text style. This opens the **Text Style** dialog box (see Figure 8.6). Text styles require that a name be assigned to the text style along with a font type. After creating a new text style or modifying an existing one, press the **C**lose button to return to the Dimension Style dialog box. Set the new or modified style in the drop-down list for the dimension style.

Field Notes

The Text Style dialog box includes a Heig**h**t: text box for controlling the height of text objects. If a height greater than 0″ is specified, all text, including text in the dimensions, reflects the height specified in the text style. It is highly recommended that the height in the Text Style dialog box be left at a value of 0″. This allows you to adjust the text height to a desired value when drawing the text.

- **Text _c_olor**   Use this drop-down list to select a color for the dimension text.

- **Text heig_h_t**   This text box is used to set the height for the text in the dimension string. This should be the actual plotted size, such as $1/8''$ for $1/8''$ high lettering on the finished plot.

- **Fraction _h_eight scale**   This text box controls the height of the individual numbers in fractions. It is common for the fraction numeral height to be the same as the whole number, but some companies use a smaller text height for the fraction numerals.

- **Draw _f_rame around text**   Check to draw a box around the dimension text.

- **Text Placement**   This area is used to place the dimension text in a location relative to the dimension and extension lines.

- **Vertical:** This drop-down list includes four options. **Centered** places the text centered on the dimension line. **Above** places the text above the dimension line. **Outside** places the text to the outside of the dimension. **JIS** Japanese Imperial Standard for text. The Above option is commonly used in Architectural drafting, but some companies prefer other standards.

- **Horizontal:** This drop down list includes five options for placing text horizontally along the dimension or extension line. **Centered** places the text centered on the dimension line. **At Ext Line 1** places the text next to the first extension line. **At Ext Line 2** places the text next to the second extension line. **Over Ext Line 1** places the text along the first extension line. **Over Ext Line 2** places the text along the second extension line.

- **Offset from dim line:** This text box specifies the distance between the text and dimension line. A distance of $1/16''$ is typical.

- **Text Alignment:** This area is used to specify whether the text is aligned with the dimension line or always horizontal in the drawing.

  - **Horizontal** This radio button specifies that the dimension text is always horizontal in the drawing, including vertical dimensions.

  - **Aligned with dimension line** This radio button aligns the text with the dimension line. This option is used in conjunction with the Vertical drop-down list. This option is commonly used for architectural drafting.

  - **ISO Standard** This radio button places the text aligned with the dimension line when the text is between the extension lines. If the text is placed outside the extension lines, the text is forced horizontally.

Exercise 8.2

1. Open EX8-1.dwg.
2. Access the Dimension Style Manager.
3. Select the Floor Plans dimension style in the Styles list. Press the Modify… button to modify the dimension style.
4. Select the Text tab.
5. Change the Text height: to $1/8''$.
6. Press the OK button to return to the Dimension Style Manager dialog box.
7. Press the Close button to return to the drawing.
8. Save the drawing as EX8-2.dwg.

The next tab in the Dimension Style dialog box is the **Fit** tab (see Figure 8.7). This tab is used to control how the dimension text fits in the dimension string, and controls the size of the dimension strings. The following settings are found in the Fit tab:

- **Fit Options** This area controls how dimension text and arrowheads are placed in the drawing. When the extension lines for a dimension are too close together, the dimension text and arrowheads for the dimension need to be placed somewhere in the

FIGURE 8.7 The Fit tab is
used to control how the
dimension text fits in the
dimension string.

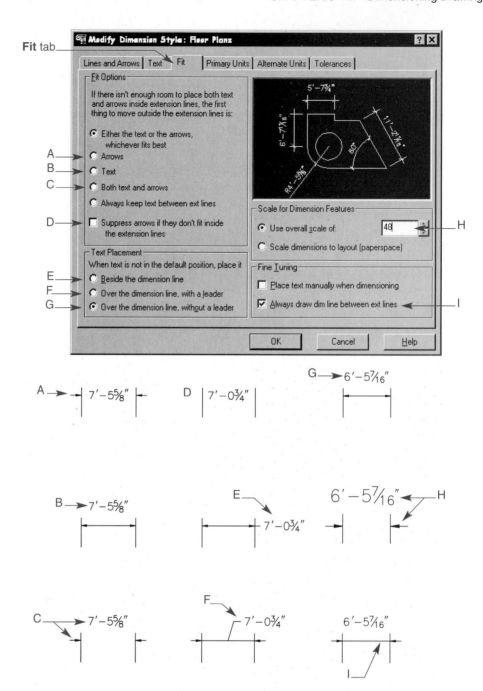

drawing. The following five radio button options plus one
check box control the placement of the text and arrowheads:

- **Either the text or the
  arrows,whichever fits best** When this radio button is selected, either the arrowheads or
  the text is moved outside the extension lines. This option is
  also known as the ***best fit***, because AutoCAD automatically
  decides how to place text, dimension lines, and arrowheads
  during tight situations.

- **Arrows**   When this option is selected, the arrowheads are the first
                                objects to move to the outside of the extension lines during a
                                crowded situation.

- **Text**   This option moves the text to the outside of the dimension
                                before moving the arrowheads outside.

- **Both text and arrows**

  This option moves both the text and the arrows outside the extension lines when the dimension space is too small.

- **Always keep text between ext lines**

  This option always forces the text to be inside the extension lines.

- **Suppress arrows if they don't fit inside the extension lines**

  If the arrowheads do not fit inside the extenstion lines, this option removes the arrows from the dimension.

- **Text Placement**

  This area specifies where the dimension text is placed if it is outside the extension lines.

- **Beside the dimension line** This option places the text beside the dimension line.

- **Over the dimension line with a leader**

  This option places the text over the dimension line with a leader line pointing back to the dimension line.

- **Over the dimension line without a leader**

  This option places the text over the dimension line without a leader line.

- **Scale for Dimension Features**

  This area is used to specify a scale for the dimension objects. This value does not scale the linear length of the dimension, only the sizes of the text and arrowheads and any other values that are set:

- **Use overall scale of:**

  This text box allows you to set the scale factor for the drawing as the scale factor value. If the drawing scale is set to $^1/_4''=1'-0''$, the scale factor equals 48. Enter 48 in this text box. This setting scales the real plotted paper dimension text size and arrowhead size up 48 times in model space. This way, when the drawing is plotted to scale the dimension text and arrowheads appear to be the correct size.

- **Scale dimensions to layout (paperspace)**

  This option is used when dimensioning the drawing through a floating viewport in layout space (paper space). The scale factor of the viewport is used as a multiplication factor for the dimension objects. Layout space and floating viewports are discussed later in this text.

- **Fine Tuning**

  This area is used to set two miscellaneous values for the text and dimensions:

- **Place text manually when dimensioning**

  Typically, the placement of a dimension requires three location picks in the drawing. The first and second picks designate the length being dimensioned. The third pick places the dimension line in relation to the extension lines and automatically adds the dimensional value. If this check box is selected, an additional point needs to be picked that designates the location of the text.

- **Always draw dim line**

  When a dimension is too small, and the text or arrows, or both objects, are placed outside the extension lines, this check box forces a line to be drawn between the extension lines.

Exercise 8.3

1. Open EX8-2.dwg.
2. Access the Dimension Style Manager.
3. Select the Floor Plans dimension style in the <u>S</u>tyles list. Press the <u>M</u>odify... button to modify the dimension style.
4. Select the Fit tab.
5. Change the following options:
   - Select the Arrows option.
   - Over the dimension line, without a leader.
   - Set the overall scale to 48 ($^1/_4''$=1'-0''=48).
   - Select Always draw dim line between ext lines.
6. Press the OK button to return to the Dimension Style Manager dialog box.
7. Press the Close button to return to the drawing.
8. Save the drawing as EX8-3.dwg.

When you have finished configuring the fit values for the placement of text and arrowheads, select the **Primary Units** tab (see Figure 8.8). This tab is used to specify the type of formatting that is used by the dimension text. The following are values that are found in the Primary Units tab:

■ **Linear Dimensions**	This area is used to specify the type of units the dimension text uses as well to control the appearance of the text in the dimension.
■ **<u>U</u>nit format**	This drop-down list controls the type of units that are displayed in the dimension text. Options include Scientific, Decimal, Engineering, Architectural, Fractional, and Windows Desktop.
■ **<u>P</u>recision**	This drop-down list controls the how accurately the dimension values are rounded when measured between two points.
■ **Fraction for<u>ma</u>t:**	This drop-down list controls how fractions appear when Architectural or Fractional units are used. The options include Horizontal, Diagonal, and Not Stacked. A *not-stacked* fraction is a fraction that appears along the same line as the main dimension text.
■ **De<u>c</u>imal Separator:**	This option is used with the Decimal units. This drop-down list is used to control the separator between the whole number and the decimal number. Options include period, comma, or space.
■ **Round off:**	This text box allows the dimension to be rounded up or down.
■ **Pre<u>fi</u>x:**	This value is placed in front of the dimension text.
■ **<u>S</u>uffix:**	This value is placed at the end of the dimension text.
■ **Measurement Scale**	This area is used to scale the dimension size. For example, a value of 4 makes a 2'-0''-long dimension read as 8'-0''. It is important to note that this is not the drawing scale factor and does not scale the text height or arrowhead length. The **Sca<u>l</u>e factor** text box is used to enter a scale to be used as the multiplication factor. The **Apply to layout dimensions only** option scales the length of the dimensions when the dimensions are placed in the layout.
■ **Zero Suppression**	This area is used to suppress zeros in the dimension text. The **<u>L</u>eading** and **<u>T</u>railing** check boxes are used for decimal units, and the **<u>F</u>eet** and **<u>I</u>nches** check boxes are used for engineering and architectural units. When a check is placed in the <u>L</u>eading or <u>F</u>eet check box, any zeros in the whole number or the feet column

Primary Units tab

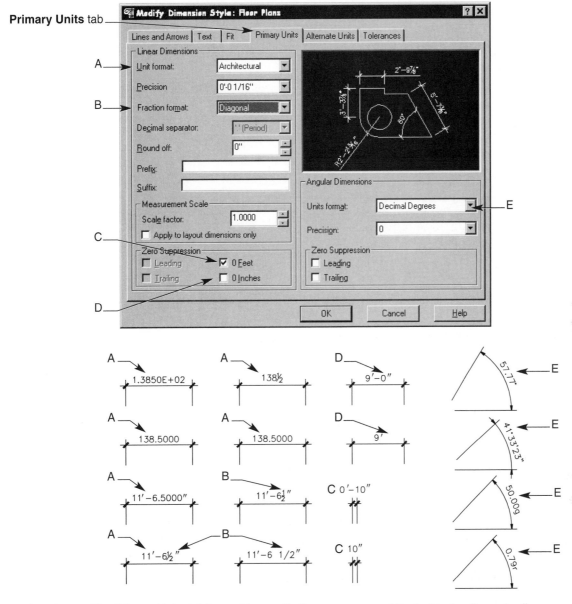

FIGURE 8.8   The Primary Units tab is used to specify the appearance of the text in the dimension line.

are not displayed on the dimension. If the Trailing or Inches is checked, the zeros at the end of the dimension are suppressed.

- **Angular Dimensions**   This area controls the appearance of the text in an angular dimension.

- **Units format:**   This selects the types of units to display for angular dimensions: Options include Decimal degrees, Degrees Minutes Seconds, Gradians, and Radians.

- **Precision:**   This area controls how angular dimension text is rounded.

- **Zero Suppression**   Similar to the Zero Suppression for linear dimensions, the **Leading** and **Trailing** check boxes are used to remove zeros in the dimension text.

The **Primary Units** tab sets up the primary or main dimension text for the dimensions. The **Alternate Units** tab is used to add another element of text in the dimension that is scaled differently. For example,

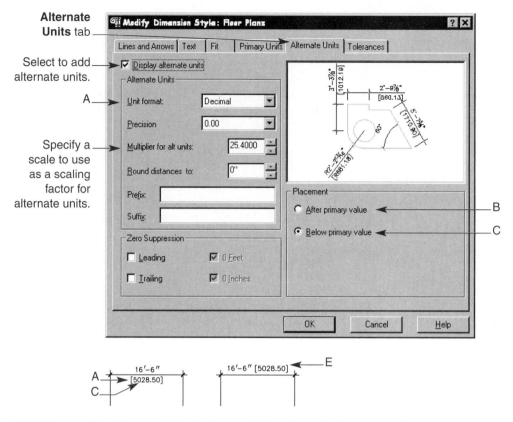

FIGURE 8.9   The Alternate Units tab places additional text that is set to a different scale next to the primary unit text.

suppose the Primary Units is set to Architectural units dimensioning in feet and inches. Alternate Units can be used to display the metric equivalent of the feet-and-inches dimension. Alternate units are specified with squared brackets around the dimensional value (see Figure 8.9). The following options are used to set the alternate units if used:

- **Display alternate units**   Select this check box to turn on alternate units.

- **Alternate Units**   This area is used to set the parameters for alternate text in the dimension:

  - **Units format:**   This drop-down list includes the unit formats for the alternate dimension text.

  - **Precision:**   This drop-down list specifies how alternate unit text will be rounded.

  - **Multiplier for alt units:**   This text box is used as a multiplier for the resulting alternate text. For example, given a dimension of a 12'-0" length and a multiplier of 2, the resulting alternate text will be 24'-0". By default the multiplier 25.4 is used to convert inches to millimeters (25.4 × inches = millimeters). When this multiplier is used on alternate text, the resulting text is the metric equivalent of the primary units.

  - **Round Distances:**   This text box value rounds the alternate unit text to the nearest value specified.

  - **Prefix:**   This value is placed in front of the alternate units.

  - **Suffix:**   This value is displayed at the end of all alternate unit text.

- **Zero Suppression**   As in the Primary Units tab, this area specifies how zeros in a dimension are displayed or suppressed.

- **Placement**   This area includes two options for the placement of the alternate text in the dimension. The **<u>A</u>fter primary value** radio button places the alternate unit text beside the primary unit text. The **<u>B</u>elow primary value** radio button places the alternate unit text below the primary unit text.

Exercise 8.4

1. Open EX8-3.dwg.
2. Access the Dimension Style Manager.
3. Select the Floor Plans dimension style in the <u>S</u>tyles list. Press the <u>M</u>odify... button to modify the dimension style.
4. Select the Primary Units tab.
5. Change the following options:
   - <u>U</u>nit Format: Architectural
   - <u>P</u>recision: $^1/_{16}''$
   - Sca<u>l</u>e Factor: 1
6. Do not use Alternate units.
7. Close the dialog box, and save the drawing as EX8-4.dwg.

The last tab in the Dimension Style dialog box is the **Tolerances** tab. This tab is used to display a tolerance after a dimensional value. Tolerances are used primarily in the mechanical engineering field. For more information regarding tolerances see the section on *Adding Tolerances to Dimensions* in the AutoCAD 2000i Help\ User's Guide.

Setting Dimension Overrides

Once a dimension style has been set up, the dimension style is used for all types of dimensions on your drawing. For example, if architectural ticks have been specified as the arrowhead choice, then linear dimensions, angular dimensions, and radius and diameter dimensions all use architectural ticks for the arrows (see Figure 8.10). You may want to use the architectural tick for the linear and diameter dimensions but closed filled arrows for angular and radius dimensions. In order to specify how a particular type of dimension is supposed to be displayed, overrides need to be set.

Parent dimensions were set up earlier. These were a set of rules that governed how all dimensions appear. When overrides are applied to different types of dimensions, the dimensions break the rules set up by the parent dimensions. Applying the overrides is known as setting up the ***child dimensions***.

FIGURE 8.10   (A) Dimensions using the same dimension style are all formatted similarly. **(B)** Dimensions with overrides attached to give a more desirable appearance.

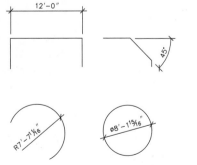

Dimensions using the same formatting

A

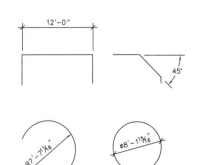

The angular and radius dimensions have had overrides attached.

B

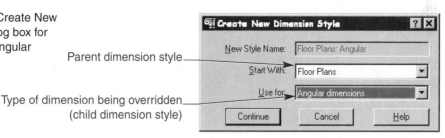

FIGURE 8.11   The Create New
Dimension Style dialog box for
setting overrides to angular
dimensions.

Parent dimension style

Type of dimension being overridden
(child dimension style)

The Dimension Style Manager is used to begin making override settings for the dimensions. To create an override for the radius dimension in Figure 8.10, select the New… button to create a new dimension style. In the **Create New Dimension Style** dialog box, select the dimension style that is being overridden in the **Start With:** drop-down list, shown in Figure 8.11. In the **Use for:** drop-down list, select the type of dimension you are setting an override to. Note that specifying a dimension type other than "All dimensions" grays out the New Style Name: text box. Once the Continue button is pressed, the New Dimension Style dialog box is displayed.

This procedure is similar to specifying the formatting for all dimensions, except that only the values specific to the type of dimension you are overriding are available. The preview pane displays only the type of dimensioning that is being overridden. Make any changes desired for the type of dimension you are overriding, then press the OK button.

The Dimension Style Manager dialog box reappears, displaying the parent dimension style with the overriding style beneath, as shown in Figure 8.12. Continue adding overrides to dimension types until you have configured all the dimension styles and overrides.

Setting the Dimension Style Current

After a dimension style is created, it must be set current before it can be used for the dimensions. To set a dimension style current, select the style from the Styles: list in the Dimension Style Manager dialog box, then press the Set Current button. Dimension styles can also be set current by selecting the dimension style from the drop-down list on the Dimension toolbar (see Figure 8.13).

Only parent styles can be set current. Any overrides set are automatically assumed to be a part of the parent style. All dimensions added to the drawing are automatically configured to reflect the current dimension settings.

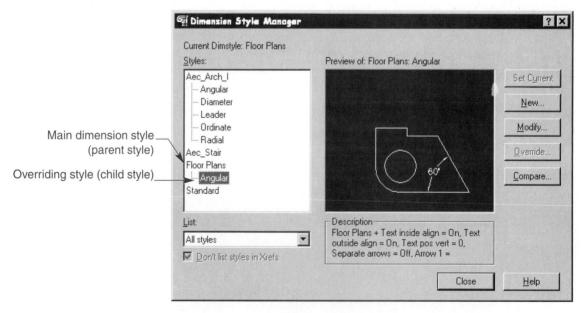

Main dimension style
(parent style)

Overriding style (child style)

FIGURE 8.12   The Dimension Style Manager displaying the dimension style
with a child overriding style applied.

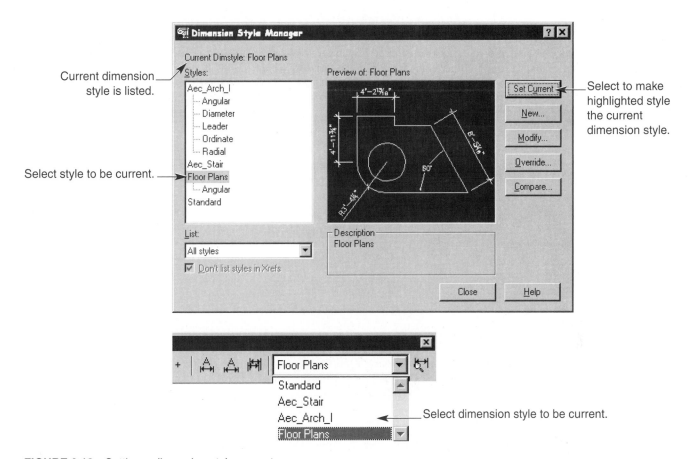

FIGURE 8.13 Setting a dimension style current.

Exercise 8.5

1. Open EX8-4.dwg.
2. Access the Dimension Style Manager.
3. Create a new dimension style starting with the Floor Plans dimension style.
4. In the Use for: drop-down list, select Angular dimensions.
5. Change the arrowheads to closed filled arrows. Change the vertical text placement of the text from Above to Centered. Change the text alignment to horizontal.
6. Create another new dimension style starting with the Floor Plans dimension style and using it for radius dimensions.
7. Change the arrowheads to closed filled arrows. Change the vertical text placement of the text from Above to Centered. Change the text alignment to horizontal.
8. Set the Floor Plans style current.
9. Save the drawing as EX8-5.dwg.

Field Notes

When the templates supplied with Architectural Desktop are used, a standard dimension style named AEC_Arch_I is available and has been set as the current dimension style. There are also overrides that have been created within the template. Save time by using this dimension style as a basis for creating your dimension standard.

Using Dimension Styles in Other Drawings

Once you have created a dimension style and any overrides, the style is part of the drawing. Dimension styles can be used in other new or existing drawings by importing the dimension style. To import a dimension style into a new or existing style, access the AutoCAD DesignCenter.

Once the DesignCenter has been opened, select the Desktop icon. Use the tree view to browse through the folders to the drawing that contains the desired dimension style. If the drawing that contains the desired dimension style is already open, use the Open Drawings icon to view the drawings contents (see Figure 8.14). With the drawing open, select the dimensions icon in the tree view or in the preview pane. A list of the dimension styles available is displayed. Drag and drop the desired style into the current drawing to import the style.

If a dimension style has overrides, then the overrides appear in the DesignCenter with the name of the dimension style followed by a dollar sign ($) and a number. If you need to use the dimension style and the overrides in another drawing, drag and drop the dimension style and overrides separately into the current drawing.

Exercise 8.6

1. Open EX7-8-Flr_Pln.dwg.
2. Use the DesignCenter to browse for the EX8-5.dwg.
3. Drag and drop the Floor Plans, Floor Plans$2, and the Floor Plans$4 dimension styles into the EX7-8-Flr-Pln.dwg. Save the EX7-8-Flr-Pln drawing as EX5-6-Flr-Pln.dwg. Close the drawing.
4. Open EX8-5.dwg.
5. Open EX7-10-Fnd.dwg
6. Use the DesignCenter's Open Drawing view to drag and drop the Floor Plans, Floor Plans$2, and the Floor Plans$4 dimension styles into the EX7-10-Fnd.dwg.
7. Save the EX7-10-Fnd drawing as EX8-5-Fnd.dwg.

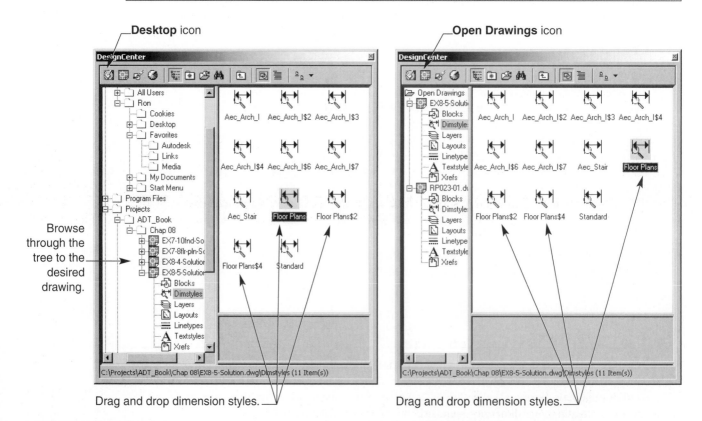

FIGURE 8.14   Drag and drop dimension styles through the AutoCAD DesignCenter into the current drawing.

Using the Dimensioning Commands

The previous section explained how to create and modify a dimension style. Once a dimension style has been created and set current, dimensions can be drawn that reflect the settings assigned by the dimension style. Standard AutoCAD has a number of dimensioning commands that can be used to draw dimensions. These commands can be found in the Dimension pull-down menu and in the Dimension toolbar (see Figure 8.15).

When a drawing is dimensioned using the commands from the pull-down menu or the toolbar, the dimension object is added to the drawing and placed on the current layer. Additionally, six of the standard AutoCAD dimensioning commands have been added to the DesignCenter. As dimensioning commands are dragged from the DesignCenter and dropped into the drawing, the dimensions are placed in the drawing with a dimensioning layer assigned to the A-Anno-Dims layer. These dimensioning commands include Linear, Continuous, Baseline, Angular, Radius, and Aligned, and they are the most common AutoCAD dimensioning commands used in architectural drawings.

Explanations of these six dimensioning commands follow. For more information on other AutoCAD dimensioning commands refer to the AutoCAD user's guide in the on-line help.

Drawing Linear Dimensions

The Linear Dimension command creates horizontal or vertical dimensions. Use one of the following options to access this command:

✓ Pick the **Dimension** pull-down menu, and select **Linear**.

✓ Pick the **Linear Dimension** button on the **Dimension** toolbar.

✓ Type **DIMLIN** or **DIMLINEAR** at the Command prompt.

All these options access the standard AutoCAD Linear Dimension command, which places the linear dimension on the current layer. Additionally, you can select the Linear Dimension from the Documentation pull-down menu, Documentation Content, and select Miscellaneous..., which opens the DesignCenter. Then, pick the Dimensions folder and select the Linear icon. You also can use the Miscellaneous button found on the Documentation-Imperial toolbar to access the Miscellaneous folder in the DesignCenter.

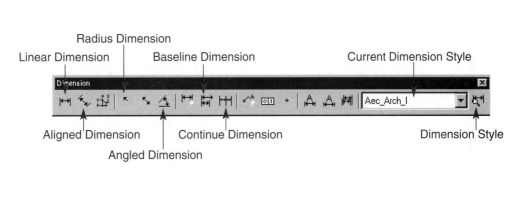

FIGURE 8.15 Standard AutoCAD dimensioning commands can be found in the Dimension pull-down menu and in the Dimension toolbar.

FIGURE 8.16 The Linear Dimension command is used to create horizontal and vertical dimensions in your drawing.

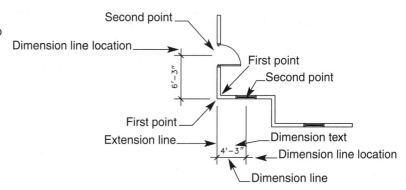

To create a linear dimension, you are prompted for two points that designate the points that are being dimensioned. Pick the first point where the first extension line will be connected. Pick the second point for the second extension line. After the second point has been selected, you are prompted for the dimension line location. The position that is picked places the dimension line at that location. When a dimension with an extension line is added to the drawing, a point is placed at the location selected. The point indicates the actual point being dimensioned, and the extension line is offset a distance away from that point. These points are placed onto a layer automatically named *defpoints*, which stands for *definition point*, as the points selected define the dimension extents. The defpoints layer is a non-plotting layer, so the points do not plot on the finished drawing. The following prompt sequences were used to create a horizontal and vertical dimension for Figure 8.16:

```
Command: DIMLIN or DIMLINEAR ↵
DIMLINEAR
Specify first extension line origin or <select object>: (pick the first point to be dimensioned)
Specify second extension line origin: (pick the second point to be dimensioned)
Specify dimension line location or
[Mtext/Text/Angle/Horizontal/Vertical/Rotated]: (pick the dimension line location)
Dimension text = 4'-3"

Command: ↵
Specify first extension line origin or <select object>: (pick the first point to be dimensioned)
Specify second extension line origin: (pick the second point to be dimensioned)
Specify dimension line location or
[Mtext/Text/Angle/Horizontal/Vertical/Rotated]: (pick the dimension line location)
Dimension text = 6'-3"
Command:
```

When you are specifying the dimension line location, pulling the mouse to the left or right turns the linear dimension into a vertical dimension; pulling the mouse up or down creates a horizontal dimension. Once the dimension line location has been specified, a dimension object is created with the appropriate text annotated in the dimension.

Before the dimension line location is picked, the following options are available for the linear dimension:

- **Mtext:**   Entering **M** at the prompt opens the **Multiline Text Editor** window, shown in Figure 8.17. This window is used to enter dimension text that is different from the default text. The default text appears in brackets (<>), which indicates that the text reflects the actual length that is being measured. If the brackets are removed, the text entered is displayed between the extension lines. You can also keep the brackets and enter additional information or values that are added to the default text.

- **Text:**   Entering **T** at the prompt displays the dimension length in brackets at the Command: prompt. Entering a different value overrides the text and displays the overridden text between the extension lines.

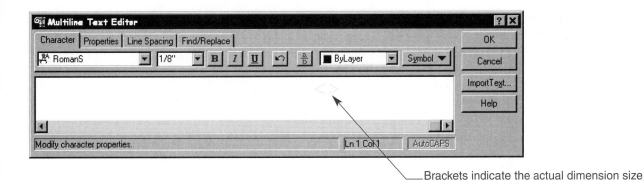

Brackets indicate the actual dimension size
that will be used in the dimension object.

FIGURE 8.17 The Multiline Text Editor allows you to make adjustments to the text in the dimension.

FIGURE 8.18 Linear
dimensions can be rotated to
dimension the length of angled
features.

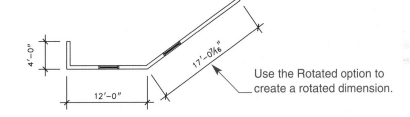

Use the Rotated option to
create a rotated dimension.

- **Angle:** Entering **A** at the prompt allows you to adjust the angle of the text between the extension lines.

- **Horizontal:** Entering **H** at the prompt creates a horizontal dimension. This overrides the default of moving the mouse up or down to create a horizontal dimension.

- **Vertical:** Entering **V** at the prompt creates a vertical dimension.

- **Rotated:** Entering **R** at the prompt rotates the angle of the dimension line. Use this option when dimensioning lengths of a building that are not horizontal or vertical (see Figure 8.18).

Field Notes

When creating dimensions in the drawing, use object snaps to snap to the geometry being dimensioned. This ensures accuracy of the dimension.

Drawing Continuous Dimensions

The **Continue Dimension** command builds on an existing linear dimension. Continuous dimensions create one "string" of dimensions in which all the dimension lines are lined up. Use one of the following options to access this command:

✓ Select the **Dimension** pull-down menu, and pick **Continue.**

✓ Pick the **Continue Dimension** button on the **Dimension** toolbar.

✓ Type **DIMCONT** or **DIMCONTINUE** at the Command prompt.

These options are used to place continuous dimensions on the current layer. In addition, you can select the Continue Dimension command from the Documentation pull-down menu, Documentation Content, and

FIGURE 8.19 Create a linear dimension before using the Continue Dimension command. Pick the second extension points for new continued dimensions.

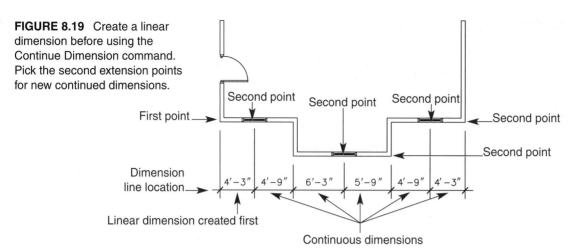

select <u>M</u>iscellaneous…, which opens the DesignCenter. Then, pick the Dimensions folder and select the Continue icon. You also can use the Miscellaneous button found on the Documentation-Imperial toolbar to access the Miscellaneous folder in the DesignCenter.

If you create a linear dimension prior to entering the Continue Dimension command, you are prompted to pick a second extension line point. The command assumes that the last point selected for the previous linear dimension is the first point of the continued dimension, so the only requirement is the second extension line point. The following prompt sequence illustrates the creation of continuous dimensions in Figure 8.19:

Command: **DIMLIN** or **DIMLINEAR** ↵
DIMLINEAR
Specify first extension line origin or <select object>: *(pick the first point to be dimensioned)*
Specify second extension line origin: *(pick the second point to be dimensioned)*
Specify dimension line location or
[Mtext/Text/Angle/Horizontal/Vertical/Rotated]: *(pick dimension line location)*
Dimension text = 4'-3"

Command: **DIMCONT** or **DIMCONTINUE** ↵
DIMCONTINUE
Specify a second extension line origin or [Undo/Select] <Select>:*(pick the second point to be dimensioned)*
Dimension text = 4'-9"
Specify a second extension line origin or [Undo/Select] <Select>:*(pick the second point to be dimensioned)*
Dimension text = 6'-3"
Specify a second extension line origin or [Undo/Select] <Select>:*(pick the second point to be dimensioned)*
Dimension text = 5'-9"
Specify a second extension line origin or [Undo/Select] <Select>:*(pick the second point to be dimensioned)*
Dimension text = 4'-9"
Specify a second extension line origin or [Undo/Select] <Select>:*(pick the second point to be dimensioned)*
Dimension text = 4'-3"
Specify a second extension line origin or [Undo/Select] <Select>:↵
Select continued dimension: ↵
Command:

FIGURE 8.20 Selecting an extension line on an existing dimension designates the extension line as the first extension line for the new continued dimension.

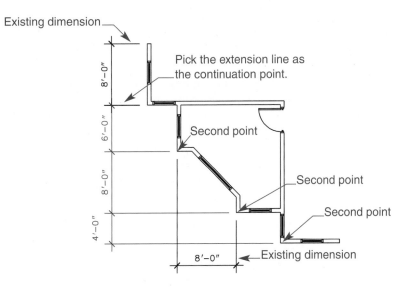

Linear dimensions do not always need to be created prior to using the Continue Dimension command. As long as there is a dimension in the drawing, the Continue Dimension command can be used. When you first access the Continue Dimension command, you are prompted to *Specify a second extension line origin or [Undo/Select] <Select>:*. Pressing the [Enter] key or right-clicking allows you to select an existing dimension's extension line. This designates it as the first extension line of the new continued dimension. Once you have selected an extension line to be used as the starting point for the new dimension, begin picking the second extension line locations for the new dimensions. The following prompt sequence is used to create continued dimensions from an existing dimension, as seen in Figure 8.20.

Command: **DIMCONT** or **DIMCONTINUE** ↵
DIMCONTINUE
Specify a second extension line origin or [Undo/Select] <Select>: ↵
Select continued dimension: *(pick an extension line to use as the continuation point)*
Specify a second extension line origin or [Undo/Select] <Select>: *(pick a second extension line point)*
Dimension text = 6′-0″
Specify a second extension line origin or [Undo/Select] <Select>:*(pick a second extension line point)*
Dimension text = 8′-0″
Specify a second extension line origin or [Undo/Select] <Select>:*(pick a second extension line point)*
Dimension text = 4′-0″
Specify a second extension line origin or [Undo/Select] <Select>:*(pick a second extension line point)*
Select continued dimension: ↵
Command:

Exercise 8.7

1. Open EX8-6-Fnd.dwg.
2. Use the Linear Dimension and Continue Dimension commands to dimension the foundation plan as illustrated in the figure.
3. Save the drawing as EX8-7-Fnd.dwg.

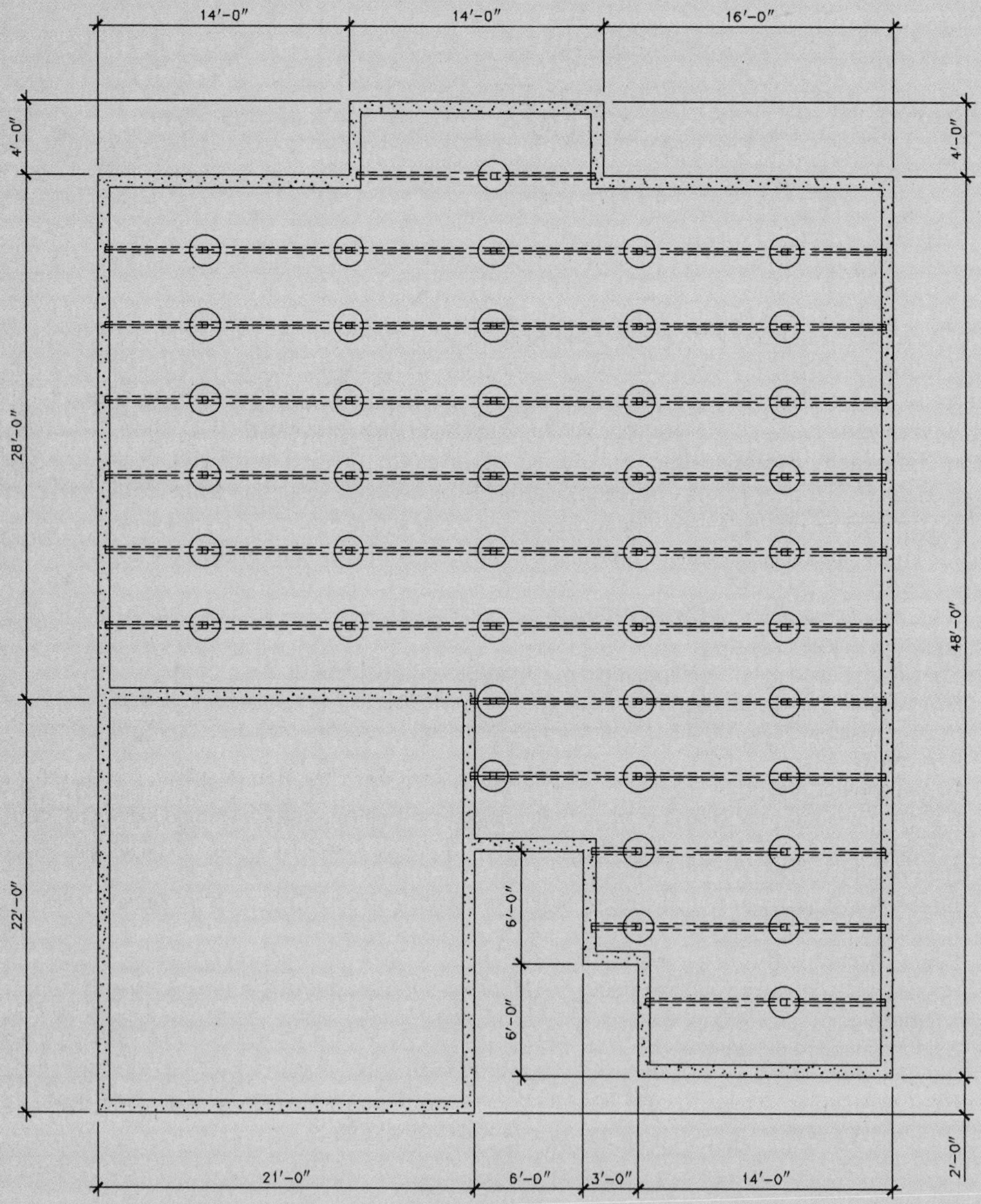

Drawing Baseline Dimensions

The term *baseline dimensions* refers to a group of dimension that all start at the same location. Baseline dimensions are also created in relation to an existing linear dimension. The Baseline Dimension command uses the first extension line of the previously created linear dimension as the first extension line for new baseline dimensions. As second extension line points are selected, new dimensions are created away from the original or previously created dimension, allowing you to create dimensions in different levels, as shown in Figure 8.21.

FIGURE 8.21 The baseline dimensions use a common extension line as a base point. New dimensions are added away from the previously created dimension line, with all the first extension line points lining up with each other.

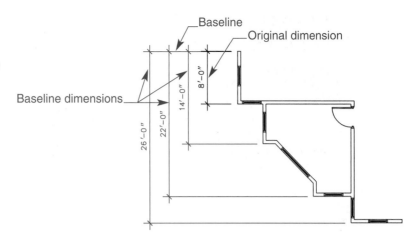

Use one of the following options to access the command:

✓ Pick the **Dimension** pull-down menu, and select **Baseline**.

✓ Pick the **Baseline Dimension** button on the **Dimension** toolbar.

✓ Type **DIMBASE** or **DIMBASELINE** at the Command prompt.

All these methods access the standard AutoCAD Baseline Dimension command, which places baseline dimensions on the current layer. In addition, you can enter the Baseline Dimension command from the Documentation pull-down menu, Documentation Content, and select Miscellaneous…, which opens the DesignCenter. Then, pick the Dimensions folder, and select the Baseline icon. You also can use the Miscellaneous button found on the Documentation-Imperial toolbar to access the Miscellaneous folder in the DesignCenter.

The following sequence is used to create the dimensions in Figure 8.22. Notice that a linear dimension is created first, then the Baseline Dimension command is used to create baseline dimensions:

Command: **DIMLIN** or **DIMLINEAR** ↵
Specify first extension line origin or <select object>: *(pick the first extension line point)*
Specify second extension line origin: *(pick the second extension line point)*
Specify dimension line location or
[Mtext/Text/Angle/Horizontal/Vertical/Rotated]: *(pick the dimension line location)*
Dimension text = 4'-3"

Command: **DIMBASE** or **DIMBASELINE** ↵
Specify a second extension line origin or [Undo/Select] <Select>: *(pick the second extension line point)*

FIGURE 8.22 Create a linear dimension first, then use the Baseline Dimension command to create baseline dimensions.

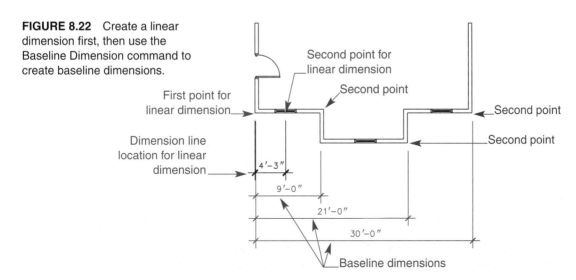

Dimension text = 9'-0"
Specify a second extension line origin or [Undo/Select] <Select>:*(pick the second extension line point)*
Dimension text = 21'-0"
Specify a second extension line origin or [Undo/Select] <Select>:*(pick the second extension line point)*
Dimension text = 30'-0"
Specify a second extension line origin or [Undo/Select] <Select>: ↵
Select base dimension: ↵
Command:

The spacing between the dimension lines is designated in the dimension style in the **Lines and Arrows** tab under the **Baseline spacing** text box. As with the Continue Dimension command, an existing extension line can be selected for use as the baseline. At the *Specify a second extension line origin or [Undo/Select] <Select>:* prompt, press the [Enter] key or right-click to enter the default option, Select. Pick an existing extension line for use as the baseline, then pick the second extension line locations. The following prompt sequence is used to create the baseline dimensions in Figure 8.23:

Command: **DIMBASE** or **DIMBASELINE** ↵
Specify a second extension line origin or [Undo/Select] <Select>: ↵
Select base dimension: *(pick an extension line as the baseline)*
Specify a second extension line origin or [Undo/Select] <Select>: *(pick a second extension point)*
Dimension text = 9'-0"
Specify a second extension line origin or [Undo/Select] <Select>: *(pick a second extension point)*
Dimension text = 21'-0"
Specify a second extension line origin or [Undo/Select] <Select>: *(pick a second extension point)*
Dimension text = 30'-0"
Specify a second extension line origin or [Undo/Select] <Select>: ↵
Select base dimension: ↵
Command:

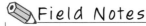

Field Notes

Continuous dimensions are the most commonly used dimension format in architectural drafting, but baseline dimensions are used in some applications. A combination of continuous and baseline dimensions can sometimes be used to lay out the dimensions in the drawing. This produces a clean, professional look to your construction documents, as shown in Figure 8.23.

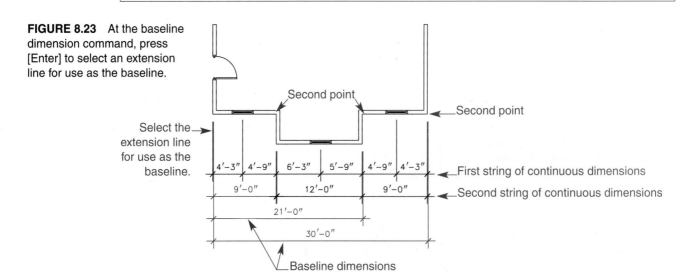

FIGURE 8.23 At the baseline dimension command, press [Enter] to select an extension line for use as the baseline.

Exercise 8.8

1. Open EX8-7-Fnd.dwg.
2. Use the Baseline Dimension command to add baseline dimensions to the drawing in the figure.
3. Save the drawing as EX8-8-Fnd.dwg.

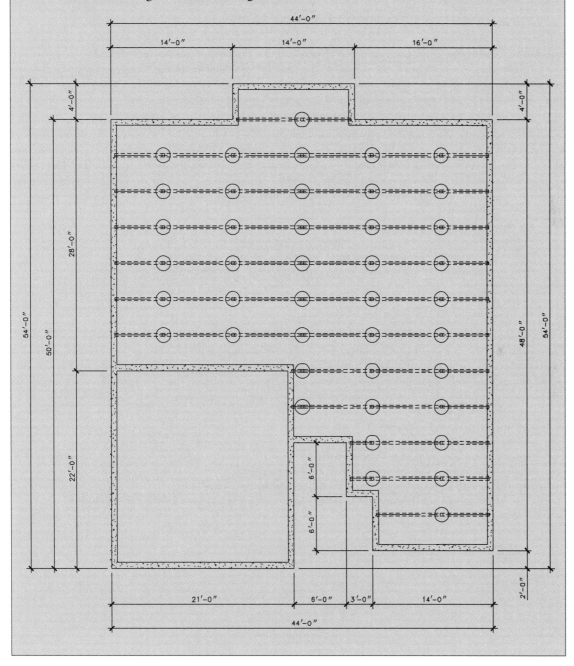

Drawing Angular Dimensions

Angular dimensions are used to dimension the angle of a feature in a drawing. The **Angular Dimension** command is used to create an angular dimension. Use one of the following options to access this command:

✓ Select the **Dimension** pull-down menu, and pick **Angular**.

✓ Pick the **Angular Dimension** button on the **Dimension** toolbar.

✓ Type **DIMANG** or **DIMANGULAR** at the Command prompt.

The AutoCAD Angular Dimension command places angular dimensions on the current layer.

In addition, you can enter the Angular Dimension command from the Documentation pull-down menu, Documentation Content, and select Miscellaneous…, which opens the DesignCenter. Then, pick the Dimensions folder, and select the Angular icon. Another option is to use the Miscellaneous button found on the Documentation-Imperial toolbar to access the Miscellaneous folder in the DesignCenter.

After entering the Angular Dimension command, you are prompted to *Select arc, circle, line, or <specify vertex>:*. Select one of the objects. You are then prompted to select a second line. Select the next object to establish the angular dimension. Finally, you must pick a place for the dimension line. Depending on how you move the cursor, one of four angles appears, indicating the sides of the objects that are dimensioned. The following prompt sequence is used to create the angular dimensions in Figure 8.24B:

Command: **DIMANG** or **DIMANGULAR** ↵
Select arc, circle, line, or <specify vertex>: *(pick one object)*
Select second line: *(pick second object)*
Specify dimension arc line location or [Mtext/Text/Angle]: *(pick dimension line location)*
Dimension text = 45
Command:

You also can specify the vertex point where the two objects cross each other to create an angular dimension. Press [Enter] at the *Select arc, circle, line, or <specify vertex>:* prompt. Pick the vertex location where the two objects cross each other, then pick a point designating the extension line location used for the first part of the angle. Next, pick a point designating the second line of the angle. The following prompt sequence is used to create an angular dimension using the vertex option shown in Figure 8.25:

Command: **DIMANG** or **DIMANGULAR** ↵
Select arc, circle, line, or <specify vertex>: ↵
Specify angle vertex: *(pick the vertex point)*
Specify first angle endpoint: *(pick a point for the first angle)*
Specify second angle endpoint: *(pick a point for the second angle)*
Specify dimension arc line location or [Mtext/Text/Angle]: *(pick a point for the dimension line location)*
Dimension text = 45
Command:

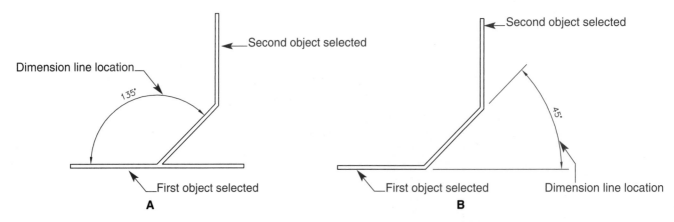

FIGURE 8.24 Selecting two objects to be dimensioned with the Angular Dimension command.

FIGURE 8.25 Press [Enter] to select a vertex point. Pick to establish the first angle, then pick a second point for the second angle.

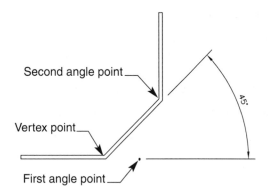

Second angle point

Vertex point

First angle point

45°

Drawing Radius Dimensions

Radial dimensions are used to dimension the radius of a curved object, such as a curved wall, pillar, or arched window. Use one of the following options to use this command:

✓ Pick the **Dimension** pull-down menu, and select **Radius**.

✓ Pick the **Radius Dimension** button on the **Dimension** toolbar.

✓ Type **DIMRAD** or **DIMRADIUS** at the Command prompt.

The Radius Dimension command places radius dimensions on the current layer.

In addition, you can enter the Radius Dimension command from the Documentation pull-down menu, Documentation <u>C</u>ontent, and select <u>M</u>iscellaneous…, which opens the DesignCenter. Then, pick the Dimensions folder and select the Radius icon. You also can use the Miscellaneous button found on the Documentation-Imperial toolbar to access the Miscellaneous folder in the DesignCenter.

Once you have entered the command, you are prompted to pick an arc or circle. Pick the arc or circle, then the dimension text location to place the dimension. The dimensional radius is displayed with the letter R, indicating that the dimension is referring to a radius. If you have a curved wall in the drawing that needs to be dimensioned, pick it as you would an arc or circle. A radius dimension will be created for the wall. The following prompt sequence is used to create a radius dimension for the curved wall in Figure 8.26:

Command: **DIMRAD** or **DIMRADIUS** ↵
Select arc or circle: *(pick an arc, circle, or curved wall)*
Dimension text = 5′-7 13/16″
Specify dimension line location or [Mtext/Text/Angle]: *(pick a point for the text placement)*
Command:

FIGURE 8.26 Using the Radius Dimension command to dimension the curved wall.

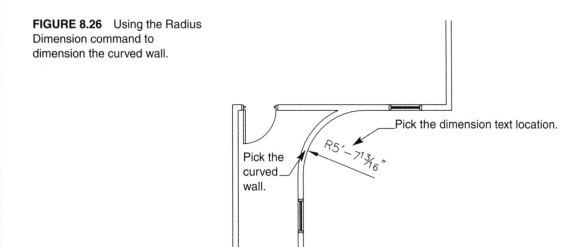

Pick the dimension text location.

Pick the curved wall.

R5′-7 13/16″

Drawing Aligned Dimensions

Aligned dimensions are used to align with an object that is not horizontal or vertical. Use one of the following options to access the Aligned Dimension command:

✓ Pick the **Dimension** pull-down menu, and select **Aligned**.

✓ Pick the **Aligned Dimension** button on the **Dimension** toolbar.

✓ Type **DIMALI** or **DIMALIGNED** at the Command prompt.

The Aligned Dimension command places aligned dimensions on the current layer.

In addition, you can enter the Aligned Dimension command from the Documentation pull-down menu, Documentation Content, and select Miscellaneous…, which opens the DesignCenter. Then, pick the Dimensions folder and select the Aligned icon. You also can use the Miscellaneous button found on the Documentation-Imperial toolbar to access the Miscellaneous folder in the DesignCenter.

The following prompt sequence is used to create the aligned dimensions displayed in Figure 8.27:

Command: **DIMALI** or **DIMALIGNED** ↵
Specify first extension line origin or <select object>: *(pick the first extension line point)*
Specify second extension line origin: *(pick the second extension line point)*
Specify dimension line location or [Mtext/Text/Angle]: *(pick the dimension line location)*
Dimension text = 5′-7 7/8″
Command: ↵
DIMALIGNED
Specify first extension line origin or <select object>: *(pick the first extension line point)*
Specify second extension line origin: *(pick the second extension line point)*
Specify dimension line location or [Mtext/Text/Angle]: *(pick the dimension line location)*
Dimension text = 5′-7 7/8″
Command:

✎Field Notes

When creating a dimension that is not horizontal or vertical it is often better to use the Linear Dimension command with the Rotated option to rotate the dimension line rather than to use the Aligned Dimension command. This allows you to pick extension line points along a different angle without having the dimension line remain parallel to the imaginary line between the two extension line points.

FIGURE 8.27   Use the Aligned Dimension command to create dimensions that are parallel to the two extension points picked.

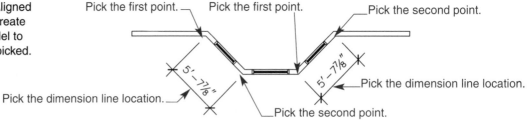

Exercise 8.9

1. Start a new drawing using the AEC Arch (Imperial) or the AEC Arch (Imperial-Int) template.
2. Draw the diagram in the figure.
3. Use the Linear, Continue, Baseline, Angular, Radius, and Aligned dimension commands as necessary to dimension the drawing.
4. Save the drawing as EX8-9.dwg.

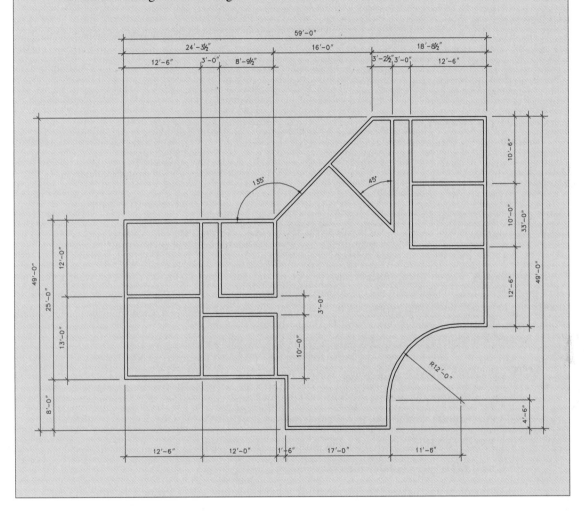

Using Grips to Modify Dimensions

There may be times when you need to modify a dimension placed in a drawing, for example, to reposition it or to move an extension line to a different point. Grips can be used to speed up the process of modifying a dimension.

To do this, first select the dimension to be modified, then select a grip to make it hot. Use the grip Stretch mode to adjust the text location, an extension line location, or a dimension line location (see Figure 8.28).

Using the Properties Command

As mentioned earlier, when dimensions are created from the Dimension pull-down menu or from the Dimension toolbar, the dimensions are placed on the current layer. In proper drafting management dimensions should be placed on their own specific layer. For example, when the dimension commands from the DesignCenter are used, the dimensions are placed on the A-Anno-Dims layer.

FIGURE 8.28   Use grips to modify a dimension.

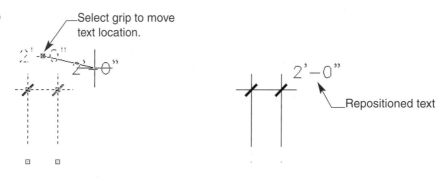

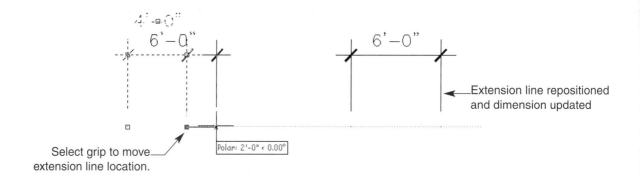

Occasionally, as you draw, objects may be placed on the wrong layer, or you may want to change other properties of the object such as color, linetype, or location. The **Properties** command can be used to modify the layer and additional properties of individual dimension objects and can be used to change the properties of all AutoCAD and Architectural Desktop objects.

Use one of the following options to access the Properties window:

✓ Pick the **Tools** pull-down menu and select **Properties**.

✓ Pick the **Properties** button on the **Standard** toolbar.

✓ Select the **Modify** toolbar and pick **Properties**.

✓ Type **PROPERTIES** at the Command: prompt.

✓ Pres the [Ctrl]+[1] key combination.

✓ Select the object(s) that need to be modified, right-click, and select **Properties** from the cursor menu.

The Properties window, shown in Figure 8.29, is displayed.

Initially, the Properties window is docked to the left side of the drawing screen, and two horizontal grab bars are displayed at the top of the window. Double-clicking on the grab bars sets the Properties window as a floating window that can be moved around the screen. The Properties window remains on the drawing screen as you work in Architectural Desktop. While the Properties window is displayed you can continue to enter commands and work on the drawing. Pick the X in the top right corner of the box to close the window.

When you open the Properties window without first selecting an object, **No selection** is displayed in the top drop-down list. This means that no objects have been selected for modification. When there are no objects selected, four categories are available: **General**, **Plot style**, **View**, and **Misc;** these categories list the current settings for the drawing. The categories can be organized alphabetically by selecting the **Alphabetic** tab or by category, by selecting the **Categorized** tab.

Beneath each category is a list of object properties. For example, in Figure 8.29 the current color is set to ByLayer. To change a property, pick the property or its current value. Depending on the property that is selected, one of the following methods is used to change the value (see Figure 8.30):

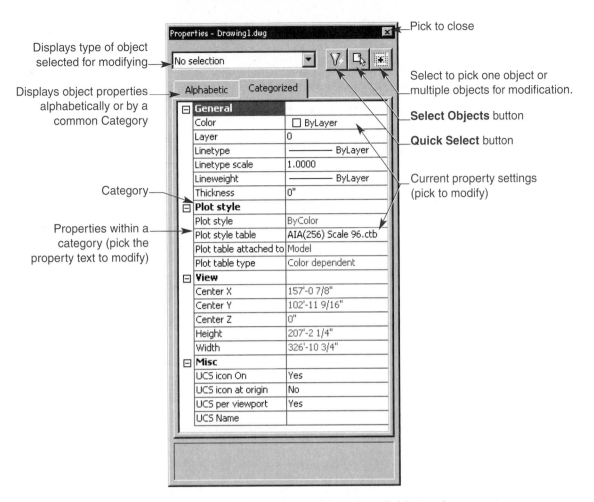

FIGURE 8.29 The Properties window lists the properties of the selected object and allows you to modify the properties.

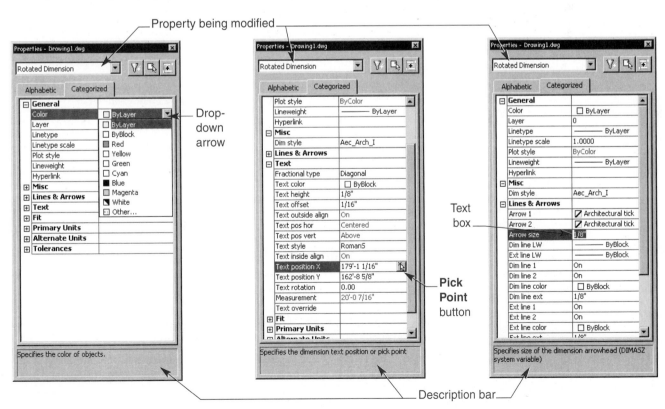

FIGURE 8.30 Different methods of changing properties. (A) A drop-down arrow contains a list of options. (B) The Pick Point button allows you to pick a coordinate location. (C) A text box allows you to enter a new value.

- A drop-down arrow button appears. Press the drop-down arrow to display a list of options. Select an option available to change to one of the values.

- A Pick Point button appears. This allows you to pick a location in the drawing to specify a new coordinate location.

- If a text box is selected, such as the arrowhead size, enter a new value to change the size.

As properties to be modified are selected, a description of what that property does is shown at the bottom of the Properties window. The upper right portion of the Properties window includes the following three buttons:

- **Quick Select**   This button opens the **Quick Select** command, which allows you to filter for objects that meet specific criteria. The selected objects can then be modified in the Properties window.

- **Select Objects**   This button allows you to pick objects that are to be modified.

- **Toggle value of**   Selecting this button allows you to pick only one object at a time to be
  PICKADD sysvar   modified. The default allows multiple objects to be modified.

Once an object has been selected for modification, the properties specific to the type of object are displayed in the Properties window. The drop-down list at the top of the window indicates the type and the number of objects being modified. For example, in Figure 8.30 a Rotated Dimension (linear dimension) has been selected. The properties of the one dimension can be modified to override the properties set from the dimension style.

If many different objects are selected, the drop-down list displays the word All and the number of objects selected in parenthesis. The only properties available are the properties that are common to all the selected objects. The drop-down list at the top of window displays all the objects that have been selected and the total number of each type of object (see Figure 8.31). Selecting an object type from the drop-down list allows the properties available for the type of object to be modified.

FIGURE 8.31   The Properties window with multiple objects selected. Select an object type to modify all the objects of one type.

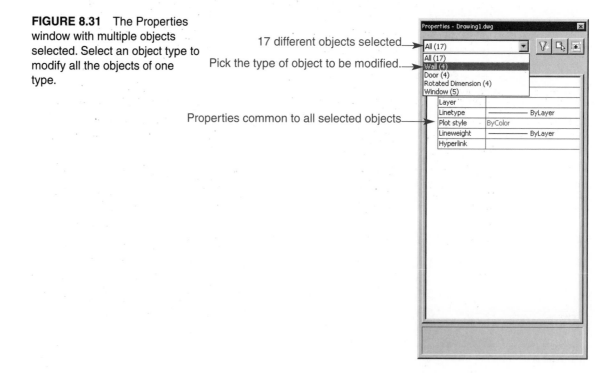

17 different objects selected

Pick the type of object to be modified.

Properties common to all selected objects

Exercise 8.10

1. Open EX8-9.dwg.
2. Create a layer named A-Anno-Dims using the Layer Manager.
3. Use the Properties command to change all the dimensions to the A-Anno-Dims layer.
4. Save the drawing as EX8-10.dwg.

Drawing Architectural Desktop Dimensions

So far we have discussed dimension styles and dimensioning commands. Architectural Desktop includes a dimensioning command that can be used to help automate the process of dimensioning. The automatic dimensioning feature creates linear dimensions (horizontal, vertical, and rotated), and continuous and baseline dimensions that dimension wall lengths, and window, door, and opening locations. The only minimum requirement is that there must be AEC wall objects in the drawing.

Use one of the following options to access the Architectural Desktop dimensioning command:

✓ Pick the **Design** pull-down menu, select **Wall Tools,** then **Dimension Walls.**

✓ Pick the **Dimension Walls** button in the **Wall Tools** toolbar.

✓ Right-click, and select **Design, Wall Tools, Dimension Walls.**

✓ Type **AECWALLDIM** at the command prompt.

Once the command has been entered you are prompted to *Select walls.* Select the walls you wish to dimension. If the wall includes any openings, such as doors and windows, the command automatically dimensions to the center of these objects. When you have finished selecting the walls, press the [Enter] key. You are then prompted to specify the side of the wall on which to place the dimensions. Pick a point that designates the location for the first dimension line string. The next prompt asks for a second point. Pull the mouse away from the first point that was picked. Architectural Desktop locates the endpoints of the walls and the center points of the openings and places dimensions in the drawing on the correct layer. The following prompt sequence is used to create the dimensions in Figure 8.32:

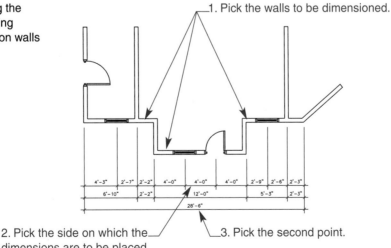

FIGURE 8.32 Using the automatic dimensioning command to dimension walls and openings.

Command: **AECWALLDIM** ↵
Select walls: 1 found
Select walls: 1 found, 2 total
Select walls: 1 found, 3 total
Select walls: ↵
Pick side to dimension: *(pick the side of the wall on which to place the first string of dimensions)*
Second point or [Parallel]: *(pull the mouse away from the first point selected and pick)*
Command:

When the second point is specified, depending on the direction the mouse is moved, the dimension strings are created perpendicular to the imaginary line between the first and second points. If it is important to have dimensions that are parallel to the walls they are dimensioning, use the Parallel option at the second prompt to ensure that the dimensions are created correctly. The following prompt sequence uses the Parallel option for Figure 8.32:

Command: **AECWALLDIM** ↵
Select walls: 1 found
Select walls: 1 found, 2 total
Select walls: 1 found, 3 total
Select walls: ↵
Pick side to dimension: *(pick the side of the wall on which to place the first string of dimensions)*
Second point or [Parallel]: **P** ↵
Command:

The dimensions created are based on the current dimension style and are individual AutoCAD dimensions. The justification of the selected walls is automatically used to place the dimensions. Any intersecting walls are also dimensioned to their wall justification. As the dimensions are standard AutoCAD linear dimensions, they can be grip stretched for fine-tuning. Duplicate dimensions can be removed with the ERASE command.

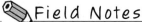 Field Notes

> The simplest way of dimensioning walls is to first select the walls to be dimensioned, right-click, select the Plan Tools cascade menu, and pick Dimension from the cursor menu.

Exercise 8.11

1. Open EX8-6-Flr-Pln.dwg.
2. Set the Floor Plans dimension style current.
3. Use the automatic dimensioning feature to dimension the floor plan as shown in the figure.
4. Use grip editing as required to adjust the dimensions. Erase any duplicate dimensions.
5. For dimensions that are too narrow to fit between extension lines, use the PROPERTIES command to adjust the text height property to $^3/_{32}''$.
6. Save the drawing as EX8-11-Flr-Pln.dwg.

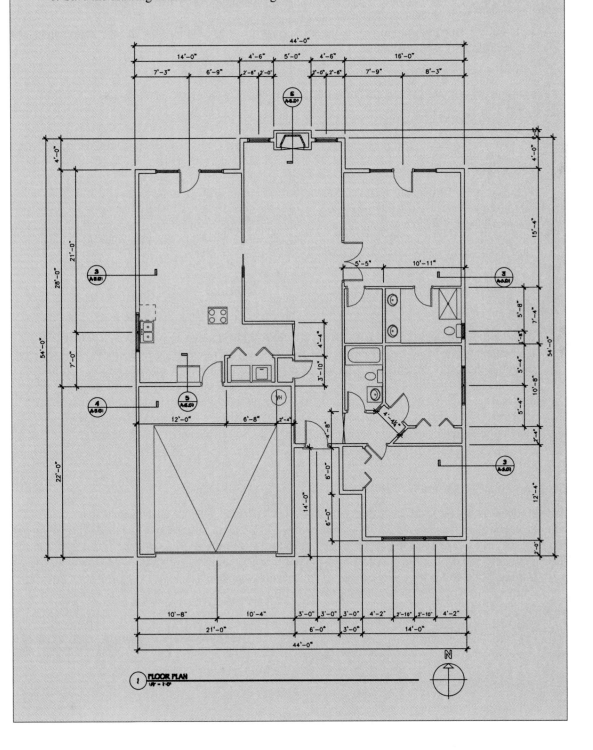

Creating AEC Dimensions

Some installation tips were presented at the beginning of this text. One of these included the installation of the International Extensions. If the International Extensions have been installed on your system, the remainder of this chapter covers some of these features.

One of the features included with the International Extensions is the AEC Dimensions command. This command creates dimensions that are attached to the objects they are dimensioning. If the wall, door, or window that has an AEC dimension attached is moved, stretched, or modified, the AEC dimension automatically adjusts to reflect the change.

Use one of the following options to add AEC Dimensions to the drawing:

✓ Select the **Documentation** pull-down menu, pick the **AEC Dimension** cascade menu, and pick **Add AEC Dimension....**

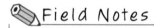

✓ Pick the **Add AEC Dimension** button from the **AEC Dimensions** toolbar.

✓ Right-click, and select **Documentation, AEC Dimension, Add AEC Dimension....**

✓ Type **AECDIMADD** at the Command: prompt.

✓ Select the objects to be dimensioned, right-click, and pick **AEC Dimension** from the cursor menu.

The **Add AEC Dimension** modeless dialog box where an AEC dimension style can be selected is displayed. Once you have selected the style to use, pick the side of the wall where you want the first string of dimensions placed. Pick a second point to establish the angle of the strings, or enter **P** for the **Perp** option to make the second point perpendicular to the first point selected. This makes the AEC dimension strings parallel to the walls (see Figure 8.33).

✎ Field Notes

> If the AEC Dimensions are not displayed, there may be two reasons why even if the International Extensions are installed. The first is that you may not be using the AEC Arch (Imperial-Int) template. This template has preestablished the display configurations so that the AEC dimensions appear in a plan view. The second is that the current display configuration has not been set up to view AEC Dimensions. To fix this problem, enter the Display Manager, and add the AEC dimension plan representation to any of the display sets desired. See Chapter 5 for more information regarding the Architectural Desktop Display System.

Creating an AEC Dimension Style

An AEC dimension style is different from an AutoCAD dimension style. *AEC dimension styles* control the number of strings used and how each string dimensions AEC objects. For example, the first string of dimensions can be set to dimension the center of any openings in a wall, the second string can dimension the lengths of walls, and the third string can dimension the overall outer limits of walls.

Use one of the following options to create an AEC dimension style:

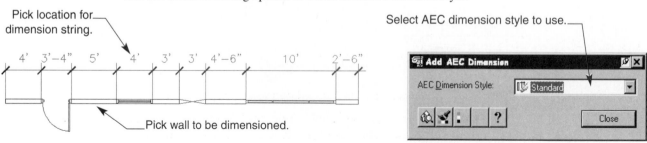

FIGURE 8.33 AEC dimensions are similar to the automatic dimensions in appearance, but they are "attached" to the objects they are dimensioning.

Edit Style button_____   _____New Style button

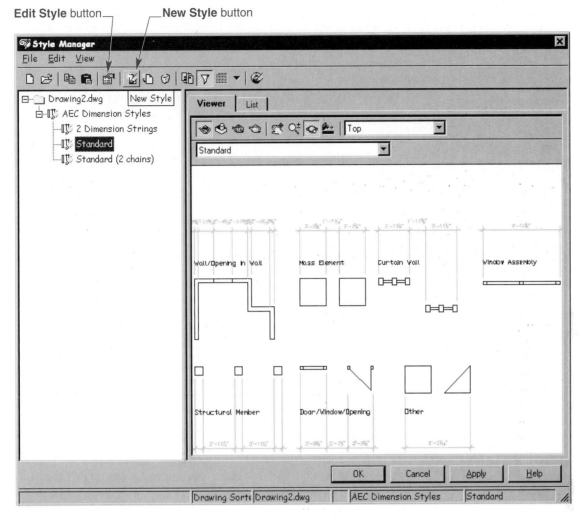

FIGURE 8.34   Create a new AEC dimension style, then edit it.

✓ Select the **Documentation** pull-down menu, pick the **AEC D_imension** cascade menu, then select **AEC D_imension Styles....**

✓ Pick the **AEC Dimension Styles** button in the **AEC Dimension** toolbar.

✓ Right-click, and select **Doc_umentation, AEC D_imension, AEC D_imension Styles....**

✓ Type **AECDIMSTYLE** at the Command: prompt.

The Style Manager dialog box, shown in Figure 8.34, is displayed. Pick the New Style button to create a new AEC dimension style, and enter a name for the style. Select the new style from the tree list on the left, then pick the Edit Style button. The **AEC Dimension Style Properties** dialog box, shown in Figure 8.35, is displayed.

The following tabs are available in the AEC Dimension Style Properties dialog box:

- **General**   This tab is used to rename and provide a description of the AEC dimension style.

- **Chains**   This tab is used to specify the number of dimension strings (chains).

- **Display Props** This tab is used to set up the appearance of the AEC dimensions in the different display representations. Note that you should set an override using the **AEC Dimension Style Property Source**, then edit the display of the AEC dimensions.

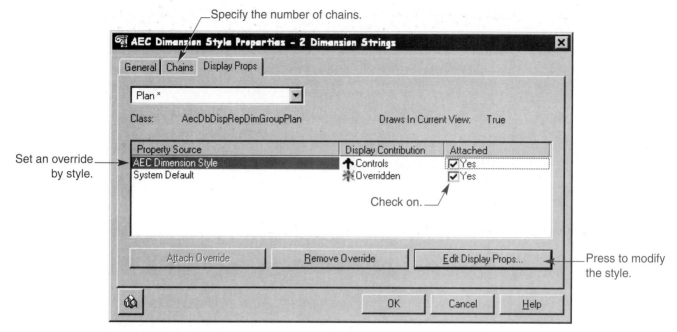

FIGURE 8.35   The AEC Dimension Style Properties dialog box is used to configure the AEC dimension style.

Determine the number of strings or chains to use for the AEC dimension style. In the Display Props tab, select the Attached check box for the AEC Dimension Style Property Source. Pick the Edit Display Props… button to edit the way the dimensions appear. The Entity Properties dialog box for the AEC dimensions is displayed (see Figure 8.36). The following tabs are available:

- **Layer/Color/Linetype**   This tab controls the visibility and appearance of the components of the AEC dimension.

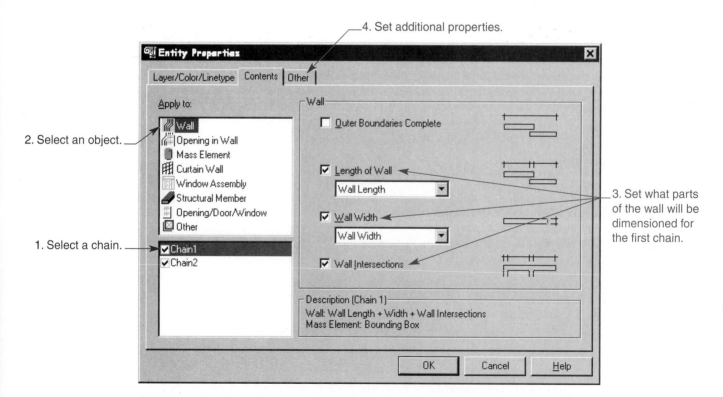

FIGURE 8.36   The Entity Properties dialog box for AEC dimensions.

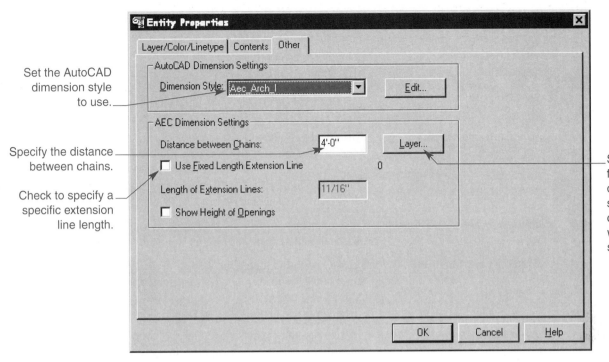

FIGURE 8.37

The image above contains the following annotations:
- Set the AutoCAD dimension style to use.
- Specify the distance between chains.
- Check to specify a specific extension line length.
- Specify the layer for the AEC dimensions. If set to 0, the current layer key will be used to set the layer.

■ **Contents**   This tab displays the number of chains specified in the AEC Dimension Style Properties dialog box, the types of objects that can be dimensioned, and how they are dimensioned.

■ **Other**   This tab contains options to control what AutoCAD dimension style to reference and how the AEC dimensions are created.

The Contents tab is divided into three areas. The **Apply to:** area lists the different AEC objects associated with the AEC dimensions. The Other icon is used for AEC objects that are not specified in the list. Below the Apply to: area is a list of the number of chains selected. To use this option, first highlight a chain, then select an object from the list above to assign how the object should be dimensioned for that string. The Other area provides a list of options for each AEC object in the Apply to: list. Select the check boxes desired to configure what will be dimensioned for each string.

When you are finished in the Contents tab, select the Other tab to control additional display settings (see Figure 8.37). The AEC dimension styles use the properties of an existing AutoCAD dimension style. Select the style to use for the AEC dimensions. Once the AEC dimension style has been configured, new AEC dimensions can be created in the drawing using the style.

Using Additional AEC Dimension Commands

There are additional AEC dimensioning commands that can be used to create and modify AEC dimensions in the drawing. The commands can be found in the AEC Dimensions cascade menu under the Documentation pull-down menu, and on the AEC Dimension toolbar:

■ **Add Manual AEC Dimension**   This command is used to pick extension line points for AEC dimensions. When this command is used the AEC dimensions are not associated with the objects, but they are considered one complete object instead of individual dimensions.

■ **Modify AEC Dimension**   Use this command to modify an existing AEC dimension object to use a different AEC dimension style.

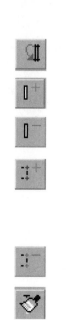

- **Convert to AEC Dimension**   Use this command to convert individual AutoCAD dimensions into one AEC dimension object.

- **Attach Objects**   Use this command to attach additional AEC objects to the AEC dimension object.

- **Detach Objects**   Use this command to disassociate of an AEC object from an AEC dimension object.

- **Add Dimension Points**   This command allows you to add additional points for an AEC dimension. To do this, first pick the points to be dimensioned, then pick the dimension chain to be used to dimension the new points. The new points are not associated with any AEC geometry.

- **Remove Dimension Points**   This command can be used to remove unwanted extension lines and points on the AEC dimension object.

- **Match AEC Dimension**   Use this button to match the display properties of an AEC dimension style to AutoCAD dimensions. Display settings such as the extension line length and text display are applied to the standard AutoCAD dimensions.

- **Activate Dim Text Grip Points**   Use this button to turn on the grips for the text in an AEC dimension. This allows you to grip edit the text location.

- **AEC Dimension Wizard**   This button starts a wizard that walks you through the process of adjusting the display settings for an AEC dimension style.

Field Notes

The use of AEC dimensions can be helpful during the design phase of the drawing. Use AEC dimensions when laying out walls and openings. As you modify walls and openings the dimension automatically updates so you can keep track of measurements. AEC dimensions can also be used alongside standard AutoCAD dimensions to aid in preparation of the finished construction documents.

|||||||||||| CHAPTER REVIEW

Use the CD-ROM to test your knowledge and skills.

Chapter Test

To check your understanding of the content provided in this chapter, access the Test file in the CH08 folder of the CD-ROM that accompanies this text.

Chapter Project

To practice the Architectural Desktop skills presented in this chapter, access the Project files in the CH08 folder of the CD-ROM that accompanies this text. The project files are in pdf format and include sample drawings and instructions for completing each project.

|||||||||||||||||||||||||||||||||||||||||||||||||||||||||||||||||||||||||||||||||||||||||||||||||||

CHAPTER 9

Editing and Modifying Walls

LEARNING GOALS

After completing this chapter, you will be able to:

◎ Use wall entity display.

◎ Create wall interfaces.

◎ Use wall modifiers.

◎ Create wall sweeps.

◎ Adjust the floor and roof line of a wall.

◎ Add an object to a wall.

◎ Use wall properties.

◎ Join, merge, and anchor walls.

In Chapter 3 of this text you were introduced to the process of adding wall objects to your drawing. In addition to adding and modifying walls, there are a number of other ways of enhancing the model walls and construction documents.

The **Modify Walls** command is used to make minor adjustments to wall objects such as the size of the wall and style being used. Additional commands can be used to enhance the look of wall objects in both plan and model representations. These commands include adding or subtracting shapes, columns, or pilasters to or from walls, and modifying the shape and controlling additional display properties of a wall. This chapter explains how to use wall modeling tools and tools to improve the use and look of walls in the construction documents.

Using Wall Entity Display

Chapter 5 explained how the display system within Architectural Desktop works. Basically, there are three main display representations used to control how walls appear in the drawing: plan, model, and reflected. These are the most often used and configured display representations in Architectural Desktop. Additional display representations are included for use with other display configurations. This section deals with the most common display representations.

As discussed in Chapter 5, display representations can be controlled globally for all the walls, by style, or by individual wall. To change the display of a wall, select the wall or wall using a style to be modified, right-click, and select Entity Display… from the cursor menu. The Entity Display dialog box, shown in Figure 9.1, is displayed.

First, select the appropriate display representation to be modified from the drop-down list in the upper left corner of the dialog box, then determine which level of control you want to modify. The **Property Source** column displays the three levels of control: Wall, Wall Style, and System Default. Selecting a property source from the list highlights the level so an override can be attached. Press the Attach Override button to attach a display override to the property source. A check mark appears in the corresponding attached check box. Finally, pick the Edit Display Props… button to modify the display for the display representation and the level of control that you have specified. The Entity Properties dialog box, shown in Figure 9.2, is displayed.

327

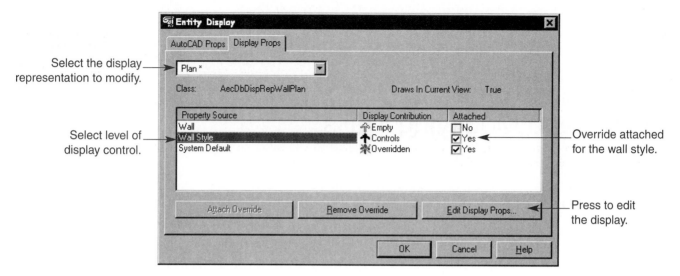

Select the display representation to modify.

Select level of display control.

Override attached for the wall style.

Press to edit the display.

FIGURE 9.1 The Entity Display dialog box for wall objects.

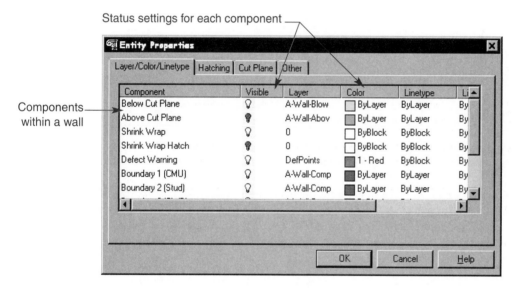

Status settings for each component

Components within a wall

FIGURE 9.2 The Entity Properties dialog box for wall objects.

Depending on the display representation chosen for modification, the number of tabs available for display control can vary. If you are modifying the plan and reflected display representations, the four tabs shown in Figure 9.2 are available. If the model display representation is selected, only the Layer/Color/Linetype tab is available. The four tabs and the display information they control are as follows:

■ **Layer/Color/Linetype** This tab lists the components of a wall object. Plan and reflected wall components can be set differently than model wall components. Settings such as component visibility, layer, color, and linetype are controlled in this tab (see Figure 9.2). To adjust a setting, pick the word or symbol next to the component that is to be modified. For example, if a hatch component is not supposed to be displayed, pick the lightbulb symbol. If the color of the shrink wrap is to be displayed as a different color, pick the swatch box to select a different color. Each component within a wall can have its own display setting to include linetype and lineweight. The following are the components included for wall objects:

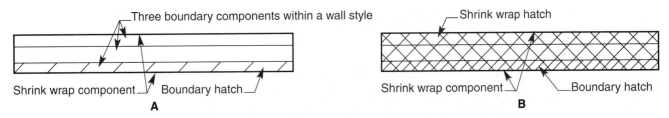

FIGURE 9.3 (A) Shrink wrap is a component applied to the outside lines of a wall. Use a wide lineweight for the shrink wrap component. (B) The shrink wrap hatch is applied between the two outermost lines of a wall instead of between individual boundaries lines.

■ **Below Cut Plane**	This component is used to control the display of items below the cut plane. The *cut plane* is where Architectural Desktop cuts through wall objects for display in the top view. By default, Architectural Desktop cuts wall objects 3′-6″ from the bottom of the wall for display in the top view.
■ **Above Cut Plane**	This component is used to control the display of items above the cut plane.
■ **Shrink Wrap**	*Shrink wrap* is a component applied to the outermost edges of the walls that allows the outside lines of a multicomponent wall to use a different color and lineweight for plotting purposes (see Figure 9.3A).
■ **Shrink Wrap Hatch**	*Shrink wrap hatch* applies a hatch pattern between the two outside lines of a wall, as shown in Figure 9.3B.
■ **Defect Warning**	This component controls the display of the defect warning symbol in a wall when there is a problem with the wall.
■ **Boundaries**	A wall style can include up to 20 individual boundaries. A *wall style boundary* is the object that contains the wall style, such as the outline. If you are modifying the wall style or an individual wall display, the number of boundaries within the wall style is displayed. If you are controlling the display of system default walls, all 20 boundaries are available for modification. Each boundary is listed with a number. If you are modifying a specific wall style, the name of the boundary is displayed in parenthesis.
■ **Hatches**	Each boundary component is assigned a hatch component that can be displayed in a plan or reflected view. A *hatch* is a pattern that is used to fill in an area. Like the boundary components, each hatch component is assigned a number. The number coincides with the boundary it is assigned to. For example, the Hatch 2 component is the hatch component assigned to the Boundary 2 component.
■ **Hatching**	This tab controls the hatches used by the hatch components in the Layer/Color/Linetype tab (see Figure 9.4). The hatch components are listed in the Component column. The hatch pattern, scale, rotation angle, and orientation can be controlled for each hatch component. As in the Layer/Color/Linetype tab, picking a symbol or word in each column presents you with options for controlling how the hatch component is displayed. The following columns are available for modification:
■ **Pattern**	Pick the swatch next to the desired hatch component to select a hatch pattern other than the default diagonal lines. The **Hatch Pattern** dialog box is shown in Figure 9.5. The **Type** drop-down

Pick to assign a
hatch pattern to the   Set the scale   Specify the rotation
hatch component.   for the hatch.   angle of the hatch.

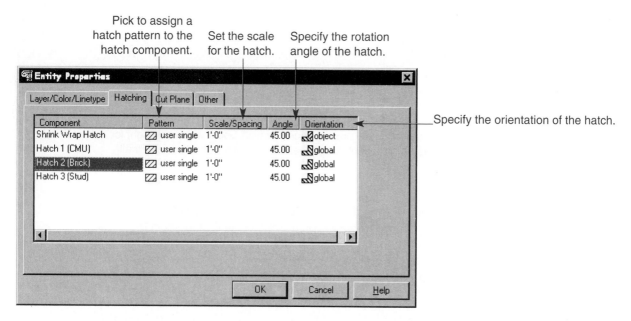

Specify the orientation of the hatch.

FIGURE 9.4   The Hatching tab controls how the hatch patterns are displayed in a wall object.

Select the type of hatch to use.

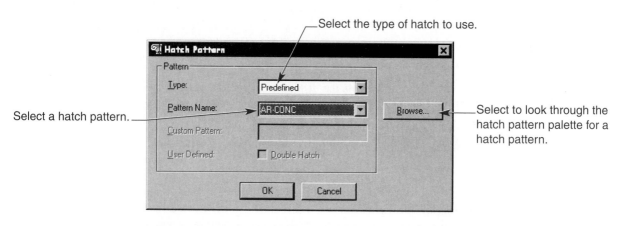

Select a hatch pattern.

Select to look through the
hatch pattern palette for a
hatch pattern.

FIGURE 9.5   The Hatch Pattern dialog box is used to assign a hatch pattern to a hatch component.

list includes options for the type of hatch pattern to use. When the Predefined type is selected, the **Pattern Name:** drop-down list becomes available with a list of AutoCAD hatch patterns that can be applied to the hatch component. Picking the <u>B</u>rowse… button opens the **Hatch Pattern Palette** dialog box, which displays a preview of the hatches available.

- **Scale/Spacing**   Pick a scale for the selected hatch pattern.

- **Angle**   Select the angle to enter a new rotation angle for the hatch pattern.

- **Orientation**   There are two options for the orientation of a hatch pattern: **Object** or **Global**. The Object option rotates the hatch relative to the angle of the wall. The Global option ensures that the hatch always remains at the same rotation angle regardless of the angle of the wall (see Figure 9.6).

- **Cut Plane**   This tab establishes the cutting plane for wall objects (see Figure 9.7). The cut plane establishes where the shrink wrap, component boundaries, and hatching will take effect. In addition to the cut

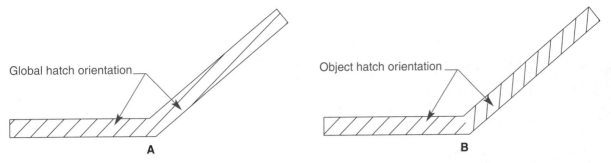

FIGURE 9.6 (A) Walls hatched using a global orientation. (B) Walls using the object orientation.

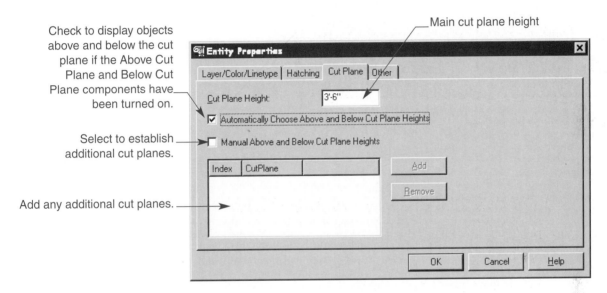

FIGURE 9.7 The Cut Plane tab is used to control where the cut plane for wall objects is set.

plane, additional cut planes can be established that cut through other components within a wall. The tab includes the following four areas for display control:

- **Cut plane Height**

  This is the main cut plane height. By default it is set to 3'-6" above the baseline of the wall. The cut plane establishes the components of the wall that will be cut and how the shrink wrap is applied to the wall components. Hatch patterns are also displayed if the cut plane cuts through the boundary where the hatch component has been assigned.

- **Automatically Choose Above and Below Cut Plane Heights**

  This check box, when checked on, determines where the Above and Below Cut Plane components are located in the wall. If the Above Cut plane and the Below Cut Plane components have been turned on, the components display the objects where the cut plane is located.

- **Manual Above and Below Cut Plane Heights**

  Check this box on to establish additional cut planes in a wall. This makes the **Add** and **Remove** buttons available so additional cut planes can be established. Press the Add button to add another cut plane. The table in the lower left is used to establish the

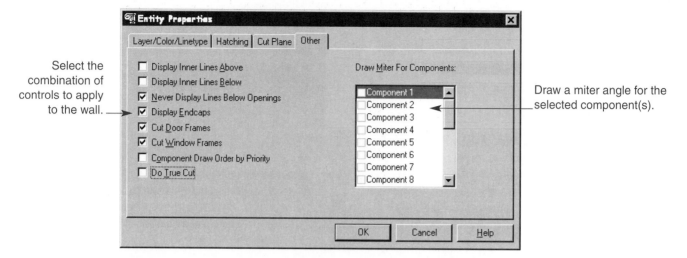

FIGURE 9.8   The Other tab is used to set additional display control for wall components.

heights of additional cut planes. If the additional cut plane is placed lower than the cut plane height, the wall components are displayed using the properties established in the Below Cut Plane component in the Layer/Color/Linetype tab. If the additional cut plane is placed higher than the cut plane height, the wall components are displayed using the properties established in the Above Cut plane component in the Layer/Color/Linetype tab.

- **Other**

  This tab controls additional display properties for wall components (see Figure 9.8). It contains the following options:

  - **Display Inner Lines Above**

    This check box turns on any component lines above the cut plane through an opening in a wall.

  - **Display Inner Lines Below**

    This check box turns on any component lines below the cut plane through an opening in a wall.

  - **Never Display lines Below Openings**

    Select this option to never show lines below the cut plane of an opening such as a door or window.

  - **Display Endcaps**

    This check mark turns on the endcaps at the ends of walls and openings. An *endcap* is a special shape to describe how the wall is to be built. Endcaps are discussed later in this text.

  - **Cut Door Frames**

    This check mark adjusts the shrink wrap of the wall to be flush with the door panel instead of with the outermost part of the door frame (see Figure 9.9A).

  - **Cut Window Frames**

    This check mark adjusts the shrink wrap of the wall to be flush with the window sash instead of with the window frame (see Figure 9.9B).

  - **Component Draw Order by Priority**

    Check this box on if a wall component is to be plotted over another wall component.

  - **Do True Cut**

    Select this check box to cut the 3D model at each of the defined cut planes. This feature is helpful when using wall sweeps and wall modifiers, as the plan view is accurately displayed. Wall sweeps and wall modifiers are described later in this chapter.

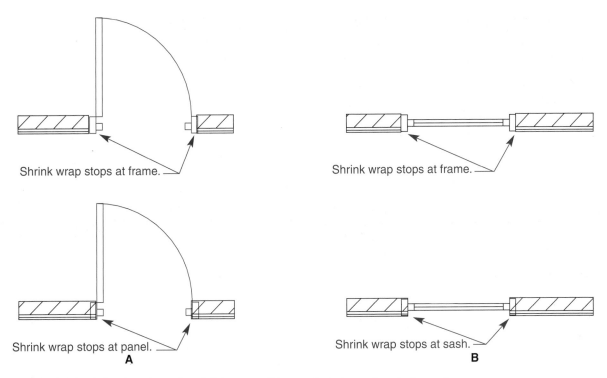

FIGURE 9.9 Adjusting where the shrink wrap will be applied at openings in the wall.

■ **Draw <u>M</u>iter for Components** — This drop-downlist displays all the possible boundaries in a wall object. Select the component(s) to have a miter line drawn where two walls cross each other or join at a corner.

Once you have finished configuring the display for the wall style or individual wall, press the OK button to make the changes appear in the drawing. If the changes do not appear, try entering the **REGEN**, **REGENALL**, or **OBJRELUPDATE** commands at the Command: prompt to update the display in the drawing.

Field Notes

Set colors for plan and reflected display representations different from those for the model display representations. Doing this gives you more control over how the drawing will appear when plotted, and you can assign renderable materials to the colors used in the model view for presentation drawings.

Creating Wall Interferences

Often when a floor plan is created, part of the design can include columns within a wall, pilasters applied to the side of a wall, or a niche carved out of the wall. All these instances can be re-created with Architectural Desktop through a system known as an *interference condition* (see Figure 9.10). AEC objects can interact with the wall object by having the shrink wrap component wrap around the AEC object that is interfering with the wall or by having the AEC object subtract out a part of the wall.

AEC objects such as columns and mass elements are typically used to interfere with a wall, although any AEC object can be used as an interference. Mass elements are discussed later in this text but can be used much like column objects. In order to use the **Interference Condition** command, a wall and an object for use as the interference must be present in the drawing. Once the AEC object has been added to the wall, three different options are available to control how the shrink wrap component of the wall interacts with the

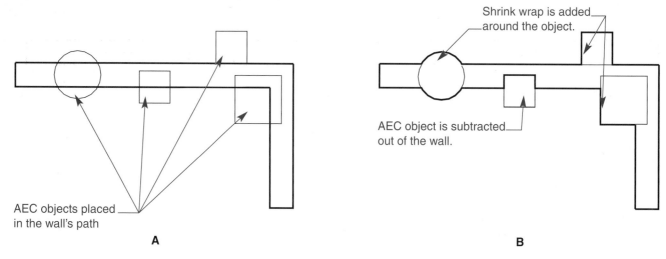

FIGURE 9.10 (A) A few different AEC objects placed along the wall. (B) The same objects after a wall interference has been applied to the walls.

AEC object: the shrink wrap can be added around the AEC object, the object can be used as a subtraction from the wall with the shrink wrap being applied to the "hole" in the wall, or the shrink wrap can be ignored.

Use one of the following options to add an interference condition to a wall:

✓ Select the **Design** pull-down menu, pick **Wall Tools,** then **Interference Condition.**

✓ Pick the **Interference Condition** button from the **Wall Tools** toolbar.

✓ Right-click, and select **Design, Wall Tools, Interference Condition.**

✓ Type **AECWALLINTERFERENCE** at the Command: prompt.

✓ Alternatively, select the wall that will have an interference applied, then right-click, and pick the **Model Tools** cascade menu, then **Interference.**

The following prompt sequence is used to add the interference to the wall in Figure 9.11 and have the shrink wrap run around the column object:

> Command: **AECWALLINTERFERENCE** ↵
> Wall interference [Add/Remove]: **A** ↵
> Select walls: *(pick the wall)*
> 1 found
> Select walls: ↵
> Select AEC entities to add: *(pick an AEC object)*
> 1 found
> Select AEC entities to add: *(pick an AEC object)*
> 1 found, 2 total

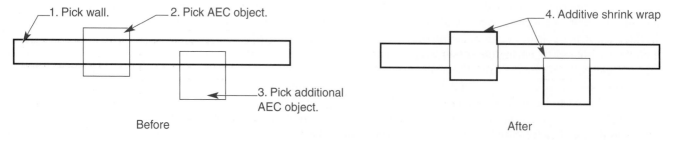

FIGURE 9.11   Adding an interference condition with additive shrink wrap.

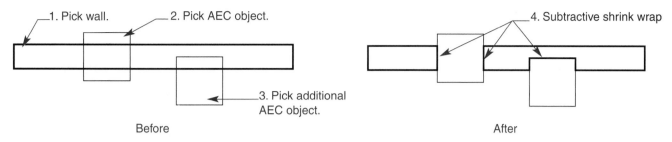

FIGURE 9.12   Adding an inference condition with subtractive shrink wrap.

```
Select AEC entities to add: ↵
Enter shrinkwrap plan effect [Additive/Subtractive/Ignore]: A
2 object(s) added to wall 1825.
Wall interference [Add/Remove]: ↵
Command:
```

Subtracting an AEC object from a wall is similar to adding it. The following prompt sequence is used to add the interference to the wall in Figure 9.12 and make the shrink wrap subtractive:

```
Command: AECWALLINTERFERENCE ↵
Wall interference [Add/Remove]: A ↵
Select walls: (pick the wall)
1 found
Select walls: ↵
Select AEC entities to add: (pick an AEC object)
1 found
Select AEC entities to add: (pick an AEC object)
1 found, 2 total
Select AEC entities to add: ↵
Enter shrinkwrap plan effect [Additive/Subtractive/Ignore]: S ↵
2 object(s) added to wall 1815.
Wall interference [Add/Remove]: ↵
Command:
```

When the Ignore shrink wrap option is used, the AEC object is added to the wall, but the shrink ignores the AEC object and remains at the exterior edges of the wall. The following prompt sequence is used to apply an interference to the wall in Figure 9.13 with ignored shrink wrap:

```
Command: AECWALLINTERFERENCE ↵
Wall interference [Add/Remove]: A↵
Select walls: (pick the wall)
1 found
Select walls: ↵
Select AEC entities to add: (pick an AEC object)
1 found
Select AEC entities to add: (pick an AEC object)
1 found, 2 total
```

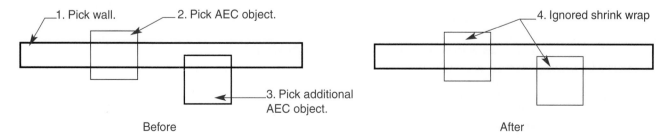

FIGURE 9.13   Adding an interference condition with ignored shrink wrap.

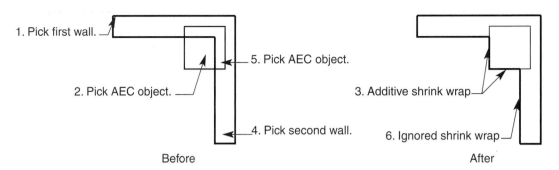

1. Pick first wall.

5. Pick AEC object.

2. Pick AEC object.

3. Additive shrink wrap

4. Pick second wall.

6. Ignored shrink wrap

Before

After

FIGURE 9.14   Adding an interference to two walls using additive shrink wrap.

```
Select AEC entities to add: ↵
Enter shrinkwrap plan effect [Additive/Subtractive/Ignore]: I ↵
2 object(s) added to wall 182B.
Wall interference [Add/Remove]: ↵
Command:
```

Occasionally, an interference needs to be placed in the corner of two walls. When this happens the interference must be applied to the two walls individually. When an additive shrink wrap is applied, one of the walls must use the additive shrink wrap, and the other wall must use the ignored shrink wrap (see Figure 9.14). The following is the command sequence:

```
Command: AECWALLINTERFERENCE ↵
Wall interference [Add/Remove]: A ↵
Select walls: (pick first wall)
1 found
Select walls: ↵
Select AEC entities to add: (pick AEC object)
1 found
Select AEC entities to add: ↵
Enter shrinkwrap plan effect [Additive/Subtractive/Ignore]: A ↵
1 object(s) added to wall 1835.
Wall interference [Add/Remove]: A ↵
Select walls: (pick second wall)
1 found
Select walls: ↵
Select AEC entities to add: (pick AEC object)
1 found
Select AEC entities to add: ↵
Enter shrinkwrap plan effect [Additive/Subtractive/Ignore]: I ↵
1 object(s) added to wall 1834.
Wall interference [Add/Remove]: ↵
Command:
```

Creating an interference in a corner with a subtractive shrink is similar to the method using additive shrink wrap, except that the interference can be added to the two walls at the same time and the subtractive shrink wrap applied to both walls, as shown in Figure 9.15. The following is the command sequence for this application:

```
Command: AECWALLINTERFERENCE ↵
Wall interference [Add/Remove]: A ↵
Select walls: Specify opposite corner: (select both walls)
2 found
Select walls: ↵
Select AEC entities to add: (select AEC object)
1 found
Select AEC entities to add: ↵
Enter shrinkwrap plan effect [Additive/Subtractive/Ignore]: S ↵
1 object(s) added to wall 1838.
```

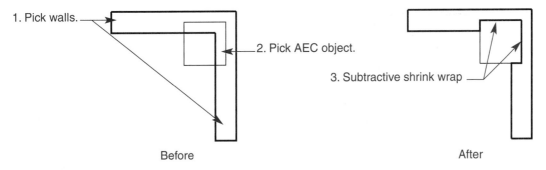

1. Pick walls.

2. Pick AEC object.

3. Subtractive shrink wrap

Before

After

FIGURE 9.15 Adding an interference to two walls with subtractive shrink wrap.

1 object(s) added to wall 1837.
Wall interference [Add/Remove]: ↵
Command:

The **Remove** option in the Interference Condition command is used to remove an interference from a wall. When you use this option you are prompted to select a wall with an interference, then select the objects to be removed from the wall.

Exercise 9.1

1. Open EX8-11-Flr-Pln.dwg.
2. Use the Structural Member catalog to add a 12 × 12 piece of rough-cut lumber to the drawing.
3. Use the Add Column command to add two columns 8′-0″ in length to the drawing, as indicated in the figure.
4. Use the Interference Condition command to add the columns to the two walls using additive shrink wrap.
5. Save the drawing as EX9-1-Flr-Pln.dwg.

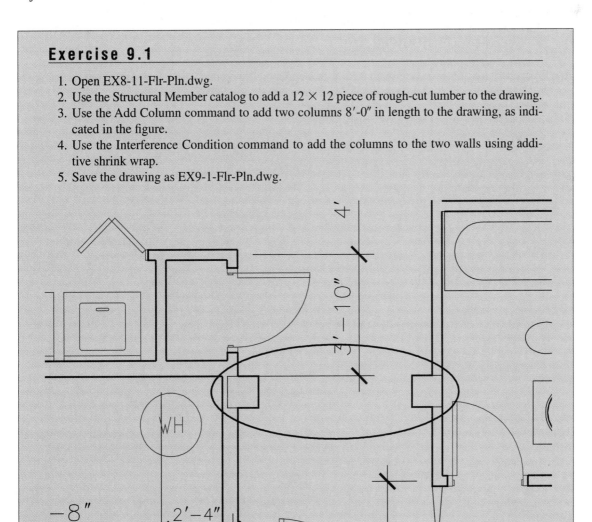

Using Wall Modifiers

Wall modifiers are custom shapes that can be added to the surface of a wall object to enhance the design or look of the wall. Wall modifiers can be used to create a wide footing in a foundation wall under a fireplace or can be applied to an interior wall to represent a custom shelf or decorative feature. Once a wall modifier has been created, it can be added to either side of a wall and at any location, as shown in Figure 9.16.

Wall modifiers are similar to an interference condition, because they are added to a wall to enhance the look of the wall. A wall modifier begins with a polyline shape that is turned into a wall modifier style, which in turn is added to the wall. Typically, the wall modifier is a vertical element that is added to the drawing. Architectural Desktop includes a simple wall modifier style shaped like a square that can be added to a wall. Each style that is created is given a name. The default wall modifier is named Standard.

Adding Wall Modifiers

Use one of the following options to add a wall modifier:

- ✓ Pick the **Design** pull-down menu, select the **Wall Tools** cascade menu, and pick **Add Wall Modifier…**.

- ✓ Pick the **Add Wall Modifier** from the **Wall Tools** toolbar.

- ✓ Right-click, and select **Design, Wall Tools, Add Wall Modifier…**.

- ✓ Type **AECWALLMODIFIERADD**.

A series of prompt sequences are displayed in the Command line area. The following prompt sequence is used to add the Standard wall modifier to the wall in Figure 9.17.

Command: **AECWALLMODIFIERADD** ⏎
Select a wall: *(select a wall)*
Select start point: *(use tracking to track from an endpoint)* **24** ⏎
Select end point: *(specify the length of the modifier)* **36** ⏎

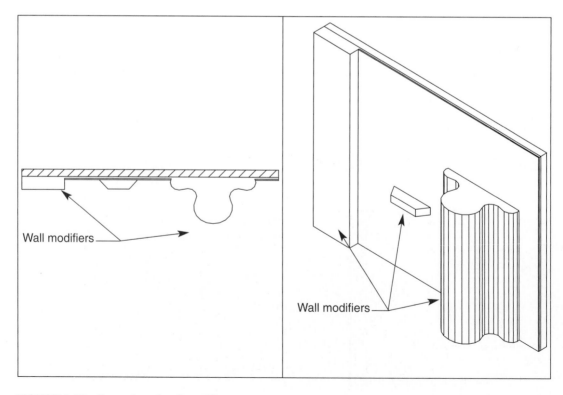

FIGURE 9.16   Examples of wall modifiers.

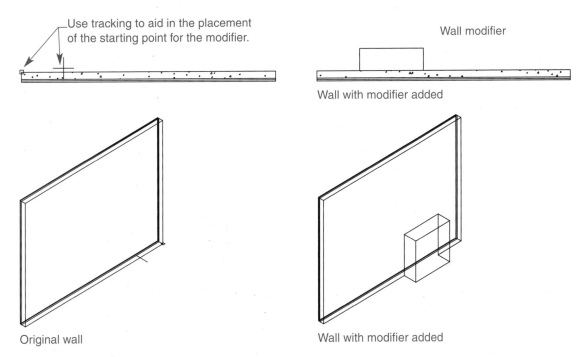

Use tracking to aid in the placement of the starting point for the modifier.

Wall modifier

Wall with modifier added

Original wall

Wall with modifier added

FIGURE 9.17 The Standard wall modifier is a square shape that is added to a wall. The modifier can be added anywhere along a wall and the size of the shape determined as the modifier is added.

> Select the side to draw the modifier: *(pick the side of the wall to place the modifier)*
> Enter wall modifier depth<6 7/8">: *(specify the depth of the modifier)* **12** ↵
> *(attach the modifier to a component and set its height)*
> Command:

After the depth for the modifier has been specified, the **Add Wall Modifier** dialog box is displayed (see Figure 9.18). This dialog box is used to select a modifier to be assigned to a wall component and to specify the height of the modifier.

Removing a Wall Modifier

Wall modifiers can be removed from a wall by using the **Remove Wall Modifier** command. Use one of the following options to access this command:

✓ Select the **Design** pull-down menu, pick the **Wall Tools** cascade menu, and select **Remove Wall Modifier....**

✓ Pick the **Remove Wall Modifier** command from the **Wall Tools** toolbar.

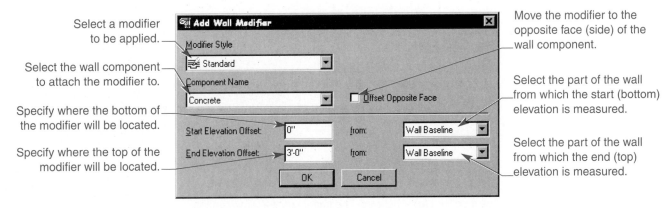

Select a modifier to be applied.

Select the wall component to attach the modifier to.

Specify where the bottom of the modifier will be located.

Specify where the top of the modifier will be located.

Move the modifier to the opposite face (side) of the wall component.

Select the part of the wall from which the start (bottom) elevation is measured.

Select the part of the wall from which the end (top) elevation is measured.

FIGURE 9.18 The Add Wall Modifier dialog box is the final step in adding a wall modifier to a wall.

✓ Right-click, and select **Design, Wall Tools, Remove Wall Modifier...** from the cursor menu.

✓ Type **AECWALLMODIFIERREMOVE** at the Command prompt.

The following prompt sequence is used to remove a wall modifier:

Command: **AECWALLMODIFIERREMOVE** ↵
Select a wall: *(pick a wall with a wall modifier)*
Select a modifier: *(pick the wall modifier)*
Convert removed modifier to a polyline? [Yes/No] <N>: *(enter yes to convert the modifier to a polyline or no to completely remove the modifier)* ↵
Command:

Wall Modifier Styles

So far you have added a wall modifier from the Standard wall modifier style. Custom polyline shapes can be created and applied to walls as modifiers. To create a custom wall modifier style, first draw the shape using a polyline. When the polyline is drawn, the starting and ending points form an imaginary plane that establishes how the modifier is placed along the wall (see Figure 9.19).

Use one of the following options for creating the style after the polyline for the modifier has been drawn:

✓ Pick the **Design** pull-down menu, access the **Wall Tools** cascade menu, and select the **Wall Modifier Styles...** command.

✓ Pick the **Wall Modifier Styles** button in the **Wall Tools** toolbar.

✓ Right-click, and select **Design, Wall Tools, Wall Modifier Styles...** from the cursor menu.

✓ Type **AECWALLMODIFIERSTYLE** at the Command prompt.

The Style Manager dialog box is displayed, opened to the **Wall Modifier Styles** section shown in Figure 9.20.

To create a new style, pick the New Style button in the Style Manager or right-click in the right pane, and select New. Type a name for the modifier and press the [Enter] key to create the named style. Next, pick the new style in the right plane, and select the Set From button, or right click and select Set From... to return you to the drawing screen where the polyline can be selected. Once the polyline has been picked, the Style Manager is redisplayed. The polyline has now been created as a wall modifier style and is ready to be applied to a wall. When a wall modifier is added, the new modifier can be selected from the **Add Wall Modifier** dialog box (see Figure 9.18).

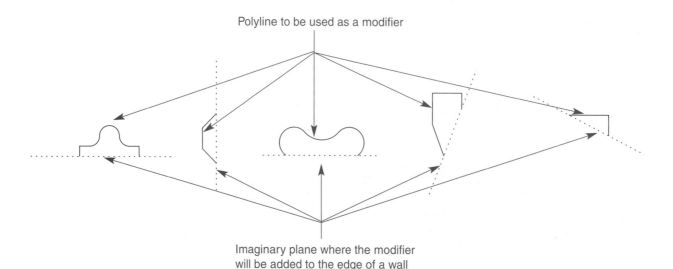

FIGURE 9.19 Polylines for use as modifiers are displayed. The imaginary line between the starting and ending points forms a plane where the modifier is attached to a wall.

New Style button ———   ——**Set From** button

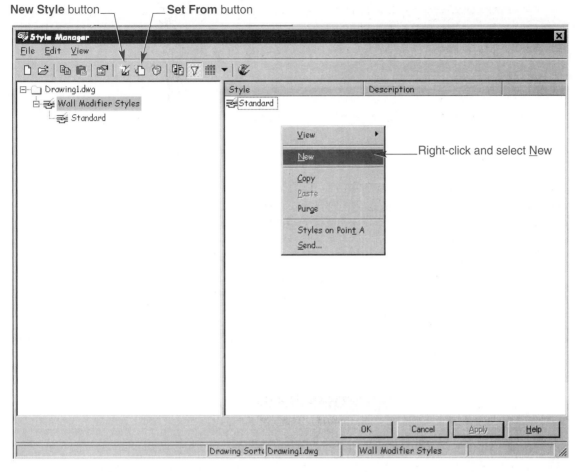

FIGURE 9.20   The Wall Modifier Styles section within the Style Manager is used to create a modifier style.

Converting a Polyline into a Wall Modifier

Another way of creating a modifier style and applying it to a wall is to convert a polyline into a modifier. To do this, draw a polyline in the shape desired along the wall where you want the modifier attached, as shown in Figure 9.21, then use one of the following options to access the **Convert Polyline to Wall Modifier** command:

✓ Pick the **Design** pull-down menu, access the **Wall Tools** cascade menu, then **Convert Polyline to Wall Modifier....**

✓ Select the **Convert Polyline to Wall Modifier** button in the **Wall Tools** toolbar.

✓ Right-click, and select **Design, Wall Tools, Convert Polyline to Wall Modifier...** from the cursor menu.

✓ Type **AECWALLMODIFIERCONVERT** at the Command prompt.

The following sequence is used to convert the polyline in Figure 9.21 into a wall modifier style and add it to the wall:

Command: **AECWALLMODIFIERCONVERT** ↵
Select a wall: *(select the wall)*
Select a polyline: *(select the polyline)*
Erase layout geometry? [Yes/No] <N>: **Y** ↵
(enter a new name for the style in the New Wall Modifier Style Name dialog box)
Command:

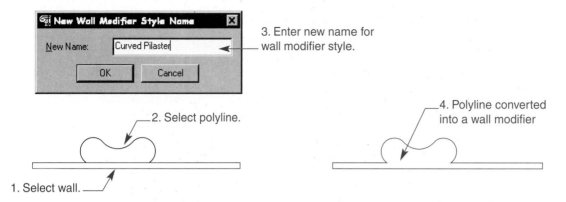

3. Enter new name for
wall modifier style.

2. Select polyline.

4. Polyline converted
into a wall modifier

1. Select wall.

FIGURE 9.21   (A) Draw a polyline with the shape desired at the location where the wall
modifier will be created. (B) The finished wall modifier after being converted and applied.

Changing a Wall Modifier

There may be times when you need to edit a wall modifier that has been added to a wall. The **Wall Properties**
dialog box can be used to make adjustments to a wall modifier. To do this, first select the wall with a modifier,
then right-click and select Wall Properties… from the cursor menu. The Wall Properties dialog box is displayed.
The **Wall Modifiers** tab is used to make any adjustments to the wall modifier (see Figure 9.22).

The top of the tab includes a list of wall modifiers attached to the selected wall and their parameters.
The following settings are available in the Wall Modifier tab:

- **Add**   Pick this button to add a wall modifier to a wall.

- **Remove**   Pick this button to remove a wall modifier from a wall.

- **Modifier Style:**   This drop-down list includes a list of wall modifier styles in the
                        drawing. Select a style from the list to apply the modifier to the wall.

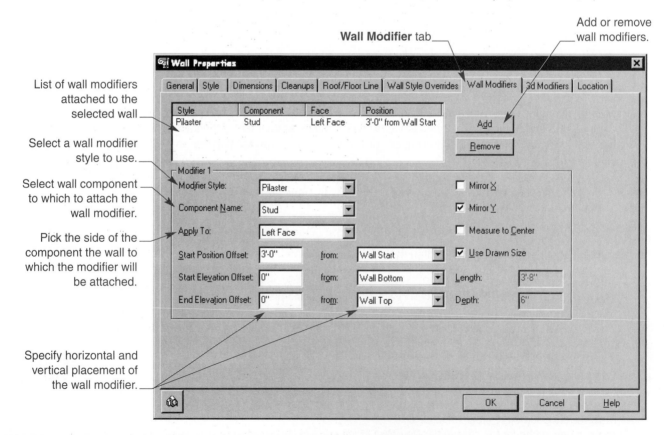

FIGURE 9.22   The Wall Modifiers tab in the Wall Properties dialog box is used to control
and adjust wall modifiers attached to a wall.

- **Component Name:** Use this drop-down list to specify a wall component to which to attach the modifier.

- **Apply To:** This drop-down list assigns the modifier to the left or right side of the selected wall component.

- **Start Position Offset:** This section includes a text box and drop-down list. Specify a distance in the text box that determines how far the modifier is placed from the start, end, or center of the wall. Use the drop-down list to determine if the modifier is placed from the starting or ending point of the wall or from the center of the wall.

- **Start Elevation Offset** This section also includes a text box and drop-down list. Use the text box to determine the distance the bottom of the modifier will be placed vertically in the wall. The drop-down list allows you to measure the vertical offset of the modifier from the wall bottom, baseline, base height, or wall top.

- **End Elevation Offset** This section is similar to the Start Elevation Offset section. This option specifies how the top of the modifier is offset from the wall bottom, baseline, base height, or wall top.

- **Mirror X** Selecting this check box mirrors the wall modifier in the X direction of the wall. The X direction of any AEC object is determined from the starting and ending points of the object. For example, a wall object's X axis is determined from the start point to the endpoint. The positive X direction is measured from the start point to the endpoint.

- **Mirror Y** Selecting this check box mirrors the wall modifier in the Y direction of the wall. The Y direction of an object is 90° from the X direction. This is typically the width of the object.

- **Measure to Center** Selecting this check box measures the Start Position Offset to the center of the modifier instead of to the first (starting) edge of the modifier.

- **Use Drawn Size** This check box uses the actual size of the polyline as the size of the wall modifier.

- **Length** When the Use Drawn Size check box is unchecked, the polyline modifier can be sized along its length by entering a length value.

- **Depth:** When the Use Drawn Size check box is unchecked, the depth (or width) of the polyline modifier can be sized by entering a value in the text box.

Use the Wall Properties dialog box to manage any wall modifiers you have placed in the drawing. It may be easier to draw the polyline where the modifier needs to be placed, then convert it to a wall modifier, then use the Wall Properties dialog box to make any adjustments to the wall modifier.

Insert Modifier as a Polyline

If you have created several wall modifier styles, they can be inserted into the drawing as a polyline, then modified. When the polyline is modified, the wall modifier style needs to be redefined, or a new wall modifier can be created. Use one of the following options to insert a wall modifier into the drawing as a polyline:

✓ Select the **Design** pull-down menu, pick the **Wall Tools** cascade menu, and select **Insert Modifier Style as Polyline.**

✓ Pick the **Insert Modifier Style as Polyline** button in the **Wall Tools** toolbar.

✓ Right-click, and pick **Design, Wall Tools, Insert Modifier Style as Polyline.**

✓ Type **AECWALLMODIFIER** at the Command prompt, then enter **P**.

The Modifier Styles dialog box, shown in Figure 9.23, is displayed. Select a modifier to insert and press the OK button.

FIGURE 9.23 The Modifier Styles dialog box is used to insert a modifier style into the drawing as a polyline.

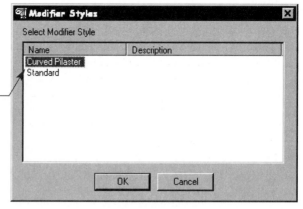

Select a modifier style to be inserted as a polyline.

Exercise 9.2

1. Open EX9-1-Flr-Pln.dwg.
2. Zoom into the fireplace area.
3. Draw a polyline 6″ × 2″ on one side of the fireplace opening to represent a pilaster for a mantle. See diagram A in the figure.
4. Name the modifier FP-Sides. Attach the modifier to the Stud component and specify the starting offset 0″ from the baseline, and the ending offset 4′-0″ from the baseline.
5. Use the Wall Properties dialog box to add the modifier style to the other side of the fireplace. See diagram B in the figure.
6. Draw a polyline to represent a mantle over the fireplace opening. Make the polyline "overhang" 2″ past the two modifiers you just created. See diagram C in the figure.
7. Convert the polyline into a modifier. Name the modifier Mantle, and set the start elevation offset 4′-0″ from the baseline, and the end elevation offset 4′-2″ from the baseline. The fireplace should look like diagram D in the figure.
8. Save the drawing as EX9-2-Flr-Pln.dwg.

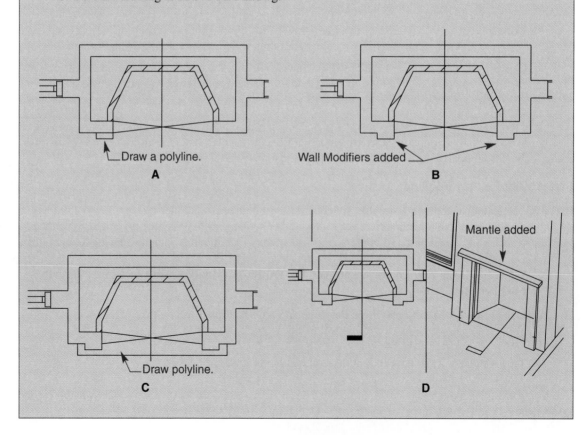

Creating Wall Sweeps

The Wall Modifier is primarily used to add vertical shapes to a wall. A *wall sweep* is used to create horizontal shapes along the length of a wall. For example, a piece of wainscot or molding can be "swept" along a wall object. Another use for wall sweeps is to create rounded horizontal edges that cannot be created in a wall style. Log homes are examples of curved shapes that can be swept horizontally along a wall (see Figure 9.24). Using a custom AEC Profile and applying it to a wall component creates the wall sweep. Multiple AEC profiles can be swept along a multicomponent wall, with one profile being swept along one component at a time. AEC Profiles were discussed in Chapter 7.

   When an AEC Profile is swept along a wall, the profile is inserted at the lower left corner of the component that is being swept and is swept from the start point to the end point of the wall (see Figure 9.25). When the profile is swept along the wall, the wall component maintains the original profile shape and size and is essentially converted to the shape of the AEC profile.

Field Notes

> When creating a polyline for conversion to an AEC profile, it is important to draw the polyline at the size and shape that is to be swept along a wall component. It is also critical to establish an insertion point for the profile at the lowest left corner of the polyline in order for the profile to appear correctly.

   Once an AEC profile has been created, it can be swept along a wall. Use one of the following options to sweep a profile along a wall component:

✓ Select the **Design** pull-down menu, pick the **Wall Tools** cascade menu, then select **Sweep Profile.**

✓ Pick the **Sweep Profile** button in the **Wall Tools** toolbar.

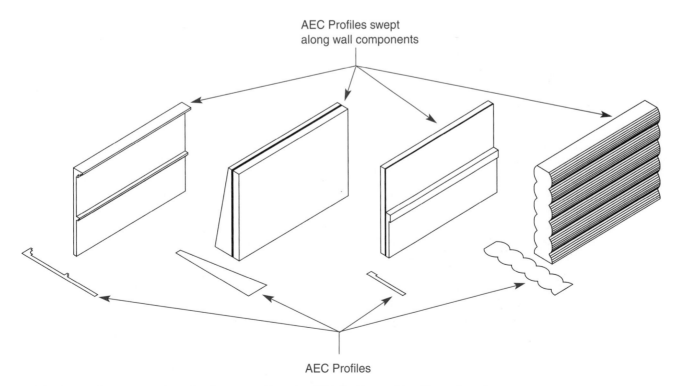

FIGURE 9.24 Some examples of wall sweeps. Note the AEC Profile used beside each wall.

FIGURE 9.25 The lower left corner of the profile is attached to the lower left corner of a wall component. The profile is then swept from the starting point to the ending point.

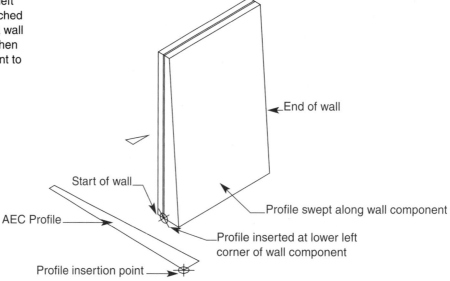

✓ Right-click, and pick the **Design, Wall Tools, Sweep Profile** command from the cursor menu.

✓ Type **AECWALLSWEEP** at the command prompt.

✓ First, pick the wall to be swept, right-click, and pick the **Model Tools** cascade menu, then Sweep Profile from the cursor menu.

The following command sequence is used to sweep the profile along the Brick-X Furring style wall in Figure 9.25:

```
Command: AECWALLSWEEP ↵
Select walls: (pick wall to be swept)
1 found
Select walls: ↵
(pick an AEC profile from the Profile Definitions dialog box)
Available Wall Style Components
1. Brick Veneer
2. Stud
3. GWB
Enter wall style component index<1>: (select a component number from the list above to apply
    the profile to) 3 ↵
Command:
```

After you have selected the wall(s) to be swept, the **Profile Definitions** dialog box appears, allowing you to select the profile that is to be used along a wall component. As indicated earlier, multiple profiles can be swept along a multicomponent wall. Each profile must be swept individually and assigned to a different component using the preceding steps outlined. It is important to remember that only one profile can be assigned to each component of a wall. If a different profile is swept along a component that is already assigned to a profile, the new profile is used along the wall component. Figure 9.26 provides an example of three separate profiles being swept along three different wall components.

Two walls that have been swept with the same profiles do not clean up with each other by default. To get the two walls to clean up, the ends of the walls must be mitered. A *miter* is the junction of two walls that are extended and joined together to form a continuous wall at a corner. Use one of the following options to miter the ends of two walls that have been swept:

FIGURE 9.26 Multiple profiles may be swept along different components of a wall.

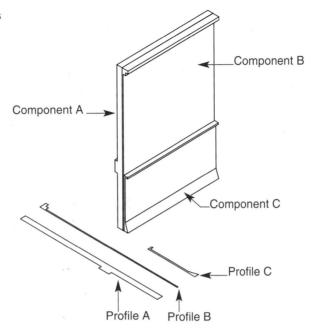

Component B

Component A

Component C

Profile C

Profile A   Profile B

✓ Pick the **Design** pull-down menu, select the **Wall Tools** cascade menu, then pick **Sweep Profile Miter Angles.**

✓ Pick the **Sweep Profile Miter Angles** button from the **Wall Tools** toolbar.

✓ Right-click, and select **Design, Wall Tools, Sweep Profile Miter Angles.**

✓ Type **AECWALLSWEEPMITERANGLES** at the command prompt.

✓ Select one of the walls first, then right-click and pick the **Model Tools** cascade menu, then **Sweep Profile Miter Angles** from the cursor menu.

The following sequence is used to miter to walls that have been swept in Figure 9.27:

Command: **AECWALLSWEEPMITERANGLES** ↵
Select the first wall: *(select one wall)*
Select the second wall: *(select second wall)*
Command:

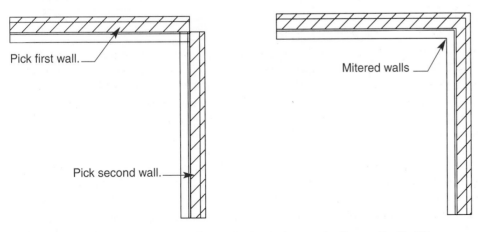

Pick first wall.

Pick second wall.

Mitered walls

FIGURE 9.27 To miter two walls that have been swept use the Sweep Profile Miter Angles command.

Exercise 9.3

1. Start a new drawing using the AEC Arch (imperial) template or the AEC Arch (Imperial-Intl) template.
2. Draw two walls 12'-0" long, 1'-0" wide, and 10'-0" high using the Stud-X wall style. Ensure that the walls form a 90° angle.
3. Create the AEC Profile displayed in the figure. Use the dimensions shown to create the polyline. Name the polyline as in the figure below and select the insertion point as displayed.
4. Sweep the wall profile along both walls.
5. Miter the angle between the two walls.
6. Use the Opening command to add Gothic openings in the swept walls.
7. Save the drawing as P9-3.dwg.

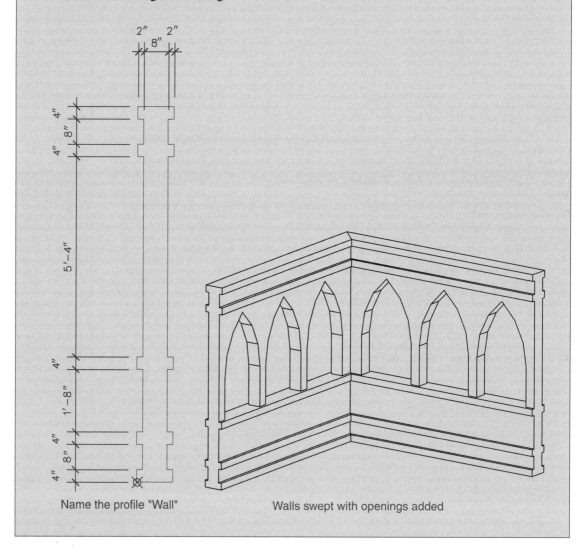

Name the profile "Wall"   Walls swept with openings added

Adjusting the Floor and Roof Line of a Wall

As you gain more experience with the modeling aspects of Architectural Desktop you may find situations in which the top of a wall needs to be projected up to the underside of a roof, or the bottom of a wall needs to be projected down to reflect a sloping site. Architectural Desktop includes two tools that project the top or bottom of a wall to a polyline or to another AEC object: the **Roof Line** and the **Floor Line** (see Figure 9.28).

The roof line has been modified.

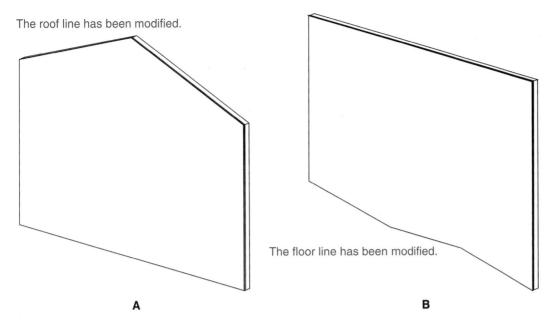

The floor line has been modified.

A   B

FIGURE 9.28 (A) The top of a wall can be projected up to any shaped polyline or AEC object. (B) The bottom of a wall can be projected down to a polyline or AEC object.

Modifying the Top of a Wall

When a wall is drawn and a roof is added over the wall, it is desirable to have the wall meet the underside of roof object. The Roof Line command is used to do this. The following discussion concentrates on projecting the top of the wall to the underside of a polyline, because roof objects have not been discussed yet. Roofs are discussed in a later chapter.

Before you can project the top of a wall up, you must have an AEC object or polyline to project to. Polylines are always drawn in the current drawing plane. For example, if you are working in the top view (floor-plan view), the polyline is drawn flat along the ground (Z axis = 0″). In order to project a wall to a polyline, the polyline must be parallel to the wall(s) that are being projected. To create a polyline that is drawn in the correct plane, use one of the preset view commands to establish an elevation view. When an elevation view has been established, adjust the XY drawing plane so it is perpendicular to your eye. Polylines can now be drawn in this plane (see Figure 9.29).

Field Notes

> After an elevation view is selected, the drawing plane is adjusted. If an isometric view is established, the drawing plane remains in the last orthographic plane that was established before changing to an isometric view. Any new objects drawn are now placed in this drawing plane. To change the drawing plane back to the bottom of the walls or the "floor," change to a top view, and the drawing plane changes to a standard flat drawing plane.

Access the polyline command and draw a polyline above the wall that is to be projected. The polyline can be drawn in any shape using polyline line segments, arced segments, or a combination of the two (see Figure 9.29).

Once the polyline has been drawn, walls that are parallel to the polyline can be projected up to the polyline. To project the top (roof line) of the wall use the Roof Line command by accessing it one of the following ways:

✓ Pick the **Design** pull-down menu, select **Wall Tools,** then pick **Roof Line.**

✓ Select the **Roof Line** button on the **Wall Tools** toolbar.

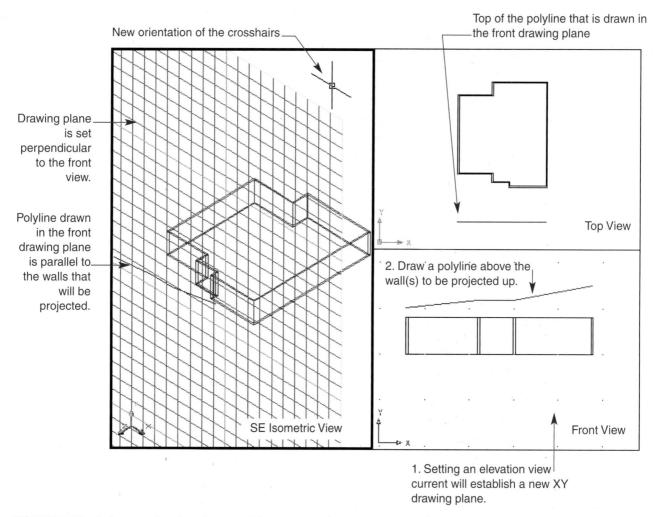

New orientation of the crosshairs

Top of the polyline that is drawn in the front drawing plane

Drawing plane is set perpendicular to the front view.

Polyline drawn in the front drawing plane is parallel to the walls that will be projected.

Top View

2. Draw a polyline above the wall(s) to be projected up.

SE Isometric View

Front View

1. Setting an elevation view current will establish a new XY drawing plane.

FIGURE 9.29 Access an elevation view to adjust the drawing plane so that a polyline can be drawn above a wall to be projected.

✓ Right-click, and select **Design, Wall Tools, Roof Line.**

✓ Type **AECROOFLINE** at the Command: prompt.

✓ Select the wall to be projected, right-click, and select **Model Tools,** then the **Roof Line** command from the cursor menu.

The following sequence is used to adjust the roof line of the wall displayed in Figure 9.30:

```
Command: AECROOFLINE ↵
RoofLine [Offset/Project/Generate polyline/Auto project/Reset]: P ↵
Select walls: (pick the wall to project up)
1 found
Select walls: ↵
Select polyline: (select the polyline to project to)
[1] Wall cut line(s) converted.
RoofLine [Offset/Project/Generate polyline/Auto project/Reset]: ↵
Command:
```

Notice that the Project option is used to project the roof line of the wall up to a polyline. The following are additional options in the Roof Line command to modify the roof line of a wall:

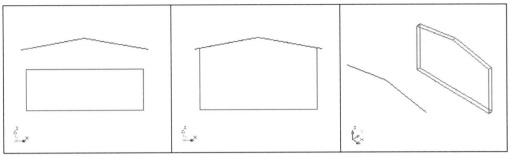

1. Establish the elevation view, then draw the polyline.

2. Use the Roof Line command to project the top (roof line) of the wall to the polyline.

3. Resulting wall with projected roof line

FIGURE 9.30   Projecting the roof line of a wall to a polyline.

- **Offset**   This option offsets the roof line of a wall up a specified distance.

- **Project**   This option, as previously illustrated, is used to project the roof line of a wall up to a polyline.

- **Generate polyline**   This option creates a polyline from the current shape of the roof line of a wall. This new polyline can be moved around the drawing and made parallel to other walls in order for the other walls to be projected.

- **Auto project**   This option projects the roof line of a wall to the underside of an AEC object such as a roof, stair, mass element, slab, or another wall.

- **Reset**   This option resets the roof line of the wall back to a flat-topped roof at the height the original wall was drawn.

Modifying the Bottom of a Wall

The Floor Line command is similar to the Roof Line command except that it affects the bottom (floor line) of a wall. The process is the same as for the roof line. First, create a polyline that is parallel to and below a wall, then project the wall down to the polyline.

Use one of the following options to access the Floor Line command:

✓ Pick the **Design** pull-down menu, select **Wall Tools,** then pick **Floor Line.**

✓ Select the **Floor Line** button on the **Wall Tools** toolbar.

✓ Right-click, and select **Design, Wall Tools, Floor Line.**

✓ Type **AECFLOORLINE** at the Command: prompt.

✓ Select the wall to be projected, right-click, then select **Model Tools,** then the **Floor Line** command from the cursor menu.

The options and the prompts are essentially the same as they were for the Roof Line command. The difference is that the bottom (floor line) of the wall projects down.

Exercise 9.4

1. Open the EX8-8-Fnd.dwg from Chapter 8.
2. Split the drawing screen into two viewports. Assign one viewport to a front view and the other to an isometric view.
3. Draw a horizontal polyline 14″ below the bottom of the walls.
4. Use the Floor Line command to project all the front and rear walls down to the polyline. Note that by projecting the foundation walls down, the footing on the wall is displayed.
5. Change the front elevation view to the left side viewing direction.
6. Draw a horizontal polyline 14″ below the bottom of the walls.
7. Use the Floor Line command to project all the side walls down to the polyline.
8. Change the left elevation view to the top view.
9. Save the drawing as EX9-4-Fnd.dwg.

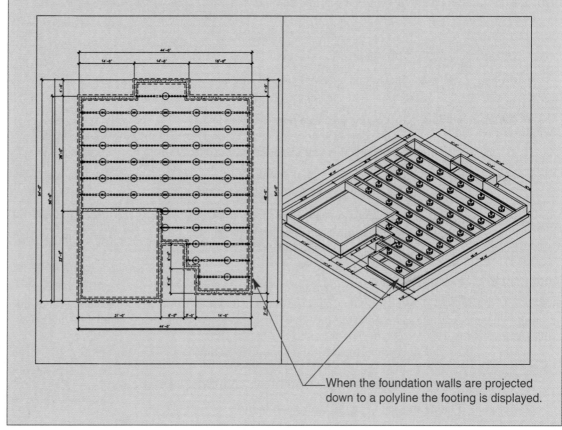

When the foundation walls are projected down to a polyline the footing is displayed.

Adding an Object to a Wall

Wall interferences and wall modifiers were discussed earlier in this chapter. Both of these commands allow you to add a shape to a wall. In the case of the interference condition, the interference "cut" a part of the wall away and filled the gap with the interfering object. The wall modifier added shapes to a wall, defining the shape from a polyline. This section explains another way of adding an object to wall.

Objects that include any mass can be added to a wall. For example, a column object includes mass and is interfered with a wall object. AutoCAD solid objects include mass, and Architectural Desktop mass elements include mass. Both of these types of objects can be added to a wall without interfering with the wall. For more information on creating an AutoCAD solid object, look up *Creating Solids* in the *AutoCAD 2000i Help*, and the *Creating Three Dimensional Objects* chapters in the on-line help within Architectural Desktop.

Mass elements are AutoCAD solids. Mass elements are thoroughly discussed in Chapter 22. For the purposes of this explanation, a simple mass element will be created and then added to a wall. To create a

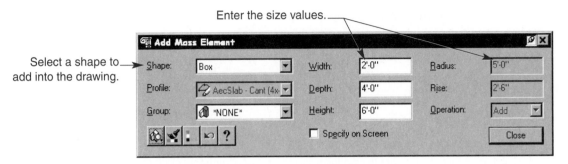

Enter the size values.

Select a shape to
add into the drawing.

FIGURE 9.31   The Add Mass Element dialog box is used to add mass elements into the drawing.

mass element, select the Concept pull-down menu, pick the Mass Element cascade menu, and select Add Mass Element…. The **Add Mass Element** dialog box, shown in Figure 9.31, is displayed.

The **Shape** drop-down menu includes many shapes that can be added to the drawing. Depending on the shape selected, different size text boxes become available. Fill out the desired size for the selected shape, and then pick a location for the mass in the drawing. After the mass element has been added to the drawing, the rotation angle for the element needs to be specified. Typically, if the desired effect is to add the mass element as part of a wall, then a wall should be drawn first and the mass element(s) added along the wall where desired. Figure 9.32 shows three different mass elements within a wall.

After the mass elements have been added along a wall at the correct positions, the object(s) can be included as part of the wall. When an object with mass is added to a wall, the function is called a *body modifier* and is accessed one of the following ways:

✓ Pick the **Design** pull-down menu, select the **Wall Tools** cascade menu, then pick **Body Modifier.**

✓ Pick the **Body Modifier** button in the **Wall Tools** toolbar.

✓ Right-click, and select **Design, Wall Tools, Body Modifier** from the cursor menu.

FIGURE 9.32   Mass elements in the shape of a cone, box, and cylinder are added along a wall.

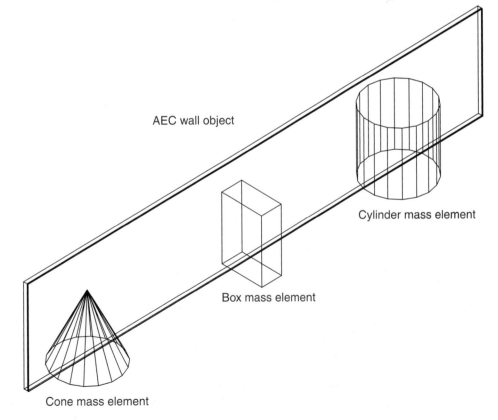

AEC wall object

Cylinder mass element

Box mass element

Cone mass element

✓ Type **AECWALLBODY** at the Command: prompt.

✓ Select the wall to be modified, right-click, and select **Model Tools, Body Modifier** from the cursor menu.

The following sequence is used to add the box object in Figure 9.32 to the wall:

Command: **AECWALLBODY** ↵
Select a wall: *(select the wall)*
Select a body: *(select the box mass element)*
Available Wall Style Components *(the following is a list of components in the wall you are adding the body modifier to)*
1. Brick Veneer
2. Stud
3. GWB
Enter component index<1>: *(enter a component number to add the mass element to)* **3** ↵
Operation [Subtractive/Replacement]<Additive>: *(make the body modifier an additive or subtractive piece in relation to the selected component)* **A** ↵
Description: *(enter a description for the body modifier)* **Partial Height Wall** ↵
Erase layout geometry? [Yes/No] <N>: Y
Command:

During the process of adding a body modifier, you are prompted to select the wall to be modified, then the body (mass object). Choose a wall component to attach the body modifier to, then choose the type of operation to perform. The body modifier can be added to or subtracted from the wall component, or it can replace the wall component. Figure 9.33 shows each of the operation options for a simple single component wall. After selecting the type of operation to use on the wall component, you can erase the original mass object from the drawing by entering a Yes at the *Erase layout geometry? [Yes/No] <N>:* prompt.

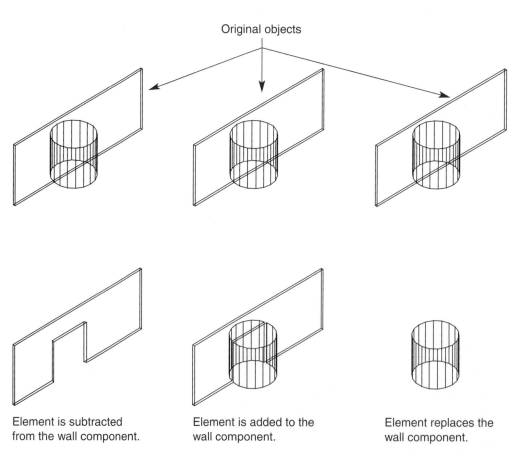

Original objects

Element is subtracted from the wall component.

Element is added to the wall component.

Element replaces the wall component.

FIGURE 9.33   Mass elements or AutoCAD solids can be subtracted from or added to a wall component or can replace the wall component.

Exercise 9.5

1. Start a new drawing using the AEC Arch (Imperial) or AEC Arch (Imperial-Int) template.
2. Create three walls 12'-0" long and 4" wide as shown in the figure.
3. Add a pyramid mass element 6'-0" wide, 2'-0" deep, and 6' high in the middle of each wall.
4. On the first wall, add the mass element as a body modifier using the Additive option. For the description enter Additive.
5. On the second wall, add the mass element as a body modifier using the Subtractive option. For the description enter Subtractive.
6. On the third wall, add the mass element as a body modifier using the Replacement option. For the description enter Replacement.
7. Save the drawing as EX9-5.dwg.

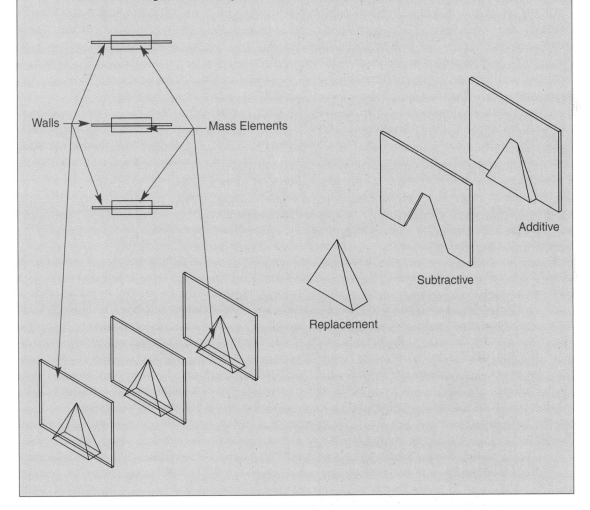

Using Wall Properties

The previous section described tools that can be used to enhance plan-view and model-view wall objects. Chapter 3 also described modifying walls by using the Modify Walls dialog box. This section explains how to use the **Wall Properties** dialog box as yet another way of managing wall objects.

Use one of the following options to access the Wall Properties dialog box:

✓ Select a wall, then right-click, and select **Wall Properties...** from the cursor menu.

✓ Type **AECWALLPROPS** at the command: prompt.

One wall or multiple walls can be selected to be modified by the Wall Properties dialog box. If more than one wall is selected, the options available in the dialog box are limited. The Wall Properties dialog box is divided into nine tabs. Each tab controls a different aspect of the select wall. The tabs available for a single wall are described next.

General Tab

The **General** tab includes an area that can be used to enter a description for the wall that is selected (see Figure 9.34). Additionally two buttons are included along the left edge to add other information.

The **Notes** button can be used to enter a series of notes regarding the wall—for example, the type of wall (fire wall, demo wall), the colors being used in Entity Display, or any additional information that may be pertinent to the selected wall.

The **Property Sets…** button is used to assign a property set definition to the wall. A *property set definition* is special information that can be assigned to an AEC object that provides information about the object such as its size, area, or style. Information available in a property set definition can be pulled from an object and assembled into a schedule. Assigning a property set to a wall can give you information about the type of finishes used, so a Wall Finish Schedule can be generated. Property set definitions and their uses are discussed in the scheduling chapter later in this text.

Style Tab

The **Style** tab indicates the type of style being used by the selected wall by highlighting the style in the list. If the wall needs to be changed to a different wall style, select a different style from the list (see Figure 9.35).

If the wall style is changed, some of the values for the wall may be changed in the other tabs within the Wall Properties dialog box. Pressing the **Reset to Style Defaults** button resets these changed values to the default properties that were set up in the wall style. The lower left corner of the dialog box also includes the floating viewer button. Selecting this button opens the viewer window so you can preview how the selected wall looks when you change different values in the Wall Properties dialog box, whether it is the wall style or a dimensional value.

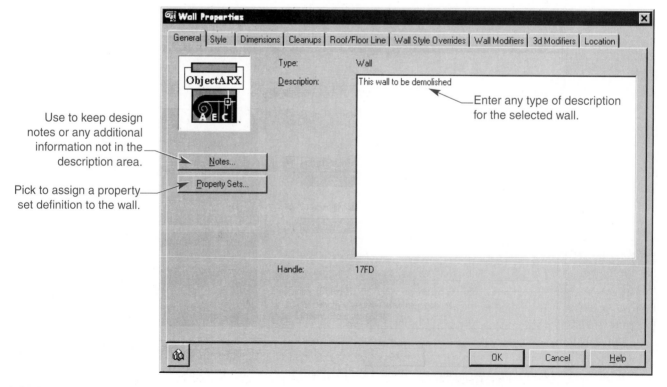

FIGURE 9.34   The General tab is used to enter information about the wall.

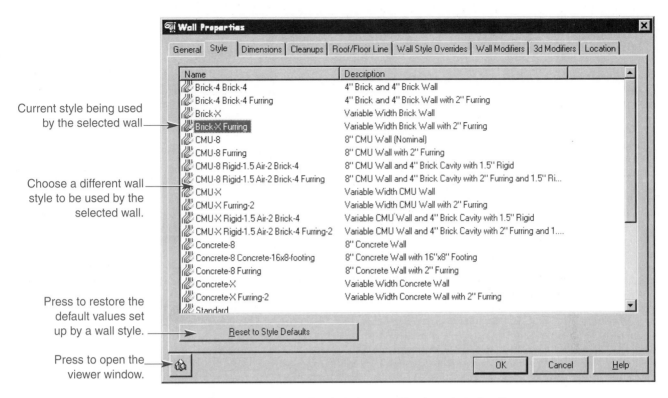

Current style being used by the selected wall

Choose a different wall style to be used by the selected wall.

Press to restore the default values set up by a wall style.

Press to open the viewer window.

FIGURE 9.35 The Style tab allows you to change the wall style being used by the selected wall.

Dimensions Tab

The **Dimensions** tab is used to control the dimensional values for the selected wall (see Figure 9.36). If a wall style has been selected that maintains a fixed width, such as a 4″ stud wall, then the **Wall Width:** text box will be grayed out. The **Base Height** text box indicates the height of the wall. If the roof line of the wall has been modified, the height text box indicates the drawn height, not the base height plus the projected wall height.

The **Length:** text box indicates the length of the wall. If a curved wall is selected, the Length text box is grayed out and the **Radius:** check box is available for modification. Entering a new value for any of the size text boxes changes the wall(s) to reflect the new values.

The **Justify** drop-down list allows you to specify a different justification to use for the selected wall.

Cleanups Tab

The **Cleanups** tab is used to control how the wall cleans up with other walls (see Figure 9.37). At the top of the tab is a check box that allows this wall to cleanup with other walls. If the check box is unchecked, the wall will not attempt to clean up when it crosses or touches another wall.

The **Cleanup Settings** area is used to control the rules for cleanup with other walls. The **Cleanup Group Definitions** drop-down list is used to assign the selected wall to a different wall cleanup group. Walls that belong to the same cleanup group will cleanup with each other. See Chapter 3 for more information regarding wall cleanup groups.

The **Cleanup Radius:** text box is used to specify the default wall cleanup radius for the selected wall. The initial default radius is set to 0″. This requires that the justification lines for walls touch or cross each other in order to cleanup. Specifying a different value can aid in the cleanup of walls. It is strongly recommended that you set your walls cleanup radius equal to the width of the wall. This helps in proper cleanup between walls. The cleanup radius, however, does not have to reflect the width of the wall. Increase or decrease the value in order to cleanup your walls. See Chapter 3 for complete information regarding the wall cleanup radius.

The **Cleanup Radius Overrides** area specifies different cleanup radii at the start and end of a wall. If a starting or ending cleanup radius is specified, this radius overrides the default cleanup radius. The **Graph**

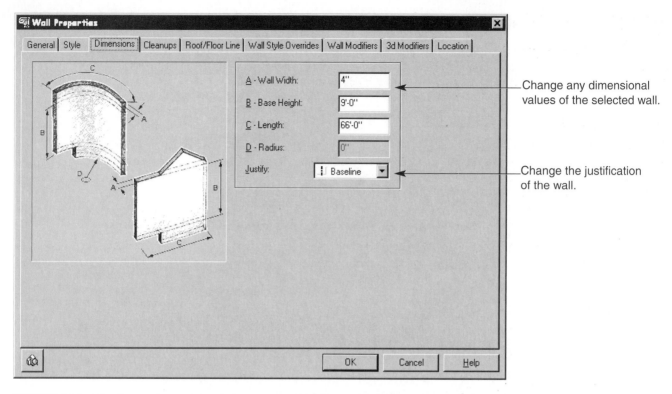

FIGURE 9.36   The Dimensions tab is used to control the dimensions of the selected wall.

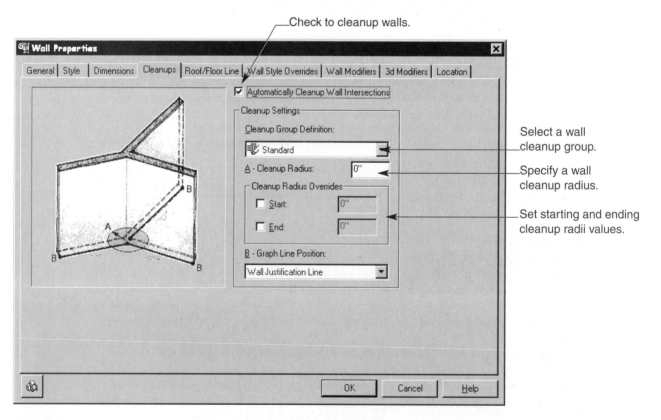

FIGURE 9.37   The Cleanups tab controls how the selected wall will cleanup with other walls.

Line Position drop-down list includes two options: **Wall Justification Line** and **Wall Center Line**, which control the actual point from which the cleanup radius is measured. For example, if the Wall Justification Line is selected, the cleanup radius originates from the wall justification point. If the Wall Center Line is selected, the cleanup radius originates from the center of the line of the wall. If you create walls using a baseline justification it is recommended that you use the Wall Justification Line option.

Roof/Floor Line Tab

The **Roof/Floor Line** tab controls how the top and bottom of the selected wall appears in elevation. If the Roof Line or Floor Line commands have been used prior to entering the Wall Properties dialog box the preview window displays the wall shape in an elevation view. Numerical information is also displayed in the table at the top of the tab (see Figure 9.38).

At the bottom of this tab are two radio buttons: **Edit Roof Line** and **Edit Floor Line**. Selecting one of these options allows you to modify the points along the top or the bottom of the wall. The preview window places vertex point boxes at all the vertex locations. The filled box indicates the vertex point that is being modified. Use the mouse to select a vertex point to be modified. The buttons in this tab have the following different functions related to modifying or adding vertex points:

- **Add Gable...** This button adds a vertex point from the top middle or bottom middle of the wall and projects it up or down depending on whether the roof line or floor line points are being modified. This button is available if the starting vertex point is highlighted.

- **Add Step...** This button adds a step in the wall either up or down depending on whether the roof line or the floor line is being edited.

- **Insert Vertex...** This button adds a vertex point to the wall. When this button is selected the Wall Roof/Floor Line Vertex dialog box appears (see Figure 9.39). This

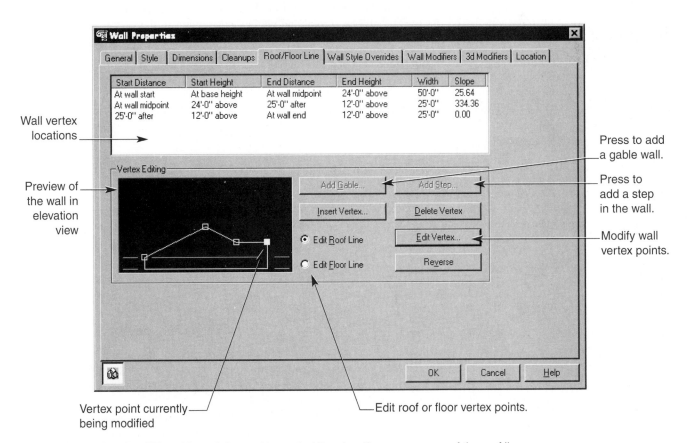

FIGURE 9.38 The Roof/Floor Line tab is used to control the elevation appearance of the roof line and floor line of the selected wall.

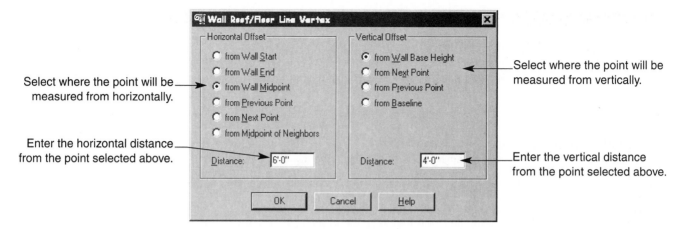

Select where the point will be
measured from horizontally.

Enter the horizontal distance
from the point selected above.

Select where the point will be
measured from vertically.

Enter the vertical distance
from the point selected above.

FIGURE 9.39   Use the Wall Roof/Floor Line Vertex dialog box to place a new vertex
horizontally and vertically in the wall.

dialog box allows you to specify horizontally where the new point will be
placed in relation to an existing point on the wall, and where the point will
be placed vertically in the wall in relation to the projected bottom of the
wall, the wall baseline, the base height of the wall, or the projected top of
the wall.

- **Delete Vertex**   This button deletes the highlighted vertex point in the preview window.

- **Edit Vertex**   Pick a vertex to modify before selecting this button. When the button is
  picked, the Wall Roof/Floor Line Vertex dialog box appears, allowing you
  to readjust the horizontal and vertical positioning of the vertex, as shown
  in Figure 9.39.

- **Reverse**   This button reverses the starting and ending points of the wall, mirroring
  your edited vertex points.

As new points are added, deleted, and modified, the table at the top of the tab displays the location
information for the vertex points. Picking a point from the table highlights the point in the preview pane at
the bottom of the tab.

Wall Style Overrides Tab

The **Overrides** tab is used to modify the starting and ending priorities of the selected wall as well as the
endcaps used on the wall (see Figure 9.40). Each component within a wall is assigned a priority. Wall components with the same priority clean up with each other. *Endcaps* are a type of graphic that controls the
appearance of the end of a wall.

Wall Priorities
As mentioned previously each component of a wall is assigned a priority. A component with a higher priority "cuts through" a wall component with a low priority, as shown in Figure 9.41. The highest priority a
wall component can have is 1. Wall components that represent different types of building materials are
assigned priorities that provide the desired display. For example, in Figure 9.41, a wall with a brick veneer
component can include two wall components. The wall stud is assigned a higher priority than the brick
veneer; therefore, the stud component cuts through the brick veneer component. Wall priorities and how
they are assigned within a wall style are discussed later in this text.

To assign a different priority to a component in a wall press the Add Override... button. The **Priority
Override** dialog box, shown in Figure 9.42, is displayed. Select where the priority for the component will
be overridden. The starting point or ending point of the wall can be selected. Enter the priority number in
the Priority text box, then select the component to assign the priority to in the Component Name text box.

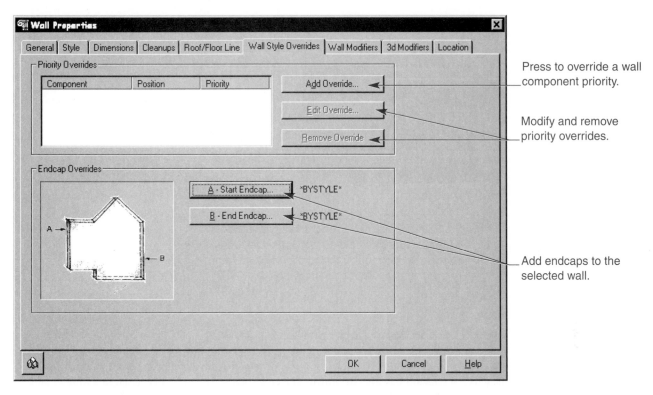

FIGURE 9.40 The Overrides tab is used to add priority overrides to wall components and endcaps to the end of walls.

FIGURE 9.41 A wall component with a higher priority will cut through a component with a lower priority.

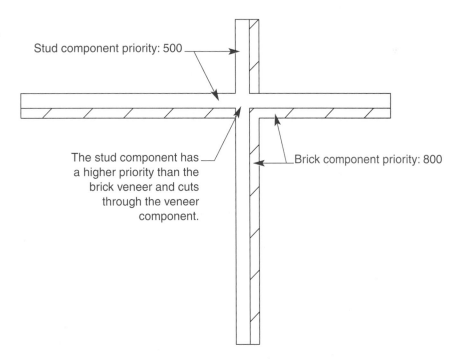

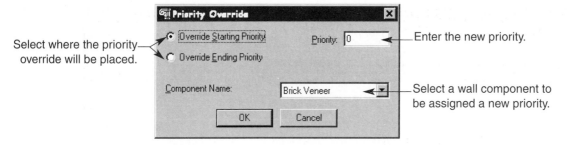

Select where the priority override will be placed.

Enter the new priority.

Select a wall component to be assigned a new priority.

FIGURE 9.42   The Priority Override dialog box assigns a different priority to the starting or ending point of a wall component.

FIGURE 9.43   Different endcaps can be assigned to the ends of walls.

Starting endcap   Ending endcap

Wall Endcaps

As described earlier, a wall endcap is a graphic that can be assigned to the end of a wall to enhance the desired display of the wall. For example, a wall endcap can be created to wrap around the end of a stud component (see Figure 9.43). Template drawings include several types of endcaps that can be assigned to different walls. Custom endcaps can also be drawn. The creation of endcaps is discussed in a future chapter.

The A - Start Endcap… and B - End Endcap buttons are used to assign an endcap to the end of a wall.

Wall Modifiers Tab

The **Wall Modifiers** tab controls any wall modifiers that have been placed in the selected wall (see Figure 9.44). The addition of wall modifiers was discussed earlier in this chapter. Existing wall modifiers can be modified in this tab, or new wall modifiers can be applied to the wall from this tab.

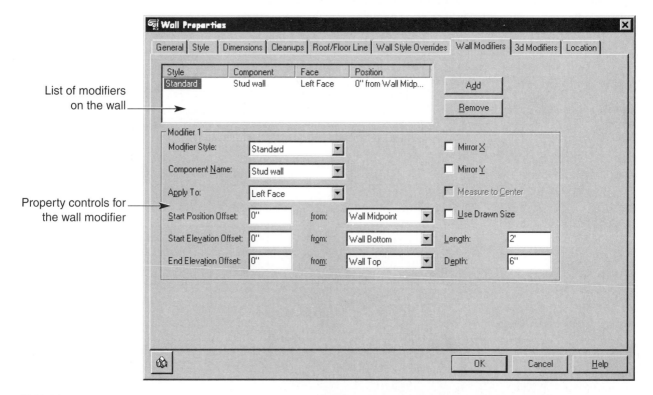

List of modifiers on the wall

Property controls for the wall modifier

FIGURE 9.44   The Wall Modifiers tab is used to manage, modify, or add wall modifiers to the selected wall.

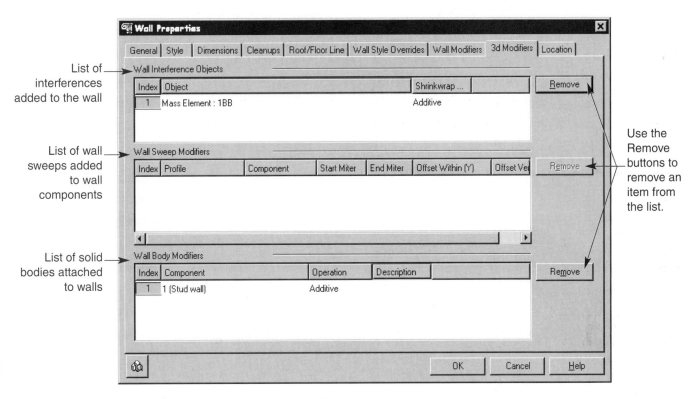

List of interferences added to the wall

List of wall sweeps added to wall components

List of solid bodies attached to walls

Use the Remove buttons to remove an item from the list.

FIGURE 9.45   The 3D Modifiers tab is used to list, modify, and remove any customization added to a wall object.

3D Modifiers Tab

The **3D Modifiers** tab is used to list any wall interferences, sweeps, or body modifiers that have been added to the wall (see Figure 9.45). Interferences, sweeps, and body modifiers can be removed from the wall, or their values can be changed by selecting the appropriate value and choosing from a list the different values that can be assigned.

Location Tab

The **Location** tab is used to identify the selected wall's location relative to a user coordinate system (see Figure 9.46). Two options at the top indicate the wall's location relative to the World coordinate system (the standard AutoCAD drawing plane) or the current User coordinate system.

When the World coordinate system is used, the Z-axis value indicates where the bottom of the wall is located. If you have walls that need to be added into the drawing at a different height (such as a second floor), adjust the wall's Z-axis value in the World coordinate system. Entering a positive value places the wall "up" in elevation.

Exercise 9.6

1. Open EX9-5.dwg.
2. Select the additive wall, right-click, and select Wall Properties... from the cursor menu.
3. Change the wall to a 12″-wide wall. Adjust the height to 6′-0″.
4. Select the subtractive wall, right-click, and select Wall Properties... from the cursor menu.
5. Change the roof line of the wall by adding a gable 4′-0″ from the base height of the wall.
6. Select the replacement wall, right-click, and select Wall Properties... from the cursor menu.
7. Change the replacement modifier to additive, then change the Z-axis location to 6′-0″.
8. Save the drawing as EX9-6.dwg.

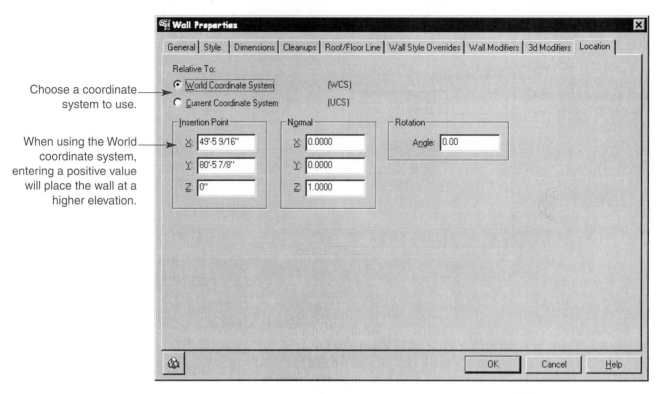

Choose a coordinate system to use.

When using the World coordinate system, entering a positive value will place the wall at a higher elevation.

FIGURE 9.46   The Location tab is used to indicate the location of the wall.

Additional Wall Modifying Commands

Three additional commands can be used to control walls created in the drawing. Walls can be joined to form one wall, difficult wall cleanups can be fixed, and objects such as doors, windows, window assemblies, and openings independent of a wall can be "anchored" or attached to a wall. Descriptions of each of these commands follow.

Joining Walls

When floor plans are created many walls are added to the drawing. Occasionally, you may want to add two walls together to form one wall. The **Join Walls** command can be used for this application. When two walls are joined together, any modifications made to one wall is transferred to the other wall. For example, wall modifiers in one wall maintain their location and are added within the new single wall.

In order for two walls to be joined together the following few simple rules must be followed:

1. Straight wall segments must have justification lines that are colinear and must touch at one endpoint.

2. Arced walls must have the same center point and radius. These walls must also touch at one endpoint.

3. Walls must use the same wall style, width, and must be in the same wall cleanup group.

Use one of the following options to activate the Join Walls command:

✓ Select the **Design** pull-down menu, pick the **Wall Tools** cascade menu, and select the **Join Walls** command.

✓ Pick the **Join Walls** button on the **Wall Tools** toolbar.

✓ Right-click, and select **Design, Wall Tools, Join Walls** from the cursor menu.

✓ Type **AECWALLJOIN** at the Command: prompt.

✓ Select a wall to be joined, right-click, and select **Plan Tools, Join** from the cursor menu.

Select the first wall to be joined, then select the second wall. The two walls are joined together and form one wall. The following sequence is used to join two walls:

Command: **AECWALLJOIN** ⏎
Select first wall: *(select the first wall)*
Select second wall: *(select the second wall)*
Merged walls successfully
Command:

Merge Walls

The **Merge Walls** command is designed to provide a means of cleaning up walls when adjusting the wall cleanup radius fails. This command should be used as a last resort, as any walls that cross through or intersect a merged wall must also be merged together.

Use one of the folloiwng options to access the Merge Walls command:

✓ Pick the **Design** pull-down menu, select the **Wall Tools** cascade menu, and pick the **Merge Walls** command.

✓ Pick the **Merge Walls** button on the **Wall Tools** toolbar.

✓ Right-click, and select **Design, Wall Tools, Merge Walls** from the cursor menu.

✓ Type **AECWALLMERGE** at the Command: prompt.

✓ Alternatively, access this command by first selecting the wall to be merged, right-clicking, and picking the **Plan Tools** cascade menu, then picking **Merge** from the cursor menu.

You are prompted to **Add** or **Remove**. Enter **A** for Add to merge another wall into the selected wall, or enter **R** for Remove to remove the merging effect. The following prompt sequence merges two walls together:

Command: **AECWALLMERGE** ⏎
Wall Merge [Add/Remove]: **A** ⏎
Select a wall: *(select the first wall to merge)*
Select walls to merge with: *(select the second wall to merge)*
1 found
Select walls to merge with: ⏎
Merged with 1 wall(s).
Wall Merge [Add/Remove]: ⏎
Command:

Anchoring to Walls

Chapter 3 described how to add a door, window, window assembly, and opening independent of a wall. During the insertion process for one of these objects, instead of selecting a wall into which to insert the object, pressing [Enter] adds the object without attaching it to the wall. If you have added any of these objects without attaching them to a wall, the **Anchor to Wall** command can be used to attach the independent object to a wall. Use one of the following options to access this command:

✓ Select the **Design** pull-down menu, pick the **Wall Tools** cascade menu, and select the **Anchor to Wall** command.

✓ Pick the **Anchor to Wall** button on the **Wall Tools** toolbar.

✓ Right-click and select **Design, Wall Tools, Anchor to Wall** in the cursor menu.

✓ Type **AECWALLANCHOR** at the Command: prompt.

Upon entering the command you are prompted to **Attach** or **Free** an object. Attaching an object attaches the AEC object to the wall. Entering the Free option removes the wall anchor from the object, releasing the object from the wall and making it independent of the wall. The following prompt sequence attaches an independent door object to a wall:

Command: **AECWALLANCHOR** ↵
Wall anchor [Attach object/Free object]: **A** ↵
Select object to be anchored: *(select the object to attach to a wall)*
Select a wall: *(select the wall to which to anchor the object)*
Command:

Field Notes

If doors or windows have been inserted into a drawing where AutoCAD lines are being used to represent walls, you may decide later to convert one of the lines into an AEC wall where the Anchor to Wall command can be used to attach the door or window to the newly converted wall.

||||||||||| CHAPTER REVIEW

 Use the CD-ROM to test your knowledge and skills.

Chapter Test

To check your understanding of the content provided in this chapter, access the Test file in the CH09 folder of the CD-ROM that accompanies this text.

Chapter Project

To practice the Architectural Desktop skills presented in this chapter, access the Project files in the CH09 folder of the CD-ROM that accompanies this text. The project files are in pdf format and include sample drawings and instructions for completing each project.

|||||||||||||||||||||||||||||||||||||||||||||||||||||||||||||||||||||||||||||||||||||||||||||||||||||||||||||

CHAPTER 10

Editing and Modifying Wall Openings

LEARNING GOALS

After completing this chapter, you will be able to:

◎ Control door, window, and opening objects.

◎ Attach a custom block to a door, window, and opening.

◎ Anchor door, window, and opening objects within a wall.

◎ Change door, window, and opening properties.

◎ Use window assembly tools.

In Chapter 3 you were introduced to the process of adding and modifying openings in a wall such as doors, windows, window assemblies, and AEC openings. Architectural Desktop includes a number of additional tools that can be used to enhance the look of these objects and prepare them for use in construction documents or in the model.

Chapter 9 introduced many additional wall commands that can be used to model 3D walls or to adjust the display of wall objects. This chapter covers the tools such as entity display available for modifying doors, windows, window assemblies, and openings and also discusses how to add custom shapes to these objects to enhance their appearance, and how to control the properties of each type of opening object.

Controlling Door Object Display

Earlier in this text we discussed the process of adding and modifying a door object. These commands provided the basic information for controlling how a door appears in a drawing. For example, when a door is added, the size, style, and opening percentage are specified, controlling the look of the door object. If the door needs to be adjusted or changed, the Modify Door command can be used to adjust the size, style, and opening percentage of the door, and grips can be used to control where the door leaf and swing is placed in relation to the wall object. Architectural Desktop includes additional tools to control many properties of a door to configure the appearance of the object. The following discussion provides many ways of controlling how doors behave in the drawing.

Door Entity Display

As mentioned in previous chapters, the entity display for any AEC Object is very important for processing how an object is to appear in a particular view. Doors and walls include their own display settings that can be managed and controlled through entity display. Chapter 5 discussed the use of the Architectural Desktop display system and described what a display representation is and how to modify it at the default level. To change the display of a door, select the door or a door using a style that is to be modified, right-click, and select Entity Display… from the cursor menu. The **Entity Display** dialog box for door objects is displayed. Select the **Display Props** tab to adjust the display properties for the selected door, as shown in Figure 10.1.

Once the Display Props tab has been accessed, you need to determine in which display representation the door needs to be adjusted. Select the representation to be modified from the drop-down list in the upper left corner of the dialog box. Next, determine the level of display control for the door.

367

Select to adjust the display
properties for the door.

Select the display
representation to be
modified from the list.

Levels of display
control (the plan
representation for
a door is being
controlled at the
door style level)

Select to attach
the override.

Pick to edit the display
of doors (in this case
the plan view door at
the style level).

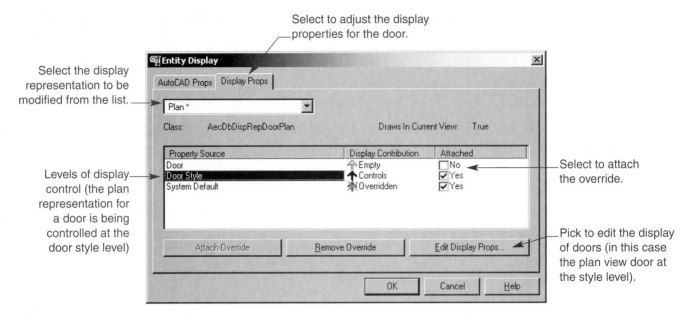

FIGURE 10.1   The Entity Display dialog box for doors is similar to the dialog box for walls.

The **Property Source** column lists the three levels of display control for the door object. The **System Default** level controls the display of doors globally unless the other doors are overridden. The **Door Style** level controls the display of doors based on the door style used by the selected door. The **Door** level controls the display of the single selected door. Pick the level of control to highlight it. Once you have selected the level of control, place a check mark in the associated box under the **Attached** column, which attaches an override to the previous level of display control. Figure 10.1 displays an override set at the style level for a door plan display representation. Press the Edit Display Props… button to adjust the display of the door. The **Entity Properties** dialog box is displayed (see Figure 10.2).

Depending on the display representation being adjusted, the Entity Properties dialog box contains different options. All the Entity Properties dialog boxes include the **Layer/Color/Linetype** tab shown in Figure 10.2. This tab controls how the components for a door object are displayed in the selected display

Controls the layer, color, and linetype
properties for door components

Controls additional display
properties for a door

Door components

Display properties for
each door component

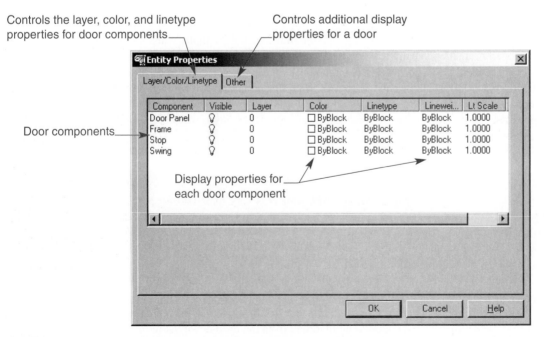

FIGURE 10.2   The Entity Properties dialog box is used to control the display of the components of door objects. In this diagram the plan representation for door objects is displayed.

representation. Doors typically include a door panel, frame, stop, and swing component. The components can be turned on or off by selecting the lightbulb symbol. Layers, Colors, Linetypes, Lineweights, Lt Scale, and Plot Style for each component can be adjusted by picking on top of the appropriate symbol or text.

Many of the display representations also include a tab named **Other**. Depending on the representation selected, the Other tab displays different options (see Figure 10.3). The Other tab displayed in Figure 10.3A is used to add custom blocks to the door object. The Other tab displayed in Figure 10.3B is used to adjust door threshold sizes. These topics are discussed later in this section.

When you have finished adjusting the display properties for door objects, press the OK button to return to the Entity Display dialog box, where you can select another display representation and override. Press the Edit Display Props… button to adjust the display properties of the newly selected display representation.

When selecting a display representation, notice that there are a few different representations available for doors that are not available for walls. Two of these representations are the elevation and the threshold plan. The *elevation representation* controls how the door appears in an elevation view. Generally, in an ele-

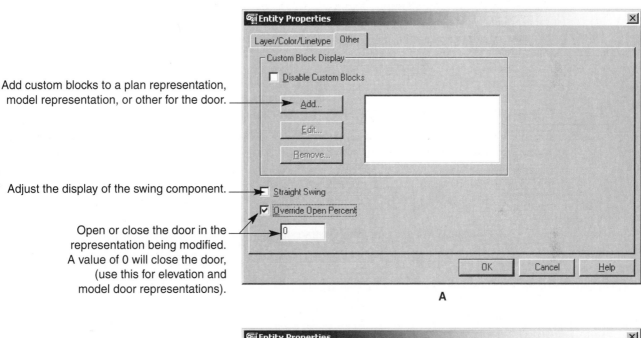

Add custom blocks to a plan representation, model representation, or other for the door.

Adjust the display of the swing component.

Open or close the door in the representation being modified. A value of 0 will close the door, (use this for elevation and model door representations).

A

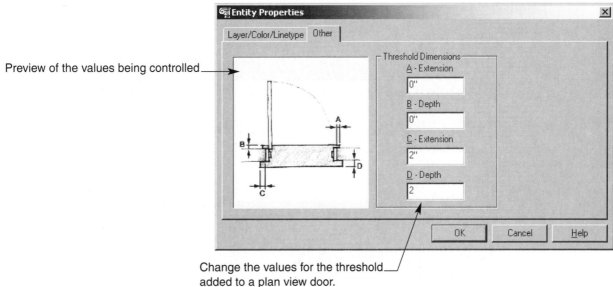

Preview of the values being controlled

Change the values for the threshold added to a plan view door.

B

FIGURE 10.3   (A) The Other tab for most of the door display representations. (B) The Other tab used for the door threshold display representations.

vation plan, door objects are displayed closed. The display representation for a door in an elevation view can be modified so the door is closed in elevation yet remains open in the plan view. To close the door, select the Elevation representation in the Entity Display dialog box, set the level of override, then pick the Edit Display Props... button. Select the Other tab in the Entity Properties dialog box. Notice that the last option in the tab overrides the opening percentage of the door. If the value is set to 0, the door appears closed in the representation being modified yet open in a plan view (see Figure 10.3A).

Threshold Plan Representation

The ***threshold plan representation*** allows you to add an inside threshold, an outside threshold, or both in a plan view. The plan representation and the threshold plan representation can be displayed at the same time for a door while looking in a top view. To display a door with a threshold, first adjust the display in the plan representation for the door, then adjust the threshold plan display representation.

The threshold plan representation includes two components: threshold A and threshold B. The *threshold A* component is a threshold that is added to the swing side of the door. *Threshold B* is added to the opposite side of the door, as shown in Figure 10.4. Either or both thresholds can be turned on or off for a single plan view door or door style.

The threshold plan representation also includes an Other tab, as shown in Figure 10.5. This tab controls the size of the thresholds on plan view doors. Values A and B control the size of threshold A. The values C and D control the sizes for threshold B.

Field Notes

> Thresholds are visible only in a top view. When assigning thresholds it is usually a good idea to set the override by individual door, because you may have only a few doors that need the threshold. If you have a door style that represents an exterior door, then you may want to set the threshold by style.

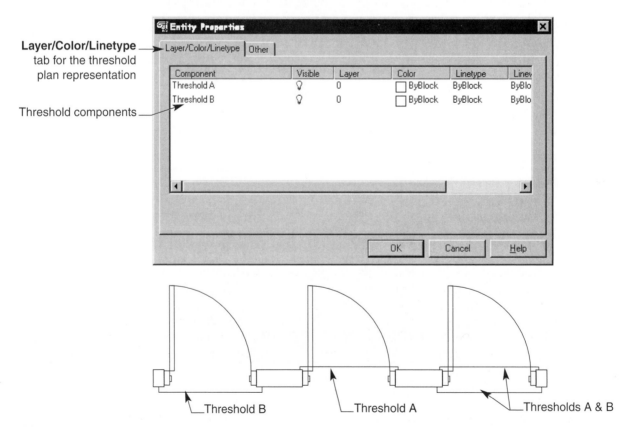

FIGURE 10.4   The threshold plan representation includes two threshold components that can be viewed in a top view.

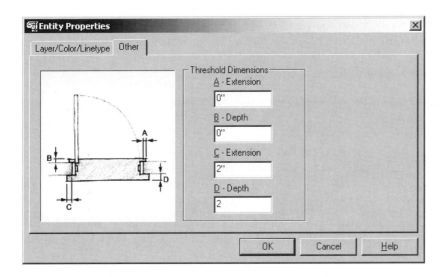

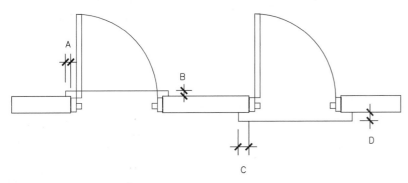

FIGURE 10.5 The Other tab for the threshold plan representation controls the size of the thresholds.

Exercise 10.1

1. Open the EX9-2-Flr-Pln.dwg.
2. Add a threshold to all the exterior doors. Set the override by individual door.
3. Close the doors in the model representation by style.
4. Change the plan representation door swing's color to 140.
5. Save the drawing as EX10-1-Flr-Pln.dwg.

Attaching Custom Blocks to a Door

Custom blocks can be attached to a door representation through entity display. All the door representations except the threshold plan representations can have a custom block attached to the door by style or individual door. Custom blocks can include a door handle or custom moulding in a model and elevation view, or a dashed line through the door representing the header lines in plan view.

The first step in attaching a block to a door is to have a block definition in the drawing. The block needs to be created on layer 0, using ByBlock colors, linetypes, and lineweights. See Chapter 7 for creating blocks. Once the block is created it can be attached to the door. If you are creating blocks that are to be used in a model view, then the blocks should be 3D. Blocks attached to an elevation or plan representation can be drawn as 2D blocks.

Creating a Door Handle Block

This section walks you through the creation of a simple 3D door handle that can be attached to a door. Start a new drawing and split the screen into four equal viewports using a 3D setup to provide you with a top, front, right, and SE isometric view.

Pick inside the front viewport to make the viewport active. Draw a circle 1″ in diameter. Use the **EXTRUDE** command to turn the circle into a 1½″-long cylinder. The following prompt sequence is used to extrude the circle:

Command: **EXTRUDE** ↵
Current wire frame density: ISOLINES=4
Select objects: *(pick the circle)*
1 found
Select objects: ↵
Specify height of extrusion or [Path]: **1-1/2** ↵
Specify angle of taper for extrusion <0>: **0** ↵
Command:

The shaft for the door handle is created (see Figure 10.6). The next step is to create the handle. Change to the isometric viewport. Use the **SPHERE** command to create a sphere and place it at the "front" of the cylinder. The following prompt sequence is used to create the handle:

Command: **SPHERE** ↵
Current wire frame density: ISOLINES=4
Specify center of sphere <0,0,0>: *(use a center object snap to snap to the end of the cylinder)*
    CEN ↵
Of *(pick the center of the front part of the cylinder)*
Specify radius of sphere or [Diameter]: **D** ↵
Specify diameter: **2** ↵
Command:

The sphere for the door handle, shown in Figure 10.7 is created.

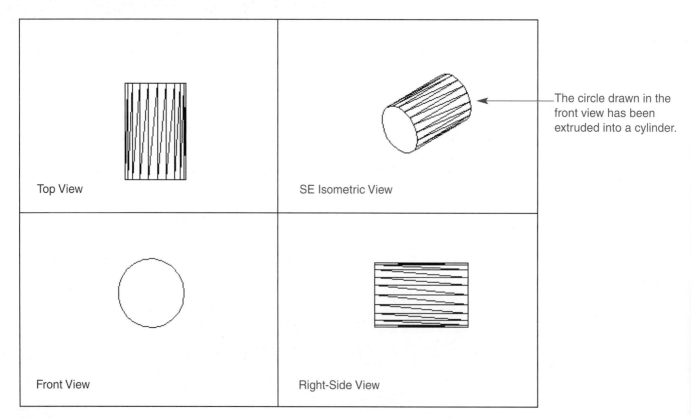

Top View	SE Isometric View
Front View	Right-Side View

The circle drawn in the front view has been extruded into a cylinder.

FIGURE 10.6   The EXTRUDE command is used to take a 2D shape and "stretch" it into a 3D shape. In this case the circle was extruded into a cylinder shape.

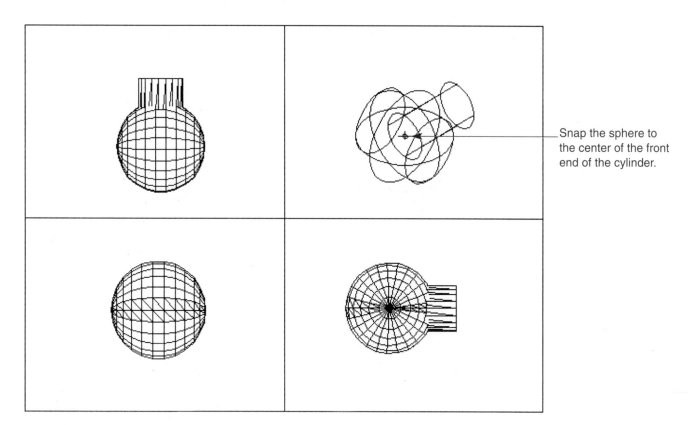

FIGURE 10.7   The SPHERE command is used to create 3D spheres. Place the sphere at the end of the cylinder using the CENTER osnap.

The final step is to join the two pieces together to form one object. Use the **UNION** command to do this. The following prompt sequence is used to join union the two pieces together.

Command: **UNION** ↵
Select objects: *(select one of the objects)* ↵
1 found
Select objects: *(select the other object)* ↵
1 found, 2 total
Select objects: ↵
Command:

The next step is to ensure that the door handle is on layer 0 and using ByBlock colors, linetypes, and lineweights. Pick the door handle shape, right-click, and select Entity Display (see Figure 10.8). Notice that the door handle does not include a Display Props tab. This is because the object is not an Architectural Desktop object but an AutoCAD solid object. Change the layer to 0, and the color, linetype, and lineweight to ByBlock. Press OK when finished.

The final step is to turn the handle into a block. Use the **BLOCK** command to turn the door handle into a block. When creating a block to be attached to a door, set the orientation of the crosshairs parallel to the floor. In the top viewport, change the view to the SE isometric view. The crosshairs are oriented correctly for selecting the insertion point of the block.

The insertion point for the block should be located at the end of the cylinder opposite the end that includes the sphere, as shown in Figure 10.9. Use CENter osnap to snap to the center of the cylinder face. Once the block is created in this drawing, it can be dragged and dropped into other drawings using the AutoCAD DesignCenter.

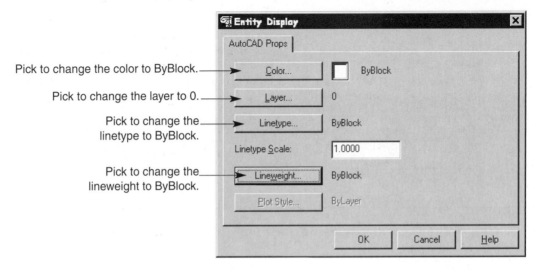

Pick to change the color to ByBlock.

Pick to change the layer to 0.

Pick to change the linetype to ByBlock.

Pick to change the lineweight to ByBlock.

FIGURE 10.8   Entity Display can be used to change AutoCAD properties for AutoCAD objects. Note that AutoCAD objects do not have a Display Props tab. This is available only to AEC Objects.

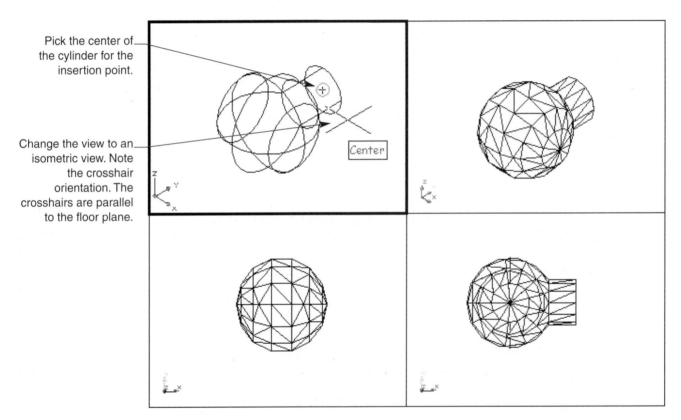

Pick the center of the cylinder for the insertion point.

Change the view to an isometric view. Note the crosshair orientation. The crosshairs are parallel to the floor plane.

FIGURE 10.9   Change the view appropriately to adjust the orientation of the crosshairs before creating the block. Use CENter osnap to specify the insertion point for the block.

Field Notes

3D blocks can be constructed from AutoCAD solids or surface objects. 3D blocks also can be constructed from AEC mass elements. The blocks do not have to be created in a different drawing file but can be created in the drawing where they will be used. AEC mass elements are discussed in Chapter 22. Additionally, knowing how to manipulate the User coordinate system can be a great help in creating 3D blocks for use in an AEC object.

Adding the Door Handle to a Door

Once the door handle block has been defined in the drawing where it will be used, it can be added to a single door or to a door style through entity display. To do this, select a door, right-click, and pick Entity Display… from the cursor menu. Select the Model representation, and set the override by style. Pick the Edit Display Props… button to modify the model view door. Select the Other tab to add the block to the door style (see Figure 10.3A). Selecting the Add… button displays the **Custom Block** dialog box, shown in Figure 10.10.

The dialog box is divided into two parts: the left side controls how the block is inserted, and the right side provides a preview of how the block appears. If you are adding blocks to a model representation, you may want to select an isometric view angle from the drop-down list in the upper right corner of the dialog box.

To add the block, pick the Select Block… button to select the door handle block. Once the door handle block is selected, it is inserted at the bottom left side of the framing component. This is the X=0, Y=0, Z=0 point for the individual door. Each AEC object includes its own coordinate locating system. Typically the *starting* point of the object establishes the 0,0,0 point for the object (see Figure 10.11).

The block can be attached to one of three door components. At the bottom of the dialog box, the block can be attached to the frame, leaf, or glass component (see Figure 10.10). As you are attaching a door handle select the Leaf Component radio button. This places the block on the door leaf component in the preview. Use the Insertion Point drop-down lists to position the block along the door's X, Y, and Z axis. The Insertion Offset text boxes are used to offset the insertion point of the block in relation to the insertion point for the door. Adjust the values until the door handle appears correctly in the preview window. The values specified for attaching the door handle to the door style are shown in Figure 10.12.

When you have finished, press the OK button to return to the Entity Properties dialog box, where you can attach additional blocks to the door. If a door handle is to be attached to both sides of the door leaf, then add two blocks and adjust their values accordingly. When blocks are added to an AEC object, they become components of the door, and their display can be controlled in the Layer/Color/Linetype tab. This is why it is important to create the original block with ByBlock properties on layer 0.

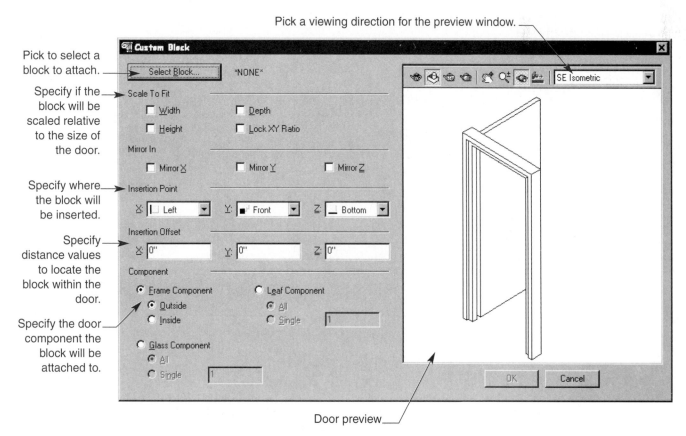

FIGURE 10.10   The Custom Block dialog box is used to add custom blocks to a door.

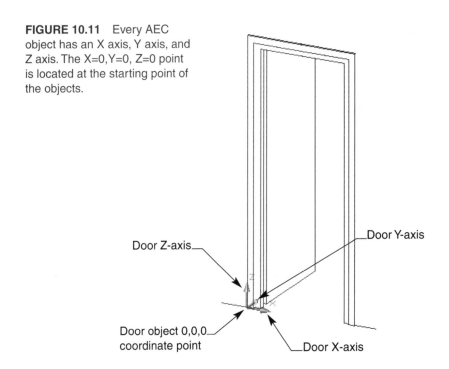

FIGURE 10.11 Every AEC object has an X axis, Y axis, and Z axis. The X=0,Y=0, Z=0 point is located at the starting point of the objects.

Door Z-axis

Door Y-axis

Door object 0,0,0 coordinate point

Door X-axis

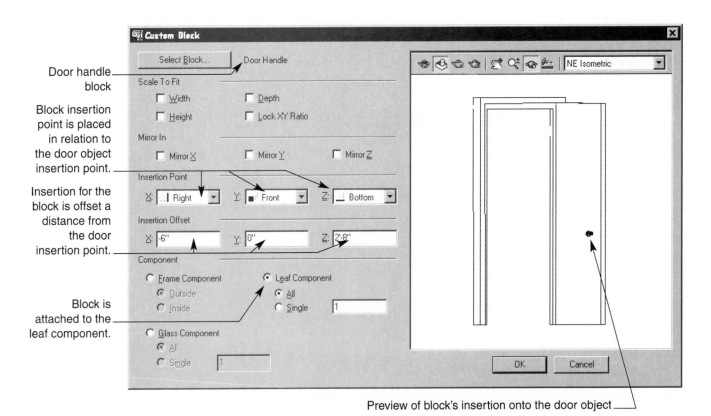

Door handle block

Block insertion point is placed in relation to the door object insertion point.

Insertion for the block is offset a distance from the door insertion point.

Block is attached to the leaf component.

Preview of block's insertion onto the door object

FIGURE 10.12 The door handle is attached to the door object.

Exercise 10.2

1. Open EX10-1-Flr-Pln.dwg.
2. Create a door handle block similar to the one described in the preceding section.
3. Attach the block to the model representation for each door style used in the drawing except the garage door.
4. Change the color of the door handle components to a gold color.
5. Save the drawing as EX10-2-Flr-Pln.dwg.

Setting Door Anchors within a Wall

As doors are added to walls they are *anchored* to a position within the wall. For example, if a door is placed in a wall that is 12″ from a wall intersection, the door is anchored 12″ from the corner and centered within the wall. The door can still be moved along the wall, but it remains centered (anchored) within the wall (see Figure 10.13).

Notice in Figure 10.13 that the door frame is centered along the wall's centerline. In many cases this may be acceptable. In this situation the frame of the door extends between the stud and brick components. In actuality, the frame of the door should probably be inset in the wall so the frame remains along the stud wall component. To adjust the door within the wall, first pick the door, right-click, and select **Reposition Within Wall**. You are prompted to select a side to offset. Pick a point on either side of the wall to reposition the door so the frame is flush with the side of the wall, as shown in Figure 10.14.

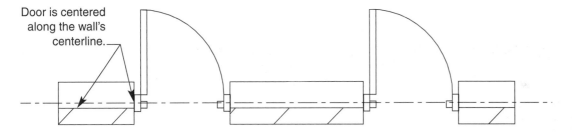

FIGURE 10.13   When doors are first placed in a wall, they are anchored to the center of the wall.

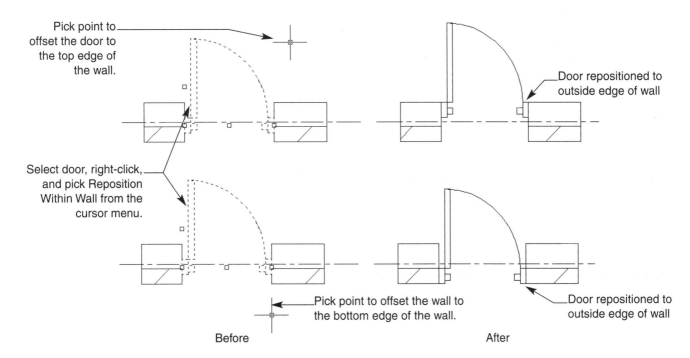

FIGURE 10.14   The Reposition Within Wall command adjusts the position of the door within the wall.

The Reposition Within Wall command includes two options. Instead of picking the side to offset the door, the Offset option can be entered by typing **O** at the prompt. Enter a distance to offset the door, then pick the side of wall to which to offset the door. This allows you to place door frames on the outside of a wall, as with garage or overhead doors that are mounted on the edge of a wall. The Center option centers the door on the centerline of the wall.

Another anchor option is to reposition a door along a wall. This option allows you to offset the door a specified distance along the face of the wall. To use this command, select the door, right-click, and select **Reposition Along Wall**. As with the Reposition Within Wall command, pick a point to the left or right of the door to position the door at either end of the wall, or use the Offset option to specify a distance to move the door.

 Field Notes

> The Reposition Along Wall option is rarely used as the MOVE command can easily be used to reposition (move) the door a specified distance.

Additional anchoring commands can be accessed by selecting the door, right-clicking, and selecting **Wall Anchor**. A cascading menu with the following series of options for controlling how a door is anchored within a wall is displayed (see Figure 10.15):

■ **Set Wall**	This option allows you to move a door into a different wall. Pick the wall where the door needs to be moved, and the door is anchored to the new wall.
■ **Release**	This option releases a door from a wall and fills in the opening within the wall. Once a door is released, it cannot be placed back into the wall.
■ **Set Position Along**	This option is similar to the Reposition Along Wall option. A series of prompts allows you to specify where a door should be placed along a wall.
■ **Set Position Within**	This option is similar to the Reposition Within Wall option. A series of prompts specifies how the door fits within the wall.
■ **Set Vertical Position**	This option controls how the door is placed vertically within the wall. This option can also be accessed through the Door Modify dialog box and by specifying the Vertical Alignment value.
■ **Set Anchored End**	This option sets where the door position is measured within the wall. This option does not make any display changes to the object but sets the distance from the end of the wall to the beginning of the door for Architectural Desktop to calculate its position.

FIGURE 10.15 The Wall Anchor cascade menu provides options for controlling the anchored position of a door within a wall.

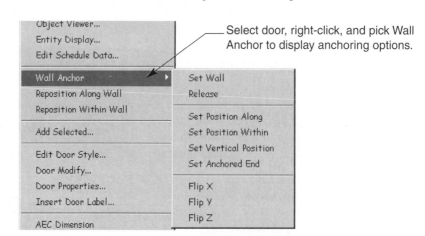

Select door, right-click, and pick Wall Anchor to display anchoring options.

- **Flip X**   This option mirrors the door leaf along the door's X axis. This is similar to modifying the door leaf location with grips. See Chapter 3 for changing the side of a wall on which the door leaf is positioned.

- **Flip Y**   This option mirrors the swing of the door along the door's Y axis. This change also can be accomplished by grip editing.

- **Flip Z**   This option mirrors the door vertically within a wall, producing a door that is upside-down.

Changing Door Properties

This section discusses how to modify the display of doors, how to add blocks to a door representation, and how to reposition doors in walls. Chapter 3 also discussed how to add and modify a door. Door objects can be manipulated in a number of ways, yet there are some things that cannot be controlled with the options mentioned earlier. The **Door Properties** dialog box can be used to round out all the modifying options available to you for adding and modifying doors.

To access the Door Properties dialog box, select the door, right-click, and pick Door Properties... from the cursor menu. The Door Properties dialog box, shown in Figure 10.16, is displayed. The dialog box contains the following five tabs that control the behavior of the selected door:

General Tab
The **General** tab is used to add a description for the selected door in the description area. The **Notes...** button can be used to enter any design criteria or attach any external documents, such as a text document or spreadsheet. The **Property Sets...** button is used to attach property set definitions to the door. As defined in Chapter 9, a property set definition is a group of smart information that can be pulled from the door object and inserted into a schedule. Property set definitions are discussed in a later chapter.

Style Tab
The **Style** tab is used to change the selected door to a different style. The door styles available in the drawing are listed in this tab. The floating viewer can be used to preview the styles before the change is made (see Figure 10.17).

Dimensions Tab
The **Dimensions** tab is used to control the dimensions of the selected door (see Figure 10.18). If a door style includes standard sizes, then the sizes are available in the **Standard Sizes** list area. Select a standard

FIGURE 10.16   The Door Properties dialog box includes settings to control the behavior of the selected door.

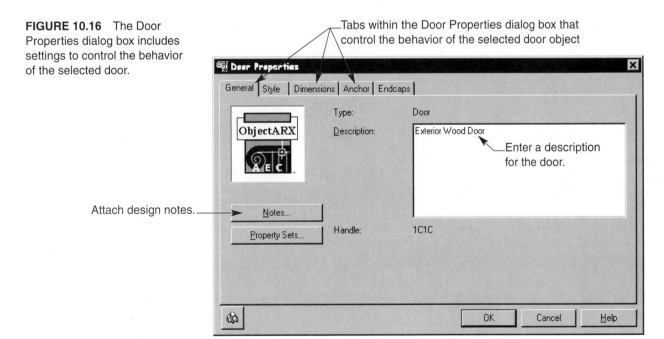

Tabs within the Door Properties dialog box that control the behavior of the selected door object

Style tab

Pick a different style to use for the selected door.

Floating Viewer button

Floating Viewer window

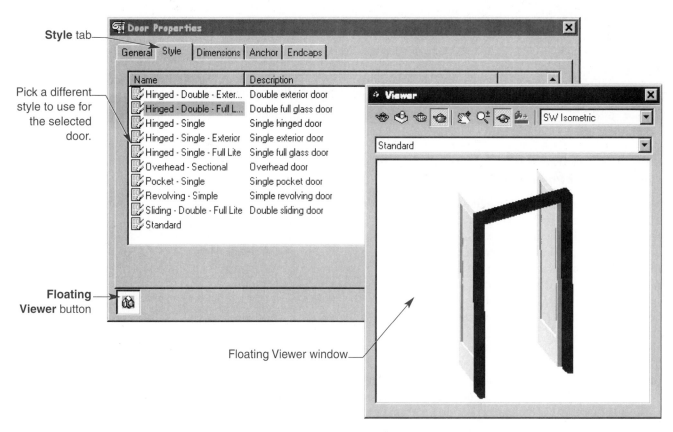

FIGURE 10.17 The Style tab is used to list the door styles available in the drawing and to choose a different style for use.

Dimensions tab

Select a new door size if available from the list.

Check to measure the rough opening for a door.

Enter custom door sizes.

Adjust the opening percentage of the door.

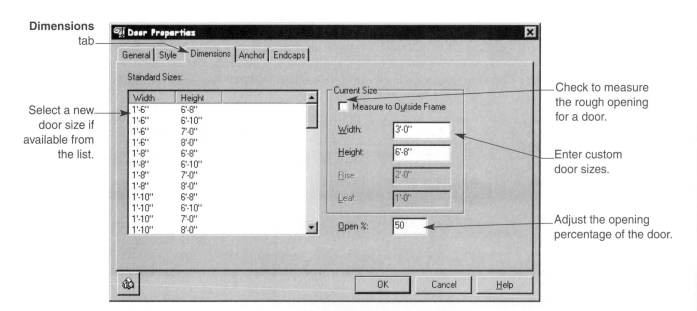

FIGURE 10.18 The Dimensions tab controls the dimensions of the selected door. The size of the door can be modified by selecting different sizes or entering a custom size.

size from the list to change the size of the door. If there are not any sizes listed or a custom door size needs to be entered, then enter the new size in the Current Size text boxes, to change the size of the door.

The door width and height can be measured to the outside of the frame to provide a rough opening size by selecting the **Measure to O̲utside Frame** check box. The opening percentage of the door can also be adjusted in this tab. If the opening percentage is being overridden through the Entity Display dialog box, changing the opening percentage here has no effect on the override.

Anchor Tab

The **Anchor** tab controls the placement of the door in a wall. This tab is divided into the following four areas. Each of these areas adjusts the door within the wall object's X, Y, and Z axes (see Figure 10.19).

- **Position Along (X)** — This area controls the placement of the door along the X axis of a wall. The positive X axis of a wall is designated from the starting point of the wall to the ending point of the wall. The following settings control the door placement along the wall:

  - **From:** — This drop-down list specifies where along the wall the placement of the door is measured. The term *Curve* is used in a general sense and refers to a straight or curved wall. Options include **Start of Curve, Midpoint of Curve,** and **End of Curve**.

  - **Distance:** — Enter a distance from one of the points selected in the F̲rom: drop-down list to a point on the door.

  - **To:** — This drop-down list specifies what point on the door the distance value is measured to. Options include **Start Edge, Center,** and **End Edge** of the door.

- **Position W̲ithin (Y)** — This area controls the placement of the door within the wall.

  - **From:** — This drop-down list specifies where the placement of the door is measured from within the wall. Options include **Left Edge of Curve Width, Center of Curve Width,** and **Right Edge of Curve Width**.

  - **Distance:** — Enter a distance from one of the points selected in the F̲rom: drop-down list to a point on the door.

  - **To:** — This drop-down list specifies what point on the door the distance value is measured to. Options include **Front, Center,** and **Back**.

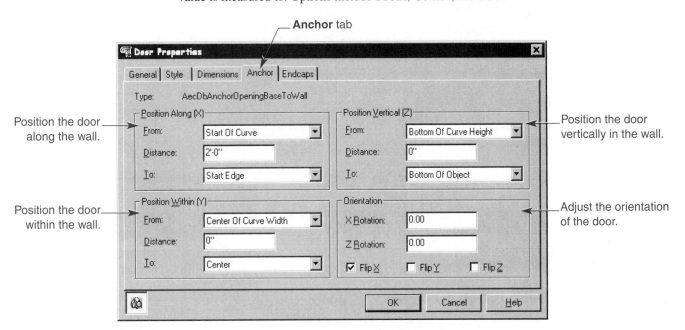

FIGURE 10.19   The Anchor tab is used to anchor the door in a wall at a specific position.

- **Position Vertical (Z)**   This area is used to place the door vertically within the wall.

  - **From:**   This drop-down list specifies where the placement of the door is measured from vertically within the wall. Options include **Bottom Of Curve Height**, **Center Of Curve Height**, and **Top Of Curve Height**.

  - **Distance:**   Enter a distance from one of the points selected in the From: drop-down list to a point on the door.

  - **To:**   This drop-down list specifies what point on the door the distance value is measured to. Options include: **Bottom Of Object**, **Center Of Object**, and **Top Of Object**.

- **Orientation:**   This area controls the direction of the door in the wall.

  - **X Rotation:**   Entering a value in this text box rotates the door in the wall's X axis, which produces an angled door.

  - **Z Rotation:**   Entering a value in this text box rotates the door along the wall's Z axis.

  - **Flip X**   This check box mirrors the door in the wall's X axis.

  - **Flip Y**   This check box mirrors the hinge/swing of the door in the wall's Y axis.

  - **Flip Z**   This check box mirrors the door in the wall's Z axis, producing an upside-down door.

Endcaps Tab

The **Endcaps** tab is used to assign a wall endcap to each side of a door. The **Starting Endcap** is assigned to the starting point of the door and the **Ending Endcap** is assigned to the ending point of a door, as shown in Figure 10.20. The creation of endcaps is covered later in this text.

Exercise 10.3

1. Open EX10-2-Flr-Pln.dwg.
2. Use the Door Properties command to change the style of the front door to Hinged - Single - Exterior.
3. Enter the description Oak Wood Door in the General tab.
4. Change the plan representation of the swing to color 140, through the Entity Display command.
5. Change the model representation for the front door so that it is closed in a model view.
6. Save the drawing as EX 10-3-Flr-Pln.dwg.

FIGURE 10.20 (A) This door is using the Standard Endcap style to cap the ends of the wall through which the door is cutting. (B) This door is using an endcap style that wraps the brick component of the wall around the stud compoent.

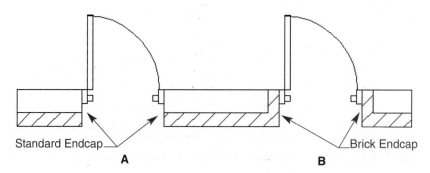

Standard Endcap   Brick Endcap

A   **B**

Controlling Window Objects

Window objects like door and wall objects include a number of tools that can be used to enhance their behavior in the construction documents as well as in the CAD model. Many of the tools are similar to the tools used for door objects. This section explains the tools used to manipulate the look and behavior of window objects.

Window Entity Display

The window display can be controlled through the use of entity display much like doors are controlled. To access the entity display for a window, first select the window, right-click, and select Entity Display…, then select the Display Props tab. Select a representation that needs to have its display adjusted, then select the level of control at which the window will be overridden. After the level of override has been determined and set, pick the Edit Display Props… button to adjust the window display. Figure 10.21 shows the components for a plan representation window.

As it is for doors, the Other tab is used to add custom blocks to a window. This topic is discussed in the next section. Window objects also include an *elevation* display representation and a *sill plan* representation. The elevation display representation controls how the window appears in an elevation view.

Sill Plan Representation

The sill plan representation creates an inside and outside windowsill and is used in conjunction with the window plan representation. The sill plan representation includes two components, the inside sill and the outside sill (see Figure 10.22). The inside of a window is determined by the location of the grip that is set off the wall when the window is selected. The *sill A* component is considered to be the inside sill, and the *sill B* is the outside component. One, both, or neither of these sills can be turned on for the top view.

The Other tab in the sill plan representation includes the dimensional values for the inside and outside sills. Values **A** and **B** control the dimensions for **sill A** (inside sill), and values C and D control the dimensions for **sill B** (outside sill) (see Figure 10.23).

Adding Muntins to Windows

Many residential and some commercial buildings include muntins in the windows as part of the design. A **muntin** is a vertical, horizontal, or angled piece between two panels or glass panes. Preconfigured muntins such as diamond or rectangular grid patterns can be added to the glazing component of a window. The elevation and model representations include a tab named **Muntins,** as shown in Figure 10.24. This tab is used to place muntin configurations within an individual window or window style. The tab looks similar to the Other tab, which adds custom blocks to the window.

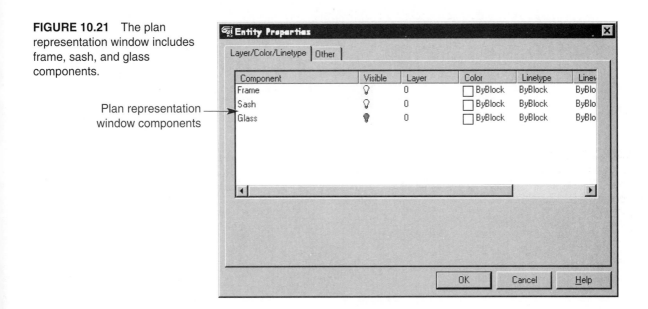

FIGURE 10.21   The plan representation window includes frame, sash, and glass components.

Plan representation window components

FIGURE 10.22 The Layer/Color/Linetype tab for the sill plan representation is used to control the graphical display of windowsill in a plan representation window.

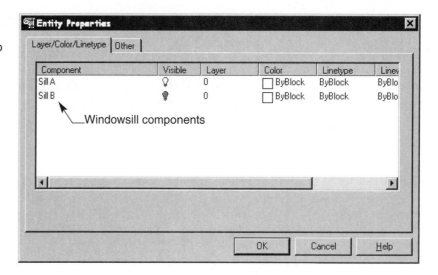

The inside grip determines the inside of the window.

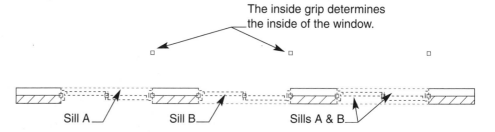

Sill A   Sill B   Sills A & B

FIGURE 10.23 Adjust the values of the windowsills in the sill plan representation.

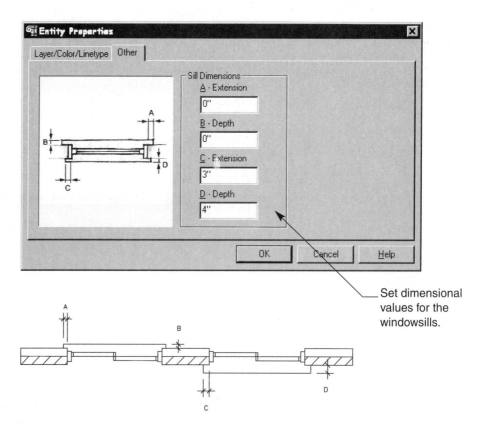

Set dimensional values for the windowsills.

Muntins tab

Press to add muntins.

Press to edit muntins.

Press to remove muntins.

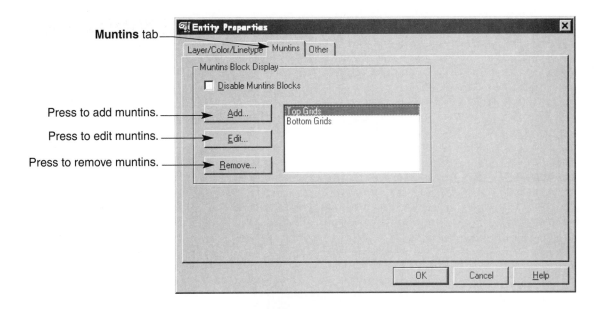

Different muntin patterns placed
in different types of windows

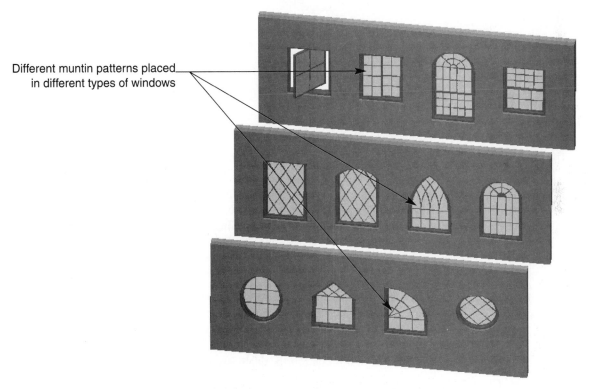

FIGURE 10.24   The Muntins tab is used to place grid patterns within walls.

To add muntins to a window, select the elevation of one of the model representations, set the level of override, and press the Edit Display Props... button. Select the Muntins tab, then press the Add... button to add a muntin configuration. The **Muntins Block** dialog box, shown in Figure 10.25, is displayed. The following settings are available in this dialog box:

- **Name:**   Enter a name for the muntin configuration. This name appears in the
                          Layer/Color/Linetype tab, where display properties can be set.

- **Window Pane**   This area specifies where the muntins are added.

  - **Top**   This radio button adds the muntins to the top glazing pane.

Specify a name for the muntin component.

Specify where the muntin will be added.

Control the size of the muntins.

Cleanup the grid intersections, and convert them to 3D objects.

Specify the grid pattern.

Specify the number of lights (glazing panels).

Use for half-round windows.

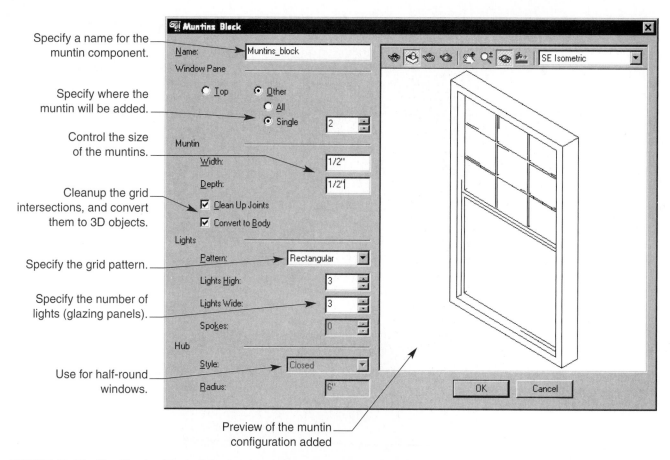

Preview of the muntin configuration added

FIGURE 10.25   The Muntins Block dialog box is used to add muntin configurations to windows.

- **Other**   This radio button provides two options. The **All** option adds muntins to all the glazing panels. The **Single** option places the muntin on a specific glazing panel. Enter a number to add the muntins to a specific window pane. Generally, for a window with multiple panes, number 1 is the bottom pane, and number 2 is the top pane.

- **Muntin**   This area controls the size of the muntins, how the grids cleanup, and whether it is a 2D or 3D object:

- **Width**   This text box controls the width size of an individual muntin.

- **Depth:**   This text box controls the depth size of an individual muntin.

- **Clean Up Joints** This option cleans up the joints between individual muntins.

- **Convert to Body** This option converts the muntins into a 3D object embedded within the window. Check this box on if you are using a rendering program to render the model.

- **Lights**   This area controls the muntin pattern and the number of lights (window panes) in the window:

- **Pattern:**   This drop-down list includes different muntin patterns that can be applied to windows. Depending on the shape of the window, such as rectangular, half-round, or Gothic, different muntin patterns are available. Typically, rectangular and diamond patterns are available for most window shapes.

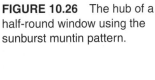

FIGURE 10.26   The hub of a half-round window using the sunburst muntin pattern.

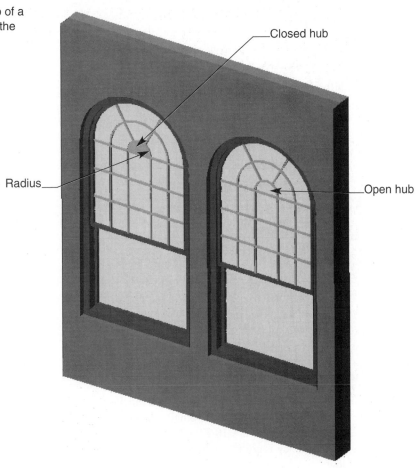

- **Lights High:**   The number in this text box determines how many window panes are created vertically by adding horizontal muntins.

- **Lights Wide:**   The number in this text box determines how many window panes are created horizontally by adding vertical muntins.

- **Spokes:**   This option is available if a starburst or sunburst muntin pattern is added. These two patterns are available for half-round or quarter-round–shaped windows only.

- **Hub**   This area becomes available when the sunburst muntin pattern is used. The *hub* is where arched muntins begin (see Figure 10.26).

    - **Style:**   This option opens or closes the hub in the muntin pattern.

    - **Radius:**   This option specifies the inside radius of the sunburst pattern.

- **Preview**   This area allows you to preview the muntin patterns added to the selected window.

After adding a muntin, press the OK button to return to the Muntins tab. You can add additional muntin configurations by pressing the Add… button again. If a window has multiple window panes, such as a double-hung or glider window, each pane can have its own muntin pattern applied. As the muntin configurations are added to the window they are also added to the Layer/Color/Linetype tab for additional display control. You may want to make the muntin components the same color as the frame and sash or pick your own color.

✎ Field Notes

- Muntins cannot be added to window assemblies; however, if a window assembly includes AEC windows, the AEC windows can have muntins applied.

- You may want to add custom muntin configurations to the window objects that are not included in the Muntins tab. To do this, create 3D objects in the pattern desired, turn them into a block, and attach the block to the window through the Other tab.

Exercise 10.4

1. Open EX10-3-Flr-Pln.dwg.
2. Add muntins to the front windows of the building in both the elevation and the model representations.
3. Adjust the color of the muntin components to match the frame and sash components for the elevation and the model representations.
4. Save the drawing as EX10-4-Flr-Pln.dwg.

Attaching Custom Blocks to a Window

Attaching a block to a window is similar to attaching a block to a door. First, create the block(s) that will be attached to the window, ensuring that the original geometry was created on layer 0 using ByBlock colors, linetypes, and lineweights (see Figure 10.27). The insertion points for the blocks should be logical places on the geometry, as the block is added to the window and assigned a location measured from the insertion point of the window.

Select the window and enter the Entity Display dialog box. Pick the desired display representation to attach the block to and the level of display control. Once in the Entity Properties dialog box, select the Other tab. Press the Add... button to add a block to the window. The **Custom Block** dialog box for windows, shown in Figure 10.28, is displayed. The following settings are available to control the addition of a block to a window:

- **Select <u>B</u>lock...** Press this button to choose the block to be attached to the window.

- **Scale To Fit** This area includes options for scaling the inserted block. The **<u>W</u>idth**, **<u>H</u>eight**, and **<u>D</u>epth** of the block can be scaled to fit the size of the window. The **Lock XY Ratio** check box locks the scale of the width and depth options.

- **Mirror In** This area controls whether the block is mirrored in the windows X, Y, or Z axis.

- **Insertion Point** This area is used to specify where the block is inserted in relation to the window. The **<u>X</u>:** drop-down list places the block along the window's left, center, or right, edge. The **<u>Y</u>:** drop-down list adjusts the block to the front, center, or back edge. The **<u>Z</u>:** drop-down list adjusts the block to the top, center, or bottom of the window's edge.

- **Insertion Offset** This area allows you to specify an offset distance for the insertion of the block relative to the points specified in the insertion point area above.

- **Component** Blocks that are placed in a window are placed relative to the **Frame** component or to the **Window** component. The window component is measured from the edge of the sash to the opposite edge of the sash.

Many blocks can be added to a window at the style or individual window level. As each block is attached it is added to the component list in the Layer/Color/Linetype tab, where display controls can be

FIGURE 10.27 Four blocks have been created for attachment to a window. Note the insertion points.

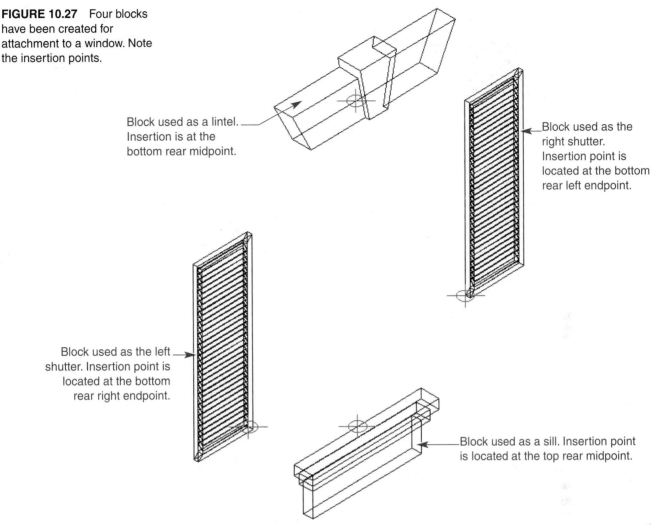

Block used as a lintel. Insertion is at the bottom rear midpoint.

Block used as the right shutter. Insertion point is located at the bottom rear left endpoint.

Block used as the left shutter. Insertion point is located at the bottom rear right endpoint.

Block used as a sill. Insertion point is located at the top rear midpoint.

applied. You can have custom blocks that are drawn at different sizes so that when the window is added to a wall, the Entity Display dialog box can be used to turn the appropriately sized block component on or off. Figure 10.29 shows the finished custom window.

 Field Notes

It is a good idea to attach blocks at the style level, as the window can be used many times throughout the drawing with the custom blocks applied. If the blocks are not to be shown on a section of the building, simply turn off the block components for that one window.

Anchoring a Window in a Wall

Anchoring doors was discussed earlier in this chapter. The same commands used to control where the door is in relation to a wall are available for windows. Use one of the following options to access the **Reposition Within Wall** command to reposition a window within a wall:

✓ Pick the **Design** pull-down menu, select the **Windows** cascade menu, then pick **Reposition Within Wall.**

✓ Right-click in the drawing, and select **Design, Windows, Reposition Within Wall.**

✓ Type **REPOSITIONWITHIN** at the Command: prompt.

✓ Select the window to be repositioned, right-click, and select **Reposition Within Wall**.

Sill block is being adjusted onto the window.

Pick to add a custom block.

Specify how the block will be scaled against the component it is assigned to.

Mirror the block along the window's X, Y, or Z axis.

Specify where the insertion point of the block will be inserted in relation to the window's insertion point.

Specify an insertion offset from the values chosen above.

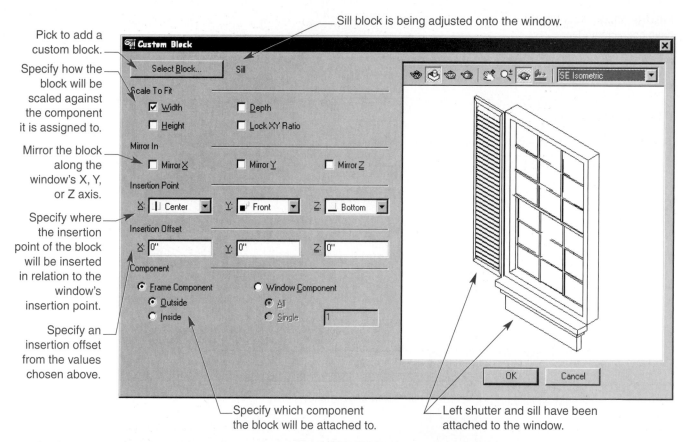

Custom Block

Select Block...   Sill

Scale To Fit
☑ Width   ☐ Depth
☐ Height   ☐ Lock XY Ratio

Mirror In
☐ Mirror X   ☐ Mirror Y   ☐ Mirror Z

Insertion Point
X: Center ▾   Y: Front ▾   Z: Bottom ▾

Insertion Offset
X: 0"   Y: 0"   Z: 0"

Component
◉ Frame Component   ○ Window Component
  ◉ Outside   ◉ All
  ○ Inside   ○ Single   1

SE Isometric ▾

OK   Cancel

Specify which component the block will be attached to.

Left shutter and sill have been attached to the window.

FIGURE 10.28   The Custom Block dialog box is used to attach custom blocks to the window object. This window has two blocks attached, the left shutter and the sill blocks. The sill block is currently being manipulated. Note where the block is inserted in relation to the window insertion point.

FIGURE 10.29   The window with all the blocks attached and muntins added.

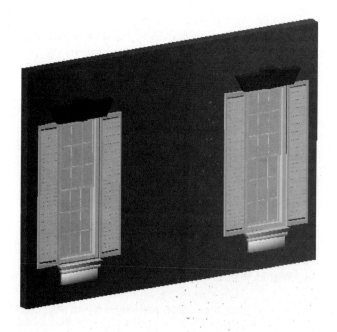

You are prompted to select the side of the wall to offset to. Pick a side of the wall, and the window is positioned along that edge of the wall.

Windows can also be repositioned along a wall using the **Reposition Along Wall** command. Use one of the following options to access this command:

- ✓ Pick the **Design** pull-down menu, select the **Windows** cascade menu, then pick **Reposition Along Wall.**

- ✓ Right-click in the drawing and select **Design, Windows, Reposition Along Wall.**

- ✓ Type **REPOSITIONALONG** at the Command: prompt.

- ✓ Select the window to be repositioned, right-click, and select **Reposition Along Wall.**

Using Window Properties

The **Window Properties** dialog box is used to control additional properties for a single window. Use one of the following options to access this dialog box:

- ✓ Select the window to be modified, right-click, and select **Window Properties…** from the cursor menu.

- ✓ Type **WINDOWPROPS** at the Command: prompt, then select the window to be modified.

As with the door object, five tabs are available for modifying the selected window.

General Tab

The **General** tab is used to add a description for the selected window in the description area (see Figure 10.30). Notes and external documents such as HTML pages, spreadsheets, or text files can be added by picking the Notes… button.

Property set definitions can be added to the window by selecting the Property Sets button. As mentioned previously, a property set includes information about the window such as width, height, manufacturer, and type that can be pulled into a schedule. Creating schedules is discussed later in this text.

Style Tab

The **Style** tab is used to select a different style for the window being modified. Available window styles in the drawing are listed in this tab. Use the floating viewer to look for the desired window style. Select the new style from the list. Press the OK button to change the window to the newly selected style (see Figure 10.31).

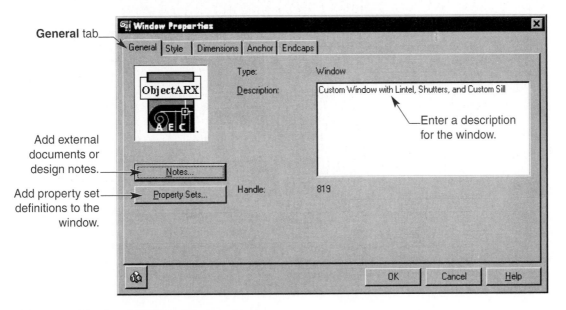

FIGURE 10.30 The General tab for Window objects.

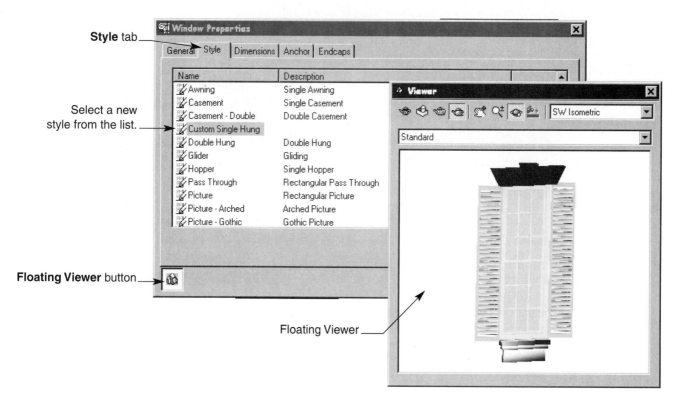

FIGURE 10.31   The Style tab lists available window styles in the drawing. Select a new style for the window being modified.

Dimensions Tab

The **Dimensions** tab controls the size of the window being modified. New sizes can be selected from the **Standard Sizes:** list on the left (see Figure 10.32). If the window style does not have a list of standard sizes, a custom size can be entered in the text boxes to the right. Window dimensions can also be set to measure to the outside of the frame if the **Measure to Outside Frame** check box is selected. If the check-box is unselected, the dimensions are measured to the inside of the frame.

Windows that can be opened such as casement windows, awning windows, gliders, and double-hung windows can be opened or closed by entering a value in the **Open %** text box.

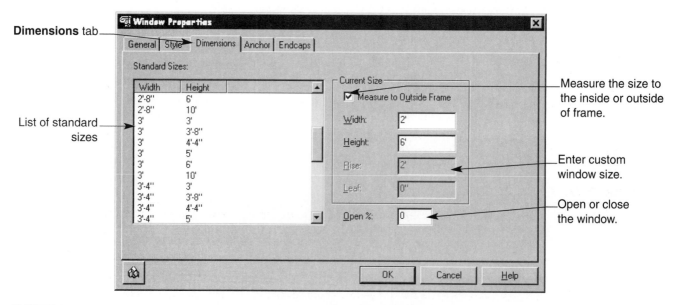

FIGURE 10.32   The Dimensions tab controls the size of the window being modified.

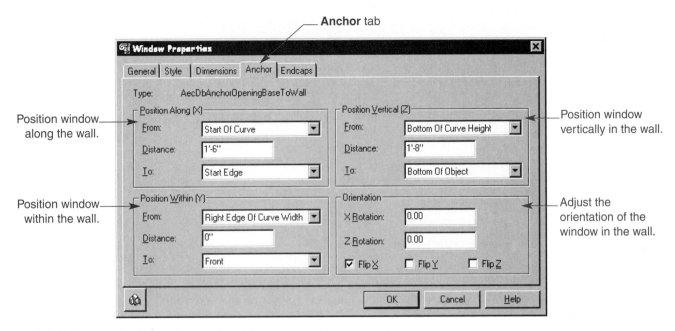

FIGURE 10.33   The Anchor tab is used to set the position of the window in a wall.

Anchor Tab

The **Anchor** tab is used to adjust where the window is placed in the wall. It is divided into four separate areas. The term *Curve* in the **From:** drop-down lists refers to the edge of a wall whether it is a straight wall or a curved wall. The **Distance:** text boxes are used to enter a distance value when the placement of the window is measured from the wall (curve) to the window. The **To:** drop-down lists are used when the distance from a point on a wall (curve) to a point on the window is measured (see Figure 10.33).

The **Position Along (X)** area is used to reposition the window along the length of the wall. The **Position Within (Y)** area is used to adjust the window within the wall. The **Position Vertical (Z)** area is used to set the wall vertically within the height of the wall. The **Orientation** area is used to orient the rotation of the window within a wall.

Endcaps Tab

The **Endcaps** tab is used to assign an endcap for the wall at each side of the window opening. Endcaps are discussed in detail in Chapter 13. Figure 10.34 displays an endcap assigned to a window.

Controlling Opening Objects

The addition of opening objects was discussed in Chapter 3. In that chapter you learned how to add and modify openings. In addition to modifying an opening, there are a few tools that can be used to enhance the look of the opening object. Opening objects have properties similar to those of doors and windows and behave in a similar manner. The following discussion explains the tools available for modifying an AEC opening object.

The Opening Entity Display

When an opening is added to a wall, it cuts a hole in a wall similar to a door or window. The opening object displays an X through the hole in a plan view to indicate that it is a void in the wall and is not filled with a door or window. The model view opening displays an edge around the hole (see Figure 10.35).

The display of properties for opening objects is controlled through Entity Display. To access the Entity Display for AEC openings, select the opening, right-click, and select Entity Display... from the cursor menu. The Entity Display dialog box, shown in Figure 10.36, is displayed. As you did for the door and window objects, select the Display Props tab to modify the display properties.

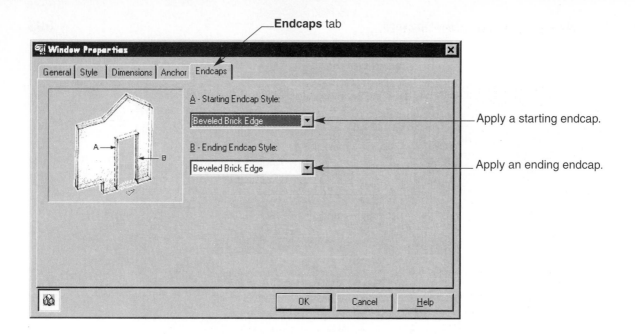

Endcaps tab

Apply a starting endcap.

Apply an ending endcap.

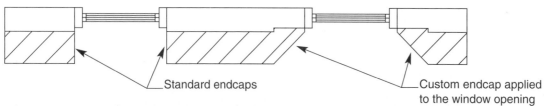

Standard endcaps

Custom endcap applied to the window opening

FIGURE 10.34 The Endcaps tab is used to assign an endcap to each side of a window that controls how the wall will terminate at the opening.

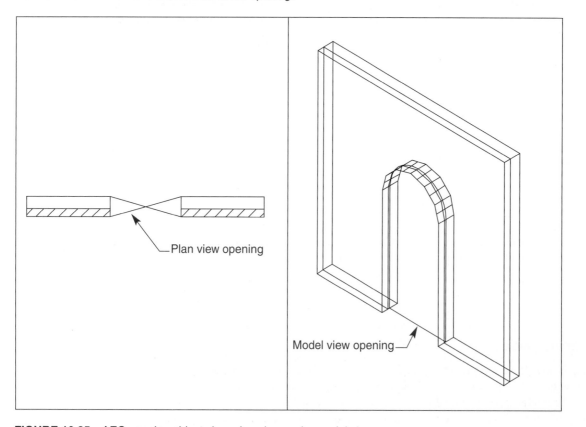

Plan view opening

Model view opening

FIGURE 10.35 AEC opening objects in a plan view and a model view.

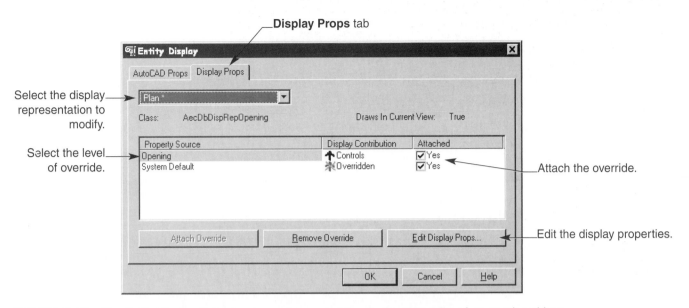

FIGURE 10.36   The Entity Display dialog box is used to control the display properties of an opening object.

AEC opening objects include considerably fewer display representations than a door or window. The Plan 1-100, Plan 1-50, and Sill plan representations are initially not configured for use with the default display configurations in the AEC Arch (Imperial) or AEC Arch (Imperial-Int) templates. To incorporate the use of these display representations, you must create a display configuration. See Chapter 5 for more information regarding display configurations. The Plan 1-100 and Plan 1-50 are simplified display representations for doors, windows, and openings.

To adjust an opening's display representation, select it from the display representations drop-down list (see Figure 10.36). Next, you must select a Property Source override. Note that for openings there is not an opening style property source available. This is because there are no opening styles in Architectural Desktop. Openings can be controlled at the system default level or at the individual opening level. Attach an override, then select the Edit Display Props... button to modify the display properties.

The plan and model representations for openings include only one component listed in the Layer/Color/Linetype tab. If the opening is turned off in one of these representations, the hole in the wall remains, but the opening cannot be selected in that view. The viewing direction must be changed, or a different display configuration must be selected in order to display the opening. Select the opening again, right-click, and pick Entity Display to turn the component back on for the representation where it was turned off.

Attaching Custom Blocks to an Opening

Like doors and windows, openings can have custom blocks attached to them. The block must be attached at the system default level or at the individual object's level. The model and plan representations include the Other tab, which is used to attach custom blocks.

A block that is commonly used for openings in a plan representation is two parallel lines that represent a dashed opening instead of the default X through the opening. Draw two parallel lines 1″ in length and 1″ apart to create the block. The insertion point can be at the end of one of the lines or at the center point between the two lines. Figure 10.37 displays the two lines as a custom block. Remember to create the two lines on layer 0, using ByBlock colors, linetypes, and lineweights.

Access the Entity Display dialog box. Select the Plan representation, and select the System Default property source. Attach the override and select the Edit Display Props... button. Pick the Other tab, then press the Add... button to add the custom block. The **Custom Block** dialog box, shown in Figure 10.38, is displayed. Pick the Select Block button to add the custom block to the opening. Scale the width and depth, and set the insertion point relative to the block's insertion point. See Figure 10.38 for the settings. When you are finished, press the OK button to return to the Entity Properties dialog box. Select the Layer/Color/Linetype tab to turn off the opening component, and set the block component to a hidden linetype. The finished opening with the block is shown in Figure 10.38.

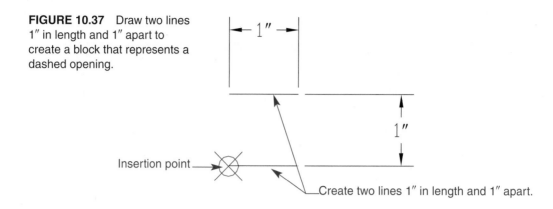

FIGURE 10.37   Draw two lines 1″ in length and 1″ apart to create a block that represents a dashed opening.

Insertion point

Create two lines 1″ in length and 1″ apart.

Attach the block from Figure 10.37.

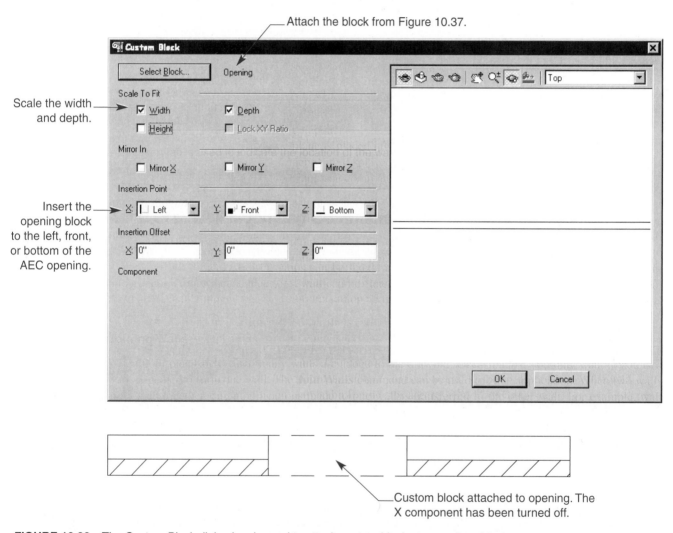

Scale the width and depth.

Insert the opening block to the left, front, or bottom of the AEC opening.

Custom block attached to opening. The X component has been turned off.

FIGURE 10.38   The Custom Block dialog box is used to attach custom blocks to opening objects.

Exercise 10.5

1. Open EX10-4-Flr-Pln.dwg.
2. Create the block shown in Figure 10.37.
3. Attach the block to the opening's plan representation.
4. Set the override to system default.
5. Turn off the AEC opening component and change the linetype of the custom block to Hidden through the Layer/Color/Linetype tab.
6. Turn the opening off in the model representation.
7. Save the drawing as EX10-5-Flr-Pln.dwg.

Anchoring an Opening in a Wall

Anchors for doors and windows were discussed earlier in this chapter. Openings also are anchored to a wall object; however, there is not a reposition *within* wall command as there is with doors and windows. The reposition *along* wall command can be used to adjust the opening along a length of wall. This command can be accessed by selecting the opening, right-clicking, and picking Reposition Along Wall from the cursor menu. Like doors, the opening can be offset a distance for repositioning along a wall, but using the MOVE command accomplishes the same task.

Alternatively, select the opening object, then right-click to access the Wall Anchor cascade menu, which contains additional options for controlling how the opening is anchored in a wall.

Using Opening Properties

The **Opening Properties** dialog box provides additional tools for modifying AEC openings. Use one of the following options to access this dialog box:

✓ Select the opening to be modified, right-click, and pick **Opening Properties...** from the cursor menu.

✓ Type **OPENINGPROPS** at the Command: prompt.

The Opening Properties dialog box includes only four tabs that can be used to adjust the selected opening. The General, Anchor, and Endcaps tabs are similar to the tabs for the Door Properties or the Window Properties dialog boxes, except they are used for AEC opening objects.

The Dimensions tab controls the dimensions of the opening as well as the shape that is used to cut a hole in the wall (see Figure 10.39). Make any adjustments, then press the OK button to accept the changes and return to the drawing.

FIGURE 10.39 The Dimensions tab in the Opening Properties dialog box controls different aspects of opening objects.

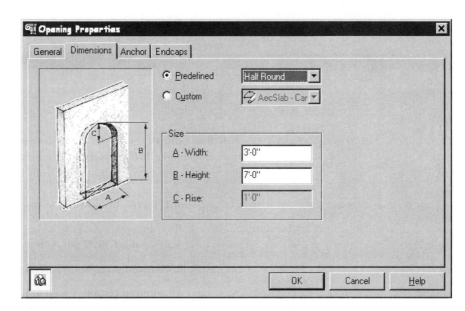

Using Window Assembly Tools

Window assemblies contain a group of windows, or windows and doors, that act as one object. The AEC Arch (Imperial) and AEC Arch (Imperial-Int) templates include a few window assembly styles that have a combination of doors and glazing components. The Standard window assembly is configured to create a group of windows that when stretched larger, automatically cause more windows to be added. A number of additional window assemblies are included with Architectural Desktop. These are discussed in a later chapter.

Using Window Assembly Entity Display

The display of window assemblies, like other AEC objects, is controlled through entity display. As with other AEC objects, the Entity Display dialog box can be accessed by first selecting the window assembly, right-clicking, then selecting Entity Display… from the cursor menu. Select the Display Props tab to modify the display properties for a window assembly. Window assemblies do not include an elevation representation. Instead, an elevation viewing direction uses the model representation.

Adjusting the Plan Representation

There are a number of different plan representations that can be used within different display configurations. The plan display representations include many components that can be turned on or off. Select the plan representation that is to be modified, then select the appropriate level of display control, attach the override, then select the Edit Display Props… button. The Entity Properties dialog box for the plan representation, shown in Figure 10.40, is displayed.

The following four tabs are included for the plan representation:

- **Layer/Color/Linetype**   This tab lists the components used for plan view window assemblies. The **Default Infill**, **Default Frame**, and **Default Mullion** are standard components on a window assembly. The hatches for plan view glazing (infills), frames, and mullions can be manipulated in this tab. The *frame* is the outer part of the assembly, the *mullion component* divides the frame into smaller sections, and the *infill component* is the piece that is filled inside the frame and mullions (see Figure 10.41). *Infill* is the component name for glazing. Window assemblies include a cut plane, hence the plan view components **Below** and **Above**, which display components above and below the cut plane. The plan view infills, frames, and mullions can be displayed with a hatch pattern. The hatch pattern components are tied to their corresponding component.

Layer/Color/Linetype tab

Components for a plan representation window assembly

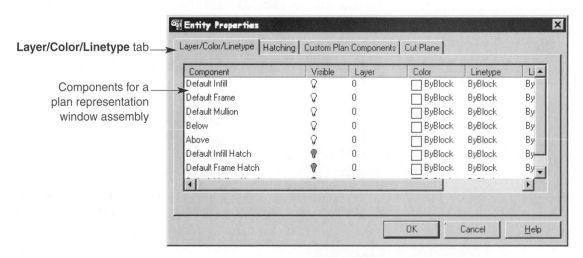

FIGURE 10.40   The Entity Properties dialog box for plan representation window assemblies.

FIGURE 10.41 The parts of a window assembly.

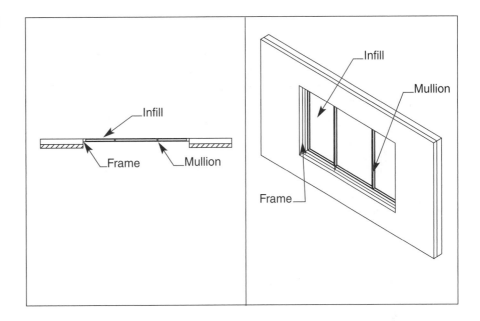

Turn components on or off as desired and set any of the display properties such as color or linetype as needed. If you are using a window assembly with a door, you may wish to turn off the Above component.

■ **Hatching** The hatch components of a window assembly can be turned on in the Layer/color/Linetype tab, then adjusted in the Hatching tab. The hatches for plan view infills, frames, and mullions can be manipulated in this tab.

■ **Custom Plan Components** This tab is similar to the Other tab in the door or window Entity Display dialog box. This allows you to add custom blocks to infills, frames, or mullions.

■ **Cut Plane** This tab designates the cutting plane for plan view window assemblies. Enter a value for the cut plane height.

The model representation is similar but includes fewer components and has only the Layer/Color/Linetype and Custom Model Components tabs.

Exercise 10.6

1. Open EX10-5-Flr-Pln.dwg.
2. Adjust the window assembly's model representations so that the color of the frame and mullion components is color 251.
3. Save the drawing as EX10-6-Flr-Pln.dwg.

Anchoring a Window Assembly in a Wall

Anchors are another tool that can be used to control window assembly behavior. Window assemblies contain anchoring tools for repositioning the assembly within the wall or along the wall similar to the anchoring tools for doors or windows. Select a window assembly and right-click to access the Reposition Within Wall and the Reposition Along Wall commands. Use these anchor tools to reposition the window assembly.

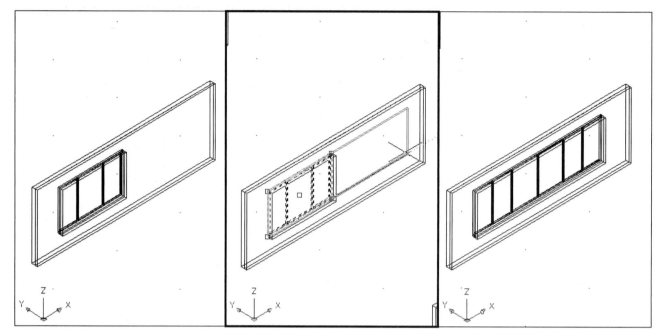

Original window assembly
with divided panels

Select the window assembly,
pick the end grip, and stretch
the assembly larger.

Window assembly after stretching.
Note the additional panels.

FIGURE 10.42   The Standard window assembly divides the object into a series of infill
panels. As the object is stretched larger, additional panels are added to the assembly.

Editing Cells

As indicated earlier, a window assembly is an object that represents windows, or windows and doors that
are grouped together to form one object. For example, when the Standard window assembly is first inserted
into a wall, it contains a series of glazing panels. As the window assembly is made wider, through the
Window Assembly modify command or by stretching the end grips, additional glazing panels are automatically added to the assembly, as shown in Figure 10.42.

The infill panels are also called *cells*. An *infill panel* is the component of a curtain wall or window
assembly that defines the "fill" between frames and mullions. These fills can be glazing (as in a window
assembly), or glazing and spandrel components for an office building curtain wall. *Spandrels* are the spaces
between series of arches. A *curtain wall* is an exterior wall that provides no structural support. The cells
(infill panels) of a window assembly can be modified to reflect a custom window assembly. Cells can be
merged or filled with a different type of infill such as a door or window. To modify the cells within a window assembly, the cell marker needs to be turned on. A *cell marker* is a symbol that identifies an individual
cell within a window assembly. To turn on the cell markers, select the window assembly, right-click, and
pick the Cell Marker cascade menu. Select the All Visible option to turn on all the markers within the window assembly. This turns on a series of markers within each cell (infill). Two cells can be merged together
by applying an override. Pick the window assembly to highlight it, right-click, and select the Overrides cascade menu and select the Merge option. This command can also be accessed by typing **GRIDASSEMBLY-
MERGECELLS** at the command prompt. You are prompted to select cell A. Select the first cell marker
that will be merged, then select the next cell marker that will be added. This merges the two cells, forming
one cell (see Figure 10.43).

Three additional override commands can be used to modify the look of the window assembly. The
Override Cell Assignment option changes the infill cell to a different style of infill such as a door. The
Override Edge Assignment option changes the frame edges or mullions to a different style, and the
Override Edge Profile option changes the shape of the frame and mullions used in the window assembly.
These options are discussed later in this textbook when infill styles and frame styles are created.

To turn off the cell markers, select the window assembly, right-click, pick the Cell Marker cascade
menu, and select Off. This turns off the editing cell markers.

FIGURE 10.43   Window
assembly cells can be merged
by using the Merge Cells
command.

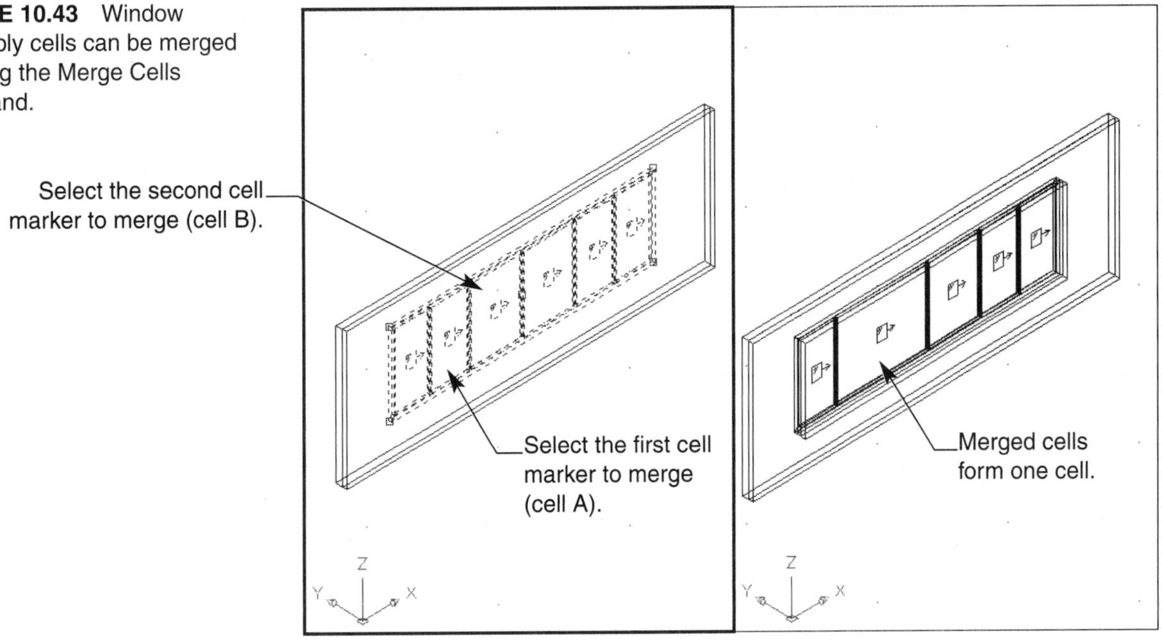

Select the second cell
marker to merge (cell B).

Select the first cell
marker to merge
(cell A).

Merged cells
form one cell.

Window Assembly Properties

Window assemblies can also be controlled through the **Window Assembly Properties** dialog box. To do this, select the window assembly, right-click, and select Window Assembly Properties... from the cursor menu, or type **WINASSEMBLYPROPS** at the command: prompt. The **Window Assembly Properties** dialog box is displayed. The **General**, **Style**, **Anchor**, and **Endcaps** tabs perform the same functions as the Door or Window Properties tabs, except the former values affect window assemblies.

The Dimensions tab controls the size of the window assembly and miter angles at the end of the assembly (see Figure 10.44). The outside edges of the assembly can be mitered to create a corner window.

The **Overrides** tab is used to manage any overrides attached to a window assembly. Figure 10.45 shows an override attached in the Cell Merges area, which indicates that cell 2 has been merged with cell 3. The overrides can be removed by selecting the override and pressing the <u>R</u>emove button.

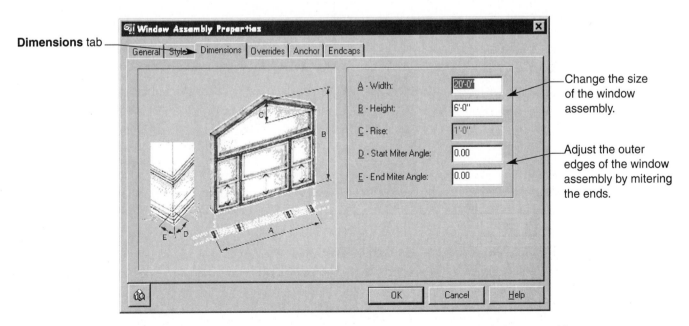

Dimensions tab

Change the size
of the window
assembly.

Adjust the outer
edges of the window
assembly by mitering
the ends.

FIGURE 10.44   The Dimensions tab controls the size and mitering angles for the selected window assembly.

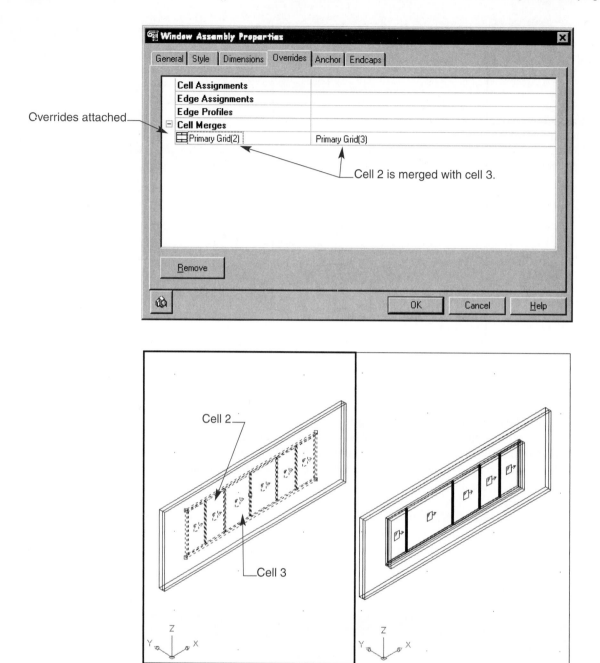

FIGURE 10.45   The override attached in the Cell Merges area indicate that cell 2 has been merged with cell 3.

CHAPTER REVIEW

Use the CD-ROM to test your knowledge and skills.

Chapter Test

To check your understanding of the content provided in this chapter, access the Test file in the CH10 folder of the CD-ROM that accompanies this text.

Chapter Project

To practice the Architectural Desktop skills presented in this chapter, access the Project files in the CH10 folder of the CD-ROM that accompanies this text. The project files are in pdf format and include sample drawings and instructions for completing each project.

| |

CHAPTER 11

Creating Stairs and Railings

LEARNING GOALS

After completing this chapter, you will be able to:

◎ Create and modify stair objects.

◎ Adjust the edge of stairs.

◎ Change the stair display.

◎ Use stair properties.

◎ Create and use stair styles.

◎ Create and modify railings.

◎ Control railing display.

◎ Use railing properties.

◎ Create and modify railing styles.

The previous chapters concentrated on creating and modifying the important Architectural Desktop objects walls, doors, and windows. These three types of objects provide the basis for construction documents. Another important object in the creation of a floor plan is the stair. The addition of a staircase is very important for any building with more than one floor. The addition of stairs also normally requires the use of railings. This chapter is devoted to the stair and railing objects included in Architectural Desktop.

Creating Stair Objects

The stair object in Architectural Desktop can be used to create several different types of stairs. Straight staircases, U-shaped stairs, L-shaped stairs, angled stairs, and curved stairs are all options in the stair design. The addition of stairs requires two basic sizes: the width of the stair runs and the floor-to-floor height. The floor-to-floor height determines the number of riser and treads used in the stair object. Stair types and components are displayed in Figure 11.1.

Like other AEC objects, stairs are displayed as 2D objects in a top view and as 3D blocks in a model view. Use one of the following options to add stairs to the drawing:

✓ Select the **Design** pull-down menu, pick the **Stairs** cascade menu, and select **Add Stairs….**

✓ Pick the **Add Stair** button in the Stairs - Railing toolbar.

✓ Right-click, and select **Design, Stairs, Add Stairs…** from the cursor menu.

✓ Type **AECSTAIRADD** at the Command: prompt.

The **Add Stairs** dialog box, shown in Figure 11.2, is displayed.

The following basic settings can be controlled for all stairs:

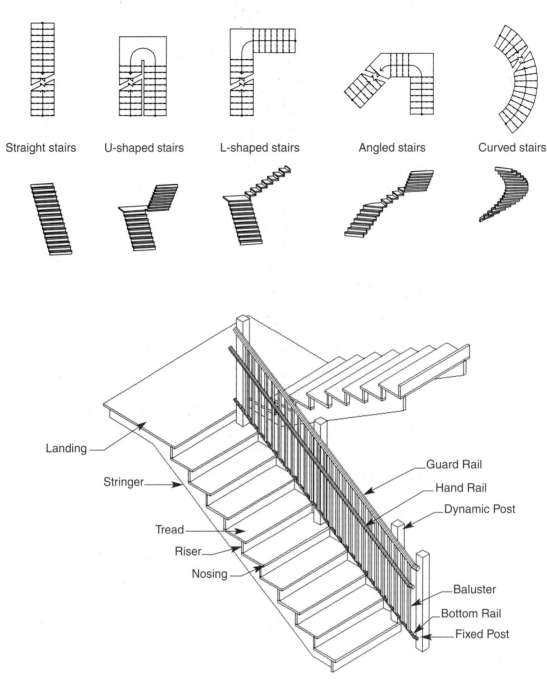

FIGURE 11.1   A sampling of the stairs that can be created using the stair object.

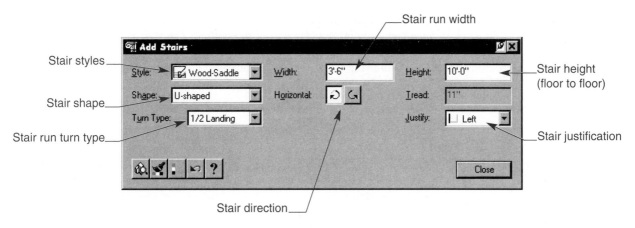

FIGURE 11.2 The Add Stair dialog box is used to add stairs into the drawing.

- **Style:** This drop-down list includes a list of stair styles available for use in the drawing. The styles list includes different versions of concrete, steel, and wood stairs.

- **Width:** This text box is used to specify a stair run width.

- **Height:** This text box is used to specify the total height of the stair object. This value should reflect the distance between floors.

- **Tread:** This text box remains grayed out but indicates the maximum tread depth based on the style for the stair selected.

- **Justify:** This drop-down list specifies how the stair will be drawn. The options include **Left**, **Center**, and **Right**. For example, the staircase can be drawn in relation to the left or right edges of the stair or from the center of the stair.

- **Shape:** This drop-down list specifies the shape or type of stair that will be drawn. Options include **U-shaped, Multi-landing, Spiral,** and **Straight**.

Depending on the shape selected, additional text and drop-down boxes appear in the dialog box. Once the stair properties have been determined, pick a point in the drawing where the first riser is to be located. After the first point has been picked, an outline of the stair is displayed, as shown in Figure 11.3. This outline indicates where the left and right boundaries of the stair will be placed in the drawing as well as the last

FIGURE 11.3 To add a stair, select the first riser location, then pick a point past the outline to add the stair.

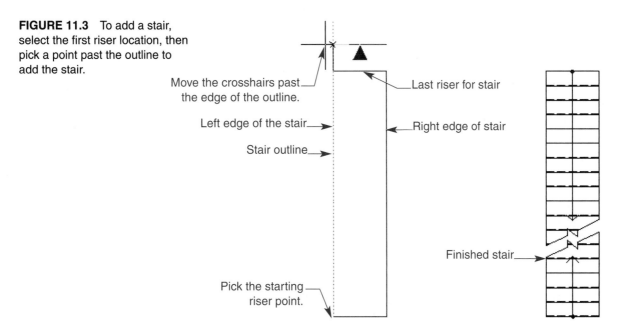

riser. The last riser location varies depending on the height specified. If a straight run of risers is being added to the drawing, pick an ending point past the end of the outline. The stair is then displayed in the drawing.

Place additional stairs in the drawing as necessary, then press the Close button or the [Enter] key.

Creating a Straight Stair

A straight stair is created by selecting the **Straight** option from the Shape: drop-down list. Select a style to use for the staircase from the Style: drop-down list. When creating a straight stair enter the width and height values to control the size of the staircase. The value in the Height: text box determines the total number of risers required to build the staircase. Select a justification for the stair as it is added to the drawing. The justification chosen determines how the stair is laid out initially in relation to the crosshair points picked (see Figure 11.4).

A left justification draws the stair along the left edge. The center justification draws the stair along the center of the staircase, and the right justification draws the stair along the right edge of the stair. When the values for the stair have been determined, select a starting point in the drawing. This establishes the first riser location. Select an ending point and angle for the stair by moving the crosshairs past the end of the outline. A straight stair is added with the number of risers automatically determined by the height value.

Exercise 11.1

1. Start a new drawing using the AEC Arch (Imperial) or AEC Arch (Imperial-Int) template.
2. Draw a straight staircase 10'-0" high. Set the justification to left, and use the Steel-Open style.
3. Draw a 5'-0"-high, 8'-0"-wide straight stair. Use the center justification and Concrete stair style.
4. Draw a straight staircase 9'-1^1/$_8$" high and 4'-0" wide. Use the right justification and Wood-Saddle stair style. Set the stair at a 45° angle.
5. Split the drawing screen into two vertical viewports using a 3D setup.
6. Save the drawing as EX11-1.dwg.

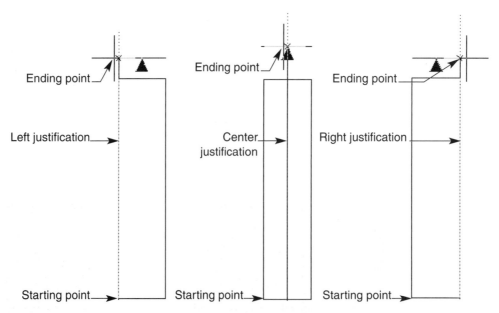

FIGURE 11.4   Selecting a justification determines the side of the stair that is used when selecting starting and ending points for the stair.

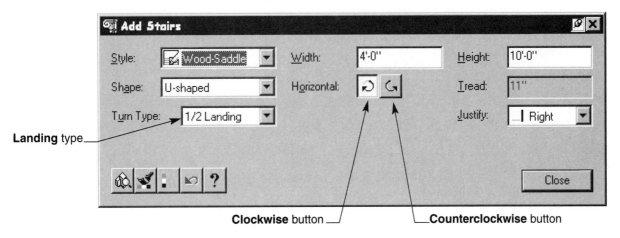

FIGURE 11.5   Selecting a U-shaped stair provides additional variables for further control of the U-shaped stair.

Creating a U-Shaped Staircase

Creating a U-shaped staircase is similar to creating a straight stair. Select the **U-Shape** option from the Sha̲pe: drop-down list. Two additional variables appear in the stair dialog box that control how the stair is to be created (see Figure 11.5).

The **Ho̲rizontal** area includes two buttons: **Clockwise** and **Counterclockwise**. These buttons control the direction the risers turn as the stair goes up in elevation (see Figure 11.6). The **Tu̲rn Type** drop-down list controls how the landing between the two runs of stairs appears. The **1/2 Landing** places a landing between the two stair runs. The **1/2 Turn** option does not add a landing; instead, the risers wind around the corners of the staircase.

The justification chosen along with the clockwise or counter clockwise settings affects how the stairs are drawn (see Figure 11.7). As with the straight stair, the starting riser location is selected first, then the ending riser. Architectural Desktop determines how the rest of the stair is added. The number of risers is

FIGURE 11.6   The U-shaped stair includes two types of turns, 1/2 Landing and 1/2 Turn. The direction of the risers moving up the stair is controlled by the Clockwise and Counterclockwise buttons.

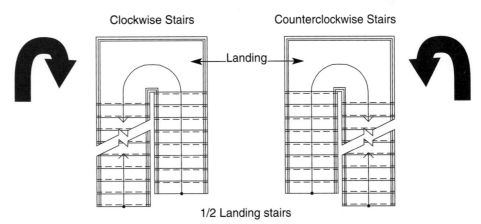

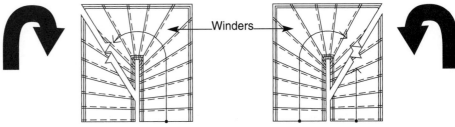

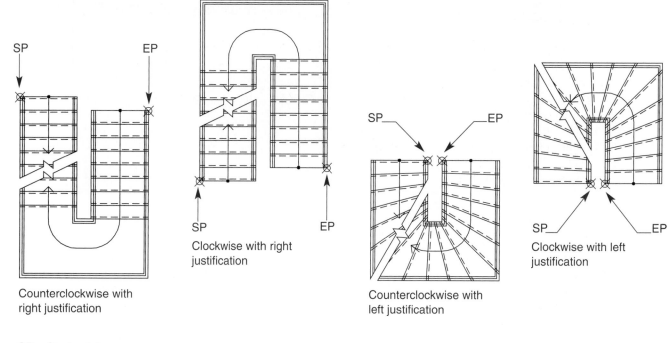

SP = Start point
EP = End point

FIGURE 11.7   Justification along with the direction of the stair determines the orientation of the staircase.

determined by the height specified. Architectural Desktop divides the total number of risers in half and creates two runs of stairs, side by side. If you specify the stair from the outside edge, be sure that the ending point selected allows enough room for both stair runs, as calculated by the formula stair run width + stair run width + a gap between the two runs. If creating the stair from the inside edge, make sure that there is a gap between the starting edge and ending edge of the stair runs.

Field Notes

If the stair runs overlap each other, the stair is invalid. Architectural Desktop displays invalid stairs with the following warning sign.

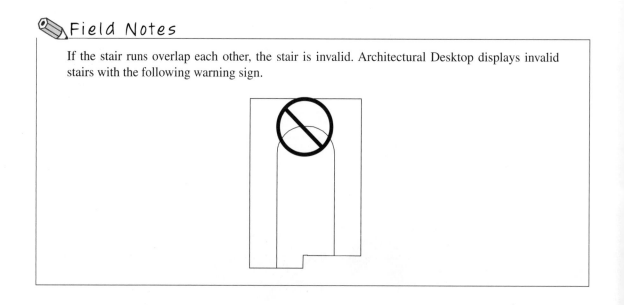

Exercise 11.2

1. Start a new drawing using the AEC Arch (Imperial) or AEC Arch (Imperial-Int) template.
2. Add a U-shaped stair 5'-0" wide and 10'-0" high. Use the 1/2 Landing turn type and set to a clockwise direction.
3. Add a U-shaped stair 5'-0" wide and 10'-0" high. Use the 1/2 Landing turn type and set to a counterclockwise direction.
4. Add a U-shaped stair 3'-0" wide and 8'-1" high. Use the 1/2 Turn turn type.
5. Split the drawing screen into two vertical viewports using a 3D setup.
6. Save the drawing as EX 11-2.dwg.

Creating a Multi-Landing Staircase

The multi-landing stair is perhaps the most flexible of the stair shapes. Multi-landing stairs can be used to create L-shaped stairs, U-shaped stairs, S-Shaped stairs, straight stairs with landings, and angled stairs (see Figure 11.8). To draw a multi-landing stair, select the **Multi-Landing** option from the **Shape:** drop-down list. Multi-landing stairs include four options in the Turn-Type: drop-down list: **1/4 Landing**, **1/4 Turn**, **1/2 Landing**, and **1/2 Turn**. The 1/4 and 1/2 Landing options create a landing between stair runs. The 1/4 and 1/2 Turn options create risers that wind around a corner of a stair.

When a multi-landing stair is created, the first point picked establishes the first riser location. Once the first point has been selected, the stair outline appears, indicating where the last riser location will be placed if a point is picked past the last riser. If a point is selected before the last riser outline, that point becomes the last point of the current stair run and the first point of a landing. Subsequent points picked create additional stair runs and landings (see Figure 11.9).

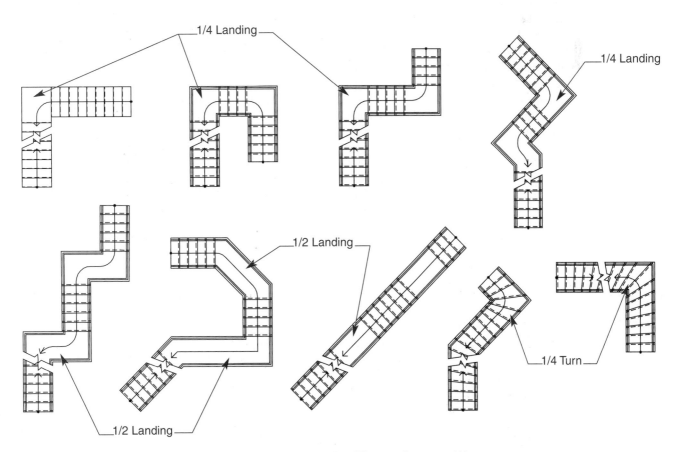

FIGURE 11.8   A few examples of multi-landing stairs. Note the difference between 1/4 Landing and 1/2 Landing.

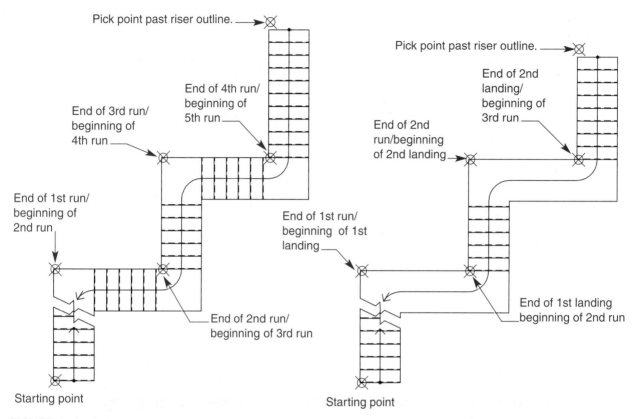

FIGURE 11.9   Selecting points before the outline of the last riser establishes landings and new stair runs.

As points are picked for a multi-landing stair the stair outline constantly readjusts to reflect the new ending riser point. As long as points are selected before the outline of the last riser, new runs and landing will be created. Selecting a point past the last riser outline creates the staircase.

After the starting point for the stair has been selected, the next point establishes where the next run will turn. Depending on how the crosshairs are moved, and the stair justification, the landing adjusts by either adding or subtracting risers from the previous run (see Figure 11.10).

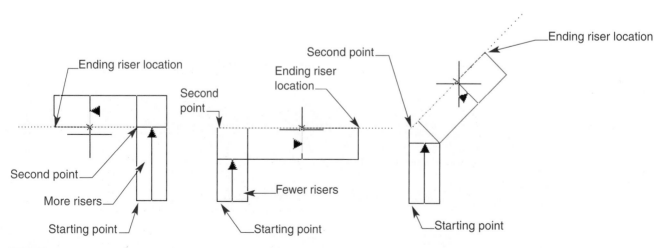

FIGURE 11.10   Establishing the next stair run. Depending on the location of the crosshairs when the next point is picked, risers are added or subtracted from the previous run.

Field Notes

- If a point is selected that causes two runs or landings to overlap each other, the stair will be created with the defect symbol. Erase the stair and draw a new stair with new points.

- To create a straight stair with landings between the runs, use the 1/2 Landing option, selecting all the points defining flight starting and ending points along the same line.

Exercise 11.3

1. Start a new drawing using the AEC Arch (Imperial) or AEC Arch (Imperial-Int) template.
2. Create several multi-landing stairs similar to Figure 11.8.
3. Try different styles, widths, and heights.
4. Save the drawing as EX 11-3.dwg.

Creating Curved Stairs

The final type of stair that can be created is a curved stair. Select the **Spiral** option from the Shape: drop-down list to create a curved stair (see Figure 11.11). Although the name *Spiral* indicates that a spiral staircase can be created, a true spiral staircase cannot be created, because the curved stair cannot create a complete circle or repeat risers on top of itself. Curved staircases can be created using one of three arc constraints plus a radius or an angle.

The **Arc Constraint:** drop-down list includes the following options for laying out a curved stair:

- The **Free** arc constraint uses the Radius: text box to create the curved stair. The radius for the curved stair is measured from the curve center point to the justification point, as shown in Figure 11.12A. To create the curved stair using the Free arc constraint, pick a point in the drawing as the center point of the curve, then pick a location for the starting riser. The Clockwise and Counterclockwise buttons determine the side of the run on which the starting riser will be placed.

- The **Total Degrees** arc constraint uses the Arc Angle: text box. The angle specified establishes the total staircase size in degrees, measured from the starting riser to the ending riser. The degree value entered must be between 10° and 355° (see Figure 11.12B). Select the center point for the stair, then the starting riser location.

- The **Degrees Per Tread** arc constraint also uses the Arc Angle: text box. The value entered in the text box becomes the total number of degrees for a single tread on the staircase. The angle for

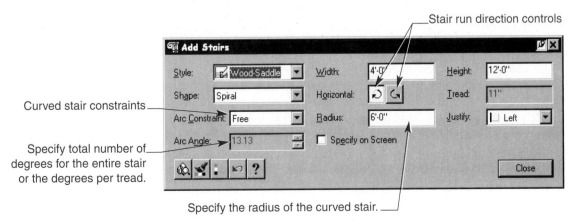

FIGURE 11.11 The Spiral shape includes many settings for creating a curved staircase. Three different methods can be used.

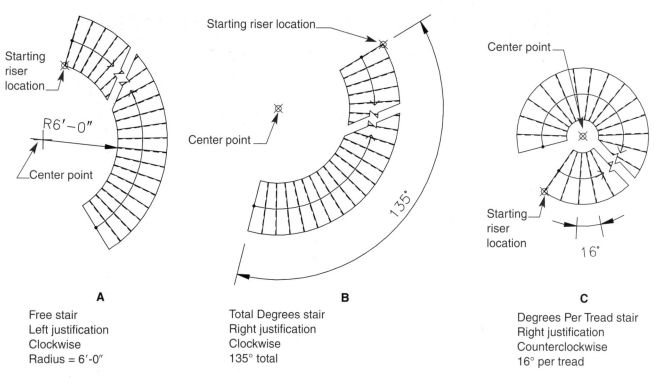

A	B	C
Free stair	Total Degrees stair	Degrees Per Tread stair
Left justification	Right justification	Right justification
Clockwise	Clockwise	Counterclockwise
Radius = 6'-0"	135° total	16° per tread

FIGURE 11.12   The Spiral stair option creates curved staircases in the drawing. Three different arc constraints can be used to aid in the curved stair layout process.

this constraint must be between 1° and 30°. To add this type of constrained stair, pick the center point location in the drawing, then the starting location for the first riser (see Figure 11.12C).

Exercise 11.4

1. Start a new drawing using the AEC Arch (Imperial) or AEC Arch (Imperial-Int) template.
2. Create each of the arc constrained curved stairs.
3. Try different styles, heights, justifications, and other settings.
4. Divide the drawing screen into two viewports with a 3D setup.
5. Save the drawing as EX11-4.dwg.

Modifying Stair Objects

Stairs that have been added to the drawing can be modified at any time. Settings such as the stair style, width, and height can be modified to reflect the design needs. Depending on the stair being modified, the clockwise or counterclockwise options also can be adjusted.

To modify a staircase, select the stair to be adjusted, right-click, and select Stair Modify... from the cursor menu. The Modify Stair dialog box is displayed. The setting content of this dialog box varies depending on the type of stair selected (see Figure 11.13). Alternatively, use one of the following options to access the Modify Stair dialog box:

✓ Pick the **Design** pull-down menu, select the **Stairs** cascade menu, then pick **Modify Stair....**

 ✓ Pick the **Modify Stair** button on the Stairs - Railings toolbar, right-click, and select **Design, Stairs, Modify Stair...** from the cursor menu.

✓ Type **AECSTAIRMODIFY** at the Command: prompt.

FIGURE 11.13 The different Modify Stair dialog boxes for the different types of stairs.

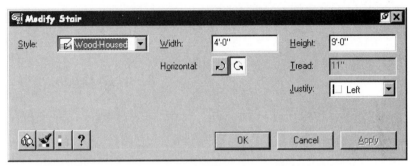

Stair Modify dialog box for multi-landing and straight stairs

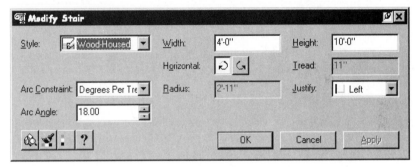

Stair Modify dialog box for U-shaped stairs

Stair Modify dialog box for spiral stairs

Make any adjustments necessary to the stair, then press the <u>A</u>pply button to apply the changes made to the stair to check the modifications. If the stair appears correct, press the OK button to apply the changes and close the dialog box.

The only modification that cannot be made is to change from one type of stair to another. For example, a multi-landing stair cannot be modified into a curved stair; however, spiral stairs have the option adjusting the Arc Constraint.

Exercise 11.5

1. Open EX11-4.dwg.
2. Use the Modify Stair command to modify the spiral stairs.
3. Make adjustments to the stairs such as style, height, width, and arc constraint.
4. Save the drawing as EX11-5.dwg.

Adjusting the Edge of Stairs

Thus far in this chapter, we have discussed the addition and modification of stairs. The creation of stairs in Architectural Desktop provides you with some common types of stairs found in many buildings. The edges

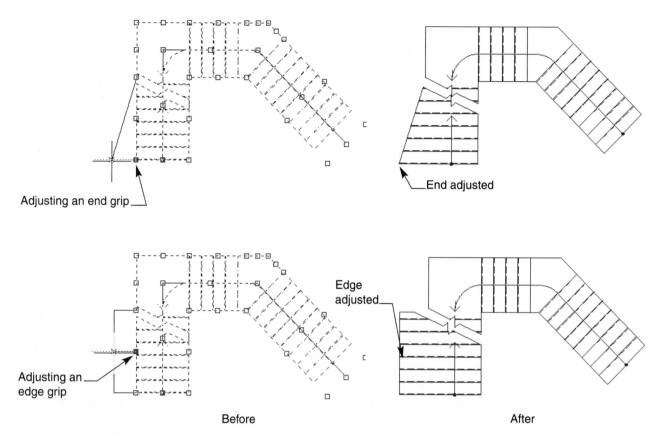

FIGURE 11.14   Grips can be used to adjust the stair runs and landings. Edges of stairs can also be projected to reflect a custom stair edge.

of standard stairs can also be modified in order to create a unique staircase. Stairs can be simply modified by stretching the grip locations around the stair, or they can have their edges modified to reflect a custom-shaped edge (see Figure 11.14).

Using Grips to Adjust Stair Edges

When an object is selected in a drawing without a command active, the object is highlighted with a series of boxes located at different vertex and midpoint locations. These boxes are known as grips. Grips were discussed in Chapter 3 for use on objects such as walls, doors, and windows. When a staircase is selected, a number of grip points become available. Typically, a grip is located at the starting, middle, ending, left, right, and center edges of stair runs (flights). If the stair includes a landing, additional points are included along the landing edges.

When a grip point is picked, that point is capable of being stretched or repositioned, as shown in Figure 11.15. If an ending grip is selected, that one point can be moved. If a grip along the middle of an edge is selected, the whole edge is repositioned. Selecting a grip on a landing can produce varying results.

Projecting Stair Edges

In addition to grip editing, the edges of stairs can be adjusted by offsetting the stair edge or projecting the edge of the stair (the right or left side) to a polyline or other AEC object. The **Customize Edges** command is used to perform these types of edge modifications.

Use one of the following options to access the Customize Edges command:

✓ Select the **Design** pull-down menu, pick the **Stairs** cascade menu, and select **Customize Edges....**

✓ Select the **Customize Edges** button from the Stairs - Railings toolbar.

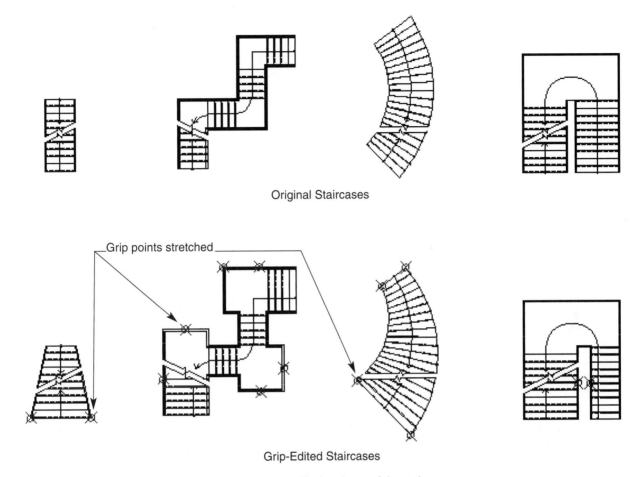

Original Staircases

Grip points stretched

Grip-Edited Staircases

FIGURE 11.15 Select a grip point along a stair to modify the shape of the stair.

✓ Right-click, and select **Design, Stairs, Customize Edges...** from the cursor menu.

✓ Type **AECSTAIRCUSTOMIZEEDGE** at the Command: prompt.

✓ Select the stair, right-click, and select **Customize Edge...** from the cursor menu.

When the command has been activated, the following prompt is issued:

Customize Stair Edge [Offset/Project/Remove customization/Generate polyline]:

Enter an option to modify the edge of the stair. After an option has been entered, you are prompted to select the edge of a stair. Select the edge to be modified and follow the remaining prompts. The following options are available for modifying stair edges:

- **Offset:**   This option offsets the selected edge the distance specified. A positive value offsets the stair edge away from the stair, and a negative value offsets the edge into the stair.

- **Project:**   This option projects the selected edge to a polyline or AEC object such as a wall. Before projecting the edge, ensure that the geometry is drawn to the desired shape and in the correct location (see Figure 11.16). Enter the Project option, select the edge, then pick the object you are projecting to.

- **Remove customization:** This option removes any edge or point modifications performed on the stair.

FIGURE 11.16 Stair edges can be projected to a polyline or to another AEC object.

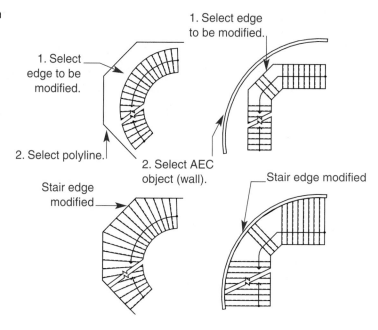

1. Select edge to be modified.

1. Select edge to be modified.

2. Select polyline.

2. Select AEC object (wall).

Stair edge modified

Stair edge modified

- **Generate polyline:**   This option creates a polyline along the edge of the selected stair edge.

Field Notes

- When the edge of a stair is projected, if the edges being projected intersect, the stair yields a default symbol. Remove the customization, adjust the object you are projecting to, and try projecting the edge again.

- If the width of the stair is modified through the Modify Stair dialog box, any shape adjustments made by grip editing or projecting are lost.

Exercise 11.6

1. Open EX11-3.dwg.
2. Experiment using the grips to modify stair edges.
3. Draw polyline shapes and walls away from different stairs for projecting to.
4. Try projecting the edge of stairs to the polylines and wall objects. (Remember, if portions of the stair will intersect when projecting, the stair is invalid.)
5. Save the drawing as EX11-6.dwg.

Changing the Stair Display

Through out this text, Entity Display has been used to modify how plan view objects appear compared with the model view object. Stair objects also react to Entity Display settings. Notice that plan view stairs display a break line that divides the stair into a flight going "up" and a flight coming "down." When the same stair is viewed in a model view, the full stair is displayed (see Figure 11.17).

To change the display of the stair, select the stair object, right-click, and pick Entity Display... from the cursor menu. Like previously discussed AEC objects, stairs include a model and a plan representation.

FIGURE 11.17 The plan view stair compared with the model view stair.

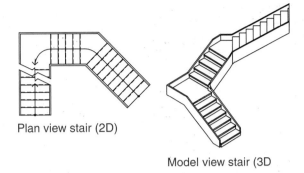

Plan view stair (2D)

Model view stair (3D

Select a display representation to modify, select the level of display control from the Property Source column, attach the override, and select the Edit Display Props button to make the modifications.

Adjusting Plan View Stairs

The **Entity Properties** dialog box for plan representation stairs includes two tabs: **Layer/Color/Linetype** and **Other**. The Layer/Color/Linetype tab includes a list of components that are displayed when the stair is viewed in a top view (see Figure 11.18). Typical components include Stringer, Riser, Nosing, Path, and Outline. These are represented as components that are going up the stair and coming down the stair. The properties of these components can be readjusted as desired by picking the symbol or text associated in each column. For example, if you do not need to see the components coming down the stair, then turn off all of the "down" components.

The Other tab includes options for adjusting additional display options (see Figure 11.19). The **Cut Plane** area is used to specify the elevation at which the break line cuts through the stair, the distance between the break lines(s), and the angle of the break line.

The **Stair Line** area controls how the path lines appear on the selected stair or stair style. The *path lines* or *walk path* are lines that have an arrow pointing up or down showing the direction of the stairs. The **Shape:** drop-down list creates a curved or straight path when the path line crosses over a landing. The **Apply to:** drop-down list applies the path in one of three ways. The **Entire** stair option applies the path to the whole stair (stair flights on each side of the break line). The **Cut Plane - Parallel** option places a walk path in the "up" direction anywhere a flight of stairs appears in the stair. The **Cut Plane - Opposite** is the default, which places the walk path going up a flight of stairs and a walk path coming down a flight of stairs.

The **Arrow** area controls how the arrowhead appears on the walk path. The size of the area and the offset from the break line are controlled in this area. The **Dim Style** drop-down list specifies the type of dimension style the walk path uses for display of the arrowheads. This value defaults to the AEC_Stair style. The **Break Mark** area specifies the type of break line used for the stairs. Custom blocks can be used for a break mark.

Adjusting Model View Stairs

Model representation for stairs includes only the Layer/Color/Linetype tab. Components of a model stair include Stringer, Tread, and Riser. The properties can be set for each component and affect how the model representation for the stair appears.

Exercise 11.7

1. Open EX11-6.dwg.
2. Select a stair, set the override for the individual stair, and turn off all the "down" components.
3. Select another stair, and set the override for the individual stair. Change the "Outline down" component to color 3 and a hidden linetype. Turn off the other "down" components. Adjust the cut plane to 7'-0".
4. Select a stair with multiple landings. Change the stair line to the Cut Plane - Opposite option and place a check in the Dra<u>w</u> for Each Flight check box.
5. Save the drawing as EX11-7.dwg.

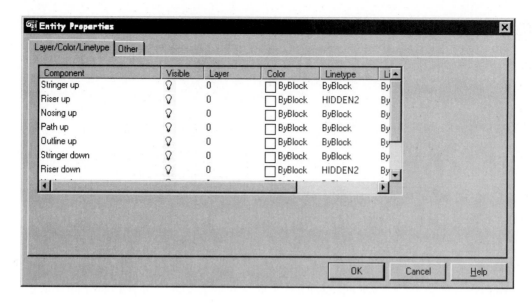

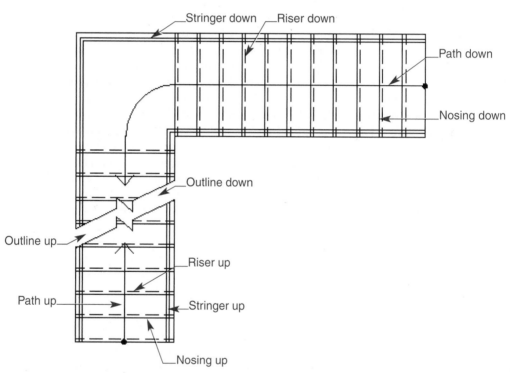

FIGURE 11.18 The Layer/Color/Linetype tab for plan view stairs controls the display of the stair when viewed in the top view.

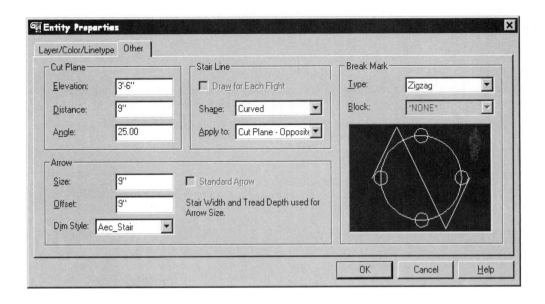

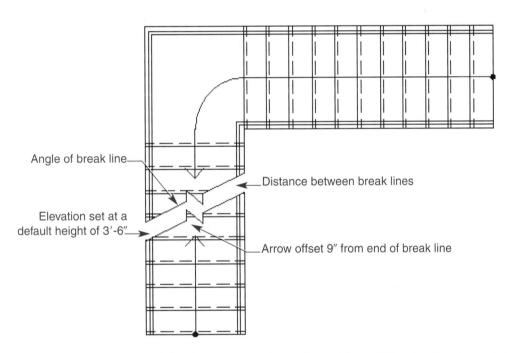

FIGURE 11.19   The Other tab for plan representation stairs contains additional options for controlling the display of plan view stairs.

Using Stair Properties

The Modify Stair and Entity Display commands provide many options for controlling how stairs appear in the drawing. In addition to these commands, the **Stair Properties** dialog box can be used to control individual settings and sizes for a selected stair. To use this command, select a stair to modify, right-click, and select Stair Properties… from the cursor menu. The Stair Properties dialog box is displayed. The dialog box includes the following seven tabs for controlling the selected stair:

- **General**   Use this tab to add a description to the stair, add notes, or attach property set definitions.

- **Style**   This tab lists the available stair styles in the drawing. Use the Floating Viewer button to preview what the style looks like and how the components have been combined for the style.

- **Dimensions**   This tab is used to control the sizes used in a stair object (see Figure 11.20). The **Width:** and **Height:** text boxes display the stair's current settings. Adjust the values to modify the selected stair. The **Interacting Dimensions** area includes a few values that can be adjusted. Select the **Padlock** button to unlock the value, and enter a new value as needed.

 Field Notes

- Changing the width of the stair removes any edge customization on the stair object.

- The values in the Interacting Dimensions area are governed by the stair style. Changing a value may result in a warning message. In this case, press the Padlock button to accept the default sizes, then modify the stair style rules. Finally, enter the Stair Properties dialog box for the individual desired settings for the selected stair. The rules for the style are discussed in the next section.

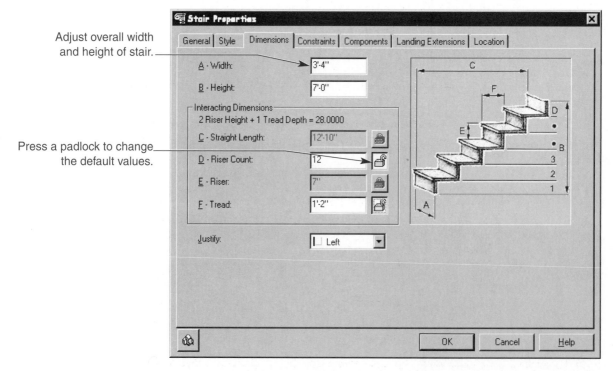

FIGURE 11.20   The Dimensions tab can be used to adjust the sizes of the stair and individual component sizes.

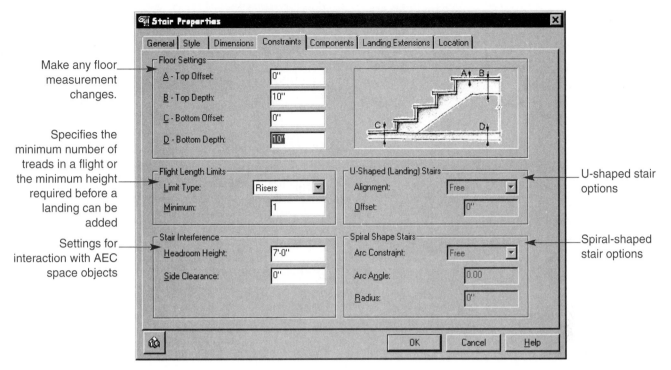

Make any floor measurement changes.

Specifies the minimum number of treads in a flight or the minimum height required before a landing can be added

Settings for interaction with AEC space objects

U-shaped stair options

Spiral-shaped stair options

FIGURE 11.21 The Constraints tab is used to set specific settings governing the creation of the stair.

- **Constraints**

This tab is used to adjust any constraints placed on the stairs, as shown in Figure 11.21. The **Floor Settings** area is used to set top and bottom settings for the stairs in relation to a floor. The **Flight Length Limits** area specifies a minimum number of treads to create a flight of stairs, or a minimum height before a landing can be added to the stair. The **Stair Interference** area is used when placing a stair within a space object. A *space object* is an AEC object found in the Concept pull-down menu. These objects are used to define the interior area of a room, to create a finished floor and finished ceiling component. The addition of these objects within a room is discussed in a later chapter. The space object produces a finished floor and ceiling for modeling purposes. Entering any values in this area directly affects where the stair cuts through the space object.

The **U-shaped (Landing) Stairs** area is used to line up risers in a U-shaped stair. Options include **Free**, which causes the risers to be added into the stair whereever treads can be fit. The **Tread to Tread** option lines up the nosing of two opposing stair flights. The *nosing* is the normally rounded front edge of a tread that extends past the riser (see Figure 11.1). The **Tread to Riser** option lines up a riser on one flight of stairs with the nosing of the second flight. The **Riser to Riser** option lines up the risers of two opposing stair flights. The **Spiral Shape Stairs** option controls how a spiral (curved) stair is constructed.

- **Components**

This tab is used to control the risers and treads created in the stair object. If these values are grayed out, the stair style governs these rules. Adjust the stair style as necessary so changes can be made in the Stair Properties dialog box. See the next section for more information regarding this tab.

- **Landing Extensions**

This tab controls how the risers and landings determine their placement in the stair object. If this tab is grayed out, the stair style governs these rules.

■ **Location**   This tab is used to readjust the stair location in the drawing. The most valuable option in this tab is the readjustment of the Z-axis (floor) variable, which places the stair vertically in the drawing.

Creating Stair Styles

When stairs are added to the drawing, one of the options available is the type or style of stair to be used. Some of the settings found in the Stair Properties dialog box are directly affected by the stair style and may or may not allow you to make adjustments to the stair. Stair styles can be customized in Architectural Desktop and rules can be set covering how stair components are created.

Stair styles can be created or modified by accessing the Stair Styles command using one of the following options:

✓ Select the **Design** pull-down menu, pick the **Stairs** cascade menu, then select **Stair Styles....**

✓ Pick the **Stair Styles** button in the Stairs - Railings toolbar.

✓ Right-click, and select **Design, Stairs, Stair Styles....**

✓ Type **AECSTAIRSTYLE** at the Command: prompt.

The Style Manager dialog box, opened to the **Stair Styles** section, as shown in Figure 11.22, is displayed.

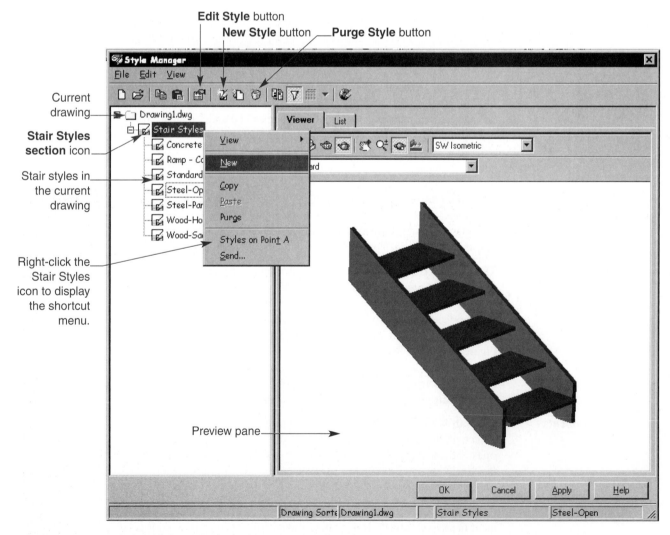

FIGURE 11.22   The Style Manager dialog box displaying the Stair Styles section.

The Style Manager dialog box is divided into two windowpanes. The left pane displays any drawings open in AutoCAD. The open folder icon in the left pane notes the current drawing. Below the current drawing file is the **Stair Styles section** icon. Pick on top of this icon to display a list of stair styles in the right pane, or press the + symbol beside the Stair Styles section icon to list the styles in the drawing in a tree view in the left pane. If a style is selected in the left pane, the preview window on the right displays the components of the selected stair style and their arrangement in the stair.

New stair styles can be created by selecting the New Style button or by highlighting the Stair Styles section icon, right-clicking, then selecting <u>N</u>ew from the shortcut menu. Enter a new name for the stair style, then press the [Enter] key. Existing stair styles or new stair styles can be edited by selecting the style from the left pane to highlight it, then picking the Edit Style button. Styles can also be edited by selecting the desired style, then right-clicking to display a shortcut menu. Select the <u>E</u>dit... option to modify the new or existing stair.

The **Stair Styles** dialog box is displayed when a stair style is edited (see Figure 11.23). This dialog box is used to control all the settings for the selected stair style. The following six tabs are available for creating the stair style:

- **General** Similar to the General tab in the Stair Properties dialog box, this tab is used to rename the style name and provide a description for the style. The **N<u>o</u>tes...** button can be used to add design notes and specs or to attach an external file such as a text document or spreadsheet to the style. The **Property Sets...** button is used to attach a Property Set Definition to the style. Property set definitions are discussed in a later chapter.

- **Design Rules** This tab is used to establish rules governing riser and tread sizes for a stair (see Figure 11.24). The **Code Limits** area allows you to set up maximum riser and tread dimensions for stairs at different slopes. The length drawn when adding a stair to the drawing determines the *slope* of a stair. Architectural Desktop automatically determines the riser and tread sizes based on the rules established in the Code Limits area. When a stair is modified through the Stair Properties dialog box, some settings such as the number of risers, riser heights, and tread depths are determined by the values set in this tab.

FIGURE 11.23 The Stair Styles dialog box is used to edit a stair style.

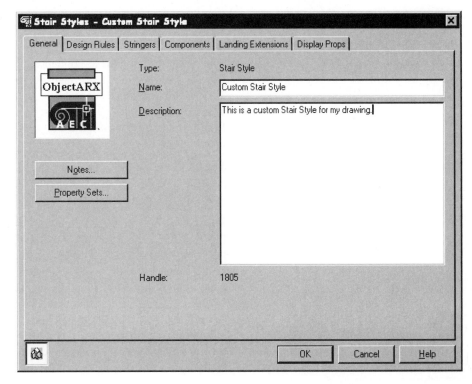

The Code Limits area specifies maximum heights and depths at different stair slopes.

The Calculator Rule applies a constraint to the stair slopes.

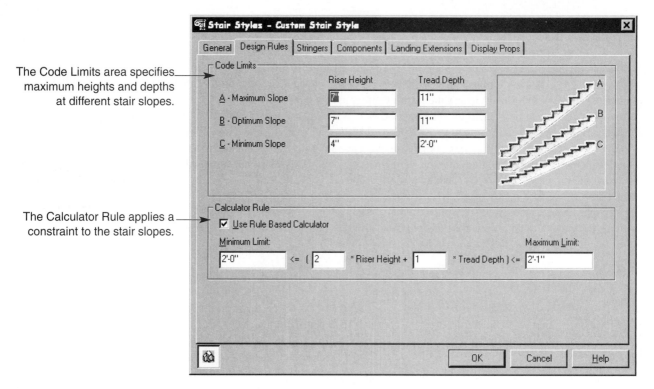

FIGURE 11.24   The Design Rules tab establishes rules for riser and tread sizes.

The **Calculator Rule** area can be used to set constraints for the calculations required to create the proper number of risers as close as possible to the **Optimum Slope**. These rules can be adjusted as needed. In many cases, this calculator can be turned off for the style, which causes Architectural Desktop to create a stair based on the Code Limits area only.

- **Stringers**

This tab specifies whether the stair is constructed with stringers, as in a steel or wood stair, or as a slab for concrete stairs. First, determine if the stair is a stringer or slab type. Select the appropriate radio button, as shown in Figure 11.25. If the **Stringers** radio button is selected, the table at the top is used to configure any stringers added to the stair. Use the **A**dd button to add stringers to the style. As stringers are added the stringer variables can be adjusted by selecting the text in the table. Some of the variables require sizes, and others include drop-down lists. Use the Floating Viewer button to preview the style and placement of the stringers.

The **Solid Slab** radio button uses the **Slab Settings** area to determine how the stair slab appears. Adjust the **Waist at Flights** and the **Waist at Landings** variables.

- **Components**

This tab is used to specify sizes for risers and treads in a stair (see Figure 11.26). The **A**l**low Each Stair to Vary** check box in the upper left corner allows you to adjust these same properties for an individual stair through the Stair Properties dialog box. Treads and risers can be added or removed by selecting the **Tread** or **Riser** check boxes. Nosings can be added to the tread by specifying a size in the **Nosing Length** text box. The **Sloping Riser** check box bevels the riser from the end of the nosing to the bottom of the riser. Use the Floating Viewer to check on the progress of the stair style. The bottom of the tab is devoted to controlling the sizes at the landing areas.

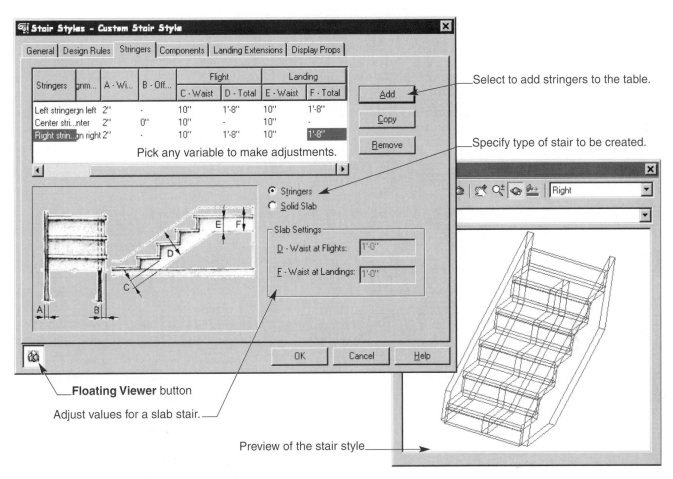

FIGURE 11.25 The Stringers tab creates a stair with stringers or a slab stair.

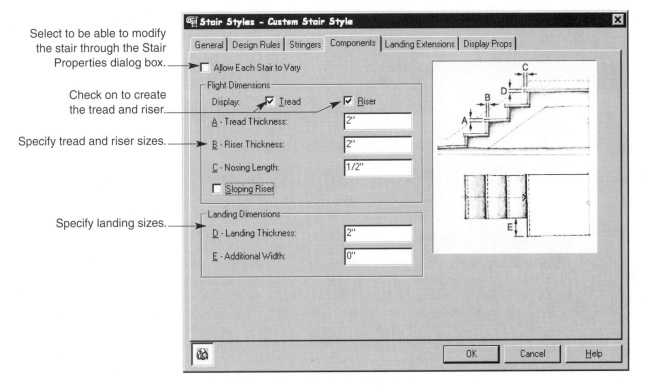

FIGURE 11.26 Use the Components tab to adjust the riser and tread sizes.

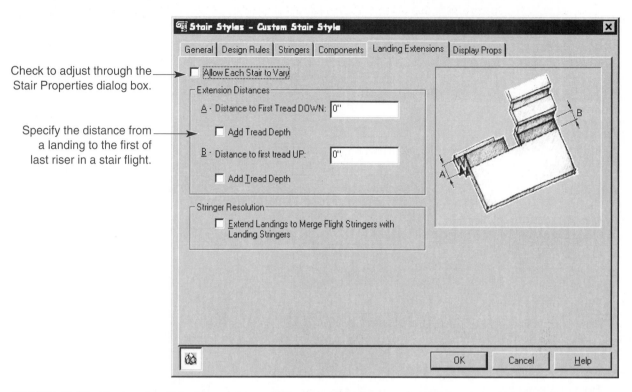

Check to adjust through the
Stair Properties dialog box.

Specify the distance from
a landing to the first of
last riser in a stair flight.

FIGURE 11.27   The Landing Extensions tab is used to control the placement of the riser in
relation to the beginning of a landing.

- **Landing Extensions** Like the Components tab, this tab includes the Allow Each Stair to
Vary check box. Check this on to be able to adjust landing extensions
per stair through the Stair Properties dialog box. The **Extension
Distances** area is used to control distances from the beginning of the
landing to the last or first riser in a flight of stairs. The **Stringer
Resolution** area includes a check box that merges stair flight stringers
with stringers under a landing (see Figure 11.27).

- **Display Props** Similar to the Entity Display dialog box, this tab controls the display
of the stairs in different display representations. The difference
between this tab and the standard Entity Display is that there are only
two property sources: **System Default** and **Stair Style**. If you are creating a stair style from scratch, you may want to also control its display settings by style (see Figure 11.28).

When you have finished setting the rules and configuring the components of the stair, press the OK button
to accept the modifications to the style and to return to the Style Manager dialog box. Press the OK button
to exit and accept the changes to the style. The new stair style is now ready for use in the drawing.

Using Stair Styles

The Style Manager is used to organize styles in any drawings that are open. If a custom stair style has been
created, it can be used in another drawing. To do this, open the drawing that contains the desired stair style
and the drawing in which the style will be used, and access the Style Manager. The Style Manager displays
both drawings in the left pane (see Figure 11.29). Select the Stair Style Section icon that contains the
desired stair style. Highlight the desired stair style from the right pane, then press and hold the left mouse
button to drag the style to the desired drawing in the left pane. When the cursor is over the desired drawing
in the tree-view pane (left pane), let go of the left mouse button. The style is added to the new drawing.

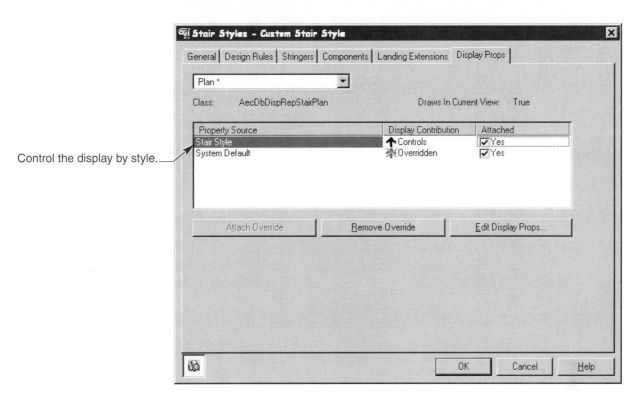

FIGURE 11.28   The Display Props tab is used to control the display properties of the stair by style.

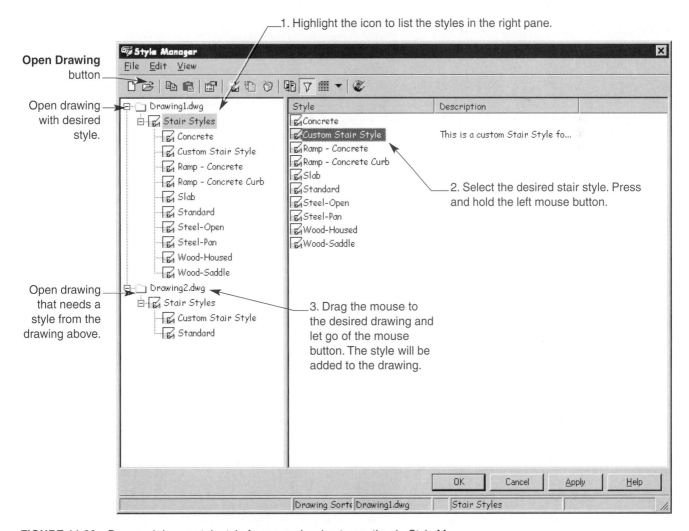

FIGURE 11.29   Drag and drop a stair style from one drawing to another in Style Manager.

Styles folder

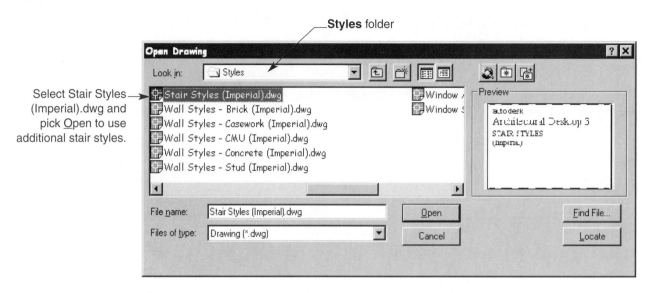

Select Stair Styles (Imperial).dwg and pick Open to use additional stair styles.

FIGURE 11.30   Use the Open Drawing button to browse to the \\Program Files\ Autodesk Architectural Desktop 3\Content\Imperial\Styles folder to use additional stair styles included with Architectural Desktop.

Additional stair styles are included in Architectural Desktop. To access these styles, select the Open Drawing button in Style Manager, as shown in Figure 11.29. The Open Drawing dialog box is displayed. Browse to the **\\Program Files\ Autodesk Architectural Desktop 3\Content\Imperial\Styles** folder, shown in Figure 11.30. Select the Stair Styles (Imperial).dwg file, then press the Open button. The drawing file is opened in the Style Manager and not in AutoCAD. Drag and drop the desired styles into your drawing following the procedures previously outlined. When you have finished accessing the Stair Styles drawing, highlight it in the left pane, right-click, and select the Close option to close the drawing.

Stair styles that are not used in the drawing take up valuable drawing file size. These unused styles can be purged from the drawing by highlighting the Stair Style Section icon, right-clicking, and selecting the **Purge** option or by pressing the Purge button. The Select Stair Styles dialog box appears with a list of the styles in the drawing and a check box beside each style (see Figure 11.31). The check mark indicates styles that will be purged. Press the OK button to purge the unused styles from the drawing. Only styles that are not being used are purged from the drawing. Individual styles can be purged by first highlighting the style in the left pane, then selecting the Purge button or right-clicking to select the Purge option.

Styles that will be purged from the drawing

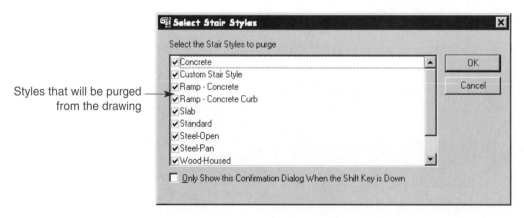

FIGURE 11.31   Check the styles to purge from the drawing.

Exercise 11.8

1. Start a new drawing using the AEC Arch (Imperial) or AEC Arch (Imperial-Int).dwt.
2. Create a new stair style.
3. Name the style Custom Stair Style.
4. Uncheck the Use Rule Based Calculator check box.
5. Create the style using three stringers 2″ wide. Place the stringers on the left, center, and right sides of the stair.
6. Turn on the Allow Each Stair to Vary check box in the Components and Landing Extensions tabs.
7. Set up any additional values as desired.
8. Save the drawing as EX 11-8.dwg.

Creating Railings

Railings are AEC objects that represent a handrail or guardrail. Railings can be freestanding at balconies or decks, added to a flight of stairs, or added to all the flights and landings of a stair object.

Use one of the following options to add railings to the drawing:

✓ Pick the **Design** pull-down menu, select the **Railings** cascade menu, and pick the **Add Railing...** command.

✓ Pick the **Add Railing** button from the Stairs - Railings toolbar, right-click, and select **Design, Railings, Add Railing...** from the cursor menu.

✓ Type **AECRAILINGADD** at the Command: prompt.

The **Add Railing** dialog box, shown in Figure 11.32, is displayed.

The Add Railing dialog box includes a list of railing styles in the drawing. Select the desired style. Use the Floating Viewer to preview the railing style. The **Attached To:** drop-down list creates a freestanding railing, a railing that is added to stair flights, or a railing that is added to the entire stair object. Additional check boxes may or may not be available depending on the railing style selected. If these check boxes are grayed out, the railing style governs their usage.

Once the options have been selected, the railing can be added to the drawing. If the **None** option has been selected from the Attached To: drop-down list, the railing can be added to the drawing as a freestanding railing. Pick a location for the starting point of the railing. As more points are picked, the railing is added to the drawing. Once the ending point of the railing has been selected, press the [Enter] key to close the Add Railings dialog box (see Figure 11.33).

If a railing is to be added to a stair, use the **Stair** or **Stair Flight** options in the Attached To: drop-down list. When these options are used, the **Offset** text box and the **Automatic** check box become available. The Offset text box specifies a distance from the edge of a stair to the railing. The Automatic checkbox automatically determines the side of the stair to which the railing is to be added. Once the railing values have been set, you are prompted to select a stair. Select the side of the stair where you would like to

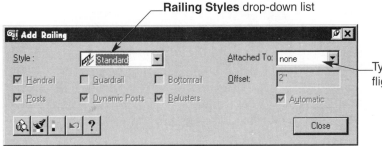

FIGURE 11.32 The Add Railing dialog box is used to add railings to the drawing.

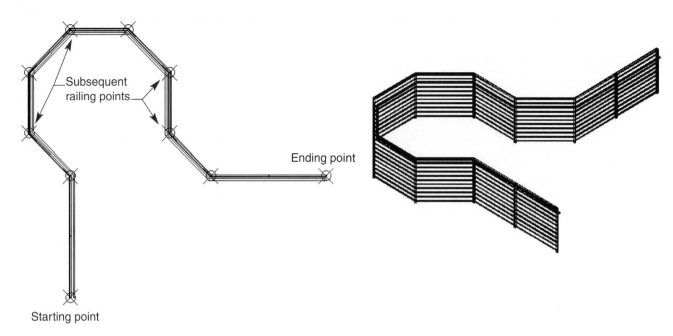

FIGURE 11.33   A freestanding railing is created using the None option from the Attached To: drop-down list. Pick points to establish the bends in the rail.

place the railing. If the Automatic check box is selected, the railing is applied to the side of the stair you selected, offset the distance specified (see Figure 11.34).

Polylines also can be converted into railings. The polyline drawn can be composed of straight segments, curved segments, or both. This is the only way to create a curved railing except for railings that are attached to a spiral stair or attached to a stair that has had its edge modified. First, draw the polyline in the desired shape, then use the **Convert to Railing** command. Use one of the following options to access this command:

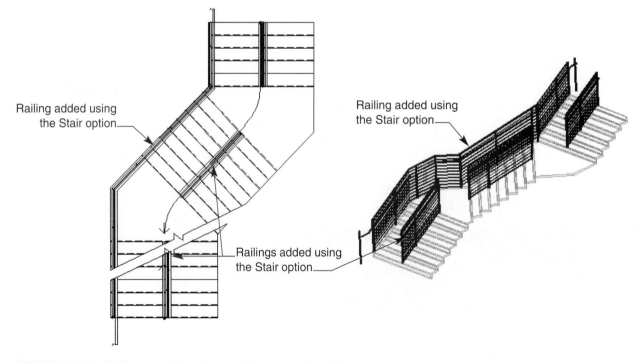

FIGURE 11.34   Railings applied using the Stair and the Stair Flight options.

✓ Pick the **Design** pull-down menu, select the **Railings** cascade menu, then pick **Convert to Railings....**

✓ Pick the **Convert to Railing** button on the Stairs - Railings toolbar.

✓ Right-click and select **Desi̲gn, Railings, C̲onvert to Railings...** from the cursor menu.

✓ Type **AECRAILINGCONVERT** at the Command: prompt.

As when converting polylines into walls, you are prompted to select a polyline, erase the layout geometry, then select the style from the Railing Properties dialog box.

Exercise 11.9

1. Open EX11-8.dwg.
2. Add a Multi-landing stair 8′-0″ wide and 12′-0″ high to the drawing.
3. Add a spiral stair with your custom railing style. Stretch the starting riser grips to make the spiral stair wider at the beginning than it is at the end.
4. Add railings to the multi-landing stair. Create railings using the Stair option at the left and right sides of the stair. Create railings for each flight using the Stair Flight option.
5. Add Railings to either side of the spiral stair.
6. Create a Railing using the None option.
7. Draw a polyline with straight and curved segments and convert it into a railing.
8. Save the drawing as EX11-9.dwg.

Modifying Railings

If you have added a railing to your drawing and determine that you used the wrong style or that some components need to be turned on or off, you can use the **Modify Railing** command to change the existing railing. To access the command, select the railing to be modified, right-click, and pick Railing Modify... from the cursor menu. The Modify Railing dialog box is displayed. This dialog box looks similar to the Add Railings dialog box (see Figure 11.35).

Alternatively, you can use one of the following options to access the Modify Railing dialog box:

✓ Pick the **Design** pull-down menu, select the **Railings** cascade menu, and pick **Modify Railing....**

✓ Select the **Modify Railing** button on the Stairs - Railings toolbar.

✓ Right-click, and pick **Desi̲gn, Railings, M̲odify Railings...** from the cursor menu.

✓ Type **AECRAILINGMODIFY** at the Command: prompt.

Change the railing style. ⎯⎯

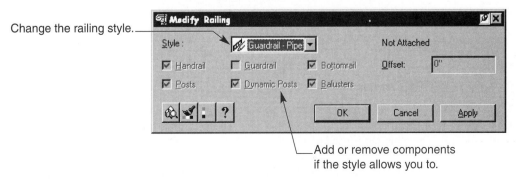

Add or remove components
if the style allows you to.

FIGURE 11.35   The Modify Railing dialog box is used to modify a railing object.

FIGURE 11.36   Select a railing, then right-click to display the railing menu. Select the Post Placement cascade menu to add new posts.

Select to control new railing posts.

Redistribute new and existing posts evenly along a railing.

Add new posts.

Remove new posts.

Hide new posts.

Show new posts.

Select a different style, or adjust any other values for the railing. Press the <u>A</u>pply button to apply the changes to the railing before exiting the Modify Railing dialog box. If the railing style permits it, you can turn railing components on or off by selecting different combinations of the check boxes in the Modify Railing dialog box. Railing styles can be modified and newly created in a drawing. Railing styles are discussed later in this chapter.

Railings can also be modified with grips. When a railing is selected, grip points are located at the corners and ends of a railing. Selecting one of these grips allows you to reposition the post or vertex location and the guard and handrails. Pick a new point to reposition the railing post. If a railing is placed along a stair, adjusting the grip points on the railing repositions the post, but not the guard or handrail.

In addition to using grips to adjust post locations, you can use the right-click cursor menu to access additional options for adding and removing posts not initially created when the railing was drawn. To access these commands, select a railing, right-click, and pick the **Post Placement** cascade menu, shown in Figure 11.36. The commands in this menu allow you to add and remove new posts, and hide or show newly added posts. The right-click menu also includes a **Redistribute Posts** option. This command can be used to redistribute existing and new posts evenly along a railing.

Controlling Railing Display

As with other AEC objects, the display properties of railings are controlled through entity display. Plan view railings can be configured to look one way in a plan representation (top view) and a different way in a model representation. To access the entity display for railings, pick a railing to display the grip boxes, right-click, and select Entity Display... from the cursor menu. The Entity Display dialog box for railings appears. Select the display representation to be modified from the drop-down list. Attach an override for the desired property source or level of display control and pick the <u>E</u>dit Display Props... button to begin editing the display.

Adjusting Plan Representation Railings

The plan representation includes two tabs. The Layer/Color/Linetype tab includes railing components that are going up the stair, and components that are coming down a stair. If a freestanding railing has been selected, the "up" components are used. Make any display changes to the components for the plan view railings.

The Other tab is used to apply an AEC profile to the different components of the railing in a plan view. Select the <u>A</u>dd... button in the Other tab to add a profile. The Custom Profile dialog box, shown in Figure 11.37, is displayed. Select a profile for use by picking the <u>S</u>elect Profile... button.

Select an AEC profile by picking the <u>S</u>elect Profile... button. Place a check in the component box that uses the custom profile, and specify any insertion criteria. When you have finished, pick the OK button to return to the Entity Properties dialog box. Any custom profiles added to plan view railings appear as components in the Layer/Color/Linetype tab.

Adjusting Model Representation Railings

The model representation railing also includes the same two tabs. Use the Layer/Color/Linetype tab to make display adjustments to the model components of the stair. These components include balusters, posts, guardrails, and handrails. The other tab is used to add custom blocks to the railing much like adding custom blocks to doors or windows.

The Other tab can be useful if you are attaching a custom post or baluster block to the railing for modeling or detailing purposes. First, create the desired 3D geometry. Remember to create the geometry on

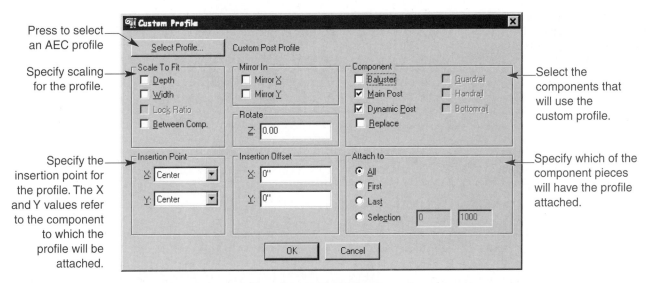

Press to select an AEC profile

Specify scaling for the profile.

Specify the insertion point for the profile. The X and Y values refer to the component to which the profile will be attached.

Select the components that will use the custom profile.

Specify which of the component pieces will have the profile attached.

FIGURE 11.37   Attach a custom AEC profile to plan representation railing components through the Other tab. Select the <u>A</u>dd... button to display the Custom Profile dialog box.

layer 0, using ByBlock colors, linetypes, and lineweights. Next, turn the geometry into a block. Select the railing to attach the block to, right-click, and select Entity Display... from the cursor menu. Select the Model representation, select the appropriate Property Source, and press the <u>E</u>dit Display Props... button.

To attach the block(s), select the Other tab, then pick the Add... button. The **Custom Block** dialog box, shown in Figure 11.38, is displayed. Select a block from the <u>S</u>elect Block... button. In the Components area, select the railing component(s) the block will be attached to. Adjust any scaling factors or insertion point locations, and press the OK button when you have finished. In the Entity Properties dialog box, make any adjustments to the new block components.

Using Railing Properties

The Railing Properties dialog box is similar to other properties dialog boxes in that it allows additional control on how the railing is constructed. Use one of the following options to access this dialog box:

✓ Select the railing that will be modified, right-click, and select **Railing Properties...** from the cursor menu.

✓ Type **RAILINGPROPS** at the Command: prompt.

The Railing Properties dialog box for freestanding railings includes the following five tabs. A railing that is attached to a stair includes two additional tabs: Extensions and Anchor.

- **General**   As with other General tabs, use this tab to enter a description, notes, or property set definitions for the selected railing.

- **Style**   This tab lists the various railing styles available in the current drawing. Pick a different style to change the selected railing's style. Use the floating viewer to preview the railing style.

- **Rail Locations** This tab is used to control the railing components of a railing object. Guardrails, handrails, and bottom rails can be turned on and their properties set in this tab (see Figure 11.39). If the values are grayed out, the railing style controls these properties.

- **Post Locations** This tab is used to control the post and baluster components of a railing. As with the Railing Locations tab, if this tab is grayed out, the railing style controls these properties (see Figure 11.40).

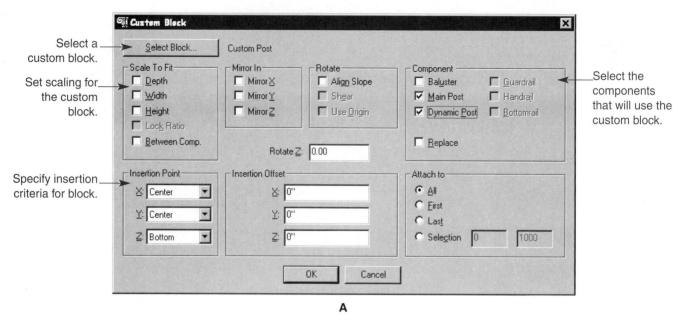

Select a custom block.

Set scaling for the custom block.

Specify insertion criteria for block.

Select the components that will use the custom block.

A

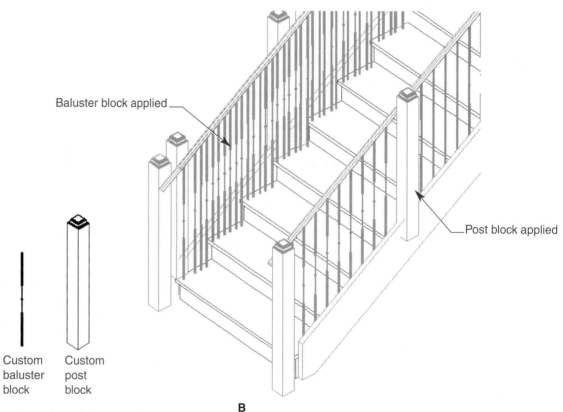

Baluster block applied

Post block applied

Custom baluster block

Custom post block

B

FIGURE 11.38   (A) The Custom Block dialog box is displayed when custom blocks that are attached to railing objects are added or edited. (B) After custom blocks are added to a railing, the railing can be used in the drawing.

Rail Locations tab ____   ____Turn components on or off

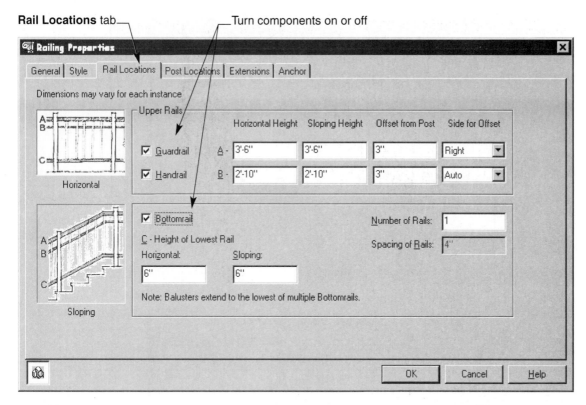

FIGURE 11.39   The Rail Locations tab is used to control the properties of the rail components.

Post Location tab ____   ____Turn posts and balusters on or off

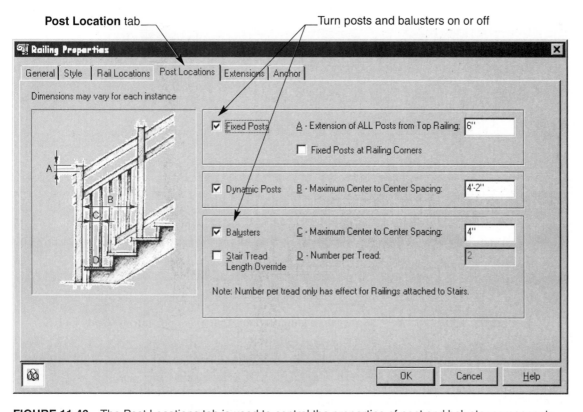

FIGURE 11.40   The Post Locations tab is used to control the properties of post and baluster components.

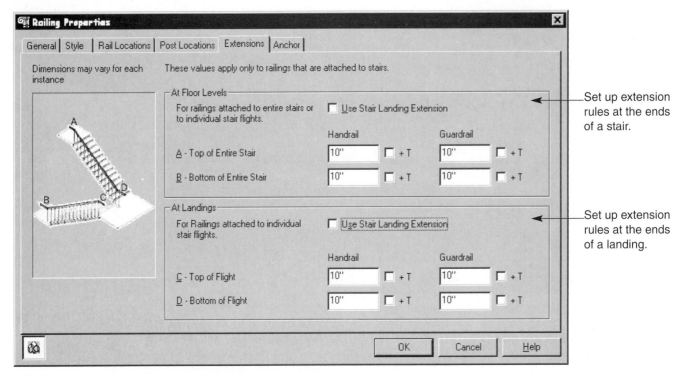

Set up extension rules at the ends of a stair.

Set up extension rules at the ends of a landing.

FIGURE 11.41   The Extensions tab is used to specify how far the guardrail or handrail will extend past the end of a flight of stairs.

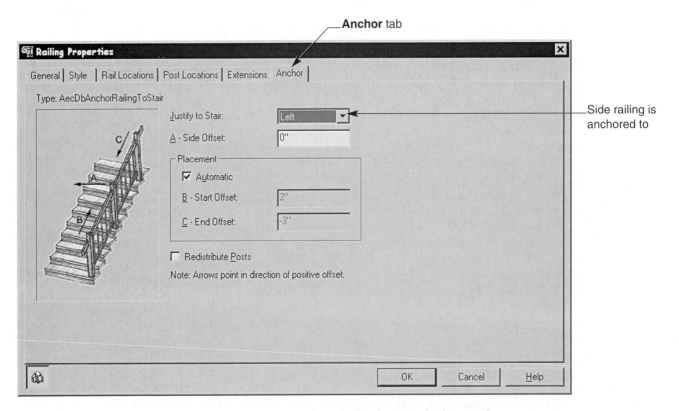

Anchor tab

Side railing is anchored to

FIGURE 11.42   The Anchor tab is used to specify anchoring criteria when attached to a stair.

■ **Location**   This tab is displayed only when a freestanding railing object is modified, and is used to specify the location of the railing.

■ **Extensions**   This tab is used to specify how guardrails and handrails extend at the ends of a stair and landings. This tab is typically used when applying ADA (American Disabilities Act) rules to a stair railing. This tab is also unavailable if controlled by the railing style (see Figure 11.41).

■ **Anchor**   This tab is available only if the railing selected is attached to a stair. The anchor controls where the railing is located on the stair. Modify the values to adjust the anchor of the railing (see Figure 11.42).

Creating and Modifying Railing Styles

Styles play a big part in the design of an architectural project. Wall styles that represent brick, concrete, and wood have their own display properties. Door and window styles create specific types of openings in a wall. Stair styles are used to represent wood, steel, and concrete stairs. Railings, too, use styles. Many of the styles included in the templates represent commercial and industrial railing types. Custom railings can be created, existing railings can be modified, and additional railings can be imported through the Railing Styles command.

Use one of the following options to access this command in order to create, modify or import railing styles:

✓ Select the **Design** pull-down menu, pick the **Railings** cascade menu, and select the **Railing Styles…** command.

✓ Select the **Railing Styles** button in the Stairs - Railings toolbar.

✓ Right-click, and select **Design, Railings, Railing Styles…** from the cursor menu.

✓ Type **AECRAILINGSTYLE** at the Command: prompt.

The **Style Manager,** open to the **Railing Styles** section, as shown in Figure 11.43, is displayed.

The Style Manager is divided into two panes, the tree view on the left and the list or preview pane on the right. Select the Railing Styles section icon to display a list of styles in the right pane. Press the + symbol beside the Railing Styles section icon to display a list of railing styles in the drawing. If a style is selected in the tree view, the preview pane shows you what the railing style looks like.

As with stair styles, pick the Open Drawings button and browse to the \\Autodesk Architectural Desktop 3\Content\Imperial\Styles folder to import additional railings. Open the railing Styles (Imperial).dwg file to access additional styles. This file contains many commercial, residential, and industrial railings. Drag and drop the desired styles from this drawing into the drawing folder icons in the Style Manager. The additional styles are added to your drawings.

To create a new railing style, press the New style button, or right-click on top of the Railing Styles section icon, and select New from the cursor menu. Type a new name for the railing style, then press [Enter] to accept the new style name. Once the style has been created, highlight the new style in the tree view, and pick the Edit Style button, or right-click and select Edit… from the cursor menu. The Railing Styles dialog box, shown in Figure 11.44, is displayed.

The Railing Styles dialog box appears similar to the Railing Properties dialog box. The former dialog box, however, is used to create a style and configure all the properties for a railing. The following six tabs are available for configuring the railing style:

■ **General**   Like previous General tabs, this tab is used to describe railing styles or to rename them. Attach notes through the N_o_tes… button, and add property set definitions to the railing style through the _P_roperty Sets… button.

■ **Rail Locations**   This tab is used to configure the railing components used in the railing style (see Figure 11.45). At the top left side of the tab is the **Allow Each Railing to Vary** check box. Placing a check in this box makes this tab available in the Railing Properties dialog box.

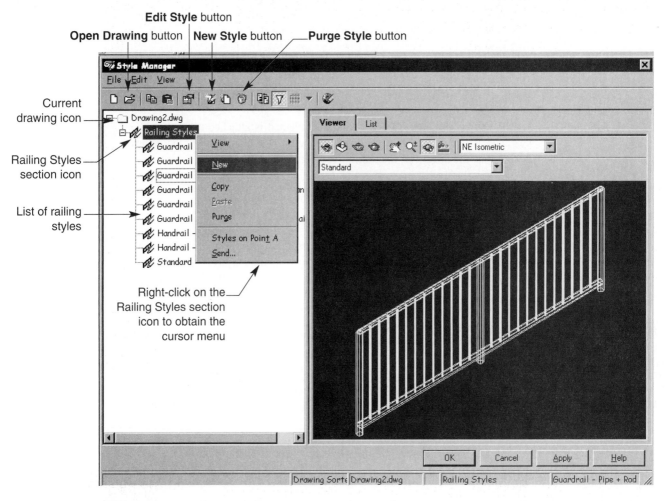

FIGURE 11.43   The Style Manager opened to the Railing Styles section.

The **Upper Rails** area is used to add guardrails, handrails or both to the railing style. Place a check in the appropriate check boxes to add any upper rails. The **Horizontal Height** and **Sloping Height** columns are used to specify the heights of the rails measured from the ground or the slope of a stair. The **Offset from Post** column is used to place the rails a distance from a post. A value of 0 places the rails centered on the posts. The **Side to Offset** column is used to specify the side of the posts where the upper railings are placed if a value greater than 0 is entered in the Offset from Post column. The **Auto** option automatically places the railing toward the center of a stair if the offset value is positive. If the railing is freestanding, the Auto option places the rails on the right side of the railing.

The bottom part of this tab controls the addition of a bottom rail(s). Check the **Bottomrail** check box to add a bottom rail to the style. Set the heights in the Horizontal and Sloping text boxes. If more than one bottom rail is to be added, enter the total number of bottom rails in the **Number of Rails** text box. When more than one bottom rail is created the **Spacing of Rails** text box is used to specify the spacing between all the bottom rails.

■ **Post Locations**   This tab is used to specify the addition and configuration of posts in the railing style, as shown in Figure 11.46. Place a check in the Allow Each Railing to Vary check box, to make this tab available in the Railing Properties dialog box. The tab has the following three areas: The **Fixed**

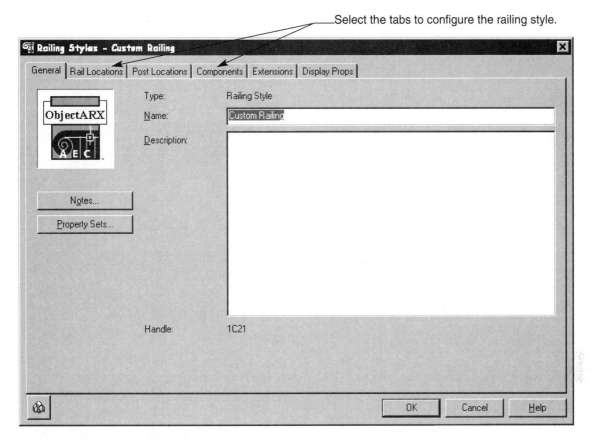

Select the tabs to configure the railing style.

FIGURE 11.44 The Railing Styles dialog box is used to specify the components and their properties for the railing style.

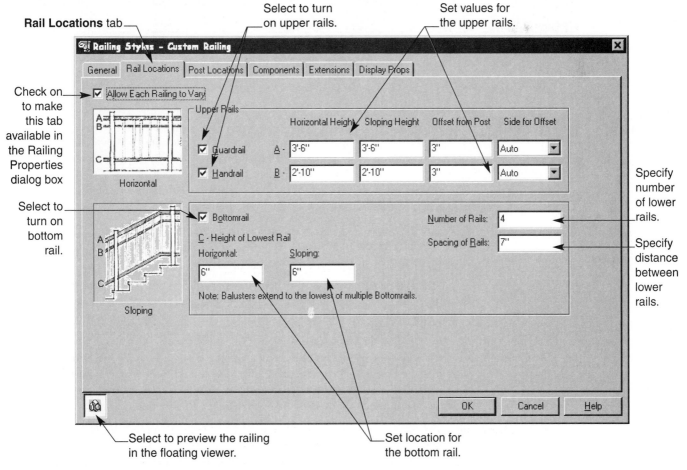

Rail Locations tab

Select to turn on upper rails.

Set values for the upper rails.

Check on to make this tab available in the Railing Properties dialog box

Select to turn on bottom rail.

Specify number of lower rails.

Specify distance between lower rails.

Select to preview the railing in the floating viewer.

Set location for the bottom rail.

FIGURE 11.45 The Rail Locations tab is used to configure railings in a railing style.

441

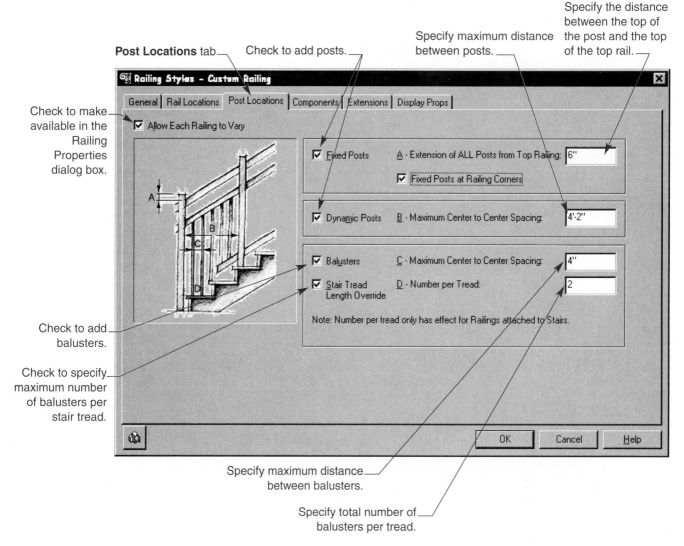

FIGURE 11.46 The Post Locations tab is used to configure posts and balusters in the railing style.

Posts area controls the addition of *fixed posts,* which are located at the starting and ending of a railing and at any corners of the railing.

The **Dynamic Posts** area is used to specify the addition of *dynamic posts,* which are added between fixed posts and spaced evenly along the railing. Specify the maximum distance between dynamic posts.

The **Balusters** area is used to configure *balusters* in the railing, which are the small posts that support the upper railing. Specify the maximum distance between balusters, and the maximum number of balusters to use on a stair tread. Figure 11.47 shows all the components of a railing.

- **Components**   This tab is used to control the shape and the size of the individual components of the railing (see Figure 11.48). Select the Profile Name column to assign an AEC profile to a component to give it a shape. When the profile text is selected, a drop-down list with a list of AEC profiles in the drawing appears. Select the values in the Width, Depth, and Rotation columns to size the components.

- **Extensions**   This tab is used to control guardrail and handrail extensions at the top and bottom of stairs or landings (see Figure 11.49). Enter the appropriate val-

FIGURE 11.47   The components of a railing.

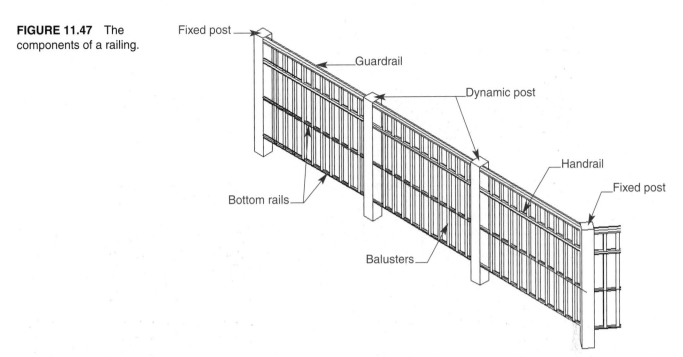

Fixed post

Guardrail

Dynamic post

Handrail

Fixed post

Bottom rails

Balusters

Select to assign a different AEC profile to use as a component shape.

Components tab

Select to specify sizes for the components.

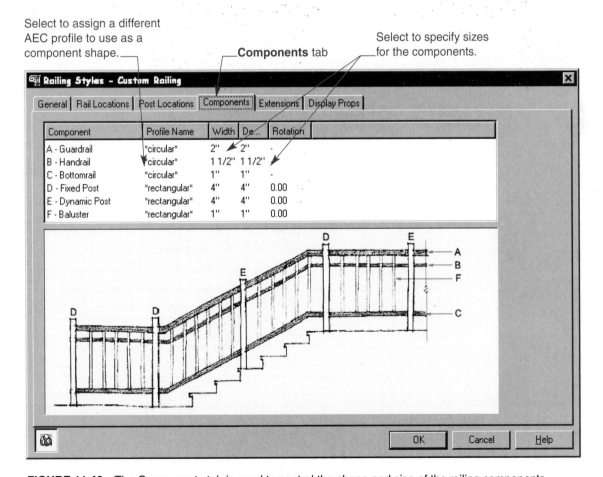

FIGURE 11.48   The Components tab is used to control the shape and size of the railing components.

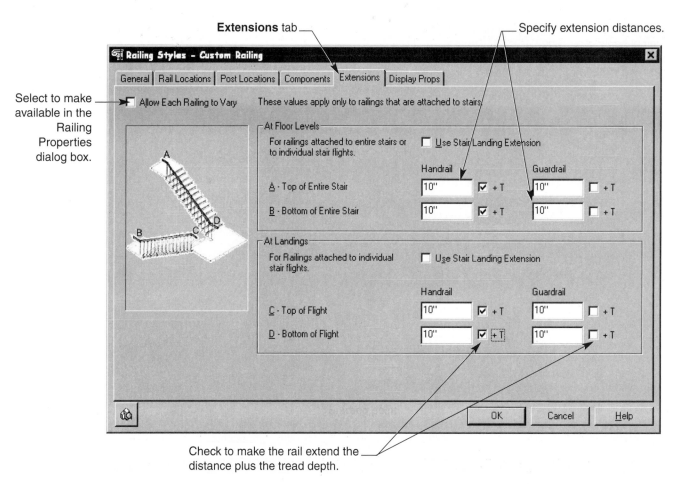

FIGURE 11.49   The Extensions tab is used to control the extension of upper rails past the ends of stairs and landings.

ues for the railing extensions in the text boxes. If building codes dictate that the railing needs to extend a distance plus the tread depth, then check on the appropriate **+T** check boxes. As with the Rail and Post Locations tabs, place a check in the Allow Each Railing to Vary check box to make this tab available in the Railing Properties dialog box.

- **Display Props**   This tab is used to control the display properties of the railing style. Only two property sources are available: System Default and Railing Style. Select the appropriate representation to modify, attach the override to the property source, and select the Edit Display Props… button to modify the display.

✎Field Notes

Most of the default railing styles do not have the Allow Each Railing to Vary check box selected. This makes it difficult to modify the railing style through the Railing Properties dialog box. Edit the railing styles so that each railing can be edited individually, for more control of the railings in the drawing.

Exercise 11.10

1. Start a new drawing using the AEC Arch (Imperial) or the AEC Arch (Imperial-Int).dwt.
2. Access the Railing Styles command. Use the Open Drawing button to browse to the \\Autodesk Architectural Desktop3\Content\Imperial\Styles folder. Select and open Railing Styles (Imperial).dwg.
3. Drag and drop the following styles into your current drawing:
   - Guardrail - Cable
   - Guardrail - Pipe + Handrail + Returns
   - Guardrail - Wood Balusters 02
4. Close Railing Styles (Imperial).dwg.
5. Edit the Guardrail - Cable style.
6. Use the Display Props tab to adjust the model colors of the railing components to color 250. Change the Post component colors to color 252.
7. Add railings to the drawing using each of the styles.
8. Save the drawing as EX11-10.dwg.

CHAPTER REVIEW

Use the CD-ROM to test your knowledge and skills.

Chapter Test

To check your understanding of the content provided in this chapter, access the Test file in the CH11 folder of the CD-ROM that accompanies this text.

Chapter Project

To practice the Architectural Desktop skills presented in this chapter, access the Project files in the CH11 folder of the CD-ROM that accompanies this text. The project files are in pdf format and include sample drawings and instructions for completing each project.

CHAPTER 12

Creating Curtain Walls

LEARNING GOALS

After completing this chapter, you will be able to:

◎ Create and modify curtain walls and curtain wall units.

◎ Override curtain wall features.

◎ Use curtain wall tools and properties.

◎ Control curtain wall display.

◎ Create curtain wall unit styles.

◎ Create window assembly styles.

Curtain walls are an architectural feature typically used on commercial buildings. Similar to the window assemblies in Architectural Desktop, curtain walls work on a grander scale. Window assemblies are typically groups of windows used on a smaller commercial building and are often inserted into a wall object. The ***curtain wall object*** is usually the entire wall with glazing panels built in.

Curtain wall objects provide a framework in which windows, doors, and window assemblies can be inserted. The framework or grid controls the spacing between infill panels (windows, doors, and window assemblies) (see Figure 12.1). Curtain walls are similar to standard AEC walls in that they include a baseline, a roofline, and a floor line, and are drawn by specifying a starting and ending point.

Curtain walls are different from standard AEC walls in that they are made up of a framework or ***grid***. Between the grid is an area called an ***infill panel***. The infill panel can be "filled in" with other AEC objects such as windows and doors. Grids contain rules that specify how infill panels are multiplied or divided as the curtain wall is made larger or smaller. A curtain wall grid is divided into a horizontal or vertical division. Divisions (grids) can be nested inside each other to create a variety of patterns for the curtain wall.

Creating Curtain Walls

Curtain walls are drawn using the **Add Curtain Wall** command, which is accessed with one of the following options:

✓ Select the **Design** pull-down menu, pick the **Curtain Walls** cascade menu, and select **Add Curtain Wall....**

✓ Pick the **Add Curtain Wall** button in the **Curtain Walls** toolbar.

✓ Right-click, and pick **Design, Curtain Walls, Add Curtain Wall...** from the cursor menu.

✓ Type **AECCURTAINWALLADD** at the Command: prompt.

The Add Curtain Walls dialog box, shown in Figure 12.2, is displayed.

The Add Curtain Walls dialog box is similar to the Add Walls dialog box. Select a curtain wall style to add into the drawing, specify the height, and determine if a straight or curved curtain wall will be creat-

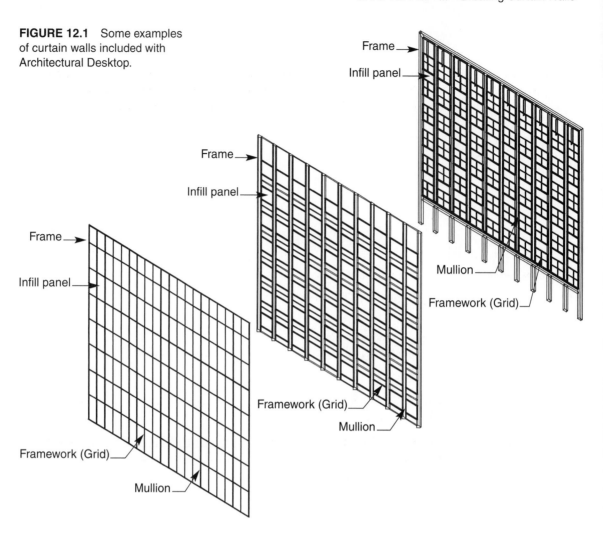

FIGURE 12.1   Some examples of curtain walls included with Architectural Desktop.

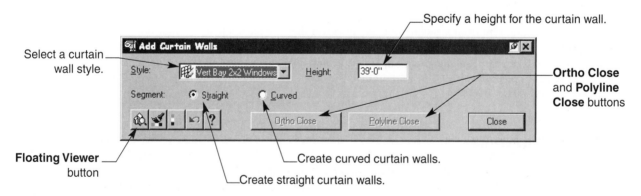

FIGURE 12.2   The Add Curtain Walls dialog box is used to add curtain walls to the drawing.

ed. Notice that there is not a width or justify option. The curtain wall width is determined within the curtain wall style and is always drawn using a baseline justification.

To draw a curtain wall, pick a point on the drawing screen for the staring point, than pick an ending point. As curtain walls are drawn they do not cleanup with themselves, as do AEC walls. You can manually cleanup the intersections of curtain walls with grip editing or the miter angles tool. When two or more curtain wall segments have been added, use the Ortho Close or Polyline Close buttons to close the building. When you have finished adding curtain walls to the drawing, press the Close button to exit the command and return to the drawing screen.

Field Notes

> If the floating viewer is used to preview the curtain wall style, it displays a preview of how the curtain wall is divided into grids without infill patterns. If you are unsure how a curtain wall will appear after using the floating viewer, draw a sample segment in the drawing to see what the style looks like.

Exercise 12.1

1. Begin a new drawing using the AEC Arch (Imperial) or AEC Arch (Imperial-Int).dwt.
2. Spilt the drawing screen into two vertical viewports with a 3D setup.
3. Draw one 50'-0"-long curtain wall for each style available. Set the height to 39'-0".
4. Use the Spandrel 13' curtain wall style to create a building 120'-0" long and 60'-0" wide. Set the height to 78'-0".
5. Save the drawing as EX11-1.dwg.

In addition to drawing curtain walls from scratch, you can convert standard wall objects into curtain walls. Use one of the following options to convert walls into curtain walls:

✓ Access the **Design** pull-down menu, select the **Curtain Walls** cascade menu, and pick the **Convert Wall to Curtain Wall** command.

✓ Pick the **Convert Wall to Curtain Wall** button in the **Curtain Walls** toolbar, right-click, and select **Design, Curtain Walls, Convert Wall to Curtain Wall** from the cursor menu.

✓ Type **AECCURTAINWALLCONVERTWALL** at the Command: prompt.

The following prompt sequence converts a wall into a curtain wall:

```
Command: AECCURTAINWALLCONVERTWALL ↵
Select walls: (select wall(s) to be converted to curtain walls)
1 found
Select walls: ↵
Curtain wall baseline alignment [Left/Right/Center] <Baseline>: ↵
Erase layout geometry? [Yes/No] <N>:Y ↵
1 new Curtain Wall(s) created.
Command:
```

When walls are converted into curtain walls, the baseline (center) of the curtain wall is added to the left, right, center, or baseline of the wall being converted, as shown in Figure 12.3. Enter the appropriate location for the curtain wall at the *Curtain wall baseline alignment [Left/Right/Center] <Baseline>:* prompt.

Another tool that can be used to add curtain walls to a drawing is the **Reference Curtain Wall** command. This command is similar to the Convert to Wall command, in that layout lines, arcs, polylines, circles, and splines are "referenced," and a curtain wall is added in their place. The layout geometry is also known as a *curve* whether the geometry is a straight or arced segment. The curtain wall references the curve, so any changes made to the shape of the curve update the curtain wall to reflect the new shape.

Use one of the following options to reference a curve:

✓ Select the **Design** pull-down menu, move to the **Curtain Wall** cascade menu, and pick the **Reference Curtain Wall** command.

✓ Pick the **Reference Curtain Wall** button on the **Curtain Walls** toolbar.

✓ Right-click, and select **Design, Curtain Wall, Reference Curtain Wall** from the cursor menu.

✓ Type **AECCURTAINWALLREFERENCE** at the command: prompt.

FIGURE 12.3 When walls are converted into curtain walls, the baseline alignment (center of curtain wall) can be placed along the converted wall's left, right, center, or baseline justification points.

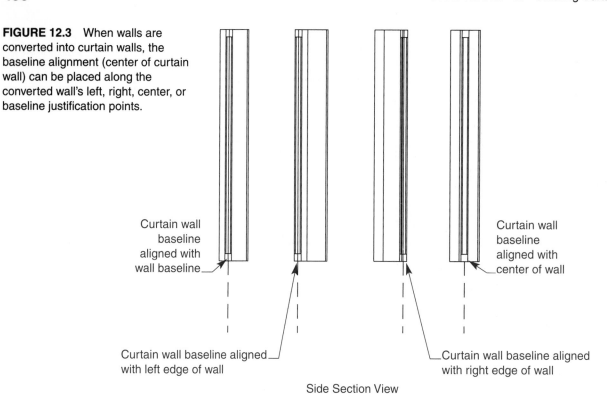

Curtain wall baseline aligned with wall baseline

Curtain wall baseline aligned with center of wall

Curtain wall baseline aligned with left edge of wall

Curtain wall baseline aligned with right edge of wall

Side Section View

The following prompt sequence is used to reference an arced polyline:

Command: **AECCURTAINWALLREFERENCE** ⏎
Select a curve: *(select a layout line, arc, circle, polyline, or spline)*
(Select the curtain wall style to use from the Curtain Wall Styles dialog box)
Command:

When using the Reference Curtain Wall command, you are prompted to select a curve. Pick one of the layout objects to reference. Once the object has been selected, the Curtain Wall Style dialog box is displayed so a curtain wall style can be selected. If the layout geometry is modified in any way, such as by being stretched with grips, the curtain wall adjusts with the geometry.

Exercise 12.2

1. Begin a new drawing using the AEC Arch (Imperial) or AEC Arch (Imperial-Int).dwt.
2. Split the drawing screen into two vertical viewports with a 3D setup.
3. Draw a wall using the Concrete-8 Furring wall style. Make the wall 32'-0″ high and 40'-0″ long.
4. Convert the wall into a curtain wall using the Ribbon curtain wall style.
5. Draw a polyline with a mixture of straight and arced segments.
6. Use the Reference Curtain Wall command to reference the polyline. Use the Square Grid 5′ × 5′ curtain wall style.
7. Select the polyline and stretch the grip points.
8. Save the drawing as EX12-2.dwg.

Modifying Curtain Walls

Most of the curtain wall styles included in the AEC templates are built with custom infill panels known as *curtain wall units*. When selecting a curtain wall for modification, you may find that the curtain wall unit is selected instead of the curtain wall framework. Selecting the actual frame edge of the curtain wall or a mullion grid selects the entire curtain wall, so it can be modified.

The **Modify Curtain Wall** command can be used to adjust curtain wall style and height. Additional properties of a curtain wall can be modified through the **Curtain Wall Properties** dialog box. In addition to these two options, Architectural Desktop includes other tools for refining the appearance of the curtain walls.

Using the Modify Curtain Walls Command

If you wish to change only the style or height of a curtain wall, then use the **Modify Current Wall.** Use one of the following options to access this command:

✓ Select the **Design** pull-down menu, pick the **Curtain Walls** cascade menu, and select **Modify Curtain Wall....**

✓ Pick the **Modify Curtain Wall** button on the **Curtain Walls** toolbar.

✓ Right-click, and select **Design, Curtain Walls, Modify Curtain Wall...** from the cursor menu.

✓ Type **AECCURTAINWALLMODIFY** at the Command: prompt.

✓ Alternatively, select the wall, right-click, and select **Curtain Wall Modify...** from the cursor menu.

The **Modify Curtain Walls** dialog box, shown in Figure 12.4, is displayed. Notice that it is similar to the Add Curtain Walls dialog box. Change the selected curtain wall to a new style, or adjust the height. When you have finished making adjustments to the curtain wall, press the Apply button to preview the changes before accepting the changes by pressing the OK button.

Field Notes

Some of the curtain walls include separate infill panels that can cause the infill panel to get selected, which may be undesirable. If you have difficulty selecting the curtain wall grid, use the following procedure to pick the desired object:

Before selecting the object, press and hold the [Ctrl] key. Move the mouse cursor over the area that includes two or more objects, and pick with the left mouse button to highlight one of the objects. Let go of the [Ctrl] key and pick with the left mouse until the desired object is highlighted. When the desired object is highlighted press the [Enter] key to select the highlighted object. This procedure is known as *object cycling*. You can use this method to cycle through the objects until the desired object is highlighted.

Modifying Elements Within a Curtain Wall

The individual elements within a curtain wall (frame, mullions, and infills) can be overridden for the individual wall selected. By using override options, infill cells can be merged or assigned a different infill definition. Frame and mullion edges can be assigned to a custom profile or definition. Overrides do not affect the curtain wall style but affect only the curtain wall being modified.

Before a cell (infill) can be merged or the cell (infill) assignment changed, the cell markers need to be turned on. To turn on the cell markers, select the curtain wall to be modified, right-click, and pick the **Cell**

FIGURE 12.4   The Modify Curtain Walls dialog box is used to adjust the style and height of a curtain wall.

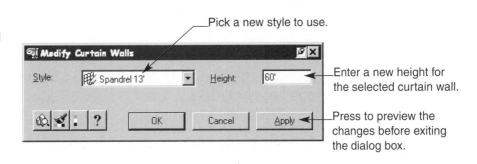

Cell Marker Option	Grid Level Displayed In	Turns cell markers...
Off	Markers turned off in all cells	Off
All Visible	Markers displayed in all cells at the lowest level	On
1st Grid	Markers displayed in the primary grid level (first set of grids)	On
2nd Grid	Markers displayed in all the cells of the secondary grid level (first nested set of grids within the primary grid level)	On
3rd Grid	Markers displayed in all the cells of the third grid level (second nested set of grids within the second grid level)	On
Other	Markers displayed in all the cells of the level that you specify on the command line: 4, 5, etc.	On

FIGURE 12.5   This table indicates the cell markers that are displayed depending on the cell marker option selected.

Markers cascade menu, then select one of the cell marker options. Curtain walls can be composed of several "nested" grids within a larger grid pattern, so you have the option of choosing which cell markers are displayed. Figure 12.5 indicates the cell markers that are displayed based on the option selected.

Once a cell marker option has been selected, a cell marker is displayed in the center of each cell (infill) and is a selection point for the cell. The cell marker changes depending on the direction of the grid and the cell assignment. For horizontal grids, the marker points up, indicating the cells are numbered from bottom to top beginning with cell number 1 (see Figure 12.6A). Markers displayed in vertical grids point to the right, indicating the cells are numbered from left to right beginning with cell number 1 (see Figure 12.6B).

After the cell markers have been turned on, the individual cells can be overridden. The curtain wall overrides are available only from the right-click shortcut menu. First, select the curtain wall that is to be edited, right-click, then select an override command from the **Overrides** cascade menu, shown in Figure 12.7.

Merging Two Cells

The first override option is the **Merge Cells** command. This override merges two cells into one. When this command is used, you are prompted to select Cell A then Cell B, where Cell A is the first cell that will be merged, and Cell B is the second cell that will be merged (see Figure 12.8). Only cells that are next to one another (horizontally or vertically) can be merged.

Depending on the cell marker level displayed, two cells (or sets of cells if at the primary grid level), can be merged.

Overriding Cell Assignments

Each cell in a curtain wall is assigned an infill. The infill assigned to a cell can be overridden with a different infill element. Some of the curtain walls include multiple infill elements, whereas others include only one infill element. Additional infills can be created through the **Curtain Wall Styles** dialog box. The creation of additional infills will be discussed later in this chapter.

When the **Override Cell Assignment** command is entered, you are prompted to select a cell. Select the cell that will have a different cell assignment. A list of available infill elements is displayed in the Command: prompt area. If you cannot see the entire list of infill elements, press the [F2] key to change to the text screen window, which provides you with the list of infill panels. Enter the number associated with the desired infill element to override the current infill cell. Once a new infill element has been selected, the curtain wall updates with the new cell (see Figure 12.9).

✎ Field Notes

In most of the curtain wall styles, only one or two infill elements are available. Additional infills can be created in a style for use as overrides. If a curtain wall has a list of infill elements use the [F2] key to switch to the text screen window to list the available infill elements. Press [F2] again to return to the drawing screen.

FIGURE 12.6 (A) Cell markers for a horizontal grid point up. (B) Cell Markers for a vertical grid point to the right.

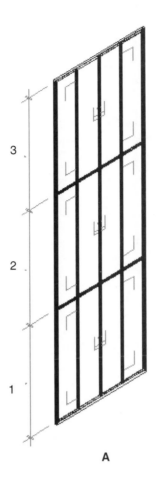

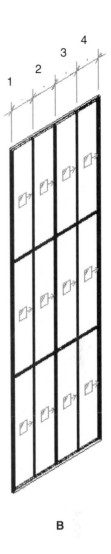

A

B

FIGURE 12.7 The Overrides cascade menu includes four overriding commands.

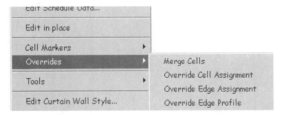

Overriding Edge Assignments

An *edge* in a curtain wall is the outer frame or the inner mullion divisions. The elements of these edges also can be overridden. The process is the same as previously discussed. Enter the **Override Edge Assignments** command, then select a frame edge or a mullion edge. After selecting the edge to be overridden, a list of frame or mullion elements is displayed in the Command: prompt area. Enter the number associated with the desired edge element (see Figure 12.10). Once the desired edge has been selected, the curtain wall updates the edge in the drawing.

Field Notes

By default there are no additional edge element definitions for mullions or frames in the curtain wall styles in the AEC templates. You can custom create your own edge definitions for frames and mullions. This process is discussed later in this chapter.

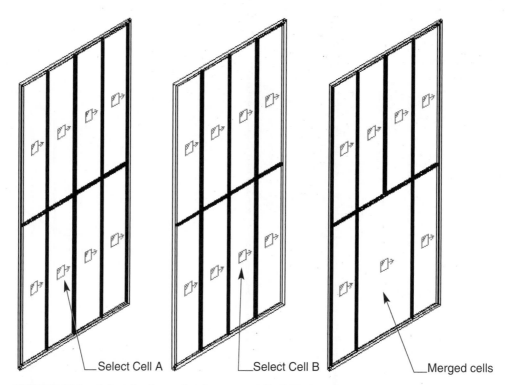

Select Cell A   Select Cell B   Merged cells

FIGURE 12.8   Select the first cell to be merged (Cell A), then the second cell (Cell B).

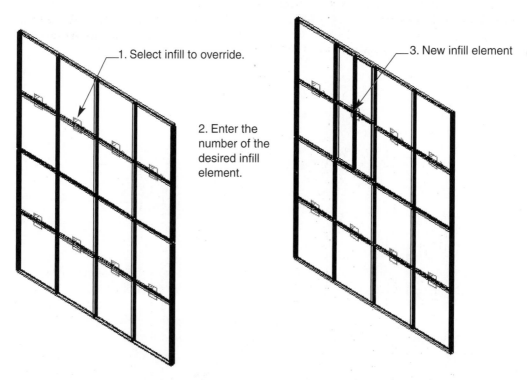

1. Select infill to override.

2. Enter the number of the desired infill element.

3. New infill element

FIGURE 12.9   Overriding a cell infill assignment with another infill element.

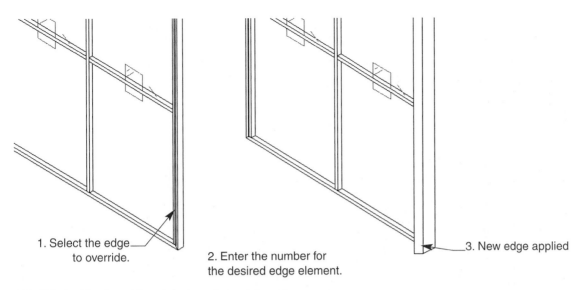

1. Select the edge
   to override.

2. Enter the number for
   the desired edge element.

3. New edge applied

FIGURE 12.10   Overriding the edge of a curtain wall.

Overriding the Edge Profile

Although the default curtain wall styles do not contain additional edge definitions that can be used as overrides, a polyline profile can be used as an override to an edge shape. First, draw a closed polyline in the shape desired for the frame. Be sure that the polyline is drawn at the bottom of the frame component, because the polyline will be converted into a frame edge at its drawn location.

Enter the **Override Edge Profile** command. You are prompted to select a vertical edge. Select a vertical frame or mullion. The next prompt asks you to select your polyline profile. Select the polyline and give it a name. The polyline is converted into the edge shape in the curtain wall (see Figure 12.11).

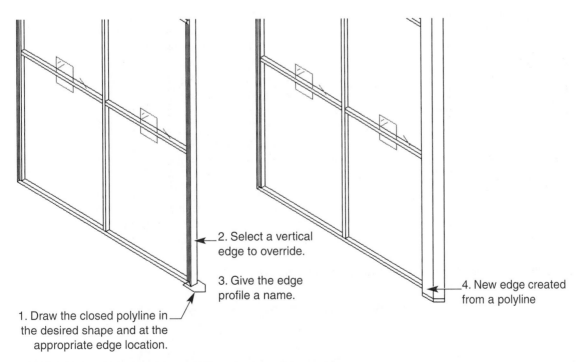

2. Select a vertical
   edge to override.

3. Give the edge
   profile a name.

4. New edge created
   from a polyline

1. Draw the closed polyline in
   the desired shape and at the
   appropriate edge location.

FIGURE 12.11   Using a custom profile as a shape for an edge.

Field Notes

When overriding an edge profile you have the option of selecting a polyline or pressing Return to choose from a list of existing profiles in the drawing to use as an edge.

Exercise 12.3

1. Begin a new drawing using the AEC Arch (Imperial) or AEC Arch (Imperial-Int).dwt.
2. Spilt the drawing screen into two vertical viewports with a 3D setup.
3. Draw a curtain wall using the Spandrel 13′ curtain wall style. Make the curtain wall 39′-0″ high and 50′-0″ long.
4. Turn all the cell markers on.
5. Use the Merge Cells command to merge a few cells. Merge some cells horizontally and some vertically.
6. Use the Override Cell Assignment command to override a few infills with the (2×2) Glazing infill element.
7. Draw a closed polyline at the ends of the curtain wall. Use the Override Edge Profile command to override the vertical edges of the curtain wall.
8. Save the drawing as EX 12-3.dwg.

Using Curtain Wall Tools

In addition to being able to modify the individual elements of a curtain wall, you can modify the appearance of curtain walls with some additional tools. To access these commands first select a curtain wall, right-click, then select the Tools cascade menu, shown in Figure 12.12. This menu contains the following five commands:

- **Reverse**   This command reverses the starting and ending points of the curtain wall. Use this command when the curtain wall appears on the wrong side.

- **Roof Line**   This command is similar to the Roof Line command used on AEC walls. Use the **Project** option to project the top of the wall (roofline) to a polyline. The polyline that is being projected to must be parallel to the curtain wall being projected. Use the **Auto project** option to project the top of the curtain wall to an AEC object such as a roof, slab, wall, or curtain wall.

- **Floor Line**   This command is similar to the Floor Line command used on AEC walls. Use the Project option to project the bottom of the curtain wall to a polyline that is parallel to the wall. Use the Auto project option to project the bottom of the wall to an AEC object, such as a slab, wall, or curtain wall.

- **Interference**   This command is used to apply an interference to a curtain wall. An interference can be a structural member, mass element, wall, or other AEC object that runs into the curtain wall. Adding an interference stops the curtain wall to either side of the interfering object.

FIGURE 12.12 The Tools cascade menu, found on the right-click shortcut menu, includes tools for modifying the curtain wall.

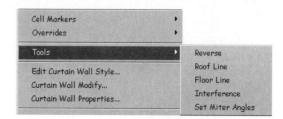

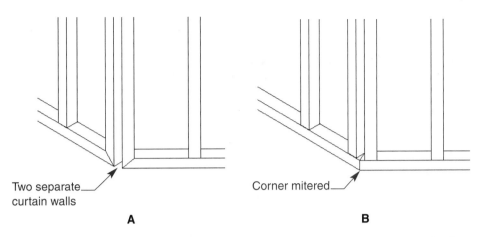

Two separate
curtain walls

Corner mitered

A   **B**

FIGURE 12.13   (A) Two curtain walls meeting at a corner. (B) The two curtain walls after mitering.

■ **Set Miter Angles**   This command is used to miter the ends of two curtain walls meeting at a corner. After entering the Set Miter Angles command, you are prompted to *Select the Second grid Assembly*. Select the second curtain wall to miter the two edges together, as shown in Figure 12.13.

Field Notes

The mitering of two curtain walls applies to the walls' horizontal components only. If the corner of the wall is to have a special type of corner, or in a plan view if the walls need to appear as if they cleanup, apply an interference condition to the walls at the corner. Another alternative is to use the Override Edge Profile as the edge profile applies to vertical edges.

Exercise 12.4

1. Open EX12-1.dwg.
2. Miter the corners for the four walls of the building.
3. Change to a front view and draw a polyline at an angle above the top of the curtain wall building.
4. Use the Roof Line command to project the front and back walls to the polyline.
5. Save the drawing as EX12-4.dwg.

Using Curtain Wall Properties

The **Curtain Wall Properties** dialog box includes additional options for modifying curtain walls. To access this dialog box, first select the curtain wall to be modified, right-click, and select Curtain Wall Properties… from the cursor menu, or type **CURTAINWALLPROPS** at the Command: prompt. The Curtain Wall Properties dialog box is displayed. The dialog box includes the following six tabs for managing and modifying the selected curtain wall:

■ **General**   This tab is used to assign a description to the selected curtain wall, to attach notes, and to add property set definitions.

■ **Style**   This tab is used to change the selected curtain wall into a different style. The floating viewer can be used to preview the appearance of the grid within the curtain wall.

Dimensions tab

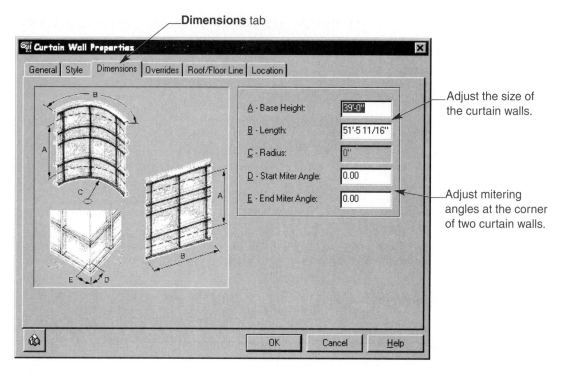

FIGURE 12.14   The Dimensions tab is used to manage the size of the curtain wall.

■ **Dimensions**   This tab controls the length and height of the curtain wall. If the curtain wall is curved, the Radius option is available. Mitering angles at the corner of two curtain walls can also be controlled from this tab (see Figure 12.14).

■ **Overrides**   This tab is used to manage any overrides applied to the curtain wall (see Figure 12.15). The overrides are listed in a table under the appropriate override heading. The overrides can be removed by highlighting an override, right-clicking, and selecting Remove from the shortcut menu or by selecting the Remove button at the bottom of the tab. The table lists the cells and edges in relation to the grid divisions in the curtain wall, which aids in locating the appropriate cell or edge for modification.

■ **Roof/Floor Line**   This tab is used to manage the roofline and floor lines of the curtain walls. The table at the top of the tab lists the vertices for the roof and the floor. Use the buttons to the lower right of the tab to add, edit, and delete vertices in the curtain wall (see Figure 12.16).

■ **Location**   Use this tab to adjust the location of the curtain wall. The most common modification made in this tab is the adjustment of the Z-axis insertion point in the World Coordinate System. Adjustments made in this text box control where the bottom of the curtain wall is placed in the XY plane (Z axis).

FIGURE 12.15 The Overrides tab is used to list and/or remove overrides applied to a curtain wall.

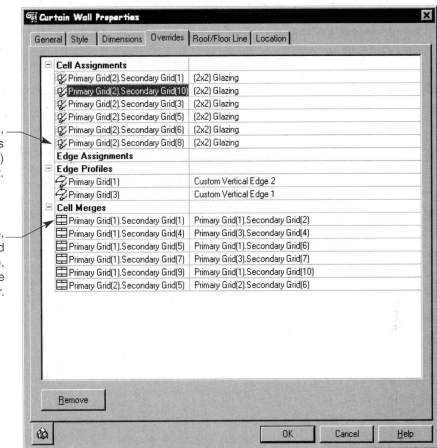

Primary Grid(2), Secondary Grid(8) is assigned the (2x2) Glazing infill element.

Primary Grid(1), Secondary Grid(1) and Primary Grid(1), Secondary Grid(2) are merged together.

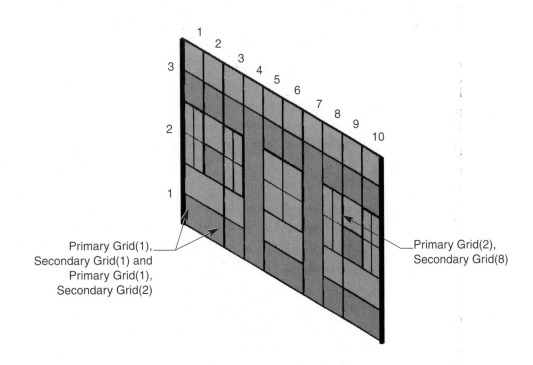

Primary Grid(1), Secondary Grid(1) and Primary Grid(1), Secondary Grid(2)

Primary Grid(2), Secondary Grid(8)

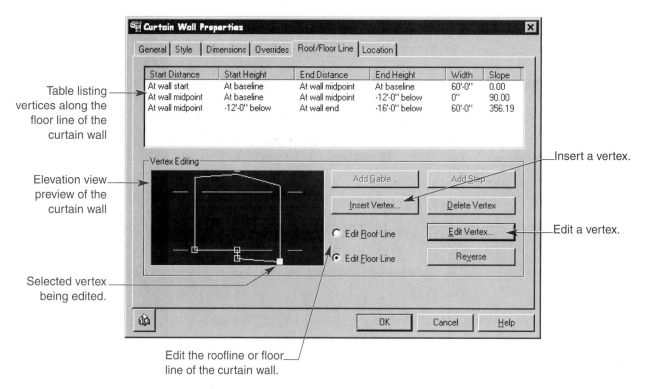

Table listing vertices along the floor line of the curtain wall

Elevation view preview of the curtain wall

Selected vertex being edited.

Insert a vertex.

Edit a vertex.

Edit the roofline or floor line of the curtain wall.

FIGURE 12.16   The Roof/Floor Line tab is used to manage locations of vertices along the roofline or floor line of a curtain wall.

Controlling Curtain Wall Display

The display properties for curtain walls, much like for other AEC objects, are controlled through **Entity Display**. To control the display properties of a curtain wall, select the curtain wall, right-click, and pick Entity Display… from the shortcut menu. As with previously discussed objects, pick the **Display Props** tab to modify the display.

In the Display Props tab, select the display representation to be modified from the drop-down list at the upper left corner of the tab (see Figure 12.17). In the Property Source column, select the level of display

Display Props tab

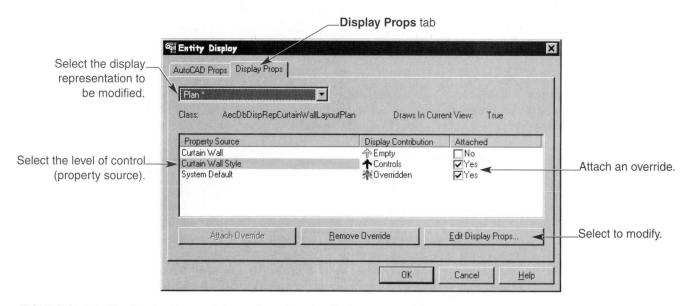

Select the display representation to be modified.

Select the level of control (property source).

Attach an override.

Select to modify.

FIGURE 12.17   The Display Props tab is used to select the display representation and the level of display control that will be edited.

control, then attach an override. After the override has been attached, press the Edit Display Props button to modify the display settings. Curtain walls contain plan and model representations that can be edited.

Modifying Plan Representation Curtain Walls

The **Entity Properties** dialog box for plan representation curtain walls includes the following four tabs. Each of these tabs controls a different aspect for the display of plan-viewed curtain walls.

- **Layer/Color/Linetype**   This tab is used to control the color, layer, and linetype assigned to plan view curtain walls. The curtain wall includes three main components: Default Infill, Default Mullion, and Default Frame. Each of these components also includes a hatching component. In addition to these components, an Above Wall and Below Wall component is included to display curtain walls above or below the curtain wall cut plane. Select any of the symbols or text to modify the values assigned to the components.

- **Hatching**   This tab is used to assign the style of hatch pattern used for a component hatch. The three hatch components are listed here.

- **Custom Plan Components**   This tab controls new components that can replace one of the three default components: Default Infill, Default Mullion, and Default Frame, similar to adding blocks to a door or window. When the Add... or Edit... button is selected, the **Custom Display Component** dialog box appears, as shown in Figure 12.18. To replace a default component, select the component that will be replaced in the Component Type drop-down list. Press the Select Element... button to assign a new component to the default component. The Draw Custom Graphic check box can be used to attach a custom block to the curtain wall, similar to the way you would attach a block to a door. When a

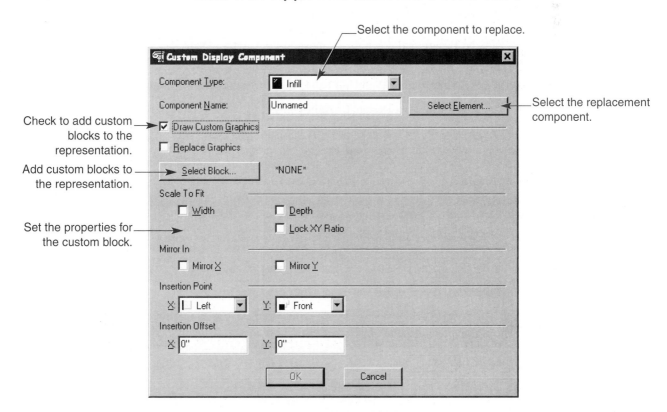

FIGURE 12.18   The Custom Display Component dialog box is used to replace a plan view component with another component or to add a custom block to the curtain wall representation.

new component type or a block is added to the plan representation curtain wall, the component is added to the Layer/Color/
Linetype tab.

■ **Cut Plane**   This tab controls the cut plane for the selected curtain wall.

Modifying Model Representation Curtain Walls

The model representation curtain wall includes two tabs: **Layer/Color/Linetype** and **Custom Model
Components**. The Layer/Color/Linetype tab lists the three components of a model view curtain wall plus
the cell marker component. Adjust any of the properties for the model-viewed curtain walls.

The Custom Model Components tab is used to assign a replacement component or block to the curtain
wall, as with the plan representation. This tab also includes a text box for controlling the size of the cell markers.

Creating Curtain Wall Units

Curtain walls are a series of elements that work together to form a wall that is divided into a number of grid
patterns. Each curtain wall includes an infill (cell), mullion, and frame. The frames and mullions are used to
divide the wall into a series of divisions. Between each mullion and frame is an infill panel. Some of the
curtain wall styles use an infill that is defined within the curtain wall style. Other curtain walls use curtain
wall units as the infill between the frames and mullions.

The *curtain wall unit* is a separate object type whose only purpose is to be used as the infill within a
curtain wall. Curtain wall units can be used to create a complex series of divisions and grids without
having to nest the complex divisions directly in the curtain wall. This allows more flexibility in overriding
the infill panel of the curtain wall.

Adding Curtain Wall Units

Curtain wall units can be added to the drawing for review and design purposes, but they cannot be added to
a wall, as can window assemblies. Use one of thefollowing options to add a curtain wall unit:

✓ Pick the **Design** pull-down menu, move to the **Curtain Walls** cascade menu, and select the
   Add Curtain Wall Unit command.

✓ Pick the **Add Curtain Wall Unit** button in the **Curtain Walls** toolbar.

✓ Right-click, and select **Design, Curtain Walls, Add Curtain Wall Unit** from the cursor menu.

✓ Type **AECCWUNITADD** at the Command: prompt.

Once the command has been entered, pick a starting location, ending location, and a height for the
curtain wall unit. After selecting the height, choose the desired curtain wall unit from the **Curtain Wall
Unit Styles** dialog box. A curtain wall unit is added to the drawing. The benefit of adding the curtain wall
unit to the drawing in this manner allows you to preview what the unit looks like within a curtain wall and
allows you to edit the unit in place, in turn updating any other units in the drawing with the changes made.

Exercise 12.5

1. Begin a new drawing using the AEC Arch (Imperial) or AEC Arch (Imperial-Int).dwt.
2. Add a curtain wall unit to the drawing.
3. Select a starting point for the curtain wall unit, pull the cursor to the right, and enter 16′-0″
   at the End Point prompt. Make the unit 12′-0″ high.
4. Select the Precast Panel + Windows curtain wall unit style.
5. Divide the drawing screen into two vertical viewports with a 3D setup.
6. Save the drawing as EX12-5.dwg.

Modifying Curtain Wall Units

The elements within a curtain wall unit can be modified with the same commands used to modify curtain walls. Before the cells of a curtain wall unit can be modified, the cell markers must be turned on. First, select the curtain wall unit, right-click, and select the Cell Markers cascade menu. Pick the level at which the cells will be modified.

Field Notes

Usually, when a curtain wall unit is modified the All Visible cell marker option is selected so all the lowest level cells can be modified.

After the cell markers are turned on, the curtain wall unit is ready to be modified with the Override commands. At this point applying overrides to the individual curtain wall unit is not very valuable, because changes made affect only the individual unit and not other units of the same style. If you wish to modify the curtain wall unit so the changes made affect the curtain wall style globally in the drawing, then you need to edit the curtain wall unit in place.

Editing a Curtain Wall Unit in Place

To edit the curtain wall unit in place, select the unit, right-click, and pick the **Edit in place** command to open the curtain wall unit so changes made to the individual curtain wall unit can be saved back to the unit style. Select the curtain wall unit, right-click, and select any of the overriding commands in the Overrides cascade menu to modify the unit. Set any overrides as required for the curtain wall unit style.

When you have finished making adjustments to the curtain wall unit, you can save the changes back to the unit style. Select the curtain wall unit, right-click, and select the Edit in place cascade menu. Pick the **Save changes** command. The **Save Changes** dialog box, shown in Figure 12.19, is displayed.

The changes made while editing the unit style in place can be saved to a unit style found in the drop-down list. New curtain wall unit styles can also be created from the changes by selecting the <u>N</u>ew... button. If the changes are to be saved in an existing style or a new style, then select the appropriate check boxes at the bottom of the dialog box. As changes are made to the curtain wall unit, the different check boxes in the **Save Changes** dialog box become available.

Field Notes

The Edit in place command is a simple way of taking an existing curtain wall unit style, applying any overrides to the unit, and saving the changes as a new style. This process creates additional styles that can be overridden in a curtain wall style.

FIGURE 12.19 The Save Changes dialog box is used to save any changes made to a curtain wall unit back to the unit style or into a new unit style.

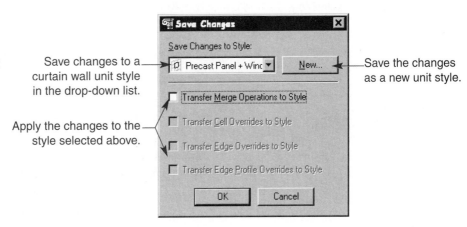

Exercise 12.6

1. Open EX12-5.dwg.
2. Turn on all the cell markers.
3. Edit the curtain wall unit in place.
4. Merge the windows indicated in the figure.
5. Save the changes as a new curtain wall unit style named Precast Panel + 2 Windows.
6. Turn off the cell markers.
7. Add the new curtain wall unit into the drawing.
8. Save the drawing as EX12-6.dwg.

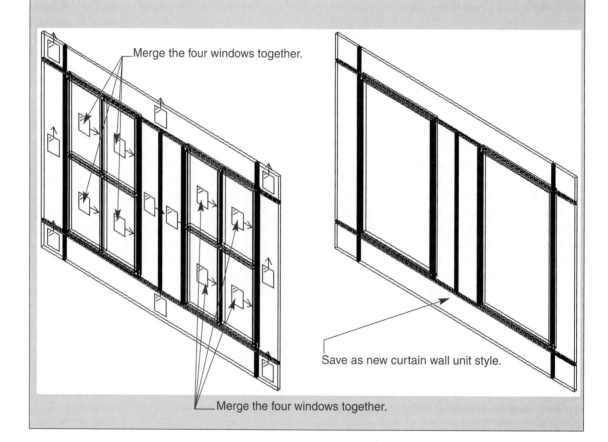

Merge the four windows together.

Save as new curtain wall unit style.

Merge the four windows together.

Curtain Wall Unit Properties

Curtain wall unit properties can be adjusted and managed through the **Curtain Wall Unit Properties** dialog box. This dialog box can be accessed by selecting a curtain wall unit, right-clicking, and selecting Curtain Wall Unit Properties… from the shortcut menu. The dialog box contains the following five tabs for unit control:

- **General** This tab is used to enter a description for the curtain wall unit, to attach notes, and to add property set definitions.

- **Style** This tab is used to change the selected curtain wall unit to a different unit style.

- **Dimensions** This tab is used to control the width and height of the unit and to set mitering angles for the unit.

- **Overrides** This tab is used to remove curtain wall unit overrides, which are listed in this tab. Select an override, then pick the Remove button to remove the override, if desired.

- **Location** This tab is used to adjust the location of the curtain wall unit in the drawing.

Curtain Wall Unit Styles

Curtain wall unit styles, curtain wall styles, and window assembly styles are all based on the same creation methods. These objects are based on four parts that when assembled together form the style. *Divisions* are rules that determine how a curtain wall, curtain wall unit, or window assembly is divided into cells. *Infills* are components that are placed in cells. The *frame* establishes the outer boundaries of the object, and the *mullions* are the boundaries within the frame that divide the infills.

As mentioned earlier, the curtain wall unit is an AEC object that is used as an infill within a curtain wall or window assembly. This allows you to create a curtain wall or window assembly with divisions separate from the details of the infill. Three methods are available for custom creating a curtain wall unit style: create a unit from scratch, convert linework into a unit, or convert a layout grid into a unit. These three methods are discussed in this section and are similar to creating curtain walls and window assemblies.

Creating a New Curtain Wall Unit Style from Scratch

Use one of the following options to create a curtain wall unit from scratch:

✓ Select the **Design** pull-down menu, pick the **Curtain Walls** cascade menu, then select **Curtain Wall Unit Styles…**.

✓ Pick the **Curtain Wall Unit Styles** button in the **Curtain Walls** toolbar.

✓ Right-click, and select **Design, Curtain Walls, Curtain Wall Unit Styles…** from the cursor menu.

✓ Type **AECCWUNITSTYLE** at the Command: prompt.

The Style Manager dialog box, shown in Figure 12.20 is displayed.

In the Style Manager dialog box, pick the New Style button, or right-click the Curtain Wall Unit Styles section icon, and select New to create a new curtain wall unit style. Enter a new name for the style next to the new icon created to create a new style. The next step is to edit the new curtain wall unit. Highlight the new style and select the Edit Style button, right-click over the new style, and pick Edit… from the shortcut menu. The **Curtain Wall Unit Style Properties** dialog box, shown in Figure 12.21, is displayed.

The dialog box is divided into four tabs. The **General** tab is used to rename the style and provide a description for the style. The **Design Rules** is the main tab and is used to define the rules for the curtain wall unit. The **Overrides** tab displays any overrides attached to the style when editing in place, and the **Display Props** tab is used to control the display of the elements within the curtain wall unit style.

Before configuring the style, you need to have an idea of how the curtain wall unit will appear. Notice in Figure 12.21 the icons in the Element Definitions area in the left part of the tab. As mentioned earlier, curtain wall units are composed of four parts: divisions, infills, frames, and mullions. When creating the style, you must create element definitions for these definitions in order to assemble them together into one unit.

Divisions establish how the cells within a unit are divided. There are two types of divisions: horizontal and vertical. Multiple divisions can be created and nested within one another to create a pattern of divisions (see Figure 12.22). A *primary grid* is always required and serves as the first level of division. As additional divisions are added they become nested in the cells of the previous division.

Infill definitions can be created from window, door, curtain wall unit, and AEC polygon styles. AEC polygons are a type of hatch pattern object. These are discussed in a later chapter. Infills can also be created from a default element within the Curtain Wall Unit Style Properties dialog box. Frame and mullion definitions by default take on a rectangular shape but can be defined by using an AEC profile definition.

For the purposes of this discussion the curtain wall unit that will be created appears in Figure 12.23. The following section explains how to create the individual definitions, then finally how to assemble the pieces into one unit.

Establishing Division Definitions

Refer to Figure 12.23, where you can see four sets of divisions used to create the curtain wall unit; therefore, four division definitions need to be created. Figure 12.24 displays the criteria for each of the divisions.

Edit Style button

Open Drawing button　　　**New Style** button　　**Purge Styles** button

Current
drawing

Right-click on
the Curtain
Wall Unit Style
section icon.

List of curtain
wall unit styles
in the current
drawing

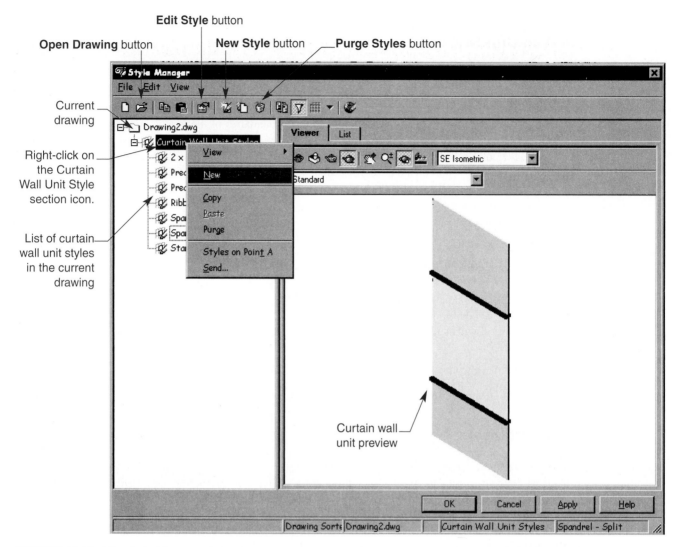

Curtain wall
unit preview

FIGURE 12.20　The Style Manager dialog box is used to organize styles in the drawing. Accessing the Curtain
Wall Unit Styles command displays the Curtain Wall Unit Styles section in Style Manager.

To create the division definitions, make the **Design Rules** tab active, and select the **Divisions** icon under the
Element Definitions folder icon. The table view changes to the division definition view (see Figure 12.25).

Select the New Division button to create a new definition. Highlight the new definition to configure it.
The table at the bottom of this view allows you to name the definition and to control the orientation and the
type of division used. In the Name: text box enter Primary Vertical Division. Select the Vertical orientation
button. In the Division Type drop-down list select Manual to allow you to manually set the vertical divisions within the definition. This definition is to have two vertical grid lines 12″ from each edge, so use the
table in the lower right corner to specify the division settings.

The table in the lower right corner lists the number of gridlines used in the definition. The Offset column establishes the measured distance from a specified point. The From column establishes where the gridlines are being measured from. The division orientation is vertical, so the From column includes the Grid
Start, Grid Middle, and Grid End options. Pick the values in the columns to make the modifications. You
can add additional gridlines by picking the Add Gridline button.

The next definition will be nested in the three cells of the primary vertical division. Create a new division definition named Secondary Horizontal Division. Set the orientation to horizontal and the division type
to manual. Set the grid lines 12″ from the grid bottom and 12″ from the grid top. These are the two main
divisions in the style.

The next two divisions are to be used for the glazing part of the style. There are two definitions, one vertical and one horizontal. The divisions in each of these definitions are to have an equal spacing. Create another

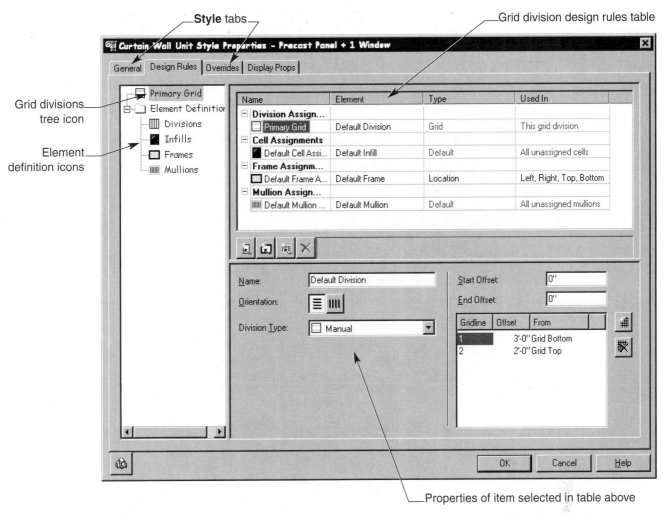

FIGURE 12.21 The Curtain Wall Unit Style Properties dialog box is used to configure the appearance of the curtain wall unit.

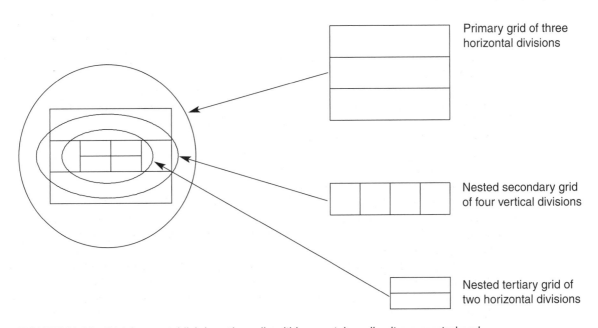

FIGURE 12.22 Divisions establish how the cells within a curtain wall unit are created and organized.

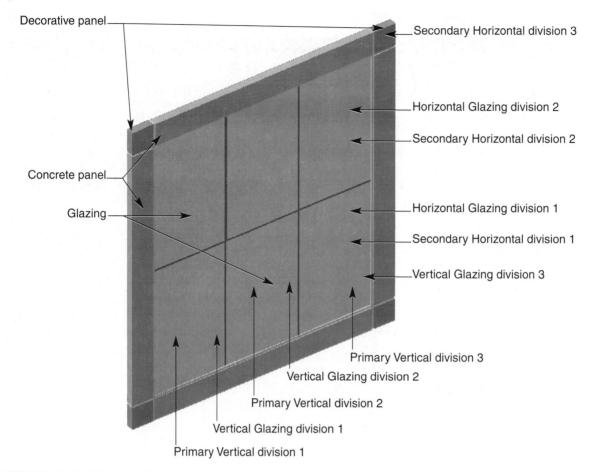

Decorative panel —
Concrete panel —
Glazing —

Secondary Horizontal division 3
Horizontal Glazing division 2
Secondary Horizontal division 2
Horizontal Glazing division 1
Secondary Horizontal division 1
Vertical Glazing division 3

Primary Vertical division 3
Vertical Glazing division 2
Primary Vertical division 2
Vertical Glazing division 1
Primary Vertical division 1

FIGURE 12.23   The desired curtain wall unit.

FIGURE 12.24   Four division definitions to be used in the creation of the curtain wall unit style.

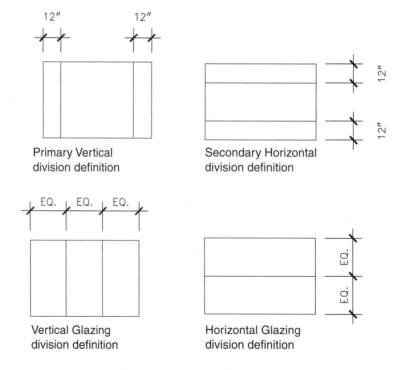

Primary Vertical division definition

Secondary Horizontal division definition

Vertical Glazing division definition

Horizontal Glazing division definition

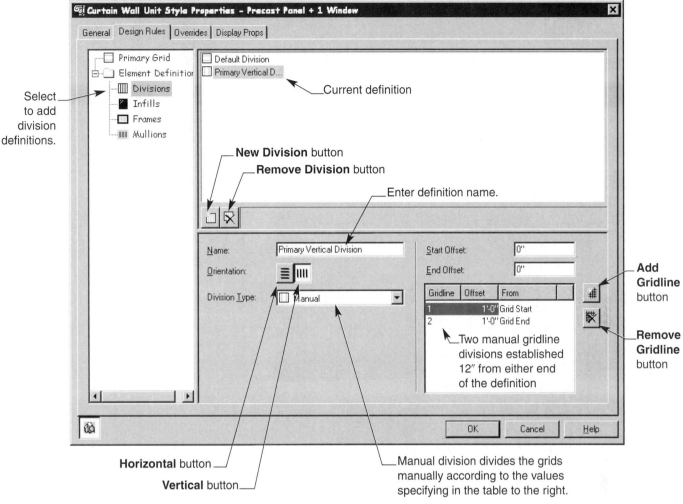

Select to add division definitions.

New Division button

Remove Division button

Enter definition name.

Current definition

Add Gridline button

Remove Gridline button

Two manual gridline divisions established 12″ from either end of the definition

Horizontal button

Vertical button

Manual division divides the grids manually according to the values specifying in the table to the right.

FIGURE 12.25 Select the Divisions icon on the left to access the division definition view.

definition named Vertical Glazing Division. Select the Vertical orientation button. In the Division Type drop-down list select the Fixed Number of Cells option to change the bottom of the tab to reflect the settings made for this type of division (see Figure 12.26). Enter a value of 3 in the Number of Cells text box. Add one final definition named Horizontal Glazing Division. Set the orientation to horizontal and the fixed number of cells to 2.

Establishing Infill Definitions

The **Infills** definitions icon is used to create infill definitions that are used within each cell of the curtain wall unit. Figure 12.23 indicates that three separate infills will be used: Decorative Panel, Concrete Panel, and a Glazing Panel. Select the Infills definition icon on the left to create the infill definitions. The Infill definitions view is displayed (see Figure 12.27). Press the New Infill button to create a new definition.

Highlight the new definition, and enter the name Decorative Panel in the Name text box. Enter a Panel Thickness of 4″. Create another infill named Concrete Panel, and set the thickness to 4″. Create the last panel, named Glazing, and set the thickness to 1″. This creates the three infills required for the curtain wall unit.

Establishing Mullion Definitions

This curtain wall unit will be used as an infill to a curtain wall, so a frame is not required; however, if you are creating a single panel of glazing as a curtain wall unit, you may want to create a frame on the panel. Two types of mullions will be used in the style, one for the glazing divisions and one for the concrete panel divisions. Select the Mullions definition icon. Again, the tab view changes to reflect mullion definitions (see Figure 12.28).

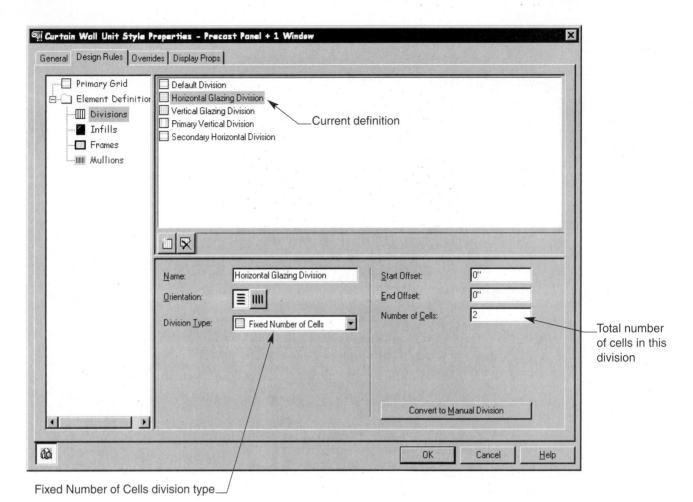

Fixed Number of Cells division type

Current definition

Total number of cells in this division

FIGURE 12.26   When Division Type is set to Fixed Number of Cells, the bottom of the tab changes so that a specific number of divisions can be specified.

Select the New Mullion button to create a new mullion definition. Highlight and name the new definition Concrete Mullion. Set the width to 1″ and the depth to 3″. You can use a custom profile as a special shape for the mullion by selecting the Use Profile check box and selecting an AEC profile from the list. Create a second mullion definition named Glazing Mullion, and set the width to 1″ and the depth to 2″. The two mullion definitions required for the style are created. You are now prepared to assemble the elements to create a custom curtain wall unit.

Adjusting Entity Display

Once the element definitions have been created, it is a good idea to adjust the definition colors in the Display Props tab to make it easier to view what you are configuring for the style. In the Curtain Wall Unit Style Properties dialog box, select the **Display Props** tab. Select the model display representation from the drop-down list and set the override to the curtain wall unit style. Select the Edit Display Props… button to edit the display properties for the model representation.

In the Entity Properties dialog box, notice that the Layer/Color/Linetype tab includes the three default components: Default Infill, Default Frame, and Default Mullion. Select the **Custom Model** Components tab. Select the Add… button to begin adding the new definitions to the model representation. The **Custom Display Component dialog box**, shown in Figure 12.29, is opened. Select the Infill option from the Component Type drop-down list, then pick the Select Element… button to select one of the new definitions. The new definition is added to the Layer/Color/Linetype tab. Press the OK button to return to the Custom Display Component dialog box. Continue to add components until all the custom definitions have been added to the Layer/Color/Linetype tab.

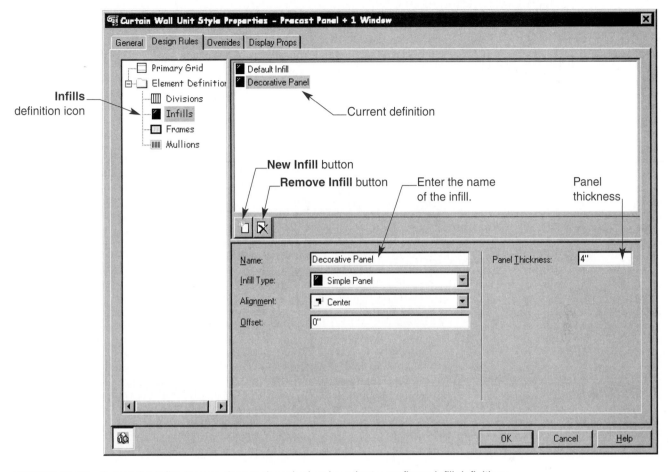

FIGURE 12.27   Select the Infills icon to change the tab view in order to configure infill definitions.

Once the new definitions have been added to the Layer/Color/Linetype tab, adjust the element colors (see Figure 12.30). You may wish to assign a different color to the three different panels for easy recognition when viewing the style in the floating viewer window. When you have finished press the OK button to return to the Entity Display tab. You can continue to adjust the display properties for the other representations, or you can select the Design Rules tab to assemble the components into a new style.

Assembling the Curtain Wall Unit Style

Now that the elements required for the style have been established and their display properties set, the style can be created. In the tree view on the left edge of the Design Rules tab, select the **Primary Grid** icon. The tab view is changed to a table display. The table gives the settings used for the primary grid, also known as the *main division*.

The primary grid includes the four elements for a curtain wall unit: division assignment, cell assignments, frame assignments, and mullion assignments (see Figure 12.31). Across the top of the table are four columns: Name, Element, Type, and Used In. At the top of the table, Primary Grid is the name of the division assignment. This grid is using the division element definition of Default Division. Pick the Default Division text under the Element column to display a drop-down list of division definitions you previously created. Select the Primary Vertical Division definition from the list.

The primary grid is then set to use your custom division that establishes grid lines 12″ from the start and end of the unit. The primary grid is now set with three separate cells. Moving down the table, note that Cell Assignments for the primary grid is set to use the Default Infill definition for all unassigned cells, as shown in Figure 12.31. Select the Default Infill text in the Element column, and select the Concrete Panel infill. The primary grid will thus use the concrete panel definition for each of the three cells.

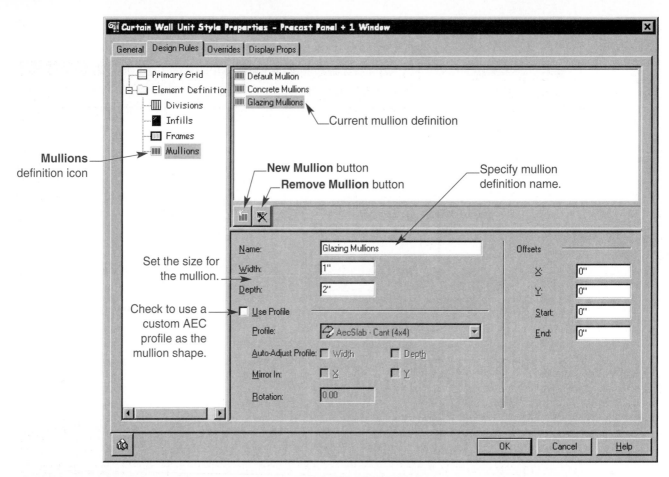

FIGURE 12.28 The Mullion definition view is used to create custom mullion definitions.

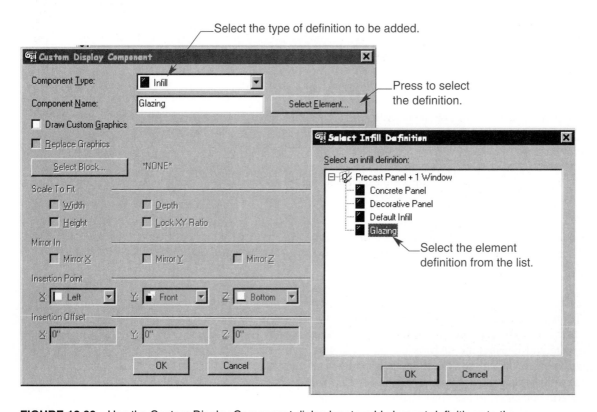

FIGURE 12.29 Use the Custom Display Component dialog box to add element definitions to the Layer/Color/Linetype tab.

FIGURE 12.30   The five new definitions created have now been added to the model display representation and assigned a different color.

The new element definitions have been added and assigned a color.

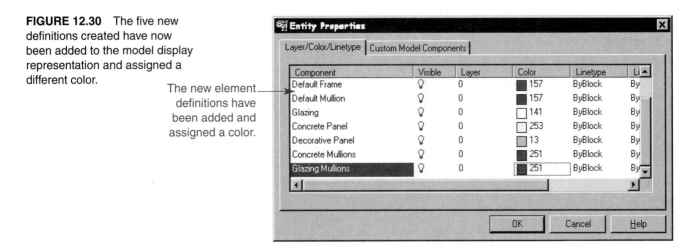

Select default division to set new division definition.

Default infill definition

Default frame definition

Default mullion definition

Select the **Primary Grid** icon.

New Cell Assignment button

New Frame Assignment button

New Mullion Assignment button

Remove Assignment button

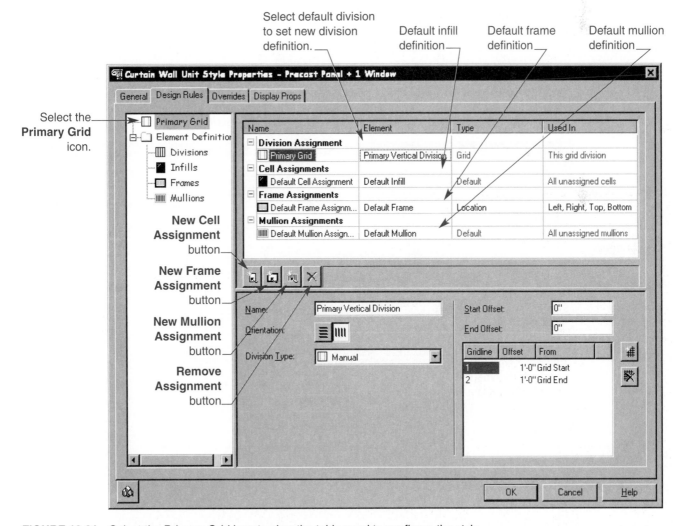

FIGURE 12.31   Select the Primary Grid icon to view the table used to configure the style.

Referring to Figure 12.23, notice that the three vertical cells need to be subdivided into three horizontal divisions, totaling nine individual cells without the glazing divisions. In other words, three horizontal divisions need to be nested within each of the three vertical divisions.

In the **Cell Assignments** area press the **New Cell Assignment** button to add another cell assignment to the primary grid. A new cell named New Cell Assignment is added. In the Element column beside the new cell, pick the text to produce a drop-down list. Pick *Nested Grid* from the drop-down list. The cell assignments name is renamed to New Nested Grid. Pick the text in the Name column to rename the New Nested Grid, Secondary Grid. The Type column should reflect the Location option. In the Used In column, pick the text to reveal an ellipsis button […]. Pick the ellipsis button to select the primary grid cells where the nested grid will be located. Make sure that the Start and End check boxes have been selected, and clear the check mark in the Middle checkbox to specify that the starting and ending cells will be nested. Notice that the tree to the left of the Primary Grid icon now includes a new icon below it named Secondary Grid, as shown in Figure 12.32.

The primary grid now includes two types of cell assignments: Default Infill and Secondary Grid. The primary grid middle cell currently uses the default infill, because it has not been assigned an infill in the Used In column. Pick the New Cell Assignment button again to add another cell assignment. Set this cell assignment to *Nested Grid*, and set the nesting to be used in the middle primary grid cell. Rename the New Nested Grid, Middle Grid. The primary grid now uses a nested grid in the starting and ending cells, and a separate nested grid in the middle cell.

Currently, the style uses the Default Frame definition around the curtain wall unit. A frame is not needed in this style, so you can turn this off. In the Default Frame Assignment element, move to the Used

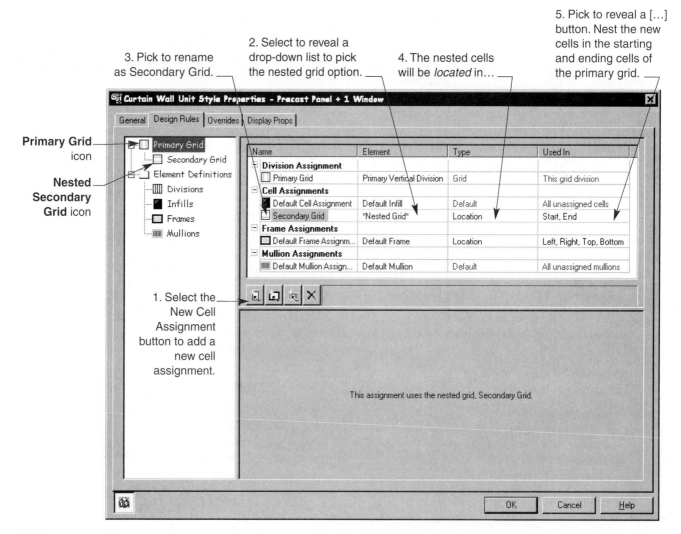

FIGURE 12.32   The primary grid is being configured to use the Concrete Panel definition for any unassigned cells. The starting and ending cells have been assigned a nested grid.

New subdivisions are
set to be nested grids.

Location for each
of the nested grids

Two new
subdivisions
to the
Primary
Grid

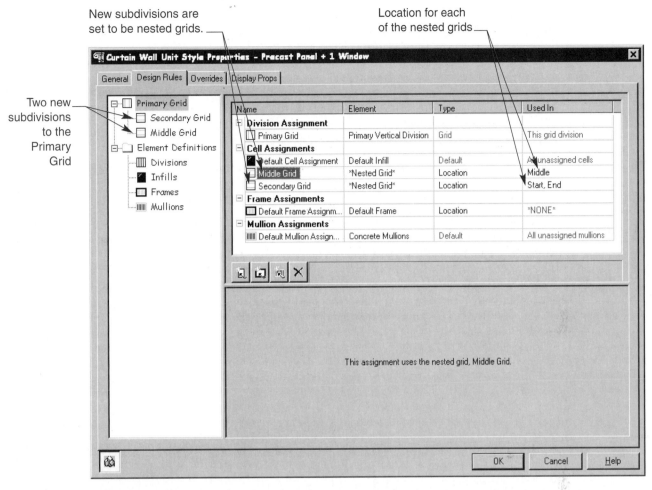

FIGURE 12.33 The primary grid assignments should appear as in the diagram. Two nested grid assignments, no framework, and concrete mullions define the primary grid.

In column. Notice that a frame is being used at the left, right, top, and bottom locations. These values reflect the four sides of the primary grid. Select the text Left, Right, Top, Bottom to reveal the ellipsis [...] button. Uncheck each of the values to turn off the frame around the primary grid.

The **Mullion Assignments** area is currently using the Default Mullion definition as the cell divider. Pick the Default Mullion text to display a drop-down list, and select the Concrete Mullions definition. This step completes the primary grid assignments, as shown in Figure 12.33. The next step is to configure the nested grid (secondary grid and middle grid) assignments.

Once you have configured the primary grid, begin making assignments for the nested grids. The secondary grid that controls the starting and ending cells of the primary grid will not be subdivided any further, but it does include two types of panels: Decorative Panel and Concrete Panel. Select the Secondary Grid icon in the tree on the left below the Primary Grid icon so that you can assign definitions to the secondary grid level.

A horizontal grid division definition was created earlier, so you can assign this division definition to the secondary grid. In the Division Assignment area for the secondary grid, select the Default Division element, and pick the Secondary Horizontal Division definition from the list to divide the starting and ending cells of the primary grid into three horizontal cells with gridlines 12″ from the top and bottom.

The secondary grid is now divided into three horizontal cells. The cell assignment is currently set to Default Infill. The top and bottom cells need to be assigned the Decorative Panel infill definition, and the middle cell needs to be assigned the Concrete Panel infill definition.

Pick the New Cell Assignment button to add a new cell assignment. In the Element column beside the new cell, select the Decorative Panel definition. Rename the New Cell Assignment text in the Name column, Top and Bottom Cells. In the Used In column, make sure the decorative panel is being used in the Top and Bottom cells. Select the New Cell Assignment button again to add another new cell assignment.

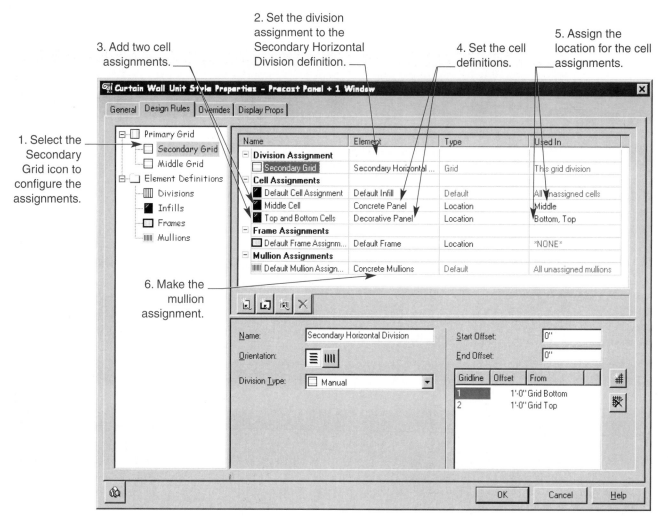

FIGURE 12.34 The secondary grid uses the Secondary Horizontal Division definition to divide the starting and ending cells of the primary grid into three cells. The top and bottom cells are assigned the decorative panel definition, and the middle cell is assigned the concrete panel definition.

Change this assignment to the Concrete Panel definition, and use this definition in the middle cell. Rename this new cell assignment Middle Cell.

The final assignment to the secondary grid is the mullion assignment. Select the Default Mullions text in the Element column, and select the Concrete Mullions definition. The settings for the secondary grid are established. Use the floating viewer to preview the results so far. Figure 12.34 displays the settings for the secondary grid.

Now that the primary and secondary grids have been defined, you can concentrate on the middle primary cell. Select the **Middle Grid** icon below the Primary Grid icon from the tree on the left. The middle grid also needs to be divided into three horizontal cells. The top and bottom cells use the Concrete Panel definition, and the middle cell becomes a nested grid for the glazing.

Select the Default Division element for the middle grid division assignment. Set this assignment to the Secondary Horizontal Division definition. In the Cell Assignments area, select the New Cell Assignment button to add a new cell assignment. Change the new cell assignment element to *Nested Grid*. Rename the New Nested Grid, Primary Glazing. Set the location for the nested grid to the middle cell, which indicates that the middle cell of the Middle Grid division will be subdivided.

Change the Default Cell Assignment to Concrete Panel, which indicates that any cells not defined (the top and bottom cells of the middle grid) will use the Concrete Panel infill definition. The last task is to set the mullion assignments. Change the Default Mullion to the Glazing Mullions definition. The assignments for the Middle Grid level are displayed in Figure 12.35.

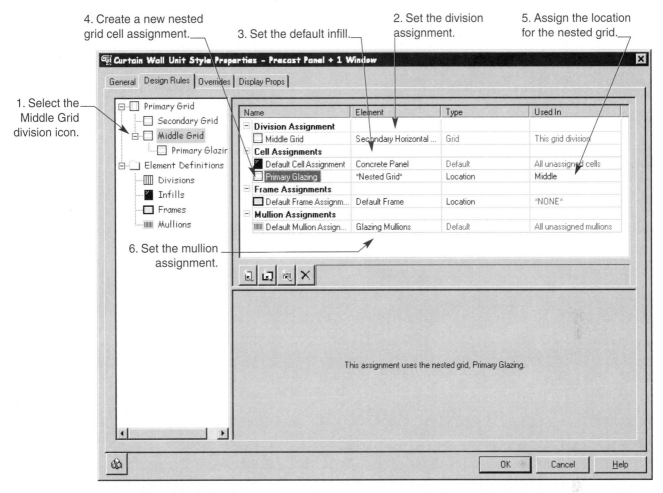

4. Create a new nested grid cell assignment.

3. Set the default infill.

2. Set the division assignment.

5. Assign the location for the nested grid.

1. Select the Middle Grid division icon.

6. Set the mullion assignment.

FIGURE 12.35   The Middle Grid division controls the middle cell of the primary grid. Assign the division element to use, followed by the cell assignments.

Now that the Middle Grid division has been defined, the Primary Glazing division needs to be defined. This division controls the middle cell of the Middle Grid division and needs to be broken down into three equal vertical glazing pieces. The three equal glazing pieces then need to be broken down further into two equal horizontal pieces.

Select the **Primary Glazing** icon from the left side of the tab. Set the Division Assignment element to the Vertical Glazing Division. This division definition splits the middle cell of the Middle Grid division into three equal cells. In the Cell Assignments area, set the Default Infill to the Glazing definition. Create a new cell assignment and make the assignment a nested grid. Name the New Nested Grid, Secondary Glazing. In the Use In column for the Secondary Glazing cell assignment, set the location to Start, Middle, and End.

Finally, set the default mullion to the Glazing Mullions definition. Figure 12.36 displays the settings for the Primary Glazing division.

The final part that needs to be configured is the last division. Select the Secondary Glazing division icon from the tree on the left. Set the Division Assignment to use the Horizontal Glazing Division definition. Set the Default Infill to the Glazing definition and the Default Mullion to the Glazing Mullions definition (see Figure 12.37). The curtain wall unit style is now complete. Use the floating viewer to preview the new style. The style is now ready to be incorporated into a curtain wall or a window assembly.

4. Create a new nested grid cell assignment.

3. Set the default infill definition.

2. Assign the division to be used.

5. Set the location for the nested grid.

1. Select the Primary Glazing division icon.

6. Set the mullion definition to be used.

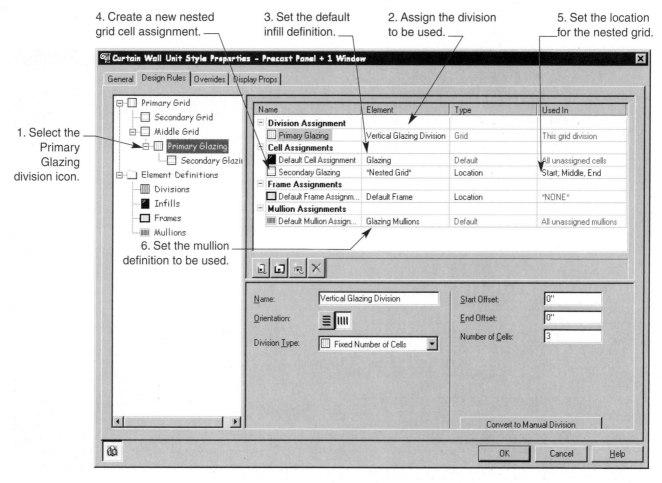

FIGURE 12.36   The settings for the Primary Glazing division control the rules for the middle cell of the Middle Grid division.

Exercise 12.7

1. Begin a new drawing using the AEC Arch (Imperial) or AEC Arch (Imperial-Int).dwt.
2. Access the Curtain Wall Unit Styles command, and create a new style named Precast Panel + 1 Window.
3. Follow the steps outlined in the previous section to create a curtain wall unit style.
4. Save the drawing as EX12-7.dwg.

Converting Linework into a Curtain Wall Unit Style

The previous section outlined how to create a curtain wall unit style from scratch. Another method for creating a curtain wall unit style is to draw the layout of the curtain wall unit using lines and polylines, then convert them into a style.

To do this, first draw the shape and gridline configuration desired for the curtain wall unit, then use the **Convert Linework to Curtain Wall Unit** command. Use one of the following options to access this command:

✓ Pick the **Design** pull-down menu, select the **Curtain Walls** cascade menu, then pick the **Convert Linework to Curtain Wall Unit** command.

✓ Pick the **Convert Linework to Curtain Wall Unit** button in the **Curtain Walls** toolbar.

3. Set the default
infill definition.

2. Set the division definition.

1. Select the
Secondary
Glazing
division icon.

4. Set the default
mullion definition.

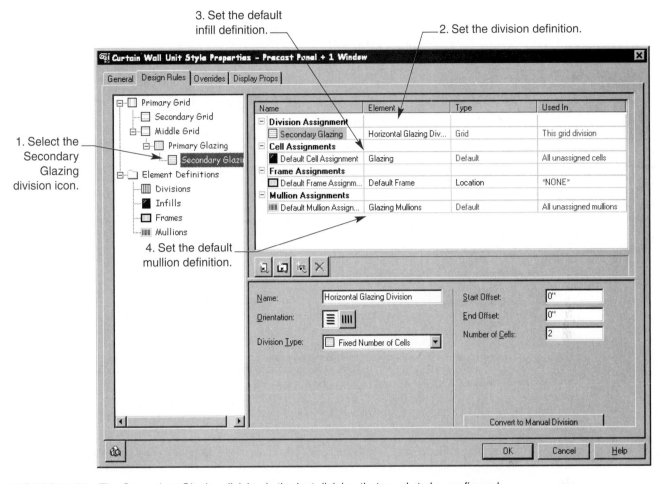

FIGURE 12.37   The Secondary Glazing division is the last division that needs to be configured.

✓ Right-click, and select **Design, Curtain Walls, Convert Linework to Curtain Wall Unit** from the cursor menu.

✓ Type **AECCWUNITCONVERT** at the Command: prompt.

The following prompt sequence is used to convert the linework in Figure 12.38 into a curtain wall unit:

```
Command: AECCWUNITCONVERT ↵
Select linework: (pick the first corner for a crossing window)
Specify opposite corner: (drag the cursor through the linework and pick a second point)
4 found
Select linework: ↵
Select baseline or RETURN for default: (pick the bottom edge of the linework to be used as the
    baseline)
Erase layout geometry? [Yes/No] <N>: ↵
Edit in place mode is active. Select Save Changes from the context menu to create a new style.
Command:
```

The linework is converted into a curtain wall unit when the command is finished. You may need to change to an isometric view if the linework was drawn in the top view. The curtain wall unit is currently in the Edit in Place mode. Select the unit, right-click, and pick the Edit in Place cascade menu. Pick the Save changes… options to save the unit as a new style.

The preceding method is a quick means of creating a curtain wall unit for use in a curtain wall; however, the drawn size of the unit is the size that will be used in the curtain wall. The curtain wall unit cannot get larger or smaller within the curtain wall as length changes are made to the curtain wall.

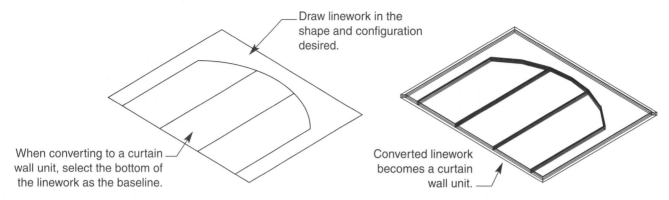

FIGURE 12.38   Draw the shape and gridlines for the curtain wall unit, then convert them into a curtain wall unit.

Exercise 12.8

1. Open EX12-7.dwg.
2. Draw the figure using a series of polylines.
3. Convert the lines to a curtain wall unit.
4. Select the unit, right-click and select Edit in Place, then Save changes….
5. Save the unit as a new style named Concrete Arched Panel.
6. Edit the style so that there is no frame on the outer boundary.
7. Create a Concrete infill and Decorative Concrete infill element definition 4″ thick.
8. Create a Glazing infill element definition 1″ thick.
9. Edit the unit in place and use overrides for the different cells, as indicated in the figure. Save the overrides back to the style when finished, ensuring that the Transfer Cell Overrides to Style is checked on.
10. Save the drawing as EX12-8.dwg.

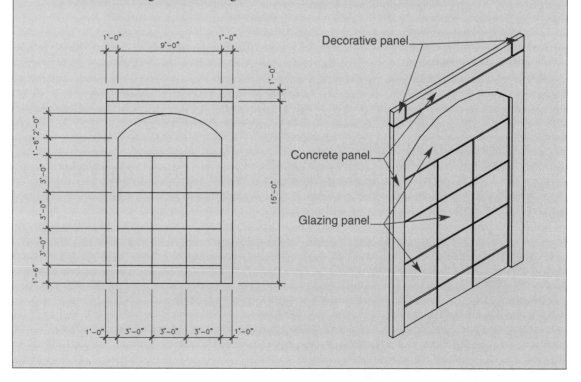

Converting a Layout Grid into a Curtain Wall Unit Style

The last method of creating a curtain wall unit is by converting it from a layout grid. A *layout grid* is a tool that can be used to lay out divisions along the X and Y axes. This tool can be found in the Desktop pull-down menu under the Layout Tools cascade menu and is the Layout Grid (2D) command. The **Add Layout Grid** dialog box allows you to specify the X and Y axes by size or by the total number of divisions to be made. Select a point in the drawing for the first corner of the grid, then select a point for the opposite corner of the grid. A new layout grid is created in the drawing screen.

After the grid is drawn, it can be converted into a curtain wall unit using one of the following options:

✓ Select the **Design** pull-down menu, move to the **Curtain Walls** cascade menu, and select the **Convert Layout Grid to Curtain Wall Unit** command.

✓ Pick the **Convert Layout Grid to Curtain Wall Unit** button from the **Curtain Walls** toolbar.

✓ Rght-click in the drawing, and select **Design, Curtain Walls, Convert Layout Grid to Curtain Wall Unit** from the cursor menu.

✓ Type **AECCWUNITCONVERTGRID** at the Command: prompt.

The following prompt sequence is used to convert a layout grid into a curtain wall unit:

Command: **AECCWUNITCONVERTGRID** ↵
Select a 2d layout grid: *(select the layout grid)*
Erase layout geometry? [Yes/No] <N>: ↵
Primary Division [Vertical] <Horizontal>: **V** ↵
(enter a new name for the curtain wall unit)
Command:

When converting a layout grid into a curtain wall unit, you must specify a primary division. Select either horizontal or vertical. After you have specified what the primary division will be, the **Curtain Wall Unit Style Name** dialog box appears so you can give the new unit a name.

Creating Window Assembly Styles

The addition of window assemblies was discussed in Chapter 3, and modification of the window assemblies was discussed in Chapter 10. Window assembly styles are discussed here because they are similar to curtain wall units and curtain wall styles. The window assembly style section also appears before the curtain wall styles discussion because curtain wall styles can use both curtain wall units and window assemblies as an infill element.

Window assemblies are created using the same types of rules as used by the curtain wall unit. The window assembly is based on a series of divisions and nested grids. In addition to the standard infill panels that the curtain wall unit uses, the window assembly can use door and window styles as infill elements. Like curtain wall units, window assemblies can be created by one of three methods: from scratch, converted from linework, or converted from a layout grid. The next section discusses how to create window assemblies.

Creating a Window Assembly from Scratch

The window assembly to be created is a door with a sidelight containing muntins as shown in Figure 12.39. Use one of the following options to create the style:

✓ Select the **Design** pull-down menu, access the **Window Assemblies** cascade menu, and pick the **Window Assembly Styles...** command.

✓ Pick the **Window Assembly Styles** button on the Doors - Windows - Openings toolbar.

✓ Right-click, and select **Design, Window Assemblies, Window Assembly Styles...** from the cursor menu.

✓ Type **AECWINASSEMBLYSTYLE** at the Command: prompt.

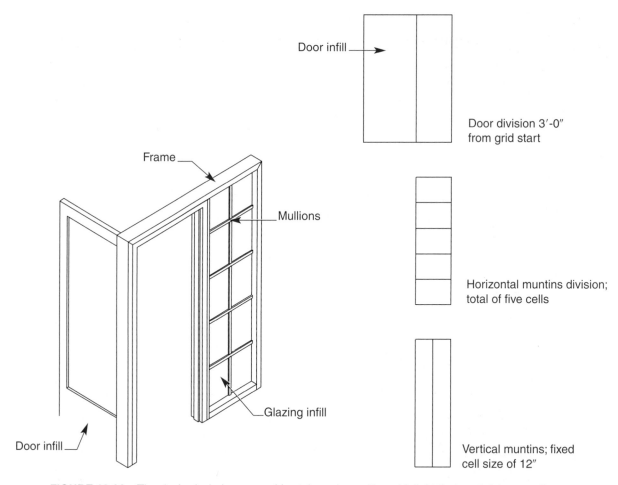

FIGURE 12.39   The desired window assembly style: a door with a sidelight that containing muntins.

The Style Manager dialog box is displayed.

The Style Manager is displayed looking at the Window Assembly section. Select the New Style button to create a new window assembly. Name the Style: Hinged Single 3-0 x 7-0 Left + Sidelight with Grids. Select the new style, right-click, and select E̲dit… from the shortcut menu. The **Window Assembly Style Properties** dialog box, shown in Figure 12.40, is displayed. This dialog box contains the following six tabs used for configuring the style.

- **General**   This tab is used to rename the style, provide a description, and add design notes and property set definitions.

- **Defaults**   This tab is used to set up any default sizes that will be filled out when the assembly style is added. The sizes act as the default size in the Add Window Assembly dialog box but can be overridden by entering a new size.

- **Shape**   This tab is used to specify a shape for the window assembly. Unlike the curtain wall unit, the window assembly can be created using different shapes such as rectangular, oval, gothic, or a custom shape determined from an AEC profile.

- **Design Rules** Similar to the tab for the curtain wall unit, this tab is used to create the division, infill, frame, and mullion definitions and assemble them into a window assembly.

- **Overrides**   This tab displays any overrides to a window assembly when you edit in place.

- **Display Props** This tab controls the display of the window assembly in different display representations.

Style configuration tabs

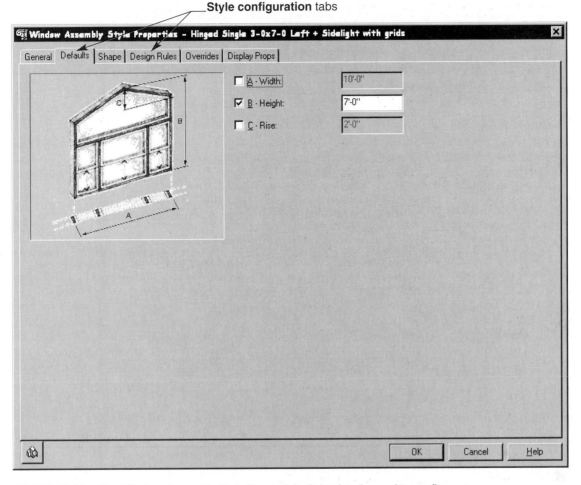

FIGURE 12.40   The Window Assembly Style Properties dialog box is used to configure a window assembly style.

   Creating the Element Definitions

Access the Design Rules tab to begin assembling the style. A review of Figure 12.39 reveals that three divisions need to be created: the primary grid places the door at the start of the grid (on the left), then a horizontal muntins division is nested to the right of the primary grid, and a vertical muntins division is nested in the five horizontal sidelight cells.

First, create the Door Division definition. Create a new division named Door Division. Use the manual division type with a vertical orientation. Add a gridline 3'-0" from the grid start.

Next, create the Horizontal division. Name the definition Horizontal Muntins. Set the division type to Fixed Number of Cells, with a total of five cells, and set the orientation to horizontal. Create one last division named Vertical Muntins. Set the Division Type to Fixed Cell Dimension, with the cell dimension set to 12" using the vertical orientation. The Fixed Cell Dimension includes an option that allows cells to grow or shrink (see Figure 12.41). When you specify whether a cell is to grow or shrink and pick the appropriate button below the drop-down list, the window assembly continues to add 12" divisions until it stops at an odd dimension. Depending on whether the starting, middle, or ending cell has been told to grow or shrink, that cell will become larger or smaller within the division. In this situation you do not want the cells to grow or shrink, so uncheck the **Auto-Adjust Cells** check box.

After you have created the divisions definitions, create the infill definitions. Two infill definitions are required: a door and the glazing. Create a new infill named Glazing, using the Simple Panel infill type. Set the panel thickness to 1". Create another infill named door. For the infill type, select Style, which displays a style box on the right side of the tab (see Figure 12.42). Browse through the Door Styles folder until you find the Hinged - Single - Full Lite style. Select the style to use for the infill.

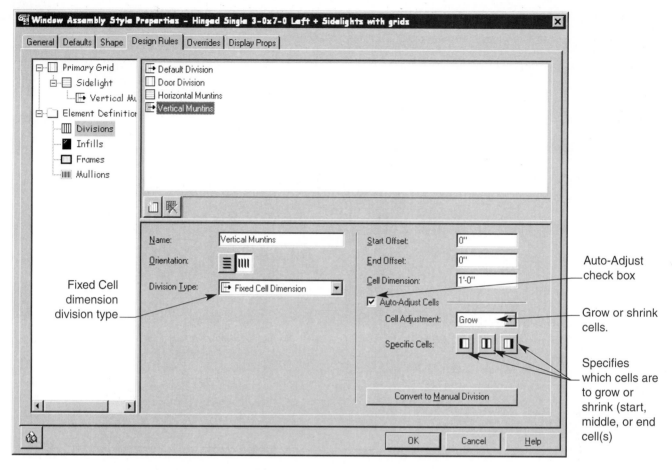

FIGURE 12.41   The Fixed Cell Dimension division type allows you to specify if cells will get larger or smaller when an odd dimensional size is entered for the window assembly.

Create a frame definition named Assembly Frame. Set the width to 2″ and the depth to 5″. Next, create two mullion definitions. One mullion definition is used between the door and the sidelight, and the other is used as the muntins in the sidelight. Create the first mullion definition, named Main Mullions, with a width of 1″ and a depth of 5″. Create the second mullion definition, named Muntins, with a width and depth of 1″. With the definitions created, the window assembly can be put together.

Creating the Rules for the Style

Select the **Primary Grid** icon in the **Design Rules** tab. Pick the Default Division text under the Element column. Change the division for the Primary Grid to the Door division. Two vertical cells are created for the primary division. Move down to the Cell Assignments area, and select the Default Cell Assignments element. Change the default infill to the Door Infill definition.

Next, add a New Cell Assignment, with the element type *Nested Grid*. The nested grid is used in the end cell. Rename the New Nested Grid, Horizontal Muntins. Change the default frame to the Assembly Frame definition. A frame is not needed below the door, so turn the frame off at the bottom of the primary grid. Select the Used In column for the frame definition, pick the [...] button, and uncheck the bottom check box.

The final step in the configuration of the primary grid is to assign the mullions dividing the door from the sidelight. Change the default mullions to the Main Mullions definition. Figure 12.43 displays the settings for the primary grid.

The next grid that needs to be configured is the Horizontal Muntins grid. Select the **Horizontal Muntins** icon from the tree on the left. Change the division type to the Horizontal Muntins definition, which divides the sidelight into five separate cells. In the Cell assignments area, set the default cell assignment to use the Glazing infill definition.

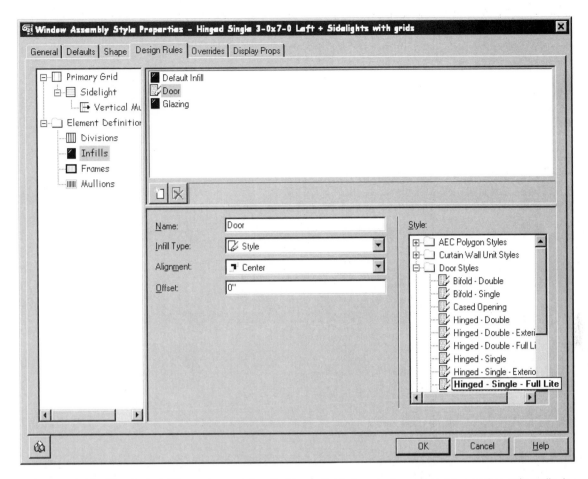

FIGURE 12.42 The Style infill type is an option that is available for window assemblies and curtain walls. It allows you to use door, window, curtain wall unit, and AEC polygon styles as infills.

Create a new cell assignment. This cell assignment determines the vertical divisions within the horizontal muntins division. Set the New Cell Assignment to *Nested Grid*. Rename the New Nested Grid, Vertical Muntins. This nested grid needs to be assigned to five cells, so change the Type column from Location to Index. Select the Location text, and change it to Index. In the Used In column, type 1,2,3,4,5, which indicates that the vertical cells need to be added to the five horizontal cells.

The frame was turned off at the bottom of the primary grid, so it is not turned on for the Horizontal Muntins division. Change the default frame to the Assembly Frame definition. In the Used In column select the *NONE* text so that you are able to place a check in the bottom check box. This turns the frame on for the bottom of the Horizontal Muntins division. Set the default mullions to the Muntins definition. Figure 12.44 displays the settings for the Horizontal Muntins division.

The final step is to configure the Vertical Muntins division. Select the Vertical Muntins division icon from the tree on the left. Change the Default Division to the Vertical Muntins definition. Change the default infill to the Glazing definition and the default mullions to the Muntins definition. This step completes the configuration of the door/sidelight window assembly. Figure 12.45 shows the settings for the Vertical Muntins division.

Exercise 12.9

1. Begin a new drawing using the AEC Arch (Imperial) or AEC Arch (Imperial-Int).dwt.
2. Create a new Window Assembly style named Door-Left with Sidelights-Right.
3. Follow the steps outlined above to create the door and sidelight window assembly.
4. Save the drawing as EX 12-9.dwg.

5. Rename the
nested grid

3. Set the Default
Cell Assignment to
the Door definition

2. Change to the
Door Division
definition

4. Create a new
cell assignment.
Set to Nested Grid
and assign to the
end cell.

1. Select the
Primary Grid
icon

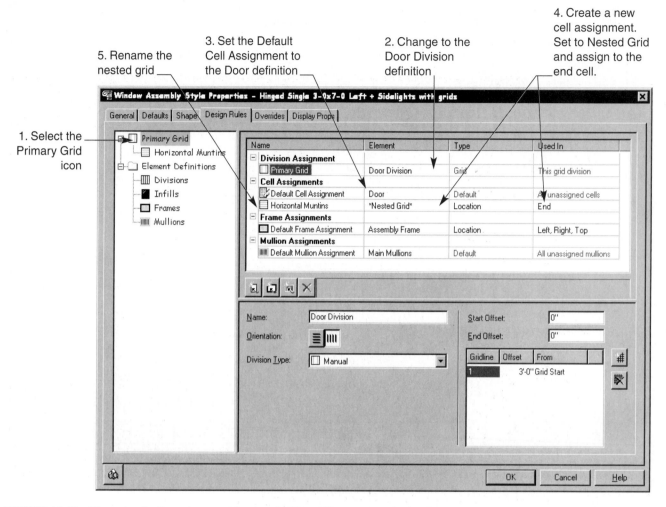

FIGURE 12.43   The rules for the primary grid separate the grid into two vertical cells, assigning the door infill and a nested infill for the sidelight.

 Converting Linework to a Window Assembly

Like curtain wall units, window assemblies can be created by converting linework into a window assembly. First, draw the desired shape and configuration for the window assembly style, then use the **Convert Linework to Window Assembly** command, which can be accessed in one of the following ways:

- ✓ From the **Design** pull-down menu, access the **Window Assemblies** cascade menu, and select the **Convert Linework to Window Assembly** command.

- ✓ Select the **Convert Linework to Window Assembly** button in the Doors - Windows - Openings toolbar.

- ✓ Right-click, and select **Design, Window Assemblies, Convert Linework to Window Assembly** from the cursor menu.

- ✓ Type **AECWINASSEMBLYCONVERT** at the Command: prompt.

As with the curtain wall unit command, select the linework that will be used to form the window assembly. When you have finished answering the prompts, a new window assembly is created. It is not a style until you select the window assembly, right-click, and select Edit in Place, then pick the Save changes… command. You can then save the window assembly as a new style for use in your drawing.

4. Add a new nested grid named Vertical Muntins. ⎯

3. Change the default cell assignment to the Glazing definition. ⎯

2. Change the division element to the Horizontal Muntins definition.

5. Set the nested grid to the Index type used in horizontal cells 1, 2, 3, 4, and 5.

1. Select the Horizontal Muntins division icon.

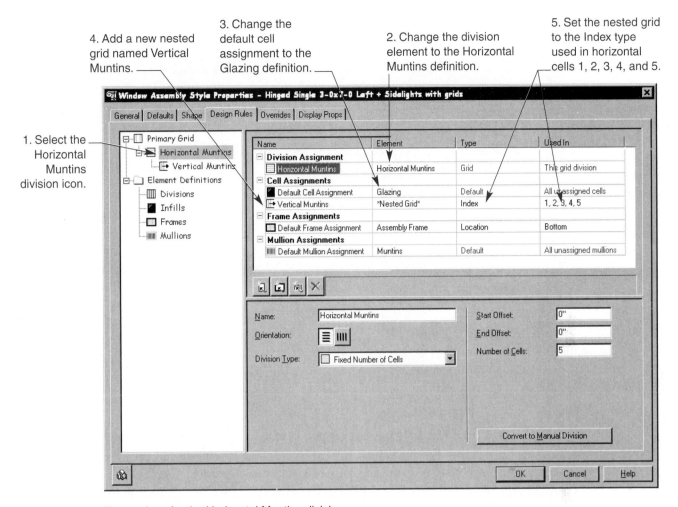

FIGURE 12.44 The settings for the Horizontal Muntins division.

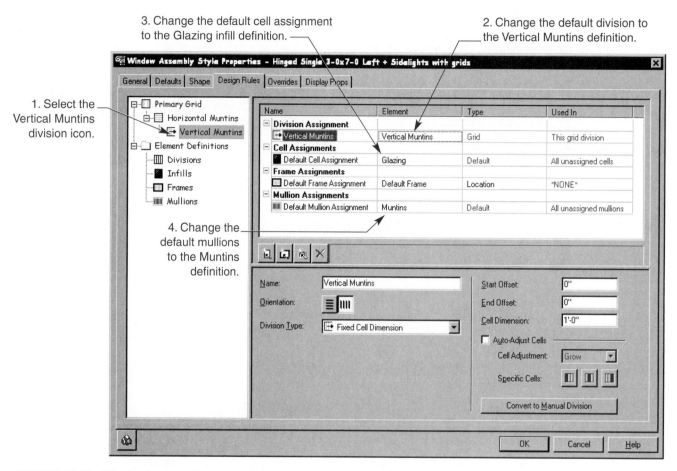

FIGURE 12.45   The final step is configuring the rules for the Vertical Muntins division.

Exercise 12.10

1. Open EX12-8.dwg.
2. Draw the figure using a series of polylines.
3. Convert the linework into a window assembly.
4. Use the Edit in Place command to save the changes as a new style. Name the style Concrete Door Panel.
5. Edit the style by creating the following four infills:
    - Concrete panel: 4″ thick
    - Decorative panel: 4″ thick
    - Glazing: 1″ thick
    - Door: Use the Double Doors door style.
6. Edit the style so that there is no frame on the outer boundary.
7. Edit the unit in place and use overrides for the different cells, as indicated in the figure. Save the overrides back to the style when finished, making sure that the Transfer Cell Overrides to Style is checked on.
8. Save the drawing as EX12-10.dwg.

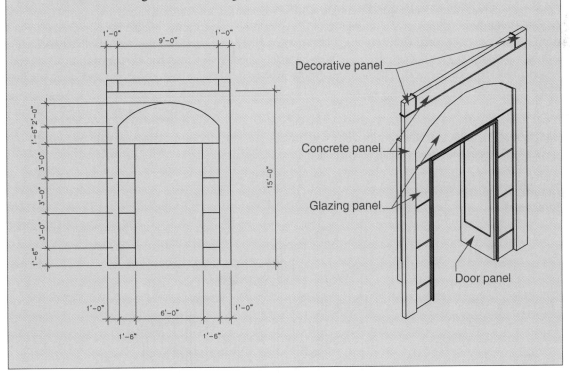

 Creating Additional Window Assemblies

You also can create window assemblies from a layout grid, by using the **Convert Layout Grid to Window Assembly** command:

✓ Select the **Design** pull-down menu, move to the **Window Assemblies** cascade menu, and select the **Convert Layout Grid to Window Assembly** command.

✓ Select the **Convert Layout Grid to Window Assembly** button in the Doors - Windows - Openings toolbar.

✓ Right-click, and select **Design, Window Assemblies, Convert Layout Grid to Window Assembly** from the cursor menu.

✓ Type **AECWINASSEMBLYCONVERTGRID** at the Command: prompt.

Answer the prompts to convert the layout grid into a window assembly.

Additional window assemblies are also included with Architectural Desktop. Use the Style Manager to open the Window Assembly Styles (Imperial).dwg, found in the \\Autodesk Architectural Desktop 3\Content\Imperial\Styles folder. Drag and drop the styles from this drawing into your current drawing in Style Manager.

Creating Curtain Wall Styles

The previous discussions have walked you through the process of creating a curtain wall unit style and a window assembly style. The techniques and dialog boxes are very similar. The process of creating the curtain wall style is the same. The only difference is that the curtain wall can use both the curtain wall unit and the window assembly style in its configuration. The goal for this section is to create a curtain wall using Exercises 12.7, 12.8, and 12.10 as infills and combine them into one curtain wall. The curtain wall will end up looking like the diagram in Figure 12.46.

Creating the Curtain Wall Style

Referring to Figure 12.47, note that the curtain wall will be broken into three separate divisions. The primary grid establishes the ends of the curtain wall 12″ from the ends, with the middle being reserved for the curtain wall units and window assemblies. The vertical bay grid establishes the vertical divisions for the curtain wall units at 11′-0″ apart. This number was established from the arched curtain wall that was converted from 11′-wide linework. The horizontal floor grid establishes the first grid line 16′-0″ from the bottom, with subsequent grid lines 13′ apart.

The infill definitions required are the Precast Panel + 1 Window (curtain wall unit), Concrete Arched Panel (curtain wall unit), Concrete Door Panel (window assembly), and an 8″-thick infill for the ends of the wall. A mullion definition between the cells will also be created 1″ wide × 3″ deep. With the parts for the curtain wall calculated, the style can be created.

FIGURE 12.46   The desired curtain wall.

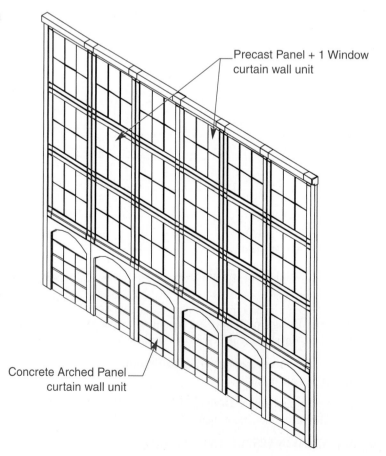

Precast Panel + 1 Window curtain wall unit

Concrete Arched Panel curtain wall unit

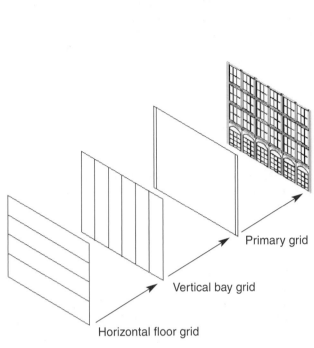

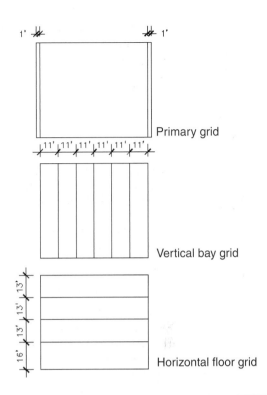

FIGURE 12.47 The curtain wall will be divided into three separate grids.

Use one of the following options to create the curtain wall style:

✓ Select the **Design** pull-down menu, pick the **Curtain Walls** cascade menu, then select the **Curtain Wall Styles...** command.

✓ Pick the **Curtain Wall Styles** button in the **Curtain Walls** toolbar.

✓ Right-click, and select **Design, Curtain Walls, Curtain Wall Styles...** from the cursor menu.

✓ Type **AECCURTAINWALLSTYLE** at the Command: prompt.

The Style Manager opened to the Curtain Wall section is displayed.

Pick the New Styles button to create a new curtain wall style. Name the style Precast Concrete Panel. Highlight the new style, right-click, and select the Edit... command from the shortcut menu. The **Curtain Wall Style Properties** dialog box appears. The dialog box is divided into the following five tabs for the configuration of the curtain wall: **General, Defaults, Design Rules, Overrides,** and **Display Props**.

A single bay in the curtain wall is 11'-0" wide, and it is desired to have a 12"-wide concrete panel at each end, so the minimum width for the curtain wall should be 13'-0". As new bays are added to the curtain wall the curtain wall increases in size in 11'-0" increments. Enter the following description in the General tab: Minimum width 13'-0", increase size in 11'-0" increments.

The Defaults tab establishes default height size for the curtain wall when it is added through the Add Curtain Wall command. The curtain wall you are creating will have four floor levels (16' + 13' + 13' + 13' = 55'), so enter the value 55'-0" for the base height. The Floor Line and Roof Line text boxes allow you enter an offset distance from the curtain wall baseline and the base height for the curtain wall.

Creating the Curtain Wall Definitions

Select the **Design Rules** tab to begin configuring the curtain wall. Begin by creating all the definitions required for the curtain wall. There are three divisions required, as noted in Figure 12.47. Select the **Divisions** icon from the Element Definitions list on the left side of the tab. Add a new division named Primary Vertical Grid. Set the orientation to vertical and the division type to Manual. Add two gridlines, one 12" from the grid start and the other 12" from the grid end.

Create the next division definition, named Vertical Bay Grid. Set the orientation to vertical and the division type to Fixed Cell Dimension, with a cell dimension of 11′-0″. Uncheck the Auto-Adjust Cells check box. Create the last division, named Horizontal Floor Grid. Set the orientation to horizontal using a Manual division type. Add four gridlines:

1. 16′-0″ from the grid bottom

2. 29′-0″ from the grid bottom

3. 42′-0″ from the grid bottom

4. 55′-0″ from the grid bottom

Next, create the four infill definitions. Select the **Infills** icon from the Element Definitions list. Add a new infill named Precast Window Panel. Set the Infill Type drop-down list to Style, then select the Precast Panel + 1 Window style from the Curtain Wall Unit Styles folder in the lower right corner of the Design Rules tab. Add another new infill named Concrete Arched Panel. Set the infill type to Style, and select the Concrete Arched Panel curtain wall unit style. Select the New Infill button, and name the new style Concrete Door Panel. Use the Concrete Door Panel style from the Window Assembly Styles folder. Create the last infill as a Simple Panel type. Name the new infill End Panels, and set the panel to be 8″ thick. This creates the four infill patterns that are used to make up the curtain wall style.

Finally, create the mullion definition. Select the **Mullions** icon from the Element Definitions list. Pick the New Mullion button. Name the new mullion Concrete Mullion, and set the width to 1″ and the depth to 3″. With all the definitions created, the curtain wall can be assembled.

 Assembling the Curtain Wall

Select the Primary Grid icon in the tree along the left of the Design Rules tab. Change the default division element to the Primary Vertical Grid definition to establish the first level of divisions for the curtain wall. Under the Cell Assignments area, select the *Nested Grid* text in the element column, and change the element type to the End Panels infill definition.

Select the New Cell Assignment button to add another cell to the Primary Vertical Grid division. Change the element of this new cell to *Nested Grid*, and rename the cell Vertical Bays. In the Used In column select the Start, End text to display the ellipsis button [...]. Select the button, and uncheck the Start and End check boxes, and place a check mark in the Middle check box to place the nested bay grid in the middle primary grid cell.

The Primary Vertical Grid is still being configured, so a frame is not required around the grid. Select the text in the Used In column for the default frame and turn off the Left, Right, Top, and Bottom frames. In the Mullion Assignments area set the default mullion to use the Concrete Mullions definition to complete the configuration for the primary grid. Figure 12.48 displays the settings for the first-level grid.

The next step is to configure the second grid, which is the first nested grid. Remember, this division is being placed in the middle cell of the Primary Vertical Grid division. Select the **Vertical Bays** icon under the Primary Grid icon in the tree along the left edge of the tab.

Change the default division to the Vertical Bay Grid definition. Add a new cell assignment to the Vertical Bay Grid and set the element to *Nested Grid*. Rename the New Nested Grid, Horizontal Floors. Currently the Vertical Bay Grid establishes 11′-0″ wide vertical bays. The Horizontal Floor nested grid needs to be told in which vertical bay cells it will be nested. In the Type column for the Horizontal Floors row, select the Location text. Change Location to Index. In the Used In column, enter: 1,2,3,4,5,6,7,8,9. This adds the horizontal floor nested grid to the first nine cells of the vertical bay division. Change the default mullion to the Concrete Mullion definition to complete the configuration of the secondary grid (Vertical Bay Grid) (see Figure 12.49).

The final step is to configure the horizontal floor cells. Select the **Horizontal Floors** icon from the tree list on the left. Change the default division to the Horizontal Floor Grid. Change the default infill to the Precast Window Panel definition. Add a new cell assignment. Change the new cell assignment to the Concrete Arched Panel definition. In the Used In column, set the panel to be placed in the bottom cell of the Horizontal Floor Grid. Finally, change the default mullions to the Concrete mullion definition to complete the curtain wall style, shown in Figure 12.50.

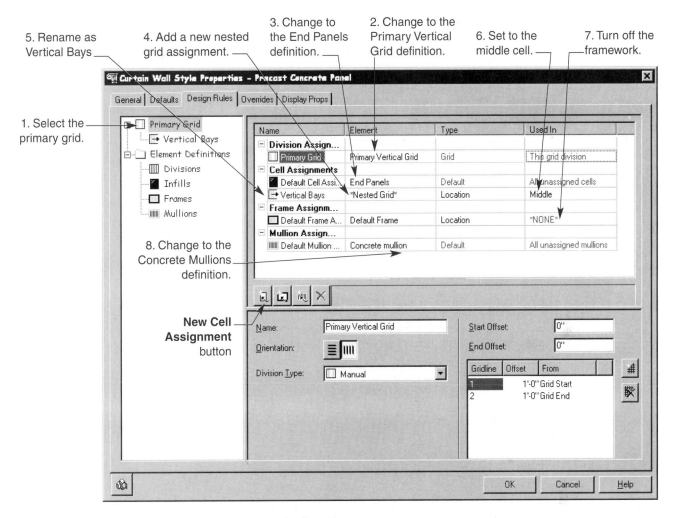

FIGURE 12.48   The primary grid settings have been configured.

Exercise 12.11

1. Open EX12-10.dwg.
2. Create a new curtain wall style named Precast Concrete Panel.
3. Follow the steps outlined in the preceding section to create the curtain wall style.
4. Add the new curtain wall into the drawing 57′-0″ long and 55′-0″ high.
5. Use the cell overrides to change two of the bottom Concrete Arched Panels to the Concrete Door Panel.
6. Save the drawing as EX12-11.dwg.

Additional Methods of Creating Curtain Wall Styles

As with the curtain wall units and the window assemblies, linework can be converted into a curtain wall style. When linework is converted to a curtain wall style then added to the drawing, the curtain wall is always drawn at the size established by the linework. There is not any flexibility in adjusting the size of the curtain wall. Use one of the following options to convert linework into a curtain wall style:

✓ Pick the **Design** pull-down menu, access the **Curtain Wall** cascade menu, then pick the **Convert Linework to Curtain Wall** command.

✓ Pick the **Convert Linework to Curtain Wall** button on the **Curtain Walls** toolbar.

1. Select the
Vertical Bays
icon.

3. Add a new nested
grid cell assignment.

2. Change the default
division to the Vertical
Bay Grid definition.

5. Change
to Index.

6. Enter the vertical bay cells to
which the nested grid will apply.
Cell numbers refer to the cells from
left to right, separated by commas.

4. Rename as
Horizontal Floors.

7. Set the default
mullions to the
Concrete mullion
definition.

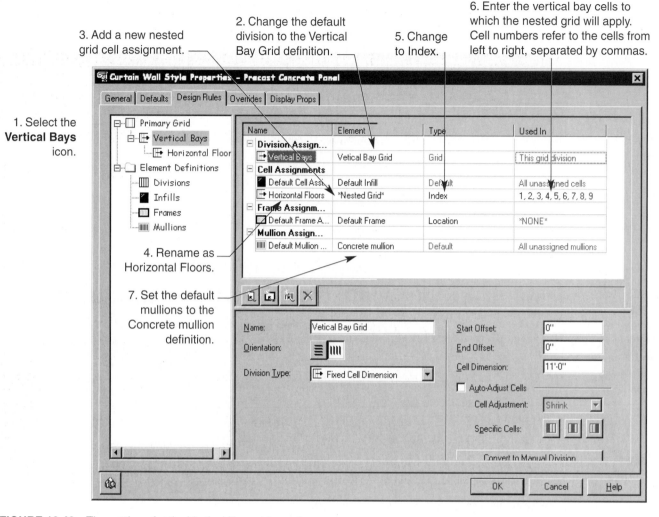

FIGURE 12.49   The settings for the Vertical Bay grid are shown.

✓ Right-click, and select **Design, Curtain Wall, Convert Linework to Curtain Wall** from the
    cursor menu.

✓ Type **AECCURTAINWALLCONVERT** at the Command: prompt.

   Follow the prompts to create the new curtain wall. After the curtain wall has been created, select the
wall, right-click, and select the Edit in place cascade menu, and pick Save Changes. Save the changes as a
new curtain wall style.

   Layout grids also can be converted into a curtain wall. First, create the layout grid using the Add
Layout Grid (2D)… command found in the Layout Tools cascade menu under the Desktop pull-down
menu. After drawing the layout grid, convert it into a curtain wall using one of the following options:

✓ Select the **Design** pull-down menu, pick the **Curtain Wall** cascade menu, then select the
    Convert Layout Grid to Curtain Wall command.

✓ Pick the **Convert Layout Grid to Curtain Wall** button on the **Curtain Walls** toolbar.

✓ Right-click, and select **Design, Curtain Wall, Convert Layout Grid to Curtain Wall** from
    the cursor menu.

✓ Type **AECCURTAINWALLCONVERTGRID** at the Command: prompt.

Follow the prompts to create the new curtain wall style.

4. Add a new cell assignment and set to the Concrete Arched Panel definition. ——

3. Change the default cell assignment to the Precast Window Panel definition.

2. Change the default division to the Horizontal Floor Grid. ——

5. Use the Concrete Arched panel in the bottom cell of the Horizontal Floor Grid.

1. Select the **Horizontal Floors** icon.

6. Change the default mullions to the Concrete mullion definition.

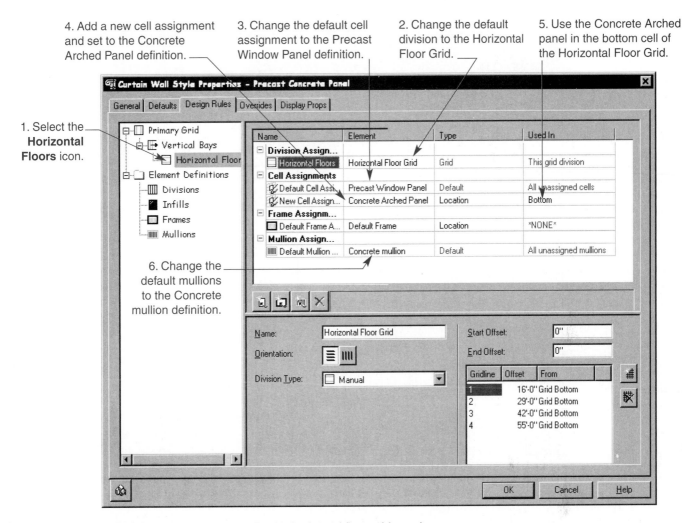

FIGURE 12.50   The settings for the horizontal floor grid are shown.

Architectural Desktop includes several additional curtain wall styles that can be used in other drawings. To use these styles, access the **Curtain Wall Styles** command. Use the Open Drawing button to browse to the \\Autodesk Architectural Desktop 3\Content\Imperial\Styles folder. Select the Curtain Wall & Curtain Wall Unit Styles (Imperial).dwg from the list of styles drawings to open the drawing in the Style Manager, as shown in Figure 12.51.

Select the Curtain Wall Styles section icon for the newly opened styles drawing. The styles are listed in the right pane. Drag and drop the styles from the right pane to the current drawing folder icon in the left pane. The additional styles are added into your current drawing for your use.

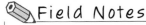 Field Notes

If you have installed the International Extensions, additional curtain wall styles can be found in the \\Autodesk Architectural Desktop 3\Content\Metric DACH\Styles folder.

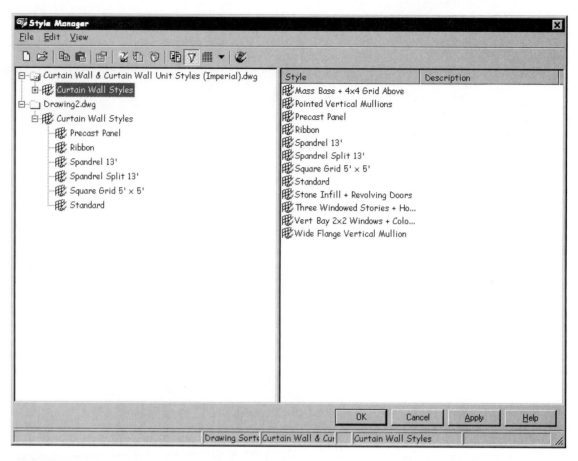

FIGURE 12.51   The Style Manager dialog box can be used to drag and drop styles from one drawing folder to another.

|||||||||||| CHAPTER REVIEW

 Use the CD-ROM to test your knowledge and skills.

Chapter Test

To check your understanding of the content provided in this chapter, access the Test file in the CH12 folder of the CD-ROM that accompanies this text.

Chapter Project

To practice the Architectural Desktop skills presented in this chapter, access the Project files in the CH12 folder of the CD-ROM that accompanies this text. The project files are in pdf format and include sample drawings and instructions for completing each project.

CHAPTER 13

Creating Styles

LEARNING GOALS

After completing this chapter, you will be able to:

◎   Use the Style Manager to manage and create styles.

◎   Import and export styles.

◎   Purge styles.

◎   Use styles over the Internet.

◎   Use standard wall, endcap, door, and window styles.

◎   Create custom wall, endcap, door, and window styles.

The AEC objects included with Architectural Desktop include many types of different styles that represent real-world objects. As you have seen so far, most of the styles available are ready to use in your construction documents. For specific types of styles, existing object styles can be modified, or new styles can be created. The **Style Manager** dialog box is used to organize all these styles in a drawing as well as to import and export styles to and from your drawing. The Style Manager can also be used to purge unused styles from the drawing, reducing the file size of the drawing.

Previous chapters introduced you to the Style Manager dialog box and showed how to create and access some object styles. This chapter builds on that knowledge and shows how the Style Manager can be used to manage all your styles. This chapter also explains how to create a custom wall and wall endcap style and how to create unique custom door and window styles that can be used in many of your future drawings.

Working with Style Manager

The Style Manager dialog box provides a central location for the management of styles within many drawings, templates, and the current drawing. This dialog box can be used to view, create, edit, and purge object styles in drawings as well as to send drawings to a consultant or to download object styles from the Internet.

The Style Manager can be accessed in a number of ways. Most commonly it is accessed from the AEC object's style command, such as the Stair Styles or Curtain Wall Styles commands. Accessing the Style Manager in this manner produces a condensed version of the dialog box. When the Style Manager is accessed from an object's style command, it filters out the other styles that are not being referenced; however, the same dialog box and command is used to manage all object styles in a drawing. The Style Manager can also be accessed to view all the object styles in a drawing.

In addition to being accessed from a particular AEC object's style command, the dialog box can be accessed from its own command. The Style Manager then appears displaying all the object style section icons. Use one of the following options to access the Style Manager in its entirety, as shown in Figure 13.1:

✓ Pick the **Desktop** pull-down menu, and pick the **Style Manager...** command.

✓ Pick the **Style Manager** button from the **AEC Setup** toolbar.

✓ Right-click, and pick **Desktop**, **Style Manager...** from the cursor menu.

✓ Type **AECSTYLEMANAGER** at the Command: prompt.

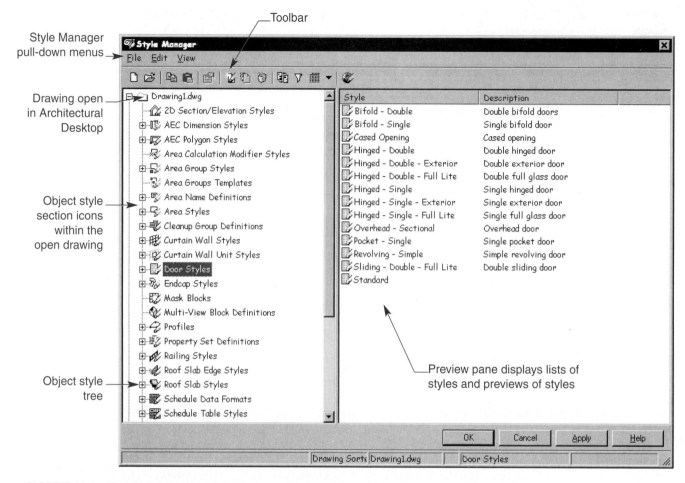

FIGURE 13.1　The Style Manager dialog box can display all the object styles in a drawing.

Across the top of the Style Manager is a menu bar of pull-down menus used to access different commands that can be executed in the dialog box. The toolbar buttons also are used to access different Style Manager commands. The Style Manager dialog box is divided into two resizable panes.

The left pane organizes object styles in any open drawings, displaying them as a series of icons and folders, branching off the drawing in which they are included. Thus, this pane is called the ***tree-view pane***. As object styles are created or added to the drawing, a + sign appears beside the style's icon, indicating that the section contains one or more object styles. Selecting the ⊞ sign expands the tree and displays a list of the styles within the style icon (see Figure 13.2).

The right pane displays different information depending on the icon selected in the left pane. If a drawing folder icon is selected, the right pane displays drawing property information such as who last saved the drawing, the name of the drawing, comments, and hyperlinks in the drawing. If a style icon is selected in the tree view, a list of object styles for that style is listed. If a style icon from the tree-view pane is selected, a preview window is displayed on the right, allowing you to preview the object style (see Figure 13.3).

In addition to accessing the commands in the toolbar or the pull-down menu area, right-clicking on different icons provides you with a list of available commands that can be selected to perform an action.

Using Toolbar Buttons

The following toolbar commands along the top of the dialog box perform a number of functions related to managing object styles in the Style Manager dialog box (see Figure 13.4).

- **New Drawing**　This button creates a new drawing folder in Style Manager to which styles can be added. When the Style Manager dialog box is closed, the new drawing is saved and can be accessed in future drawings and object styles pulled from it for use in the current drawing.

FIGURE 13.2 The left pane, also known as the tree view, lists the style icons in a particular drawing. A ⊞ sign indicates that style definitions exist for that object style.

Drawings open in Architectural Desktop

Drawing currently being managed

Object style icons

Press the ⊞ sign to expand the branch into a list of styles in the style icon

List of door styles in the drawing

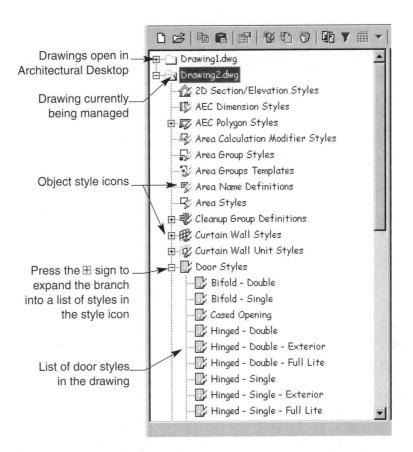

FIGURE 13.3 Selecting an icon from the tree view display different information in the right pane. An individual window style has been selected from the tree-view list. A preview of the style is displayed in the right pane.

The Window Styles icon has been expanded to display a list of object styles.

Select an individual style icon to preview it.

A preview of the selected window style.

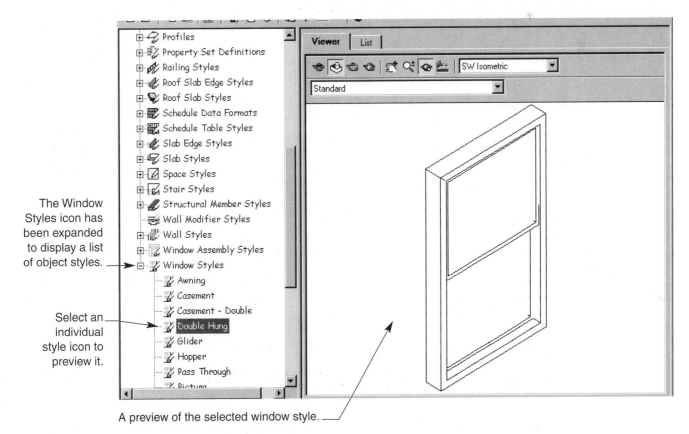

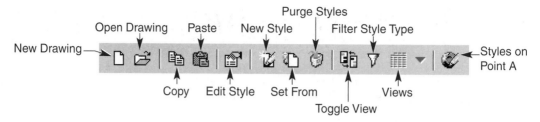

FIGURE 13.4   The toolbar buttons in the Style Manager dialog box are used to create, modify, and manage object styles.

- **Open Drawing**

This button is used to open other drawings from the hard drive or from network drives so that object styles can be accessed and dragged into the current drawing. Use this button to access additional drawings that include object styles in Architectural Desktop. These drawings can be found in the \\Autodesk ArchitecturalDesktop 3\Content\Imperial *or* Metric\Styles folders. If the International Extensions have been installed, additional styles can be found in the \Autodesk Architectural Desktop 3\Content\Metric DACH\Styles folder.

- **Copy**

This button is used to copy styles from one drawing folder in the tree view to another.

- **Paste**

This button adds object styles copied from one drawing folder to another.

- **Edit Style**

This button becomes available when an object style has been selected from either the right or the left panes in Style Manager. It is used to edit an existing style or a newly created style.

- **New Style**

This button is used to create a new object style. First, select an object style section icon from the tree view, then pick the New Style button to create the new style.

- **Set From**

This button is used for endcap styles, masking block definitions, profile definitions, and wall modifier styles. When one of these styles is high-lighted in the left or right panes, and this button is picked, the drawing window reappears, allowing you to select a polyline for use as one of these types of styles.

- **Purge Styles**

This button is used to purge the drawing of unused or unwanted object styles. Only styles that are not being used can be purged from the drawing.

- **Toggle View**

This button is used to toggle the left pane from viewing the drawing folder icons with the styles beneath, to viewing the object style icons with the drawing folders they belong to under each style icon.

- **Filter Style Type**

This button is used to look at styles of one type. For example, suppose you wanted to view only the wall styles available in a drawing folder icon and not all the styles in the folder. Select the Wall Styles icon, then pick the Filter Style Type button to display the wall styles in each of the drawing folder icons. When a particular object style is accessed from pull-down menus or from buttons on a toolbar, Style Manager automatically filters for that type of style.

- **Views**

This button contains two parts, an icon representing the type of icon view that is shown in the right pane, and a drop-down arrow with options. Use this button to toggle among small icons, large icons, list view, and the details view. The object style icons appear different depending on the view selected.

- **Styles on Point A**   This button is used to open a special view in the right pane. The view displays a section in the Point A Web site that can be used to download sample styles. The sample styles are constantly changing, so be sure to visit once in a while to obtain more object styles.

Using Views

As mentioned in the Toggle View button section, styles can be viewed under each drawing folder, or the drawing files to which the style belongs can be displayed below the style's icon. Depending on how you are viewing the object styles, the right pane displays something different when a style icon is selected from the left tree view.

If you are viewing object styles that are branched off a drawing folder, a list of individual object styles is displayed in the right pane when you pick a style icon from the left tree view (see Figure 13.5A). If you are viewing object styles in a drawing folders that is branched off a style icon, the right pane displays a table of styles in each drawing when you select the style icon, as shown in Figure 13.5B.

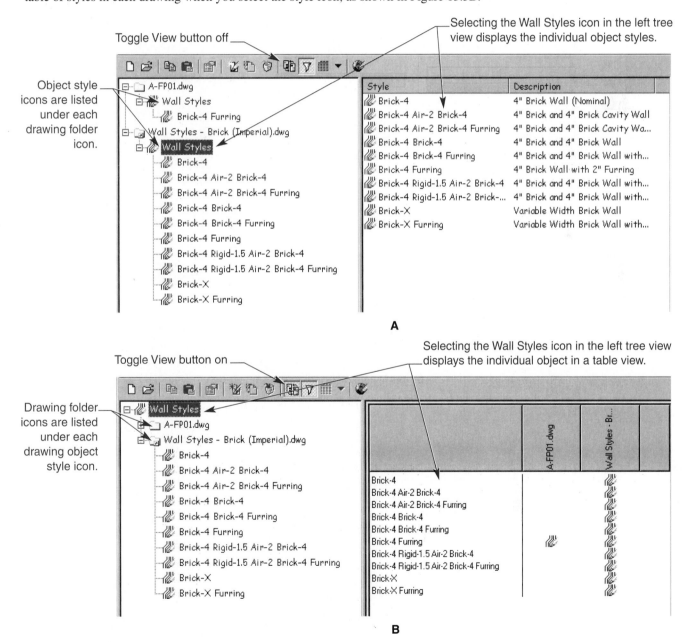

FIGURE 13.5   (A) Object styles for each drawing are listed under the drawing folder icons (Toggle View button turned off). (B) The Object Style icon lists the drawing folders open in Style Manager (Toggle View button turned on).

In either of the Toggle View modes, if an individual object style is selected from the left tree view, a preview window for that style is displayed in the right pane, as shown in Figure 13.6. The preview pane includes the **Viewer** and **List** tabs. The Viewer tab allows you to preview the style in a number of different types of views for different display configurations (see Figure 13.6A). The List tab provides a description of the style and includes a detailed list of the components and their size settings (see Figure 13.6B).

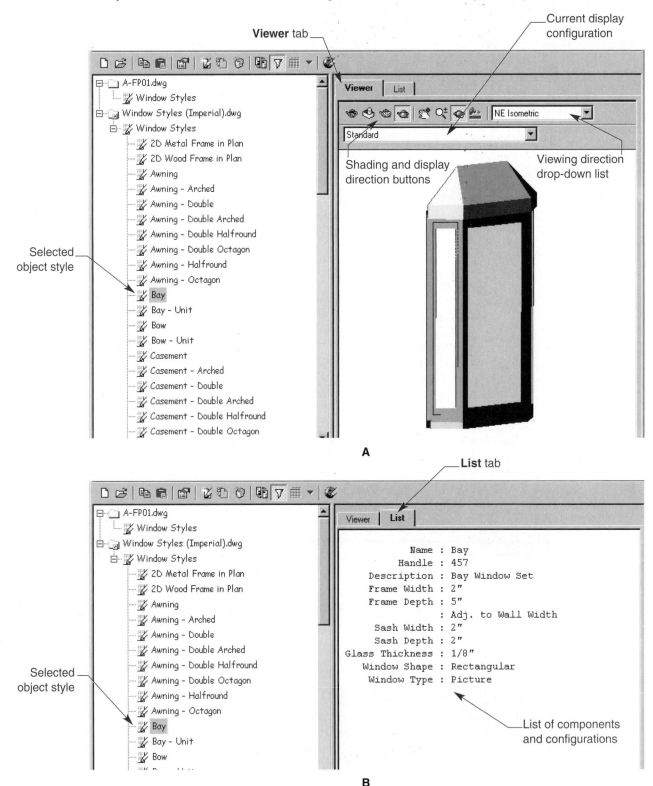

FIGURE 13.6   (A) The Viewer tab is used to preview the object type in a number of different viewing directions and display configurations. (B) The List tab provides a description of and configuration settings for the selected style.

FIGURE 13.7 (A) Drawings that are open in Architectural Desktop are represented by a standard drawing folder in the tree view. (B) Drawings that are open only in Style Manager are represented as a folder with a boxed arrow.

A   B

The drawing folder icons in the tree view also appear different. Drawings that are currently open in Architectural Desktop are displayed as a standard file folder, as shown in Figure 13.7A. As drawings are opened in Style Manager or as new drawings are created in Style Manager the folder icons are displayed with a boxed arrow (see Figure 13.7B), which indicates that the drawing is not open in Architectural Desktop but only within the Style Manager.

Drawings that are open only in the Style Manager dialog box can be closed when you have finished accessing their styles. Drawings that are open in Architectural Desktop cannot be closed from Style Manager. To close a drawing that is only in the Style Manager, first select the drawing folder with an arrow, then pick the File pull-down menu, and select Close. You can also close the folder by right-clicking over the folder icon and selecting Close.

Importing and Exporting Styles

Object styles can be imported and exported between drawings directly in the Style Manager dialog box. First, open any drawings that contain styles that you would like to use in another drawing. Use the preview pane to view the styles to ensure that you are picking the correct styles to be imported or exported.

To import object styles you must drag them from one drawing folder to another drawing folder icon or to another drawing's object style section icon. You can drag individual styles from the right-pane view to another folder or style section icon, or from the tree view to another folder (see Figure 13.8). Styles cannot be dragged from the Style Manager to the drawing screen.

Multiple styles can be selected in the right pane by holding down the [Ctrl] key when picking on the individual style icons. Many styles in a row can be selected by picking one style then holding the [Shift] key and selecting a style farther down the list. This method selects all the styles between the two picks. Press and hold the left mouse button to drag the styles to the new drawing, then let go the left mouse button to add the styles to the other drawing.

Another method of transferring styles is to cut and paste them. First, select the individual desired styles, then use the Copy button to copy the styles, then select the drawing folder to which the copied styles is to be transferred, and pick the Paste button.

Occasionally, as you copy styles between drawings you may encounter a situation in which a style already exists in the receiving drawing. In this case the Style Manager displays a warning that you have found a style with a duplicate name (see Figure 13.9). This dialog box contains the following three options for dealing with the style that is being copied over:

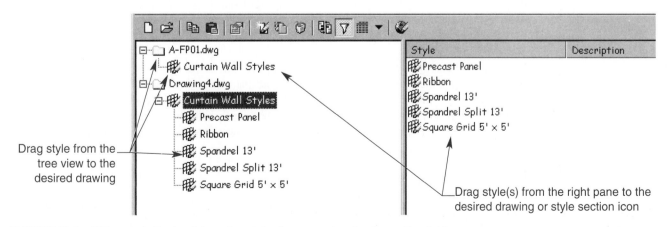

FIGURE 13.8 Different methods of dragging styles from one drawing to another folder.

FIGURE 13.9 The
Import/Export - Duplicate Names
Found dialog box is displayed
when you try to copy a style that
already exists in the receiving
drawing.

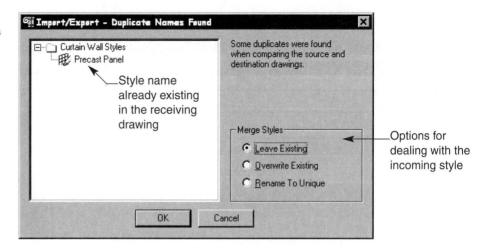

- **Leave Existing**   Use this radio button to leave the existing style in the receiving draw-
                          ing and discard the style trying to be transferred.

- **Overwrite Existing** Use this radio button to overwrite the existing style in the receiving
                          drawing with the style being transferred.

- **Rename to Unique**   Use this radio button to rename the incoming style with a number
                          after the style name. Typically if the style name already exists in the
                          receiving drawing, the newly imported drawing will have the number
                          (1) appended to its name. Additional importing of the same style will
                          continue to sequentially number the style names.

Purging Styles

The standard AEC Arch templates include many styles that can be used when creating drawings. Often, only a
few of these styles are actually used. If there are styles in the drawing that are not being used, they can be
purged from the drawing. Purging unused object styles from the drawing reduces the size of the drawing file.

Object styles can be purged from the drawing globally or individually. When you purge styles globally, all the unused styles in the drawing are purged. When you purge object styles individually, specific
unused object styles are removed.

To purge styles globally select the drawing folder icon containing the styles to be purged, then select
the Purge Styles button (see Figure 13.10). You can also right-click over the drawing folder icon and select
Purge from the cursor menu. The **Select Style Types** dialog box, shown in Figure 13.10, is displayed. The
list of items in this dialog box includes all the object style types available. A check in the box beside the
style type indicates that any unused styles of that type will be purged from the drawing. Press the OK button to purge all unused styles from the drawing globally.

To purge styles individually, select the style type icon from the list in Style Manager, then pick the
Purge Styles button, or right-click and select Purge. The **Select *(current object type)* Styles** dialog box is
displayed. Depending on the style type selected, the name of the style appears in title bar of this dialog box
(see Figure 13.11).

A list of individual styles for the selected style type is listed in the dialog box. A check mark beside
the style name indicates that the style will be purged, with the exception that if a style is currently being
used in the drawing, it will not be purged from the drawing, even with a check placed in the check box.
When you have finished purging object styles, exit the Style Manager dialog box and save the drawing.

Depending on the number of styles used and purged, the drawing file can be significantly reduced in size.

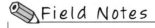 Field Notes

> It is usually good practice to purge all unused styles from the drawing at the end of a project to
> significantly reduce the file size.

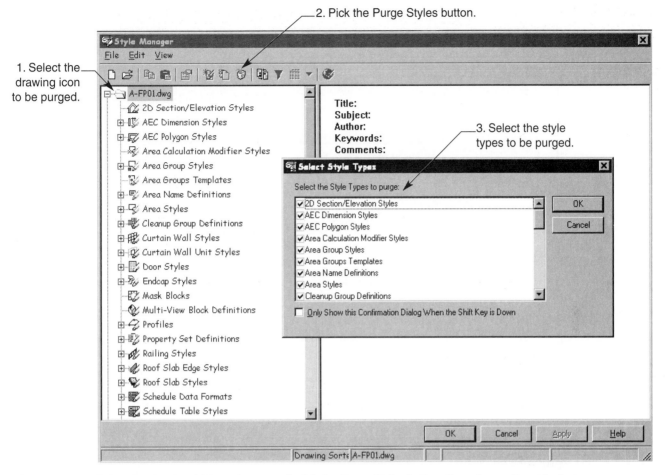

FIGURE 13.10   Select the drawing folder to be purged, pick the Purge Styles button, then choose the object style types to be purged.

Using Styles Over the Internet

In addition to importing object styles by dragging and dropping them from a drawing, you can download styles from the Internet. To do this, select the **Styles from Point A** button to access a special section of the **Point A** Web site (see Figure 13.12).

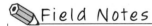 Field Notes

> At the time of this writing, object styles were accessible to everyone with a copy of Architectural Desktop 3. There are plans, however, to make this area accessible only to VIP members. To become a VIP member, contact your local Autodesk reseller, or visit the Autodesk Web site (www.autodesk.com) for more information.

Once the Style on Point A button has been selected, the right pane of the Style Manager displays available object styles for downloading. Press and hold the left mouse button over the desired style icon, drag to the left pane, and let go of the left mouse button to download the style from the Point A Web site and import it into the current drawing. The ability to drag a style from a Web site into the drawing is a new technology known as *iDrop*.

Styles can also be sent to another user via e-mail. First ensure that you have an e-mail account on your computer system, then select the desired object style to be sent. Pick the File pull-down menu in Style

2. Pick the Purge Styles button.

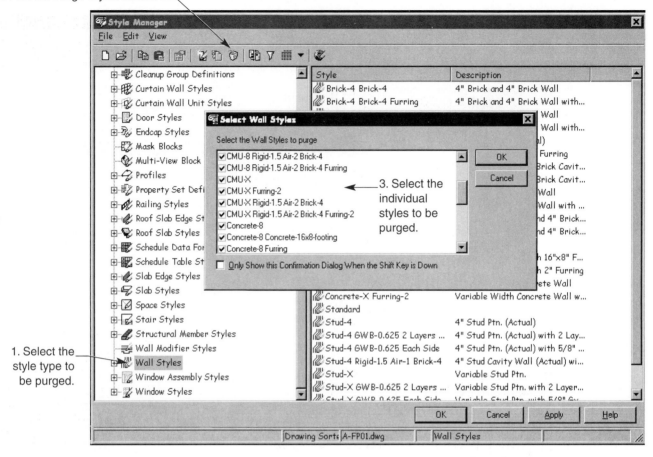

FIGURE 13.11   Highlight a style type icon, then press the Purge Styles button to display the Select *(current object type)* Styles dialog box. Select the styles in the list to be purged.

Manager, and select the <u>S</u>end... option. You can also right-click over the highlighted style, and select <u>S</u>end... from the cursor menu to attach the style within a drawing file to a new e-mail document. Address the e-mail and send the style to a coworker, client, or consultant.

Exercise 13.1

1. Start a new drawing using AEC Arch (Imperial) or AEC Arch (Imperial-Int).dwt.
2. Access the Style Manager dialog box.
3. Expand the different style section icons to list the different styles available.
4. Select different styles and preview them in the right pane.
5. Select the Door styles icon, then press the Purge Styles button. Purge all the door styles except Hinged - Single and Hinged - Double.
6. Pick the drawing folder icon, then purge all the unused object styles from the drawing except the Door style type.
7. Exit the Style Manager dialog box and save the drawing as EX13-1.dwg.

2. Pick the Styles on Point A button

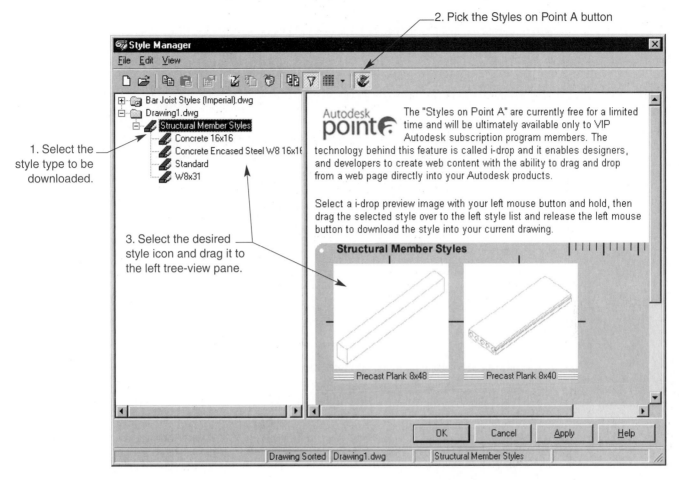

1. Select the style type to be downloaded.

3. Select the desired style icon and drag it to the left tree-view pane.

FIGURE 13.12 Select the Styles on Point A button to access downloadable object styles from the Internet.

Using and Creating Wall Styles

As mentioned in previous discussions, the AEC Arch (Imperial) and AEC Arch (Imperial-Int) templates include a general assortment of wall styles that can be used when creating a building project. Additional wall styles are included with Architectural Desktop. In addition to using wall styles provided by Architectural Desktop, you can create custom wall styles to reflect your design needs.

Importing Additional Wall Styles

Architectural Desktop includes more than 150 different wall styles. These wall styles are divided into five types of walls: Brick, Casework, Concrete, CMU, and Stud. The templates include some of these wall styles, and the remaining styles are found in the \\Content\Styles directory.

Wall styles can be used to model any type of horizontal object, such as casework. These types of walls represent cabinetry. They are considered a wall style because the wall style commands are used to create an object that looks like a cabinet with a countertop, as shown in Figure 13.13.

Additional wall styles can be imported into the current drawing through the Style Manager command or the Wall Styles command.

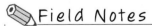

Field Notes

Keep in mind that if you access the Wall Styles command, the Style Manager is still being used, but the Filter Style Type button is on, which filters for wall styles.

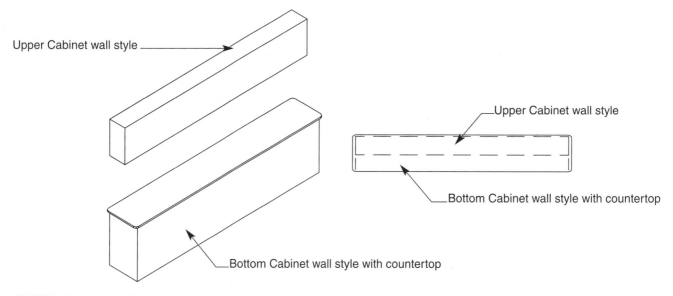

Upper Cabinet wall style

Upper Cabinet wall style

Bottom Cabinet wall style with countertop

Bottom Cabinet wall style with countertop

FIGURE 13.13   A wall style that has been configured to appear as a piece of casework.

Use one of the following options to import the additional wall styles:

✓ Select the **Design** pull-down menu, then pick the **Walls** cascade menu, and select the **Wall Styles...** command.

✓ Pick the Wall Styles button on the **Walls** toolbar.

✓ Right-click, and select **Design**, **Walls**, **Wall Styles...** from the cursor menu.

✓ Type **AECWALLSTYLE** at the Command: prompt.

The Style Manager dialog box is displayed, filtering for wall styles.

Use the Open Drawings button to open the drawings containing the additional wall styles. Browse to the \\Autodesk Architectural Desktop 3\Content\Imperial\Styles folder. There are five drawings that include additional wall styles, as shown in Figure 13.14. Pick one of the drawings to open the drawing file in Style Manager. Drag and drop the desired wall styles from the newly opened drawing file into your current drawing folder icon.

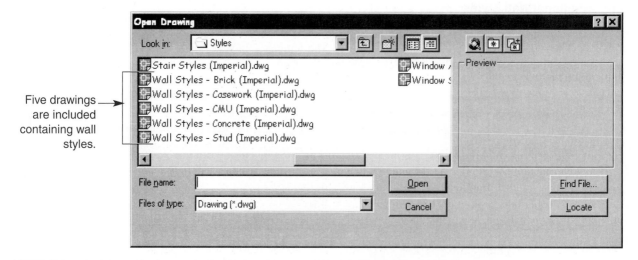

Five drawings are included containing wall styles.

FIGURE 13.14   Five drawings are included that contain additional wall styles.

Field Notes

- Additional metrically scaled wall styles can be found in the \\Autodesk Architectural Desktop 3\Content\Metric\Styles folder.
- As new styles are used, created, and downloaded you may want to store the wall styles used most often in a drawing that can be accessed quickly. Use the New Drawing button in the Style Manager to create a new blank drawing. Drag and drop the most commonly used wall styles into this new drawing. Thus, you do not have to access the different wall style drawings but, instead, only your custom wall style drawing that contains the specific styles you use on projects.

Exercise 13.2

1. Open EX10-6-Flr-Pln.dwg.
2. Access the Wall Styles command.
3. Use the Purge command to purge the unused wall styles from the drawing.
4. Press the Open Drawing button and browse to the \Autodesk Architectural Desktop 3\Content\Imperial\Styles folder. Select the Wall Styles - Casework (Imperial).dwg file and open it.
5. Drag and drop the following styles into the current drawing:
   - Casework - 36 (Base with Counter)
   - Casework - 36 (Isle)
   - Casework - 42 (Upper)
6. Close the Wall Styles - Casework (Imperial).dwg file.
7. Exit the Style Manager and apply the style changes to the drawing by pressing the OK button.
8. Save the drawing as EX13-2-Flr-Pln.dwg.

 Creating Custom Wall Styles

The preceding section explained how to import wall styles into the drawing from an external drawing. This drawing may be a previously created drawing or a wall style drawing found in the Styles folder. In addition to using preestablished wall styles, you can create custom wall styles for use in your drawings. Custom wall styles can be as complex or as simple as needed.

To create a custom wall style, access the Wall Styles command. The Style Manager dialog box is displayed, filtered to the wall styles section. Select the New Style button to create a new custom wall style. A new wall style is created in Style Manager. Enter a name for the new wall style, then select the new wall style, right-click, and select Edit.... Select the wall style, then pick the Edit Style button, or double-click on the new wall style. Any of these methods accesses the Wall Style Properties dialog box for the new wall style (see Figure 13.15).

The Wall Style Properties dialog box is divided into the following five tabs used to set the design parameters for the wall style:

- **General**   This tab is used to rename and to provide a description for the wall style to add design notes, documents, and spreadsheets to the style, and to attach property set definitions to the style (see Figure 13.15). Property set definitions are described later in this text.

- **Defaults**   This tab is used to set any defaults for the wall style (see Figure 13.16). When defaults have been set in this tab, they are applied to the Add Wall dialog box walls are added to the drawing. These settings establish only default values when the wall is added, and can be modified through the Wall Properties dialog box at any time. If a particular setting has been set as a default, it appears in the Add Wall dialog box with a blue color, which indi-

Tabs used to set wall style parameters

Name of the custom wall style

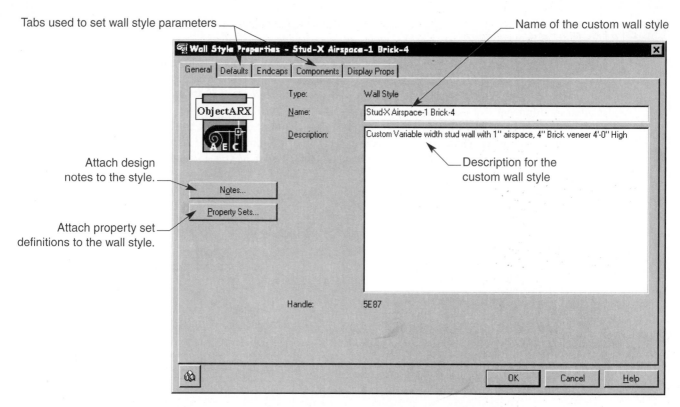

Attach design notes to the style.

Attach property set definitions to the wall style.

Description for the custom wall style

FIGURE 13.15   The Wall Style Properties dialog box is used to configure the parameters for the new custom wall style.

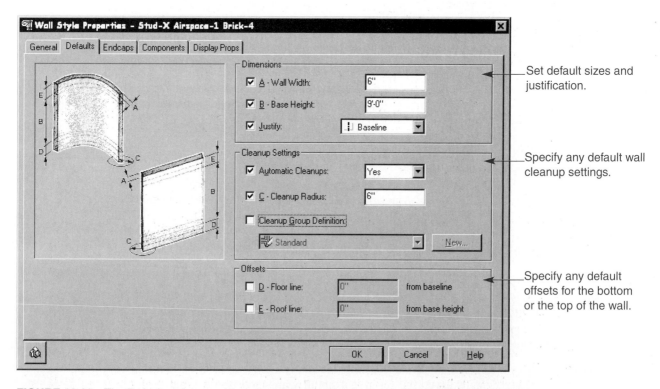

Set default sizes and justification.

Specify any default wall cleanup settings.

Specify any default offsets for the bottom or the top of the wall.

FIGURE 13.16   The Defaults tab is used to set up default values for the wall style. Whenever a wall is added to the drawing the default values appear in the Add Walls dialog box. These values can be overridden by entering a new value in the appropriate text boxes in the Add Wall dialog box.

cates that the value is a default value. If the default values are overridden in the Add Wall dialog box, they appear in a red color.

- **Endcaps**   This tab is used to place a custom shape at the end of a wall or at the edges of a wall when an opening in the wall is encountered (see Figure 13.17). The creation of endcap styles is discussed later in this chapter.

- **Components**   This tab is used to create the wall style by adding wall components to the style and setting parameters for the behavior of each component within a wall (see Figure 13.18). The Components tab is discussed in greater detail later in this section.

- **Display Props** This tab is used to configure the display properties of the wall style in plan, model, reflected, and any other display representations.

Building the Wall Style

When walls are added or modified, wall styles are used to portray how the wall appears in a top view (plan and reflected representation) and an elevation or isometric view (model representation). Depending on the wall style, one component (boundary) or multiple components are used to represent different parts of the wall. For example, whereas a simple stud wall can include only one wall component that represents the wall, a complex wall style such as a stud wall with brick veneer and an airspace uses three components to represent each part of the wall (see Figure 13.19).

Wall styles can use a total of 20 components. These components can be used to represent studs, brick, concrete, CMU block (concrete masonry unit), siding, gypsum board, wainscot, baseboards, trim boards, countertops, casework, and many other construction materials. A wall components can be created only as a rectangular shape but can be modified using a wall sweep to represent a curved shape after the wall has been added to the drawing. See Chapter 9 on how to use wall sweeps.

Configuring Wall Components

Once a wall style has been created and the Wall Style Properties dialog box has been entered, the wall style can be configured. Select the Components tab to begin adding the wall components (see Figure 13.18).

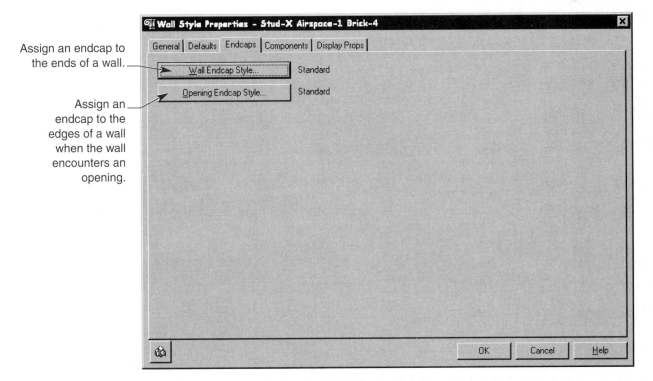

FIGURE 13.17   The Endcaps tab is used to assign an endcap to the edges of a wall or to the ends of a wall when a wall opening is encountered.

Select to add a new
wall component.

Select to remove a
wall component.

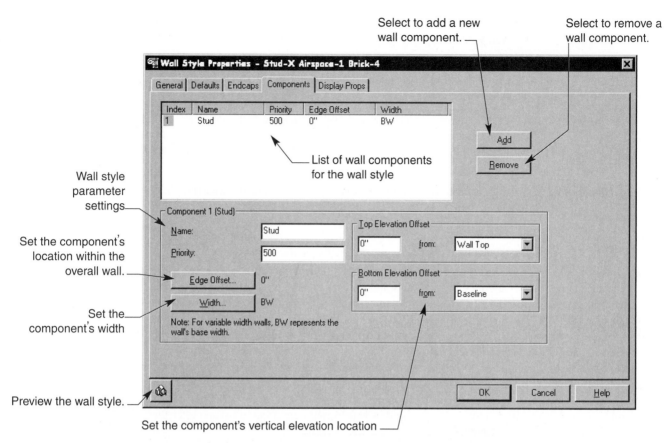

List of wall components
for the wall style

Wall style
parameter
settings

Set the component's
location within the
overall wall.

Set the
component's width

Preview the wall style.

Set the component's vertical elevation location

FIGURE 13.18   The Components tab is used to assemble the wall style and to set any rules on the behavior of the components in the drawing.

FIGURE 13.19   Wall styles are composed of wall components representing different construction materials.

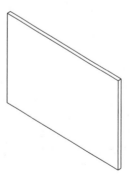

Simple wall using one component

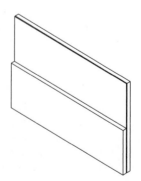

Complex wall using three components

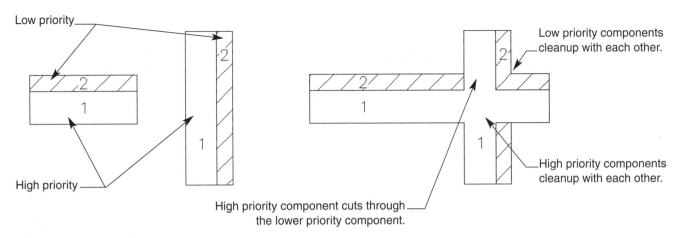

FIGURE 13.20 Priorities assigned to wall components establish how a wall will interact with other walls and components.

Initially, a new wall style includes one component with the name Unnamed. To rename the component, select the index number corresponding to the component in the list box, then enter a new name in the Name: text box.

Below the Name: text box, is the Priority: text box. Each wall component is assigned a priority. Wall components with the same priority cleanup with other walls' wall components. A higher priority wall component cuts through a wall component of a lower priority (see Figure 13.20). A value of 1 is the highest priority a wall component can be assigned. If a wall component is assigned a priority of 1, it can cleanup with only another wall component with a value of 1 and cuts through all other priority wall components.

Appendix G text (on the CD-ROM) provides a list of priorities that are assigned to different wall components within the default Architectural Desktop wall styles. For example, a stud wall component is assigned a priority of 500, and a brick veneer component is assigned a priority of 800, which indicates that a stud wall component can cut through a brick veneer component. The priorities established within the program vary, allowing you to use any priority value in between the Architectural Desktop priority values. For example, if you are creating a wall style that includes a stud and furring, you can assign the stud a priority of 500, and the furring can be assigned a value of 510.

Based on the type of wall component you are creating, assign a priority from the list in Appendix G. If a priority is not listed for the desired component, review the priority list and determine what the best priority would be. Keep in mind that higher priority components cut through lower priority wall components.

After determining the wall component name and its priority, you need to situate the wall within the wall. The Edge Offset... button is used to place the wall component in relation to the baseline. When walls are added to the drawing, one of the wall justifications is the Baseline option. This justification indicates that the wall is drawn from the baseline of the wall style. If you are looking at the wall from the right side, the wall components are added to the style by measuring a distance from the baseline to the left edge of the wall component (see Figure 13.21). A positive distance moves the component to the right, and a negative distance moves the wall component to the left.

Select the Edge Offset... button to specify a distance. The **Component Offset** dialog box is displayed, as shown in Figure 13.22. Enter the distance from the baseline to the left edge of the component in the Base Value: text box. For example, a value of 6″ places the component 6″ away from the baseline in the positive direction. A negative value places the component in the negative direction.

Below the Base Value: text box is an area designed for advanced placement of the wall component. Check on the Use Base Width: check box to configure this type of component placement. The Operator and the Operand settings become available. The Operator drop-down list contains four functions: Addition, Subtraction, Multiplication, and Division. For example, to create a wall component that is always centered on the baseline no matter what the width of the wall during the Add Wall process, use the Multiplication operator times an Operand value of −1/2. This setting indicates that when a wall is added to the drawing using a baseline justification, the component looks at the wall width specified (Base Width or BW) and multiplies the width by negative one-half, repositioning the component so the left edge is half the distance of the width from the baseline.

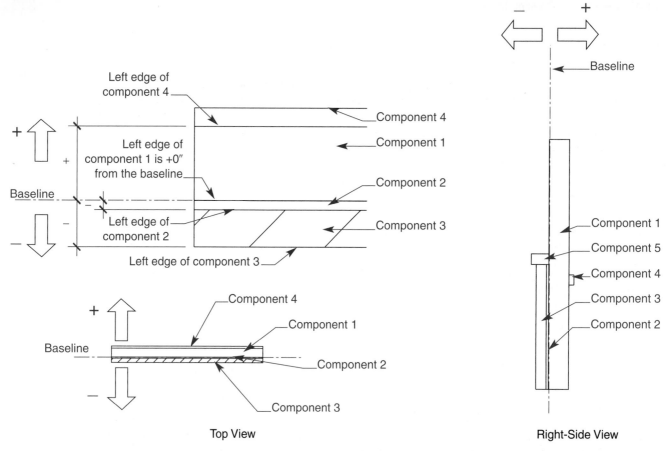

FIGURE 13.21 The placement of a wall component is determined by the distance from the baseline to the left edge of the component.

FIGURE 13.22 The Component Offset dialog box is used to place the wall component in relation to the baseline of the walls.

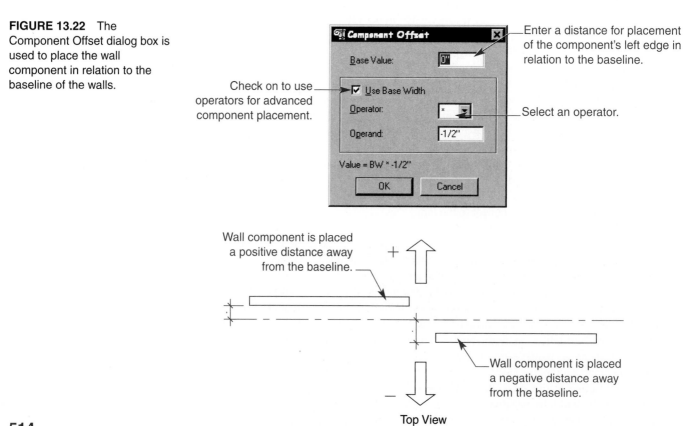

 Field Notes

In most cases wall components are placed at a specific distance from the baseline whether it is a positive or negative distance. Generally, the Use Base Width option is checked off for each of the components, unless the component is to vary in placement based on the width of the wall. This is a difficult process to master. Experiment with the operators. To make it easier to see how the components are placed in a wall style, use the floating viewer looking at the right view. You also may review an existing Architectural Desktop complex wall style that allows you to specify a wall width for ideas on how to use the operators and operands.

After determining where the wall component is to be placed in the wall style, select the <u>W</u>idth... button to determine the wall component width. The **Component Width** dialog box, shown in Figure 13.23, is displayed. Enter the width of the wall component in the <u>B</u>ase Value: text box. A value of 0″ allows you to specify the width for the wall when you use the Add Wall command. If a value other than 0″ is entered, the width is fixed to the component, and you do not have the option of changing the width of the wall when adding the wall to the drawing. Like the Component Offset dialog box, the Component Width dialog box includes an operator area. If a value greater than 0″ is entered for the base value, then uncheck the <u>U</u>se Base Width check box. If the wall is to be a variable-width wall, and the base value is set to 0″, then check on the <u>U</u>se Base Width, and set the operator to a + and the operand to 0″.

After setting the wall component edge offset and width, you need to configure vertical elevation offsets for the wall. The lower right side of the Components tab includes two areas: **<u>T</u>op Elevation Offset** and **<u>B</u>ottom Elevation Offset** (see Figure 13.18). Each of these areas sets the elevation for the wall component.

The <u>B</u>ottom Elevation Offset area sets the bottom of the wall component in relation to four vertical locations: Wall Bottom, Baseline, Base Height, and Wall Top. First set an offset distance, then select a vertical offset (from) location. For example, if the bottom of a wall component is to be located 3′-8″ off the floor, enter a value of 3′-8″ from the baseline. The four location options are described next:

- **Wall Bottom**   This location is used when the wall component has the ability to be projected downward when the Floor Line command is used on the wall. This location specifies the lowest elevation of a wall. See Chapter 9 on how to use the Floor Line command.

- **Baseline**   This location generally represents the floor plate location. The baseline of a wall is always measured to the current 0 Z-axis location. Use this option if the wall component cannot be projected down with the Floor Line command.

- **Base Height**   This location represents the height of the wall as specified in the Add Wall dialog box. The bottom of a wall can be specified in relation to the base height of the wall.

- **Wall Top**   This option is used if the wall component can be projected upward with the Roof Line Command. See Chapter 9 on how to use the Roof Line command.

FIGURE 13.23 The Component Width dialog box is used to set a width for the wall component.

Uncheck if hardwiring a width to the component. Check on if making the wall component a variable-width component.

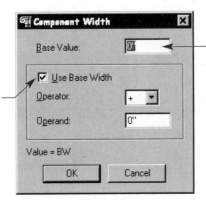

Enter a width for the wall component. If the wall style width is to vary, then set the value to 0″.

When placing the bottom of a wall component, you can place the component anywhere vertically in the wall by specifying a distance and a vertical offset location. The <u>T</u>op Elevation Offset area is used to adjust the top of the wall component. As with the <u>B</u>ottom Elevation Offset, enter a vertical distance in relation to a vertical offset location.

Once you have configured the vertical elevation offsets for the wall component, you can add and configure additional wall components by picking the A<u>d</u>d button. Figure 13.24 displays a wall style that includes a stud, airspace, brick veneer, brick veneer cap, and a wainscot (a piece of trim on the inside of the wall). The settings for the wall components are listed in the figure.

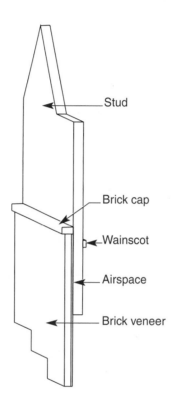

Index #	Name	Priority	Edge Offset	Width	Btm. Elev. Offset	Top Elev. Offset
1	Stud	500	BV=0″	BV=0″+0 (BW)	0″ from BL	0″ from WT
2	Airspace	700	BV=−1″	BV=1″	0″ from WB	4′-0″ from BL
3	Brick Veneer	800	BV=−5″	BV=4″	0″ from WB	4′-0″ from BL
4	Brick Cap	801	BV=−7″	BV=7″	4′-0″ from BL	4′-4″ from BL
5	Wainscot	1400	BV=0″+0 (BW)	BV=2″	3′-4″ from BL	3′-8″ from BL

BL: Baseline
BV: Base Value
BW: Base Width
WB: Wall Bottom
WT: Wall Top

FIGURE 13.24   A custom complex wall style can be created to represent any type of wall with up to 20 components.

After configuring the wall style with components, you may want to set any defaults for the wall style in the **Defaults** tab. You can also configure the display for the wall style in the Display Props tab. Set all display properties for the plan and model representations for the wall style by using the Wall Style property source.

Be creative when designing custom wall styles. The wall style can be used to create many types of walls, both detailed and schematic. Wall styles can be used to create walls with trim, veneer, casework, shelving, exterior corner quoins, custom railings, and anything you can dream up (see Figure 13.25). Remember that the components are rectangular and are represented horizontally. If you need shapes other than rectangular, then apply a wall sweep to the wall components.

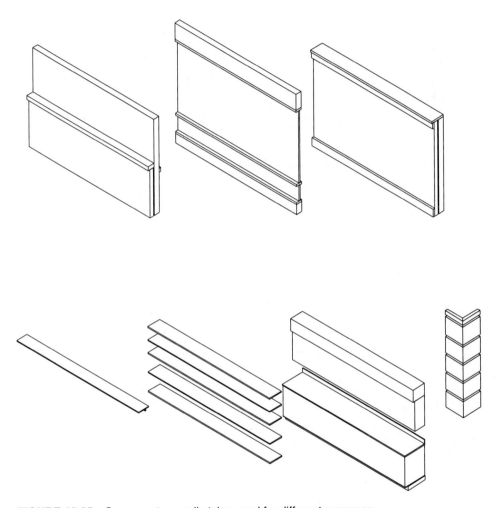

FIGURE 13.25   Some custom wall styles used for different purposes.

Exercise 13.3

1. Open EX13-2-Flr-Pln.dwg.
2. Access the Wall Styles command and edit the Casework-36 (Isle) wall style.
3. In the Defaults tab set the Casework - Top cleanup group definition current.
4. Add a component with the following properties:

Index #	Name	Priority	Edge Offset	Width	Bottom Elevation Offset	Top Elevation Offset
3	Case - Bar	2020	BV= 9″	BV= 18″	2′-6″ from BL	2′-7″ from BL

5. Modify the display properties for the model wall so the new component matches the color of the Case - Counter wall component.
6. Add a wall using the Casework-36 (Isle) wall style to the kitchen area. See the figure for placement.
7. Create a new cleanup group definition named Quoins.
8. Create a new wall style named Stucco Quoins, then edit the style.
9. In the Defaults tab, set the Quoins cleanup group definition current.
10. Create the following components and configure the settings as outlined in the table.

Index #	Name	Priority	Edge Offset	Width	Bottom Elevation Offset	Top Elevation Offset
1	Q1	1110	BV= 0″	BV=3″	0″ from WB	1′-4″ from WB
2	Q2	1110	BV= 0″	BV=3″	1′-6″ from WB	2′-10″ from WB
3	Q3	1110	BV= 0″	BV=3″	3′-0″ from WB	4′-4″ from WB
4	Q4	1110	BV= 0″	BV=3″	4′-6″ from WB	5′-10″ from WB
5	Q5	1110	BV= 0″	BV=3″	6′-0″ from WB	7′-6″ from WB
6	Q6	1110	BV= 0″	BV=3″	7′-8″ from WB	9′-2″ from WB
7	Q7	1110	BV= 0″	BV=3″	9′-4″ from WB	10′-10″ from WB
8	Q8	1110	BV= 0″	BV=3″	11′-0″ from WB	0″ from WT

11. Use the Stucco Quoins wall style to add 18″-long quoins 8′-0″ high at the front corners of the building. See the figure for placement.
12. Save the drawing as EX13-3-Flr-Pln.dwg.

Exercise continues on next page

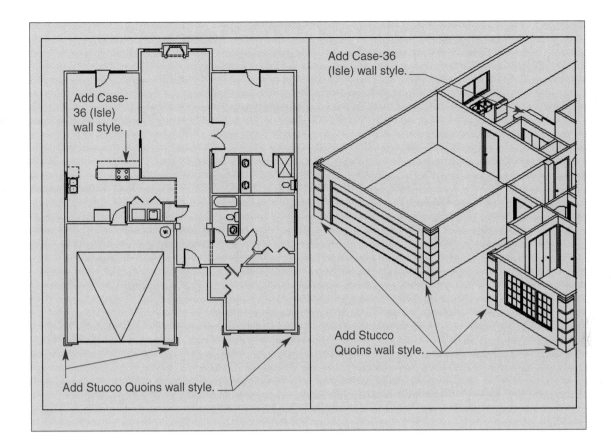

 ## Creating Wall Endcaps

Another useful tool in creating a wall or wall style is the use of endcaps in a wall. As defined earlier, endcaps are a graphic that is applied to the end of a wall or wall opening to provide a better representation of wall component construction (see Figure 13.26).

Drawing a polyline to represent the ending wall condition creates a wall endcap. A polyline should be drawn for each wall component within the wall style. The polylines are then assigned to the wall components through the **Wall Endcap Style** dialog box.

Creating Custom Endcap Styles

When a polyline is drawn for the wall components endcap, the starting and ending points of the polyline must line up in the same plane (see Figure 13.27). The polyline shape can vary depending on how the component should end.

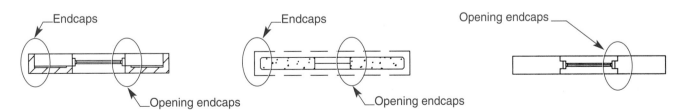

FIGURE 13.26   Some examples of endcaps applied to a wall style.

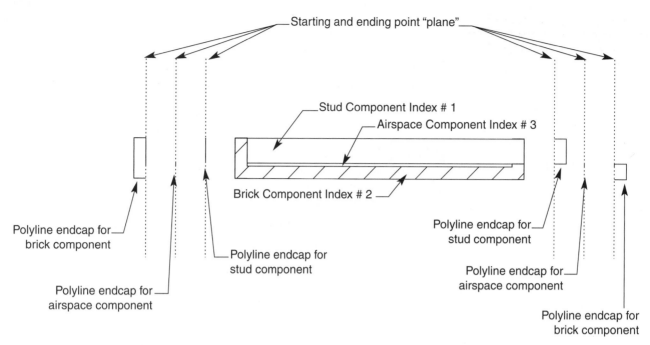

FIGURE 13.27   Draw a polyline for each wall component ensuring that the starting and ending points of the polyline line up in the same plane.

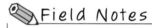Field Notes

The polylines for the endcaps do not have to be drawn separated as shown in Figure 13.27. It is often easier to draw the polyline over the end of the wall, where the endcap is to appear, to better visualize where the components will end in relation to the other wall component endcaps. Keep in mind that the starting and ending points of each of the polylines must line up in the same plane.

After drawing the polylines defining how the wall components will end, you need to group the polylines together in a wall endcap style. Use one of the following options to create an endcap for a wall style:

✓ Select the **Design** pull-down menu, pick the **Wall Tools** cascade menu, and select the **Wall Endcap Styles...** command.

✓ Pick the **Wall Endcap Styles** button from the **Wall Tools** toolbar.

✓ Right-click, and select **Design**, **Wall Tools**, **Wall Endcap Styles...** from the cursor menu.

✓ Type **AECWALLENDCAPSTYLE** at the Command: prompt.

The Style Manager dialog box is displayed, opened to the **Wall Endcap Style** section.

Select the New Style button to create a new style and give it a name. Defining an endcap style is similar to defining a profile. First, select the style name from the Style Manager dialog box, then pick the Set From button, or right-click and pick Select From... from the cursor menu. You are returned to the drawing screen, where you are prompted for a few settings. The following command sequence is used to create an endcap style for one of the ends displayed in Figure 13.27:

Command: **AECWALLENDCAPSTYLE** ↵
Select a polyline: *(select the polyline for the stud endcap)*
Enter component index for this segment <1>: *(enter a component to assign the polyline to or press [Enter])* ↵

FIGURE 13.28 Different offset values place the endcap at different points within the wall.

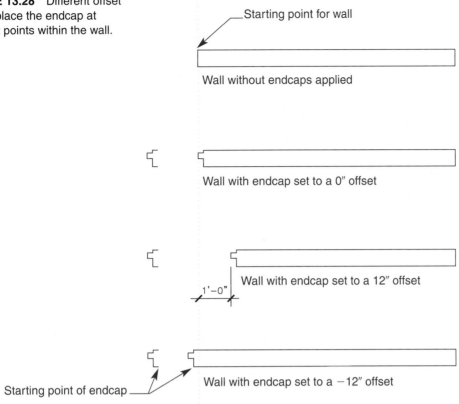

Starting point for wall

Wall without endcaps applied

Wall with endcap set to a 0″ offset

Wall with endcap set to a 12″ offset

1′–0″

Starting point of endcap

Wall with endcap set to a −12″ offset

Add another component? [Yes/No] <N>: **Y** ↵
Select a polyline: *(select the polyline for the brick endcap)*
Enter component index for this segment <2>:*(enter a component to assign the polyline to or press [Enter])* ↵
Add another component? [Yes/No] <N>: **Y** ↵
Select a polyline: *(select the polyline for the airspace endcap)*
Enter component index for this segment <3>:*(enter a component to assign the polyline to or press [Enter])* ↵
Add another component? [Yes/No] <N>: ↵
Enter return offset <0″>: ↵
Command:

During the process of defining the endcap, you must select each polyline and assign it to a wall component index number. These numbers correspond to the index numbers in the Wall Style Properties dialog box. The last prompt allows you to specify a return offset. Entering a positive value offsets the endcap into the wall the specified distance. Entering a value of 0″ places the outermost edge of the endcap at the ending point of the wall, and entering a negative value places the starting and ending points of the polyline endcap at the end of the wall (see Figure 13.28).

After the endcap style has been defined, the Style Manager is redisplayed. Press the OK button to save the style in the drawing. The endcap can now be assigned to a wall, wall style, or wall opening such as a door, window, or opening.

To assign the endcap to an individual wall, select the wall, right-click, and select Wall Properties... from the cursor menu. Pick the **Wall Style Overrides** tab and specify a starting and ending endcap for the wall (see Figure 13.29A). If the endcap is to be assigned to the wall style, access the **Wall Style Properties** dialog box, and assign the endcap in the **Endcaps** tab. This tab allows you to specify an endcap for the ends of the wall and at any openings in the wall (see Figure 13.29B). Doors, windows, and openings also can have endcaps assigned but only in their respective properties dialog boxes. Select the Endcaps tab and assign an endcap for each side of the opening (see Figure 13.29C).

FIGURE 13.29 Different
methods of assigning a wall
endcap.

Wall Style Overrides tab

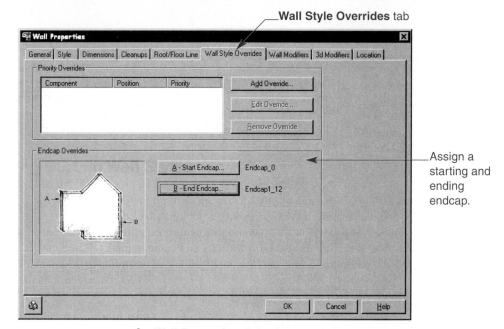

A   Wall Properties dialog box

Endcaps tab

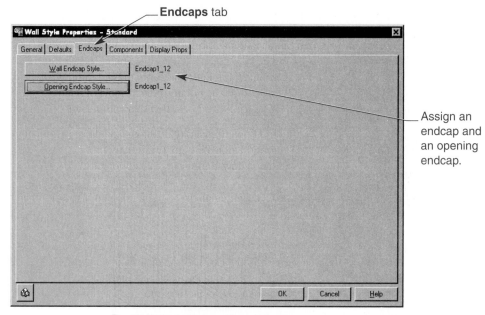

B   Wall Style Properties dialog box

Endcaps tab

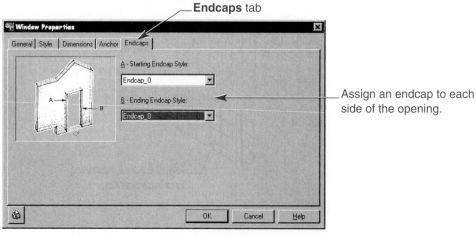

C   Window Properties dialog box

Field Notes

Endcap polylines respect the direction in which the wall is drawn. If you define an endcap, then assign it to a wall, and the endcap appears reversed or mirrored, try redrawing the polyline in the opposite direction. For example, if the polyline is drawn clockwise and it appears mirrored, try drawing the polyline in a counterclockwise direction and redefining the endcap style.

If a wall or wall style includes an endcap style, you can override the style without having to access the Wall Properties dialog box. Use one of the following options to override an endcap:

✓ Access the **Design** pull-down menu, select the **Wall Tools** cascade menu, then select the **Override Endcap Style** command.

✓ Pick the **Override Endcap Style** button in the **Wall Tools** toolbar.

✓ Right-click, and select **Design**, **Wall Tools**, **Override Endcap Style** from the cursor menu.

✓ Type **AECWALLAPPLYENDCAP** at the Command: prompt.

On entering this command, you are prompted to select a point. Select the side of the wall that includes the endcap to be overridden. After the point is selected, the **Select an Endcap Style** dialog box is displayed, allowing you to select the endcap to use at the end of the wall selected.

You can also insert a wall endcap into the drawing as a polyline for any type of modification by using one of the following options:

✓ Select the **Design** pull-down menu, access the **Wall Tools** cascade menu, then pick **Insert Endcap Style as Polyline**.

✓ Pick the Insert **Endcap Style as Polyline** button in the **Wall Tools** toolbar.

✓ Right-click, and select **Design**, **Wall Tools**, **Insert Endcap Style as Polyline** from the cascade menu.

✓ Type **AECWALLENDCAP** at the Command: prompt.

The **Endcap Styles** dialog box is displayed with a list of endcaps in the drawing. Select the endcap you would like to insert into the drawing for modification.

Exercise 13.4

1. Open EX13-3-Flr-Pln.dwg.
2. Use the Override Endcap Style command to override the end of the eating bar drawing in EX 13.3 that is crossing into the wall. Set the override to the Standard endcap style.
3. Add casework to the kitchen. See the figure for placement. Use the Casework-36 (Base with Counter). Lace the walls on the Casework - Top cleanup group.
4. Override the endcap at the refrigerator location to use the Standard endcap.
5. Draw a new endcap as shown in the figure. Create the new endcap with the name Casework-36 (Counter End1), and apply it to Index 2.
6. Use an override to apply the new endcap to the end of the casework.
7. Add upper cabinets using the Casework-42 (Upper) style. Set the upper cabinets to the Casework - Top cleanup group.
8. Save the drawing as EX13-4-Flr-Pln.dwg.

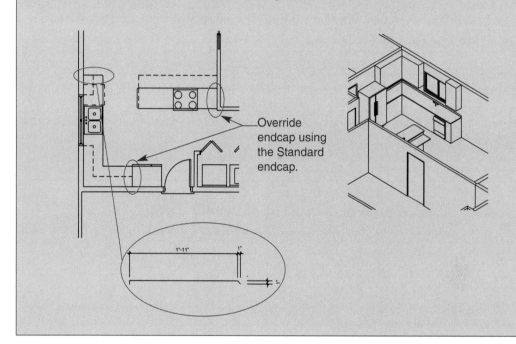

Override endcap using the Standard endcap.

Using Door Styles

As with the wall styles, Architectural Desktop includes a number of door styles and the ability to custom create door styles. Many of the door styles can be found in the AEC templates, and additional styles can be imported through the Style Manager dialog box. Custom doors can also be created using standard shapes and door types or form a custom AEC profile.

Importing Additional Door Styles

The AEC Arch (Imperial) and AEC Arch (Imperial-Int) templates contain many of the door styles needed to create a set of construction documents. Additional styles can be imported to the drawing through the use of the Style Manager dialog box. You must use the Style Manager to import or create door styles. Access the Style Manager dialog box using one of the following options:

✓ Select the **Design** pull-down menu, pick the **Doors** cascade menu, then select the **Door Styles...** command.

✓ Pick the **Door Styles** button in the **Doors - Windows - Openings** toolbar.

✓ Right-click, and select **Design**, **Doors**, **Door Styles...** from the cursor menu.

✓ Type **AECDOORSTYLE** at the Command: prompt.

The Style Manager dialog box is displayed, opened to the **Door Styles** section. To import door styles to your drawing, select the Open Drawing button. Browse to the \\Autodesk Architectural Desktop 3\Content\Imperial\Styles folder. Select the Door Styles (Imperial).dwg, then press the Open button to open the drawing file within the Style Manager. Browse through the drawing's door styles by selecting a door style in the left tree view. This allows you to preview the styles in the right pane.

Drag the desired styles from the drawing into your current drawing folder icon in the tree view to import the styles into your drawing for use. More than 40 different door styles available. The door styles range form simple door panel styles to door styles that have custom blocks and glazing applied. When you are finished importing door styles, select the Door Styles (Imperial).dwg to highlight it, right-click, and select Close from the cursor menu to close the drawing.

Additional door styles can be found and downloaded from the Internet by selecting the Styles on Point A button. Use the drag-and-drop procedure to download the styles from the Internet to the current drawing. Like wall styles, unused door styles take up space in the drawing. Use the Purge Styles button to purge any unwanted door styles from the drawing to reduce the size of the drawing.

Field Notes

> Additional door styles can be found in the \\Autodesk Architectural Desktop 3\Content\ Metric\Styles folder. If you have installed the International Extensions, look in the \\Autodesk Architectural Desktop 3\Content\Metric DACH\Styles folder for more door styles.

Exercise 13.5

1. Open EX13-4-Flr-Pln.dwg.
2. Access the Door Styles command.
3. Use the Open Drawing button to open Door Styles (Imperial).dwg.
4. Drag and drop the following styles into the drawing:
   - Bifold - Double - Louver
   - Hinged - Single - Exterior Panel
   - Overhead - 4 Window
5. Use the Door Modify command to change all the bifold doors to the newly imported style.
6. Change the front door to the Hinged - Single - Exterior Panel door style.
7. Change the garage door to the Overhead - 4 Window door style.
8. Make any desired entity display changes to the new doors.
9. Use the Purge command to purge unused door styles from the drawing.
10. Save the drawing as EX 13-5-Flr-Pln.dwg.

Creating Custom Door Styles

Custom door styles can also be created through the Style Manager dialog box. A custom door can take on any default shape such as a rectangle or arched top, or can use a custom AEC profile as its shape. In order for a custom door to be created, the definition must exist in the Style Manager. In Style Manager, select the New Style button to create a new door style. Enter a name for the door style, then highlight the name, and select the Edit Style button, or right-click and select Edit... from the cursor menu. The **Door Style Properties** dialog box, shown in Figure 13.30, is displayed.

The Door Style Properties dialog box includes the following five tabs for configuring the door style:

FIGURE 13.30 The Door Style Properties dialog box is used to create a custom door style.

Door style configuration tabs

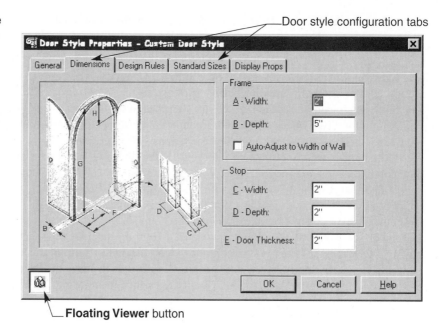

Floating Viewer button

- **General**　This tab is used to rename and to provide a description for the door style, to attach design or specification notes, and to assign property set definitions to the door style.

- **Dimensions**　This tab is divided into three separate areas (see Figure 13.30). The **Frame** area controls the size of the door frame. Enter a width and depth for the frame. Below the frame depth is a check box that adjusts the depth of the door to reflect the width of the wall where it is inserted. The **Stop** area is used to configure the size of the door stop. At the bottom of the tab is the setting that controls the thickness of the door panel.

- **Design Rules**　This tab is used to configure the shape of the door and the type of door that is being created (see Figure 13.31). The left side of the tab is used to specify the shape of the door. The **Predefined:** radio button controls the predefined shapes for the door style. The drop down-list beside the Predefined: radio button includes several different types of shapes. As you select shapes from the drop-down list use the floating viewer to preview what the door will look like. The **Custom:** radio button is tied to the drop-down list to its right. This list includes all the AEC profiles in the current drawing. Any AEC profile can be used to create a custom door shape. If a profile includes rings that are void (see Chapter 7) the voided areas become glass components in the door.

  After you have selected the shape, select the type of door to use from the **Door Type** area on the right. As with the shape drop-down lists, use the floating viewer to preview how the door will appear using the different shapes in combination with the door type.

- **Standard Sizes**　This tab is used to specify predefined sizes for the door style. These sizes are then available when the door is added into the drawing through the Add Doors dialog box (see Figure 13.32). Select the Add... button to display the Add Standard Sizes dialog box. Enter sizes that you need for this type of door.

- **Display Props**　This tab controls the display of doors. The display for the doors can be controlled at the system default level or at the door style level. Remember to control the plan, reflected, model, and elevation representations for the door. Custom blocks can also be attached for most of the representations. See Chapter 10 for adding custom blocks to the doors.

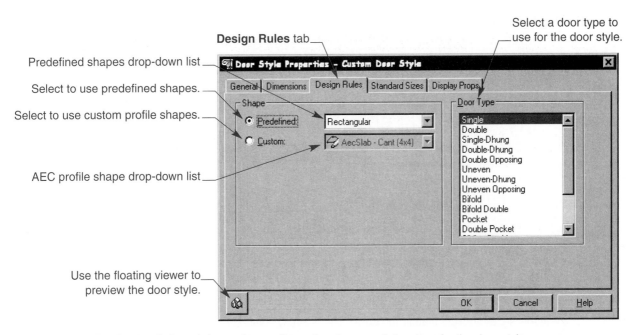

FIGURE 13.31 The Design Rules tab is used to configure the shape and door type for the door style.

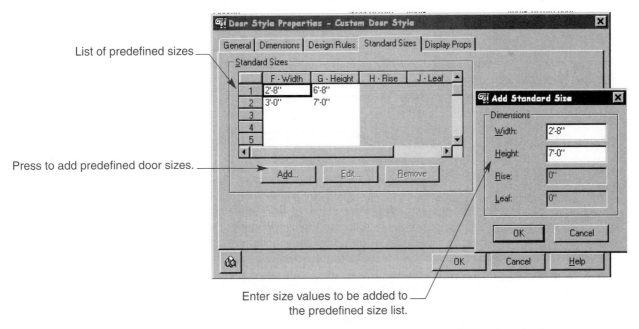

FIGURE 13.32 The Standard Sizes tab is used to define predefined sizes that are available when the door is added through the Add Doors dialog box.

✎ Field Notes

It is generally a good idea to control the display of objects at the style level, especially when customizing your own styles.

Exercise 13.6

1. Open EX13-5-Flr-Pln.dwg.
2. Access the Door Styles command.
3. Modify the Hinged - Single door style.
4. In the Dimensions tab, uncheck the Auto-Adjust to Width of Wall check box. Set the frame depth to 5″ and the stop depth to 1-1/2″.
5. In the Design Rules tab, select the <u>C</u>ustom: radio button. Select the Custom Door profile created in Chapter 7.
6. Use the floating viewer to preview the style.
7. In the Display Props tab change the model representation doors by style so that the glass component is the same color as the door panel.
8. Save the drawing as EX 13-6-Flr-Pln.dwg.

Using Window Styles

Window styles are very similar to door styles. Window styles can be imported from other drawings, or customized. Use one of the following options to access the **Window Styles** command to import or create a new window style:

✓ Select the **Des<u>i</u>gn** pull-down menu, pick the **W<u>i</u>ndows** cascade menu, then select the Window <u>S</u>tyles... command.

✓ Pick the **Window Styles** button in the **Doors - Windows - Openings** toolbar.

✓ Right-click, and select **Design, W<u>i</u>ndows, Window <u>S</u>tyles...** from the cursor menu.

✓ Type **AECWINDOWSTYLE** at the Command: prompt.

The Style Manager is displayed, where new styles can be created or imported.

Importing Additional Window Styles

To import window styles, select the Open Drawing button. Any existing drawing can be used to select available window styles. Architectural Desktop includes additional window styles found in the Window Styles (Imperial) drawing. Browse to the \\Autodesk Architectural Desktop 3\Content\Imperial\Styles folder and open the Window Styles (Imperial) drawing.

In Style Manager, browse through the list of styles until you find the desired window styles. Drag and drop the styles from the opened drawing in Style Manager to another drawing folder in Style Manager. The Window Styles (Imperial) drawing includes many window styles such as bow, bay, and shutter window styles.

When you have finished importing styles, select the drawing that was opened in Style Manager, right-click, and select the C<u>l</u>ose option. This closes the drawing and helps keep Style Manager from getting cluttered with open drawings. Like other styles, unused window styles take up room in the drawing. Use the Purge Styles button to purge any unwanted window styles from the drawing and reduce the size of the drawing file.

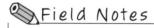

Additional window styles can be found in the \\Autodesk Architectural Desktop 3\Content\ Metric\Styles folder.

Exercise 13.7

1. Open EX13-6-Flr-Pln.dwg.
2. Use the Purge Styles command to purge any unused window styles from the drawing.
3. Use the Open Drawing button to open the Window Styles (Imperial) drawing. Import the following window styles:
   - Bow
   - Shutters - Dynamic
4. Close the Window Styles (Imperial) drawing when you have finished importing.
5. Modify the kitchen window to use the Bow window style.
6. Save the drawing as EX13-7-Flr-Pln.dwg.

 Creating Custom Window Styles

To create custom window styles, access the Window Styles command, then select the New Style button in Style Manager. Enter a new name for the style, then edit the style. The **Window Style Properties** dialog box, shown in Figure 13.33, is displayed. This dialog box is similar to the Door Style Properties dialog box, except some of the options are different. The Window Style Properties dialog box is divided into the following five tabs:

- **General** — This tab is used to rename and provide a description for the window style, to add design and specification notes, and to attach any property set definitions to the window style.

- **Dimensions** — This tab controls the sizes of the frame sash and glass (see Figure 13.33). Use the **Auto-Adjust to Width of Wall** option to make the depth of the window the same as the width of the wall.

- **Design Rules** — This tab is used to specify the shape and type of the window being created (see Figure 13.34). Unlike with door styles, the use of a custom profile

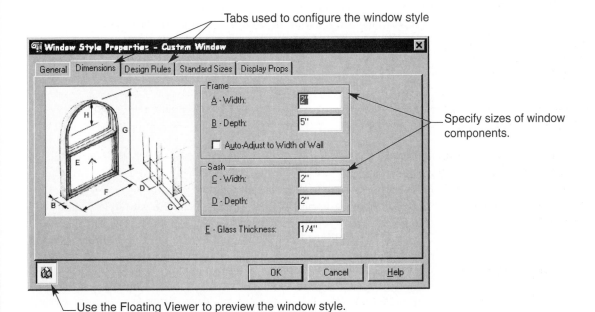

FIGURE 13.33 The Window Style Properties dialog box includes five tabs for configuring a window style.

Design Rules tab

Specify the type of window to use.

List of predefined window shapes

Select to use predefined shapes.

Select to use a custom profile shapes.

List of AEC profiles in the drawing

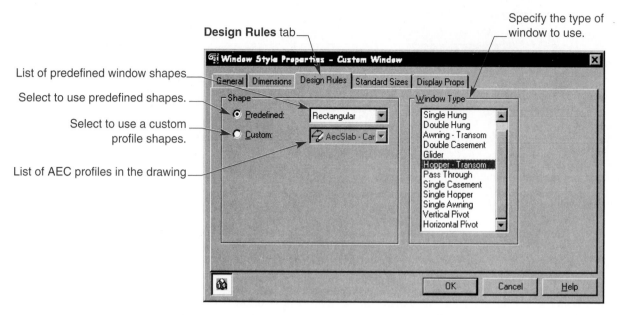

FIGURE 13.34   The Design Rules tab is used to specify a shape and type of window for the new window style.

shape with rings creates actual voids within the window. The edges of the inner rings become muntins within the window, and the void is an open hole in the window style. If you create a window style using a profile with voids, fill the voids with a block representing glazing panels, and attach the block through the Display Properties tab for the appropriate display representation.

- **Standard Sizes**   This tab is used to add any predefined window sizes for the window style. These sizes are displayed in the Add Windows dialog box when a window is added to the drawing.

- **Display Props**   This tab is used to control the display of the window in the different window representations. As with the door styles, you may need to configure the plan, model, reflected, sill plan, and elevation window representations.

When you have finished configuring the window style, the style is ready to be used.

Exercise 13.8

1. Open EX13-7-Flr-Pln.dwg.
2. Access the Window Styles command.
3. Use the Copy button to copy the Picture window style.
4. Paste the style back into the drawing using the Paste button. Name the new style Picture (2).
5. Edit the Picture (2) window style. Rename the style Picture - Arch Top. Change the description to Arched picture.
6. Select the Design Rules tab and change the shape to the Arch predefined shape.
7. Save the changes to the style by pressing the OK button. Exit of the Style Manager dialog box by pressing the OK button.
8. Modify the front middle window to use the Picture - Arch Top window style. Adjust the height to 5'-6" and the vertical alignment to 7'2".
9. Save the drawing as EX13-8.dwg.

CHAPTER REVIEW

Use the CD-ROM to test your knowledge and skills.

Chapter Test

To check your understanding of the content provided in this chapter, access the Test file in the CH13 folder of the CD-ROM that accompanies this text.

Chapter Project

To practice the Architectural Desktop skills presented in this chapter, access the Project files in the CH13 folder of the CD-ROM that accompanies this text. The project files are in pdf format and include sample drawings and instructions for completing each project.

CHAPTER 14

||||||||||||||||||||||||||||||||||||||||||||||||||||||||||||||||||||||||||||||||||

Using Slab Objects

LEARNING GOALS

After completing this chapter, you will be able to:

◎　Add slabs to a drawing.

◎　Modify slabs.

◎　Convert polylines and walls into slabs.

◎　Modify the edges of slabs.

◎　Use slab properties.

◎　Import slabs and slab edge styles.

◎　Create slabs and slab edge styles.

◎　Use slab tools.

◎　Extend slabs.

◎　Miter slabs.

◎　Cut slabs.

◎　Add and remove slab vertices.

◎　Add and remove a hole in a slab.

◎　Use Boolean operations.

So far this text has described how to create objects for use in a floor plan, such as walls doors, windows, and blocks. As you create your construction documents keep in mind that you are also creating a model of the building on the computer. One of the key elements in any real building before walls are added is a floor. Architectural Desktop includes an object known as a *slab* specifically for this purpose.

　　Slab objects do not have to be used solely as floors; they can be used to represent driveways and sidewalks on a site plan, or countertops, ceilings, and roofs (see Figure 14.1). Architectural Desktop also uses the slab technology in commands designed specifically for the design of roofs. Roofs designs are discussed in a later chapter.

　　A *slab* is an element that appears to have a thickness, similar to an AutoCAD solid, with an edge that can represent any shape. Slabs can be added by selecting vertex points, much like picking points for a polyline, or they can be converted from closed polylines or walls that form a closed room. These objects can also be carved, stretched, extended, and trimmed to form any desired shape. In addition to being able to be modifed to any shape, slab objects also can have holes cut into them.

Adding Slabs to a Drawing

Slabs can be added to the drawing with the **Add Slabs** command. Use one of the following options to access this command:

FIGURE 14.1 Slab objects are used to create floors, sidewalks, ramps, countertops, and ceilings.

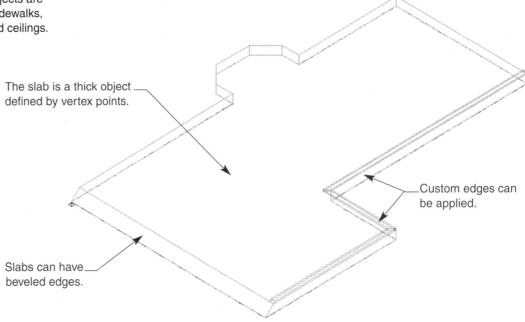

The slab is a thick object defined by vertex points.

Custom edges can be applied.

Slabs can have beveled edges.

✓ Select the **Design** pull-down menu, pick the **Slabs** cascade menu, and select the **Add Slabs…** command.

✓ Pick the **Add Slab** button from the **Slabs** toolbar.

✓ Right-click, and select **Design, Slabs, Add Slabs…** from the cursor menu.

✓ Type **AECSLABADD** at the Command: prompt.

Enter the thickness for the slab.

Enter a height for the slab if using the projected mode.

Select a slab style.

Choose the slab mode.

Enter an overhang.

Control the slab justification.

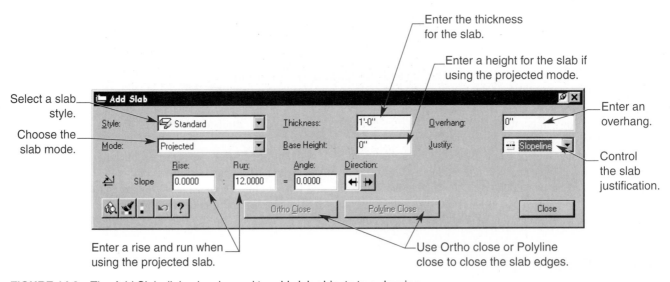

Enter a rise and run when using the projected slab.

Use Ortho close or Polyline close to close the slab edges.

FIGURE 14.2 The Add Slab dialog box is used to add slab objects to a drawing.

The Add Slab dialog box, shown in Figure 14.2, is displayed. The following settings are available in the Add Slab dialog box:

- **Style:**   Use this drop-down list to specify a slab type for use.

- **Thickness:**   Enter a thickness for the slab. If the slab is representing a 4″-thick concrete walk, then enter 4″ in this text box.

- **Overhang:**   The slab edges are defined as vertex points are selected in the drawing. If you are creating a slab that needs to overhang past the edge of the vertex points, then enter a value in this text box.

- **Mode:**   This drop-down list includes two options: **Direct** and **Projected**. The Direct mode creates a flat slab and grays out the Base Height and the Rise and Run text boxes. The Projected mode creates a slab that can be set at an angle. Selecting this mode allows you to set the Base Height (plate height) and the rise and run.

- **Base Height:**   This text box is available only if the Projected mode has been selected. Enter a value to place the slab up in elevation along the Z axis.

- **Justify:**   This drop-down list includes four justifications: **Top, Center, Bottom**, and **Slopeline.** The justification used establishes how the slab will be laid out when vertex points are selected. For example, if the slab is to be drawn from the top edge down, then select the Top justification. The Slopeline justification is used in the Projected mode. The *slopeline* is determined by the first and second vertex points and is the edge that determines the edge of the slab that projects up.

- **Slope**   This area is available when a projected slab is being created. Specify a **Rise**, **Run**, and **Angle** for the slope of the slab. The **Direction** buttons determine the direction the projected slab is created in relation to the slopeline.

- **Ortho Close**   Similar to the Ortho Close button for the Add Walls dialog box, this button is available after two vertex points have been selected and closes the slab with two edges based on a direction.

- **Polyline Close** This button closes the slab between the first and last vertex points picked. This option becomes available after three vertex locations have been selected.

After configuring the settings for the slab, pick a point in the drawing to establish the first vertex point. This point establishes the pivot point. When the slope of a projected slab is changed, the slope adjusts from the pivot point. Continue picking vertex points to establish the slab. When you have finished selecting points pick the Ortho Close or the Polyline Close buttons to close the slab. Slabs must be closed in order to be created.

Exercise 14.1

1. Open EX9-4-Fnd-Pln.dwg
2. Add a Standard style slab 4″ thick using the projected mode.
3. Set the rise to 1/2″ and the run to 12″. Set the direction to left.
4. Set the base height to 18″.
5. In the top view, add a slab to the garage area, picking the starting point at the lower left inside corner of the walls. Pick the additional vertex points in a counterclockwise direction. This will establish the slopeline at the front of the garage with the slab sloping from the rear of the garage down to the front at 1/8″ per foot.
6. Add a driveway and walks to the front of the building, as in the figure. Use the 4″ thick projected slab with a rise of 0″ and a base height of 18″.
7. Save the drawing as EX14-1-Fnd-Pln.dwg.

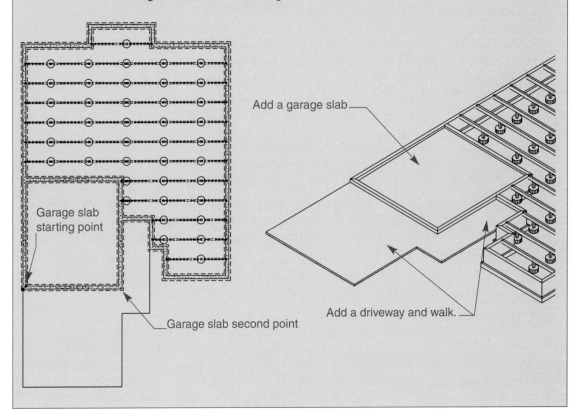

Add a garage slab

Garage slab starting point

Add a driveway and walk.

Garage slab second point

Modifying Slabs

Like other objects, slabs can be modified after they have been added to the drawing. Use one of the following options to modify an existing slab:

✓ Select the **Design** pull-down menu, access the **Slabs** cascade menu, and pick the **Modify Slabs…** command.

✓ Pick the **Modify Slabs** button from the **Slabs** toolbar.

✓ Right-click, and select **Design, Slabs, Modify Slabs…** from the cursor menu.

✓ Type **AECSLABMODIFY** at the Command: prompt.

✓ Alternatively, select the slab, right-click, and select **Slab Modify…** from the shortcut menu.

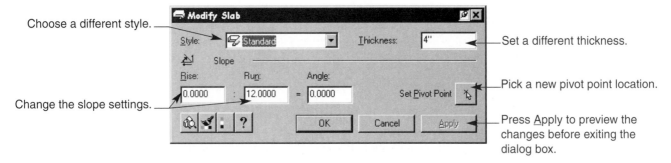

FIGURE 14.3 The Modify Slabs dialog box is used to make adjustments to slab objects.

When this command is entered and a slab has been selected, the **Modify Slab** dialog box appears, as shown in Figure 14.3. Make any adjustments to the style, thickness, or slope. In addition to making such changes, you can relocate the pivot point anywhere within the drawing by selecting the **Set Pivot Point** button. Selecting a new point establishes that point as the pivoting point for the slope of the slab.

As with other Modify dialog boxes, select the Apply button to apply the changes before exiting. This allows you to determine if your changes appear correctly or if you need to make further refinements.

Converting Polylines and Walls to Slabs

Another method of creating a slab is to convert closed polylines or walls that form a closed room into a slab. It is very important when polylines or walls are converted into slabs that the objects are closed in order for the slab to be created correctly. Use one of the following options to access the **Convert to Slabs** command:

✓ Select the **Design** pull-down menu, pick the **Slabs** cascade menu, then select the **Convert to Slabs...** command.

✓ Pick the **Convert to Slabs** button in the **Slabs** toolbar.

✓ Right-click, and select **Design, Slabs, Convert to Slabs...** from the cursor menu.

✓ Type **AECSLABCONVERT** at the Command: prompt.

The following prompt sequences are used to convert a closed polyline into a slab:

Converting to a Projected slab:

Command: **AECSLABCONVERT** ↵
Select walls or polylines: *(select closed polyline)*
1 found
Select walls or polylines: ↵
Erase layout geometry? [Yes/No] <N>: ↵
Creation mode [Direct/Projected]<Projected>: ↵
Specify base height<8'-0">: *(enter a base height for the projected slab or press [Enter] for the default height)* 3'0" ↵
Specify slab justification [Top/Center/Bottom/Slopeline]<Bottom>: *(enter the justification for the new slab)* T ↵
(select the style and thickness for the slab in the Slab Properties dialog box)

Converting to a Direct slab:

Command: **AECSLABCONVERT** ↵
Select walls or polylines: *(select closed polyline)*
1 found
Select walls or polylines: ↵

Erase layout geometry? [Yes/No] <N>: ⏎
Creation mode [Direct/Projected]<Projected>: D ⏎
Specify slab justification [Top/Center/Bottom/Slopeline]<Bottom>: *(enter the justification for the new slab)* C ⏎
(select the style and thickness for the slab in the Slab Properties dialog box)
Command:

When a polyline is converted to a slab, the command allows you to choose between a **direct** slab or a **projected** slab. When you choose a projected slab, you can specify the base height measured from the current 0 Z axis to the justification location. If you choose the direct slab, the slab justification point is placed in the current 0 Z axis.

When walls are converted into a slab, the projected option is automatically used, and the slab justification points are placed at the top of the walls. The following prompt sequence is used when converting a series of walls that form a closed room into a slab:

Command: **AECSLABCONVERT** ⏎
Select walls or polylines: *(select walls that form a closed room)*
Specify opposite corner: 8 found
Select walls or polylines: ⏎
Erase layout geometry? [Yes/No] <N>: ⏎
Specify slab justification [Top/Center/Bottom/Slopeline]<Bottom>: ⏎
Specify wall justification for edge alignment
[Left/Center/Right/Baseline]<Baseline>: *(specify where the edges of the slab will be placed in relation to the wall's justification)* ⏎
Specify slope Direction [Left/Right]<Left>: ⏎

Field Notes

Be careful when converting walls into a slab. The walls cannot go past the edge of another wall or stop short of another wall without causing an undesired slab. If the walls extend past the enclosed area or fall short of the enclosed area, then each wall is converted into a slab rather than having the slab placed within the closed room. When converting walls into a slab, it is often best to convert the exterior shell of the building into a slab.

Exercise 14.2

1. Open EX13-8-Flr-Pln.dwg.
2. Draw a polyline around the exterior walls of the living area of the residence. Do not include the garage area (see the figure).
3. Convert the polyline into a slab using the Standard style.
4. Use the Direct slab option with a justification of Top. Set the slab thickness to 2″.
5. Save the drawing as EX14-2-Flr-Pln.dwg.

Draw a polyline around the "living" areas of the house.

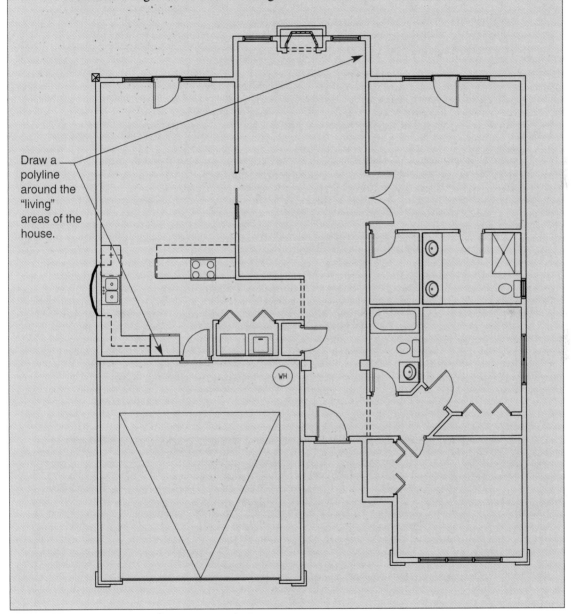

Modifying the Edges of Slabs

Depending on the style of slab selected, a custom edge may be added to the slab edges (see Figure14.4). The AEC Arch (Imperial) and the AEC Arch (Imperial-Int) templates include three slab styles: **Standard, Curb,** and **Cant**. The Standard slab creates a slab without custom edges. The Curb and Cant include custom edges in their style definition.

Slab edges can be assigned to all edges of the slab or to individual edges of a slab. Use one of the following options to access the **Edit Slab Edges** command:

✓ Select the **Design** pull-down menu, pick the **Slabs** cascade menu, and select the **Edit Slab Edges...** command.

✓ Pick the **Edit Slab Edges** button in the **Slabs** toolbar.

✓ Right-click, and select **Design, Slabs, Edit Slab Edges...** from the cursor menu.

✓ Type **AECSLABEDGEEDIT** at the Command: prompt.

✓ Select the slab, then right-click, and pick **Edit Edges...** from the shortcut menu.

You are then prompted to select edges. Select the edges of the slab to be modified. When you have finished selecting edges, press the [Enter] key, and the **Edit Slab Edges** dialog box is displayed (see Figure

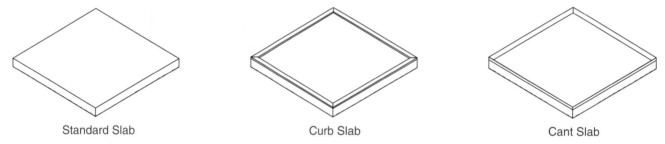

Standard Slab   Curb Slab   Cant Slab

FIGURE 14.4   Custom edges can be assigned to a style or individual edges of a slab. The AEC templates include three types of slab styles with different edge styles.

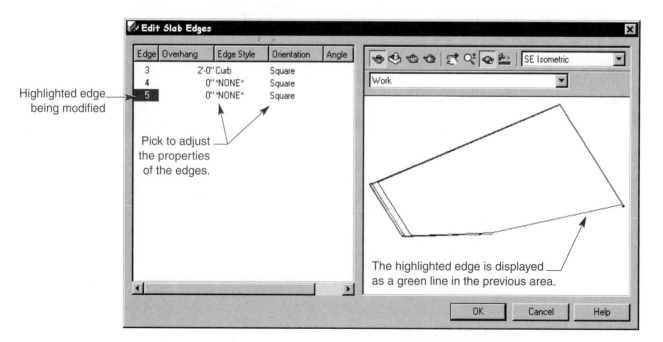

FIGURE 14.5   The Edit Slab Edges dialog box allows you to modify the edges of a slab.

14.5). This dialog box allows you to modify the edge(s) of a slab. The edge number is displayed along the left edge of the dialog box. Select any edge number to modify that edge. When an edge is highlighted in the table it is displayed as a green edge in the preview window, showing you which edge is currently being modified.

You can edit multiple edges by holding down the [Ctrl] key as you pick edges in the table. To modify the highlighted edges, pick any of the existing values in the table to change the setting. The following values can be changed:

- **Overhang**   Pick the existing value to reveal a text box. Enter the new value for the overhang, as shown in Figure 14.6A.

- **Edge Style**   Selecting the existing value displays a drop-down list with available edge styles in the drawing. Select a different edge style for the highlighted edge (see Figure 14.6B).

- **Orientation** When the existing value is selected, a drop-down list including two options becomes available. The **Plumb** option always makes the edge of the slab perpendicular to the ground. The **Square** option squares the end of the slab between the top and bottom edge of the slab, as shown in Figure 14.6C.

- **Angle**   Pick in this area to enter an angular value for the slab edge (see Figure 14.6D).

FIGURE 14.6 The different edge settings control how the edge of the slab will appear.

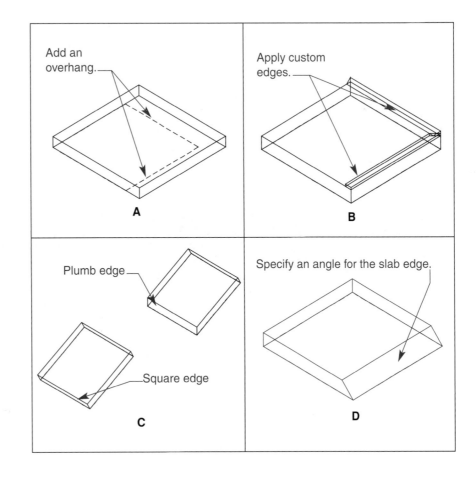

Exercise 14.3

1. Start a new drawing using the AEC Arch (Imperial) or the AEC Arch (Imperial–Int) template.
2. Create five 8'-0" x 8'-0" rectangles.
3. Use the Convert option to convert the polylines into a projected slab using the Standard slab style.
4. Use the Modify Slabs command to modify two of the slabs to have a 6" rise per 12" run.
5. Use the Edit Slab Edges command to modify the edges of each slab to create slabs similar to those in Figure 14.6.
6. Save the drawing as EX14-3.dwg.

Using Slab Properties

Another method of modifying a slab is by using the **Slab Properties** command. This command allows you to modify many of the slab settings in one dialog box. The Slab Properties command can be accessed by selecting the slab, right-clicking, and picking Slab Properties... from the shortcut menu or by typing **SLABPROPERTIES** at the Command: prompt.

The **Slab Properties** dialog box is displayed. This dialog box is divided into the following five tabs:

- **General**   This tab is used to give a description to the selected slab(s), to attach design notes, and to assign property set definitions to the selected slab(s).

- **Style**   This tab is used to change the style of the selected slab (see Figure 14.7). If a different style is selected, the slab changes to that new style; however, the properties of the new style, such as the edge style and thickness, are not applied. If the slab is being changed to a different style and you would like the properties of the new style to be applied to the selected slab, then press the **Reset to Style Defaults** button.

FIGURE 14.7 The Style tab is used to change to a different slab style for the selected slab.

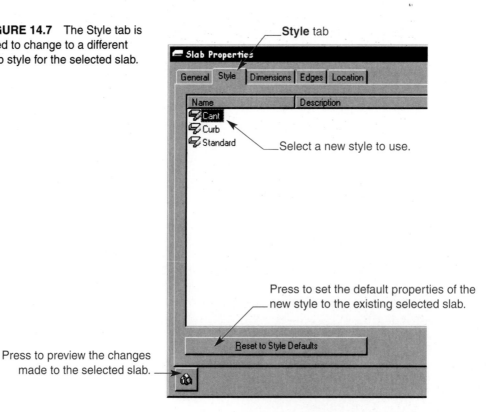

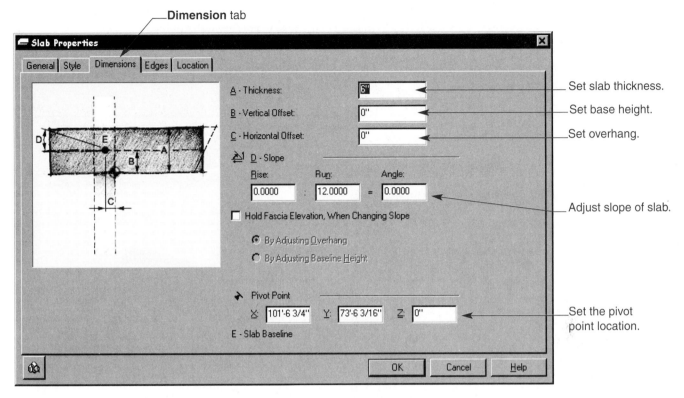

FIGURE 14.8 The Dimensions tab is used to control dimensional settings on the selected slab.

- **Dimensions** This tab is used to set the dimensions for the selected slab. The **Vertical Offset** controls the distance from the 0 Z axis to the justification of the slab. The **Horizontal Offset** controls the overhang of the slab. The slope of the slab and the pivot point location also can be controlled in this tab (see Figure 14.8).

- **Edges** This tab is the same tab used when editing slab edges (see Figure 14.5). Use this tab to adjust the edges of the selected slab.

- **Location** This tab indicates the current location of the slab in the drawing. If the slab is to placed up in elevation, readjust the Z insertion point for the World coordinate system.

Slab and Slab Edge Styles

In addition to the three slab styles and slab edge styles included in the templates, additional slabs and slab edge styles can be customized or imported from external drawings.

Importing Slab and Slab Edge Styles

Architectural Desktop includes additional styles in an external drawing that are not included in the templates. Use one of the following options to access the Slab Styles command to import these styles:

✓ Select the **Design** pull-down menu, pick the **Slabs** cascade menu, then select the **Slab Styles...** command.

✓ Pick the **Slab Styles** button in the **Slabs** toolbar.

✓ Right-click, and select **Design, Slabs, Slab Styles...** from the cursor menu.

✓ Type **AECSLABSTYLE** at the Command: prompt.

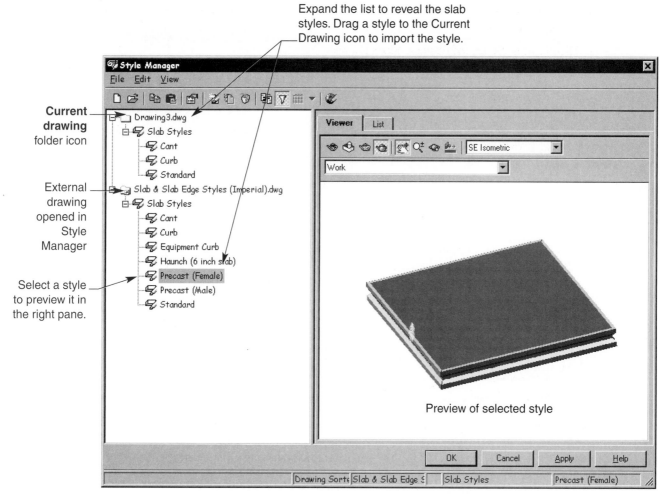

FIGURE 14.9   The Style Manager dialog box is used to import, export, and organize styles.

The Style Manager dialog box is displayed, opened to the slabs section.

Use the Open Drawing button to browse to the \\Autodesk Architectural Desktop 3\Content\Imperial\ Styles folder. This folder contains a list of object styles that can be imported into the drawing. Select the Slab and Slab Edge Styles (Imperial) drawing and press the Open button. The Slab and Slab Edge Styles (Imperial) drawing is opened as a folder icon on the left side of the Style Manager. Expand the drawing folder to reveal the additional slab styles.

Selecting a style from the left tree view allows you to preview the slab style (see Figure 14.9). If the **Slab Styles** section icon is selected, the slab styles are listed in the right side of the window. Drag the desired styles from the right or left panes into the current drawing folder icon to import them. Slab styles that include an edge automatically import the assigned slab edge style into the drawing.

✎ Field Notes

> Additional slab styles can be found in the \\Autodesk Architectural Desktop 3\Content\Metric\ Styles folder. If the International Extensions have been installed, additional slab styles can be found in the \\Autodesk Architectural Desktop 3\Content\Metric DACH\Styles folder.

If you need to import only the slab edge styles, then use one of the following options to access the **Slab Edge Styles** command:

✓ Pick the **Design** pull-down menu, select the **S̲labs** cascade menu, then pick the **Slab Edg̲e Styles...** command.

✓ Pick the **Slab Edge Styles** button in the **Slabs** toolbar.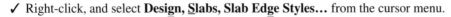

✓ Right-click, and select **Desi̲gn, S̲labs, Slab Edge Styles...** from the cursor menu.

✓ Type **AECSLABEDGESTYLE** at the Command: prompt.

To import the slab edge styles, open the Slab and Slab Edge Styles (Imperial) drawing. Drag and drop any desired edge styles into the current drawing folder icon.

Exercise 14.4

1. Open EX14-3.dwg.
2. Access the Slab Styles command.
3. Use the Open Drawing button to browse to the \\Autodesk Architectural Desktop 3\ Content\Imperial\Styles folder. Open the Slab and Slab Edge Styles (Imperial) drawing.
4. Drag and drop the following styles into the drawing:
   - Equipment Curb
   - Haunch (6 inch slab)
   - Precast (Female)
   - Precast (Male)
5. Copy the four flat slabs to the side.
6. Use the Slab Properties dialog box to change the copied slabs to the new styles.
7. Save the drawing as EX14-4.dwg.

 Creating a Slab Style

The Slab Styles command previously discussed can also be used to create custom slab styles. Once the Style Manager dialog box has been opened, select the New Style button to create a new slab style. Enter a name for the new slab style. After naming the style, highlight the style, and select the Edit button, right-click and select E̲dit..., or double-click on top of the style. The Slab Styles dialog box is opened. This dialog box includes the following four tabs used to configure the style:

- **General**   This tab is used to rename and provide a description of the slab style, to attach documents, spreadsheets, and HTML documents and to assign property set definitions to the style.

- **Defaults**   This tab is used to set up defaults that appear in the Add Slabs dialog box when a slab is drawn (see Figure 14.10). These settings specify only the default values and can be overridden at the time the slab is being created or modified. As in the Dimensions tab found in the Slab Properties dialog box, the thickness, vertical offset (base height), horizontal offset (overhang), and slope can be configured for the style.

  At the bottom of the tab is the **Perimeter Edges** check box. Select this check box to add an edge style to the slab style. Browse through the drop-down list of edge styles and select the desired one. If a slab edge does not exist, or if you want to custom create a new one, select the **N̲ew Edge Style** button to create a new slab edge style.

- **Design Rules** This tab is used to control the design rules for the slab (see Figure 14.11). If the slab has a specific thickness, then select the **Has F̲ixed Thickness** check box. The **T̲hickness** text box becomes available to set a slab thickness to the style. The **Thickness Offset** text box controls where the bottom of the slab is located in relation to the vertical offset. A negative value

Defaults tab

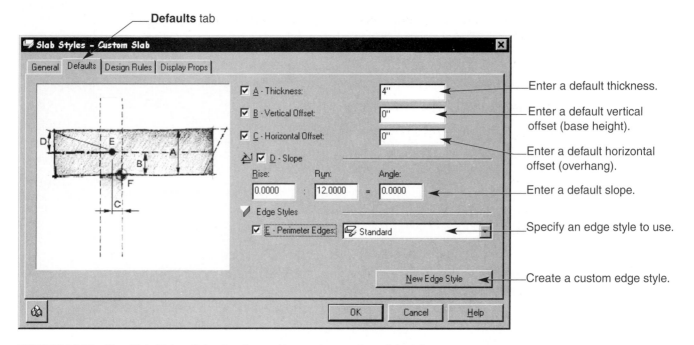

Enter a default thickness.

Enter a default vertical offset (base height).

Enter a default horizontal offset (overhang).

Enter a default slope.

Specify an edge style to use.

Create a custom edge style.

FIGURE 14.10   The Slab Styles dialog box is used to create a custom slab style.

Design Rules tab

Select to hardwire a thickness to the slab.

Enter a hardwired slab thickness.

Specify a thickness offset.

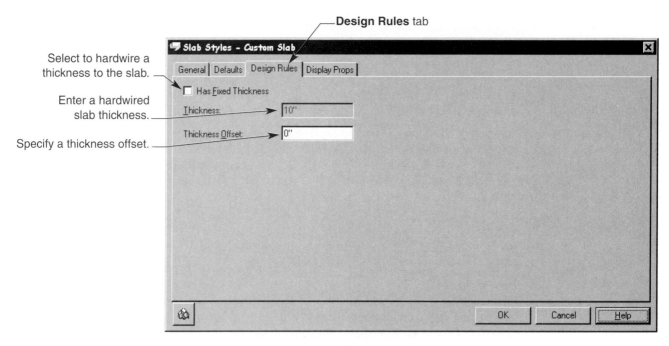

FIGURE 14.11   The Design Rules tab allows you to specify a hardwired slab thickness and offset.

places the bottom of the slab below the justification vertex points. A positive value places the bottom of the slab above the justification vertex points.

■ **Display Props**   This tab is used to control how the slab appears in plan and model representations. When modifying the display of plan representation slab components, you can specify a cut plane to cut through the slab. The cut plane can display the cut line at the perimeter edges or in the body of the slab. In addition to the cut plane, a hatch component can be turned on and configured for the slab. Both the plan and model representations include the pivot point component. You can decide to turn this component off or set it to a layer that does not plot. The component will plot if turned on.

Creating a Slab Edge Style

Often, when a custom slab style is created, an edge is associated with it. The Defaults tab in the Slab Styles dialog box allows you to create a new slab edge style. Select the <u>N</u>ew Edge Style button to display the **Slab Edge Styles** dialog box, or enter the Slab Edge Styles command, create a new style, then edit it to access this dialog box. The Slab Edge Styles dialog box includes the following three tabs:

- **General**   This tab is used to rename and provide a description for the style and to attach design notes and assign property set definitions through their respective buttons.

- **Defaults**   This tab is used to specify default values for the edge of the slab (see Figure 14.12). As with the slab styles, these settings control only the default values for the slab edge. These values can be modified at the time the slab is added or modified.

- **Design Rules** This tab is used to add a custom shape for the edge style (see Figure 14.13). Initially, two check boxes are available for customizing the edge. The **Fascia** and **Soffit** check boxes, when selected, allow you to add a custom shape for the fascia and the soffit. The *fascia* is the edge of the slab. The *soffit* is a custom shape placed below the overhang.

When these check boxes are selected, an associated drop-down list becomes available. The items included in the list are AEC profiles defined in the drawing. See Chapter 7 for information on how to create a profile. When creating a profile for use as a fascia or soffit specify the insertion points on the profile as indicated in Figure 14.14. Specify any other settings including soffit location in this tab. Refer to the diagram in this tab for value settings.

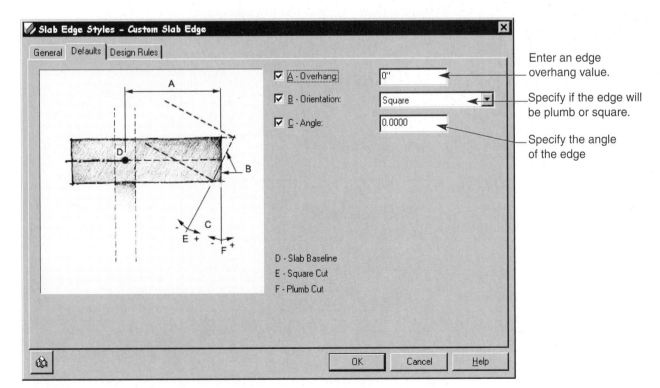

FIGURE 14.12   Set any defaults for the slab edge in the Defaults tab.

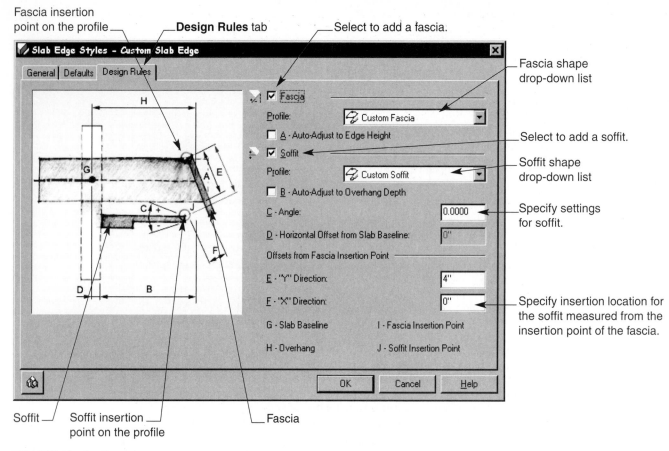

FIGURE 14.13 The Design Rules tab is used to assign a fascia and soffit to the edge of a slab.

FIGURE 14.14 When creating a fascia profile, set the insertion point at the upper left corner of the profile. If creating a profile for a soffit, specify the upper right corner as the insertion point on the profile.

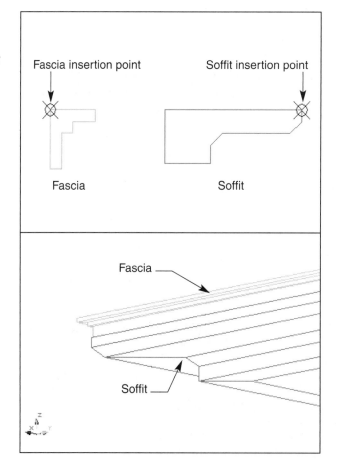

Exercise 14.5

1. Open EX14-2.dwg.
2. Create a 2″ x 4″ rectangle. Create a new profile named 2×4 and specify the insertion point as the upper left corner of the rectangle.
3. Create a new slab edge style named Trim Board. In the Design Rules tab turn on the Fascia check box and specify the 2×4 as the fascia. Clear the Auto-Adjust to Edge Height check box.
4. Create a new slab style named Subfloor. In the Design Rules tab check on the Has Fixed Thickness, and specify a thickness of 2″.
5. In the Defaults tab, check on the Perimeter Edges check box, and select the Trim Board edge style.
6. In the Display Props tab, edit the model representation by style. Change the color of the fascia component to color 7. Turn off the pivot point component.
7. Edit the plan representation to not show the fascia or pivot point components.
8. Use the Slab Properties command to change the existing floor slab to the new Subfloor style. Be sure to press the Reset to Style Default button in the Style tab.
9. Save the drawing as EX14-5-Flr-Pln.dwg.

Using Slab Tools

The previous sections covered how to add and modify slab objects and how to create slab and edge styles. In addition to these tools, there are commands for performing advanced editing on slabs. Tools similar to those used on polylines are available and are described next.

Trimming Slabs

Slabs can be trimmed in much the same way as polylines, arcs, or circles. When one of these objects is trimmed, you are able to determine the side of the cutting plane that is to be removed from the drawing. The process is similar for slabs: establish the trimming plane, then trim the slab. When a slab is trimmed, part of the slab is removed from the drawing. Objects that can be used as trimming planes include polylines, walls, or other slabs. If the slab being trimmed includes a custom edge, the fascia and/or soffit that intersects the trimming plane is trimmed at the same angle as the slab it is attached to.

If the trimming plane is a polyline, the slab is trimmed to the edge of the polyline. If the trimming object is a wall or slab, the trimming plane is the closest surface to the part of the slab that will remain. The new edge of the slab is based on the angle of the wall or slab that was used as the trimming plane.

Use one of the following options to trim a slab:

✓ Select the **Design** pull-down menu, access the **Slab Tools** cascade menu, and pick the **Trim Slab** command.

✓ Pick the **Trim Slab** button from the **Slab Tools** toolbar.

✓ Right-click, and select **Design, Slab Tools, Trim Slab** from the cursor menu.

✓ Type **AECSLABTRIM** at the Command: prompt.

✓ Select the slab, then right-click, pick the **Tools** cascade menu, and select the **Trim** command.

The following prompt sequence is used to trim the slab away from a wall in Figure 14.15:

Command: **AECSLABTRIM** ↵
Select trimming object (a slab, wall, or polyline): *(select the wall)*
Select a slab to trim: *(select the slab)*
Specify side to be trimmed: *(pick the side of the slab to be trimmed)*
Command:

FIGURE 14.15   Trimming a slab
away from a wall.

1. Select the trimming
plane (wall).

2. Select the slab
to be trimmed.

3. Pick the side of the
slab to be removed.

4. The slab is trimmed
away from the wall.

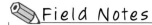

Field Notes

- If a polyline with a mixture of arcs and straight segments is used as a trimming plane, the slab follows the contours of the polyline. If a curved wall is used as a trim plane, the slab is trimmed along the straight-line angle between the starting and ending points of the arced wall.
- It is not possible to trim only an overhang. The trimming plane must intersect the defined boundary of the slab at some location.

Extending Slabs

In addition to being trimmed slabs, slabs can be extended. In order to extend a slab, the boundary being extended to must be a slab or wall object. When a slab is extended, one edge is actually moved to the boundary. Thus, you are prompted to select the adjacent edges that will be extended. Select the two adjacent edges to the edge of the slab that is to be extended to the boundary (see Figure 14.16).

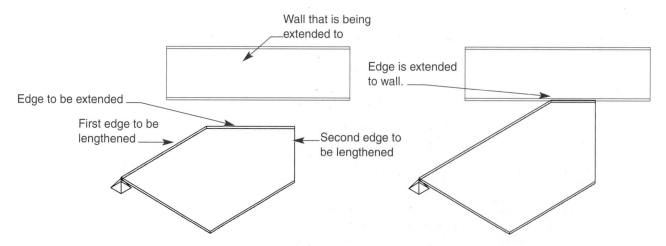

FIGURE 14.16   When extending a slab to a wall or another slab, select the adjacent edges of the slab edge being extended.

Use one of the following options to extend the edge of a slab:

✓ Pick the **Design** pull-down menu, select the **Slab Tools** cascade menu, and select the **Extend Slab** command.

✓ Pick the **Extend Slab** button in the **Slab Tools** toolbar.

✓ Right-click, and select **Design, Slab Tools, Extend Slab** from the cursor menu.

✓ Type **AECSLABEXTEND** at the Command: prompt.

✓ Select the slab, then right-click, and select the **Tools** cascade menu, then the **Extend** command from the shortcut menu.

The following prompt sequence is used to extend the edge in Figure 14.16:

Command: **AECSLABEXTEND** ↵
Select an object to extend to (a slab or wall): *(select the wall)*
Select a slab to extend: *(select the slab edge to be extended)*
Slab will be extended by lengthening two edges.
Select first edge to lengthen: *(select the first edge to be lengthened to the wall)*
Select second edge to lengthen: *(select the second edge to be lengthened to the wall)*
Command:

Exercise 14.6

1. Start a new drawing using the AEC Arch (Imperial) or the AEC Arch (Imperial–Int) template.
2. Create the figure using slabs, walls, and polylines.
3. Trim the slab way from the wall and polyline.
4. Extend the edge of the slab to the wall.
5. Save the drawing as EX14-6.dwg.

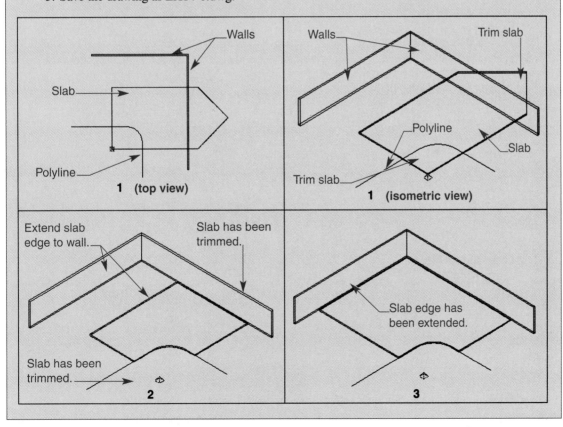

Mitering Slabs

Another command useful when working with slabs at different slopes is the **Miter Slab** command. This command allows you to miter two slabs together and join them at the points where the two slabs meet. Use one of the following options to access this command:

✓ Select the **Design** pull-down menu, pick the **Slab Tools** cascade menu, and select the **Miter Slab** command.

✓ Pick the **Miter Slab** button in the **Slab Tools** toolbar, right-click, and select **Design, Slab Tools, Miter Slab** from the cursor menu.

✓ Type **AECSLABMITER** at the Command prompt.

✓ Select the slab to be mitered, right-click, then pick the **Tools** cascade menu, then the **Miter** command from the shortcut menu.

When the Miter Slab command is used two different methods of mitering the slab are available and are described next.

Field Notes

> *Mitering* is an operation that can be performed only on two slabs of the same style regardless of the mitering method.

Miter by Intersection

The **Miter by Intersection** option is used when two slabs of the same style intersect each other (see Figure 14.17). When the two slabs are mitered using the Intersection option, the slabs are trimmed along the intersection of their boundary lines. The new edges of the two slabs are set to an angle that supports a true miter cut between the slabs. The following prompt sequence is used to miter the three slabs displayed in Figure 14.17:

```
Command: AECSLABMITER ⏎
Miter by [Intersection/Edges] <Intersection>: ⏎
Select first slab at the side to keep: (select the first slab at the side of the miter that is to be kept)
Select second slab at the side to keep: (select the second slab at the side of the miter that is to
    be kept)
Command: AECSLABMITER ⏎
Miter by [Intersection/Edges] <Intersection>: ⏎
```

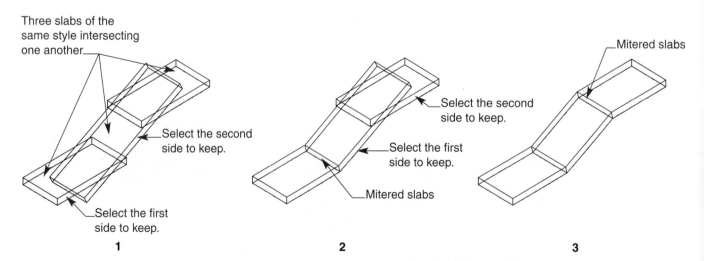

FIGURE 14.17   The Miter Slab command using the Intersection option miters two slabs that cross through each other.

Select first slab at the side to keep: *(select the first slab at the side of the miter that is to be kept)*
Select second slab at the side to keep: *(select the second slab at the side of the miter that is to be kept)*
Command:

 Field Notes

> If multiple edges on a slab need to be mitered, try stretching them using the grip boxes to overlap the other slabs, then use the Miter by Intersection option.

Miter by Edge

The Miter Slab command with the **Edges** option provides more flexibility than the Miter by Intersection option, because two particular edges are specified for mitering. This option allows you to miter the edges of two slabs without their having to intersect each other (see Figure 14.18). If the slabs do not intersect, then

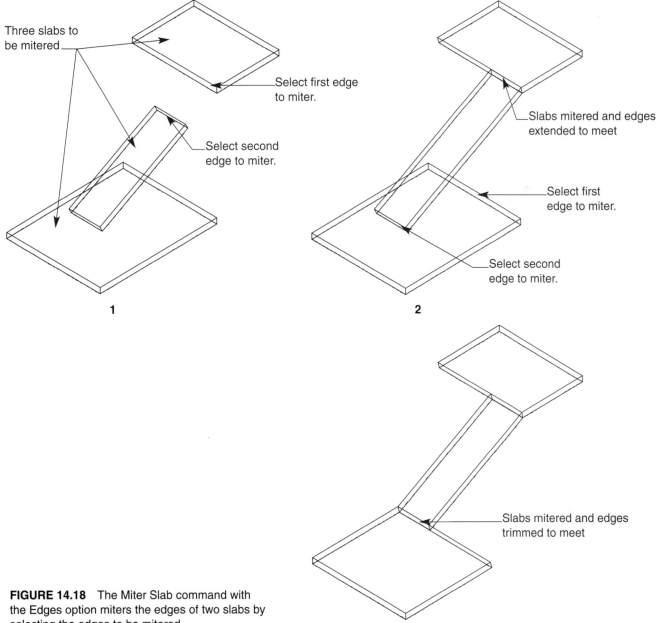

FIGURE 14.18 The Miter Slab command with the Edges option miters the edges of two slabs by selecting the edges to be mitered.

the edges are extended to the point at which they meet. If the two slabs cross through each other, then the slabs are trimmed as with the Miter by Intersection option. The following command sequence is used to miter the slabs by edge shown in Figure 14.18:

> Command: **AECSLABMITER** ↵
> Miter by [Intersection/Edges] <Intersection>: **E** ↵
> Select edge on first slab: *(select the first edge to miter)*
> Select edge on second slab: *(select the second edge to miter)*
> Command: **AECSLABMITER** ↵
> Miter by [Intersection/Edges] <Intersection>: **E** ↵
> Select edge on first slab: *(select the first edge to miter)*
> Select edge on second slab: *(select the second edge to miter)*
> Command:

Exercise 14.7

1. Start a new drawing using the AEC Arch (Imperial) or the AEC Arch (Imperial–Int) template.
2. Create the slabs displayed in Figures 14.17 and 14.18.
3. Use the Miter by Intersection and the Miter by Edge options accordingly.
4. Save the drawing as EX14-7.dwg.

Cutting Slabs

The Trim Slab command was discussed earlier. This command allowed you to trim the slab and remove a portion of the slab from the trimming line. The **Cut Slab** command can be used to split a slab apart. This command cuts the slab into two separate slabs by using a polyline or 3D object as the cutting line.

If you are using a polyline, the polyline can be a closed polyline or an open polyline that crosses through the slab. If the open polyline crosses in and out of the slab, multiple slabs are created between the polyline cut lines.

 Field Notes

> When Architectural Desktop uses a curved polyline as a cutting plane for a slab, the edges along the curve are made up of a series of straight-line segments. Adjusting the AEC-FACETDEV variable to a value smaller than 1/2″ causes more straight edges to be created, making the slab appear curved. Set this variable before cutting curved slabs.

A 3D object also can be used to cut a slab into two pieces. 3D objects include walls, slabs, mass elements, or AutoCAD solids. The only condition on using a 3D object as a cutting plane is that the object must cross through the slab being cut. When the slab is cut using a 3D object the slab is cut along the surface of the 3D object.

Use one of the following options to access the Cut Slab command:

✓ Pick the **Design** pull-down menu, then select the **Slab Tools** cascade menu, and pick the **Cut Slab** command.

✓ Pick the **Cut Slab** button in the **Slab Tools** toolbar.

✓ Right-click, and select **Design, Slab Tools, Cut Slab** from the cursor menu.

✓ Type **AECSLABCUT** at the Command prompt.

✓ Select the slab to be cut, right-click, then pick the **Tools** cascade menu, then the **Cut** command from the shortcut menu.

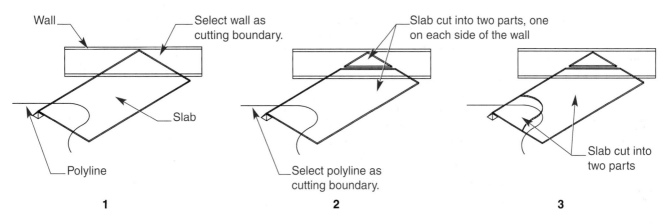

Wall — Select wall as cutting boundary. — Slab — Polyline — Slab cut into two parts, one on each side of the wall — Select polyline as cutting boundary. — Slab cut into two parts

1   **2**   **3**

FIGURE 14.19   Use the Cut Slab command to cut the slab into two pieces. Both pieces are retained in the drawing.

The following prompt sequence is used to cut the slab in Figure 14.19:

```
Command: AECFACETDEV ↵
Set new FacetDev<1/2">: 1/8 ↵
Command: AECSLABCUT ↵
Select a slab: (select the slab to cut)
Select cutting objects (a polyline or connected solid entities): (select the cutting object (wall))
1 found
Select cutting objects (a polyline or connected solid entities): ↵
Command: AECSLABCUT ↵
Select a slab: (select the slab to cut)
Select cutting objects (a polyline or connected solid entities): (select the cutting object (polyline))
1 found
Select cutting objects (a polyline or connected solid entities): ↵
Erase layout geometry [Yes/No] <N>: ↵
Command:
```

Exercise 14.8

1. Start a new drawing using the AEC Arch (Imperial) or the AEC Arch (Imperial–Int) template.
2. Create the objects displayed in the figure. Be sure to place the wall so that it cuts through the slab.
3. Use the Cut Slab command to cut the slab into separate pieces.
4. Save the drawing as EX14-8.dwg.

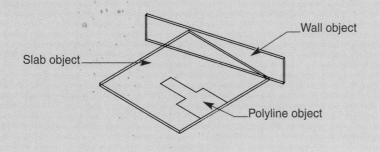

Slab object — Wall object — Polyline object

Adding Slab Vertices

Occasionally, when a slab is modified an additional edge needs to be added in order for the modifications to be correct, particularly when the slab is stretched with grips or the Slab Extend command is used. To overcome this problem, the **Add Vertex** command can be used to place a new vertex on the slab, which divides the existing edge into two edges.

Use one of the following options to add vertices to a slab:

- ✓ Select the **Design** pull-down menu, pick the **Slab Tools** cascade menu, then the **Add Slab Vertex** command.

- ✓ Pick the **Add Slab Vertex** button in the **Slab Tools** toolbar.

- ✓ Right-click, and select **Design, Slab Tools, Add Slab Vertex** from the cursor menu.

- ✓ Type **AECSLABADDVERTEX** at the Command prompt.

- ✓ Select the slab, right-click, then pick the **Tools** cascade menu, then the **Add Vertex** command from the shortcut menu.

The following command sequence is used to add vertices to the slab in Figure 14.20 so that the edges can be stretched properly:

Command: **AECSLABADDVERTEX** ↵
Select a slab: *(select the slab)*
Specify point for new vertex: **MID** ↵
of *(select the midpoint of one of the slab edges)*
Command: *(select the slab to obtain grips, then select a grip point)*
** STRETCH **
Specify stretch point or [Base point/Copy/Undo/eXit]: **END** ↵
of *(select the endpoint of a wall)*
Command: *(select the slab to obtain grips, then select a grip point)*
** STRETCH **

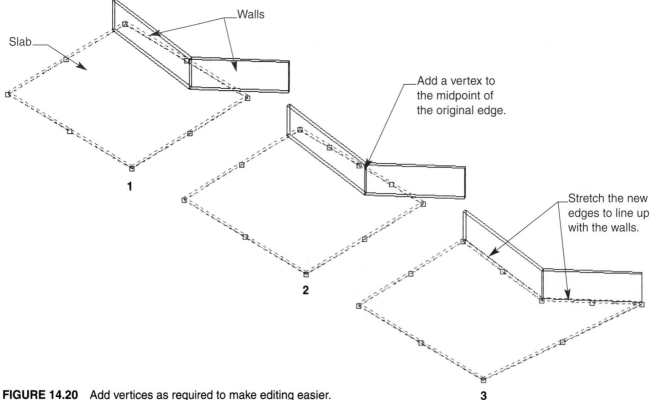

FIGURE 14.20 Add vertices as required to make editing easier.

Specify stretch point or [Base point/Copy/Undo/eXit]: **END** ↵
of *(select the endpoint of a wall)*
Command: *(select the slab to obtain grips, then select a grip point)*
** STRETCH **
Specify stretch point or [Base point/Copy/Undo/eXit]: **END** ↵
of *(select the endpoint of a wall)*
Command:

After a vertex has been added, select the object to display the grip boxes. A new grip box is located at the point you selected in addition to two grip boxes at the middle of the two edges created.

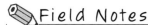
Field Notes

> Use object snaps when trying to add a vertex at a specific location. If you are trying to add a vertex at a point where two objects cross in different "Z" planes, use the Apparent Intersection object snap.

Removing Slab Vertices

The slab tools also include a option for removing excess vertices. Access the **Remove Slab Vertex** command to remove any undesired vertices by using one of the following options:

- ✓ Select the **Design** pull-down menu, access the **Slab Tools** cascade menu, and select the **Remove Slab Vertex** command.

- ✓ Pick the **Remove Slab Vertex** button in the **Slab Tools** toolbar.　

- ✓ Right-click, and select **Design, Slab Tools, Remove Slab Vertex** from the cursor menu.

- ✓ Type **AECSLABREMOVEVERTEX** at the Command: prompt.

- ✓ Select the slab, right-click, then pick the **Tools** cascade menu, then the **Remove Vertex** command.

Once the command has been entered, pick a point close to or on top of the vertex to be removed. The vertex is removed and the two adjacent edges are joined together (see Figure 14.21).

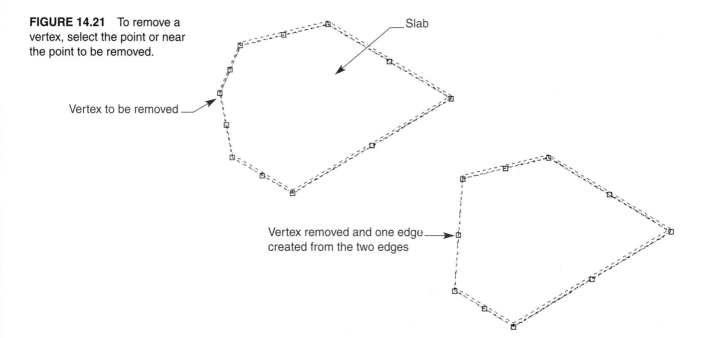

FIGURE 14.21　To remove a vertex, select the point or near the point to be removed.

Slab

Vertex to be removed

Vertex removed and one edge created from the two edges

Adding and Removing Holes in Slabs

A useful tool for cutting holes in slabs for elevator shafts, stairwells, and flues is the **Slab Hole** command. Holes can be added to a slab through the use of a polyline or a 3D object.

If a polyline is used, the polyline is projected through the thickness of the slab to create the hole. A 3D object is an object with mass such as a wall, mass element, or AutoCAD solid. The 3D object must extend through both sides of the slab in order to cut the hole.

After the hole is cut, new edges are created at the edge of the hole and can be modified using any of the commands discussed previously.

✎ Field Notes

> If a hole is cut with an object that extends outside the slab or overlaps the edge of a slab, the hole is cut only on the slab, not the fascia and soffit components. If you wish also to cut the edge away, use the Cut Slab command.

Adding a Hole
Use one of the following options to cut a hole in a slab:

✓ Select the **Design** pull-down menu, move to the **Slab Tools** cascade menu, and select the **Slab Hole** command.

✓ Pick the **Slab Hole** button in the **Slab Tools** toolbar.

✓ Right-click, and select **Design, Slab Tools, Slab Hole** from the cursor menu.

✓ Type **AECSLABHOLE** at the Command prompt.

✓ Select the slab, right-click, pick the **Tools** cascade menu, and select the **Add Hole** command from the shortcut menu.

The following prompt sequence is used to add a hole in Figure 14.22:

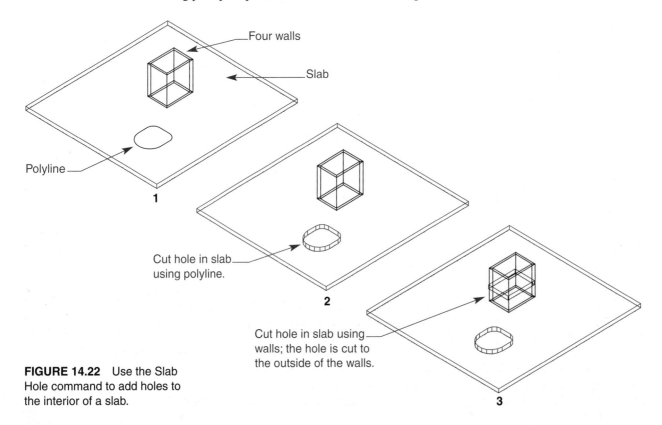

FIGURE 14.22 Use the Slab Hole command to add holes to the interior of a slab.

Command: **AECSLABHOLE** ⏎
Hole [Add/Remove] <Add>: ⏎
Select a slab: *(select the slab object)*
Select a closed polyline or connected solid entities to define a hole: *(select the polyline)*
1 found
Select a closed polyline or connected solid entities to define a hole: ⏎
Erase layout geometry? [Yes/No] <N>: **Y** ⏎
Command: **AECSLABHOLE** ⏎
Hole [Add/Remove] <Add>: ⏎
Select a slab: *(select the slab object)*
Select a closed polyline or connected solid entities to define a hole: *(select the four walls using a crossing)*
4 found
Select a closed polyline or connected solid entities to define a hole: ⏎
Erase layout geometry? [Yes/No] <N>: ⏎
Hole location relative to cutting objects [Inside/Outside] <Outside>: ⏎
Command:

Field Notes

When you use 3D objects that have an interior void, such as the walls around a room or a solid with a hole, you are prompted to cut to the inside or to the outside of the objects. If you select Inside, the hole edges are placed on the interior of the 3D object. If you select Outside, the hole is cut using the outer boundaries of the 3D object(s).

Removing a Hole from a Slab

Holes also can be removed from a slab with the **Remove** option of the Slab Hole command. Once the command has been entered, you are first prompted to Add or Remove a hole. Type **R** at the Command: prompt to remove the hole, then select the edge of the hole to be removed. The hole is removed and the void is filled.

Exercise 14.9

1. Open EX14-8.dwg.
2. Draw four walls 8'-0" high that form a closed room in the slab. Use Wall Properties to move the Z location to –36".
3. Cut a hole in the slab using the walls as the cutting objects. Cut the slab to the outside of the walls.
4. Draw two closed polylines in the slab, and cut holes using the polylines.
5. Add vertices to the edge of the slab.
6. Stretch the new vertex point(s).
7. Remove one of the holes cut by a polyline.
8. Save the drawing as EX14-9.dwg.

Using Boolean Operations

There may be times when a slab requires 3D geometry to be added or removed that cannot be defined when the slab or edge styles are added—for example, a pedestal in the middle of a slab or pit that does not cut all the way through a slab. Such geometry can be created using a Boolean operation. A **Boolean operation** is a type of command that allows you to add 3D objects to or subtract them from a slab.

The **Boolean Add** operation joins the 3D object to the slab, and the **Boolean Subtract** operation removes a portion of the slab. The original objects remain on the screen, and any modifications made to them affect the Booleaned slab. If these objects are erased from the drawing, the slab updates by removing them from the slab. These objects are generally placed on a layer that can be frozen or not plotted.

Adding 3D Objects to Slabs

Use one of the following options to add a 3D object to a slab:

- ✓ Select the **Design** pull-down menu, access the **Slab Tools** cascade menu, and pick the **Boolean Add/Subtract** command.
- ✓ Pick the **Boolean Add/Subtract** button from the **Slab Tools** toolbar.
- ✓ Right-click, and select **Design, Slab Tools, Boolean Add/Subtract** from the cursor menu.
- ✓ Type **AECSLABBOOLEAN** at the Command prompt.
- ✓ Select the slab, right-click, and select the **Tools** cascade menu, then the **Boolean** command.

The following prompt sequence is used to add the objects in Figure 14.23 to the slab:

```
Command: AECSLABBOOLEAN ↵
Select a slab: (select the slab)
Select objects: (select the first object)
1 found
Select objects: (select the second object)
1 found, 2 total
Select objects: (select the third object)
1 found, 3 total
Select objects: (select the fourth object)
1 found, 4 total
Select objects: ↵
Boolean [Additive/Subtractive/Detach] <Additive>: ↵
4 object(s) attached.
Command:
```

Subtracting 3D Objects from Slabs

3D objects also can be subtracted from slab objects using the Boolean Subtract option. Use the Boolean Add/Subtract command with the subtractive option. The following prompt sequence is used to subtract 3D objects from the slab in Figure 14.24:

```
Command: AECSLABBOOLEAN ↵
(select the slab)
Select objects: (select the first object)
```

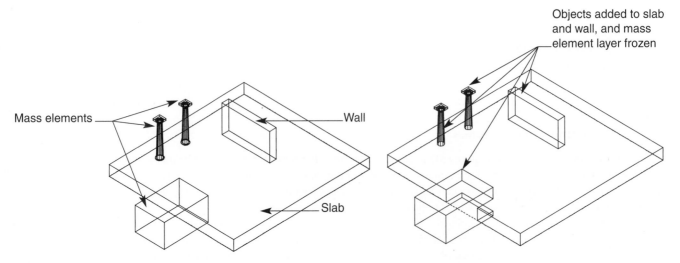

FIGURE 14.23   Use the Boolean Add option to add 3D objects to a slab.

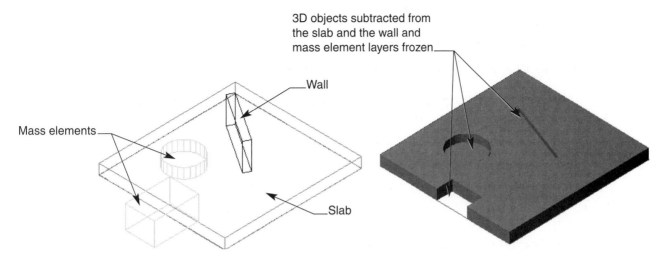

FIGURE 14.24 Use the Boolean Subtract option to subtract 3D objects from the slab.

```
1 found
Select objects: (select the second object)
1 found, 2 total
Select objects: (select the third object)
1 found, 3 total
Select objects: ↵
Boolean [Additive/Subtractive/Detach] <Additive>: S ↵
3 object(s) attached.
Command:
```

Field Notes

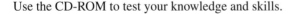

The Boolean Add/Subtract command also includes a Detach option that is used to detach 3D objects that are part of the Boolean process for the slab.

CHAPTER REVIEW

Use the CD-ROM to test your knowledge and skills.

Chapter Test

To check your understanding of the content provided in this chapter, access the Test file in the CH14 folder of the CD-ROM that accompanies this text.

Chapter Project

To practice the Architectural Desktop skills presented in this chapter, access the Project files in the CH14 folder of the CD-ROM that accompanies this text. The project files are in pdf format and include sample drawings and instructions for completing each project.

||||||||||||||||||||||||||||||||||||||||||||||||||||||||||||||||||||||||||||||||||||||||||||

CHAPTER 15

Working with Reference Files

LEARNING GOALS

After completing this chapter, you will be able to:

◎ Use reference drawings.

◎ Attach an external reference to the current drawing.

◎ Manage reference files.

◎ Use the Xref Manager.

◎ Detach, reload, and unload reference files.

◎ Bind an xref.

◎ Work with xref paths.

◎ Clip an external reference.

◎ Edit reference drawings.

◎ Set up external reference configurations.

◎ Establish project files.

◎ Transmit reference documents to other people.

◎ Work with demand load.

Architectural Desktop lends itself very well to the creation of single-level floors for a building. So far in this text you have created floor plans and foundation plans in separate files. Some offices create an entire building in one drawing file. One disadvantage to this method is that it produces a larger drawing file, because all the components of the building are included in one place. Another disadvantage is that objects at different floors must have their layers and their placement in the Z axis or elevation managed manually. For example, if a multistory building is created in one drawing file, layer overrides that specify the floor level currently being drawn must be set before the objects that belong on that floor can be drawn. When it is time to draw another floor level, the layer override and the elevation need to be reset to the next floor for the new objects. Because Architectural Desktop creates objects as 2D components in a top view, the objects are always drawn in the current Z axis.

Changing the Z axis is not the only answer, however. If the UCS (User coordinate system) is relocated at a new elevation, and trace walls are on the first floor, the new walls are snapped to the first floor. Thus, second-floor walls are created over the top of the first-floor walls. This ends up causing wall cleanup problems unless the second-floor walls were also placed in their own cleanup group. Needless to say, creating a multistory building in a single file can become a management headache, but it is possible.

AutoCAD combats this problem with the use of externally referenced files (XREFs). An *xref* is similar to the separate sheet drawn for each floor plan of a building in traditional drafting techniques. Placing one sheet on top of another causes the floor plans to stack on top of one another. Xref stands for *external reference*. An **external reference** is a source drawing outside the current drawing that AutoCAD/ Architectural Desktop refers to when information is needed (see Figure 15.1). With AutoCAD/Architectural Desktop the individual drawings or floor plans can be stacked together in an assembly file, thus stacking the floors together to create a 3D building model. Elevations and sections can be created from this assembly file. Elevations and sections are discussed in a future chapter.

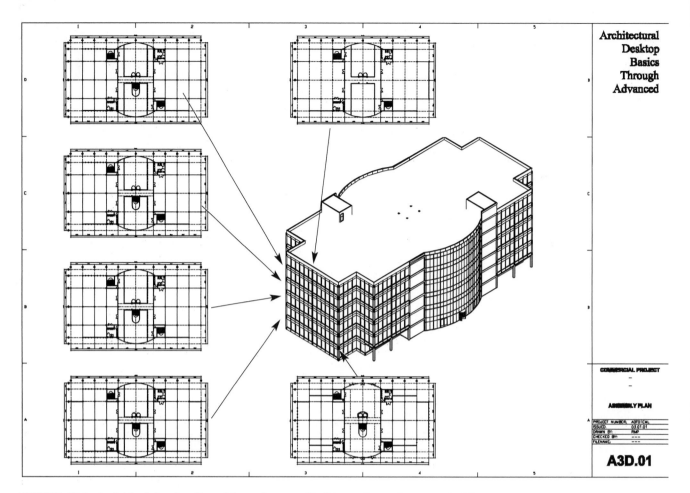

FIGURE 15.1   Individual drawing files of floor plans are Xref'd together to form one building.

The following are the advantages to using xrefs:

- A reference drawing is an external drawing file that another drafter can work on, because the file is being referenced for a separate part of the construction documents, such as sections and elevations, or another floor plan.

- Reference files are inserted much like a block is inserted, and acts like a block, with the geometry grouped together.

- Reference files can be copied, moved, and rotated like a block.

- Reference files occupy very little file space in comparison with blocks or copied geometry.

- Layers within the reference can have their colors and linetypes changed or can be frozen, thawed, or turned on or off without affecting the original drawing file.

- Referenced drawings are standard drawings that can be opened and modified.

- If a reference file has been modified, it can be updated in any drawing where it is referenced into without having to be reinserted.

This chapter discusses reference drawings and how to use them in Architectural Desktop.

Using Reference Drawings

Many architectural offices, especially large firms, assign different parts of the construction documents to different groups within the company. For example, one group is assigned to the conceptual design and development of the project, another group is assigned to laying out the floor plans, and other groups are assigned to developing the sections and elevations, details, interiors, mechanical requirements, schedules, documentation, and printing. As each group works on its own assigned tasks for the same set of construction documents, it needs to reference the other groups' work. For example, the section and elevation group needs to reference the floor plans to lay out the sections and elevations correctly. The detail group needs to reference the work of the section and elevation group to create accurate details.

Chapter 7 explained how to use blocks as standard symbols within a drawing and how to insert a drawing file by using the **INSERT** command. The section and elevation group could insert the floor plans as a block into its drawings for use as a reference; however, the disadvantage to doing this is that the inserted drawing adds physical layers, blocks, dimension styles, text styles, and linetypes to the drawing, dramatically increasing the size of the drawing file. Another disadvantage to inserting the floor plan as a reference is that it may still be under development. If the floor plans change, the inserted drawing must be erased and reinserted to reflect the updated version of the floor plans. If the floor plans are xrefed into the drawing, the reference can quickly be updated with the latest changes to the floor plans, and the file size kept small, because the xref references the original drawing layers, blocks, dimension styles, text styles, and linetypes in the external drawing.

When an external drawing is xrefed into another drawing, the external drawing geometry is not added to the receiving drawing. The only part of the external drawing that is added is the graphics displayed on the drawing screen. This keeps the receiving drawing file size small and also allows several drafters in an office to reference the same external drawing with the guarantee that any changes made to the external file will be updated the next time the receiving drawing is reopened or the reference file is reloaded in the receiving drawing.

The **XREF** command is used to reference external drawings into a receiving drawing, also called the *current drawing*. The receiving drawing can be a drawing of the next floor of the building, an assembly drawing of several references that will be used to develop elevations, or a site plan that needs to reference the "footprint" of the building. This command is a standard AutoCAD command and can be found in the AutoCAD menus.

Use one of the following options to manage reference drawings and insert new references:

✓ Select the Insert pull-down menu, then select the **X̲ref Manager...** command.

✓ Pick the **External Reference** button in the **Reference** or **Insert** toolbar.

✓ Type **XREF** or **XR** at the Command: prompt.

The **Xref Manager** dialog box is displayed, where multiple references can be managed and inserted into the current drawing (see Figure 15.2).

Attaching an XREF to the Current Drawing

Attaching an external reference is similar to inserting a block. An xref that is inserted into the current drawing is said to be *attached* to the drawing. In order to attach an xref, access the Xref Manager dialog box. Once the dialog box is displayed, select the A̲ttach... button. The **Select Reference File** dialog box is displayed, allowing you to browse the hard drive or network drives to locate the external drawing for attachment. When you have found the desired drawing, select the file in the list box to highlight it, then pick the O̲pen button to open the file.

Once the reference file has been selected and the O̲pen button pressed, the **External Reference** dialog box is displayed, as shown in Figure 15.3. This dialog box is used to specify how and where the xref is to be placed in the current drawing. The name and path to the external drawing is displayed in the upper left corner of the dialog box. The dialog box contains the following options.

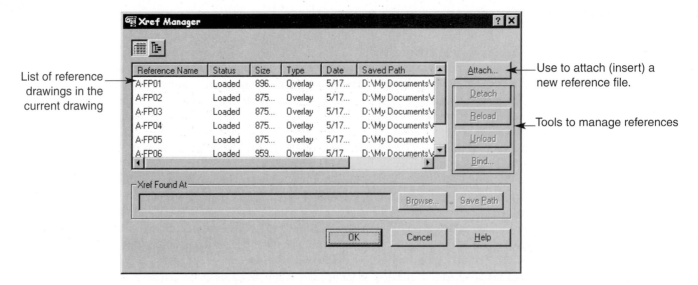

List of reference drawings in the current drawing

Use to attach (insert) a new reference file.

Tools to manage references

FIGURE 15.2   The Xref Manager dialog box lists any drawings that are in the current drawing and provides tools for managing and inserting new references.

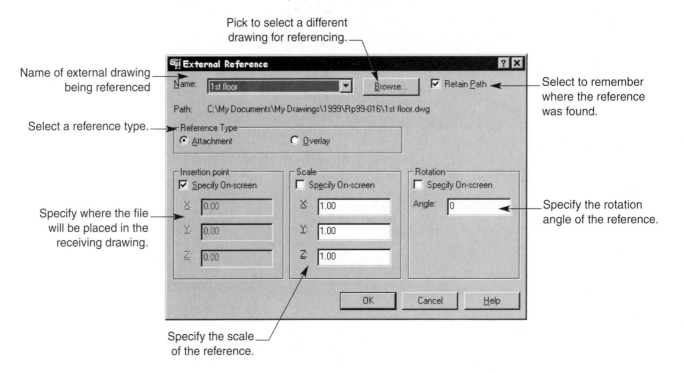

Pick to select a different drawing for referencing.

Name of external drawing being referenced

Select to remember where the reference was found.

Select a reference type.

Specify where the file will be placed in the receiving drawing.

Specify the rotation angle of the reference.

Specify the scale of the reference.

FIGURE 15.3   The External Reference dialog box is used to control how the xref will placed in the drawing.

- **Name:** This drop-down list contains the name of the reference file currently being referenced into the drawing. The drop-down list includes a list of existing references in the drawing.

- **Browse...** This button can be used to locate a different reference file than the one currently being displayed in the Name: drop-down list. Select this button to return to the Select Reference File dialog box, to browse for another file.

- **Retain Path** This check box when checked causes the Xref Manager to remember where the xref was found so it can reload the xref the next time the drawing is opened or the reference is reloaded. If the check is removed from this box, the Xref Manager looks only in the *Support File Search Path* locations specified in the Options dialog box or in the same folder in which the receiving drawing is saved, to try to locate the xref for loading.

- **Reference Type** This area is used to specify how the xref is attached to the drawing. The options include **Attachment** and **Overlay**. When an xref is attached the Attachment radio button is selected by default. Both options allow you to attach the reference to the drawing but behave in two different ways. Both of these options are described in the next section. For the purposes of this discussion, use the attachment option.

Below the Reference Type area are three areas for specifying the xref insertion point, scale, and rotation. Each area includes a **Specify On-screen** check box. If this check box is unchecked in each of the areas, the text boxes become available, allowing you to specify insertion criteria.

- **Insertion point** This area is used to specify the insertion point location for the reference file. By default all drawing files include an insertion point located at 0,0,0 (X, Y, Z). When the drawing being referenced is attached to the receiving drawing, its default insertion point is placed at an absolute coordinate in the receiving drawing. If the Specify On-screen check box is selected in this area, the referenced drawing is attached to the crosshairs, and you must pick a point in the receiving drawing to place and locate the reference file. If the box is unchecked, you must specify an absolute coordinate in the text boxes. As long as the individual floors have been drawn to line up with each other in the X and Y coordinates, the only value that may need to change when the xref is attached is the Z-axis (elevation) value.

- **Scale** This area allows you to specify the scale of the xref being attached. Initially, the Specify On-screen check box is unchecked, allowing you to specify a scale for the xref. This allows you to enter different scales for the xref along the X, Y, and Z axes. Generally, if you are referencing a floor plan in order to assemble the building, or to use one floor to lay out another floor, the scale should remain at 1.00 on all three axes.

- **Rotation** This area specifies the rotation angle of the attached xref. The Specify On-screen check box allows you to specify the rotation angle on the drawing screen after you have selected the insertion point. If this box is unchecked, you can enter the rotation angle in the text box.

When you have finished specifying the parameters, the reference drawing is attached to the current drawing (see Figure 15.4). Selecting the file highlights all the objects in the reference, because they are all grouped together like a block. The new geometry and Architectural Desktop objects can be traced over the top of the reference file, or the building can be assembled for elevations, sections, and schematic design.

If another drawing file is to be referenced, use the XREF command to attach the file. If a drawing is already attached to the drawing, the Xref Manager dialog box lists the reference file already attached and any related information about the xref file (see Figure 15.5). Use the Attach... button to attach another reference drawing.

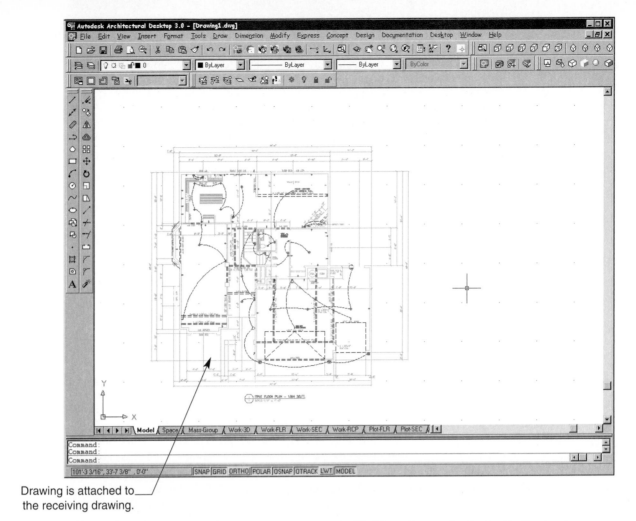

Drawing is attached to
the receiving drawing.

FIGURE 15.4 After the attachment parameters are specified the reference is attached to the drawing.

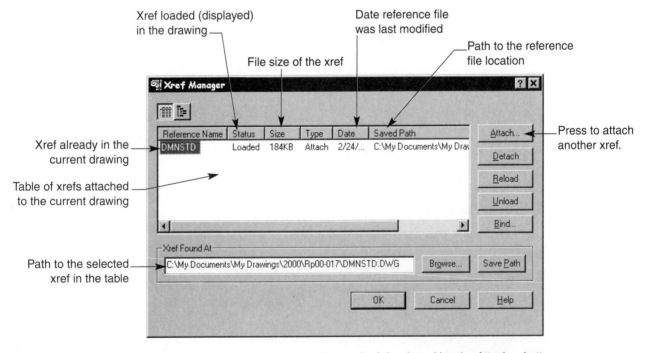

Xref loaded (displayed)
in the drawing

Date reference file
was last modified

Path to the reference
file location

File size of the xref

Xref already in the
current drawing

Table of xrefs attached
to the current drawing

Path to the selected
xref in the table

Press to attach
another xref.

FIGURE 15.5 The Xref Manager dialog box lists any currently attached drawings. Use the Attach… button
to attach another reference drawing.

Managing Reference Files

Based on the previously outlined procedures, attaching an xref is similar to inserting a block or drawing. Both commands specify the name of the file being attached, the insertion point, scale, and rotation angle. Both commands also place the file on the current layer. If the layer is frozen or turned off, the reference file, block, or inserted drawing file is not displayed in the drawing screen.

Any objects in the reference file are being borrowed by the receiving drawing and do not become a permanent part of the drawing as an inserted block or drawing would. As a result, the file size of the receiving drawing is kept to a minimum.

 Field Notes

> Because the Xref is inserted on the current layer, if you freeze or turn off the layer when making another layer current, the xref is not display in the drawing. For greater control over which xrefs are displayed in the drawing, it is advisable to create a layer for each Xref that is to be attached to the drawing. For example, you can freeze the upper floor xref layer so you can reference the main floor xref in order to create the foundation plan. If xref layer names have the prefix XREF, it becomes easier to filter for the xref layers in the Layer Manager dialog box. Examples of xref layer names for a building are XREF-1st-floor, XREF-2nd-floor, and XREF-3rd-floor.

When an xref has been attached to the receiving drawing, the objects within the xref are displayed in the drawing screen. Special xref layers are also loaded into the receiving drawing to support the display of the xref objects. The xref layers are added to the receiving drawing for the sole purpose of controlling the display of the individual xref objects.

If the Layer drop-down list is opened, it displays a list of xref layers that have been added to the receiving drawing. These layers are grayed out and cannot be set current or used by any new objects created in the drawing (see Figure 15.6); however, you do have control over the display behavior of the xref layers. You can freeze/thaw, turn on/off, lock/unlock and change the colors and linetypes of these layers. Changing any of these display settings affects the xref layers in the receiving drawing only and does not affect the display of the objects in the original reference drawing.

All these display controls can be managed from the Layer Manager dialog box, shown in Figure 15.7. This dialog box was discussed in Chapter 6. If a drawing includes an xref, the Layer Manager dialog box displays an xref group icon that lists all the layers in the xref. Select the color swatches or linetype assignments to assign new colors or linetypes to the xref layers. The xref layers can also be frozen/thawed, turned on/off, locked/unlocked, or set to a plotting/non-plotting status through this dialog box.

The xref layers that are referenced into the receiving drawing are grayed out for easy identification and are created using a unique naming convention. When a reference file has been xref'd into the current drawing the name of the reference file is added to the beginning of the layers within the reference file fol-

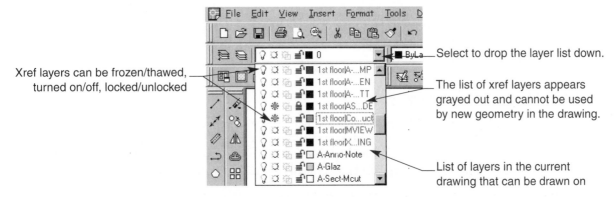

FIGURE 15.6 The Layer drop-down list displays a list of xref layers and the current drawing file layers. Xref layers can be used only by the xref objects, but their display behavior can be controlled in this drop-down list or in the Layer Manager dialog box.

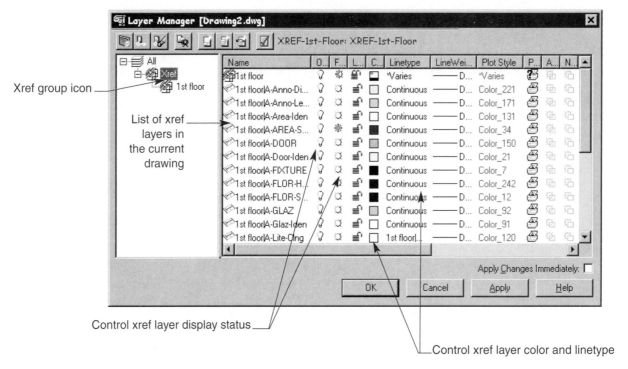

Xref group icon

List of xref layers in the current drawing

Control xref layer display status

Control xref layer color and linetype

FIGURE 15.7   Use the Layer Manager to manage layers in the current drawing or layers in reference files.

lowed by a pipe symbol, then added to the receiving drawing. In Figure 15.7, the list of xref layer names all include the prefix 1st floor followed by the name of the layer within the original reference file.

For example, a reference drawing named A-FP01 may include the layers A-Wall and A-Glaz. When the A-FP01 file is xref'd into another drawing, the layers get named A-FP01|A-Wall and A-FP01|A-Glaz. If several different xrefs of different floors are referenced into a receiving drawing, each of the layers in the original references is named with this unique naming convention. This makes searching for a particular layer easier, as a filter group can be created that searches for all the layers of a particular xref. Due to the nature of xrefs their layer names cannot be renamed, as can those of layers in the current drawing.

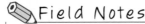Field Notes

When a drawing is referenced, the xref is attached with the same colors and display status as in the original drawing. If you are referencing a file to check the relationship of the geometry against the geometry that is drawn in the receiving drawing, or if you are planning to trace the outline of the floor plan, it is a good idea to change the colors of the xref layers to a different color. This makes it easier to differentiate between what is being drawn and what is being referenced. Remember, changing the colors of xref layers affects their color only in the receiving drawing and not in the original reference file.

Attachment versus Overlay

The previous section described how to use the Xref Manager to reference a drawing file. After a drawing has been selected for use the External Reference dialog box is used to set the parameters for attaching the xref. When an xref is attached one of two reference types can be selected: **Attachment** or **Overlay** (see Figure 15.8), which control how the Xref behaves in the receiving drawing.

Using the Attachment Option

Another benefit to using xrefs is the ability to reference the file into a drawing, which in turn is referenced into another drawing. This procedure is called ***nesting xrefs***. For example, each drawing for a set of plans is

FIGURE 15.8   When referencing a drawing select one of the reference types for use.

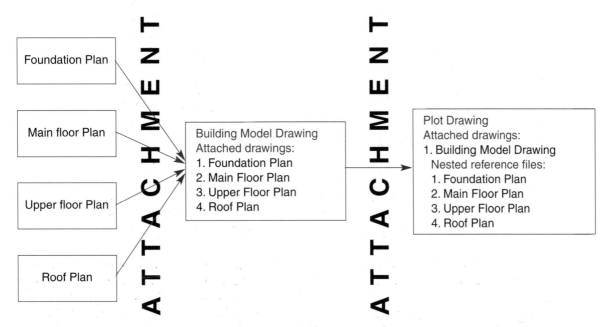

FIGURE 15.9   Using the attachment reference type causes the xref to be nested inside the receiving drawing. If the receiving drawing is then xref'd into another drawing, the next receiving drawing includes the reference files attached to the first receiving drawing.

created in individual drawing files. The individual drawing files are then xref'd into an assembly file to create the building model. The building model can then be xref'd into a drawing that is set up for plotting.

As the individual plan drawings are referenced into the building model drawing, and the building model drawing is referenced into the plot drawing, the plan drawings end up in the plot drawing because they are nested inside the building model drawing (see Figure 15.9). If changes are made to the plan drawings, the plan drawings are updated in the building model drawing and the plot drawing the next time these drawings are opened or the xrefs are reloaded.

The attachment reference type is used to nest the reference drawing(s) into the receiving drawing. If the receiving drawing is xref'd into any other drawings it carries any attached xrefs within it to the drawing(s).

Using the Overlay Option

Often when the floor plans for a project are initially laid out, one or more drawings are referenced into the file so the new plan can be traced over the reference drawing to ensure accuracy in the shape of the building and the placement of objects that interact between floors, such as stairs. In this case, the desired effect is to overlay the reference file into the drawing as another drafter or project manager xrefs the individual plans together into an assembly file to create the building model.

For example, suppose it is your job to create the upper floor plan for a set of construction documents, and the main floor has already been laid out by another drafter. The main floor contains information that governs how the upper floor is to be drawn. Thus, you can reference the main floor into your drawing so you can trace over it.

The overlay reference type also attaches an xref to the receiving drawing; however, it does not nest the xref into the receiving drawing. Instead, the overlaid xref is attached only to the receiving file and cannot be taken forward to any other drawings if the receiving drawing is xref'd into other drawings (see Figure 15.10).

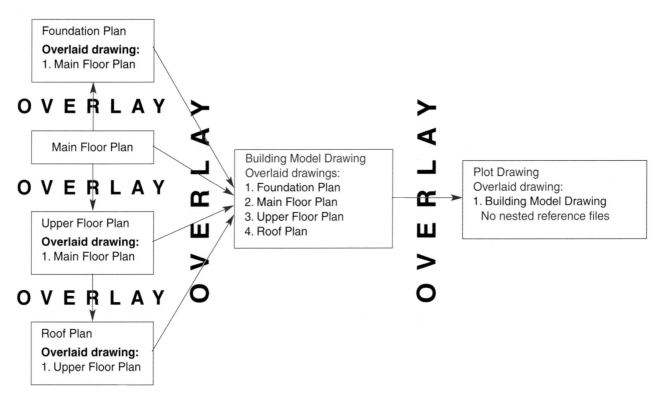

FIGURE 15.10   When the overlay reference type is used, the reference drawing can be attached only one step forward. It cannot be nested into other drawings.

The Overlay reference type aids the drafter or project manager who is assembling the plans into the building model. If the main floor is attached to the upper floor drawing using the attachment reference type, the main floor will go forward to the building model drawing. If the assembly drawing already has the main floor drawing attached, a conflict occurs, because two references with the same name are being referenced into the same drawing file (see Figure 15.11).

As long as the xref files being referenced are overlays, the main plan drawings can be assembled without causing any problems, as shown in Figure 15.12. Once a reference file has been attached to a drawing, all geometry within the xref is grouped together. The xref can then be moved, copied, rotated, and scaled within the drawing without affecting the original reference file. If the reference file needs to be removed from the drawing, use the **Detach** option to detach the xref instead of erasing it. Erasing an xref from a drawing can create some undesired results later in the project.

Field Notes

It is generally good practice to use the overlay reference type when using reference files to avoid accidentally creating duplicate references; however, office standards usually determine which drawings are overlaid and which drawings are attached. If all drawings are overlaid, the drafter or project manager assembling the files needs to xref files that are not being carried forward using the attachment reference type.

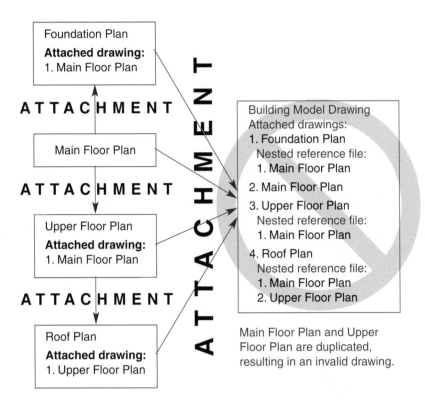

FIGURE 15.11 Attaching a nested reference into a drawing that already has the reference attached will cause a conflict.

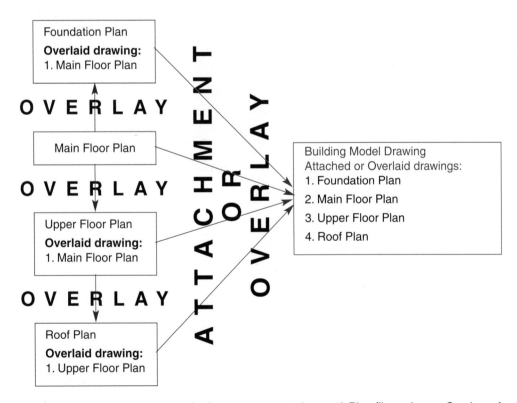

FIGURE 15.12 A combination of reference types can be used. Plan files using an Overlay reference type cannot be duplicated in the assembly drawing.

Exercise 15.1

1. Start a new drawing using the AEC Arch (Imperial) or the AEC Arch (Imperial-Int) template.
2. Create two new layers named Xref-Foundation and Xref-Main Floor.
3. Set the Xref-Main Floor layer current. Use the XREF command to attach the EX14-5-Flr-Plan.dwg. Set the reference type to Attachment and the insertion point to X:0, Y:0, and Z:0. The scale should be X:1, Y:1, Z:1, and the rotation should be set to 0. This step will xref in the main floor drawing created in the previous exercises.
4. Change to an isometric view. Set the Xref-Foundation layer current.
5. Use the XREF command to attach EX14-1-Fnd. Set the reference type to Attachment and the insertion point to X:0, Y:0, and Z:-32". The scale should be X:1, Y:1, Z:1, and the rotation should be set to 0.
   [*Note:* As the foundation plan is set below the main floor, the Z axis of the insertion point is set to 32" below the 0" Z axis. This value was determined by the height of the foundation walls measured from the foundation wall baseline to the top of the foundation wall.]
6. Save the drawing as EX15-1-Model.dwg.

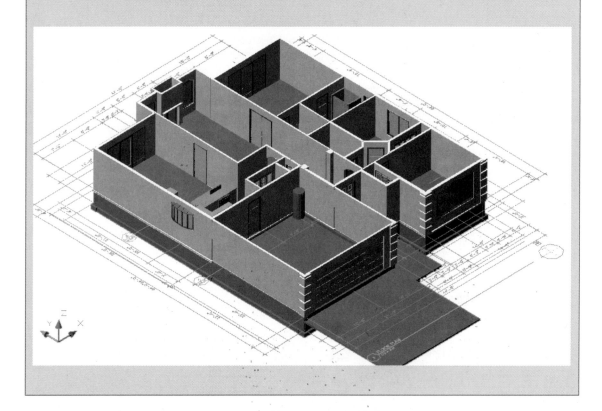

Using the Xref Manager Dialog Box

As mentioned earlier, The Xref Manager dialog box is used to attach and manage xref drawing files referenced in the current drawing. After a file has been referenced into the current drawing, the Xref Manager lists important information about any reference files in the drawing (see Figure 15.13).

Using the List View

The **Reference Name** column lists the xrefed drawings. The list of drawings can be displayed in either a list view or a tree view, by selecting the appropriate button in the upper left corner of the dialog box. By default, each time the Xref Manager dialog box is opened, the list view is displayed. If you are looking at a

FIGURE 15.13 The Xref Manager dialog box is used to attach new references and to manage existing references in the current drawing.

List of referenced files in the current drawing

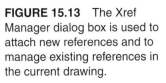
List View (F3) button
Tree View (F4) button

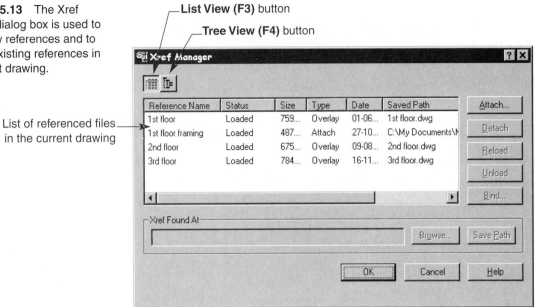

tree view of the xrefs, press the **List View (F3)** button, or the **[F3]** key to display a list of the reference files. In the list view, the following series of columns displayed across the top of the list box give specific information about the reference files in the drawing:

- **Reference Name** This column lists the names of reference files in the current drawing.

- **Status** This column describes the current status of each reference file according to the following reference classifications:

  - **Loaded** The xref file is attached to the drawing and is currently loaded within the drawing.

  - **Unloaded** The xref is attached to the drawing but has not been loaded into the drawing.

  - **Unreferenced** The xref is attached to the drawing but has been erased.

  - **Not Found** The xref was not found in the support file search paths or the directory in which the current drawing is saved.

  - **Unresolved** The xref is attached to the drawing but cannot be read by AutoCAD. This most likely means that the reference file has been corrupted.

  - **Orphaned** The xref is attached to the current drawing but is a nested xref dependent on its host drawing, which may be Unreferenced, Unresolved, Not Found, or Unloaded.

  - **Reload** This status marks the xref to be reloaded in the drawing. Press the Reload button to the right of the dialog box to reload the drawing. Once the xref is reloaded, the status changes to Loaded.

  - **Unload** This status marks the xref to be unloaded from the current drawing. The xref is not displayed even though it is still attached to the drawing. If the status is changed from Loaded to Unload, the xref is unloaded from the drawing when the OK button is selected.

- **Size** This column lists the size of each xref in the current drawing.

- **Type** This column indicates whether the xref is using the attachment or overlay reference type. Double-clicking on this value for an xref can change it from attach to overlay or vice versa. This is a good way to change the reference type if the xref was loaded with the wrong reference type; otherwise, the xref would need to be detached then reattached using the proper reference type.

- **Date**   This column lists the last saved date of the reference file.
- **Saved Path**   This column displays the saved path to the reference file. If the name of the reference file appears here without a path, the xref was attached without retaining the path.

Using the Tree View

Another way of viewing reference files in the Xref Manager is to use the tree view. Select the **Tree View (F4)** button or press the **[F4]** key to change the view from the list view to the tree view, as shown in Figure 15.14. In the tree view the current drawing is indicated at the top of the box with the AutoCAD drawing icon. Xrefs appear below the current drawing in a tree view, using one of two icons.

An icon that appears as two sheets of paper with a paperclip indicates that the drawing is using the overlay reference type. An icon that appears as a single sheet of paper with a paper indicates that the reference file is using the attachment reference type. A reference file that appears to branch under another xref indicates that the reference is a nested xref within its host drawing and has been brought forward to the current drawing.

Exercise 15.2

1. Open EX15-1-Model.dwg.
2. Access the Xref Manager dialog box.
3. View the list of information for each reference file.
4. Change the view to the tree view. Note the icons for the reference files. These should be set to the Attachment reference type.
5. Change to the list view.
6. Change the type value for each xref by double-clicking on the word Attach under the Type column to change the xref from an attachment reference type to an overlay reference type.
7. Change to the tree view. Note how the xref icons have changed.
8. Exit the Xref Manager, and save the drawing as EX15-2-Model.dwg.

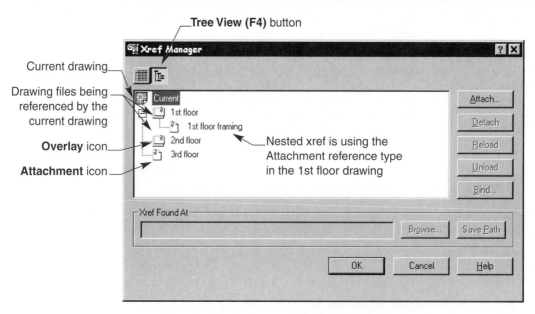

FIGURE 15.14   The tree view lists reference drawings with an icon indicating their level of nesting and placement within the current drawing.

Detaching, Reloading, and Unloading Reference Files

Once a reference file has been attached to the drawing, whether it is an attachment or overlay reference type, it is a permanent part of the drawing. Each time the drawing is opened, the reference files are reloaded displaying the most recent version of the xref(s). As long as the paths to the reference files have been saved or AutoCAD can find the xref in the support file search paths the xrefs are always reloaded.

Detaching an Xref

The xref is a permanent part of the drawing, so erasing the xref image removes the image only from the screen and not from the drawing's memory. Erasing is not a recommended method for removing the reference file. If an xref no longer needs to be referenced, it must be detached through the Xref Manager.

To detach an Xref, first access the Xref Manager, select the xref that to be detached from the list, then pick the Detach button (see Figure 15.15). This procedure detaches the reference file from the drawing and removes all layers associated with the xref. The referenced drawing image is also removed from the drawing screen. If the reference file needs to be used again, it must be attached to the drawing again.

Field Notes

> If an xref containing nested xrefs is detached from the drawing, any nested reference files are also removed from the drawing.

Reloading an Xref

Each time a drawing containing an xref is opened, AutoCAD reloads any reference files, displaying the graphics of the latest version of the reference file. Occasionally, you may need to reload the xref manually, as when another person is editing the drawing that you are referencing into your drawing.

Suppose that the other drafter has made a big design change to the floor plan drawing you are referencing for elevation drawings, and it affects how the elevations need to appear. In order to create the elevations correctly you need to relaod the floor plan with the latest changes. To do this, access the Xref Manager, highlight the reference to be reloaded, and select the Reload button. The status of the xref is changed from Loaded to Reload, as shown in Figure 15.16. The reference file is reloaded into the drawing once you exit the Xref Manager. After an xref has been reloaded, the status changes back to Loaded.

Unloading an Xref

When a drawing is being referenced it uses memory within the file to which it is attached in order to display the graphical image; however, it is still uses considerably less memory than a standard block. In some situa-

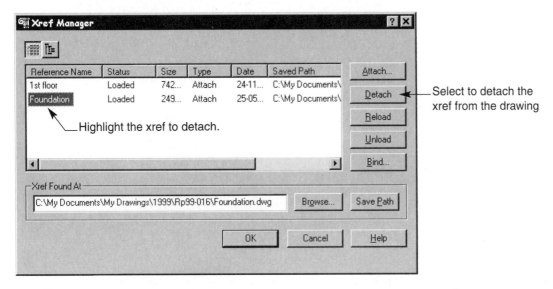

FIGURE 15.15   Use the Detach button to remove an xref from a drawing.

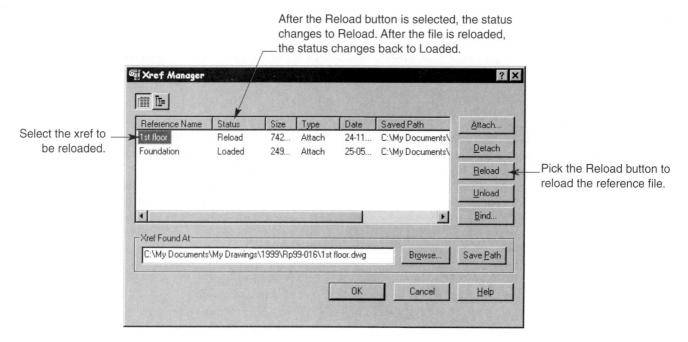

FIGURE 15.16 Use the Reload button to reload an xref into the drawing to obtain the latest version of the reference file.

tions, you may need to turn off or unload the xref. To unload an xref, select the reference file in the Xref Manager, then pick the Unload button (see Figure 15.17).

When a reference file is unloaded, the graphics and supporting layers are not displayed in the drawing. This is similar to turning off the layer where the reference was inserted. The difference is that when the reference file is unloaded, the graphics and supporting layers are not loaded into the drawing, which saves memory in the drawing causing it to use even less file space. The advantage is that the reference file is still attached to the drawing and knows where it was inserted, and the scale and rotation angle it was assigned. You can reload the reference file into the drawing at any time by using the Reload button.

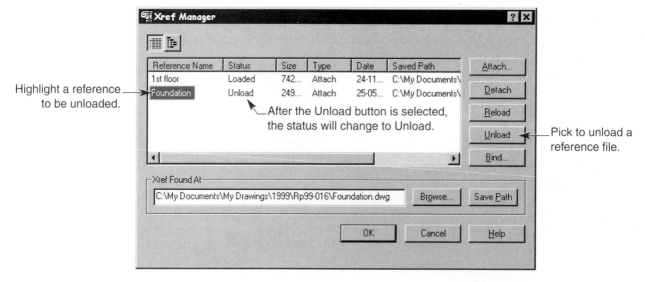

FIGURE 15.17 Use the Unload button to unload an xref but still maintain it is an xref in the drawing.

Binding an Xref

As indicated earlier, an Xref is similar to an inserted block. A reference file that is attached to a drawing groups all the objects together into one object and is placed on the current layer when attached. ***Binding*** is the process of permanently adding an xref to the current drawing. When a reference file is bound to the drawing in which it is inserted, it loses its ability to be reloaded and updated. It is also converted to a standard AutoCAD block within the drawing and increases the size of the receiving drawing. Binding is helpful if you need to send the full drawing to a consultant, another student, a plotting service, or a client.

When a drawing file is inserted into another drawing with the INSERT command, it also inserts blocks, dimension styles, layers, linetypes, text styles, and AEC object styles included in the drawing file. The xref is similar but loads only the layers that are simplified versions, because they can be used only by the xref geometry. When an xref is bound to the drawing into which it is referenced, the xref is turned into a block in the drawing and fully inserts the layers, blocks, dimensions styles, linetypes, text styles, and AEC object styles that are included in the original referenced drawing.

When a reference file is attached to a drawing, all definitions and styles that the reference is dependent on are given unique names. These styles and definitions include AEC styles, blocks, dimension styles, layers, linetypes, and text styles. The unique name is determined by name of the reference file plus the name of the style or definition. The reference file name and the style name are separated by a vertical bar symbol (|). This renaming technique allows you to easily distinguish between the style in the drawing and the style in the reference file.

For example, a drawing file with the name A-FP01 includes a layer name of A-WALL. When the drawing file is attached to another drawing the A-WALL layer is renamed (in the receiving drawing only) with the name A-FP01|A-WALL.

When a reference file is bound to the drawing to which it is attached, all dependent objects are renamed again to reflect that they have become a permanent part of the drawing. This is helpful if the receiving drawing includes a layer with a name that is also in the reference file. The renaming keeps the bound reference geometry on its own layers and the geometry in the receiving file on its own layers.

To bind an xref, access the Xref Manager, highlight the xref that is to be bound, then pick the **B**ind… button. The **Bind Xrefs** dialog box, shown in Figure 15.18, is displayed. The following two options are included: **B**ind and **I**nsert. These options are described next.

Using the Bind Option

When the **B**ind option is selected from the Bind Xrefs dialog box, the xref is converted to a block definition in the receiving drawing, with the same name as the reference file name. All dependent styles and definitions are also added to the receiving drawing and renamed. The vertical bar (|) in the style and definition

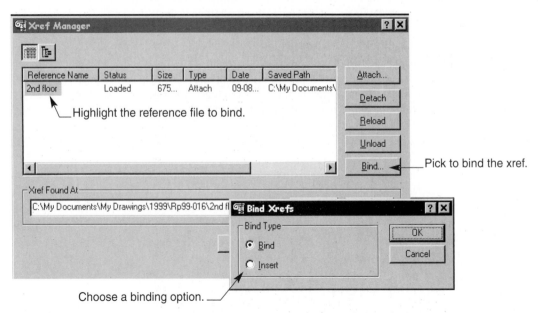

FIGURE 15.18   To bind an xref, highlight the xref in the Xref Manager, then pick the Bind… button. The Bind Xrefs dialog box is displayed with two options for binding the xref.

name is replaced with two dollar signs and a number (0). For example, the attached layer name A-FP01|A-WALL is renamed A-FP01$0$A-WALL when the Xref is bound.

The number between the dollar signs is automatically incremented if a style or definition name already exists with the same name. For example, if the A-FP01$0$A-WALL layer name already exists in the drawing, the new bound layer of A-FP01$1$A-WALL is created. The numbering system ensures unique names for xref-dependent styles and definitions that are bound into the receiving drawing.

Using the Insert Option

The Insert option treats the xref like a block that is inserted into the drawing. This option also converts the xref into a block definition with the same name as the reference file and adds the dependent styles and definitions to the receiving drawing. The difference between this option and the Bind option is the naming convention used for the dependent styles and definitions.

When an xref is bound to the drawing with the Insert option, the reference file name is stripped from the style and definition names. For example, if the xref A-EL01 includes the layer name A-DOOR, the xref-dependent layer name A-EL01|A-DOOR is renamed A-DOOR when bound to the drawing with the Insert option.

If the receiving drawing already includes styles and definitions with the same names found in the reference file, then the receiving drawings styles and definitions take precedent. Any xref geometry is then placed on the receiving drawings styles and definitions with their properties. For example, if a reference file and the file it is attached to both include a block named TOILET that appears different in each drawing file. If the reference file is bound to the drawing with the Insert option, the TOILET block from the reference file takes on the properties of the TOILET block in the receiving drawing. Any occurrences of the TOILET block in the reference drawing are updated to reflect the appearance of the TOILET block in the receiving drawing.

Binding a reference file is usually not an optimal option when using xrefs. The purpose of using xrefs is to reference a drawing. When an xref is bound, the link between the receiving drawing and reference file is broken. Any changes to the original reference file are no longer updated in the receiving drawing. Binding the xref also increases the size of the receiving drawing, because the whole drawing (geometry, styles, and definitions) is inserted into the receiving drawing.

Working with Xref Paths

When attaching a reference file, you have the option of retaining the path or not in the **External Reference** dialog box. If you select this option, the path to the reference file appears in the Saved Path column in the Xref Manager dialog box. If an xref cannot be found in the Saved Path location when the receiving drawing is opened, AutoCAD searches through a series of paths in the **Support File Search Path** folder, and in the folder from which the receiving drawing was opened, in an attempt to load the xref. The Support File Search Path folder can be found in the **Options** dialog box, under the **Files** tab.

If a drawing file matching the reference name is found, the drawing is loaded into the receiving drawing. If the drawing file is found in a different folder than was specified in the Saved Path column, the xref still loads as long as the reference file names are the same. A comparison of the path in the Saved Path column and the path in the Xref Found At area reveals where the reference file was found (see Figure 15.19). If the paths are different, but the correct xref is loaded, the path needs to be updated. Enter the Xref Manager and highlight the xref whose path need to be updated, then pick the Save Path button at the bottom of the dialog box. The new path to the xref is saved, so the next time the drawing is opened or the xref is reloaded, AutoCAD knows where to find the xref.

If a reference file is moved to a new location on the hard drive or the network drives, and the new location is not in the search paths, the next time a drawing with the xref attached is opened, the status for the reference file reads Not Found, as shown in Figure 15.20. The new path can be updated by selecting the Browse… button in the Xref Found At area which allows you to browse for the missing xref. When the xref is found, it is loaded into the drawing. Use the Save Path button to save the new path so AutoCAD can find the xref the next time the drawing is opened.

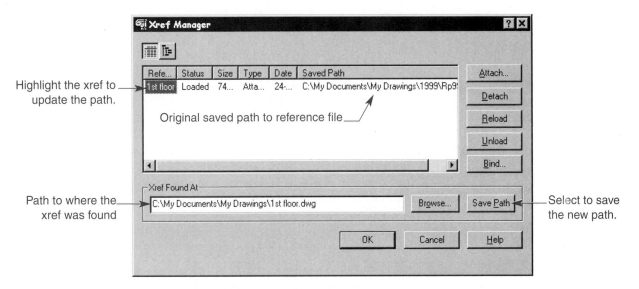

Highlight the xref to update the path.

Original saved path to reference file

Path to where the xref was found

Select to save the new path.

FIGURE 15.19   To update the saved path to an xref, highlight the xref, then pick the Save Path button.

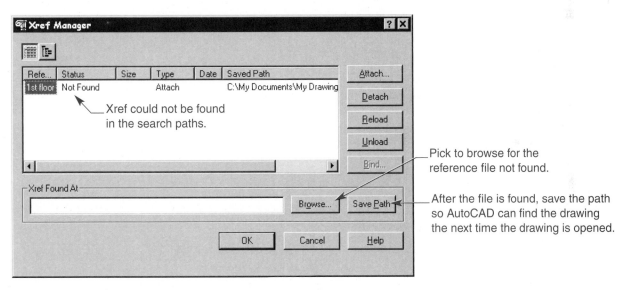

Xref could not be found in the search paths.

Pick to browse for the reference file not found.

After the file is found, save the path so AutoCAD can find the drawing the next time the drawing is opened.

FIGURE 15.20   Use the Browse… button to look for xrefs that are not found. Once the xrefs are loaded, use the Save Path button to save the new paths to the xrefs.

Exercise 15.3

1. Open EX15-2-Model.dwg.
2. Access the Xref Manager dialog box.
3. Highlight the EX14-1-Fnd.dwg, then pick the <u>U</u>nload button.
4. Close the Xref Manager and observe what happens to the foundation reference file.
5. Access the Xref Manager dialog box.
6. Highlight the EX14-1-Fnd.dwg, then pick the <u>R</u>eload button.
7. Close the Xref Manager and observe what happens to the foundation reference file.
8. Save the drawing as EX15-3-Model.dwg.

Clipping an External Reference

A set of construction documents requires several views of different portions of a building. Often, these views are detail drawings that display how building components are assembled or enlarged areas of drawings that require some special notes. In these cases, the ability to use an xref and cut away the portions not required is very important. AutoCAD includes the **XCLIP** command, which allows you to clip (cut) a portion of an xref out of a drawing.

When the XCLIP command is used, a boundary, referred to as the ***clipping boundary***, is established on the xref. Any reference geometry outside the boundary is clipped away. Reference geometry that is inside the boundary is displayed within the receiving drawing. Although the geometry appears to be trimmed away from the boundary, the original reference file remains untouched and includes all the original geometry. The XCLIP command also allows you to modify and remove clipping boundaries from an Xref.

Use one of the following options to access the XCLIP command:

✓ Select the **Modify** pull-down menu, pick the **Clip** cascading menu, and select the **Xref** command.

✓ Pick the **External Reference Clip** button on the **Reference** toolbar.

✓ Type **XCLIP** or **XC** at the Command: prompt.

On entering this command, you are prompted to select xrefs for clipping, then to select from a list of options for the method to be used to clip the reference. The following prompt sequence is used to create a rectangular clip of the floor plan in Figure 15.21:

Command: **XC** or **XCLIP**↵
Select objects: *(select the xref)*

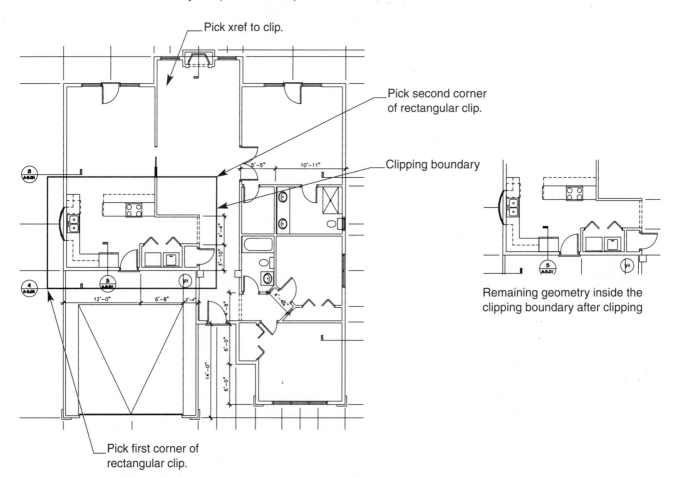

FIGURE 15.21   Use the XCLIP command to clip away a portion of an xref.

```
1 found
Select objects: ↵
Enter clipping option
[ON/OFF/Clipdepth/Delete/generate Polyline/New boundary] <New>: ↵
Specify clipping boundary:
[Select polyline/Polygonal/Rectangular] <Rectangular>: ↵
Specify first corner: (pick first corner)
Specify opposite corner: (pick second corner)
Command:
```

Once the xref has been selected and you no longer need to select xrefs for clipping, the following series of options is displayed for creating and modifying a clipped xref:

- **ON/OFF:**   After an xref is clipped the clip is turned on, clipping away undesired xref geometry, which unloads the portion of unused xref from the drawing. The OFF option turns the clip off and reloads the entire xref. The clipping boundary is still remembered by the xref, so it can be turned back on using the ON option.

- **Clipdepth:**   This option creates a front and a back clipping plane that can be defined. The *front clipping plane* cuts away the top portion of the xref. The *back clipping plane* cuts away the bottom portion of the xref.

- **Delete:**   This option removes the clipping boundary from the xref regardless of whether the clip is turned on or off.

- **Generate Polyline:** This option generates a polyline from the clipping boundary. The new polyline is drawn on the current layer.

- **New boundary:**   This option is used to create a new clipping boundary, using one of three methods: **Select polyline, Polygonal,** or **Rectangular**.

Creating a Clipping Boundary

The New boundary option is used to create a new clipping boundary within the xref. Selecting this option displays a prompt with the following three choices of clipping boundaries that can be created:

- **Select polyline:** This option requires that a polyline be drawn on top of the xref around the area that will be retained by the clipping. The polyline must be drawn with straight-line segments; otherwise, the clip will create a straight cut between the starting and ending points of the polyline arc. If the desired clipping must be curved, then first draw the polyline with straight-line segments, then use the PEDIT command to edit the polyline into a spline. If the polyline is not closed, the clip creates a boundary between the starting and ending points of the polyline.

- **Polygonal:**   This option creates a clipping boundary from several selected vertex points that establish an irregular-shaped polygon. When you are finished selecting points for the polygonal shape, press the [Enter] key to close the starting and ending points of the clip.

- **Rectangular:**   This is the default option and creates a clipping boundary by picking two opposite corners of a rectangular shape, as shown in Figure 15.21.

Once a clipping boundary has been established, any geometry outside the clipping boundary is unloaded from the drawing. Like an unclipped xref, the clipped reference can be moved, copied, scaled, and rotated within the drawing without affecting the original reference file. If the clipped xref is modified, the clipping boundary is retained and stays with the modified reference. If the reference is copied, a new clipping boundary is created for the new copy of the reference file. If an xref that contains nested xrefs is clipped, the nested xrefs are also clipped and displayed if they are inside the clipping boundary, or else they are clipped away from the xref.

Initially, when the Polygonal or Rectangular clip options are used, the clipping boundary is invisible, displaying only geometry that is within the boundary. The clipping boundary can be displayed by setting the **XCLIPFRAME** system variable to a value of 1. The variable can be changed by picking the External Reference Clip Frame button in the Reference toolbar, or by typing **XCLIPFRAME** at the Command: prompt. A value of 1 turns the frame on, which is placed on the same layer as the xref and can be plotted. A value of 0 turns the clipping frame off. The clip frame can also be turned on or off by selecting the Modify, Object, External Reference, Frame command from the pull-down menu. Figure 15.22 displays the three different types of clips with the clipping boundary turned off.

Field Notes

To create a round clipping boundary for an enlarged detail, use the **POLYGON** command to draw a polygon with many sides around the desired clipping area. Use the PEDIT command to spline the polygon to make it appear round. Use the XCLIP command with the Select Polyline option, and select the splined polygon to clip the Xref.

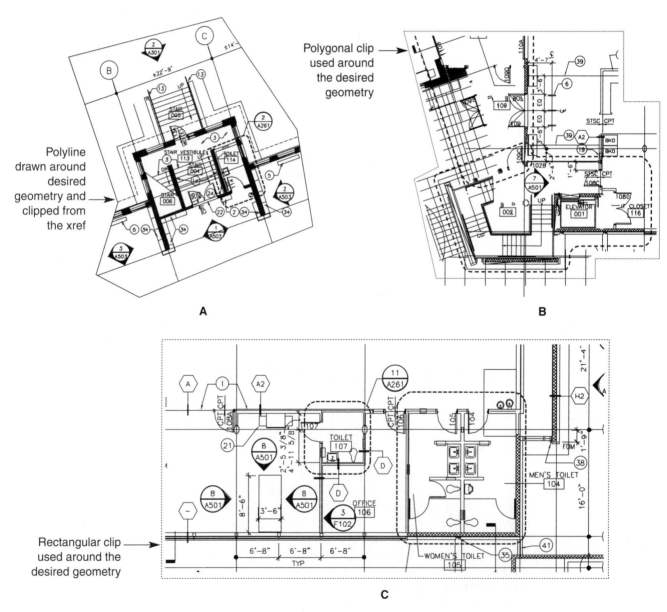

FIGURE 15.22   The three types of clipping boundaries with clip frames turned off: (A) Select polyline. (B) Polygonal. (C) Rectangular. Drawing Courtesy SERA Architects Inc., Portland, OR.

Exercise 15.4
_____

1. Open EX15-3-Model.dwg.
2. Use the XCLIP command to clip the left rear corner of the main floor xref from the drawing.
3. Use the XCLIP command to clip the front right corner of the foundation from the drawing.
4. Save the drawing as EX15-4-xclip.dwg.

Editing Reference Drawings

As you attach references into a drawing you may notice a part of the file that a drafter forgot to change or a design element that needs to be changed. In these situations you need to edit the original reference in order to complete the set of plans. AutoCAD provides a tool known as *reference editing* to edit an xref in the current drawing. This timesaving editing procedure allows you to edit a reference, a nested reference, or a block in the current drawing without having to open the original drawing file, make the changes, and reload the reference. Any changes made to the xref file are then saved back to the original reference file without your ever having to exit out of the current drawing.

Field Notes

> Reference editing is best used for minor revision work. Larger changes should be made within the original reference file.

Use one of the following options to access the **REFEDIT** command, which must be used to edit a reference file from the current drawing:

✓ Select the **Modify** pull-down menu, pick the **In-place Xref and Block Edit** cascade menu, then select the **Edit Reference** command.

✓ Pick the **Edit Block or Xref** button in the **Refedit** toolbar.

✓ Type **REFEDIT** at the Command: prompt.

Once the command has been entered, you are prompted to select a reference.

You can edit xrefs or blocks with the **REFEDIT** command. At the prompt, select the xref or block that is to be edited. The **Reference Edit** dialog box is displayed, as shown in Figure 15.23. A preview of the

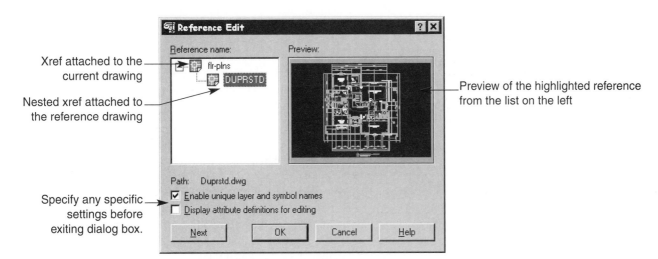

FIGURE 15.23 The Reference Edit dialog box lists the name of the reference selected plus any nested xrefs or blocks at the point at which the reference was selected.

selected xref is displayed in the Preview: panel to the right, and the name of the file being edited is high-lighted on the left. The point that is picked when a reference file is selected determines the objects that are displayed in the tree view along the left side of the dialog box. In Figure 15.23, the point selected also hap-pened to be over the top of a nested xref; thus, both the nested xref and the reference file are included in the list. Either of these references can then be edited.

Below the Reference name list a path is displayed showing you where the xref is located on the hard drive or network drives. Selecting between the xref and the nested xref changes the highlighted name, the preview, and the path in this dialog box. Two additional check boxes are included at the bottom of the dia-log box: the **Enable unique layer and symbol names** and the **Display attribute definitions for editing**.

The Enable unique layer and symbol names check box is active by default. This option renames the xref-dependent layer and block names within the selected reference during the editing process. If this box is selected, layer and symbol names are preceded by a n, where the *n* represents an incremented number. This renaming technique is used during the editing process only. When the file is finished being edited, the layer names revert to the xref naming convention *drawing name/layer name*. If the check is removed, the dependent layer names are not renamed during the editing process.

The Display attribute definitions for editing check box is used when a block that contains attributes is edited. An ***attribute*** is a piece of AutoCAD geometry that is added to a block and can provide data about the block that the drafter manually enters at the time the block is inserted. Selecting this check box allows you to edit the attributes within a block that is being edited.

If an xref is picked at a point where a dependent block or nested Xref is located, the block or nested xref appears in the Reference name list. Use the Next button to cycle through the list of xrefs and blocks in the list. When the xref or block that you wish to edit is highlighted in the list, press the OK button to edit the file. You are then prompted to *Select nested objects:* at the Command: line.

This prompt is asking you to pick geometry within the selected xref or block that is to be edited. Pick on top of any geometry that needs to be edited, then press the [Enter] key. The objects selected create what is known as a ***working set***. When you have finished selecting the geometry to be edited, press the [Enter] key. Any geometry that was not selected is grayed out. The geometry included in the working set remains the same color and shade, indicating that this is the geometry that has been selected for editing. After the geometry to be included in the working set has been selected, the Refedit toolbar is also displayed (see Figure 15.24).

The Refedit toolbar displays the name of the reference or block being edited and remains on the screen during the editing process. The toolbar can be used to add or remove objects from the working set, and save or discard changes made to the original reference file. The toolbar contains the following buttons:

- **Reference**

  This text box displays the name of the reference or block currently being edited.

- **Add objects to working set**

  This button is used to add to the current working set any geometry that is faded out. Any new geometry drawn is automatically added to the working set.

- **Remove objects from working set**

  This button removes objects from the working set and fades the color of the geometry.

- **Discard changes to reference**

  This button exits the reference editing session without saving the changes to the original reference file. Any changes, additions, or modifications made to the work-ing set are discarded.

- **Save back changes to reference**

  This button saves any changes made to the original ref-erence or block being edited when you have finished editing the geometry in the working set.

Once the working set has been created, use any AutoCAD or Architectural Desktop commands to modify existing geometry and to create new geometry. Figure 15.24 shows some wall objects and multi-view blocks being edited. The Wall Modify command can be used to adjust the properties of the selected walls, and the MOVE command can be used to reposition the multi-view blocks. If new AEC objects or polylines need to be added to the reference file, then draw them in at this point.

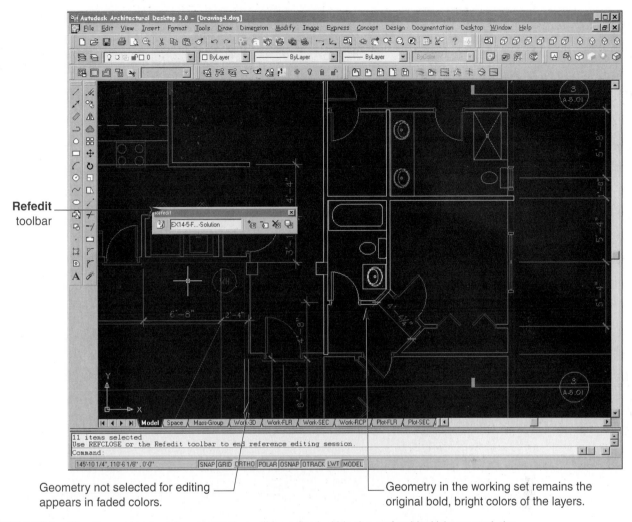

Refedit toolbar

Geometry not selected for editing appears in faded colors.

Geometry in the working set remains the original bold, bright colors of the layers.

FIGURE 15.24 After the working set (geometry that will be edited within the xref or block) is created, the geometry not selected is faded out, since it is not being modified. The Refedit toolbar is also displayed during the editing process.

After making the modifications to reference file, pick the Save back changes to reference button to save the changes to the reference file, or pick the Discard changes to reference button to discard the changes made. If the Save back changes to reference button is selected, a warning box appears, indicating that the changes made will be saved in the original reference file. Pick the OK button in the warning box to apply the changes.

Field Notes

When the REFEDIT command is used to change the geometry in an Xref, the changes are saved to the original file, which may affect other drawings that are also referencing the file being edited. Any changes made through the REFEDIT command should be carefully evaluated, as they may affect other drawings down the line. If the command is used on blocks, any blocks in the reference file using the block definition being edited are also updated with the changes.

Exercise 15.5

1. Open EX15-3-Model.dwg.
2. Use the REFEDIT command to edit the main floor reference file.
3. Edit the garage door's vertical alignment to be −8″ from the threshold.
4. Save the changes to the main floor Xref.
5. Use the REFEDIT command to modify the foundation reference file.
6. Add a 16′-0″ × 8″ rectangular opening to the front garage wall.
7. Set the vertical offset such that it cuts out the front foundation wall allowing the garage door to fit into the hole.
8. Save the changes to the foundation Xref.
9. Save the drawing as EX15-5-Model.dwg

Move the door 8″ below
the bottom of the wall.

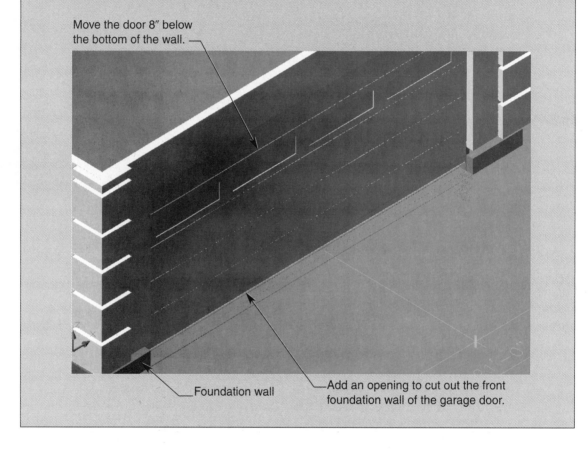

Foundation wall

Add an opening to cut out the front
foundation wall of the garage door.

Setting Up Xref Configurations

When you work with reference files, it is important that you understand how AutoCAD/Architectural Desktop finds and uses xrefs. The following discussion explains configuration options that you should explore before referencing drawing files.

Finding Reference Files

When a drawing containing an xref is opened, AutoCAD searches through folders (directories) on the disk drives and/or network drives in order to locate, load, and display the reference file. If the xref was attached to the drawing with the Retain Path check box selected, then AutoCAD looks in the saved path for the reference file (see Figure 15.25). If the reference file has been moved from the directory to which the path is pointing, then AutoCAD looks through the Support File Search Path folder looking for the reference. If the reference is found in one of these paths, then the reference file is loaded into the drawing. If the file cannot be found, the reference is not loaded, and the status in the Xref Manager indicates that the xref is Not Found.

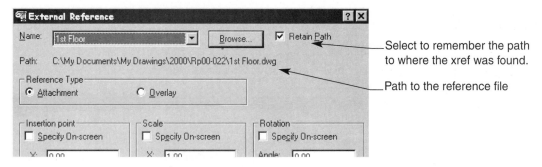

FIGURE 15.25   Select the Retain Path option if the path to the reference is to be remembered.

Using the Support File Search Path

In order to review the paths that AutoCAD is looking through in the Support File Search Path, use one of the following methods to access the Options dialog box:

✓ Select the **Tools** pull-down menu, and pick **Options....**

✓ Right-click in the drawing window or in the Command: prompt area and select **Options...** from the cursor menu.

✓ Type **OPTIONS** or **OP** at the Command: prompt.

The **Options** dialog box, shown in Figure 15.26, is displayed.

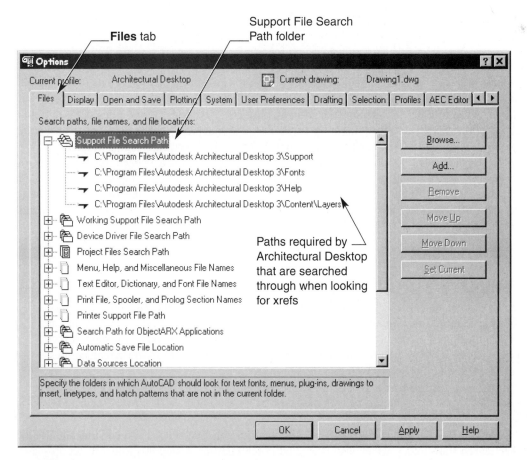

FIGURE 15.26   The Files tab in the Options dialog box includes the Support Files Search path as well as additional paths required by Architectural Desktop.

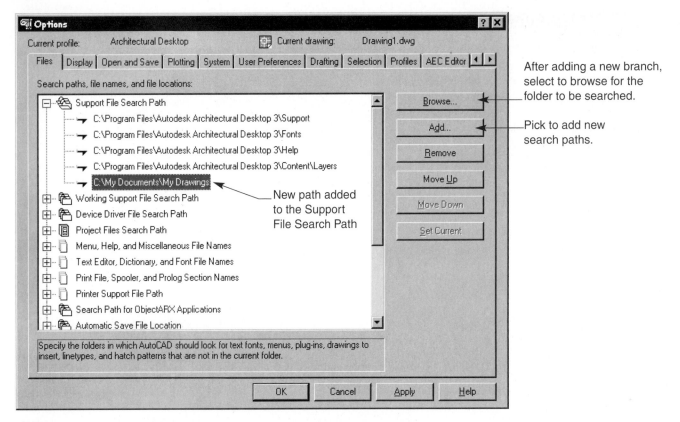

After adding a new branch, select to browse for the folder to be searched.

Pick to add new search paths.

New path added to the Support File Search Path

FIGURE 15.27 The new search path is added to the Support File Search Path folder.

The Options dialog box is used to set parameters and variables that tell Architectural Desktop how to work and behave. A number of tabs are included in the dialog box, controlling different sets of parameters. The Files tab is the first tab in the list and includes several search paths for Architectural Desktop to find required files in which to run. The Support File Search Path is the first folder in the Files tab (see Figure 15.26). Pick the plus + sign next to the folder to expand a list of search paths. These paths are created at the time Architectural Desktop is installed and are required in order for Architectural Desktop to function.

If a reference file cannot be found when a drawing containing an Xref is opened, then Architectural Desktop searches through these paths. You can add additional search paths to the Support File Search Path folder by picking the Add... button. A new branch is added to the tree. Select the Browse... button to browse through a list of folders to create a search path. Once the path has been entered, Architectural Desktop searches through this path looking for any missing xrefs (see Figure 15.27).

Field Notes

Be cautious about adding too many paths in the Support File Search Path. The more paths that are added, the more time Architectural Desktop takes to access files it needs to function as searches through all the search paths. This area is usually used to locate paths for support files used by Architectural Desktop. If multiple search paths are required to locate reference files, then use the Project Files Search Path folder, as discussed next.

Using the Project Files Search Path

If multiple folders need to be searched through in order to locate reference files it is a good idea to use the **Project Files Search Path** folder, which also is located in the Files tab. Using this folder to set up search paths helps increase performance, as Architectural Desktop does not have to search through multiple paths in the Support File Search Path.

The Project Files Search Path folder is used to set up different project folders, each with its own set of search paths. The idea is to create a project folder for each of your projects and set up search paths to any folders containing files such as drawings, documents, spreadsheets, and images required for the project. You can create as many project folders as needed, but only one project folder can be active at any given time.

When a drawing file containing xrefs is opened, Architectural Desktop first searches for the xref's saved path if it was retained. If the reference file cannot be found, Architectural Desktop searches through the active project folder and its search paths looking for the xref. If the xref still cannot be found, then Architectural Desktop looks through the Support File Search Path folder. If the xref is still missing, then the reference status in the drawing is listed as Not Found.

To create a project folder, locate the Project Files Search Path folder in the Files tab within the Options dialog box, and expand the tree as shown in Figure 15.28. Press the Add... button to create a project name folder. Enter a name for the new project folder. Expand the new project folder by pressing the + sign. Add any search paths that are to be used by the project folder by pressing the Add... button, then pressing the Browse... button to browse for the folder to be searched. When you have finished creating the search paths for the project folder, highlight the project folder and press the Set Current button to set the project folder as the active project folder to search through. Setting a project folder active instructs Architectural Desktop to search through this folder first if the reference file cannot be found. A small white symbol is added to the project folder, indicating it is the active project folder.

If only one project folder is created with a number of search paths, Architectural Desktop always searches here first for any missing reference file. In many offices, drafters work on several different projects throughout the life cycle of each project. Set up a project folder with the appropriate search paths for each of these projects.

Throughout this text you have been working on one residential project and several end-of-chapter problems. Multiple project folders can be created in such situations, one for each drawing project, with each project folder containing search paths to the drawing files required for each of the residential and commercial projects. Remember that only one project folder can be active at a time. Therefore, if you are work-

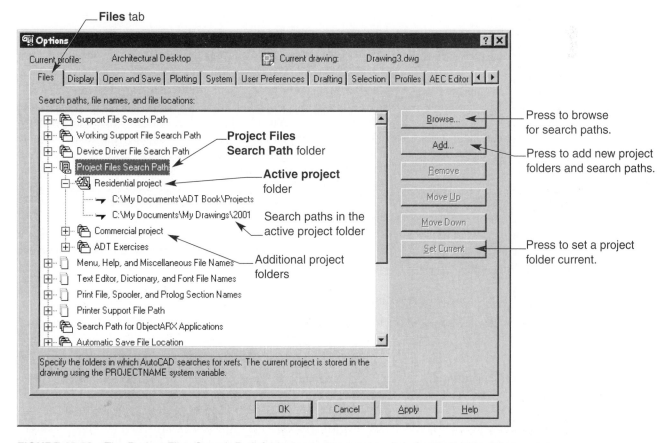

FIGURE 15.28   The Project Files Search Path folder is used to create a list of project folders that contain search paths to files required by a project.

ing on a commercial project, set the commercial project folder current before you begin working. If you later need to work on a residential project, then set a residential project folder current before working.

When working with project folders it is also a good idea to set up a drawing variable in each drawing of a project that points back to the project folder to which it belongs. For example, suppose that a project folder is created for the residence used in the exercises in this book. Each drawing created for this project, such as the latest floor plan and latest foundation plan, needs to point back to the project folder to which it belongs. Use the **PROJECTNAME** variable to do this. Within each drawing, enter PROJECTNAME at the Command: prompt. You are then prompted to enter the name of the project folder.

The following prompt sequence is used to set up the PROJECTNAME variable for a drawing:

Command: **PROJECTNAME** ⏎
Enter new value for PROJECTNAME, or . for none <"active project folder">: *(enter the project folder name to which to point this drawing)* ⏎

The advantage of setting this variable in each drawing is that every time this drawing is opened, it looks in the project folder specified in the PROJECTNAME variable for any reference files, regardless of the active project folder.

 Field Notes

> If you are sharing drawing files among offices, consultants, or contractors outside your office, do not retain the path to a reference file. If you do retain the xref path and send the drawings outside the office, Architectural Desktop may not be able to find the reference files, because the path locations are pointing to the folders on your computer or in your office. Set up project folders and search paths so AutoCAD always looks in the project folders for the reference files. Thus, consultants can place the reference files in any folder they have on their own systems and use their own project folders to find the required reference files.

Working with Other People

Construction documents for architectural projects are generally established from several sources with different parts of the plans outsourced to different offices specializing in a specific part of the project. For example, the design, floor plans, and elevations are generated by the architect, structural plans are generated by the structural engineer, mechanical plans are generated by a mechanical engineer, electrical plans are generated by an electrical engineer, and site plans are generated by a civil engineer. Each of these sources requires some of the same drawings that the others are creating. When drawings must be shared among offices or contractors, it is recommended that the path locations to the reference drawings not be retained, because each office has a different directory (folder) structure. Instead the project folders previously discussed should be used to locate reference files.

If drawings need to be outsourced to another office or consultant, Architectural Desktop provides a tool, known as **eTransmit,** to aid you in finding all the required drawing files that need to be sent. Use one of the following options to acess this tool:

✓ Select the **File** pull-down menu, then pick the **eTransmit...** command.

 ✓ Pick the **eTransmit** button from the **Standard** toolbar.

✓ Type **ETRANSMIT** at the Command: prompt.

The **Create Transmittal** dialog box, shown in Figure 15.29, is displayed. If you receive a warning message, you must save the drawing before entering the eTransmit command.

The dialog box is divided into three tabs: **General, Files,** and **Report.** The General tab contains options for the drawing that is being transmitted. The Files tab lists any dependent files on which the drawing relies. The Report tab is used to generate a report of the dependent files and instructions where to place some of them. An expanded discussion of the tabs follows:

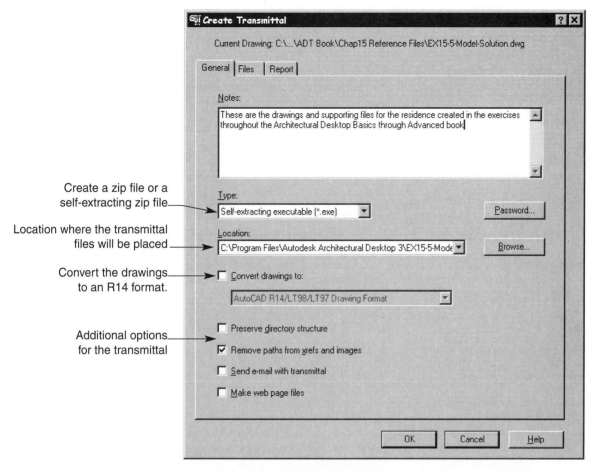

Create a zip file or a
self-extracting zip file

Location where the transmittal
files will be placed

Convert the drawings
to an R14 format.

Additional options
for the transmittal

FIGURE 15.29 The Create Transmittal dialog box enables you to locate all required files for a drawing and send them to another location.

■ **General**	This tab includes options for dealing with the transmittal files. The current drawing and any supporting files are placed together so they can be sent to a consultant or to another office.
■ **Notes:**	This area is used to enter any notes to the transmittal.
■ **Type:**	This list provides three different options for dealing with the drawing file and any required files. The files can be placed in a folder on the hard drive or on network drives, placed in a self-extracting zip file, or combined in a zip file. The last two options require that zip decompression software such as PKUNZIP or WINZIP be installed on your system.
■ **Password...:**	This button is used to assign a password to the zipped transmittal set. Others will need to provide the password in order unzip the file and access the contents of the transmittal set. A password can be assigned only to the Self-extracting executable or the zip-type transmittal set.
■ **Location:**	This text box is used to specify where the previous type is to be placed.
■ **Browse...:**	This button is used to browse for a folder location.
■ **Convert drawings to:**	This check box allows you to save the drawing and any dependent drawings (reference files) as an AutoCAD R14 drawing or

an AutoCAD 2000 drawing. Select the type of format from the drop-down list.

- **Preserve <u>d</u>irectory structure:**   This option saves the directory information for all the files in the transmittal set when they are sent to the consultant. Clearing this box places all the files in the target folder when it is unzipped.

- **Remove paths from <u>x</u>refs and images:**   This check box removes the saved path of xrefs and image files in the drawing so the consultant can place the reference file in one location and use a project folder to find the files.

- **<u>S</u>end email with transmittal:**   This option creates a new e-mail and attaches the files to the e-mail for transmittal to the consultant.

- **<u>M</u>ake web page files:**   This option creates a Web page with a link to the dependent files.

- **Files**   This tab lists all dependent files in the current drawing. All reference files, font types, plot styles, and images are listed in this tab. A check box beside each file allows you to choose whether to send the file in the transmittal (see Figure 15.30).

- **List View (F3)**   This button displays the files in a list.

- **Tree View (F4)**   This button lists the files in a tree view showing you the source of the dependent files.

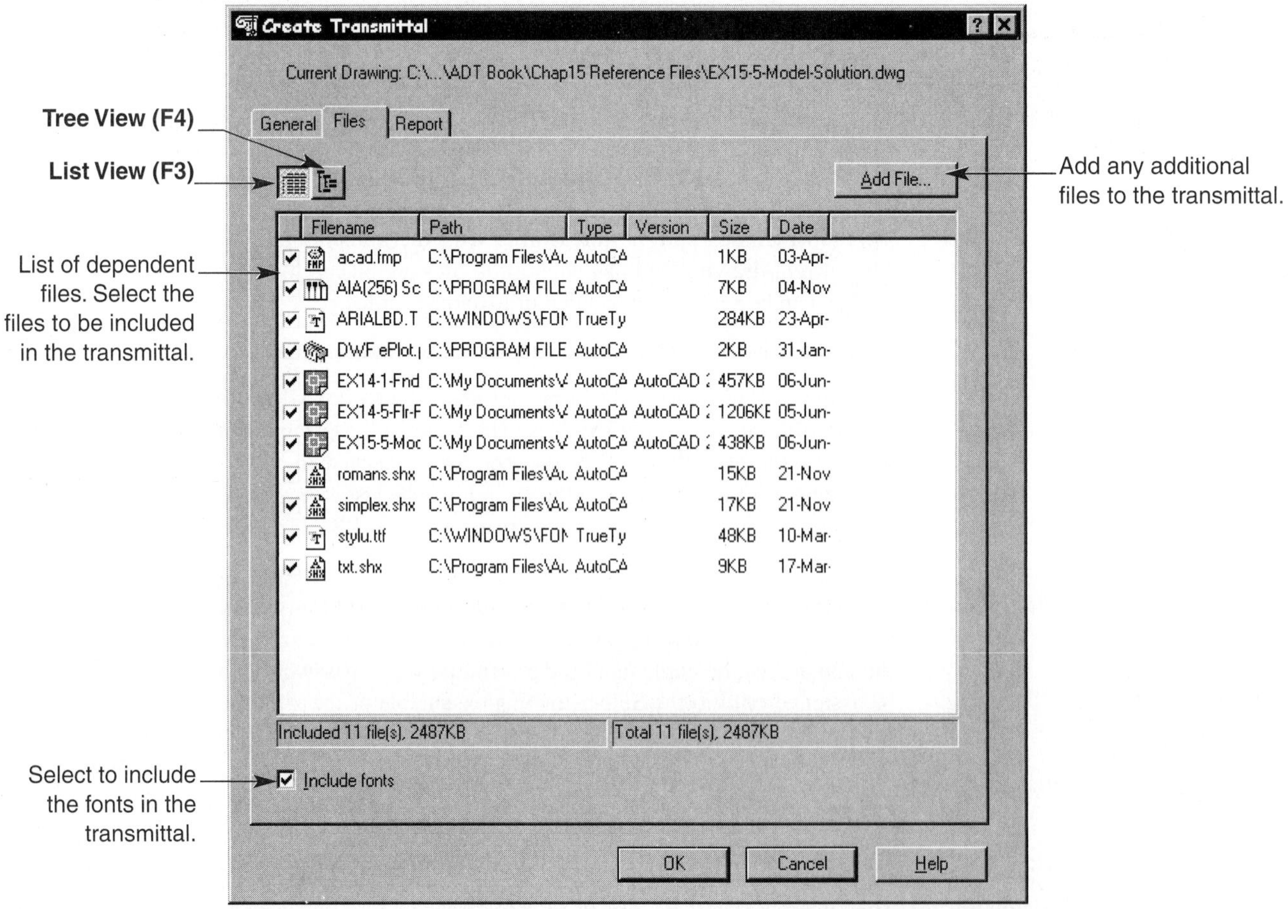

FIGURE 15.30   Use the Files tab to list the files that will be added to the transmittal.

- **Add File...**   This button allows you to add additional files to the transmittal.

- **Include fonts**   This option allows you to include the fonts in the transmittal.

- **Report**   This tab lists the report that is to be sent in the transmittal. This
                           tab also includes a **Save As** button to save a copy of the transmit-
                           tal for your own records.

When you have finished in this dialog box, press the OK button to generate the transmittal.

Exercise 15.6

1. Open EX15-5-Model.dwg.
2. Use the eTransmit command to create a transmittal.
3. If you have a zip program installed, create a zip file. If not, transmit the files to a folder on your hard drive.
4. Press the OK button when you have finished to process the transmittal.
5. Locate the transmitted files and ensure that the files were transmitted to the zip file or to the folder.
6. Exit the drawing without saving.

Understanding Demand Load

Demand load controls how much of a reference file is loaded when it is attached to the receiving drawing. When demand load is enabled, Architectural Desktop loads only the parts of the reference file required to generate the image of the xref as it is seen in the original reference file. This improves performance and saves disk space, because the entire reference file is not loaded. For example, any objects on frozen layers, or geometry outside clipping boundaries is not loaded.

Demand load is enabled by default. To control this setting, select the Open and Save tab in the Options dialog box, as shown in Figure 15.31. The following three demand load options are available in the Demand load Xrefs: drop-down list in the External References (Xrefs) area:

- **Enabled**   This option turns on demand loading in the current drawing. If a
                           drawing is being referenced in the currently open drawing, then
                           the reference file becomes a *read-only* file, and no other users can
                           open and edit the reference file, although they can xref the file
                           into another drawing.

- **Disabled**   This option turns demand loading off in the current drawing. If a
                           drawing is being referenced into the currently open drawing, it
                           remains open so other users can open and edit the reference file.

- **Enabled with copy**   This option turns off demand loading in the current drawing, but
                           the xref file can still be opened and edited by another user, as in
                           the Disabled option. AutoCAD copies the xref file to a temporary

FIGURE 15.31 The Open and Save tab in the Option dialog box contains the External Reference (xrefs) area that controls how xrefs are loaded into the drawing.

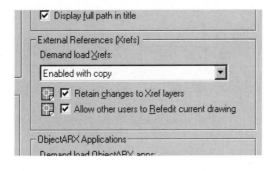

folder and treats it as a completely separate file from the original xref. Each time the reference file is reloaded in the current drawing, a new copy of the reference is placed in the temporary folder.

In addition to the demand load options, the following two additional options are included in the External References (Xrefs) area. These options have a red AutoCAD icon beside them, which indicates that these settings are remembered within the drawing file.

- **Retain changes to Xref layers**

    This option remembers the state of the reference layers in the active drawing. For example, if a few reference layers are frozen and others have their colors changed, the drawing remembers these changes the next time the drawing is opened, rather than reloading the reference with the original layer state from the original reference drawing.

- **Allow other users to Refedit current drawing**

    This option allows other drafters to be able to use the **REFEDIT** command on this drawing while it is being referenced by another drawing file.

|||||||||| CHAPTER REVIEW

 Use the CD-ROM to test your knowledge and skills.

Chapter Test

To check your understanding of the content provided in this chapter, access the Test file in the CH15 folder of the CD-ROM that accompanies this text.

Chapter Project

To practice the Architectural Desktop skills presented in this chapter, access the Project files in the CH15 folder of the CD-ROM that accompanies this text. The project files are in pdf format and include sample drawings and instructions for completing each project.

CHAPTER 16

Creating Roofs

LEARNING GOALS

After completing this chapter, you will be able to:

◎ Create a single- and a double-sloped roof.

◎ Convert polylines and walls to roofs.

◎ Modify roofs.

◎ Use roof and roof slab properties.

◎ Use grips to modify roofs.

◎ Edit edges and faces.

◎ Adjust the roof entity display.

◎ Create roof slabs.

◎ Convert a roof into a roof slab.

◎ Modify roof slabs and roof slab edges.

◎ Import roof slabs and roof slab edge styles.

◎ Trim, extend, miter, and cut roof slabs.

◎ Add and remove roof slab vertices.

◎ Add and remove holes.

◎ Use Boolean operations on a roof slab.

◎ Create a roof dormer.

Thus far you have explored the process of using Architectural Desktop objects to create a building model. Within the building model, walls, doors, windows, stairs, slabs, and many other AEC objects all contribute to the creation of 2D construction documents and a 3D model. Roofs also can be created using Architectural Desktop. Two types of AEC objects can be used to draw a roof: the AEC roof object and AEC roof slabs.

The AEC roof object is used to model an entire roof of a building. Roofs can be created by selecting points much like picking points for a slab or can be converted from polylines or walls enclosing a room or area. The roof pitch or slope, overhang, and roof plate height are controlled through a dialog box, which can be edited at any time.

For more modeling control over the appearance of a roof, the roof object can be converted to roof slabs. These objects are similar to slab objects and provide control over individual faces and the edge style of a roof. Custom fascias and soffits can be added to the edge of a roof slab to control the detail.

Creating a Roof

Roofs objects are created in one of two ways, by picking points in the drawing to establish vertex (corner) locations for the roof or by converting polylines or walls into roofs. When points are picked to draw a roof, the points are usually picked around the outside edges of the walls. As the corner points are picked the roof object begins taking shape as the points become hips and valleys for the roof (see Figure 16.1).

597

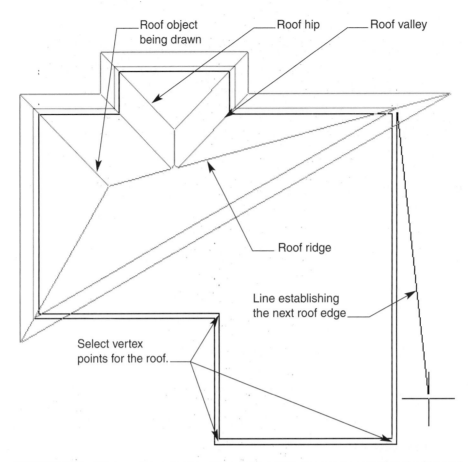

FIGURE 16.1   When points are picked to create a roof, the points establish the hip and valley lines of the roof.

Use one of the following options to add a roof by selecting points:

✓ Pick the **Design** pull-down menu, select the **Roofs** cascade menu, then pick the **Add Roof...** command.

✓ Pick the **Add Roof** button in the **Roofs - Roof Slabs** toolbar.

✓ Right-click, and select **Design, Roofs, Add Roof...** from the cursor menu.

✓ Type **AECROOFADD** at the Command: prompt.

The Add Roof dialog box, shown in Figure 16.2, is displayed.

The dialog box includes the following settings to control how the roof will be created as it is drawn on the screen.

- **Shape:**   This drop-down list includes two options, SingleSlope and DoubleSlope. The **SingleSlope** option creates a roof with one slope or pitch around the whole roof (see Figure 16.3A). This is the most common roof shape. The **Doubleslope** option creates a roof using two different slopes. This option is used to create a gambrel or a steeple roof (see Figure 16.3B). Selection of the DoubleSlope shape makes more text boxes become available.

- **Plate Height:** This text box specifies where in the Z axis the roof will be drawn. The corner points of the roof are placed at this height, which represents where the roof sits on top of a wall (see Figure 16.4).

- **Upper Height:** This text box becomes available when a DoubleSlope roof is being created. This value is measured from the bottom of the wall to the point where the

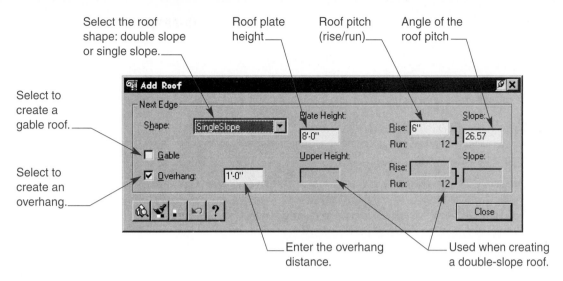

Select the roof shape: double slope or single slope.

Roof plate height

Roof pitch (rise/run)

Angle of the roof pitch

Select to create a gable roof.

Select to create an overhang.

Enter the overhang distance.

Used when creating a double-slope roof.

FIGURE 16.2 The Add Roof dialog box is used to add roofs to the drawing.

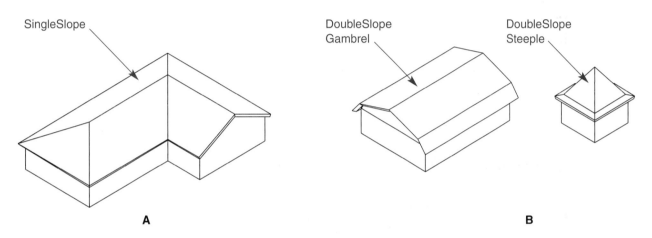

SingleSlope

DoubleSlope Gambrel

DoubleSlope Steeple

A

B

FIGURE 16.3 The roof object is used to create one of two shapes. (A) The SingleSlope roof. (B) The DoubleSlope roof.

FIGURE 16.4 The Plate Height establishes where the roof will sit on top of a wall.

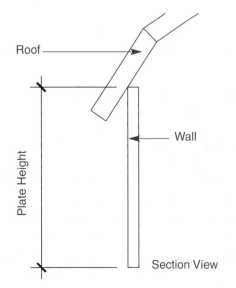

Roof

Wall

Plate Height

Section View

FIGURE 16.5 The DoubleSlope roof shape uses both the plate height and the upper height values. The Rise text boxes control how much the roof rises per 12″ of run for each roof slope.

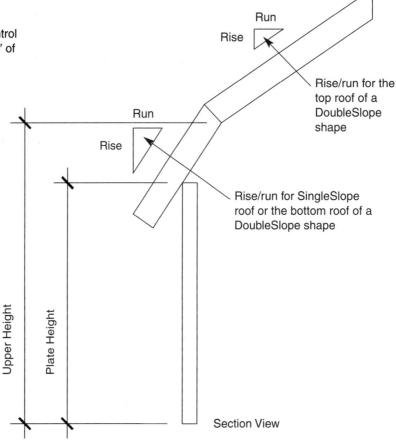

two roofs of a double roof meet (see Figure 16.5). Adjusting this text box causes the lower roof to become taller or shorter.

- **Rise and Run:**   The <u>R</u>ise text box at the top of the dialog box is always available when a roof is being created. When the SingleSlope shape is being created, this text box controls the rise over an automatic 12″ run. The rise (pitch) entered establishes the sloping angle of the roof. If a DoubleSlope shape is being created, the top <u>R</u>ise text box controls the bottom sloped roof. The bottom <u>R</u>ise text becomes available and controls the slope of the upper sloped roof (see Figure 16.5).

- **Slope:**   Like the Rise text boxes, the top <u>S</u>lope text box controls the angle of the roof for a SingleSlope roof or the bottom roof of a DoubleSlope roof. If a DoubleSlope roof is used, the bottom S<u>l</u>ope text box becomes available and controls the slope of the upper roof. These values are automatically adjusted as a different rise is entered.

- **<u>G</u>able**   This check box is used to create a gable roof end. A ***gable roof*** is a roof that does not slope on an end but has sloping roofs to each side (see Figure 16.6). To use this option when creating a gable roof, pick the first corner for the gable roof, then select the check box before selecting the second point of the gable end. After selecting the second point of the gable end of the roof uncheck the gable text box to resume creating sloped roofs. There are

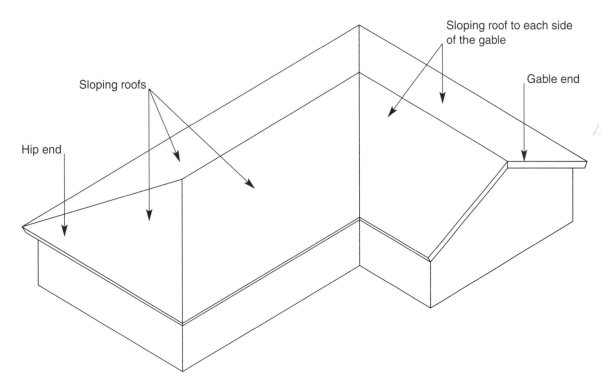

FIGURE 16.6   The difference between a gable and a hip roof.

also other methods of creating a gable end that are discussed later in this chapter.

- ■ **Overhang:**   This check box is used to add an overhang to the roof. The overhang text box becomes available so that an overhang distance can be entered. The overhang is automatically added as the roof points are selected .

Drawing a Single-Sloped roof

The SingleSlope shape roof can be used to create a hip roof, a gable roof, or a combination of hip and gable roofs. After entering the Add Roof dialog box, select the **SingleSlope** shape and determine the plate height for the roof. This is usually established by the height of the wall where the roof sits. If you are creating a roof with multiple plate heights, use the lowest plate to draw the roof first, then you can change the plate heights of the different edges later. Multiple-plate roofs are discussed later in this chapter.

Next, determine the rise or pitch of the roof. This is automatically established by the slope of the roof sides. Any rise height can be entered. If you are adjusting the slope value, the maximum slope that can be used when initially creating a roof is 89°. Creating a roof with a slope of 0° establishes a flat roof. An individual slope edge can be set to any angle between 0° and 180°. It is easiest to modify the slope edges of a roof after the initial shape has been drawn. Determine if the roof includes an overhang, and if so, specify the overhang depth in the Overhang text box.

Once the parameters for the roof have been determined, the roof can be drawn. Pick the first point for the roof. Initially nothing appears until the second and third points of the roof have been picked. Use the Endpoint object snap to select exterior corners of the building. Begin selecting all the exterior points of the building. As the points are picked the roof begins to appear. When you have picked the last point (the last point before the first point selected), press the [Enter] key to close and accept the new roof. Figure 16.7 displays the points picked to create a roof.

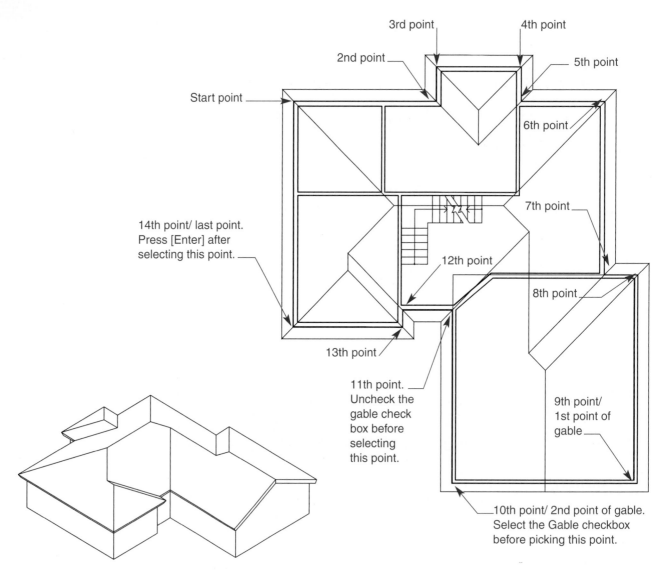

FIGURE 16.7   Select all the exterior wall corners to use for the vertex locations for the roof.

Exercise 16.1

1. Open EX14-5-Flr-Pln.dwg
2. Add a SingleSlope roof to the floor plan. Set the plate height to 8′-0″, the rise to 4″, and the overhang to 1′-0″.
3. See the figure for the roof corner points. [*Hint*: At the entry area try using Otracking, which was discussed in Chapter 2.]
4. Save the drawing as EX16-1.dwg.

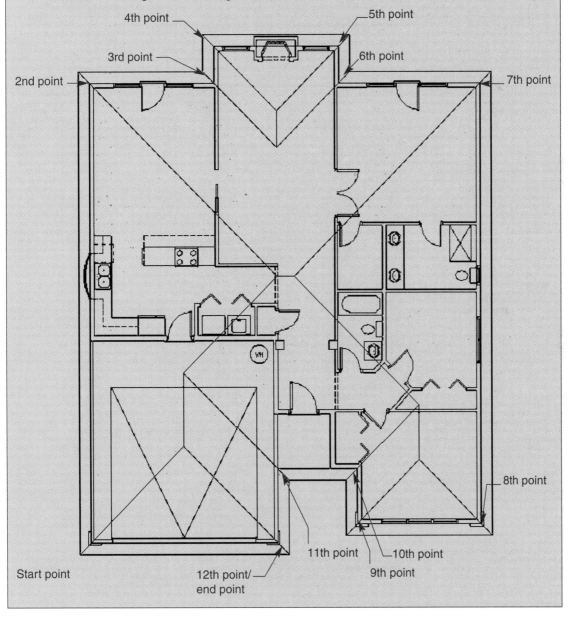

Drawing a Double-Sloped Roof

The DoubleSlope shape is used to create a roof that contains two roof slopes on each side of the ridge. In the Add Roof dialog box, select the DoubleSlope shape. Determine the plate height for the roof, then determine the upper height. This value specifies the point at which the lower roof slope meets the upper roof slope (see Figure 16.5). Next, specify the rise values for each slope. If you use an overhang, then specify the overhang depth.

After determining the parameters for the double-sloped roof, begin picking the points for the roof. Figure 16.8 shows the points picked to create a double-sloped roof.

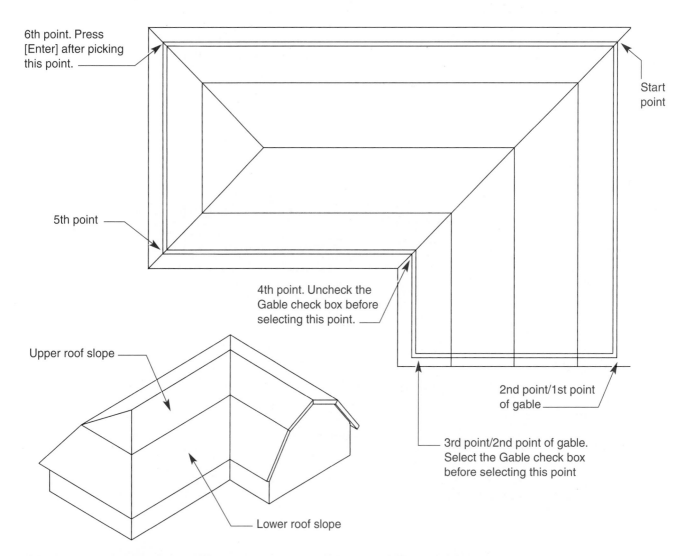

FIGURE 16.8   Creating a DoubleSlope roof is similar to creating a SingleSlope roof. The difference is the use of two roof slopes per side of roof.

Exercise 16.2

1. Start a new drawing using the AEC Arch (Imperial) or the AEC Arch (Imperial-Int) template.
2. Draw 8'-0" walls high, using the sizes indicated in the figure.
3. Add a DoubleSlope roof using the following parameters:
   - Plate height: 8'-0"; lower roof rise: 16"
   - Upper height: 16'-0"; upper roof rise: 6"
   - Overhang: 12"
4. Save the drawing as EX16-2.dwg.

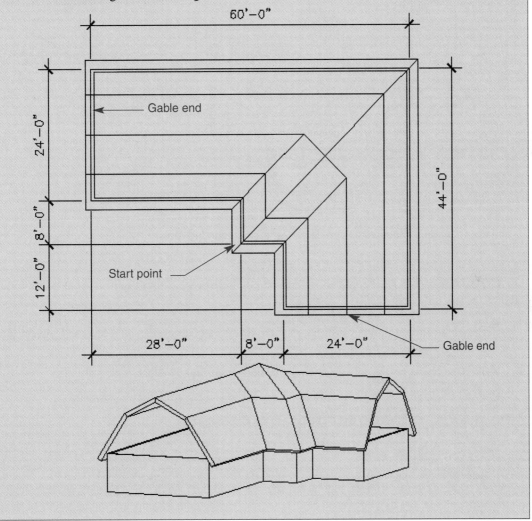

Field Notes

You can create roofs in a separate drawing file, by referencing the building floor that has a roof added above it. Using the reference, select the exterior points of the walls. After the roof is created, it can then be referenced into an assembly drawing for elevations and sections.

Converting Polylines and Walls to Roofs

In addition to drawing roofs manually, you can create roofs by converting polylines and walls into roofs. The ability to convert these objects into roofs provides the opportunity to create rounded edges for a roof (see Figure 16.9).

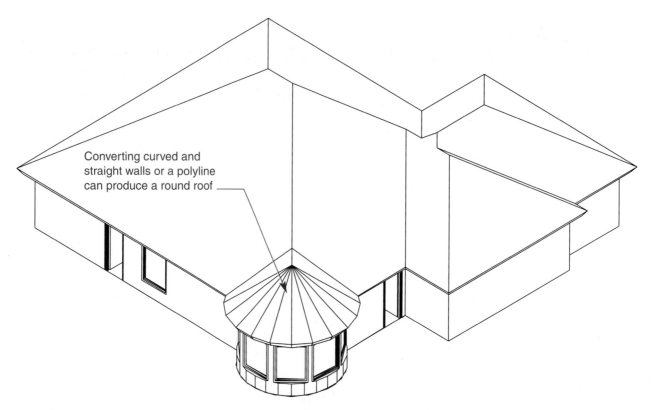

Converting curved and
straight walls or a polyline
can produce a round roof

FIGURE 16.9   Walls and polylines with arc segments can be converted into rounded roofs.

Use the **Convert to Roof** command to convert polylines or walls into a roof. Use one of the following options to access this command:

✓ Select the **Design** pull-down menu, pick the **Roofs** cascade menu, and select the **Convert to Roof...** command.

✓ Pick the **Convert to Roof** button on the **Roofs - Roof Slabs** toolbar.

✓ Right-click, and select **Design, Roofs, Convert to Roof...** from the cursor menu.

✓ Type **AECROOFCONVERT** at the Command: prompt.

Converting Polylines into Roofs

Converting a polyline into a roof is the most flexible method of converting an object to a roof. The polyline can be any combination of straight and curved segments and does not have to be a closed polyline. If the polyline is not closed, the roof command closes the roof between the starting and ending points of the polyline, adding a sloped side between the two points. It is usually best to draw a polyline around the outer edges of your building model and then convert the polyline into a roof. The following prompt sequence is used to convert a polyline into a roof:

> Command: **AECROOFCONVERT** ↵
> Choose walls or polylines to create roof profile:
> Select objects: *(select the polyline)*
> 1 found
> Select objects: ↵
> Erase layout geometry? [Yes/No] <N>: ↵
> Roof modify
> [Shape/overhangValue/Overhang/PHeight/PRise/PSlope/UHeight/URise/USlope/Match]: *(the modify roof dialog box appears and allows you to adjust the parameters of the new roof object)*

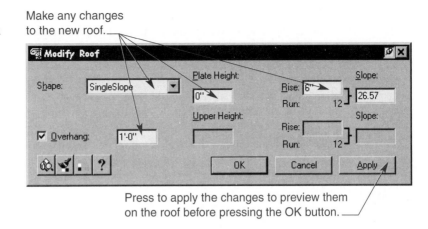

FIGURE 16.10 The Modify Roof dialog box appears after a polyline or walls are converted into roofs. Make any parameter changes to the newly created roof.

Make any changes to the new roof.

Press to apply the changes to preview them on the roof before pressing the OK button.

After you select the polyline to be converted, and choose whether to erase the pline, the roof is displayed on the drawing screen, and the **Modify Roof** dialog box appears (see Figure 16.10). This dialog box is similar to the Add Roof dialog box and allows you to modify any settings for the new roof. Adjust any of the settings, and press the <u>A</u>pply button to see the changes made to the roof. When you have finished adjusting the roof, press the OK button to finish the command.

Field Notes

If a polyline that includes curved segments is converted, the segments initially appear as a number of straight sloped segments, making the curved roof appear "blocky." This appearance can be adjusted later by modifying the number of segments used by the curve. This technique is discussed later in this chapter.

Converting Walls into Roofs
Converting a series of walls is the same process as converting a polyline; however, when walls are converted into a roof, the walls must form a closed building or room (see Figure 16.11A). If the walls being converted do

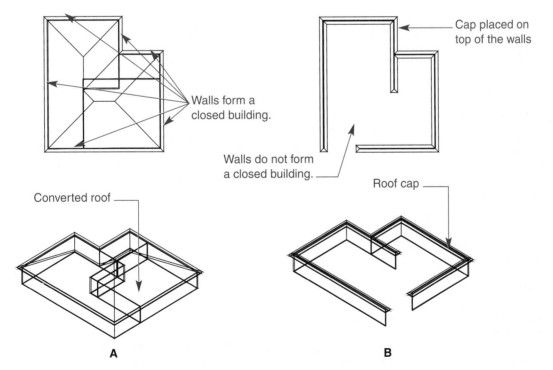

FIGURE 16.11   (A) Walls that form an enclosed area can be converted into a roof. (B) Walls that do not form an enclosed area are converted to a roof cap over the individual walls.

not form a closed building, then the roof is added on top of the walls to form a "cap," as shown in Figure 16.11B. The slightest opening between two walls forms the roof on top of the individual walls being converted.

The prompt sequence for converting walls is the same as it is for converting polylines. Select the walls to be converted, ensuring that the walls selected form a closed building. If you have walls that enclose the entire building, you can use a crossing window to select all the walls, because Architectural Desktop looks only at the outer boundary of walls for the conversion process to a roof.

When the walls are converted, the plate height is automatically filled in the Modify Roof dialog box. If the walls being converted include several different heights, then the plate height reads *VARIES* in the text box. Converted polylines and walls create hip roofs only. If a gable is desired on an end, or if the entire roof needs to be composed of gables, you can make these changes later by modifying the roof edges.

 Field Notes

> If walls with multiple heights are converted, the roof object is created using the different plates properly adjusting the roof slopes and valley and hip intersections.

Exercise 16.3

1. Start a new drawing using the AEC Arch (Imperial) or the AEC Arch (Imperial-Int) template.
2. Draw the polyline in the figure.
3. Convert the polyline into walls. Use the Modify Walls command to adjust the heights of the walls as indicated.
4. Convert the walls into a single-sloped roof, with a 3″ rise and 24″ overhang.
5. Save the drawing as EX16-3.dwg.

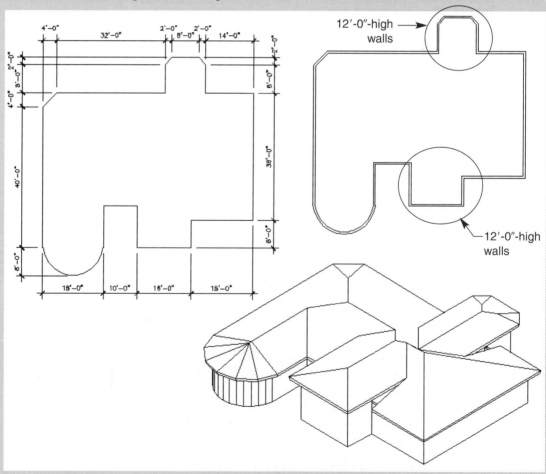

Modifying Roofs

After a roof has been added to the drawing it can be modified by several different methods to reflect any design changes The **Modify Roof** command can be used to change the shape between SingleSlope and DoubleSlope, and the plate height and the rise. **Roof Properties** allows you to modify the edges and slopes of a roof, change the thickness of the roof, and adjust the fascia edge of the roof. Roofs can also be modified by using grips or by editing individual faces. All these methods are described in this section.

Using the Modify Roofs Command

The Modify Roof command is used to change general properties of a roof. For example, the overall slope, plate height, overhang, and shape are adjusted in this dialog box. Use one of the following options to access this command:

✓ Open the **Design** pull-down menu, select the **Roofs** cascade menu, and pick the **Modify Roof...** command.

✓ Pick the **Modify Roof** button from the **Roofs - Roof Slabs** toolbar.

✓ Right-click, and select **Design, Roofs, Modify Roof...** from the cursor menu.

✓ Type **AECROOFMODIFY** at the Command: prompt.

✓ Select the roof, then right-click, and pick **Roof Modify...** from the shortcut menu.

The Modify Roof dialog box, shown in Figure 16.10, is displayed.

This is the same dialog box that is used to convert polylines or walls into roofs. Change any of the parameters to modify the existing roof. Use the <u>A</u>pply button to apply the changes before exiting the dialog box.

Use Roof Properties

The **Roof Properties** command gives you advanced control over the behavior of roofs. This command can be activated by selecting a roof to be modified, right-clicking, and selecting Roof Properties... from the shortcut menu, or by typing **ROOFPROPS** at the Command: prompt. The **Roof Properties** dialog box, shown in Figure 16.12 is displayed.

The Roof Properties dialog box is made up of the following three tabs:

- **General**   This tab is used to provide a description for the selected roof, to attach design notes, and to assign property set definitions.

- **Dimensions**   This tab is the heart of the Roof Properties dialog box. The tab is divided into two separate areas: **<u>S</u>elected Roof Edges** and **Roof <u>F</u>aces (by Edge).** The <u>S</u>elected Roof Edges area displays all the edges on the selected roof. The **Edge** column lists the different edges within the roof. For example, edge 0 is the edge between the first and second points picked when the roof is created. The values that can be changed in this table include the **(A) Height** column, the **(B) Overhang** column, and the **Segments** column. The Segments column can be changed only if the edge happens to be a curved edge. Pick on top of the values in the table to adjust them.

    The **Roof <u>F</u>aces (by Edge)** area controls the face or sloping side of the roof belonging to the edge that is highlighted in the <u>S</u>elected Roof Edges area. For example, in Figure 16.12 edge 0 is highlighted. The face information for edge 0 is displayed in the Roof <u>F</u>aces (by Edge) table. Each edge of a roof includes a face numbered 0. The information that can be adjusted for a face is found in the **Slope** column. To change the slope for a face, pick on the slope value and enter a new angle. Remember, the slope is automatically determined when a new roof is added by specifying a rise value.

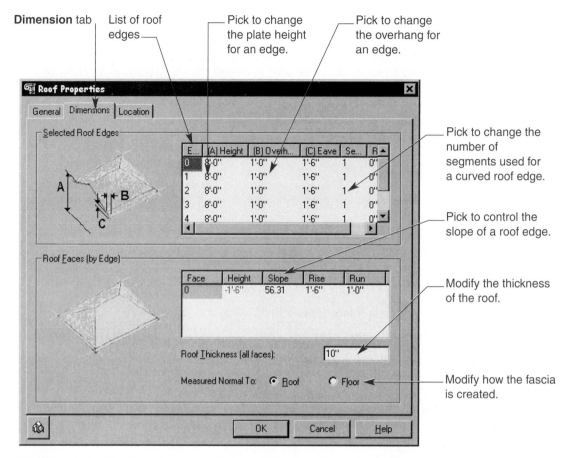

FIGURE 16.12   The Roof Properties dialog box allows you to control the parameters of each edge of the roof.

Below the Roof Faces (by Edge) table the is the **Roof Thickness (all faces):** text box. The value in the text box represents the thickness of the roof. The roof object is considered to be one entire piece that represents the rafter and the sheathing thickness. Enter a new value for the thickness of the roof.

At the bottom of the tab is the **Measured Normal To:** area. This setting contains two radio buttons that control how the fascia for a roof object appears. The default for all fascia edges of a roof is the **Roof** radio button, which is similar to the Square option for a slab object. When the **Roof** option is selected, the edges of the roof are cut at an angle determined by the slope of the face. The **Floor** option is similar to the Plumb option for slab objects. This option creates roof edges that always appear vertical. Figure 16.13 demonstrates the difference between the two options.

■ **Location**   This tab includes information regarding the position of the roof object in the drawing. The **Insertion Point** area indicates where in the World Coordinate System the roof is located. Adjusting the Z insertion point is similar to changing the plate height in the Modify Roof dialog box.

🖊️**Field Notes**

Although the roof edges and face slopes can be adjusted in this dialog box, it is difficult to know which edge and face is actually being adjusted. To specify the edge and face that is being modified use the Edit Edges/Faces command. This command is discussed later in this chapter.

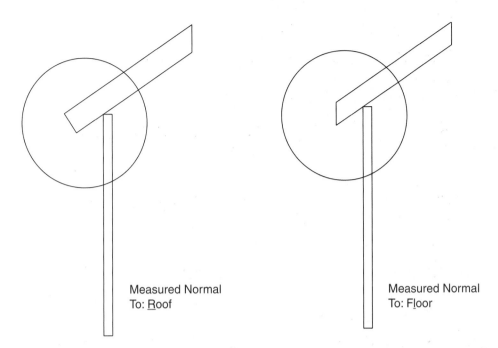

FIGURE 16.13   The Measured Normal To: area creates edges that are square to the slope of the face or plumb to the ground.

Exercise 16.4

1. Open EX16-1-Flr-Pln.dwg.
2. Change the overhangs of all the edges to 1'-6".
3. Use the Roof Properties command to change the roof thickness to 8".
4. Change the Measured Normal To: to the Floor option.
5. Save the drawing as EX16-4-Flr-Pln.dwg.

Using Grips to Modify Roofs

When a roof is selected, all the edges of the roof become highlighted. Grip boxes are also placed around the roof at all roof vertex corners, intersections of ridges, hips, and valleys. Picking a grip and stretching its location allows you to modify the roof in some way. Selecting a grip between two corner grips adjusts the roof plate location. Stretching a corner grip adjusts the two slopes on either side of the corner.

Gable end roofs can also be created with grips. To do this, select a grip at the intersection of two hips to highlight the grip (see Figure 16.14A). Stretch the grip point past the edge of the roof and pick with the left mouse button, as shown in Figure 16.14B. The ridge is extended to the edge of the eave and the sloping face is turned into a gable end.

Using grips is often easier than manually creating a gable by placing a check in the Gable check box each time a gable is to be added when you are creating a roof. Use this method also after converting poly-lines or walls into a roof, as the roof is initially created as a hip roof.

Sometimes, stretching a grip to create a gable end produces undesired results. When the face of a roof being turned into a gable happens to be on the same plane as another face, stretching the grip causes both faces to be turned into gable roofs, as shown in Figure 16.15. In this situation use the Edit Edges/Faces command to edit the slope for the face that is to be turned into a gable. This procedure is discussed in the next section.

Grips can also be used on a **DoubleSlope** roof to create a Dutch hip roof (see Figure 16.16). If the Dutch hip is to have two different slopes, make the adjustments to the rise/slope as necessary. If the Dutch hip is to have the same slope, use the same rise/slope for the upper and lower roofs. In either case a DoubleSlope shape needs to be used. After the roof is drawn select the grip where the two hips intersect, and pull the grip point past the edge between both sloped roofs.

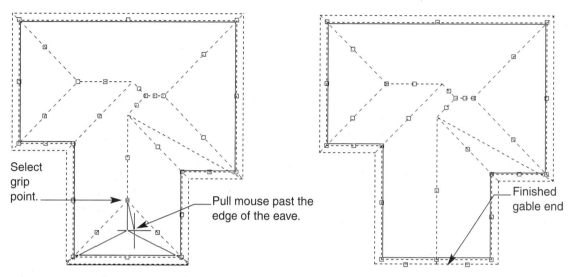

Select
grip
point.

Pull mouse past the
edge of the eave.

Finished
gable end

FIGURE 16.14   Use grips to create gable ends in a roof.

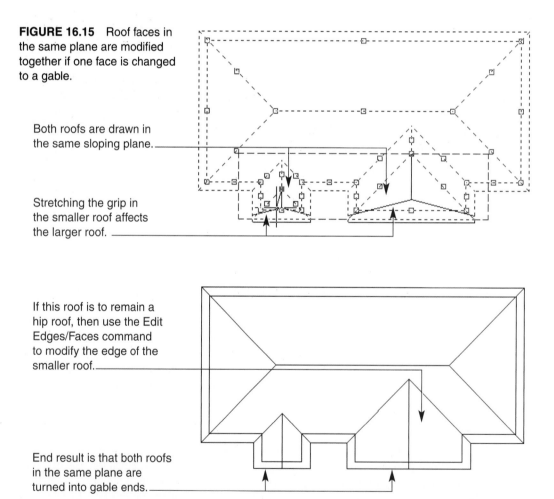

FIGURE 16.15   Roof faces in
the same plane are modified
together if one face is changed
to a gable.

Both roofs are drawn in
the same sloping plane.

Stretching the grip in
the smaller roof affects
the larger roof.

If this roof is to remain a
hip roof, then use the Edit
Edges/Faces command
to modify the edge of the
smaller roof.

End result is that both roofs
in the same plane are
turned into gable ends.

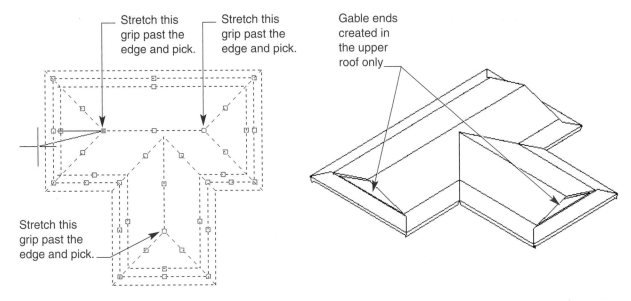

Stretch this grip past the edge and pick.

Stretch this grip past the edge and pick.

Gable ends created in the upper roof only

Stretch this grip past the edge and pick.

FIGURE 16.16 Creating a Dutch hip roof.

 Field Notes

When gable end roofs are created, an open hole is displayed between the top of the wall and the bottom of the roof. Use the Roof Line command with the Auto project option to automatically project the walls up to the bottom of the roof. With the Auto project option, the wall can be projected to any AEC object that has an edge that can be projected to. Walls can also be projected to roofs that are in an xref. To do this, first select the walls to be projected, then select the roof within the xref. The walls are projected up to the roof.

Exercise 16.5

1. Start a new drawing using the AEC Arch (Imperial) or the AEC Arch (Imperial-Int) template.
2. Draw a 32′ × 18′ rectangle.
3. Convert the polyline to a roof.
4. Use grips to turn the two shorter ends of the roof into gables.
5. Save the drawing as EX16-5.dwg.

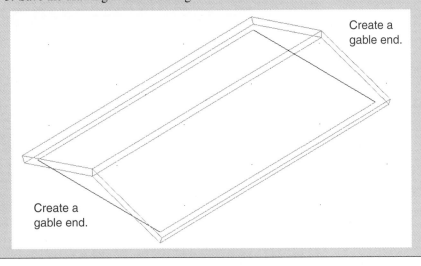

Create a gable end.

Create a gable end.

Editing Edges and Faces

The Roof Properties dialog box in which all the edges of a roof can be modified was discussed earlier. The difficulty with using the Roof Properties dialog box to modify edges is that it is hard to tell which edges are being modified. The Edit Edges/Faces command makes this task simpler because you are able to select the edge that needs to be modified.

Use one of the following options to access this command:

- ✓ Select the **Design** pull-down menu, pick the **Roofs** cascade menu, then select **Edit Roof Edges/Faces….**

- ✓ Pick the **Edit Roof Edges/Faces** button from the **Roofs - Roof Slabs** toolbar.

- ✓ Right-click, and select **Design, Roofs, Edit Roof Edges/Faces…** from the cursor menu.

- ✓ Type **AECROOFEDITEDGES** at the Command: prompt.

- ✓ Select the roof to be modified, right-click, and select **Edit Edges/Faces…** from the shortcut menu.

Once the command has been accessed, you are prompted to select edges. Select any roof edge or eave to modify it. When you have finished selecting the edges that you want to modify, press the [Enter] key. The **Edit Edges/Faces** dialog box, shown in Figure 16.17, is displayed. This dialog box looks similar to the Roof Properties dialog box. The difference is that only the edge(s) selected are displayed in the Selected Roof Edges area.

If you need to readjust the plate heights for different edges of the roof, select the edges that need to be modified, then change the values in the (A) Height column. As in the Roof Properties dialog box, the only edge values that can be modified are the height, overhang, and segments if the edge is curved.

FIGURE 16.17 The Edit Edges/Faces dialog box allows you to specify the edges to be modified.

One edge has been selected to be modified.

Adjust the height and overhang for the edge.

If the edge is curved, increase the number of segments to make the roof appear smoother.

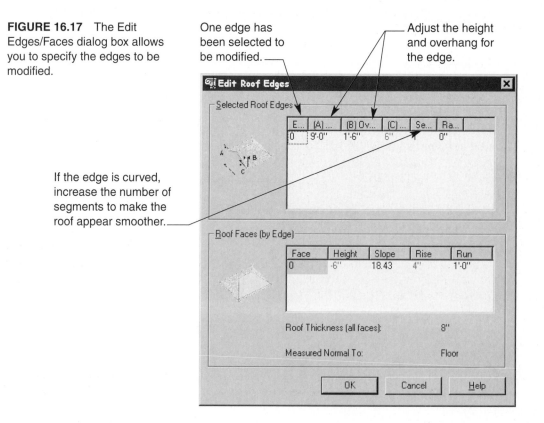

Field Notes

If you use the Edit Roof Edges/Faces command to edit a double-sloped roof, the selected edge displays two faces. Change both faces to reflect a 90° slope to create a complete gable end.

Exercise 16.6

1. Open EX16-4-Flr-Pln.dwg.
2. Change the height of the specified edges to 9'-0".
3. Use the Roof Line command to project the walls under the 9'-0" plate roofs to the roof.
4. Save the drawing as EX16-6-Flr-Pln.dwg.

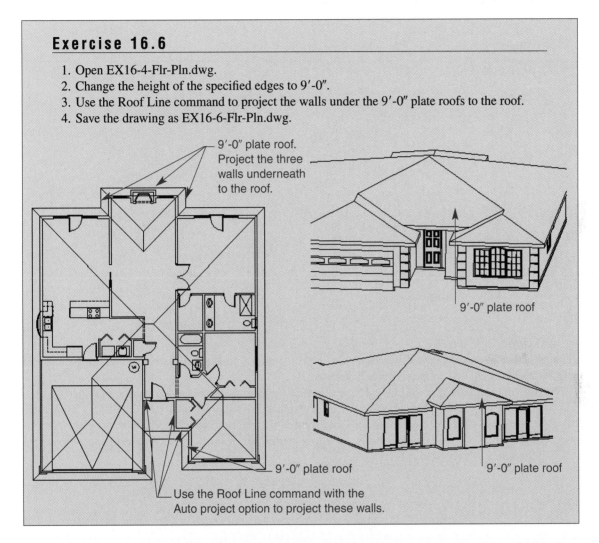

9'-0" plate roof. Project the three walls underneath to the roof.

9'-0" plate roof

9'-0" plate roof

9'-0" plate roof

9'-0" plate roof

Use the Roof Line command with the Auto project option to project these walls.

Grip editing was discussed in the previous section. A problem was encountered in Figure 16.15 where both ends of the roof in the same plane were being modified into a gable end. With the Edit Roof Edges/Faces command the one edge that needed to be turned into a gable end can be modified. To change a sloped roof into a gable end, adjust the slope to a 90° angle. This "tips" the slope straight up vertically, turning the slope into a gable end. The slope for an edge must be between 0° and 180° (see Figure 16.18).

Field Notes

The Edit Edges/Faces dialog box can also be used to create a shed roof. First, draw a rectangular roof, then use Edit Edges/Faces to modify the slope of three sides of the roof to 90°.

Adjusting the Roof Entity Display

The addition of AEC roofs produces an orange-colored roof in the model and plan representations. Like other objects, AEC roofs have the ability to appear different depending on the viewing direction. To modify

FIGURE 16.18 Changing the
slope of a face to a 90° angle
will tip the slope on end
vertically, creating a gable end.

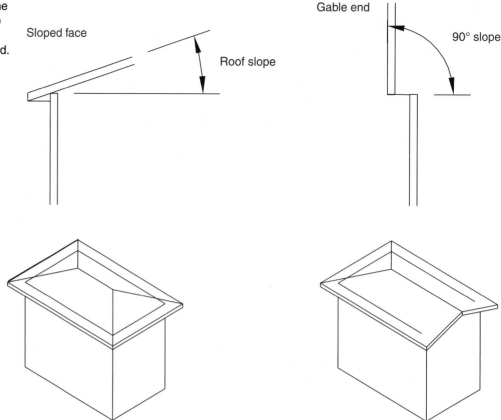

the appearance of a roof in the different display representations, first select the roof, then right-click, and select Entity Display… from the shortcut menu.

AEC roofs include plan and model representations. Select the representation to be modified and the property source (level of display control), attach an override if required, then edit the display properties. Both the plan and model representations for a roof object include roof and eave components. These components can be turned on or off, have layers assigned, or colors adjusted. Make any display changes as needed and exit the Entity Display dialog box.

 Field Notes

In many construction document sets that include roofs, the roof outline is shown in the upper floor as a dashed or dotted line. Use the Entity Display to change the plan view roofs so that they use a dashed or dotted linetype. You can create a custom display configuration that includes the dashed plan roof and another that shows the roof as a solid line for a roof plan.

Creating Roof Slabs

Slab objects were discussed in Chapter 14. These objects are used to create floors and solid shapes with mass. The use of a projected slab allows you to slope the slab at an angle to create ramps, driveways, and other sloping surfaces. The roof slab is a similar type of object using the same techniques and type of commands as available for the standard slab objects.

Roof slabs are used to model a single face or side of a roof. The roof slab, like a standard slab object, is composed of a thick face or body surrounded by edges. Each edge can be assigned its own edge style to create fascia boards and soffits. As with the standard slabs, selecting several points to define the shape of the slab creates roofs slabs. The first two points selected establish the slope line or edge that determines the beginning of the slope or angle of the roof. Typically, the fascia and soffit are created along this first edge (see Figure 16.19).

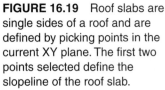

FIGURE 16.19 Roof slabs are single sides of a roof and are defined by picking points in the current XY plane. The first two points selected define the slopeline of the roof slab.

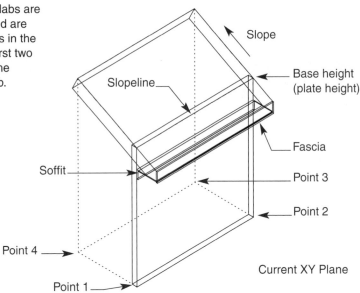

The difference between a roof slab and an AEC roof object is that the roof slab is a separate part or side of a roof. Multiple roofs slabs placed together can be used to model an entire roof but remain a series of individual sides that can be edited separately. Thus, roof editing tools are required to get the hip, valley, and ridge edges to miter and clean up properly.

 Field Notes

> When you are designing complex roofs with many hips and valleys, it is suggested that you initially create the roof using the AEC roof object to create as much of the roof as possible. When the roof is significantly complete but needs further editing that cannot be accomplished through standard roof tools, convert the roof into roof slabs that can be edited individually.

In addition to manually creating roof slabs, you can convert roof slabs from AEC roofs, walls, and polylines. Although the individual sides do not interact with each other as they do with an AEC roof, the roof edges can be trimmed, extended, and mitered as an AEC slab can. Holes for skylights can be cut in a roof slab, and any type of edge style can be applied to the different edges of a roof slab.

Adding Roof Slabs

Use one of the following options to access the **Add Roof Slab** command to add roof slabs to the drawing:

✓ Select the **Design** pull-down menu, pick the **Roofs** cascade menu, and select the **Add Roof Slabs...** command.

✓ Pick the **Add Roof Slab** button from the **Roofs - Roof Slabs** toolbar.

✓ Right-click, and select **Design, Roofs, Add Roof Slabs...** from the cursor menu.

✓ Type **AECROOFSLABADD** at the Command: prompt.

The **Add Roof Slab** dialog box, shown in Figure 16.20, is displayed. The following settings are available in the Add Roof Slab dialog box:

- **Style:** If you are using one of the AEC templates, a number of roof slab styles are available for use when creating a roof. These different styles include fascia and soffit boards attached to the slopeline edge.

Select a roof slab   Enter a base height   Specify the thickness   Specify an   Select the
style for use. ——   (plate height). ——   of the "rafters" and   overhang.   justification for
                                                                 "sheathing."   the roof slab.

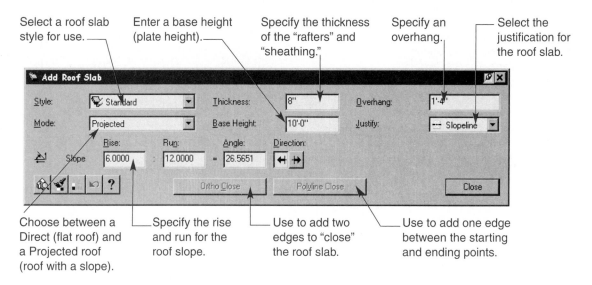

Choose between a   └─ Specify the rise   └─ Use to add two   └─ Use to add one edge
Direct (flat roof) and   and run for the   edges to "close"   between the starting
a Projected roof   roof slope.   the roof slab.   and ending points.
(roof with a slope).

FIGURE 16.20   The Add Roof Slab dialog box is used to add roof slab objects to the drawing.

- **Thickness:**   This text box is used to specify a roof slab thickness. The thickness represents the total width of the rafter and sheathing over the rafters.

- **Overhang:**   This text box specifies an overhang. The overhang is automatically assigned to the slopeline edge of the roof slab. The other edges of the roof can be edited through the Roof Slab Properties dialog box.

- **Mode:**   Two options are included in this drop-down list. The **Direct** mode creates a flat roof slab and is used to create flat roofs at the top of a building. The **Projected** mode creates a roof slab that can be set at any slope.

- **Base Height:**   This text box is used to specify plate height for a roof slab. The base height is applied only to the slopeline of the roof slab. The first two points selected specify the slopeline and the roof plate location.

- **Justify:**   This drop-down list includes four justifications. The **Slopeline** justification is used with a Projected mode. The slopeline is determined by the first and second points picked. Unless you need to specify the creation of the roof slab from a different justification value, use the Slopeline option.

- **Slope:**   This area is available when a projected roof slab is being created. Adjust the **Rise** and **Run** values, which in turn adjust the angle of the slope. The **Direction** buttons determine the direction in which the projected slab is created in relation to the slopeline.

- **Ortho Close**   This button is available after two vertex points have been selected and closes the slab with two edges.

- **Polyline Close**   This button closes the slab between the first and last vertex points picked. This option becomes available after three vertex locations have been selected.

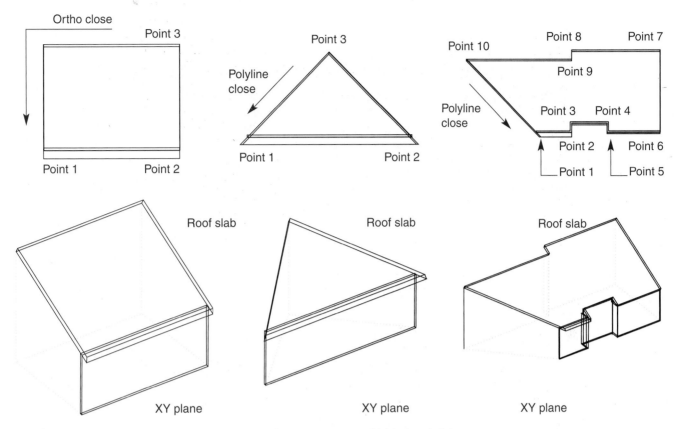

FIGURE 16.21 Selecting different combinations of points creates multisided roof slabs.

Once the parameters for the roof slab have been determined, pick a point on the drawing to establish the first vertex point. This point also establishes the pivot point location. Selecting the second point of the roof slab then determines the slopeline, which is the edge that remains along the roof plate. The remainder of the slab is sloped at the angle specified in the Rise and Run text boxes. Continue picking vertex points to establish the slab. The remaining selected points are picked in the current XY plane but are projected up in the Z axis to determine the proper slope. Once you have picked the last point, use the Ortho Close or the Polyline Close buttons to close the roof slab (see Figure 16.21). Roof slabs must be closed in order to be created.

 Field Notes

Points selection to establish roof slabs can be difficult to visualize initially. To make this process easier, sketch the roof plan using polylines to establish a 2D plan with hip, valley, and ridge lines. Use the Add Roof Slab command to trace over the points of the 2D geometry, to place the roof slabs in the proper locations. Adjust the plate heights and slopes as necessary to create the 3D roof model.

Exercise 16.7

1. Start a new drawing using the AEC Arch (Imperial) or the AEC Arch (Imperial-Int) template.
2. Use polylines to sketch out the roof plan in the figure.
3. Change the view to an isometric view, to make tracing the points easier.
4. Use the Add Roof Slab command to trace over all the points for each side of the roof. Set the following parameters:
   - Style: 10-1 × 8 Fascia+Frieze
   - Overhang: 1'-4"
   - Base height: 9'-0"
   - Rise: 6"
   - Run: 12"
5. Create each side of the roof by tracing the points designating the sides. [*Note:* Some of the edges and intersections at the fascias may not appear as desired. These can be corrected through the Roof Slab Properties dialog box, discussed next.]
6. Save the drawing as EX16-7.dwg.

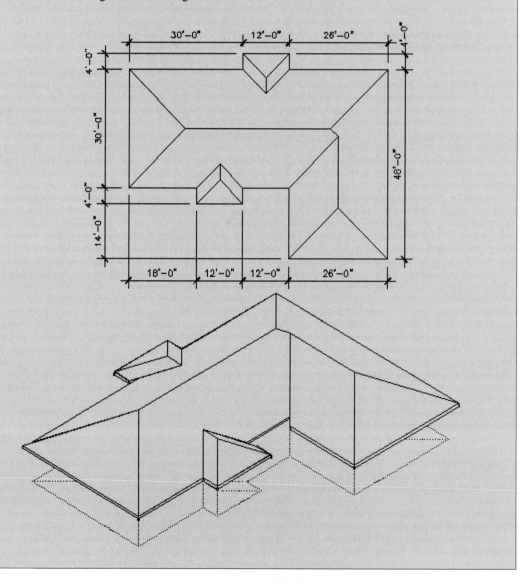

Converting a Roof into Roof Slabs

The standard AEC roof object is very good to use for laying out the basic shape of a roof. By modifying the plate heights and the overhangs, you can utilize this type of roof to model many different roof situations. Sometimes, however, the roof object is limited in its modeling capabilities, most notably when several different plate heights are used. Roofs at a lower plate height do not "fill in" under the eave of a roof at a higher plate height. Also, when a dormer or a skylight is added, holes cannot be cut into a standard roof object, as they can into a roof slab.

Both AEC roof objects and roof slabs have positive and negative features. Roof objects can be quickly laid out and the faces or sides of the roof generated automatically. Gable ends can quickly be modeled and double-sloped roofs can be simply created; however, advanced modeling techniques such as adding fascias and soffits, adjusting a roof edge, or cutting holes in the roof cannot be used on the standard roof object.

The use of roof slabs facilitates the addition of custom fascia and soffit edges, dormers, and holes for skylights; however, each side of the roof needs to be added, and the sides do not automatically cleanup or miter at the edges. Thus, due to the advantages and disadvantages of each of these objects, both are used in certain situations to create a roof.

As indicated earlier in this chapter, it is highly recommended that you initially draw the basic roof shape with the standard roof object. Make all adjustments to the overhangs and the plate heights for each of the eave edges that you can, then convert the roof object into roof slabs to take advantage of all the advanced editing capabilities of the standard slab object. Use one of the following options to convert a roof into roof slabs:

- ✓ Select the **Design** pull-down menu, pick the **Roofs** cascade menu, then select the **Convert to Roof Slabs...** command.

- ✓ Select the **Convert to Roof Slabs** button in the **Roofs - Roof Slabs** toolbar.

- ✓ Right-click, and pick **Design, Roofs, Convert to Roof Slabs...** from the cursor menu.

- ✓ Type **AECROOFSLABCONVERT** at the Command: prompt.

- ✓ Select the roof, right-click, and select **Convert to Roof Slabs** from the shortcut menu.

The following prompt sequence is used to convert a roof object into roof slabs:

```
Command: AECROOFSLABCONVERT ↵
Select roof, walls or polylines: (pick the roof object)
1 found
Select roof, walls or polylines: ↵
Erase layout geometry? [Yes/No] <N>: Y ↵
Specify roof slab style name or [?] <Standard>: (press [Enter] to accept the default roof slab
   style or enter a ? to list the styles available) ? ↵
Defined styles.
 04 - 1x8 Fascia
 04 - 1x8 Fascia + Frieze
 04 - 1x8 Fascia + Soffit
 10 - 1x8 Fascia
 10 - 1x8 Fascia + Frieze
 10 - 1x8 Fascia + Soffit
 Standard
Specify roof slab style name or [?] <Standard>: 10 - 1x8 Fascia + Soffit↵
Command:
```

The preceding sequence erased the original roof object and converted it into roof slabs. If you think that you may need to make adjustments to the original roof, you can choose not to erase the original object. The default roof slab style is the *Standard* style. Entering the ? lists the styles available in the drawing. Use the [F2] key to switch to the **Text Screen** window to see a list of the roof slab styles. Choose a style and enter the name at the prompt. Enter the roof slab style as it is displayed in the **Text Screen** window with the appropriate spaces and hyphens.

Once this process has been completed the roof object is converted into individual roof slab edges and can be modified with any of the roof slab editing tools.

Exercise 16.8

1. Open EX16-6-Flr-Pln.dwg.
2. Convert the roof into roof slabs.
3. Erase the original roof object, and use the 10 - 1x8 Fascia + Soffit style.
4. Save the drawing as EX16-8-Flr-Pln.dwg.

The Convert to Roof Slabs command can also be used to convert walls and polylines into roofs. Converting these two objects may yield unexpected results. When walls are converted into a roof slab, each wall is used to create an individual square-shaped slab (see Figure 16.22). The wall is used as the slopeline, and the starting and ending points of the wall are used as the first and second points of the roof slab; however, this conversion does not miter the edges between two sides of the building, which must be done manually.

Conversion of a polyline changes the shape of the polyline into one roof slab, as shown in Figure 16.23. The first and second points of the polyline are used as the first and second points for the roof slab and establish the slopeline.

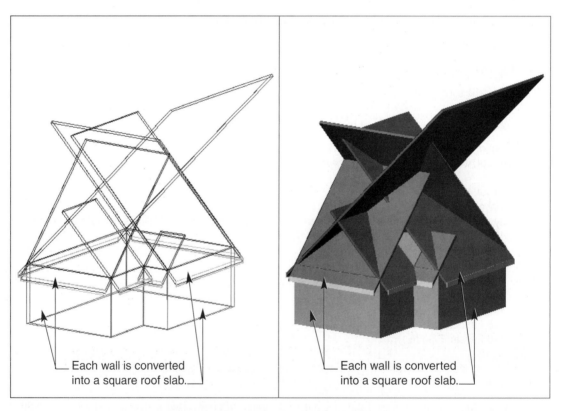

FIGURE 16.22   Conversion of walls into roof slabs converts each wall into its own roof slab. Additional editing is required to miter the edges of each side of the roof.

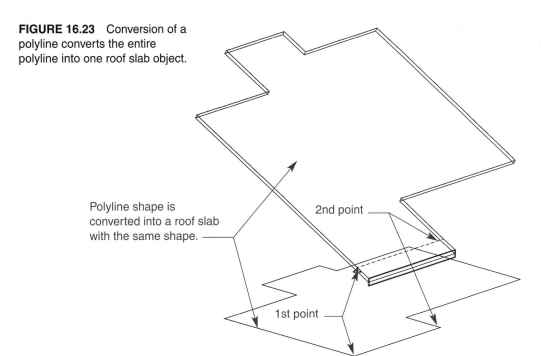

FIGURE 16.23 Conversion of a polyline converts the entire polyline into one roof slab object.

Polyline shape is converted into a roof slab with the same shape.

2nd point

1st point

Modifying Roof Slabs

Simple modifications can be made to the roof slab after it has been drawn. These modifications include the roof slab style, thickness, rise and run, and the pivot point location. Use one of the following options to modify an existing roof slab:

✓ Select the **Design** pull-down menu, access the **Roofs** cascade menu, and pick the **Modify Roof Slab...** command.

✓ Pick the **Modify Roof Slab** button from the **Roofs - Roof Slabs** toolbar.

✓ Right-click, and select **Design, Roofs, Modify Roof Slab...** from the cursor menu.

✓ Type **AECROOFSLABMODIFY** at the Command: prompt.

✓ Select the roof slab, right-click, and select **Roof Slab Modify...** from the shortcut menu.

After this command is entered and a roof slab is selected, the Modify Roof Slab dialog box appears, as shown in Figure 16.24. Make any adjustments to the style, thickness, or the rise and run of the roof slab. You can relocate the pivot point anywhere within the drawing or on the roof slab by selecting the **Set Pivot Point** button. Picking a new point in the drawing or on the roof slab establishes that point as the pivot point for the slope. Any future changes to the rise and run are adjusted from the new pivot point.

FIGURE 16.24 The Modify Roof Slab dialog box is used to make adjustments to roof slab objects.

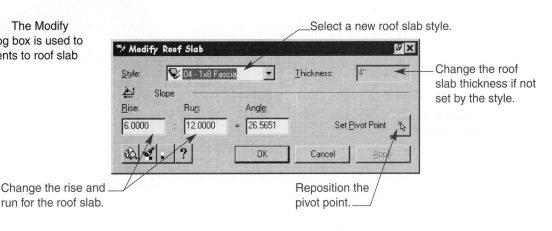

Select a new roof slab style.

Change the roof slab thickness if not set by the style.

Change the rise and run for the roof slab.

Reposition the pivot point.

Modifying the Roof Slab Edges

The roof slab styles found in the Architectural Desktop templates include several different types of roof slab edge styles. Some of the styles include edges that look like fascias, friezes, soffits, and a combination of these items. When the roof slab is first drawn, the roof slab edge style is applied only to the first edge drawn. This first edge is typically the overhang side of the roof.

As with the standard slab object and the slab edge objects, roof slab edge styles can be assigned to any roof slab edge. Individual roof edges can use different combinations of roof slab edge styles.

Use one of the following options to modify the edges of a roof slab:

- ✓ Select the **Design** pull-down menu, pick the **Roofs** cascade menu, and select the **Edit Roof Slab Edges...** command.

- ✓ Pick the **Edit Roof Slab Edges** button in the **Roofs - Roof Slabs** toolbar.

- ✓ Right-click, and pick **Design, Roofs, Edit Roof Slab Edges...** from the cursor menu.

- ✓ Type **AECROOFSLABEDGEEDIT** at the Command: prompt.

- ✓ Select the roof slab, then right-click, and pick **Edit Edges...** from the shortcut menu.

Once the command has been entered, you are prompted to select the edges. Select the edges of the roof slab to be modified or added. When you have finished selecting the roof edges, press the [Enter] key to display the **Edit Roof Slab Edges** dialog box, shown in Figure 16.25. This dialog box is used to modify the edge(s) of a roof slab. The edge being modified is listed along the left side of the dialog box. Edge 1 is always the slopeline edge of the roof slab. Select any of the edges to be modified by holding the [Ctrl] key and selecting an edge number to highlight it. When an edge has been selected in the list box, the edge is highlighted in the preview window, showing you which edge (s) are being modified.

The following values in the Edit Roof Slab Edges dialog box can be modified by picking on top of their existing values:

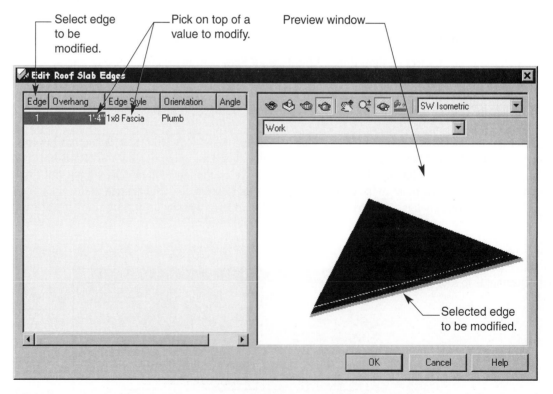

FIGURE 16.25   The Edit Roof Slab Edges dialog box is used to modify the edges of a roof slab.

- **Overhang** This column is used to specify the overhang distance for the roof slab edge. Pick on top of the existing value to reveal a text box. Enter the new overhang distance in the text box.

- **Edge Style** This column is used to assign a roof slab edge style to the selected edge. Picking on top of the existing value displays a drop-down list with available roof slab edge styles in the drawing. Select a different edge style from the list to assign to the highlighted roof edge.

- **Orientation** This column controls whether the roof edge is cut at an angle (square) or plumb to the ground. Selecting the existing value provides a drop-down list with the **Plumb** and **Square** options. The Plumb option always makes the edge of the slab perpendicular to the ground. The Square option "squares" the end of the slab between the top and bottom edge of the slab.

- **Angle** This column is used to specify a bevel angle for the roof edge. Pick in the area below the column to enter an angular value for the slab edge.

Exercise 16.9

1. Open EX16-8-Flr-Pln.dwg.
2. Adjust the overhang edges for all the roof slabs in the figure.
3. Set the overhang to 2'-0", and use the 1x8 Fascia + Soffit roof slab edge style.
4. Set the overhang for the left and right side roof slabs in the rear raised roof to 6".
5. Save the drawing as EX16-9-Flr-Pln.dwg.

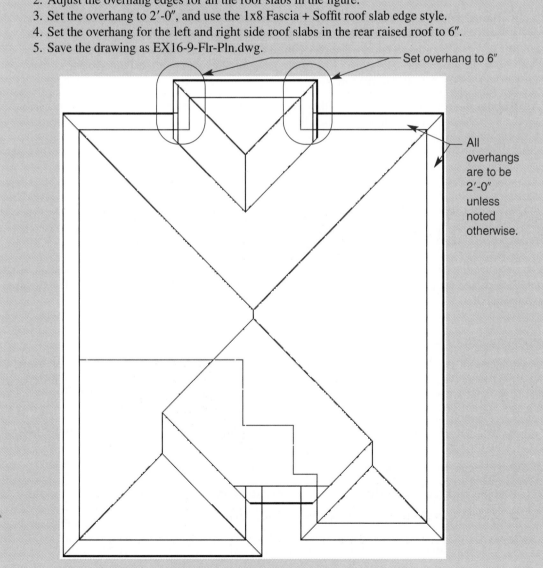

Set overhang to 6"

All overhangs are to be 2'-0" unless noted otherwise.

Using Roof Slab Properties

Roof slabs can also be modified through the **Roof Slab Properties** dialog box. This command provides many of the roof slab settings in one dialog box. Use one of the following options to access the Roof Slab Properties command:

> ✓ Select the roof slab, right-click, and pick **Roof Slab Properties...** from the shortcut menu.

> ✓ Type **ROOFSLABPROPERTIES** at the Command: prompt.

The Roof Slab Properties dialog box is displayed on the drawing screen. The dialog box is divided into the following five tabs:

- **General**   This tab is used to provide a description for the selected roof slab(s), to attach design notes, and to assign property set definitions to the selected roof slab(s).

- **Style**   This tab is used to change the selected roof slab into a different roof slab style. If a different roof slab style is selected, the roof slab is changed to the new style, but the properties of the new style, such as the edge style and thickness, are not applied. In order for the edge style and thickness to be applied to the roof slab, pick the **Reset to Style Defaults** button.

- **Dimensions**   This tab is used to set the dimensional values for the selected roof slab. The **Vertical Offset** controls the distance from the 0 Z axis to the justification of the roof slab. The **Horizontal Offset** controls the overhang used for the roof slab. The slope of the roof slab and the pivot point location also can be controlled in this tab.

- **Edges**   This tab contains the same information as specified in the Edit Roof Slab Edges dialog box (see Figure 16.25). Use this tab to make any adjustments to the edges of the selected slab.

- **Location**   This tab indicates the current location of the roof slab in the drawing.

Roof Slab and Roof Slab Edge Styles

When you use the AEC Arch templates, a few roof slab styles are available for use in the creation of your roofs. You can create custom roof slab styles and roof slab edge styles or import them to reflect the design of your roof.

Importing Roof Slab and Roof Slab Edge Styles

As with many of the other style-based objects, Architectural Desktop includes additional roof slab styles and roof slab edge styles that can be imported into the drawing. Use one of the following options to access these styles through the Style Manager dialog box:

> ✓ Select the **Design** pull-down menu, pick the **Roofs** cascade menu, then select the **Roof Slab Styles...** command.

> ✓ Select the **Roof Slab Styles** button found in the **Roofs - Roof Slabs** toolbar.

> ✓ Right-click, and select **Design, Roofs, Roof Slab Styles...** from the cursor menu.

> ✓ Type **AECROOFSLABSTYLE** at the Command: prompt.

The Style Manager dialog box is displayed, opened to the Roof Slab Styles section, as shown in Figure 16.26.

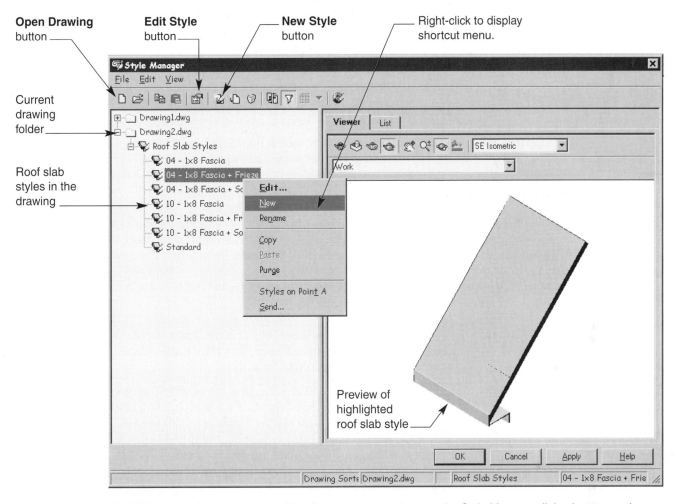

FIGURE 16.26 Access the Roof Slab Styles command to open the Style Manager dialog box opened to the Roof Slab Styles section.

Pick the Open Drawing button to browse to the \\Autodesk Architectural Desktop 3\Content\ Imperial\Styles folder. Along with additional AEC object styles, this folder includes roof slab and roof slab edge styles. Select the **Roof Slab & Roof Slab Edge Styles (Imperial)** drawing, and press the Open button. The Roof Slab and Roof Slab Edge Styles (Imperial) drawing is opened and displayed as a folder icon on the left side of the Style Manager. Expand the drawing folder to reveal the additional roof slab styles.

Picking on top of a style from the left tree view displays a preview of the roof slab style (see Figure 16.26). If the Roof Slab Styles section icon is selected, the roof slab styles are listed on the right side of the window. Drag the desired styles from the right or left panes into the current drawing folder icon to import them into a drawing. Roof slab styles that include an edge automatically import the assigned roof slab edge style into the drawing.

✎ Field Notes

Additional slab styles can be found in the \\Autodesk Architectural Desktop 3\Content\Metric\Styles folder. If the International Extensions have been installed, additional slab styles can be found in the \\Autodesk Architectural Desktop 3\Content\Metric DACH\ Styles folder.

If only roof slab edge styles need to be imported, then use one of the following options to access the Roof Slab Edge Styles command:

✓ Pick the **Design** pull-down menu, select the **Roofs** cascade menu, then pick the **Roof Slab Edge Styles...** command.

✓ Pick the **Roof Slab Edge Styles** button in the **Roofs - Roof Slabs** toolbar.

✓ Right-click, and select **Design, Roofs, Roof Slab Edge Styles...** from the cursor menu.

✓ Type **AECROOFSLABEDGESTYLE** at the Command: prompt.

To import the slab edge styles, open the Roof Slab & Roof Slab Edge Styles (Imperial) drawing. Drag and drop any desired edge styles into the current drawing folder icon.

Creating a Roof Slab Style

In addition to importing roof slab styles, you can create custom roof slab styles by accessing the Roof Slab Styles command previously outlined. Once you have entered the Style Manager dialog box, select the New Style button to create a new roof slab style, then enter a name for the new roof slab style. After naming the style, highlight the style and select the Edit Style button, right-click, and select Edit..., or double click on top of the style. The Roof Slab Styles dialog box, shown in Figure 16.27, is displayed. This dialog box includes the following four tabs used to configure the style:

- **General**   This tab is used to rename and provide a description for the roof slab style, to attach documents, spreadsheets, and HTML files to the style, and to assign a property set definition to the style.

- **Defaults**   This tab is used to set default values that appear in the Add Roof Slab dialog box when a roof slab is being drawn (see Figure 16.27). These settings designate the default values for new roof slabs but can be overridden at the time the roof slab is being created or modified. At the bottom of the tab are the **Baseline Edge** and **Perimeter Edges** check boxes. The Baseline Edge option assigns a roof slab edge style to the baseline (slopeline) of the roof slab. The Perimeter Edges option assigns a roof slab edge style to all the roof slab edges except the slopeline edge. If a roof slab edge does not exist, or if you want to create a new custom roof slab edge style, select the New Edge Style button to display the Roof Slab Edge Styles dialog box.

FIGURE 16.27 The Roof Slab Styles dialog box is used to create a new or modify an existing roof slab style.

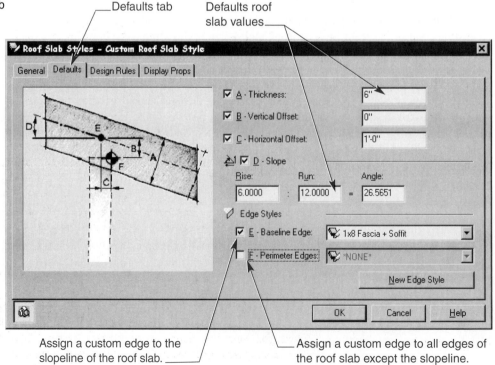

Defaults tab

Defaults roof slab values

Assign a custom edge to the slopeline of the roof slab.

Assign a custom edge to all edges of the roof slab except the slopeline.

- **Design Rules** This tab is used to set the design rules for the roof slab. If the roof slab is to be created with a set thickness, select the **Has Fixed Thickness** check box. The **Thickness** text box becomes available, so a constant thickness can be applied to the style. The **Thickness Offset** text box controls where the bottom of the roof slab is located in relation to the vertical offset (base height). A negative value places the bottom of the roof slab below the justification vertex points. A positive value places the bottom of the slab above the justification vertex points.

- **Display Props** This tab is used to control the appearance of the slab appears in plan and model representations. The plan representation roof slab includes a cut plane height that can be specified to cut through the roof slab. In addition to the cut plane, a hatch component can be turned on for the slab and configured. Both the plan and model representations include the pivot point component. You may decide to turn this component off or set it to a layer that will not plot, because the component will plot if turned on.

Exercise 16.10

1. Open Ex16-7.dwg.
2. Create a new roof slab style named 10in Tile Roof.
3. Set the thickness to always be 10″.
4. Set the default slope to an 8″ rise.
5. Create a new edge style named Gutter and Frieze. Assign the edge style to the baseline of the roof slab. Do not set any parameters to the edge style.
6. Save the drawing as EX16-10.dwg.

Creating a Roof Slab Edge Style

Usually when a roof slab is created, an edge is associated with it. The Defaults tab in the Roof Slab Styles dialog box allows you to create a new slab edge style. Select the New Edge Style button to display the **Roof Slab Edge Styles** dialog box, or enter the Roof Slab Edge Styles command, create a new style, then edit it to access this dialog box. The Slab Edge Styles dialog box includes the following three tabs:

- **General** This tab is used to rename the style and enter a description for the style.

- **Defaults** This tab specifies default values that control the edge of the roof slab (see Figure 16.28). As with the roof slab styles, these settings control only the default values for the slab edge. The values can be modified when the roof slab is added or modified.

- **Design Rules** This tab is used to add a custom shape for the roof slab edge style (see Figure 16.29). Initially, two check boxes are available to customize the edge. The **Fascia** and **Soffit** check boxes allow you to add a custom shape for the fascia and the soffit.

  A drop-down list is associated with each of these check boxes. The items included in the lists are AEC profiles defined in the drawing. See Chapter 7 for information on how to create a profile. When creating a profile for use as a fascia or soffit specify the insertion points on the profile, as indicated in Figure 16.30.

  The fascia option allows you to adjust the height of the profile to equal the height or thickness of the slab. Select the check box to automatically adjust the size of the fascia. The Soffit command also provides an option for adjusting the length to equal the width of the overhang by selecting the check box. Additionally, the Soffit option includes settings to locate the soffit in relation to the fascia profile.

Defaults tab

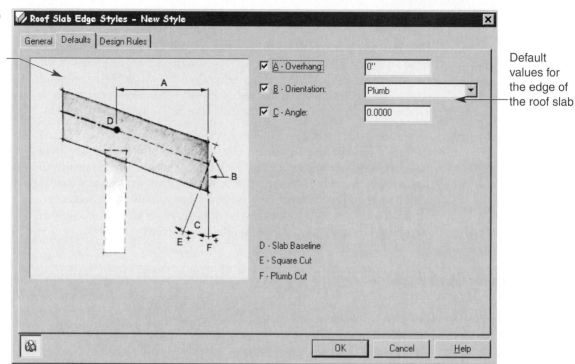

FIGURE 16.28   Set any defaults for the roof slab edge in the Defaults tab.

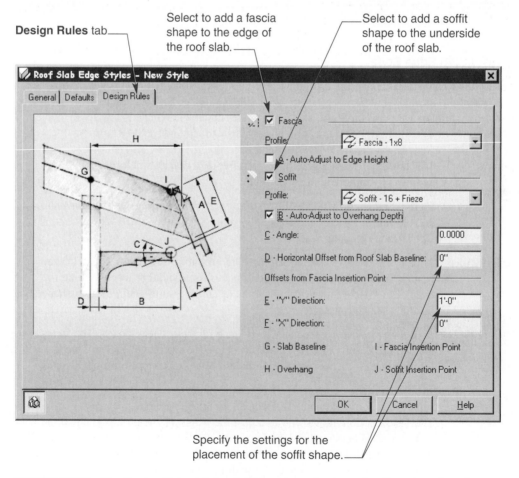

FIGURE 16.29   The Design Rules tab is used to assign a fascia and soffit to the edge of a roof slab.

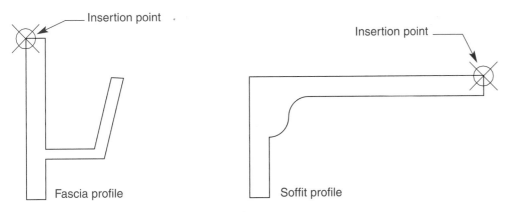

FIGURE 16.30   Use the upper right edge of the profile as the insertion point for the fascia profile. Use the upper right corner of the profile as the insertion point for the soffit profile.

When you have finished specifying the roof slab edge style, the style is ready to be assigned to a roof slab style or to be placed on the edge of an existing roof slab.

Exercise 16.11

1. Open EX16-10.dwg.
2. Draw the fascia and soffit profiles shown in the figure using a closed polyline. Turn the polylines into profiles with the names and insertion points displayed.
3. Access the Roof Slab Edge Styles dialog box. Edit the Gutter and Frieze edge style.
4. Set the default overhang to 18″, the orientation to Plumb, and the angle to 0.
5. Set the fascia to the Gutter profile. Auto-adjust the height of the fascia.
6. Set the soffit to the Frieze profile. Auto-adjust the overhang depth and adjust the Y direction to 10″.
7. Use the Roof Slab Properties dialog box to change the existing slabs to the 10in tile Roof roof slab style and an 8″ rise. Assign the Gutter and Frieze edge style to all of the overhangs in the roof slabs.
8. Make sure that all the overhang edges are 16″, plumb, and assigned to the Gutter and Frieze edge style.
9. Save the drawing as EX16-11.dwg.

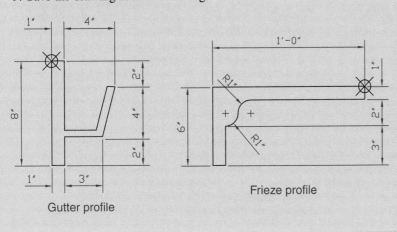

Using Roof Slab Tools

Roof slabs, which are similar to standard slab objects, include the following tools that can be used to modify the shape of the slab.

Trimming Roof Slabs

Roof slab objects include a trim command that can be used to trim a roof slab against a trimming plane. The trimming plane can be a polyline, slab, or wall. If a polyline is used as a trimming plane, the polyline must extend to the edge or past the edge of the roof slab that is to be trimmed.

Once the trimming planes have been established in the drawing, the roof slab can be trimmed away from the trimming plane. When a roof slab is trimmed, part of the roof slab is removed from the drawing. If the part of the roof slab being trimmed includes a custom roof slab edge, that will also be trimmed. The edge is trimmed at the point it intersects the trimming plane and is cut at the same angle as the trimming plane.

Use one of the following options to trim a slab:

✓ Select the **Design** pull-down menu, access the **Roof Slab Tools** cascade menu, and pick the **Trim Roof Slab** command.

✓ Pick the **Trim Roof Slabs** button from the **Roof Slab Tools** toolbar.

✓ Right-click and select **Design, Roof Slab Tools, Trim Roof Slabs** from the cursor menu.

✓ Type **AECROOFSLABTRIM** at the Command: prompt.

✓ Select, the roof slab, then right-click, and pick the **Tools** cascade menu, and select the **Trim** command.

The following prompt sequence is used to trim the slab away from a polyline in Figure 16.31:

Command: **AECROOFSLABTRIM** ↵
Select trimming object (a slab, wall, or polyline): *(select the polyline)*
Select a slab to trim: *(select the roof slab)*
Specify side to be trimmed: *(pick the side of the roof slab to be trimmed)*
Command:

✏️Field Notes

It is not possible to trim only an overhang. The trimming plane must intersect the defined boundary of the slab at some location in order to trim the overhang.

FIGURE 16.31　Trimming a roof slab away from a polyline.

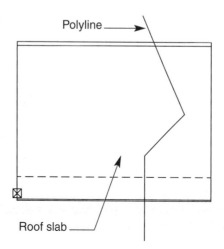
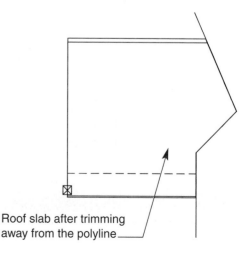

Polyline

Roof slab

Roof slab after trimming away from the polyline

Extending Slabs

In addition to being trimmed, roof slabs edges can be extended to a wall, to another slab, or to a roof slab object. When a roof slab is extended, the edge that is being extended is being moved to the face of the wall or other slab. The sides next to the edge being moved are the actual edges being extended (see Figure 16.32A). Use one of the following options to access the Extend Roof Slab command to extend the edges of a roof slab:

✓ Pick the **Design** pull-down menu, select the **Roof Slab Tools** cascade menu, and select the **Extend Roof Slab** command.

✓ Pick the **Extend Roof Slab** button from the **Roof Slab Tools** toolbar.

✓ Right-click, and select **Design, Roof Slab Tools, Extend Roof Slab** from the cursor menu.

✓ Type **AECROOFSLABEXTEND** at the Command: prompt.

✓ Select the roof slab, then right-click, select the **Tools** cascade menu, then the **Extend** command from the shortcut menu.

The following prompt sequence is used to extend the edge in Figure 16.32:

Command: **AECROOFSLABEXTEND** ↵
Select an object to extend to (a slab or wall): *(select the dormer wall)*
Select a slab to extend: *(select the roof slab to be extended)*
Slab will be extended by lengthening two edges.
Select first edge to lengthen: *(pick one edge of the roof slab to be extended)*
Select second edge to lengthen: *(pick the second edge of the roof slab to be extended)*
Command:

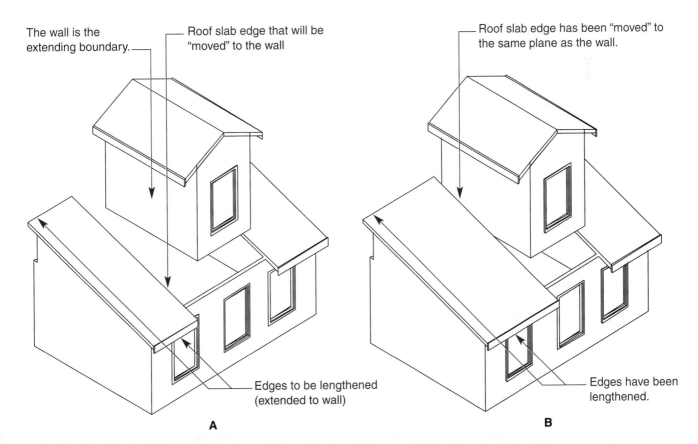

FIGURE 16.32 (A) The roof slab needs to be extended to the dormer wall. The side edge needs to be moved, and the top and bottom edges need to be extended. (B) The roof slab is extended to the plane of the dormer wall.

Mitering Roof Slabs

Occasionally, when roof slabs are modified, the edges of the roof slab overlap or do not touch each other to form a "clean" edge. This clean edge can be attained by mitering the edges of the roof slabs. When two roof slabs are mitered, the edge of each roof slab is joined together with a miter angle applied to each edge where the two slabs meet. Use one of the following options to access this command:

✓ Select the **Design** pull-down menu, pick the **Roof Slab Tools** cascade menu, and select the **Miter Roof Slab** command.

✓ Pick the **Miter Roof Slab** button in the **Roofs - Roof Slabs** toolbar.

✓ Right-click, and select **Design, Roof Slab Tools, Miter Roof Slab** from the cursor menu.

✓ Type **AECROOFSLABMITER** at the Command: prompt.

✓ Select the slab to be mitered, right-click, pick the **Tools** cascade menu, then the **Miter** command from the shortcut menu.

The Miter Roof Slab command includes two methods of mitering the roof slab. These options are described next.

Miter by Intersection

The **Miter by Intersection** option is used when two slabs intersect each other, as at the ridge, valley, or hip lines as the roof slabs are being modified (see Figure 16.33A). The intersection option miters the two slabs at the intersection of the two edges. The new edges of the two slabs are then set to an angle that creates a true miter between the roof slabs. Mitering by intersection does not miter the angle between overhangs. The following prompt sequence is used to miter the two roof slabs displayed in Figure 16.33:

Command: **AECROOFSLABMITER** ↵
Miter by [Intersection/Edges] <Intersection>: ↵
Select first roof slab at the side to keep: *(select an edge of a roof slab on the side of the roof slab that will be kept)*
Select second roof slab at the side to keep: *(select an edge of the other roof slab on the side of the roof slab that will be kept)*
Command:

 Field Notes

> If you are mitering two slabs that include overhangs, ensure that the overhangs overlap each other before mitering to cause the edges at the overhangs to be mitered as well. Use grips to stretch the edge of the roof slab past the overhang of the other slab that will be mitered, then miter the two roof slabs.

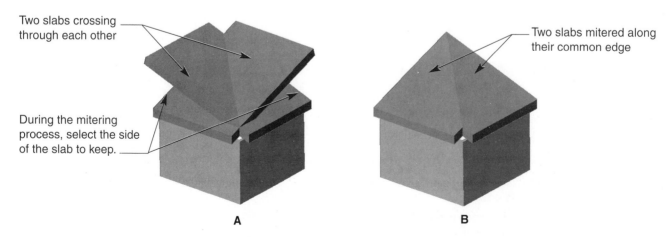

FIGURE 16.33 (A) Two roof slabs cross through each other at the hip line. (B) The two slabs after mitering.

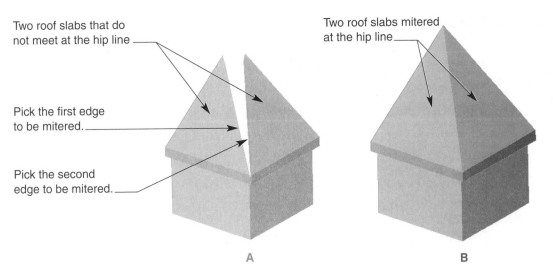

Two roof slabs that do not meet at the hip line

Two roof slabs mitered at the hip line

Pick the first edge to be mitered.

Pick the second edge to be mitered.

A   B

FIGURE 16.34   (A) Two roof slabs do not meet at the hip line. Select the edge of one roof slab to use as the first edge to be mitered, then pick the second edge to use as the second mitered edge.

Miter by Edge

Another way of mitering the edges of two roof slabs is by edge. This option allows you to miter the edges of two slabs even when they do not intersect each other (see Figure 16.34A). If the slabs do not intersect, then the edges are extended to the point at which they meet. If the two slabs cross through each other, then the slabs are trimmed similar as by the intersection option. The following command sequence is used to miter the slabs shown in Figure 16.34 by edge:

Command: **AECROOFSLABMITER** ⏎
Miter by [Intersection/Edges] <Intersection>: **E** ⏎
Select edge on first slab: *(select the edge of the first roof slab to miter)*
Select edge on second slab: *(select the edge of the second roof slab to miter)*
Command:

Exercise 16.12

1. Open EX16-11.dwg.
2. Use the Miter Roof Slab command to miter the roof slab edges.
3. Use the Miter by Edge option.
4. Save the drawing as EX16-12.dwg.

Cutting Roof Slabs

The **Cut Roof Slab** command is similar to the Trim Roof Slab command in that it also cuts the slab but retains both portions of the roof slab. A polyline or a 3D object with mass can be used as a cutting tool to cut a roof slab.

If a polyline is used as the cutting plane, it can be a closed polyline or an open polyline that crosses through the roof slab. If the open polyline crosses in and out of the roof slab, multiple roof slabs are created.

3D objects that can be used to cut a roof slab include walls, slabs, roof slabs, mass elements, or AutoCAD solids. When a 3D object is used as a cutting tool, the object must extend through the roof slab at both the top and bottom sides. If the object is to create two individual roof slabs, the 3D object must also extend through the edges of the roof slab. If the 3D object does not extend through the edges of the roof slab, then a hole is cut out of the roof slab.

Use one of the following options to access the Cut Roof Slab command:

✓ Pick the **Design** pull-down menu, select the **Roof Slab Tools** cascade menu, and pick the **Cut Roof Slab** command.

✓ Pick the **Cut Roof Slab** button in the **Roofs - Roof Slabs** toolbar.

✓ Right-click, and select **Design, Roof Slab Tools, Cut Roof Slab** from the cursor menu.

✓ Type **AECROOFSLABCUT** at the Command prompt.

✓ Select the roof slab to be cut, right-click, pick the **Tools** cascade menu, then the **Cut** command from the shortcut menu.

The following prompt sequence is used to cut the slab in Figure 16.35:

Command: **AECROOFSLABCUT** ↵
Select a slab: *(select the slab to cut)*
Select cutting objects (a polyline or connected solid entities): *(select the cutting object (polyline))*
1 found

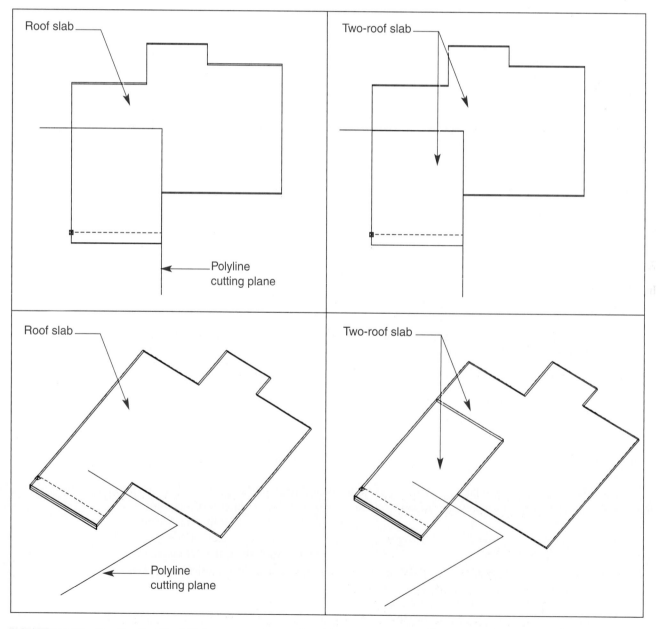

FIGURE 16.35   Use the Cut Roof Slab command to cut the slab into two pieces. Both pieces are retained in the drawing.

Select cutting objects (a polyline or connected solid entities): ↵
Erase layout geometry [Yes/No] <N>: ↵
Command:

Adding Roof Slab Vertices

When roof slabs are initially created, the points selected become the vertex points for the roof slab. Occasionally, when a roof slab is modified an additional edge or vertex needs to be placed. Adding a vertex along the edge of a roof slab creates two new edges. Adding a new vertex is useful when an edge is not available to be used in the Extend Roof Slab command.

Use one of the following options to add vertices to a slab:

✓ Select the **Design** pull-down menu, pick the **Roof Slab Tools** cascade menu, then the **Add Roof Slab Vertex** command.

✓ Pick the **Add Roof Slab Vertex** button in the **Slab Tools** toolbar.

✓ Right-click, and select **Design, Roof Slab Tools, Add Roof Slab Vertex** from the cursor menu.

✓ Type **AECROOFSLABADDVERTEX** at the Command: prompt.

✓ Select the slab, right-click, pick the **Tools** cascade menu, then the **Add Vertex** command from the shortcut menu.

The following command sequence is used to add vertices to a slab edge:

Command: **AECROOFSLABADDVERTEX** ↵
Select a roof slab: *(select the slab)*
Specify point for new vertex: **MID** ↵
of *(select the mdpoint of one of the slab edges)*
Command:

Once you have added a vertex, pick on top of the roof slab to display the grip boxes. A new grip box is located at the point you selected, dividing the edge into two new edges.

Remove Slab Vertices

In addition to adding vertices to a slab edge, you can remove them. Use one of the following options to access the **Remove Slab Vertex** command:

✓ Select the **Design** pull-down menu, access the **Roof Slab Tools** cascade menu, and select the **Remove Roof Slab Vertex** command.

✓ Pick the **Remove Roof Slab Vertex** button in the **Roofs - Roof Slabs** toolbar.

✓ Right-click, and select **Design, Roof Slab Tools, Remove Roof Slab Vertex** from the cursor menu.

✓ Type **AECROOFSLABREMOVEVERTEX** at the Command: prompt.

✓ Select the slab, right-click, pick the **Tools** cascade menu, then the **Remove Vertex** command.

Once you have entered the command, pick a point on top of the vertex to be removed. This removes the vertex and joins the two adjacent edges together.

Adding Holes to Roof Slabs

The addition of skylights and dormers to a roof generally requires that a hole be placed in the roof. The **Roof Slab Hole** command can be used to accomplish this. When a hole is added to a roof slab, the new vertices and edges are created at the perimeter of the hole.

Holes can be generated from a closed polyline or from a 3D object such as a wall, mass element, slab, or AutoCAD solid. When a polyline is used to create a hole, it must be within the boundary edge of the slab and is projected through the roof slab. If a 3D object is used, the object must physically extend through both the top and bottom faces of the roof slab. If the solid objects create a void in the slab, such as the four walls of a chimney, you are prompted to cut the hole on the inside or outside of the 3D objects (see Figure 16.36).

Use one of the following options to access the Roof Slab Hole command:

✓ Pick the **Design** pull-down menu, select the **Roof Slab Tools** cascade menu, and pick the **Roof Slab Hole** command.

✓ Pick the **Roof Slab Hole** button from the **Roof Slab Tools** toolbar.

✓ Right-click, and select **Design, Roof Slab Tools, Roof Slab Hole** from the cursor menu.

✓ Type **AECROOFSLABHOLE** at the Command: prompt.

✓ Select the roof slab, right-click, and pick the **Tools** cascade menu, then select the **Add Hole** command to add a hole, or the **Remove Hole** command to remove a hole.

When you enter the command, you are prompted to add or remove a hole. The following prompt sequences are used to add a hole and remove a hole from a roof slab:

Add Hole

Command: **AECROOFSLABHOLE** ↵
Hole [Add/Remove] <Add>:↵
Select a roof slab: *(select a roof slab that will have a hole cut through)*
Select a closed polyline or connected solid entities to define a hole: *(select a polyline or a 3D object(s) for use as a hole)*
1 found
Select a closed polyline or connected solid entities to define a hole: ↵
Erase layout geometry? [Yes/No] <N>: **Y** ↵
Command:

Remove Hole

Command: **AECROOFSLABHOLE** ↵
Hole [Add/Remove] <Add>: **R** ↵
Select an edge of hole to remove: *(select the edge of an existing hole)*
Command:

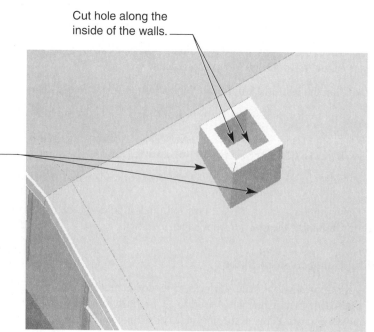

FIGURE 16.36 Four walls are being used to create a chimney. The hole can be cut on the inside or on the outside of the four walls.

Cut hole along the inside of the walls.

Cut hole along the outside of the walls.

Exercise 16.13

1. Open EX16-9-Flr-Pln.dwg.
2. Modify the chimney walls to be 12'-0" high.
3. Use the Roof Slab Hole command to add a hole into the rear roof slab through which the chimney extends. Cut the hole to the outside of the chimney walls.
4. Save the drawing as EX16-13-Flr-Pln.dwg.

Using Boolean Operations on a Roof Slab

Occasionally, when creating a roof you may need to add additional detail. For example, the addition joists and cornice details cannot be added to the roof through any of the roof commands; however, these types of details can be incorporated into the roof slab after they have been created. The **Boolean Add/Subtract** command allows you to join additional objects into the roof slab to form one object. The advantage of doing this is that when a section is cut through the roof and the detail, the section formed is displayed as a single outline.

Use one of the following options to access this command:

✓ Select the **Design** pull-down menu, pick the **Roof Slab Tools** cascade menu, and select the **Boolean Add/Subtract** command.

✓ Pick the **Boolean Add/Subtract** button from the **Roof Slab Tools** toolbar.

✓ Right-click, and select **Design, Roof Slab Tools, Boolean Add/Subtract** from the cursor menu.

✓ Type **AECROOFSLABBOOLEAN** at the Command: prompt.

✓ Select the roof slab to display the grip boxes, right-click, and pick the **Tools** cascade menu, then select the **Boolean** command.

When you use the Boolean Add/Subtract command you are first prompted to select a roof slab, then to select objects. Select any 3D objects to be joined to the roof slab. After selecting these objects, you have the option of adding the objects to the roof slab, subtracting their mass away from the roof slab, or detaching them from a roof slab.

If the 3D objects are added to the roof slab, the roof slab is adjusted to include these objects. The original objects remain in the drawing, so they can be modified if required. Any modifications to the 3D objects are reflected in the roof slab where they have been added.

 Field Notes

When objects are added to or subtracted from a roof slab, the original entities remain in the drawing on their original layers. It is common to move these objects to a layer that is set to not plot or that is frozen so the objects do not appear in the finished plot but are incorporated into the roof slab, which plots and displays the appropriate profiles.

Creating a Roof Dormer

The **Roof Dormer** command is used to cut a hole in a roof slab for a dormer and then to miter the edges of the dormer roof with the face of the main roof slab.

To create a roof dormer, you must create all the parts of the dormer first. Construct the dormer as a minihouse, with all the walls that create a room plus the roof on top of the dormer walls. Include the back wall, because all the walls are used to cut a hole in the main roof slab (see Figure 16.37).

Ensure that the rear and side dormer walls extend past the ridge of the roof slab.

Add roof slabs to the top of the dormer walls.

Construct all four sides of the dormer, with the side and rear walls extending past the roof slab ridgeline.

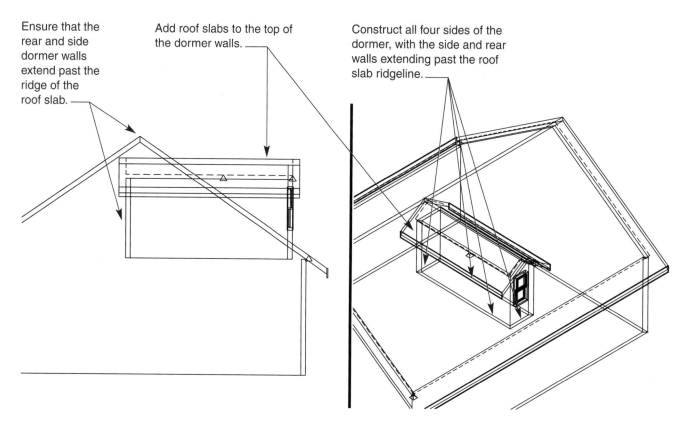

FIGURE 16.37   In order to create a dormer, first construct the dormer as a small house with all the sides and roof slabs.

 Field Notes

In order for the dormer to be constructed properly, the walls must extend past the end (ridgeline) of the roof slab.

Once you have created the dormer, enter the Roof Dormer command:

✓ Select the **Design** pull-down menu, pick the **Roof Slab Tools** cascade menu, then select the **Roof Dormer** command.

✓ Pick the **Roof Dormer** button in the **Roof Slab Tools** toolbar.

✓ Right-click, and select **Design, Roof Slab Tools, Roof Dormer** from the cursor menu.

✓ Type **AECROOFSLABDORMER** at the Command: prompt.

The following prompt sequence is used to create the dormer in Figure 16.38:

Command: **AECROOFSLABDORMER** ↵
Select a roof slab: *(select the main roof slab that the dormer will cut through)*
Select objects that form the dormer: *(select the first wall)*
1 found
Select objects that form the dormer: *(select the second wall)*
1 found, 2 total
Select objects that form the dormer: *(select the third wall)*
1 found, 3 total
Select objects that form the dormer: *(select the fourth wall)*
1 found, 4 total

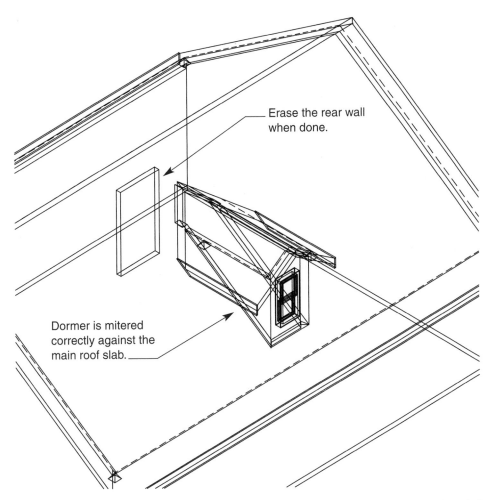

Erase the rear wall when done.

Dormer is mitered correctly against the main roof slab.

FIGURE 16.38 The dormer has cut through the main slab, mitered the dormer roof, and sliced the dormer walls against the main roof slab.

Select objects that form the dormer: *(select a dormer roof slab)*
1 found, 5 total
Select objects that form the dormer: *(select the other dormer roof slab)*
1 found, 6 total
Select objects that form the dormer:
Slice wall with roof slab [Yes/No] <Y>: *(enter Y to slice the walls against the top of the main roof slab)* **Y** ↵
Command:

After the dormer walls are sliced, the dormer roof slabs and the main roof slabs are mitered, the dormer walls are projected to the top face of the main roof slab, and a hole is cut in the main roof slab. The rear wall can now be erased, as it is no longer required.

 Field Notes

The Roof Dormer command is good for creating standard roof dormers; however, you may run into situations in which you have to use standard roof slab editing tools to cut the hole for the dormer, miter the edges of the dormer roof slabs with the main roof slab, then use the Floor Line command to project the walls to the main roof dormer.

|||||||||| CHAPTER REVIEW

 Use the CD-ROM to test your knowledge and skills.

Chapter Test

To check your understanding of the content provided in this chapter, access the Test file in the CH16 folder of the CD-ROM that accompanies this text.

Chapter Project

To practice the Architectural Desktop skills presented in this chapter, access the Project files in the CH16 folder of the CD-ROM that accompanies this text. The project files are in pdf format and include sample drawings and instructions for completing each project.

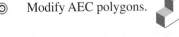

‖‖

CHAPTER 17

Creating Elevations

LEARNING GOALS

After completing this chapter, you will be able to:

◎   Create exterior 2D elevations.

◎   Adjust elevation line properties.

◎   Create and use elevation styles.

◎   Establish rules for elevation styles.

◎   Set up the entity display.

◎   Edit elevations.

◎   Generate interior elevations.

◎   Edit elevation marks.

◎   Create 3D elevations.

◎   Use the Hidden Line Projection command.

◎   Add AEC polygons.

◎   Use arc polygons.

◎   Create AEC polygon styles.

◎   Convert polylines to AEC polygons.

◎   Modify AEC polygons.

So far Architectural Desktop tools that aid in the development of drawing plans have been discussed. Floor plans, foundation plans, and roof plans are used in a set of construction documents to describe the layout of a building. These drawings show the horizontal plane of the building and explain any detail required to construct the building. Another important part of the construction document set are the elevations and sections.

Elevations show how the vertical face of a building is to appear. Through the use of exterior and interior elevations, information not found in plan sheets can be added and documented. Elevations include information such as the building materials, distances between floors, and horizontal and vertical dimensions that cannot be found elsewhere. Similar to plan sheets, elevations are 2D projections of the 3D model, as shown in Figure 17.1.

This chapter discusses the tools included in Architectural Desktop for the creation of elevations.

Creating Elevations

Once the building model has been created, elevations can be established from it. Architectural Desktop creates elevations from the model representation of the objects. You can create the elevations in the same drawing where you draw the walls, doors, and windows, or you can create the elevations from a series of referenced drawings.

Before creating an elevation, you must draw an *elevation line*, which determines the extents of the elevation view that is used to generate an elevation. A 2D elevation is created from the elevation line by orthographic projection. *Orthographic projection* is any projection of the features of an object onto an

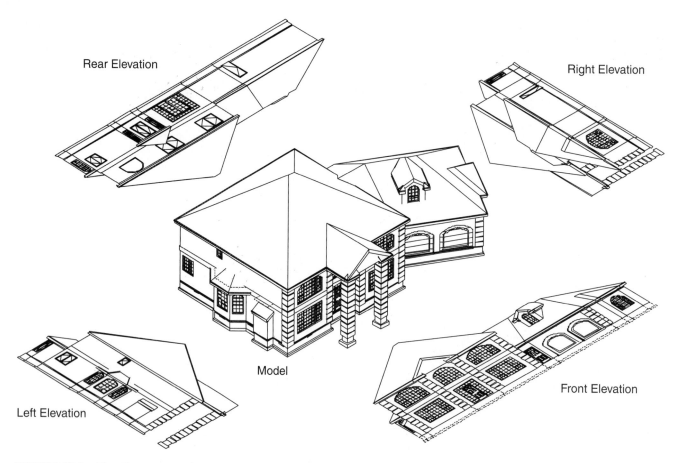

Rear Elevation

Right Elevation

Model

Left Elevation

Front Elevation

FIGURE 17.1   Elevations are typically two-dimensional drawings of a three-dimensional building.

imaginary plane called a ***plane of projection***. The features of the object are projected by lines of sight that are perpendicular to the plane of projection.

Use one of the following options to create an elevation line in the drawing:

✓ Select the **Documentation** pull-down menu, access the **Elevations** cascade menu, and pick the **Add Elevation Line** command.

✓ Pick the **Add Elevation Line** button from the **Elevations** toolbar.

✓ Right-click, and select **Documentation**, **Elevations**, **Add Elevation Line** from the cursor menu.

✓ Type **AECBLDGELEVATIONLINEADD** at the Command: prompt.

After you have entered the Add Elevation Line you are prompted for a starting point and an ending point. These points specify the outer boundaries for the elevation. The following prompt sequence is used to create the elevation line in Figure 17.2:

Command: **AECBLDGELEVATIONLINEADD** ↵
Elevation line start point: *(pick a point to start the elevation line)*
Elevation line end point: *(pick a point to end the elevation line)*
Command:

When selecting points to establish an elevation line, pick the points in a counterclockwise direction around the building. See Figure 17.2. If you are creating multiple elevations, then you need an elevation line for each elevation. Make sure that the points you select are beyond the edge of the building being considered for the elevation.

When you have finished creating the elevation line, a box is added to the drawing, as shown in Figure 17.2. This box determines the parts of the building model that are available for the elevation object. For a 2D elevation the box needs to wrap only around the part of the building that will be shown in the elevation. For a

FIGURE 17.2 Pick a starting point then an ending point to establish an elevation line.

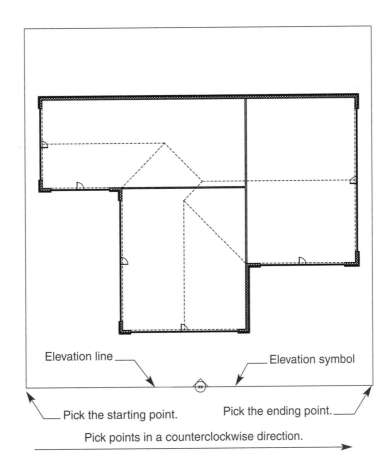

3D elevation, the box should wrap around the entire building model. The box can be enlarged or shrunk by selecting the box to display the grips, picking a grip point, and respecifying the location of the point.

The elevation line box appears if you are using the Work display configuration. If you are using the Plot display configuration, the box disappears, leaving only the elevation mark. The display of these items can be controlled with entity display.

Exercise 17.1

1. Start a new drawing using the AEC Arch (Imperial) or the AEC Arch (Imperial-Int) template.
2. Create two xref layers: Xref-flr-pln and Xref-fnd-pln.
3. Set the Xref-fnd-pln layer current. Use the XREF command to attach EX14-1-Fnd.dwg. Use the Overlay type and insert at X=0, Y=0, Z=0.
4. Set the Xref-flr-pln layer current. Use the XREF command to attach EX16-9-flr-pln.dwg. Use the Overlay type and insert at X=0, Y=0, Z=2′-10″.
5. Freeze the dimension layers for both referenced drawings.
6. Create an elevation line that will be used to generate a front elevation.
7. Draw an elevation line at the front of the building in front of and to the left of the garage, and an ending point in front of and to the right of the front bedroom.
8. Use grips to stretch the box completely around the building.
9. Save the drawing as EX17-1-Elev.dwg.

Creating the 2D Elevation

Once you have placed the elevation line in the drawing, use one of the following options to create the elevation:

✓ Select the **Documentation** pull-down menu, pick the **Elevations** cascade menu, then select the **Create Elevation...** command.

✓ Pick the **Create Elevation** button from the **Elevations** toolbar.

✓ Right-click, and select **Documentation, Elevations, Create Elevation...** from the cursor menu.

✓ Type **AECBLDGELEVATIONLINEGENERATE** at the Command: prompt.

✓ Select the elevation line, right-click, and pick **Generate Elevation...** from the shortcut menu.

After entering the command, select the elevation line. The Generate Section/Elevation dialog box, shown in Figure 17.3, is displayed. The Generate Section/Elevation dialog box is divided into the following four areas:

■ **Result Type** This area is used to determine the type of elevation to be created. The default is the 2D elevation. The 2D elevation creates a two-dimensional, flat elevation drawing, placing it in the current XY plane. The 3D elevation creates a three-dimensional elevation that is virtually a copy of the building model with one color for plotting.

The Style to Generate drop-down list includes a list of elevation and section styles that can be used to generate the elevation or section. The Standard elevation style generates the 2D and 3D elevation using one color for plotting purposes. Creating an elevation style is described later in this chapter.

■ **Selection Set** This area is used to select the objects to be added to the elevation. Pick the **Select Objects** button to return the drawing screen. Select any objects to be

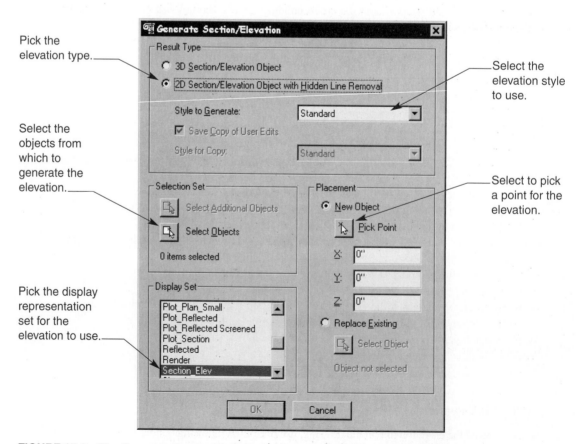

FIGURE 17.3 The Generate Section/Elevation dialog box is used to create an elevation.

displayed in the elevation. If you select an object by accident, hold the Shift key and reselect the object to remove it from the selection set.

- **Placement**   This area is used to pick a point in the drawing to place the 2D or 3D elevation. Use the **Pick Point** button to pick a location on the drawing screen, or enter an absolute coordinate in the X, Y, and Z text boxes.

- **Display Set**   This area includes a list of the available display representation sets in the drawing. Selecting a display set from the list assigns the new elevation to that display set, so it can be displayed only in a display configuration that is using that display set.

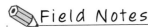Field Notes

It is recommended that the **Section_Elev** display set be used when creating an elevation. This displays the elevation correctly in the Work or Plot display configurations. If a different display set is picked, you may not see the elevation after it is created, unless you select a display configuration that uses the display set.

When you have finished specifying elevation criteria, press the OK button to create the elevation. Depending on the complexity of the building model and the number of items selected for use in the elevation, Architectural Desktop generates the elevation at the point specified in the Generate Section/Elevation dialog box (see Figure 17.4).

Once the elevation is finished being generated, it is displayed on the drawing screen. Selecting the elevation object highlights all the objects (walls, doors, windows). When the elevation is generated all the selected objects are placed together into a one-color elevation object that acts like a block. As long as the elevation remains in this state and is not exploded, it can be updated to reflect the latest changes to the plan drawings. Once the elevation object is displayed in the drawing, exploding the object breaks the link to the original building model.

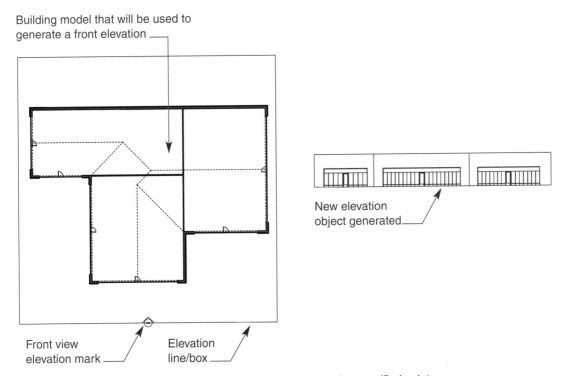

FIGURE 17.4   The new elevation is generated and placed at the specified point.

Exercise 17.2

1. Open EX17-1-Elev.dwg.
2. Select the elevation line, right-click, and pick the Generate Elevation... command from the shortcut menu.
3. Create a 2D elevation using the Standard elevation style.
4. Select all the objects used in the building model.
5. Pick a point to the right side of the xref'd floor plans.
6. Press the OK button to generate the elevation. An elevation object is created of the side of the building specified by the elevation line. The object is displayed using a single color.
7. Save the drawing as EX17-2-Elev.dwg.

 Adjusting the Elevation Line Properties

When an elevation is created from an elevation line, the elevation object appears as a single color. This is the color designated for the elevations when the building model is completely inside the elevation line box. Some offices prefer to create elevations using different colors that represent the depth of the building. For example, if a building has several jogs in its outline, sides of the building farther away are plotted in gray tones to represent the depth or distance away from the closest side of the building.

Color in the drawing screen is used during plotting to interpret the plotted color and grayscale of the linework. For example, objects in an elevation closest to you may be plotted in a heavy-weight black color to represent objects in front. As the building is stepped away from you the linework may be plotted in a lightweight gray color. The colors used for the linework on the drawing screen can be any color desired. For example, you may decide that the heavy-weight black line on the plotted paper should be white on the drawing screen, and the lightweight gray line on the plotted paper should be cyan on the screen. Plotting and the use of colors is discussed later in this text.

Elevations can be adjusted to reflect this type of coloring and plotting. When the elevation line is first created, the box that wraps around the building model is considered to be one subdivision. Any object within the box is generated into an elevation using the color assigned to the first subdivision. Additional subdivisions in the elevation line box can be created with their own colors assigned to each subdivision.

Use one of the following options to add additional subdivision lines:

✓ Select the **Documentation** pull-down menu, pick the **Elevations** cascade menu, then select the **Elevation Line Properties...** command.

 ✓ Pick the **Elevation Line Properties** button in the **Elevations** toolbar.

✓ Right-click, and select **Documentation, Elevations, Elevation Line Properties...** from the cursor menu.

✓ Type **AECBLDGELEVATIONLINEPROPS** at the Command: prompt.

✓ Select the elevation line, right-click, and pick **Elevation Line Properties...** from the shortcut menu.

Once the command has been activated, you are prompted to select an elevation line. Select the elevation line that will be used to add more subdivisions. The **Section/Elevation Line Properties** dialog box is displayed. This dialog box includes the following three tabs:

- **General**   This tab is used to specify a description for the elevation such as the front or right side elevation. Any design notes such as the number of subdivisions and the colors assigned to each can be added to the elevation line using the Notes... button. Property set definitions can also be added to the elevation line.

- **Dimensions**   This tab controls the dimensions and the heights used for the elevation line box (see Figure 17.5). Additional subdivisions can also be added in this tab. This tab includes the following two areas for configuring the elevation line box:

    - **Graphic**   This area controls the addition of subdivisions to the elevation line. Select
      Divisions   the Add... button to add a subdivision line. The **Add Subdivision** dialog is
      from Cut   displayed, allowing you to enter a distance measured from the elevation
      Plane   line to the new subdivision. When a subdivision is added, any object

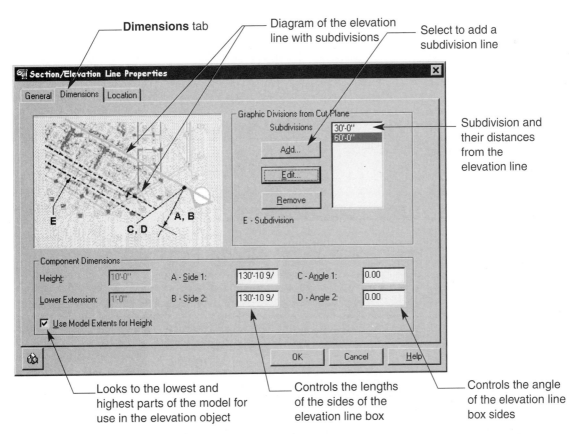

FIGURE 17.5   The Dimensions tab within the Section/Elevation Line Properties dialog box is used to add subdivisions and control the dimensions of the elevation line box.

between the elevation line and the first subdivision is created using a color. Any object between the first subdivision and the next subdivision or the end of the elevation line box is created using a different color.

You can add up to 10 subdivisions within the elevation line box. Subdivisions can be edited using the **Edit...** button or removed from the elevation line using the **Remove** button.

- **Cut Plane and Component Divisions**   This area controls the dimensions of the elevation line. The elevation line box can be used to specify a height or depth for the elevation object by entering values in the **Height** and **Lower Extension** text boxes. If a check mark is placed in the **Use Model Extents for Height** check box, the entire height and depth of the building is used for the elevation. The lengths of the sides in the elevation line box are controlled in the **Side 1** and **Side 2** text boxes. The angles of the sides can also be controlled in the **Angle 1** and **Angle 2** text boxes.

- **Location**   This tab is used to determine where the elevation line and box is placed.

After making any changes to the elevation line, pick the OK button to exit the dialog box. Any added subdivisions or adjustments made to the elevation line box are created. The subdivisions can be repositioned within the box using grips. Select the elevation line box, then select a grip on a subdivision line and reposition the line as desired. Any objects between the subdivision lines are assigned a different color for plotting.

Once a subdivision line or lines have been added and the position adjusted, the elevation object needs to be assigned colors that are used between the subdivisions. To assign colors to the different subdivisions, select the elevation object, right-click, and pick the Entity Display... command. The **Entity Display** dialog box is displayed. Select the appropriate level of display control from the Property Source column. Attach any overrides as required, then pick the Edit Display Props... button. In the **Layer/Color/Linetype** tab, assign a color to each subdivision used in the elevation line box (see Figure 17.6).

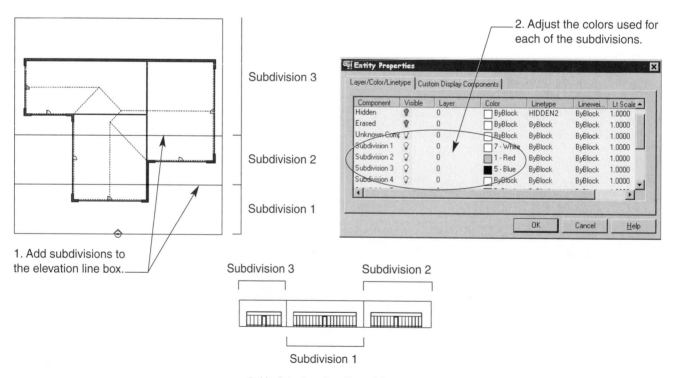

FIGURE 17.6   Add subdivisions to the elevation line box, assign colors to the subdivisions, then update the elevation.

After assigning the colors to each subdivision, use one of the following options to update the elevation object:

- Select the **Documentation** pull-down menu, access the **Elevations** cascade menu, then pick the **Update Elevation...** command.

- Pick the **Update Elevation** button in the **Elevations** toolbar.

- Right-click, and pick **Documentation, Elevations, Update Elevation...** from the cursor menu.

- Type **AECBLDGSECTIONUPDATE** at the Command: prompt.

- Select the elevation object, right-click, then select **Update...** from the shortcut menu.

The Generate Section/Elevation dialog box is again displayed. Press the OK button to update the elevation with the same settings. The elevation object is updated and is displayed with the different colors based on the location of the objects between the subdivision lines.

Exercise 17.3

1. Open EX17-2-Elev.dwg.
2. Select the elevation line, right-click, and pick Elevation Line Properties... from the short-cut menu.
3. Add two elevation lines, one at 10'-0", and the other at 20'-0". Press the OK button to exit.
4. Use the grip boxes to adjust the subdivision lines so that the first subdivision includes the garage and the front bedroom, the second subdivision includes the closet in the front bedroom, and the third subdivision includes the front entryway. See the figure for placement.
5. Select the elevation object, right-click, and select Entity Display...from the shortcut menu.
6. Edit the display properties for the System Default property source.
7. Change the colors for the following components:
   - Subdivision 1: Color 140
   - Subdivision 2: Color 211
   - Subdivision 3: Color 120
8. Exit the Entity Display dialog box, then update the elevation object.
9. Save the drawing as EX17-3-Elev.dwg.

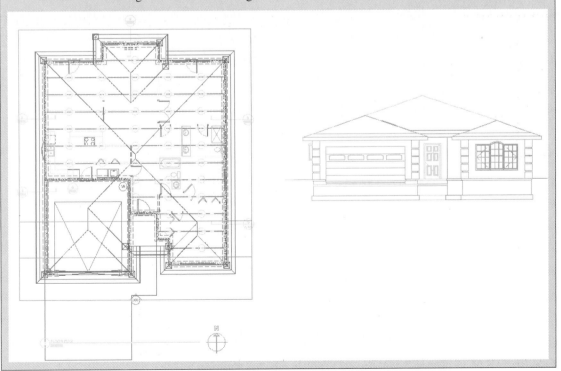

Creating and Using Elevation Styles

When a 2D elevation object is generated, a style is used to control the colors of the geometry between the subdivision lines. New styles can be generated with rules that intercept the color of AEC objects and assign them to a color to be used in the elevation object. For example, a rule can be created to intercept the window color in the building model and assign the windows a red color in the elevation object.

Use one of the following options to create an elevation style:

- Select the **Documentation** pull-down menu, pick the **Elevations** cascade menu, then select **Elevation Styles….**

- Pick the **Elevation Styles** button in the **Elevation** toolbar.

- Right-click, and select **Documentation, Elevations, Elevation Styles...** from the cursor menu.

- Type **AEC2DSECTIONSTYLE** at the Command: prompt.

The Style Manager appears, opened to the 2D Section/Elevation Styles section, as shown in Figure 17.7.

To create a new elevation style pick the New Style button or right-click over the 2D Section/Elevation icon and select New from the shortcut menu. See Figure 17.7. Enter a new name for the style in the text box provided to create a new elevation style. Highlight the new style and pick the Edit Style button, right-click over the new style, and select Edit…, or double-click on top of the new style. The **2D Section Elevation Styles** dialog box is displayed. This dialog box includes the following four tabs:

- **General**   This tab is used to provide a description for the elevation style. Construction notes and documents can be attached to the style through the Notes… button. Property set definitions can also be attached to the elevation style.

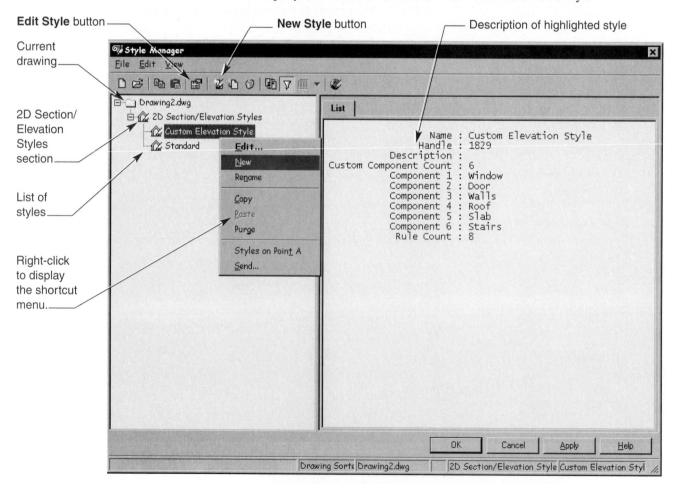

FIGURE 17.7   The Style Manager is used to create and edit elevation styles.

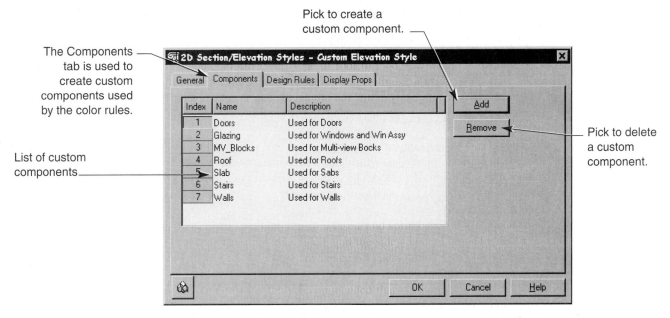

FIGURE 17.8 Add custom components that will be used by the color rules in the Design Rules tab.

- **Components** This tab allows you to create components for the elevation object that are used by the color rules. See Figure 17.8. For example, when subdivision lines are added in an elevation line box, the components used are Subdivision 1, Subdivision 2, Subdivision 3, and so on. Custom components can be added here for the color rules section in the Design Rules tab. You may decide to create components that reflect the type of linework that is assigned to the component. Some examples of components are Walls, Glazing, Doors, Roofs, Slabs, and Stairs.

  To add custom components pick the <u>A</u>dd button. A new custom component named *Unnamed* is created in the Name column. Pick the component to display a text box in which a new name can be entered. Add as many custom components as needed for the color rules. You can edit this tab as the style is developed by selecting a component and removing it or renaming it.

 Field Notes

Generally, when creating an elevation style, you should create custom components that reflect each type of object in the drawing. You can always edit these lists can to reflect office standards or when a color rule is changed.

- **Design Rules** This tab is used to create rules on how the elevation object interprets the colors of the building model and transfers them to a particular component within the elevation style (see Figure 17.9). Use the <u>A</u>dd button to add a rule. Creating the rules is discussed in the next section.

- **Display Props** This tab is used to configure the colors of the components (both default and custom) within the elevation style. This procedure is described in greater detail in the next section.

The **Design Rules** tab is used to create rules on how building model colors are interpreted and transferred to a display component in the elevation style.

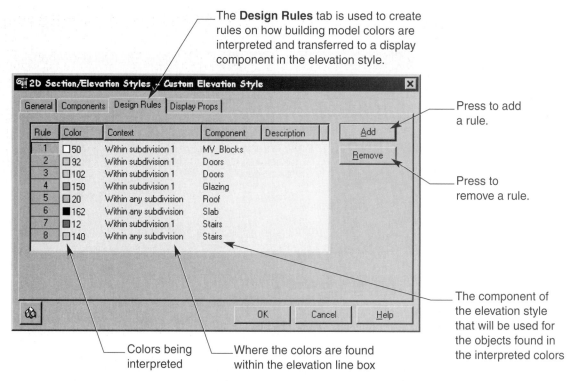

Press to add a rule.

Press to remove a rule.

The component of the elevation style that will be used for the objects found in the interpreted colors

Colors being interpreted

Where the colors are found within the elevation line box

FIGURE 17.9 The Design Rules tab is used to create rules about how the building model colors are interpreted and sent to a display component within the elevation style.

Exercise 17.4

1. Open EX17-3-Elev.dwg.
2. Create a new elevation style named Residential Elevations.
3. Edit the style and access the Components tab.
4. Create the following components with descriptions:

Index	Name	Description
1	Doors	Use for Door objects
2	Roof	Use for Roof objects
3	Slabs	Use for Slab objects
4	Walls	Use for Wall objects
5	Windows	Use for Window objects

5. Exit the Elevation Styles dialog box and save the drawing as EX17-4-Elev.dwg.

  Creating Rules for the Elevation Style

Before creating a rule, it is necessary to understand what the elevation style is looking for. When the elevation object is first created, a display set is used. In Figure 17.10 the Section_Elev display set is used. This display set (display representation set) groups together a series of display representations to be used when this set is viewed. Display representations and display sets were discussed in Chapter 5.

The Section_Elev display set looks to the Elevation display representation for windows, doors, and structural members, to the Model ByBlock display representation for the majority of the other AEC objects, and to the General display representations for AEC objects that do not have an Elevation or Model ByBlock

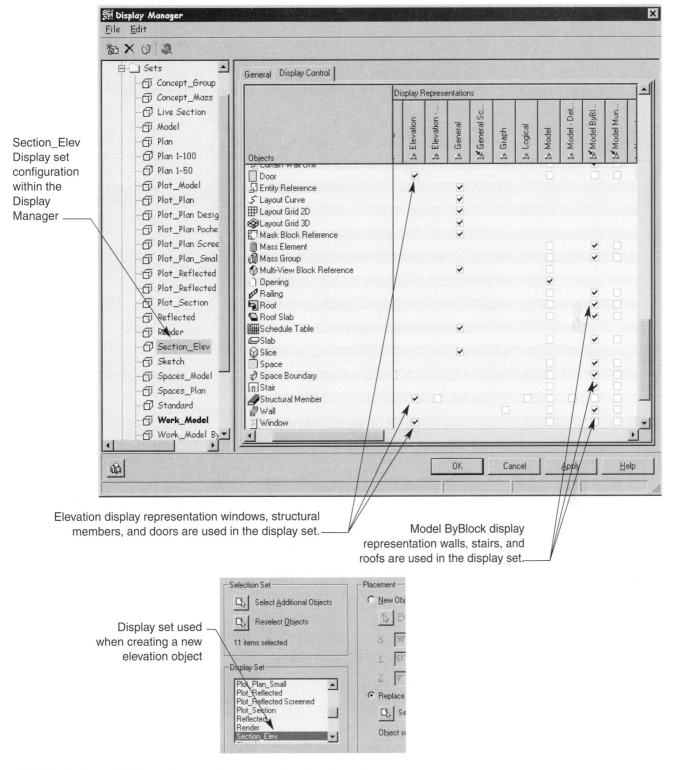

Section_Elev Display set configuration within the Display Manager

Elevation display representation windows, structural members, and doors are used in the display set.

Model ByBlock display representation walls, stairs, and roofs are used in the display set.

Display set used when creating a new elevation object

FIGURE 17.10   A display set is used when a new elevation object is created.

representation. The Design Rules tab looks to the colors used in these representations for interpretation. If another display set is used when the elevation is created, then the representations interpreted are different. Use the Display Manager to see the display set being used by a display set. See Figure 17.10.

Understanding that the elevation door, window, and structural members are being interpreted, you can make rules to intercept these elevation colors and translate the colors to a different component within the elevation style. To create the rules, access the elevation style and pick the Design Rules tab. Use the Add button to add a rule to the elevation style (see Figure 17.11).

FIGURE 17.11   Use the <u>A</u>dd button to add a new rule to the elevation style.

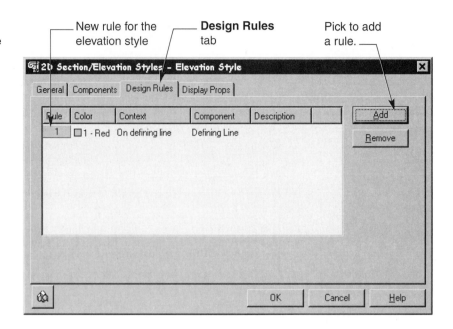

Once a new rule has been added, it can be configured. The Color column includes a color swatch representing the color of the display representation that the elevation style is looking for in the building model. Pick the swatch to select a color from the **Select Color** dialog box.

The **Context** column includes the first part of the rule that defines where the color being looked for can be found. Selecting the text in this column displays a drop-down list of options. For example, in Figure 17.11, in the Color column, the color red is being looked for on the defining line (elevation line). If a red color is found for one of the display representations and it is on the elevation line, then it is placed on the Defining Line elevation style component. Colors can be looked for in a particular subdivision within the elevation line box, on any component that is visible in the drawing, or on the defining line.

The **Component** column is the second part of the rule that defines which elevation style component the color being looked for is to be assigned. Select the text under this column to see a list of components. The following default components are listed: subdivisions, a hidden component, and the defining line component. Any custom components added in the Components tab are also listed. Figure 17.12 is an example of several rules in the elevation style. Rule 1 looks for Color 150 (the elevation door color) within the first subdivision of the elevation line box and places it on a custom component named Doors.

The **Description** column is used to enter a description for the rule. This column does not need to be filled out.

FIGURE 17.12   This rule looks for a color within a subdivision and places it on a custom component.

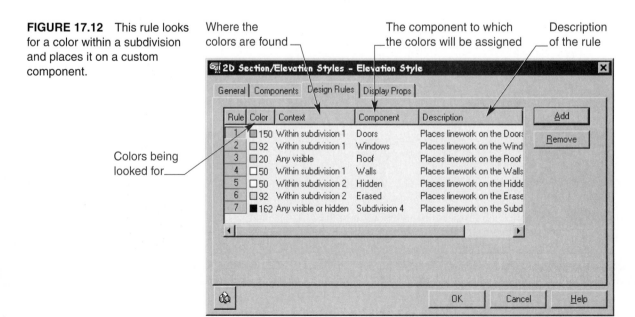

Setting Up the Entity Display

Before the rule can be set into motion, the display properties for the elevation style need to be manipulated. In the **2D Section/Elevation Styles** dialog box, pick the **Display Props** tab. Notice that the elevation object only has one display representation—General. Select the 2D Section/Elevation Style property source, attach an override, and select the Edit Display Props... button. The **Entity Properties** dialog box is displayed.

This dialog box is divided into two tabs: **Layer/Color/Linetype** and **Custom Display Components**. The Layer/Color/Linetype tab includes all the default display components within the elevation object. These components can be adjusted for visibility, layer, color, linetype, and lineweight. Notice that the custom components created in the Components tab and used in the Design Rules tab are not available.

Use the Custom Display Components tab to add the custom components defined in the Components tab and used in the Design Rules tab to the elevation object. When adding custom display components to the display properties, make sure that the name defined in the Components tab matches the new display component name added in the Custom Display Components tab. For example, if a component named *Doors* is added in the Components tab, and a rule places Color 150 on the *Doors* component, then a custom display component named *Doors* needs to be added. Figure 17.13 displays the custom display components used by the rules in Figure 17.12 and added to the display properties.

 Field Notes

> The custom display component name *must* match the component name. If the names do not match, the design rule using the component is invalid and will not create the elevation object correctly.

After the custom display components have been added, they are also placed in the Layer/Color/Linetype tab. Any display settings can now be assigned to these components. Figure 17.14 displays the Layer/Color/Linetype tab with the display settings for the elevation object.

When you have finished adjusting the display properties, press the OK button to return to the 2D Section/Elevation Styles dialog box. Press OK to return to the Style Manager and press OK to return to the drawing screen. The elevation style is now ready for use in new elevation objects or in updated elevation objects. When updating an existing elevation object, pick the style from the Style to Generate drop-down list in the **Generate Section/Elevation** dialog box shown in Figure 17.15.

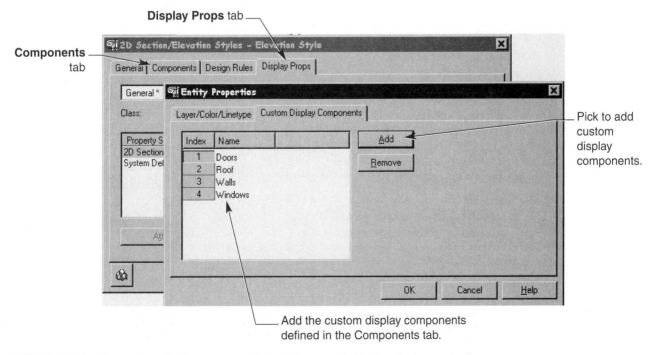

FIGURE 17.13 The custom display components have been added to the display properties.

FIGURE 17.14 Once the
display settings have been
adjusted for the elevation style,
the rules will be utilized when
elevation objects are created.

Design rules

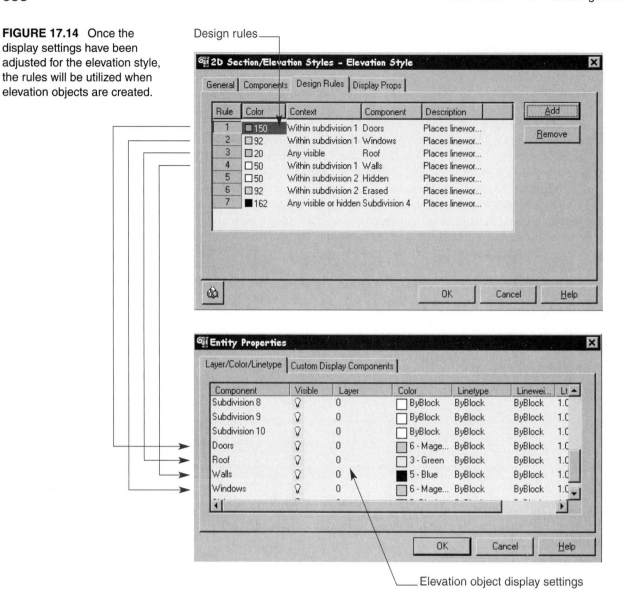

Elevation object display settings

FIGURE 17.15 Select the
appropriate elevation style to use
when creating a new or updating
an existing elevation object.

Select the elevation style to use for
the new or updated elevation object.

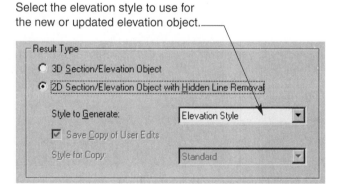

Exercise 17.5

1. Open EX17-4-Elev.dwg.
2. Access the elevation style named Residential Elevations.
3. Edit the style and access the Design Rules tab.
4. Create the following rules:

Rule	Color	Context	Component
1	150	Within subdivision 1	Doors
2	22	Within any subdivision	Roof
3	162	Any visible	Slabs
4	50	Within subdivision 1	Walls
5	92	Within subdivision 1	Windows
6	150	Within subdivision 2	Doors
7	92	Within subdivision 2	Windows
8	50	Within subdivision 2	Subdivision 2
9	150	Within subdivision 3	Within subdivision 3
10	92	Within subdivision 3	Within subdivision 3
11	50	Within subdivision 3	Within subdivision 3

5. Access the Display Props tab. Select the 2D Section/Elevation Styles property source and attach an override. Pick the Edit Display Props button.
6. Add the following display components and assign the following colors to each:

Custom Display Component	Color Assigned in Layer/Color/Linetype
Doors	Magenta (color 6)
Roof	Color 51
Slabs	Color 45
Walls	Color 60
Windows	Magenta (color 6)

7. Update the existing elevation object using the new style.
8. Save the drawing as EX17-5-Elev.dwg.

Editing the Elevation Object

Occasionally when an elevation object is created linework appears on the wrong component or is displayed in a location that should not be seen. In most situations these problems can be eliminated through the use of an elevation style and the elevation style design rules. When linework cannot be controlled through the design rules, it can be edited and placed on a particular elevation style component or removed from the elevation altogether.

In addition to editing the components that the linework is assigned to, you can merge new or existing geometry into an elevation object. Geometry such as polylines representing trim work, signage, or hatches can all be merged into the elevation object, forming one piece.

Editing Linework

Use one of the following options to begin editing linework:

✓ Select the **Documentation** pull-down menu, pick the **Elevations** cascade menu, then select the **Edit Linework** command.

✓ Pick the **Edit Linework** button from the **Elevations** toolbar.

✓ Right-click, and select **Documentation, Elevations, Edit Linework** from the shortcut menu.

✓ Type **AEC2DSECTIONRESULTEDIT** at the Command: prompt.

✓ Select the elevation object to be edited, right-click, and select **Edit Linework...** from the shortcut menu.

After entering the command you are prompted to select linework to edit. Pick the lines that need to be changed to a different elevation style component. Linework must be selected individually; crossing windows and standard windows cannot be used to select the linework. When you have finished selecting the linework to be changed, press the [Enter] key. You are then prompted to enter a component index number. Enter the number of the component that the selected linework needs to be changed to. The following prompt sequence is used to change three lines to a component named Doors:

Command: **AEC2DSECTIONRESULTEDIT** ↵
Select linework to edit: *(select a line)*
Select linework to edit: *(select a line)*
Select linework to edit: *(select a line)*
Select linework to edit: ↵
Fixed Components

1. Defining Line
2. Hidden Vectors
3. Erased Vectors
4. Subdivision 1
5. Subdivision 2
6. Subdivision 3
7. Subdivision 4
8. Subdivision 5
9. Subdivision 6
10. Subdivision 7
11. Subdivision 8
12. Subdivision 9
13. Subdivision 10

User Components

14. Doors
15. Roof
16. Slabs
17. Walls
18. Windows
Enter component index<1>: *(enter a component number)* 14↵
Command:

When you are prompted to select a component index number for the selected linework to be moved to, a list of all the elevation style components is displayed in the Command: prompt area. The list may be long, so press the [F2] key to display the Text Window. The components are listed with the fixed components (default components) first, then the **User** components. Enter the appropriate number for the linework to be assigned to the new component.

 Field Notes

Two fixed components that can be useful are the Hidden Vectors and the Erased Vectors. The Hidden Vectors component displays the edited linework as hidden lines. This feature may be helpful in an elevation when a series of lines needs to be indicated on the elevation that would normally not be seen. The Erased Vectors component initially is set to OFF. Placing edited linework on this component removes the lines from the elevation.

Exercise 17.6

1. Open EX17-5-Elev.dwg.
2. Use Entity Display to turn on the Hidden Vectors component.
3. Use the Edit Linework command to change the foundation wall and footing lines to the Hidden Vectors component.
4. Remove any unwanted lines from the elevation and place them on the Erased Vectors component.
5. Save the drawing as EX17-6-Elev.dwg.

Merging Linework

In addition to editing exiting linework in the elevation, you can merge new geometry into the elevation and have it become a part of the elevation object. Use one of the following options to merge linework:

✓ Pick the **Documentation** pull-down menu, select the **Elevations** cascade menu, then pick the **Merge Linework** command.

✓ Select the **Merge Linework** button in the **Elevations** toolbar.

✓ Right-click, and select **Documentation, Elevations, Merge Linework** from the cursor menu.

✓ Type **AEC2DSECTIONRESULTMERGE** at the Command: prompt.

✓ Select the elevation object, right-click, and select the **Merge Linework...** command from the shortcut menu.

As with the Edit Linework command, select the geometry that is to be merged into the elevation object. When you have finished selecting the geometry, press [Enter] and select the component index number where the new geometry will be placed. The following prompt sequence is used to merge new geometry to the elevation:

Command: **AEC2DSECTIONRESULTMERGE** ↵
Select a 2D Section/Elevation: *(select the elevation object)*
Select entities to merge: *(pick the first corner of a crossing window)*
Specify opposite corner: *(pick the second corner of a crossing window)*
4 found
Select entities to merge: *(pick another object)*

1 found, 5 total
Select entities to merge: ↵
Fixed Components

1. Defining Line
2. Hidden Vectors
3. Erased Vectors
4. Subdivision 1
5. Subdivision 2
6. Subdivision 3
7. Subdivision 4
8. Subdivision 5
9. Subdivision 6
10. Subdivision 7
11. Subdivision 8
12. Subdivision 9
13. Subdivision 10

User Components

14. Doors
15. Roof
16. Slabs
17. Walls
18. Windows
Enter component index for non 2D section objects<1>: *(enter the component number or press [Enter] to accept the default)*
Command:

 Field Notes

- Most AutoCAD and Architectural Desktop objects can be merged into an elevation. Lines, polylines, arcs, blocks, and hatches are examples of entities that can be merged.
- Linework should be edited and merged after the elevations have been decided on. If the elevation is updated after editing, the original linework that was edited is displayed again in the elevation object. The edited linework is also displayed but is a separate group of lines. Use the **Merge Linework** command after updating an elevation to group the edited linework back into the elevation object.

Exercise 17.7

1. Open EX17-6-Elev.dwg.
2. Add a ground line to the elevation.
3. Insert some elevation tree multi-view blocks from the DesignCenter.
4. Use the Merge Linework command to merge these items with the elevation.
5. Save the drawing as EX17-7-Elev.dwg.

Generating Interior Elevations

Interior elevations can also be created for the set of construction documents. Either the elevation line or the elevation marks found in the DesignCenter can be used (see Figure 17.16). Use one of the following options to find these DesignCenter symbols:

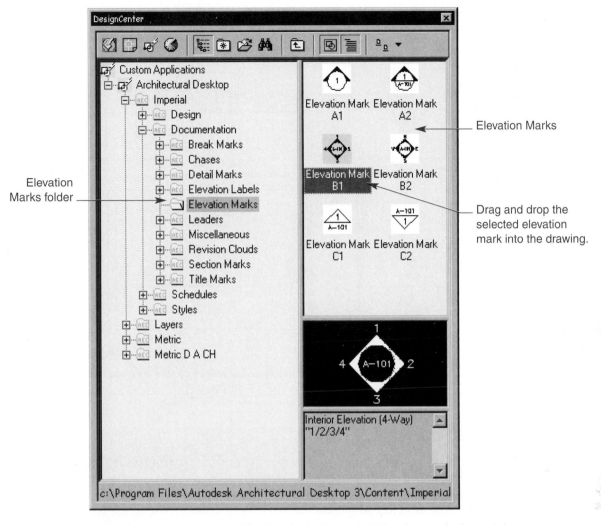

FIGURE 17.16 Architectural Desktop includes several elevation marks that can be used to develop elevatons or to document the construction documents.

✓ Select the **Documentation** pull-down menu, access the **Documentation Content** cascade menu, and pick the **Elevation Marks...** command.

✓ Pick the **Elevation Marks** button from the **Documentation - Imperial** toolbar.

✓ Right-click, and select **Documentation, Documentation Content, Elevation Marks...** from the cursor menu.

✓ Type **AECDCSETIMPELEVATIONMARKS** at the Command: prompt.

The DesignCenter is displayed, where the blocks can be dragged and dropped into the drawing.

When one of these elevation symbols is dragged into the drawing and dropped in place, several prompts appear. First, you are prompted for the location then the direction of the mark. Pick a point for the insertion of the mark, then specify where the tag will be pointing. After you have picked the direction, a dialog box is displayed, allowing you to enter text for the symbol. Next, you are asked whether an elevation line should be created.

If an elevation line is generated, it can be used to create the 2D elevation. Walk through the same process of creating the elevation by selecting the new interior elevation line, then generating the elevation. Interior elevations of the building model are produced that can be edited with any additional detail (see Figure 17.17).

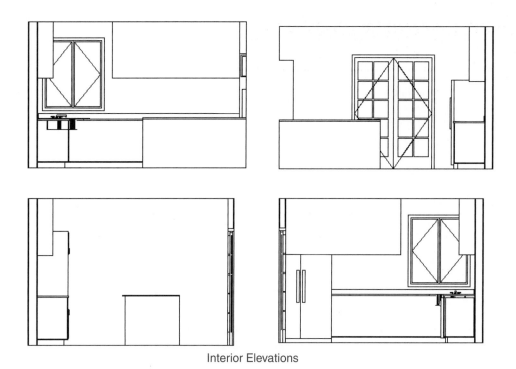

Interior Elevations

FIGURE 17.17   Interior elevations can be created if the elevation mark includes an elevation line.

Exercise 17.8

1. Open EX17-7-Elev.dwg.
2. Add the elevation mark B2 to the center of the kitchen area.
3. Follow the prompts at the Command: line and create an elevation line for each direction of the symbol.
4. Adjust the elevation line boxes for the new symbol. [*Hint:* It may be easier to freeze the foundation xref layer in order to adjust the boxes.]
5. Generate the four elevations for the kitchen.
6. Make any adjustments to the interior elevations using the Edit Linework and the Merge Linework commands.
7. Save the drawing as EX17-8-Elev.dwg.

Editing the Elevation Marks

The elevation marks included with Architectural Desktop come in two forms. The marks that are inserted from the DesignCenter are standard AutoCAD blocks with attributes attached. When the block is inserted, you are prompted for the symbol's text value in the Edit Attributes text box (see Figure 17.18A). The default elevation mark that is inserted when an elevation line is created is a multi-view block. This block is automatically placed in the drawing and displays the text "XX" (see Figure 17.18B).

Two different methods are used to modify the value within an elevation mark based on the type of mark that is being modified. For the elevation marks that are inserted from the DesignCenter, use the **Edit Attribute** command. Use one of the following options to access this command:

✓ Select the **Modify** pull-down menu, pick the **Attribute** cascade menu, and select the **Single...** command.

✓ Pick the **Edit Attribute** button in the AutoCAD Modify II toolbar.

✓ Type **ATTEDIT** at the Command: prompt.

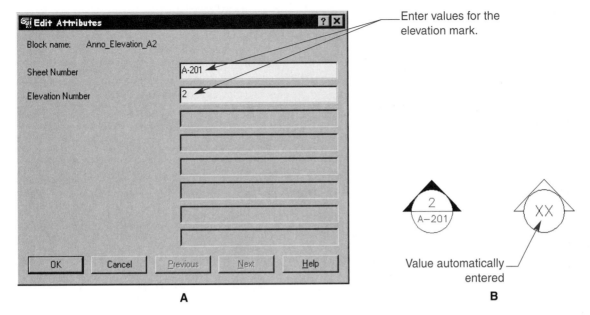

Enter values for the
elevation mark.

Value automatically
entered

A   B

FIGURE 17.18   Elevation marks come in two types: (A) Blocks. (B) Multi-view blocks.

You are prompted to select a block reference. Select the elevation mark with the text to be modified. You can also double-click on the symbol to display the Edit Attributes dialog box to change the text values.

To modify the multi-view default elevation mark:

✓ Select the symbol, right-click, and select **Multi-View Block Properties** from the shortcut menu.

✓ Type **MVBLOCKPROPS** at the Command: prompt, then pick the symbol.

The **Multi-View Block Reference Properties** dialog box, shown in Figure 17.19, is displayed. Select the **Attributes** tab. In the Attributes tab, select the value to be changed in the value column and enter a new value for the text. Press the OK button when you have finished, to change the value.

FIGURE 17.19   The Multi-View Block Reference Properties dialog box includes the Attributes tab that is used to modify attributes within the multi-view block symbol.

Attributes tab

Pick value to change
then enter a new value.

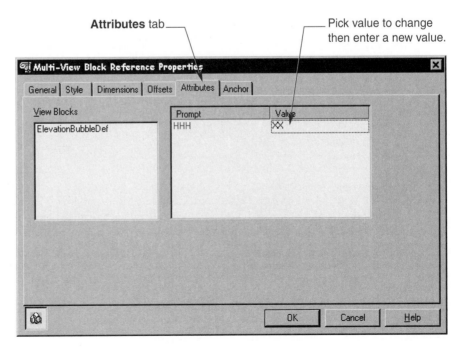

Creating a 3D Elevation

The preceding section explained how to create and manage a 2D elevation object. Another type of elevation object that can be created is the 3D elevation object, which can be used to create a perspective view of the building model. This object also uses an elevation line to generate the elevation object. It is useful to create the 3D elevation object when you need an elevation with the same subdivision and color schemes as used by the 2D elevation. The difference is that a copy of all the geometry in the building model is made (see Figure 17.20).

To create a 3D elevation, first create the elevation line as you would for a 2D elevation, and add any subdivision lines as required. Select the elevation line, right-click, and select the Generate Elevation… command from the shortcut menu. In the Generate Section/Elevation dialog box, select the 3D Section/Elevation Object radio button in the Result Type area (see Figure 17.21).

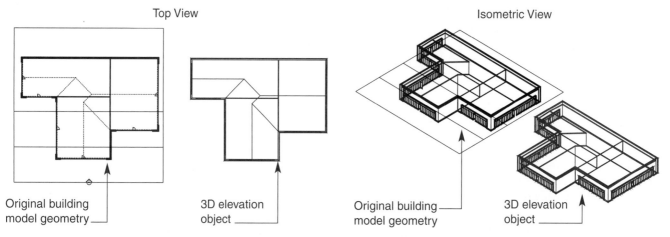

Top View   Isometric View

Original building   3D elevation   Original building   3D elevation
model geometry   object   model geometry   object

FIGURE 17.20   The 3D elevation object is used when a three-dimensional object using the same color schemes as the 2D elevation is required.

FIGURE 17.21   Select the 3D Section/Elevation Object radio button to create a three-dimensional elevation object.

Select to create a 3D elevation object.

Pick the objects to be included in the new elevation.

Select a location for the elevation.

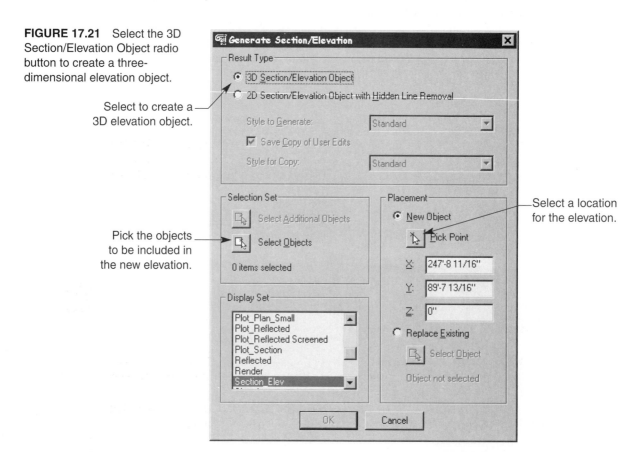

Note that selecting this option makes the elevation style to use unavailable, because 3D elevations do not use an elevation style but rely on entity display to show the subdivision colors. Complete the elevation process by selecting the objects to be included in the elevation and selecting a point for the new 3D elevation object.

This process creates a new elevation object that is three-dimensional, as shown in Figure 17.20. Select the 3D elevation object, right-click, and pick Entity Display from the shortcut menu. The subdivision's display representation is used to manage the colors of the 3D elevation object's subdivisions.

Like the 2D elevation object, the 3D elevation object maintains a link to the original building model. Any changes made to the building model can be updated in the 3D elevation. The drawback to the 3D elevation is that linework cannot be edited. What you see in the building model is duplicated in the 3D elevation.

 Field Notes

- The editing capabilities of 3D elevations are limited. The object can be updated to reflect changes made to the building model. It is recommended that you use 2D elevations so that you can modify the elevation and set rules about how colors in the building model are interpreted and used in the elevation.
- Another problem with the 3D elevation is that it consumes memory. When several 3D elevations are used, the drawing performance becomes slower as the 3D geometry is repeated in the 3D elevation.

Exercise 17.9

1. Open EX17-1-Elev.dwg.
2. Add two subdivision lines to the elevation line box. Adjust their locations so the subdivision includes the different jogs on the front of the building.
3. Generate a 3D elevation object and place it beside the building model.
4. Change the entity display of the 3D elevation object to reflect different colors for the subdivisions. Update the 3D elevation as required.
5. Save the drawing as EX17-9-3D-Elev.dwg.

 Using the Hidden Line Projection Command

Another command that is useful for creating flat-view objects of a three-dimensional building is the **Hidden Line Projection** command. This command can be used to create 2D elevations from the building model, or 2D perspectives from a perspective view. This command interprets a viewing direction of the building model and turns it into a flat block.

Use one of the following optons to access this command:

✓ Select the **Desktop** pull-down menu, pick the **Utilities** cascade menu, and select the **Hidden Line Projection** command.

✓ Pick the **Hidden Line Projection** button from the **Utilities** toolbar.

✓ Right-click, and pick **Desktop, Utilities, Hidden Line Projection** from the cursor menu.

✓ Type **AECCREATEHLR** at the Command: prompt.

The best way to create the flat block is to first obtain the viewing direction that you want to create, then run the command. If you are working with more than one viewport, you must remain in the viewport that will be used to generate the flat view throughout the process.

Once the command is entered, a series of prompts is issued to create the flat block. The following prompt sequence is used to turn the perspective in Figure 17.22 into a 2D, flat perspective block:

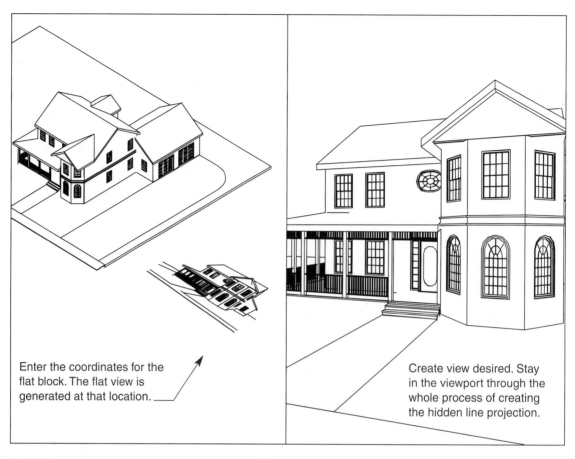

Enter the coordinates for the flat block. The flat view is generated at that location.

Create view desired. Stay in the viewport through the whole process of creating the hidden line projection.

FIGURE 17.22 The Hidden Line Projection command can be used to create a flat block of a view, whether it is an elevation, top, isometric, or perspective view.

Command: **AECCREATEHLR** ↵
Select objects: *(pick the first corner of a crossing window to select the geometry)*
Specify opposite corner: *(pick the second corner of the crossing window)*
6 found
Select objects: ↵
Block insertion point: *(enter absolute coordinates for the block insertion)* 170′,0 ↵
Insert in plan view [Yes/No] <Y>: ↵
Command:

When you enter the command initially you are prompted to select objects. You can select any objects that will be included in the flat-view block by selecting objects in any viewport. After you have finished selecting the geometry, the next prompt asks you to specify a block insertion point. You can center an absolute coordinate such as 0,0, or you can pick an insertion point on the drawing for the block. After you have specified the block insertion point, you *must* be in the viewport that contains the desired view. If you are not in the desired view's viewport, a hidden line projection will be created of the view that is currently active. Finally, you must specify whether the block will be placed in a plan view or the current view's XY plane.

Exercise 17.10

1. Open EX17-8-Elev.dwg.
2. Use the camera object to create a perspective view of the building model.
3. Use the Hidden Line Projection command to create a flat block of the perspective.
4. When prompted for an insertion point, enter 0,0.
5. Change the view back to a top view, and move the new perspective block to a blank spot in the drawing.
6. Save the drawing as EX17-10-Elev.dwg.

Using AEC Polygons

In many cases elevations are detailed with patterns of lines to show the type of finish materials to be used. Drawing these patterns by hand can be a difficult process. In standard AutoCAD, hatching is often used as a tool to fill an area with a pattern. Architectural Desktop includes a tool known as **AEC Polygons** that can be used to fill a 2D or 3D area (see Figure 17.23). AEC Polygons are multisided flat objects that can be used to fill any kind of shape with a pattern and can quickly be modified as the design of the building changes.

Creating AEC Polygon Styles

Before an AEC Polygon can be used, a style representing the type of building material needs to be created. Use one of the following options to create an AEC Polygon style:

✓ Pick the **Desktop** pull-down menu, access the **AEC Polygon** cascade menu, then select the **AEC Polygon Styles...** command.

✓ Pick the **AEC Polygon Styles** button in the **AEC Polygon** toolbar.

✓ Right-click, and select **Desktop, AEC Polygon, AEC Polygon Styles...** from the cursor menu.

✓ Type **AECPOLYGONSTYLE** at the Command prompt.

The **Style Manager** appears, open to the **AEC Polygon** section. Press the New Style button to create a new AEC Polygon style. Enter a name, then edit the style. When the style is edited, the **AEC Polygon Style Properties** dialog box is displayed. The following tabs are available in this dialog box:

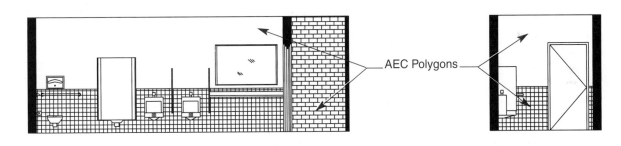

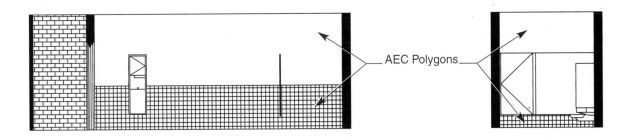

FIGURE 17.23 AEC Polygons can be used to fill an area of the drawing with a hatch pattern.

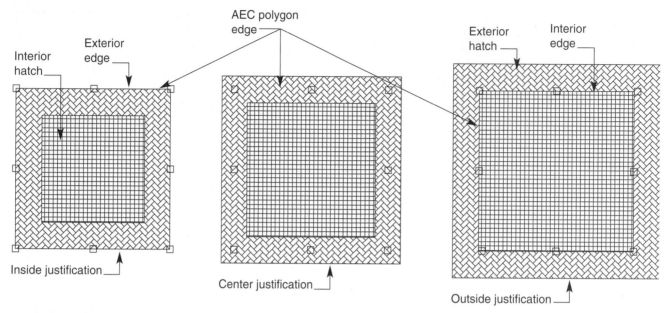

FIGURE 17.24   AEC Polygons can be created with or without an edge and with two different types of hatch patterns.

▪ **General**	This tab is used to rename the AEC Polygon style, to provide a description to the style, and to attach any design notes or property set definitions.
▪ **Dimensions**	This tab includes two options: **Edge <u>W</u>idth** and **<u>J</u>ustify**. The Edge <u>W</u>idth text box controls the width of the exterior edge of the AEC polygon (see Figure 17.24). The width can be set to 0″ if an edge is not desired. The <u>J</u>ustify drop-down list controls the justification points for the AEC polygon (see Figure 17.24).
▪ **Display Props**	This tab controls the display of AEC Polygons in the drawing. Two types of display representations are available: **Model** and **True Color**.

The Model display representation is used in the default plan and model views. This representation allows you to display a hatch pattern on the interior and exterior of the AEC Polygon (see Figure 17.24). This representation includes several components that can be turned on or off. Turn on the **Interior and Exterior Hatch** components to display a hatch pattern. In the **Hatching** tab assign a hatch to the two hatch components. Pick the swatch in the **Pattern** column to assign a predefined or user hatch to the hatch components. Make any adjustment to scale and rotation by picking the values in the two components.

The True Color representation can be incorporated into a custom display configuration to display an interior and an exterior solid fill (see Figure 17.25). The True Color representations display a solid color using any true color available in addition to the AutoCAD standard 256 colors.

Field Notes

AEC Polygon styles can be incorporated as infill panels in a curtain wall style or in a window assembly style.

FIGURE 17.25 True-color filled AEC Polygons can be used in presentation drawings.

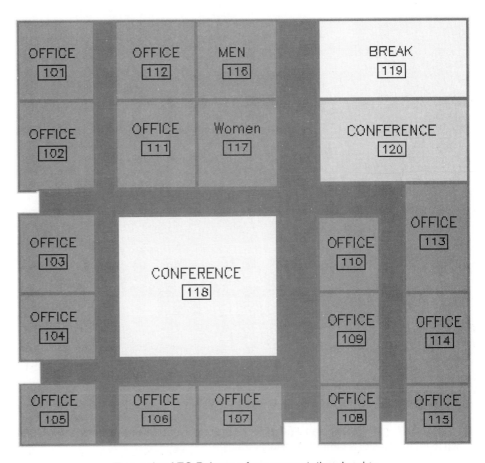

True-color AEC Polygons for a presentation drawing

 Adding AEC Polygons

Once you have created the AEC Polygon style(s), you can add an AEC polygon to the drawing. Use one of the following options to access the Add AEC Polygon command to draw a polygon:

✓ Select the **Desktop** pull-down menu, access the **AEC Polygon** cascade menu, and pick the **Add AEC Polygon...** command.

✓ Pick the **Add AEC Polygon** button from the **AEC Polygon** toolbar.

✓ Right-click, and select **Desktop, AEC Polygon, Add AEC Polygon...** from the cursor menu.

✓ Type **AECPOLYGONADD** at the Command: prompt.

The Add AEC Polygon dialog box, shown in Figure 17.26, is displayed. Select the desired AEC Polygon style from the Style: drop-down list. As with a polyline, begin picking vertex locations within the drawing for the AEC Polygon. When you have finished selecting points, pick the Polyline Close button to close the AEC Polygon. The AEC Polygon is displayed in the shape that you drew.

FIGURE 17.26 The Add AEC Polygon dialog box is used to specify the AEC Polygon style for the new AEC Polygon shape.

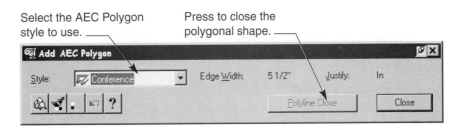

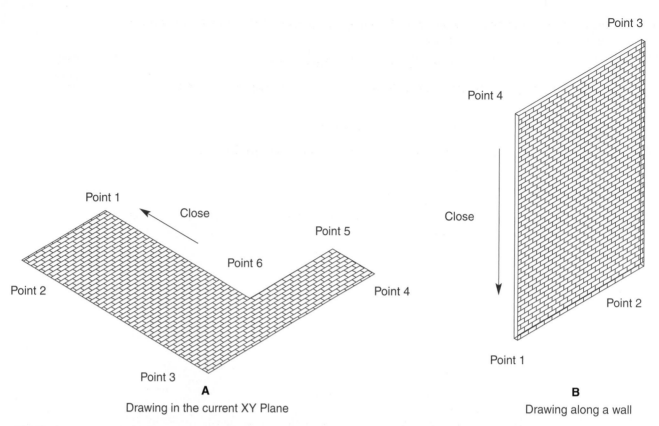

FIGURE 17.27   (A). Creating an AEC Polygon in the current XY drawing plane. (B) Creating an AEC Polygon on the face of a three-dimensional wall.

You can draw AEC Polygons in the current XY drawing plane or in any other plane by snapping the vertex point to a 3D point such as the top of a wall (see Figure 17.27). To aid in creating straight segments, use the POLAR or the ORTHO button in the menu bar at the bottom of the drawing screen.

 Converting Polylines to AEC Polygons

There may be situations in which a curved area needs to have a hatch pattern applied. The Add AEC Polygon command is limited to creating polygons as straight-line shapes. If a complex area or an area that includes curves needs to have the hatch pattern applied, use the **Convert to AEC Polygon** command. This command converts only a closed polyline into an AEC Polygon with the style of your choosing. Use one of the following options to access this command:

✓ Pick the **Desktop** pull-down menu, select the **AEC Polygon** cascade menu, then pick the **Convert to AEC Polygon...** command.

✓ Pick the **Convert to AEC Polygon** button in the **AEC Polygon** toolbar.

✓ Right-click, and select **Desktop, AEC Polygon, Convert to AEC Polygon...** from the cursor menu.

✓ Type **AECPOLYGONCONVERT** at the Command: prompt.

The command prompts you to select a polyline, then allows you to change the style to be used.

 Modifying an AEC Polygon

Once you create an AEC Polygon you can modify it by selecting the grip points along the polygon and specifying a new location for the point. In addition, you can change the style used by the AEC Polygon with the **Modify AEC Polygon** command. Use one of the following options to access this command:

✓ Select the **Desktop** pull-down menu, pick the **AEC P̲olygon** cascade menu, then select the **M̲odify AEC Polygon…** command.

✓ Pick the **Modify AEC Polygon** button in the **AEC Polygon** toolbar.

✓ Right-click, and select **Des̲ktop, AEC P̲olygon, M̲odify AEC Polygon…** from the cursor menu.

✓ Type **AECPOLYGONMODIFY** at the Command: prompt.

✓ Select the AEC Polygon, right-click, then select **AEC Polygon Modify** from the shortcut menu.

In the Modify AEC Polygon dialog box, select a different style for use.

You can also modify the AEC Polygon by selecting the polygon, right-clicking, and selecting **AEC Polygon Properties….** This command allows you to change the polygon style or the location of the polygon.

Exercise 17.11

1. Open EX17-8-Elev.dwg.
2. Create the following AEC Polygon Styles:

Dimensions Tab			Entity Properties within the Display Props Tab				
Style Name	Edge Width	Justify	Component Display	Interior Hatch Component	Interior Hatch Scale	Interior Hatch Rotation	
Roof	0″	In	Exterior Edge: Off	Predefined: AR-RSHKE	1″	0	
			Interior Hatch: On		6″	0	
Shake	0″	In	Exterior Edge: Off	Predefined: AR-SAND			
			Interior Hatch: On				

3. Use AEC Polygons to add a shake hatch pattern to the roof in the front elevation object.
4. Use AEC Polygons to add a stucco pattern to the walls in the front elevation object. Wrap the AEC Polygon completely around the front window. You will deal with this later.
5. Save the drawing as EX17-11-Elev.dwg.

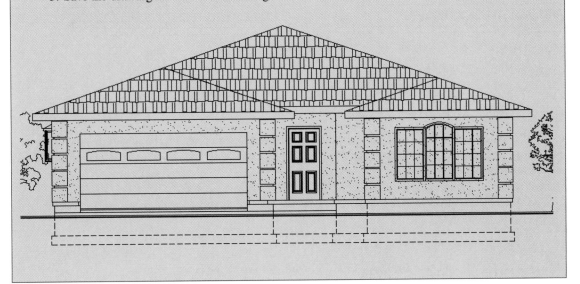

Using Modifying Tools to Adjust AEC Polygons

Like slab and roof slab objects, AEC Polygons include tools that can be used to change the shape and manner in which they are used. AEC Polygons can be divided, joined, trimmed, and subtracted from one another. The following section describes these tools and how they can be used.

Dividing an AEC Polygon

An AEC Polygon can be divided into two parts by specifying a dividing line. Use one of the following options to access the **Divide** command:

     ✓ Select the **Desktop** pull-down menu, pick the **AEC Polygon** cascade menu, and select the **Divide** command.

     ✓ Pick the **Divide** button in the **AEC Polygon** toolbar.

     ✓ Right-click, and pick **Desktop, AEC Polygon, Divide** from the cursor menu.

     ✓ Type **AECPOLYGONOPDIVIDE** at the Command: prompt.

     ✓ Select the polygon, right-click, and select the **Operation** cascade menu, then the **Divide** command from the shortcut menu.

Once the command has been activated, you are prompted to select an AEC Polygon. Select the polygon to divide. Next, pick a dividing line start point and an ending point. The selected AEC Polygon is divided into two pieces (see Figure 17.28).

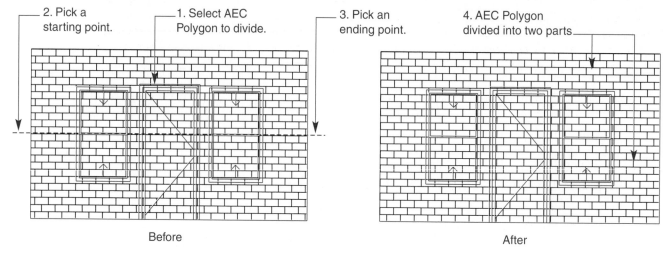

FIGURE 17.28   Use the AEC Polygon Divide command to divide a polygon into two pieces.

Joining Two Polygons

As AEC Polygons are created they may mistakenly get overlapped or need to be joined together to form a uniform pattern. The **Join** command allows you to join two AEC Polygons together. Use one of the following options to access this command:

     ✓ Pick the **Desktop** pull-down menu, access the **AEC Polygon** cascade menu, and select the **Join** command.

     ✓ Pick the **Join** button in the **AEC Polygon** toolbar.

     ✓ Right-click, and pick **Desktop, AEC Polygon, Join** from the cursor menu.

     ✓ Type **AECPOLYGONOPJOIN** at the Command: prompt.

     ✓ Select the polygon, right-click, and select the **Operation** cascade menu, then the **Join** command from the shortcut menu.

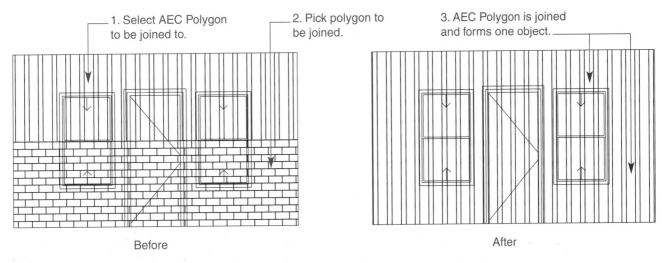

FIGURE 17.29   Use the Join option to join two AEC Polygons together.

After entering this command you are prompted to select an AEC Polygon that you want to be joined. Select the polygon that contains the hatch pattern to be used after joining another polygon, then select the polygon that will be joined. One AEC Polygon is created using the first selected polygon's hatch pattern (see Figure 17.29).

   Subtracting Polygons

AEC Polygons can be subtracted from one another to form holes in a polygon for windows or openings in an elevation, or used to take part of a polygon away from an edge. In order to use this option two AEC Polygons must exist: the main AEC Polygon that will remain, and a secondary polygon to be subtracted from the main polygon (see Figure 17.30). The two polygons must overlap each other at some point. Use one of the following options to access the command **Subtract**:

✓ Select the **Des\underline{k}top** pull-down menu, access the **AEC P\underline{o}lygon** cascade menu, and select the **Subtract** command.

✓ Pick the **Subtract** button in the **AEC Polygon** toolbar.

✓ Right-click, and pick **Des\underline{k}top, AEC P\underline{o}lygon, \underline{S}ubtract** from the cursor menu.

✓ Type **AECPOLYGONOPSUBTRACT** at the Command: prompt.

✓ Select the polygon, right-click, and select the **Operation** cascade menu, then the **Subtract** command from the shortcut menu.

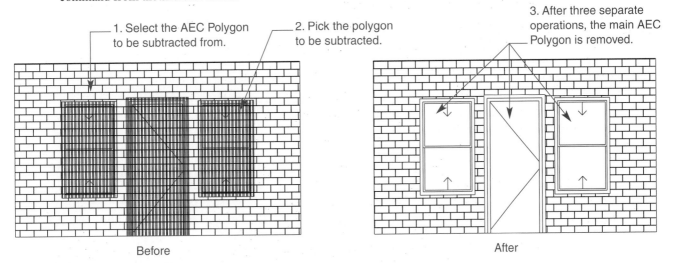

FIGURE 17.30   Use overlapping AEC Polygons to remove a portion of the main polygon with the Subtract command.

After entering the command, you are prompted to select a polygon to subtract *from*. Select the AEC Polygon that will have a portion removed, then select the AEC Polygon that is the subtracting object (see Figure 17.30).

 Creating an Intersection from Two AEC Polygons

When two objects overlap each other, an intersection is formed. The intersection between two AEC Polygons is the portion of the hatch that both polygons share. The **Intersect** command allows you to create one AEC Polygon from the shared or intersected area (see Figure 17.31). Use one of the following options to access this command:

✓ Pick the **Desktop** pull-down menu, select the **AEC Polygon** cascade menu, then pick the **Intersect** command.

 ✓ Pick the **Intersect** button in the **AEC Polygon** toolbar.

✓ Right-click, and select **Desktop, AEC Polygon, Intersect** from the cursor menu.

✓ Type **AECPOLYGONOPINTERSECT** at the Command: prompt.

✓ Select the polygon, right-click, and select the **Operation** cascade menu, then the **Intersect** command from the shortcut menu.

After activating this command, you are prompted to select an AEC Polygon. The polygon selected becomes the style that is used for the remaining AEC Polygon. Next, pick the polygon that is intersecting the previously picked polygon. You are then prompted to erase the original polygons or to keep them. An intersection of the shared areas between the two polygons is created (see Figure 17.31).

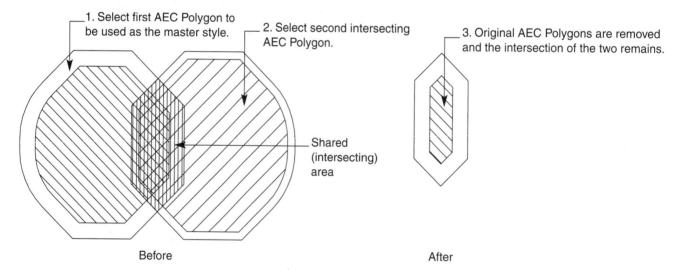

FIGURE 17.31   Use the Intersect command to create an intersecting AEC Polygon and erase the original polygons.

 Trimming an AEC Polygon

Trimming an AEC Polygon is similar to dividing except that a portion of the polygon is removed from the drawing. Use one of the following options to access the **Trim** command:

✓ Select the **Desktop** pull-down menu, access the **AEC Polygon** cascade menu, then pick the **Trim** command.

 ✓ Pick the **Trim** button in the **AEC Polygon** toolbar.

✓ Right-click, and select **Desktop, AEC Polygon, Trim** from the cursor menu.

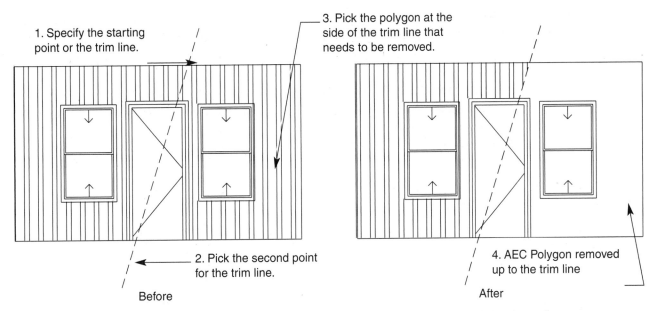

1. Specify the starting point or the trim line.

3. Pick the polygon at the side of the trim line that needs to be removed.

2. Pick the second point for the trim line.

4. AEC Polygon removed up to the trim line

Before

After

FIGURE 17.32 Unlike the Divide command, the Trim command removes a portion of the AEC Polygon from the trim line.

✓ Type **AECPOLYGONOPTRIM** at the Command: prompt.

✓ Right-click, select the **Operation** cascade menu, then the **Trim** command.

You are then prompted to select the first point of the trimming line. Pick this point, then pick the ending point. After establishing the trimming line, pick the AEC Polygon on the side of the trim line that you want to remove from the drawing (see Figure 17.32).

 Adding and Removing Vertex Points

Like the slab and roof slab objects, AEC Polygons have the capability of having vertex points added or removed, as when another point is required for a hatching pattern. Use one of the following options to access the **Add Vertex** command:

✓ Pick the **Desktop** pull-down menu, access the **AEC Polygon** cascade menu, then pick the **Add Vertex** command.

✓ Pick the **Add Vertex** button in the **AEC Polygon** toolbar.

✓ Right-click, and select **Desktop, AEC Polygon, Add Vertex** from the cursor menu.

✓ Type **AECPOLYGONOPADDVERTEX** at the Command: prompt.

✓ Right-click, then select the **Operation** cascade menu, then the **Add Vertex** command.

To add another vertex, simply pick a location along the existing AEC Polygon. To remove vertex points use the **Remove Vertex** command to pick on top of and remove a vertex located on an AEC Polygon. Use one of the following options to access this command:

✓ Select the **Desktop** pull-down menu, pick the **AEC Polygon** cascade menu, then select the **Remove Vertex** command.

✓ Pick the **Remove Vertex** button in the **AEC Polygon** toolbar.

✓ Right-click, and select **Desktop, AEC Polygon, Remove Vertex** from the cursor menu.

✓ Type **AECPOLYGONOPREMOVEVERTEX** at the Command: prompt.

✓ Right-click, then select the **Operation** cascade menu, then the **Remove Vertex** command.

Exercise 17.12

1. Open EX17-11-Elev.dwg.
2. Draw a polyline around the front windows.
3. Convert the polyline to an AEC Polygon style.
4. Use the Subtract command to remove the newly created AEC Polygon from the stucco siding polygon on the walls.
5. Save the drawing as EX17-12-Elev.

|||||||||||| CHAPTER REVIEW

 Use the CD-ROM to test your knowledge and skills.

Chapter Test

To check your understanding of the content provided in this chapter, access the Test file in the CH17 folder of the CD-ROM that accompanies this text.

Chapter Project

To practice the Architectural Desktop skills presented in this chapter, access the Project files in the CH17 folder of the CD-ROM that accompanies this text. The project files are in pdf format and include sample drawings and instructions for completing each project.

CHAPTER 18

Creating Sections

LEARNING GOALS

After completing this chapter, you will be able to:

◎ Create a 2D section line.

◎ Draw a 2D section.

◎ Update the section object.

◎ Edit and merge section linework.

◎ Use section line blocks and detail marks.

◎ Access and use the Quick Slice command.

◎ Define section styles.

◎ Adjust section line properties.

◎ Draw a 3D section.

◎ Create and modify a live section.

The previous chapter discussed Architectural Desktop elevations and the tools for creating them. Another important part of drawing construction documents is adding sections and details. Architectural Desktop also includes tools for generating section objects.

Sections are similar to elevations, as they also provide information regarding the vertical aspects of the building; however, instead of displaying how the surface of the building or room appears, sections show how the building is assembled. A *section* is a drawing that displays part of the building as if it has been cut through and a portion removed, showing the interior detail. Sections are typically created as two-dimensional drawings with additional information explaining construction techniques (see Figure 18.1).

Architectural Desktop provides utilities for creating both 2D and 3D sections. These are similar to elevation objects except they are created on their own section layer. 2D and 3D sections can be created from geometry within a drawing or from xrefs attached to the drawing. In addition to these two utilities, a display known as a Live Section can be created that physically cuts the building model at the cut line and removes the portion of the building.

Creating Sections

Before a section can be created, a section line must exist in the drawing. The *section line* is a line that is drawn through the part of the building that is to be cut through. The section line is also known as the *cutting line*, *cutting plane line*, or the *defining line*. Use one of the following options to add a section line:

✓ Select the **Documentation** pull-down menu, pick the **Sections** cascade menu, then select the **Add Section Line** command.

✓ Pick the **Add Section Line** button in the **Sections** toolbar.

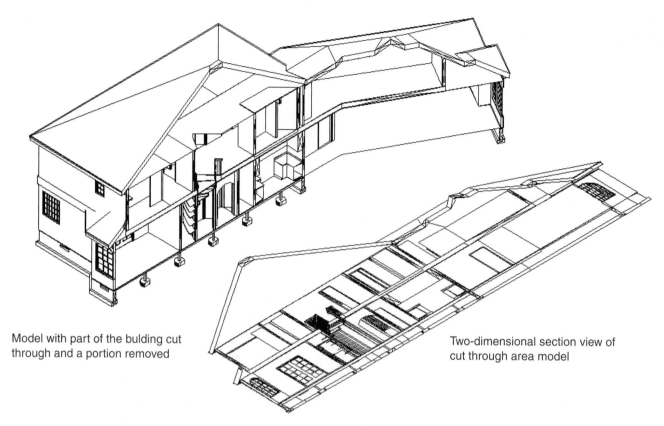

Model with part of the bulding cut
through and a portion removed

Two-dimensional section view of
cut through area model

FIGURE 18.1   A section is typically a two-dimensional drawing of the interior of a three-
dimensional building.

✓ Right-click, and select **Documentation, Sections, Add Section Line** from the cursor menu.

✓ Type **AECBLDGSECTIONLINEADD** at the Command: prompt.

A sections line can be created as a straight line, as shown in Figure 18.2A, or it can be jogged to create an irregular section cut, as shown in Figure 18.2B. Once the command has been entered, you are
prompted to select points. The first point you pick establishes the starting point of the section line.
Additional points may be selected to create either a straight or an irregular-shaped section line.

The direction in which you draw the section line determines what the section displays. Select points
in a counterclockwise direction around the building to view the interior of the building (see Figure 18.3).
The section line is used as a cutting plane, and the objects that are crossed by the section line become the
defining edge of the section. The section line arrows placed at each end of the line determine the portion of
the building that is seen in the section. The portion of the building on the other side of the section line is
removed from the view displaying the section.

Once you have determined the points establishing the section line, you are prompted to enter a length.
The length determines how far the section extends into the building; that is, it determines the *section box*.
Anything within the box is considered during the process of the section creation, as shown in Figure 18.3.

If a slice through the building is to be made, then set the length to a small value such as 1/8″. Some
offices prefer to see the interior of a room that is being cut with a section line. In these cases, set the length
to a value that places the end of the section box past the interior objects that will be included in the section.
For example, suppose a section is to cut through a building, and the section is going to display the windows
along the back wall of the rooms being cut. In this case, establish the length deep enough so the section box
encompasses the interior room windows.

The final prompt that must be answered is the height. The height determines how high the section box
cuts through for the section. If you are creating details you may determine that the section box cuts through
only the first floor instead of the entire building. If a full section is created, you may determine that the
height needs to extend through the full height of the building. The following prompt sequence is used to
determine the section line and box for the section in Figure 18.4:

FIGURE 18.2 (A) A section line can be created as a straight cut through the building. (B) A section line can be drawn using several vertex points to establish the cutting plane of the section line.

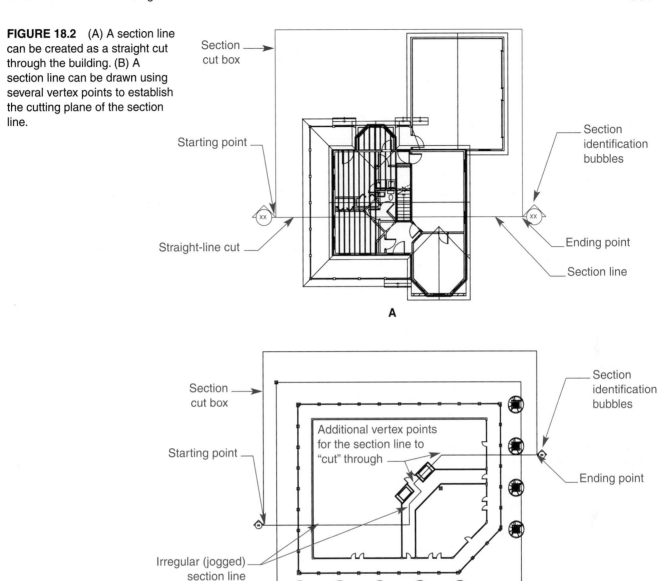

Section cut box

Starting point

Section identification bubbles

Straight-line cut

Ending point

Section line

A

Section cut box

Section identification bubbles

Starting point

Additional vertex points for the section line to "cut" through

Ending point

Irregular (jogged) section line

B

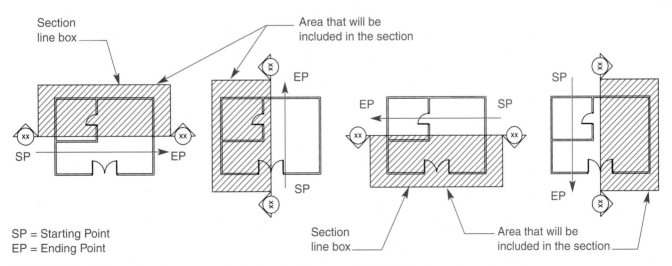

Section line box

Area that will be included in the section

SP = Starting Point
EP = Ending Point

EP

SP

EP

SP

EP

SP

SP

EP

Section line box

Area that will be included in the section

FIGURE 18.3 Select points in a counterclockwise direction around the building to create sections of the interior of the building.

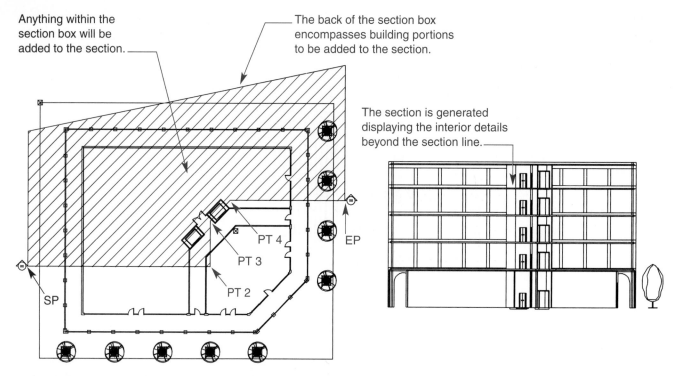

Anything within the section box will be added to the section.

The back of the section box encompasses building portions to be added to the section.

The section is generated displaying the interior details beyond the section line.

PT 4

PT 3

PT 2

EP

SP

FIGURE 18.4   The section line and box determine where the section will be cut and what will be included within the section object.

Command: **AECBLDGSECTIONLINEADD** ↵
Create polyline for section:
Specify start point: *(pick the starting point for the section line)*
Specify next point: *(pick point 2 for the section line)*
Specify next point: *(pick point 3 for the section line)*
Specify next point: *(pick point 4 point for the section line)*
Specify next point: *(pick the ending point for the section line)*
Specify next point: ↵
Enter length <20'-0">: **65'** ↵
Enter height <65'-0">: **80'** ↵
Command:

Once the section line has been determined, the 2D or 3D section object can be created. Sections can be created from physical geometry in the drawing file and from geometry that is within a referenced file attached to the drawing. If a 2D section is to be created, a section style is used to determine how the parts within the section are to appear. If a 3D section is to be created, the entity display for the section object can be adjusted to reflect how objects are going to appear.

Exercise 18.1

1. Open EX17-1-Elev.dwg.
2. Erase the elevation line.
3. Create a section that cuts through the middle of the building.
4. Set the length to 30′-0″ and the height to 24′-0″
5. Save the drawing as EX18-1-Sect.dwg.

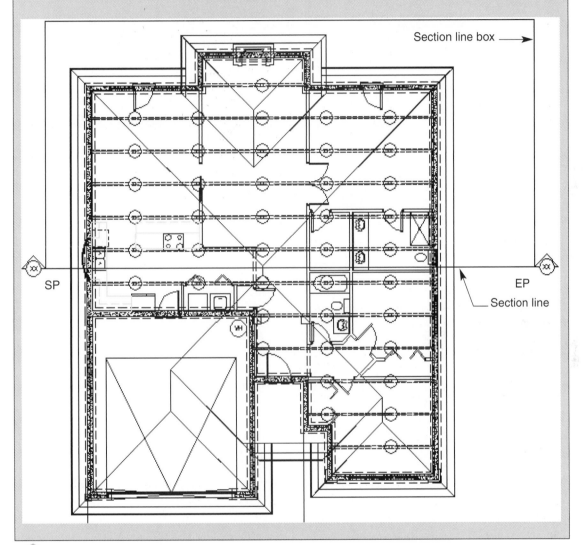

 Defining the Section Style

A section style is created by the same process used to create an elevation style. The style is used to create both the 2D elevation object and the 2D section object. Creation of elevation styles was discussed in Chapter 17. This section takes a short look at the creation of a section style. For a more in-depth explanation refer to the Elevation Styles section in Chapter 17.

Use one of the following options to enter this command:

✓ Pick the **Documentation** pull-down menu, access the **Sections** cascade menu, and pick the **Section Styles…** command.

✓ Pick the **Section Styles** button in the **Sections** toolbar.

✓ Right-click, and select **Documentation, Sections, Section Styles…** from the cursor menu.

✓ Type **AEC2DSECTIONSTYLE** at the Command: prompt.

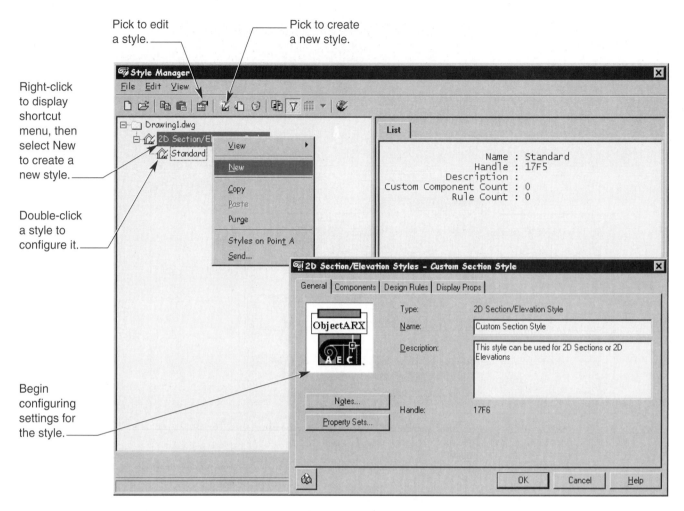

FIGURE 18.5   The Style Manager dialog box is used to configure the new section style. Once a new style has been created, double-click the new style to display the 2D Section/Elevation Styles dialog box and begin configuring the style.

When the command is entered, the **Style Manager** dialog box appears, open to the **2D Section/ Elevation Styles** section (see Figure 18.5). Select the New Style button to create a new section style, or right-click over the style section in the left pane, then select <u>N</u>ew from the shortcut menu. Enter a new name for the style, then double-click the new named style to begin configuring the style. The 2D Section/Elevation Styles dialog box, shown in Figure 18.5, is displayed.

The 2D Section/Elevation Styles dialog box contains the following four tabs:

- **General**   This tab is used to rename the section/elevation style and to provide a description for its use. Because the style can be used for either 2D sections or 2D elevations, you may decide to create only one style, or you may decide to create one style for use with sections and another style for use with elevations. You can add any notes or additional information regarding the style of an elevation by selecting the <u>N</u>otes... button, and you can add information about the style that will be pulled into a schedule by assigning a property set definition using the <u>P</u>roperty Sets... button.

- **Components**   This tab is used to create components that are used in the Design Rules tab (see Figure 18.6). The components that are created are then incorporated into a set of rules that determine how the geometry from the building model is to appear in the 2D section object. You may create components that represent the types of objects in the building model such as windows, walls, and doors or components that represent how the component will be displayed such as lines beyond, hidden lines, and cut lines.

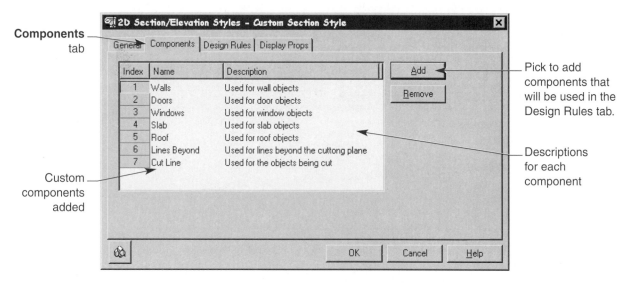

FIGURE 18.6 The Components tab is used to create custom components that will be incorporated into a design rule. The design rules are then used to intercept color from the building model and process it into colors that will be used in the 2D section object.

Use the <u>A</u>dd button to create new components for the design rules. When a component is added, the name Unnamed is placed in the Name column. Pick on top of the name to rename the component. You can add a description for each component by picking in the description area.

■ **Design Rules** This tab is used to establish rules covering how the colors of the objects used in the building model are to appear in the 2D section (see Figure 18.7).

You can add a rule by picking the <u>A</u>dd button. The **Color** column includes a color swatch. Pick the swatch to choose a color that is to be looked for when the section object is created. The color chosen should be a color that is used by the drawing in its Model ByBlock or Model display representation. For example, if looking for walls in the building model, pick color 50. The **Context** column specifies where in the building model the color is to be looked for. For example, for the color 50, a rule can be created that looks for color 50 (Color column) in the first subdivision (Context column) of the section line box. Finally, specify a component where the interpreted color will be placed in the section object. The list of components includes default components for a section object plus the components created in the components tab.

You can establish as many rules as necessary to create the section object. In Figure 18.7 several rules have been created. Some of the colors are repeated but are being looked for in a different context and sent to a different component.

■ **Display Props** This tab is used to configure how the section object is displayed. There is only one display representation, General. Select the appropriate property source (typically, 2D Section/Elevation Style if configuring a custom style), attach the override, and then press the <u>E</u>dit Display Props... button. The Entity Properties dialog box, shown in Figure 18.8, is displayed. This dialog box includes two tabs: **Layer/Color/Linetype** and **Custom Display Components**.

The Layer/Color/Linetype tab initially includes a list of default components for the section object (see Figure 18.8A). The first component is the defining line. AEC objects are placed on this component, unless a design rule specifies differently. Two components are initially turned off: Hidden and Erased. These two components are used when the linework is edited within the section object. The Hidden component displays information using a hidden line. The remaining components are used when AEC objects are found in the different subdivisions of the section line box.

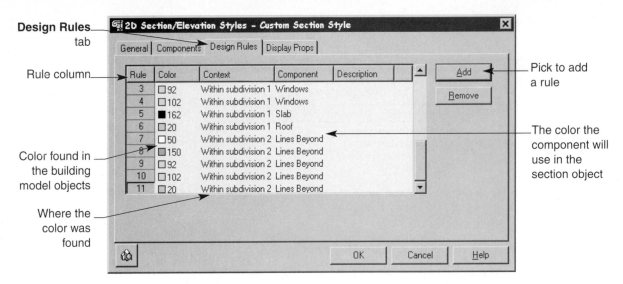

FIGURE 18.7 The Design Rules tab is used to set up rules about what is found in the building model and how it is to be processed into the 2D section object.

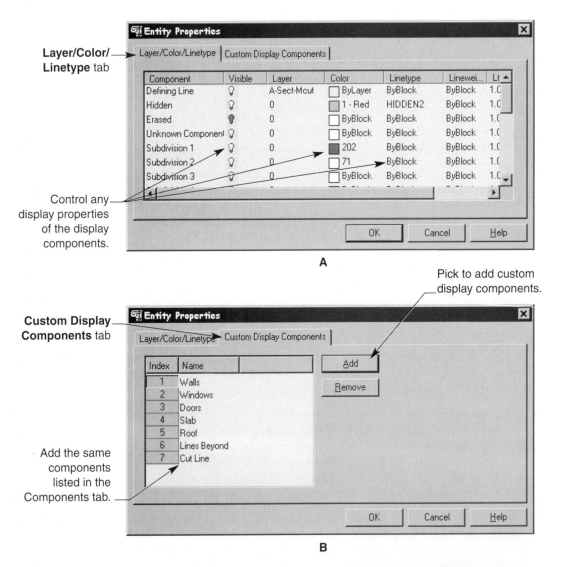

FIGURE 18.8 (A) The Layer/Color/Linetype tab includes a list of display components within a 2D section object. Specify any display properties for the components. (B) The Custom Display Components tab is used to add custom display components listed in the Components tab and place them in the Layer/Color/Linetype tab.

The Custom Display Components tab is used to add components to the Layer/Color/Linetype tab. If components were defined in the Components tab, they also need to be added to this tab and spelled the same way in both tabs. Once a custom display component is added, it becomes a component in the Layer/Color/Linetype tab and has all the same display controls as the default components.

Once you have configured the 2D Section/Elevation style press the OK button to return to the Style Manager. Press OK again to exit the Style Manager and generate the section using the newly configured style.

Exercise 18.2

1. Open EX18-1-Sect.dwg.
2. Create a new section style named Section Style.
3. Edit the style and access the Components tab.
4. Create the following components with descriptions:

Index	Name	Description
1	Doors	Use for Door objects
2	Roof	Use for Roof objects
3	Slabs	Use for Slab objects
4	Walls	Use for Wall objects
5	Windows	Use for Window objects
6	Lines Beyond	Use for linework beyond the cutting line

5. Create the following rules:

Rule	Color	Context	Component
1	150	Within subdivision 1	Doors
2	22	Within subdivision 1	Roof
3	162	Any visible	Slabs
4	50	Within subdivision 1	Walls
5	92	Within subdivision 1	Windows
6	150	Within subdivision 2	Lines Beyond
7	92	Within subdivision 2	Lines Beyond
8	50	Within subdivision 2	Lines Beyond
9	102	Within subdivision 2	Lines Beyond
10	22	Within subdivision 2	Lines Beyond

6. Access the Display Props tab. Select the 2D Section/Elevation Style property source and attach an override. Pick the Edit Display Props button.
7. Add the following display components and assign the following colors to each:

Custom Display Component	Color Assigned in Layer/Color/Linetype
Doors	Magenta (color 6)
Roof	Color 51
Slabs	Color 45
Walls	Color 60
Windows	Magenta (color 6)
Lines Beyond	Color 11

8. Save the drawing as EX18-2-Sect.dwg.

Adjusting Section Line Properties

When creating the section style you may have noticed components named Subdivision *(n)*, where *n* is a number between 1 and 10. Like the elevation line discussed in Chapter 17, the section line can be divided into subdivisions. Each subdivision then represents a different area of the building in the 2D section. For example, objects found in subdivision 1 can be a blue color in the 2D section object, and objects found in subdivision 2 can be a red color in the 2D section object (see Figure 18.9).

A total of 10 subdivisions can be added to the section line box. The subdivisions are also used in the 2D Section/Elevation style in the Design Rules tab. When rules have been established, a rule may look for a specific color in a specific subdivision.

The section line and the subdivision lines can be controlled through the Section Line Properties command. Use one of the following options to access this command:

✓ Select the **Documentation** pull-down menu, pick the **Sections** cascade menu, then pick the **Section Line Properties...** command.

✓ Pick the **Section Line Properties** button in the **Sections** toolbar.

✓ Right-click, then select **Documentation, Sections, Section Line Properties...** from the cursor menu.

✓ Type **AECBLDGSECTIONLINEPROPS** at the Command: prompt.

✓ Select the section line, right-click, then select **Section Line Properties...** in the shortcut menu.

The Section/Elevation Line Properties dialog box, shown in Figure 18.10, is displayed.
The dialog box is divided into the following three tabs:

- **General**   This tab is used to enter a description for the section line, to attach any notes, or to assign property set definitions.

- **Dimensions**   This tab is used to control the dimensional settings of the section line. This tab has two areas. The **Graphic Divisions from Cut Plane** area is used to add subdivisions to the section line. Use the **A**dd... button to add a subdivision and

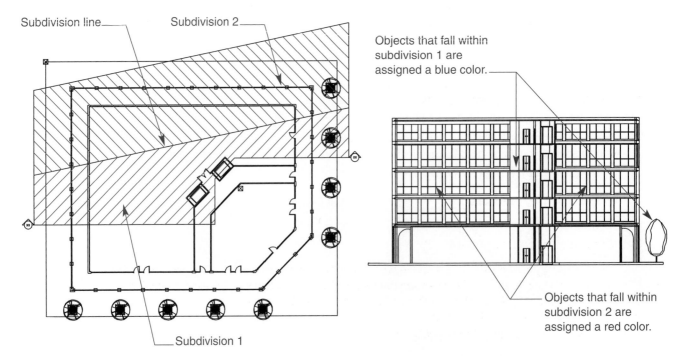

FIGURE 18.9   Subdivisions are used by the section object to find objects in these sectors and to interpret their colors in each subdivision for placement into the 2D section object.

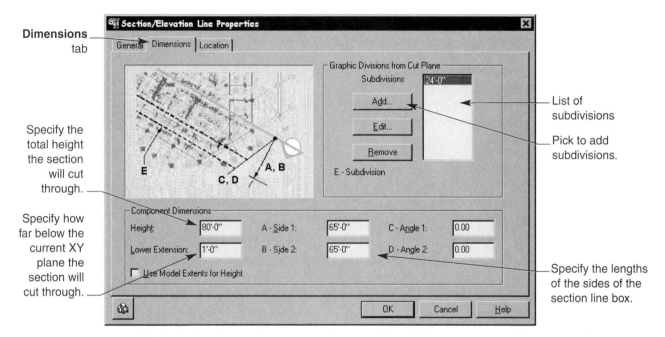

FIGURE 18.10 The Section/Elevation Line Properties dialog box is used to adjust the properties of the section line box and add subdivisions to the line.

specify its placement by a distance value measured from the defining line. The **Component Divisions** area is used to control how high the section box cuts through the building model, how far below the current XY drawing plane the section cuts through, and the lengths of the sides of the section line box.

■ **Location**   This tab can be used to control where the section line box is placed in the drawing. When a section line is initially created in the drawing, the line is placed in the current XY drawing plane. Use this tab to reestablish where on the Z axis the section line box starts.

In addition to adjusting the section line box through this dialog box, you can adjust it and the subdivisions with grips. Pick the section line box to display the grip boxes. Pick a grip box to stretch its location, and reshape the section line box or move a subdivision line.

Exercise 18.3

1. Open EX18-2-Sect.dwg.
2. Add a subdivision 16'-0" away from the defining line.
3. Edit the section style created in EX18-2-Sect so that the color used for subdivision 1 is blue, and the color for subdivision 2 is red.
4. Save the drawing as EX18-3-Sect.dwg.

Drawing a Two-Dimensional Section

After the section line has been added and configured with any subdivision lines, and a section style is available, a 2D section can be created. Like 2D elevation objects, the 2D section creates a two-dimensional section of anything visible within the section line box. Where the section line crosses through geometry, that geometry appears to be cut away in the section.

Use one of the following options to create a 2D section:

✓ Select the **Documentation** pull-down menu, access the **Sections** cascade menu, and select **Create Section....**

✓ Pick the **Create Section** button in the **Sections** toolbar.

✓ Right-click, and select **Documentation, Sections, Create Section...** from the cursor menu.

✓ Type **AECBLDGSECTIONLINEGENERATE** at the Command: prompt.

✓ Select the section line, right-click, then pick **Generate Section...** from the shortcut menu.

Once the command is entered, the Generate Section/Elevation dialog box is displayed (see Figure 18.11). The dialog box is divided into the following four main areas:

- **Result Type**   This area is used to specify the type of section that will be created. The choices are 2D Section or 3D Section. Once you have selected the type, choose the section style from the Style to Generate drop-down list.

- **Selection Set**   This area is used to select the geometry that will be used to create the section. Initially, the Select Objects button is available. Press this button to select geometry in the drawing that will be calculated for the section. After the geometry has been selected, the Generate Section/Elevation dialog box reappears to finish processing the information.

- **Placement**   This area is used to select a location for the section object. Use the Pick Point button to pick a point in the drawing to locate the section object. Once the point has been selected, the Generate Section/Elevation dialog box reappears so that the section can be generated.

- **Display Set**   This area is used to select the display set that will be used when the design rules process the section. As indicated earlier, the design rules process color and transfer the color of objects to a component within the section style. Generally, the Section_Elevation display set is used. This display set processes the Model ByBlock colors of the AEC objects and uses them in the creation of the section object.

FIGURE 18.11 The Generate Section/Elevation dialog box is used to create a section object.

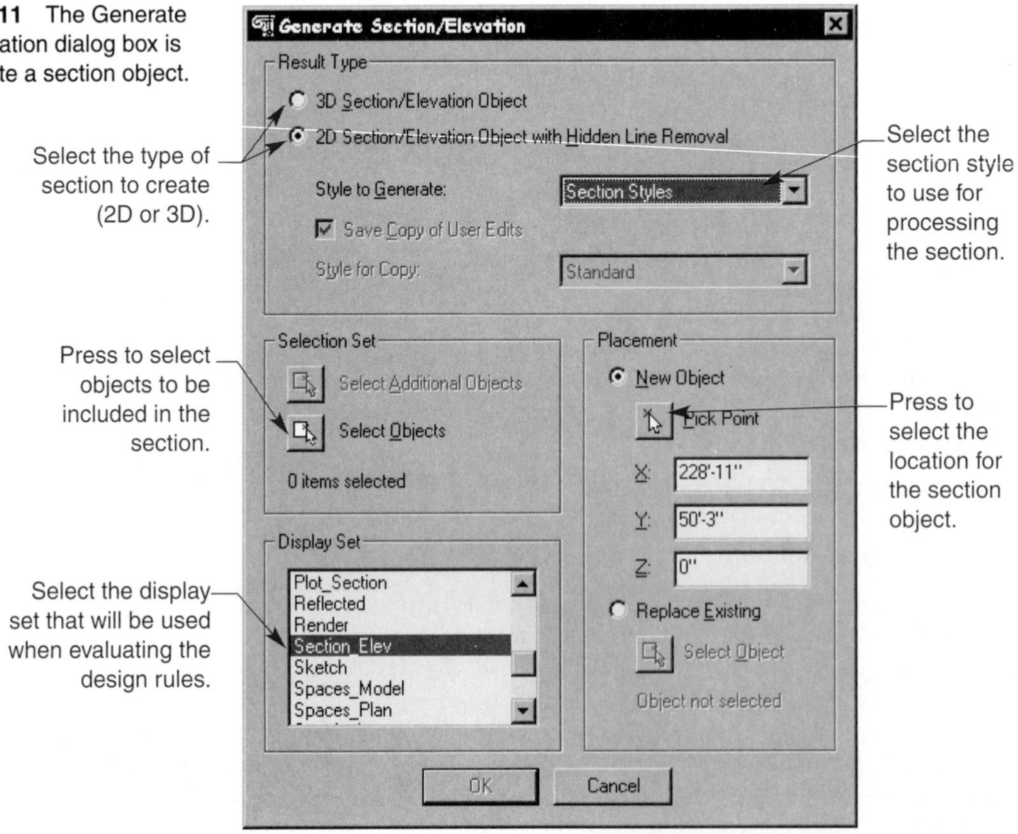

Select the type of section to create (2D or 3D).

Select the section style to use for processing the section.

Press to select objects to be included in the section.

Press to select the location for the section object.

Select the display set that will be used when evaluating the design rules.

Field Notes

> Although you can choose the display to use for processing the section object, it is recommend-ed that you use the Section_Elevation display set until you fully understand how the Architectural Desktop display settings work. You can then begin experimenting with how other display sets control how the design rules are processed.

When you have finished specifying the settings for the section object, press the OK button to begin creating the section object. When you have finished, the section object is created at the location specified with any colors used by the design rules.

Exercise 18.4

1. Open EX18-3-Sect.dwg.
2. Generate the section from the section line.
3. Create the 2D section using the section style created in EX18-2-Sect.dwg.
4. Place the section object to the side of the xref'd floor plans.
5. Use the Section_Elevation display set.
6. Save the drawing as EX18-4-Sect.dwg.

Updating the Section Object

A major benefit to using a 2D section object is that it maintains a link to the original geometry that was used to create the section. Whenever a change is made to the building model, the section can be edited. This process works well for xref'd drawings as well. Make the changes to the original reference drawing, reload the xref in the drawing in which the sections are created, and then update the section.

Use one of the following options to update the section object:

✓ Pick the **Documentation** pull-down menu, select the **Sections** cascade menu, then pick the **Update Section...** command.

✓ Pick the **Update Section** button in the **Sections** toolbar.

✓ Right-click, and select **Documentation, Sections, Update Section...** from the cursor menu.

✓ Type **AECBLDGSECTIONUPDATE** at the Command: prompt.

✓ Select the section object, right-click, and select **Update...** from the shortcut menu.

When a section is updated the **Generate Section/Elevation** dialog box is displayed, as shown in Figure 18.12. If additional geometry has been added and needs to be added to the section, pick the **Select Additional Objects** button in the Selection Set area. You can also select a different section style, as well as a different display set. Press the OK button to begin updating the section.

FIGURE 18.12 The Generate
Section/Elevation dialog box is
displayed when a section object
is updated.

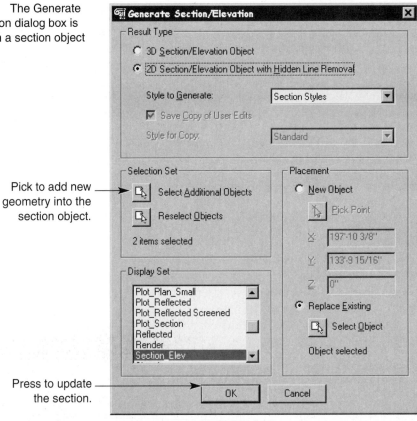

Pick to add new
geometry into the
section object.

Press to update
the section.

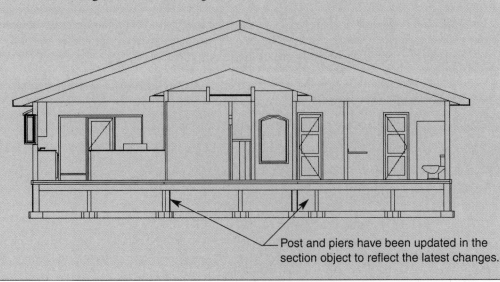

Exercise 18.5

1. Open EX18-4-Sect.dwg.
2. Notice that the post and pier structural pieces appear off the ground line.
3. Open EX14-1-Fnd.dwg. Change the post and pier structural style so the end of the 4×4
   post is set 2'-10" from the starting point of the structural style.
4. Use the Structural Properties command to adjust the length of the post and piers to 2'-10".
   Change the location of the structural members to be −12" on the Z axis. Save the drawing.
5. In the EX18-4-Sect.dwg, reload the EX14-1-Fnd.dwg.
6. Update the section object. The post and piers should appear in the proper locations and at
   the proper height.
7. Save the drawing as EX18-5-Sect.dwg.

Post and piers have been updated in the
section object to reflect the latest changes.

Editing Section Linework

The section object is only as accurate as the building model. The more detailed the model, the more detailed the section. In many cases it is not worthwhile to model everything in the building model. In these cases, the section object can be edited to display a section with the information required to create the construction documents.

Two different commands can be used to edit the section object: **Edit Linework** and **Merge Linework**.

Editing Linework

As the section is created from what the section line "sees," some undesirable linework may get included in the section object. For example, if a section line box includes a part of the roof that would be seen in the section, yet is displayed within the section, this becomes undesirable linework. Other linework may be placed on a color not desired because the design rules could not differentiate two objects using the same color. This also becomes undesirable linework.

Such lines within the section can be edited and placed on components within the section style for further display control. As long as the section remains one complete object (not exploded individual lines), the section object maintains a link back to the original geometry. The linework cannot be edited through any conventional AutoCAD commands, because the section will no longer update when the building model is changed. The **Edit Linework** command maintains the link and yet modifies how the lines appear in the section. Use one of the following options to access this command:

✓ Select the **Documentation** pull-down menu, pick the **Sections** cascade menu, then select the **Edit Linework** command.

✓ Pick the **Edit Linework** button from the **Sections** toolbar.

✓ Right-click, and select **Documentation, Sections, Edit Linework** from the cursor menu.

✓ Type **AEC2DSECTIONRESULTEDITAT** the Command prompt.

✓ Select the section object, right-click, then select **Edit Linework…** from the shortcut menu.

Once this command has been entered you are prompted to select linework to be edited. Begin selecting the linework to be changed to the same display component within the section style. For example, if several lines need to be placed on the Erased component, then select the linework that will appear to be erased from the section. The lines within the section can be selected only individually; crossing boxes and windows cannot be used to select the lines within the section.

When you have finished selecting the lines to be edited, press the [Enter] key. You are then prompted to select a component number to place the lines. Each display component within a section style is assigned a number. For example, the Defining Line component is 1, and the Erased Vectors component is 3. Enter the number of the desired component, and the selected linework is moved to that display component. The following prompt sequence is used to remove the two lines in Figure 18.13.

```
Command: AEC2DSECTIONRESULTEDITAT ⏎
Select linework to edit: (select a line in the section)
Select linework to edit: (select a line in the section)
Select linework to edit: ⏎
Fixed Components
1. Defining Line
2. Hidden Vectors
3. Erased Vectors
4. Subdivision 1
5. Subdivision 2
6. Subdivision 3
7. Subdivision 4
8. Subdivision 5
9. Subdivision 6
10. Subdivision 7
11. Subdivision 8
12. Subdivision 9
```

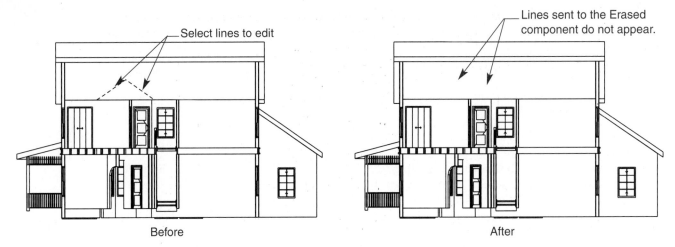

FIGURE 18.13   Linework can be edited to move lines within the section to another display component within the section style

13. Subdivision 10
User Components
No user components
Enter component index<1>: *(enter a component number from above)* 3↵
Command:

 Field Notes

■   The list of available components to which linework can be sent is displayed in the Command prompt area. If the full list cannot be displayed, press the [F2] key to display the text window. Find the component number desired, then enter it in the Command: prompt. Press the [F2] key to return to the drawing screen.

■   It is strongly recommended that lines be edited after the final design for the building has been finished. If the section object is updated, the section loses any edits performed on the lines within the section. Lines that were sent to the Erased Vectors component become visible, and lines placed on other components revert to their original display components.

Exercise 18.6

1. Open EX18-5-Sect.dwg.
2. Begin editing the linework within the section object.
3. Place lines that are to be erased on the Erased Vectors component.
4. Place any other linework on other desired display components
5. Save the drawing as EX18-6-Sect.dwg.

Merging Linework
In addition to editing existing linework, you can incorporate newly drawn lines into the section object. This routine is known as *merging linework*. As mentioned in the previous section, the section object is only as accurate as the building model. Unless structural framing has been incorporated into the building model, the section does not display framing.

In this case, structural framing can be added to the section through the use of polylines, lines, and rectangles. Once any linework has been added to the section, it can be merged into the section object and placed on a component.

Use one of the following options to merge new lines into the section:

✓ Select the **Documentation** pull-down menu, pick the **Sections** cascade menu, then select **Merge Linework**.

✓ Pick the **Merge Linework** button from the **Sections** toolbar.

✓ Right-click, and select **Documentation, Sections, Merge Linework** from the cursor menu.

✓ Type **AEC2DSECTIONRESULTMERGE** at the Command: prompt.

✓ Select the section object, right-click, then select **Merge Linework...** from the shortcut menu.

When this command is entered you are prompted to select a 2D section object. Pick the desired section. You are then prompted to select entities to be merged. Select any new or existing linework in the drawing to be merged. When you have finished selecting the lines, press the [Enter] key and specify a component for the linework to go to. The following prompt sequence is used to merge new linework into a section object:

Field Notes

When selecting linework to be merged, you can use crossing boxes and windows to select the lines for merging.

Command: **AEC2DSECTIONRESULTMERGE** ↵
Select a 2D Section/Elevation:
Select entities to merge: *(select a line or use a crossing/window to select new lines)*
1 found
Select entities to merge: *(select a line or use a crossing/window to select new lines)*
1 found, 2 total
Select entities to merge: ↵
Fixed Components
1. Defining Line
2. Hidden Vectors
3. Erased Vectors
4. Subdivision 1
5. Subdivision 2
6. Subdivision 3
7. Subdivision 4
8. Subdivision 5
9. Subdivision 6
10. Subdivision 7
11. Subdivision 8
12. Subdivision 9
13. Subdivision 10
User Components
No user components
Enter component index for non 2D section objects<1>: *(enter the component number for the newly merged lines to be assigned)* 1 ↵
Command:

Once the lines have been merged, they are displayed in the color of the component they were assigned and become a part of the section object.

Field Notes

As when linework is edited, if the section object is edited, the merged linework is no longer a part of the section object; however, the linework is grouped together as a separate section object retaining the components to which the lines were assigned. Use the Merge Linework command to remerge the lines into the section object.

Exercise 18.7

1. Open EX18-6-Sect.dwg.
2. Add linework as displayed in the figure. Draw framing members and adjust the interior room lines as necessary.
3. Merge the lines into the section object onto components of your choosing.
4. Save the drawing as EX18-7-Sect.dwg.

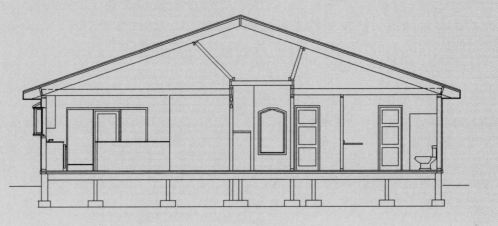

 Drawing a Three-Dimensional Section

The creation of a 3D section can be helpful when studying the three-dimensional relationships among structural parts of the building (see Figure 18.14). These 3D sections can be used to study the model or to create presentation cutaways of the model or can be turned into 2D perspective sections using the Hidden Line Projection command discussed in Chapter 17.

The 3D section is created in the same manner as the 2D section. First, create the section line and any subdivisions desired, then use the **Create Section** command to generate the section. In the **Generate Section/Elevation** dialog box, select the **3D Section/Elevation Object** radio button (see Figure 18.15). Note that the Style to Generate drop-down list is not available. 3D sections do not use a section style for the displayed colors. Instead, they rely on the entity display settings configured for the 3D section object.

After picking the 3D Section/Elevation Object radio button, pick the Select Objects button to select the objects to be included in the 3D section. After selecting all the items within the section line box, press [Enter] to return to the Generate Section/Elevation dialog box. Select the Pick Point button to select a location for the new section object. When you have finished, press the OK button to generate the section.

The 3D section is created at the location specified before the section is generated. The geometry that is created is a copy of the portion of the building model that is within the section line box. If the section line box is stretched to adjust the shape, the 3D section object can be updated. The 3D section is tied to the section line in the drawing. If the building model changes, the 3D section can be updated the same way as a 2D section. Select the 3D section, right-click, and select Update… from the shortcut menu.

The display for the 3D section can be controlled by selecting the 3D section object, right-clicking, and picking Entity Display… from the shortcut menu. When editing the display for the 3D section, the only components whose display you can control are the defining line and up to 10 subdivisions. Turn any of the components on or off and assign colors to the different subdivisions.

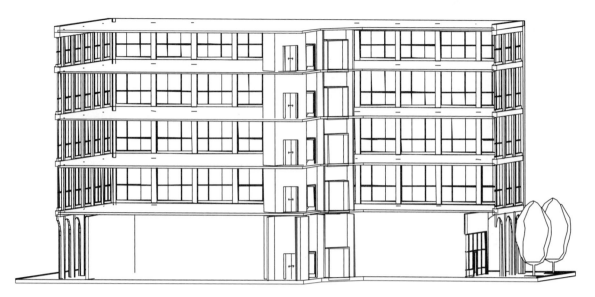

FIGURE 18.14 A 3D section can be used to study the building elements of the design or to create presentation cutaways of the building.

FIGURE 18.15 Select the 3D Section/Elevation Object radio button to create a 3D section.

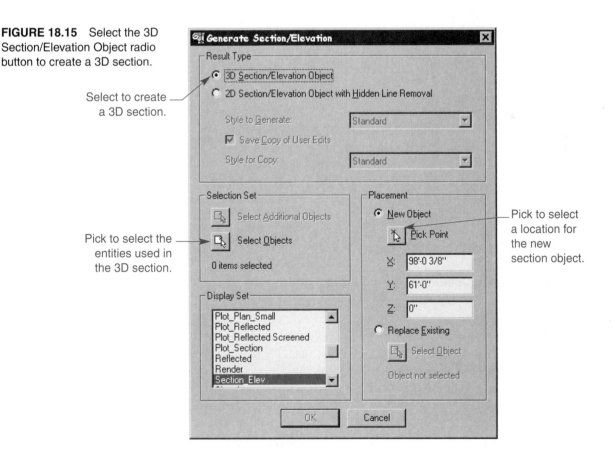

Select to create a 3D section.

Pick to select the entities used in the 3D section.

Pick to select a location for the new section object.

Field Notes

Like 3D elevation objects, 3D section objects use a lot of computer resources. If a 3D section is needed for study or presentation, you may decide to generate a 3D section, then use the Hidden Line Projection to create a flat block of the 3D section, then erase the original 3D object from the drawing.

Exercise 18.8

1. Open EX18-1-Sect.dwg.
2. Create a 3D section from the section line.
3. Place the 3D section off to the side of the building model.
4. Use a camera to establish a perspective from the front of the section object.
5. Generate the camera view.
6. Save the drawing as 18-8-Sect-3D.dwg

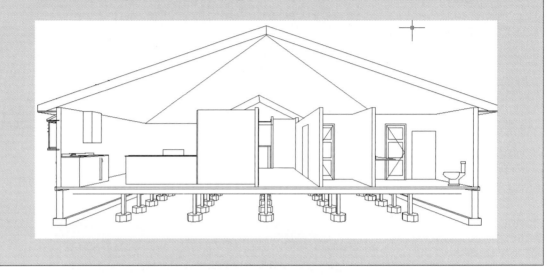

Using Section Line Blocks and Detail Marks

When a section line is first created, a section symbol is placed at the starting and ending points of the section line. The arrow on the symbol indicates the direction in which the viewer is observing the section. The section symbol typically points in the direction the section line box creates the section. Included within the symbol is text that indicates the section number. Architectural Desktop includes a few different types of section symbols that can be used to generate sections or details. This symbol is also commonly referred to as the *section bubble* or *cutting plane line symbol*.

Modifying the Default Section Symbol

When a section line is initially added to a drawing, the section symbol placed at the ends of the section line contains text that represents the section number. The text is represented by two *X*s within the section bubble. These *X*s can be modified to reflect the section number for the section line. To modify this text, select the symbol, right-click, and pick **Multi-View Block Properties** from the shortcut menu. The **Multi-View Block Reference Properties** dialog box, shown in Figure 18.16, is displayed. Select the Attributes tab to modify the text within the section bubble. The prompt column includes the attribute that controls the text value within the section bubble. Pick the associated text in the Value column to change the XX to a section number. Press the OK button to accept the change.

Using the Section Marks

You can add additional section marks to the drawing or use them to replace the default section marks placed when the section line is drawn by using one of the following options:

✓ Select the **Documentation** pull-down menu, pick the **Documentation Content** cascade menu, then select **Section Marks....**

✓ Pick the **Section Marks** button in the **Documentation - Imperial** toolbar.

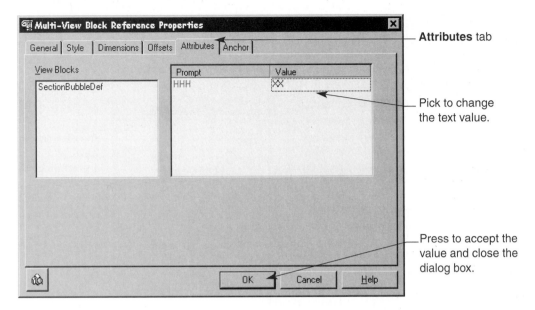

FIGURE 18.16   The Attributes tab in the Multi-View Block Reference Properties dialog box is used to change the section number in the default section bubble.

✓ Right-click, and select **Documentation, Documentation Content, Section Marks...** from the cursor menu.

✓ Type **AECDCSETIMPSECTIONMARKS** at the Command: prompt.

The DesignCenter window is displayed, open to the Section Marks folder (see Figure 18.17). Four different section marks are included. Drag and drop the symbols from the DesignCenter into the drawing. As with the section line, you are prompted to pick a starting point, then any additional points. Once you have picked the last point, press the [Enter] key. The **Edit Attributes** dialog box is displayed, where you can enter a value for the section bubble.

After specifying the section number, press OK in the Edit Attributes dialog box. You are then prompted to specify the side of the bubble on which the arrow is to be placed. Pick a point on one side of the section bubble. The final prompt asks if a section object is to be added. Entering a Y at the prompt creates a section line box the length and depth of the section line. The following prompt sequence is used to add a section mark to the drawing in Figure 18.18.

```
Command: AECANNOSECTIONMARKADD ↵
Adding: Section Mark (w/Sheet No. & Tail)
Specify first point of section line or [Symbol/Arrow/Tail]: _SYMBOL
Symbol block name or [?]<NONE>: Anno_Section_A2
Specify first point of section line or [Symbol/Arrow/Tail]: _ARROW
Arrow block name or [?]<NONE>: Anno_Arrow_A2
Specify first point of section line or [Symbol/Arrow/Tail]: _TAIL
Tail length,width<0.000,0.000>: .1875,.0625
Specify first point of section line or [Symbol/Arrow/Tail]: _Endofparam
Specify first point of section line or [Symbol/Arrow/Tail]: (pick first point for section line)
Specify next point of line or [Symbol/Arrow/Tail]: (pick second point for section line)
Specify next point of line or [Break/Symbol/Arrow/Tail]: ↵
Select block reference: (enter values in the Edit Attributes dialog box)
Command:
Specify side for Arrow: (pick the side of the bubble on which the arrow is to be placed)
Add AEC section object? [Yes/No] <N>: Y ↵

Command:
```

FIGURE 18.17   The Section Marks folder within the DesignCenter includes four different types of section marks. Drag and drop these into the drawing and answer any of the prompts in the Command: line.

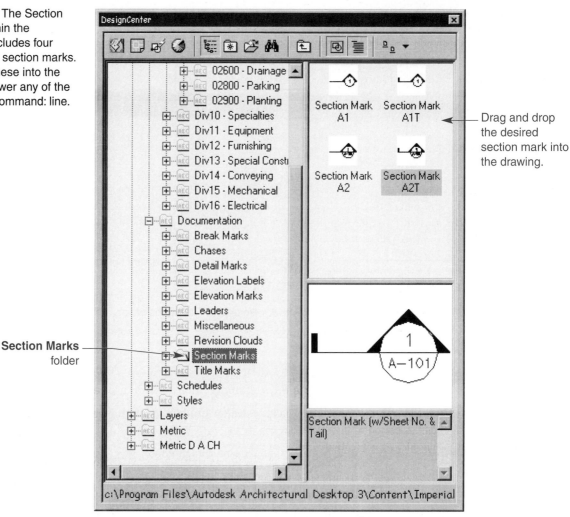

Drag and drop the desired section mark into the drawing.

Section Marks folder

FIGURE 18.18   A short section mark can be used to generate a section detail.

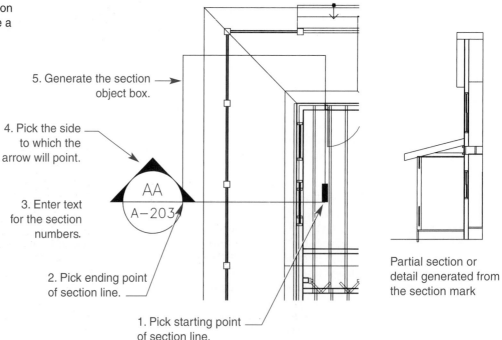

5. Generate the section object box.

4. Pick the side to which the arrow will point.

3. Enter text for the section numbers.

2. Pick ending point of section line.

Partial section or detail generated from the section mark

1. Pick starting point of section line.

If a section line box is generated, its properties can be adjusted as with the standard section line box. Use grips to increase or decrease the size of the box. Use the **Section Line Properties** command to add subdivision lines or adjust the heights for the section box. You can also modify the text within the section mark. The section marks within the DesignCenter are standard AutoCAD blocks and not multi-view blocks. Double-clicking on the symbol allows you to edit the attribute text within the bubble.

Once the section mark has been configured and the section line box properties adjusted, the section can be generated as a 2D or 3D section object.

Finding Additional Section Symbols

Along with the section marks found in the DesignCenter additional section and elevation symbols are included if the International Extensions were installed. These symbols are called *elevation labels* (see Figure 18.19). These labels can be used to document different elevation or level changes within sections, details, or elevations.

Use one of the following options to access these symbols:

✓ Select the **Documentation** pull-down menu, pick the **Documentation Content** cascade menu, then select the **Elevation Labels...** command.

✓ Pick the **Elevation Labels** button in the **Documentation - Imperial** toolbar.

✓ Right-click, and pick **Documentation, Documentation Content, Elevation Labels...** from the cursor menu.

✓ Type **AECDCSETIMPELEVATIONLABELS** at the Command: prompt.

FIGURE 18.19 The Elevation Label section is included in the Documentation Content cascade menu if the International Extensions were installed. There are three types of elevation labels that can be used.

The Elevation Labels section includes three types of symbols.

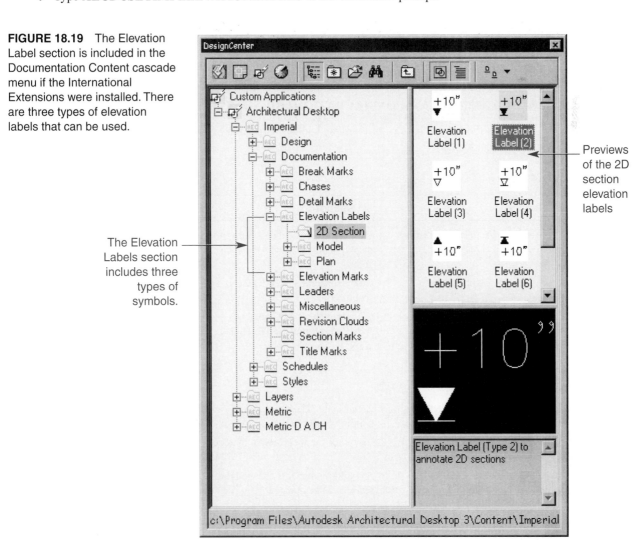

Previews of the 2D section elevation labels

Like the default section mark, these symbols are multi-view blocks. When you drag these symbols into the drawing, you are prompted to enter values for the elevation label. If you need to change the values, select the multi-view blocks, right-click, and pick **Multi-View Block Properties** from the shortcut menu, then change the attribute values in the Attributes tab.

 ## Creating Live Sections

The preceding sections have explained how to use Architectural Desktop to create sections for your construction documents. Another type of section that can be created is the live section. This tool is available only if the International Extensions are installed.

Live sections get their name from the fact that the building model is physically cut away, revealing the interior of the building. A live section is similar to a 3D section with the exception that the live section is not a three-dimensional copy but the actual AEC objects being cut. When the section is cut, a special display configuration is created that shows the building model cut away. Changing the display configuration from the newly created section configuration to a standard work, plot, or reflected ceiling configuration displays the building model in its entirety.

The AEC objects within the live section can be modified using commands you would use to modify the building model in its uncut form. Figure 18.20 tabulates the differences between standard 2D and 3D sections and live sections.

Like the standard sections, live sections use the section line to determine what will be cut away. Once the section line is created, the live section can be cut using one of the following options:

✓ Select the **Des<u>k</u>top** pull-down menu, pick the **Live S<u>e</u>ction Display** cascade menu, then select the **<u>A</u>dd Live Section Configuration** command.

✓ Pick the **Add Live Section Configuration** from the **Live Section Display** toolbar.

✓ Right-click, and select **Des<u>k</u>top, Live S<u>e</u>ction Display, <u>A</u>dd Live Section Configuration** from the cursor menu.

✓ Type **AECSECTIONCONFIGURATIONADD** at the Command: prompt.

✓ Select the section line, right-click, and select **Add Live Section Configuration...** from the shortcut menu.

2D Section	3D Section	Live Section
Uses section style to control display settings	Uses the Bldg Sections Subdivision display representation for display control	Uses the AEC objects' Section display representation for display control
Cuts a separate 2D plan view section	Cuts a separate 3D model view section	Cuts the building model into a section
Cuts through all AEC objects, AutoCAD entities, and through xrefs	Cuts through all AEC objects, AutoCAD entities, and through xrefs	Cuts through walls, doors, windows, mass elements, mass groups, stairs, railings, roofs, roof slabs, slabs, spaces, space boundaries, curtain wall layouts and units, window assemblies, structural members
Creates one 2D object copied from building model	Creates one 3D object copied from building model	Cuts the actual building model, retaining the original objects
Linework can be edited and merged into 2D object	Linework cannot be edited or added	Original AEC objects can be modified
As building model is changed the section can be updated	As building model is changed the section can be updated	Changes made to the building model are automatically adjusted

FIGURE 18.20   The differences between the standard sections and the live section.

FIGURE 18.21 The Add Live Section Configuration dialog box is used to create the new live section display configuration.

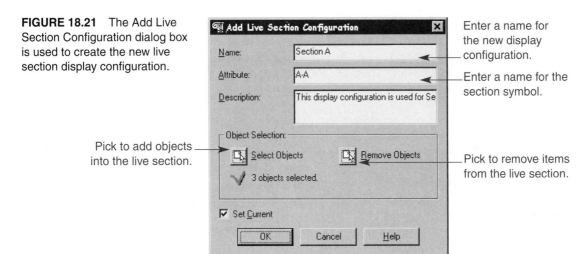

Enter a name for the new display configuration.

Enter a name for the section symbol.

Pick to add objects into the live section.

Pick to remove items from the live section.

The **Add Live Section Configuration** dialog box, shown in Figure 18.21, is displayed. Enter a name for the display configuration to be used to display the cutaway building model in the Name text box. Next, enter a value for the section symbol attribute. Finally, pick the Select Objects button to begin selecting the objects that will be included in the live section. Refer to Figure 18.20 for objects that can be added into the live section.

Once the OK button is pressed, Architectural Desktop begins processing the live section. When the processing is finished, the current display configuration is set to the new live section display configuration, and the building appears to be cut away (see Figure 18.22).

The objects that remain on the screen are the same objects that were initially drawn. The Section display representation for each object is used to display the object as a cutaway modeling component. A hatch pattern can be added to the object at the point the section line cuts the object. To add the hatch pattern, select the type of object that is to have its display altered, right-click, and select the Entity Display command.

In the Entity Display dialog box, select the Section display representation (see Figure 18.23), then select the level of display control from the Property Source column. Attach the override and begin editing the display properties of the object for the section display configuration. Two tabs are available for the section display representation: Layer/Color/Linetype and Hatching. Adjust the colors and other display properties as necessary in the Layer/Color/Linetype tab and configure the hatch used by the object in the Hatching tab.

If at any time you need to modify the whole building model, change the current display configuration to a configuration that allows you to work on the model (try Work, Work Reflected, and the Plot configurations).

FIGURE 18.22 The finished building after a new live section display configuration is created.

Select the Section display representation to modify how the cut object appears in the display configuration.

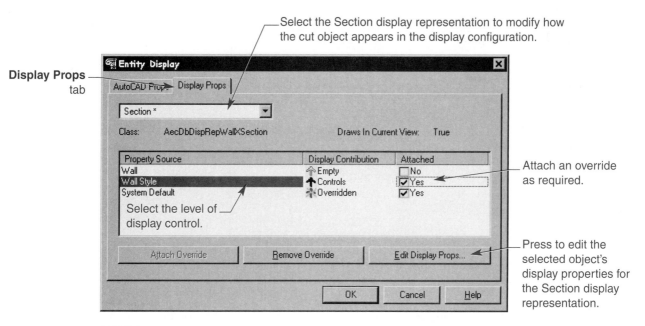

Display Props tab

Attach an override as required.

Select the level of display control.

Press to edit the selected object's display properties for the Section display representation.

FIGURE 18.23   The Entity Display dialog box for each object type that is cut by the live section includes a section display representation. Adjust the display properties of the selected object through the Edit Display Props button.

Exercise 18.9

1. Open EX16-9-Flr-Pln.dwg.
2. Create a section line across the building model.
3. Create a live section from the section line.
4. Name the configuration Section A-A. Specify the name for the section symbol in the Attributes text box.
5. After viewing the live section return to the Work display configuration. Note that the entire building is redisplayed.
6. Save the drawing as EX18-9-Flr-Pln.dwg.

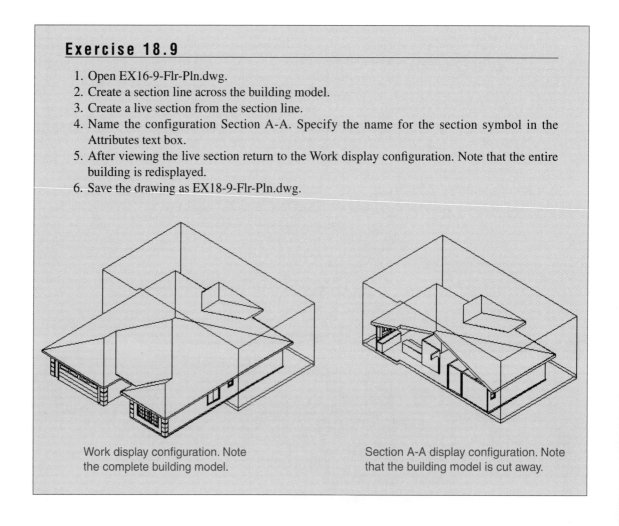

Work display configuration. Note the complete building model.

Section A-A display configuration. Note that the building model is cut away.

Modifying a Live Section

As indicated earlier, objects within the live section are the actual building objects. Any editing commands such as Entity Display, the properties dialog boxes, and AutoCAD commands can be used to modify the objects in a live section view. As the objects are modified the live section is automatically updated to reflect the changes made to the drawing.

New objects can also be added to the building model. If you are working in the live section display configuration, new objects are added but not displayed. If you are planning on adding new objects, it is best to first set the Work display configuration current, then add the new objects to the drawing.

After the new objects have been added to the building model, the live section needs to be updated with the Modify Live Section Configuration command. Use one of the following options to access this command:

✓ Select the **Desktop** pull-down menu, then the **Live Section Display**.

✓ Pick the **Modify Live Section Configuration** button on the **Live Section Configuration** toolbar.

✓ Right-click, and select **Desktop, Live Section Display, Modify Live Section Configuration** from the cursor menu.

✓ Type **AECSECTIONCONFIGURATIONMODIFY** at the Command: prompt.

✓ Select the section line, right-click, and pick **Modify Live Section Configuration...** from the shortcut menu.

When the command is first entered you are prompted to select a section line or enter one of two options. Selecting the section line displays the **Modify Live Section Configuration** dialog box. This dialog box is very similar to the Add Live Section Configuration dialog box (see Figure 18.24). You can also enter the name of the live section that is to be modified by selecting the Name option. Entering a D at the Command: prompt for the Dialog option immediately displays the dialog box.

This dialog box also allows you to add additional objects to the configuration or remove objects from the configuration. Pick the Add Objects button to add any newly created objects to the section. When you have finished adding or removing, select the OK button to close the dialog box.

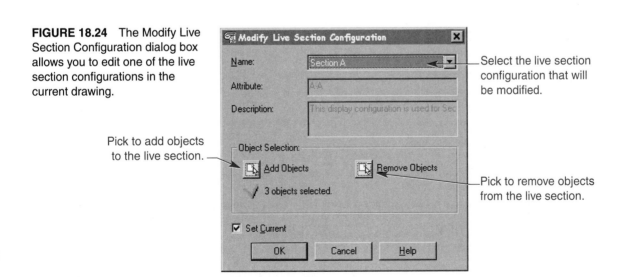

FIGURE 18.24 The Modify Live Section Configuration dialog box allows you to edit one of the live section configurations in the current drawing.

Select the live section configuration that will be modified.

Pick to add objects to the live section.

Pick to remove objects from the live section.

Using the Quick Slice Command

Another useful tool for quickly creating a 2D schematic section is the **Quick Slice** command. This tool creates a polyline around any solid shape or object it finds within the points selected.

Use one of the following options to access this command:

- ✓ Pick the **Desktop** pull-down menu, and select the **Utilities** cascade menu, then pick **Quick Slice**.

- ✓ Pick the **Quick Slice** button in the **Utilities** toolbar.

- ✓ Right-click, and select **Desktop, Utilities, Quick Slice** from the cursor menu.

- ✓ Type **AECQUICKSLICE** at the Command: prompt.

When this command is initiated, you are prompted to select the objects for the quick slice, then to pick points that define where the polyline will cut through. The following prompt sequence is used to create a quick slice of the stair and walls in Figure 18.25.

Command: **AECQUICKSLICE** ↵
Select entities to slice: *(pick objects or use a crossing box to select the objects (first point of crossing))*
Specify opposite corner: *(pick objects or use a crossing box to select the objects (second point of crossing))*
4 found
Select entities to slice: ↵
Pick first point: *(pick first point to establish the cutting line)*
Select second point: *(pick second point to establish the cutting line)*
[2] polyline(s) created.
Command:

The slice(s) that remains can be used to create a custom profile or to lay out sections manually.

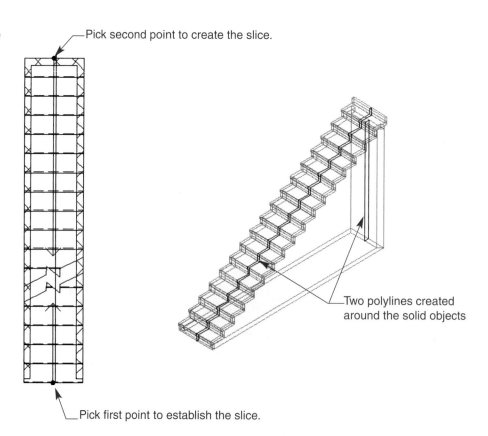

FIGURE 18.25 The Quick Slice command can be used to create a polyline slice through solid objects.

Pick second point to create the slice.

Two polylines created around the solid objects

Pick first point to establish the slice.

Field Notes

When using the Quick Slice command, create the slice in the top view to accurately pick the first and second points. This creates the quick slice. Then change to an elevation or isometric view to see the polyline slices. Freeze any of the building model layers to work on the section.

CHAPTER REVIEW

Use the CD-ROM to test your knowledge and skills.

Chapter Test

To check your understanding of the content provided in this chapter, access the Test file in the CH18 folder of the CD-ROM that accompanies this text.

Chapter Project

To practice the Architectural Desktop skills presented in this chapter, access the Project files in the CH18 folder of the CD-ROM that accompanies this text. The project files are in pdf format and include sample drawings and instructions for completing each project.

CHAPTER 19

Using Schedules

LEARNING GOALS

After completing this chapter, you will be able to:

◎ Add schedule tags to a drawing.

◎ Add room and finish tags to a drawing.

◎ Renumber schedule tags.

◎ Use property set definitions.

◎ Create a schedule.

◎ Update a schedule.

◎ Modify the properties of a schedule table.

◎ Export a schedule table.

◎ Access additional schedule table styles.

◎ Edit a schedule.

One of the key elements in producing a set of construction documents is the addition of schedules. *Schedules* provide information such as the types and sizes of doors and windows in a building, the finish materials used in different rooms, and equipment counts. The information found within a schedule is in a tabular format that is referred to as the *schedule table* (see Figure 19.1). Architectural Desktop includes several predefined schedules for use in your drawings and provides tools for customizing schedules to report on specific information required by the construction documents.

The schedules that are created maintain links back to the objects on which they are reporting. As new objects are added to the drawing or as information about the objects changes, the schedule can automatically be updated. Schedules can also report on objects that belong to a reference drawing and maintain links to the referenced geometry.

In addition to the predefined schedules, Architectural Desktop includes several notation blocks or symbols known as schedule tags. *Schedule tags* are numbers or symbols that point the person reviewing the drawing to a specific set of information within a schedule. The information found in the schedule is obtained from information attached to the object in the drawing or from the schedule tag linked to the object. This information is known as a *property*. Entire sets of properties, called a *property set,* can be attached to an object. This chapter explains the tools used to attach schedule tags to objects, fill out property set information, and to create a schedule.

Adding Schedule Tags to a Drawing

Schedule tags can be attached to AEC objects and to standard AutoCAD entities. The symbol is typically a geometric shape containing a letter or number indicating its position within a schedule. Other schedule tags may include only text or a leader line pointing to the object (see Figure 19.2).

Architectural Desktop includes several different types of schedule tags for use. Some of the tags must be attached to specific types of AEC objects. For example, the door tag must be attached to a door object. The schedule tags are divided into four types: Door & Window Tags, Object Tags, Room & Finish Tags,

DOOR AND FRAME SCHEDULE

| MARK | DOOR | | | | | | | FRAME | | | | | | FIRE RATING LABEL | HARDWARE | | | NOTES |
| | SIZE | | | MATL | GLAZING | LOUVER | | MATL | EL | DETAIL | | | | | SET NO | KEYSIDE | RM NO | |
	WD	HGT	THK			WD	HGT			HEAD	JAMB	SILL						
001	3'-0"	7'-0"	1"	WOOD	Y	0"	0"	WOOD	--	--	--	--	--	--	--	--	--
002	2'-6"	7'-0"	2"	WOOD	--	0"	0"	WOOD	--	--	--	--	--	--	--	--	--
003	2'-4"	7'-0"	2"	WOOD	--	0"	0"	WOOD	--	--	--	--	--	--	--	--	--
004	2'-8"	6'-8"	1"	WOOD	--	0"	0"	WOOD	--	--	--	--	--	--	--	--	--
005	2'-8"	7'-0"	2"	WOOD	--	0"	0"	WOOD	--	--	--	--	--	--	--	--	--
006	3'-0"	7'-0"	1"	WOOD	--	0"	0"	WOOD	--	--	--	--	--	--	--	--	--
007	3'-0"	7'-0"	1"	WOOD	--	0"	0"	WOOD	--	--	--	--	--	--	--	--	--
008	9'-0"	8'-0"	1 3/4"	WOOD	--	0"	0"	WOOD	--	--	--	--	--	--	--	--	--
009	9'-0"	8'-0"	1 3/4"	WOOD	--	0"	0"	WOOD	--	--	--	--	--	--	--	--	--

WINDOW SCHEDULE

| MARK | SIZE | | TYPE | MATERIAL | NOTES |
	WIDTH	HEIGHT			
1	5'-8"	1'-4"	TRANSOM PICTURE	WOOD	--
10	3'-0"	5'-0"	SINGLE-HUNG	WOOD	--
11	3'-0"	5'-0"	SINGLE-HUNG	WOOD	--
12	3'-0"	5'-0"	SINGLE-HUNG	WOOD	--
13	3'-0"	6'-6"	HALF ROUND PICTURE	WOOD	--
14	3'-0"	6'-6"	HALF ROUND PICTURE	WOOD	--
15	3'-0"	6'-6"	HALF ROUND PICTURE	WOOD	--
2	3'-0"	5'-0"	SINGLE-HUNG	WOOD	--
3	3'-0"	5'-0"	SINGLE-HUNG	WOOD	--
4	3'-0"	5'-0"	SINGLE-HUNG	WOOD	--
5	3'-0"	5'-0"	SINGLE-HUNG	WOOD	--
6	3'-0"	3'-6"	SINGLE-HUNG	WOOD	--
7	3'-0"	5'-0"	SINGLE-HUNG	WOOD	--
8	3'-0"	5'-0"	SINGLE-HUNG	WOOD	--
9	3'-0"	5'-0"	SINGLE-HUNG	WOOD	--

ROOM SCHEDULE

NO	NAME	LENGTH	WIDTH	HEIGHT	AREA
001	ROOM	23'-3"	27'-3"	9'-0"	633.56 SF
002	ROOM	4'-9"	17'-5"	9'-0"	15.61 SF
003	ROOM	12'-11"	14'-9"	9'-0"	174.15 SF
004	ROOM	12'-6"	10'-3"	9'-0"	105.74 SF
005	ROOM	9'-1"	15'-5"	9'-0"	79.43 SF
006	ROOM	8'-10"	5'-6"	9'-0"	48.22 SF
007	ROOM	13'-3"	19'-9"	9'-0"	257.63 SF
008	ROOM	8'-2"	14'-3"	9'-0"	89.34 SF
009	ROOM	8'-11"	9'-3"	9'-0"	75.28 SF
010	ROOM	7'-5"	7'-9"	9'-0"	26.62 SF
011	ROOM	4'-11"	7'-0"	9'-0"	22.52 SF
					1528.11 SF

FIGURE 19.1　Examples of predefined schedules.

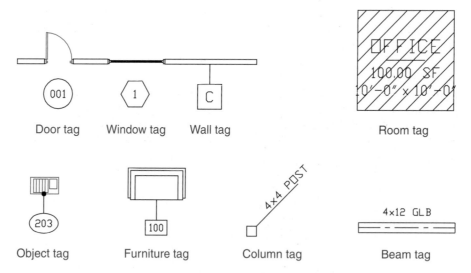

Door tag　　Window tag　　Wall tag　　　　　　　Room tag

Object tag　　Furniture tag　　Column tag　　Beam tag

FIGURE 19.2　Examples of schedule tags included with Architectural Desktop.

and Wall Tags. These schedule tags can be found in the **Documentation** pull-down menu or in the **Schedule - Imperial** or **Schedule - Metric** toolbars.

Adding Door and Window Tags

The Door & Window Tags section includes one door tag and one window tag. These tags must be attached to the type of AEC object on which they are reporting.

Use one of the following options to access the Door & Window Tag section:

✓ Select the **Documentation** pull-down menu, pick the **Schedule Tags** cascade menu, then select **D̲oor & Window Tags....**

✓ Pick the **Door & Window** tags button from the Schedule - Imperial (or Metric) toolbar.

✓ Right-click, and select **Doc̲umentation, Schedule Tag̲s, D̲oor & Window Tags...** from the cursor menu.

✓ Type **AECDCSETIMPDOORWINDOWTAGS** at the Command: prompt.

The DesignCenter is displayed, opened to the Door & Window Tags section, as shown in Figure 19.3. The Door & Windows Tags folder is displayed along the left side of the window in a tree view, with previews of the tags on the right. To insert a door or window tag into the drawing, drag the symbol from the DesignCenter into the drawing screen, or double-click the symbol. In either case, the following prompt is displayed at the Command prompt:

Select object to tag [Symbol/Leader/Dimstyle]:

Depending on whether you have selected a door tag or a window tag, pick a door or window object to link the tag with the AEC object. After selecting the door or window, pick a location for the tag block. The drawing scale selected in the Drawing Setup dialog box determines the size of the tag block.

After a location for the tag is picked, the **Edit Schedule Data** dialog box is displayed (see Figure 19.4). Depending on the object (door or window), this dialog box contains property set definitions, containing individual properties. Any property values can be edited for the door or window selected by picking the value. When you have finished editing the properties press the OK button to close the dialog box. Property sets and properties are discussed in greater detail later in this chapter.

FIGURE 19.3 The Door & Window Tags command displays the DesignCenter window, opened to the Door & Window Tag blocks.

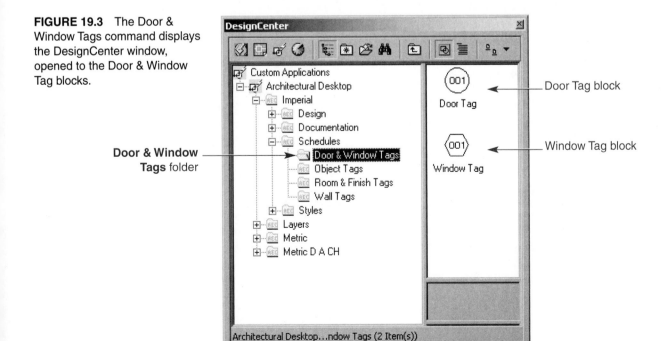

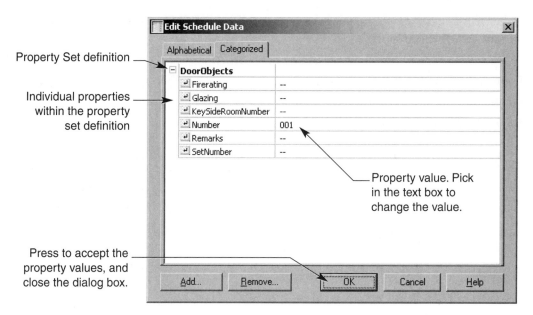

FIGURE 19.4   The Edit Schedule Data dialog box contains property information that is assigned to the door. This information can then be assembled into a schedule.

Once the first door or window tag is inserted, successive tags are sequentially numbered. In the Edit Schedule Data dialog box, the Number property can be used to adjust the door or window number as shown in Figure 19.4. If the number value is changed, tag numbers following are sequenced from the new number. For example, if the door number is changed to 101, then the next tags are numbered 102, 103, and so on.

Exercise 19.1

1. Open EX 16-9-Flr-Pln.dwg.
2. Add door tags to all the doors in the drawing.
3. Add window tags to all the windows in the drawing, including the window assemblies.
4. Save the drawing as EX19-1-Flr-Pln.dwg.

Field Notes

The tags included in Architectural Desktop are multi-view blocks. They can be displayed in the top view using the Work, Plot, and Reflected display configurations. These blocks are not displayed in an isometric or model view.

Adding Object Tags

Object tags are tags that are attached to standard AutoCAD block references, multi-view blocks, or structural members. These tags are typically used in schedules that must report on the types of equipment, building material, or furniture counts, and in schedules that reflect the total number, type, and cost of a series of objects.

Use one of the following options to add object tags:

✓ Select the **Documentation** pull-down menu, access the **Schedule Tags** cascade menu, then select the **Object Tags...** command.

✓ Pick the **Object Tags** button in the Schedule - Imperial (or Metric) toolbar.

✓ Right-click, and pick **Documentation, Schedule Tags, Object Tags...** from the cursor menu.

✓ Type **AECDCSETIMPOBJECTTAGS** at the Command prompt.

The DesignCenter is displayed, opened to the Object Tags section folder, shown in Figure 19.5.

The Object Tags folder includes a number of tags that can be attached to different objects in the drawing. Drag and drop the desired symbol from the right minipreview pane into the drawing, or double-click to insert the tag. Once a tag has been selected, the following prompt is displayed at the Command line:

Select object to tag [Symbol/Leader/Dimstyle]:

Select the block, multi-view block, or structural member to which the tag is to be attached. The following prompt is displayed next:

Specify location of tag <Centered>:

Pick a location in the drawing where the tag is to be inserted. After the tag location is selected, the Edit Schedule Data dialog box is displayed, allowing you to specify any data for the properties in the property set definition that is being attached to the object. Press the OK button to complete the process. You can continue to select blocks, multi-view blocks, or structural members to which to attach object tags. When you have finished press the [Enter] key to complete the process.

FIGURE 19.5 The Object Tags folder includes several object tags that can be attached to blocks, multi-view blocks, and structural members.

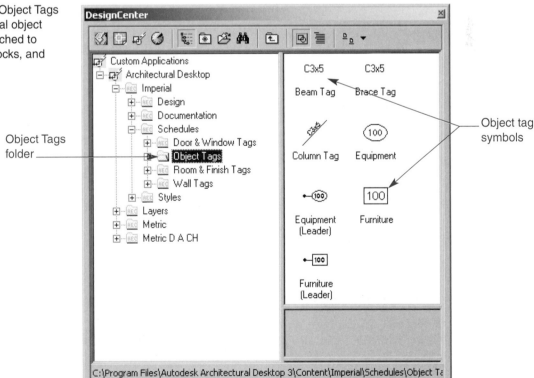

Exercise 19.2

1. Open EX 19-1-Flr-Pln.dwg.
2. Add object tags to the kitchen appliances. See the figure.
3. Use the Edit Schedule Data dialog box to number the tags.
4. Save the drawing as EX19-2-Flr-Pln.dwg.

Object tags

Adding Room & Finish Tags

Room and finish tags are tags added to a drawing to label a room name or number. Information can be attached to the room to indicate the square footage, length, and width and the finish materials. These tags can be attached only to an AEC object known as a *space*. *Spaces* are AEC objects that include a boundary defining the room size and shape, and a hatch pattern within the boundary (see Figure 19.6). Selecting the space and picking an insertion point location adds the room tags. Spaces are discussed in Chapter 23.

After spaces have been added to the drawing, use one of the following options to add the room and finish tags:

✓ Select the **Documentation** pull-down menu, pick the **Schedule Tags** cascade menu, then select **Room & Finish Tags....**

✓ Pick the **Room & Finish Tags** button from the Schedule - Imperial (or Metric) toolbar.

✓ Right-click, and pick **Documentation, Schedule Tags, Room & Finish Tags...** from the cursor menu.

✓ Type **AECDCSETIMPROOMANDFINISHTAGS** at the Command prompt.

The DesignCenter is displayed, opened to the Room & Finish Tags folder (see Figure 19.7). Three tags are available for use. Drag and drop the tags into the drawing, and select a space object, or double-click the minipreview icon to add the tags to a space.

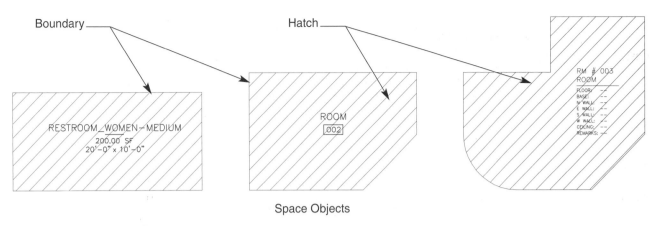

FIGURE 19.6 Spaces are AEC objects that include a boundary and a hatch pattern. Room tags must be attached to this type of object.

FIGURE 19.7 The Room & Finish Tags folder includes three room tags that can be attached to space objects.

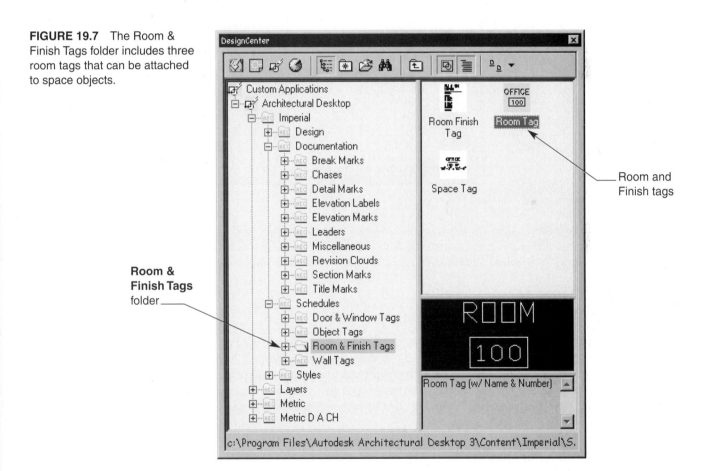

Adding Wall Tags

Wall tags are used to mark the different types of walls in the drawing. These tags must be attached to a wall object. Use one of the following options to access the wall tags:

✓ Pick the **Documentation** pull-down menu, select the **Schedule Tags** cascade menu, then pick **Wall Tags....**

✓ Pick on the **Wall Tags...** button from the Schedule - Imperial (or Metric) toolbar.

✓ Right-click, and pick **Documentation, Schedule Tags, Wall Tags...** from the cursor menu.

✓ Type **AECDCSETIMPWALLTAGS** at the Command prompt.

FIGURE 19.8 The Wall Tags folder includes two types of wall tags.

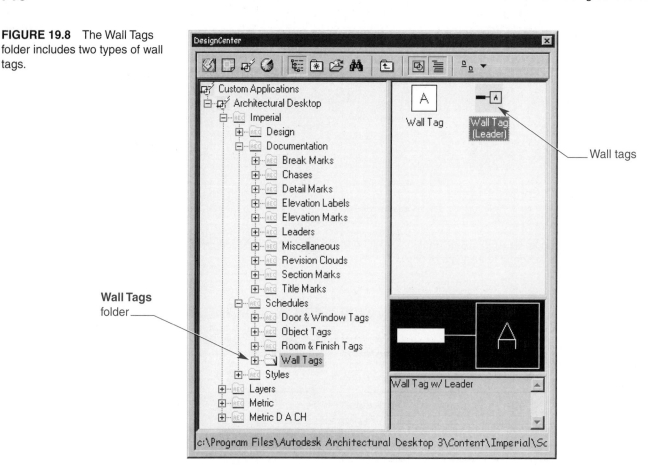

The DesignCenter is displayed, opened to the Wall Tags folder (see Figure 19.8). Two types of tags are available. Drag and drop the tag into the drawing, or double-click the block icon to insert the tag. Select a wall to which to attach the tag, then fill out the schedule information in the Edit Schedule Data dialog box.

Anchoring Tags to AEC Objects

You can copy schedule tags around the drawing so that you do not have to add all the tags into the drawing during the same operation or continue dragging the block icons from the DesignCenter. When a tag is copied in the drawing, its property set information remains linked to the original object to which it was attached. Any changes made to a property value are reflected in the original tag and the copied tag.

In these cases the copied tag must be anchored to the new object using one of the following options:

✓ Select the **Documentation** pull-down menu, pick the **Schedule Tags** cascade menu, and pick **Anchor Tag to Object.**

✓ Right-click, and pick **Documentation, Schedule Tags, Anchor Tag to Object** from the cursor menu.

✓ Type **AECTAGANCHORADD** at the Command: prompt.

✓ Select the copied tag, right-click, pick the **Tag Anchor** cascade menu, and select the **Set Object** command from the shortcut menu.

You are prompted to select the tag, then the new object to which the tag will be anchored. Once the copied tag has been anchored to the new AEC object, the property values within the property set definition are then unique to the object to which the tag is anchored and can be edited in the Edit Schedule Data dialog box.

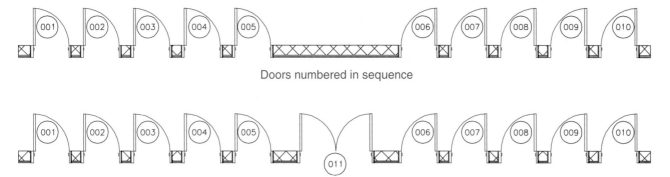

Doors numbered in sequence

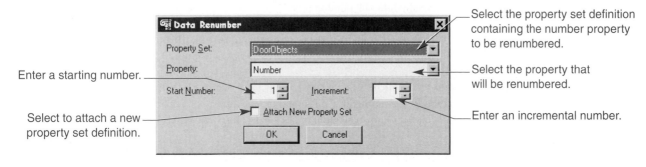

Doors numbered out of sequence

FIGURE 19.9   A new door is added to a wall, causing the door numbering to be out of sequence.

Renumbering Schedule Tags

As the construction documents are created a new door, window, or piece of equipment may be added to the drawing, which can disrupt the numbering scheme for the tags. For example, suppose 10 doors have been added to the drawing with 10 door tags attached. Another door is added between door number 5 and door number 6 with a door tag numbered 11 (see Figure 19.9). This causes the doors to be numbered out of sequence.

Fortunately, a command is available for renumbering schedule tags without having to individually modify each of the tags' schedule data. Use one of the following options to access this command:

- ✓ Pick the **Documentation** pull-down menu, select the **Schedule Data** cascade menu, then pick the **Renumber Data...** command.

- ✓ Pick the **Renumber Data** button from the Schedule - Imperial (or Metric) toolbar.

- ✓ Right-click, and select **Documentation, Schedule Data, Renumber Data...** from the cursor menu.

- ✓ Type **AECPROPERTYRENUMBERDATA** at the Command: prompt.

The **Data Renumber** dialog box, shown in Figure 19.10, is displayed. First, select the property set definition that contains the property that needs to be renumbered. Next, select the property to be renumbered. Enter a starting number to be used to start renumbering the schedule tags, then enter a number for incrementing successive tag numbers. For example, if renumbering odd-numbered doors, you may begin the starting number at 5 and increment the numbers by 2.

When you have finished setting the parameters for the renumbering of the tags, press the OK button to begin renumbering. Select the schedule tag to be numbered with the start number specified. Continue to pick the tags that need to be renumbered. These tags are modified based on the incremental number specified. When you have finished, press the [Enter] key to exit the command. The schedule tags should be renumbered, and the property set definitions attached to the tags should be updated.

Select the property set definition containing the number property to be renumbered.

Enter a starting number.

Select to attach a new property set definition.

Select the property that will be renumbered.

Enter an incremental number.

Data Renumber

Property Set: DoorObjects
Property: Number
Start Number: 1   Increment: 1
☐ Attach New Property Set
OK   Cancel

FIGURE 19.10   The Data Renumber dialog box is used to set up how the schedule tags will be renumbered.

Understanding Property Set Definitions

Property set definitions are automatically assigned to AEC objects when a tag is attached to an object. When a tag is inserted and the object to which it is attached is selected, the **Edit Schedule Data** dialog box appears (see Figure 19.11). The properties available vary depending on the type of tag being inserted and the type of object selected.

Property set definitions are a collection of properties assigned to an object and reported in a schedule. The schedule in turn queries the objects in the drawing, looking for properties and property set definitions to assemble the information into a logical table. The property set definitions available for use are those that can be assigned individually to AEC objects or assigned to an object style.

FIGURE 19.11 The Edit Schedule Data dialog box includes a list of properties available for the type of object selected. This dialog box displays the properties available for door objects.

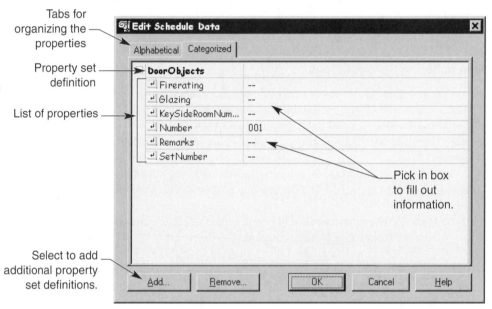

Property Set Definitions by Object

Each tag that is inserted and attached to an object has its own property set definition assigned to the object. This ensures that each AEC object in the drawing contains unique information. For example, when door tags are attached to door objects each door has a unique number so you can keep track of it.

When the Edit Schedule Data dialog box is displayed, the property set definition for that type of object appears. If additional property sets are available, the <u>A</u>dd... button is highlighted, allowing you to attach additional property set definitions to the object (see Figure 19.12). Selecting the <u>A</u>dd... button displays the **Add Property Sets** dialog box, shown in Figure 19.12. A list of available property set definitions that can be assigned to the objects is displayed. Placing a check mark in the box beside the property set name adds the property set definition to the object. Press the OK button to add the property set definition to the AEC object and the Edit Schedule Data dialog box.

If you need to change the property information within a property set definition, or if you need to add additional property set definitions to the object, you need to access the Edit Schedule Data dialog box again. Use one of the following options to access this dialog box after the tag is added:

✓ Select the **Docu<u>m</u>entation** pull-down menu, select the **Schedule <u>D</u>ata** cascade menu, and pick **<u>A</u>ttach/Edit Schedule Data....**

✓ Pick the **Attach/Edit Schedule Data** button in the Schedule - Imperial (or Metric) toolbar.

✓ Right-click, and select **Docu<u>m</u>entation, Schedule <u>D</u>ata, <u>A</u>ttach/Edit Schedule Data...** from the cursor menu.

✓ Type **AECPROPERTYDATAEDIT** at the Command: prompt.

FIGURE 19.12 The Add Property Sets dialog box is displayed when the <u>A</u>dd... button in the Edit Schedule Data dialog box is picked. A check next to the property set adds the definition to the object.

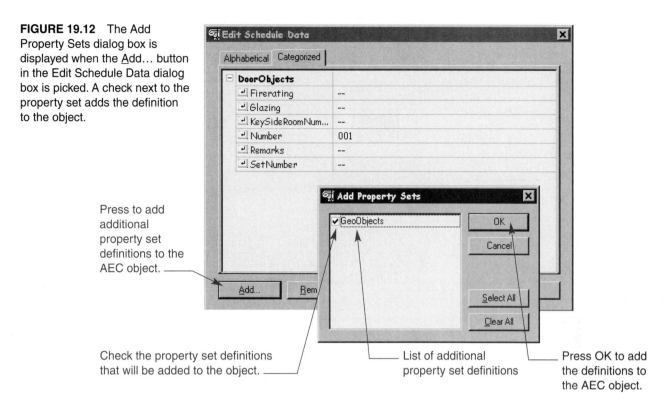

Press to add additional property set definitions to the AEC object. ——

Check the property set definitions that will be added to the object. ——

List of additional property set definitions

Press OK to add the definitions to the AEC object.

You are prompted to select objects. Select any objects with property set definitions that need to be edited or added. When you have finished selecting objects to modify their properties, press the [Enter] key to display the Edit Schedule Data dialog box. Begin editing any properties and attaching new property set definitions to the object(s). You also can enter this dialog box by first selecting an object(s) then right-clicking and selecting Edit Schedule Data... from the shortcut menu. When you have finished editing the properties, press the OK button to exit the dialog box.

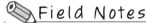 Field Notes

- If several objects are selected that have differing property values, a *VARIES* value is displayed in the text box beside the property value that is different in each object.

- Attach property set definitions individually to objects when the objects need to contain unique information such as an itemized number, separate notes, or a specific description.

Property Set Definitions by Style

The creation of object styles was discussed in previous chapters. In each of the object style dialog boxes the General tab included a Property Sets button (see Figure 19.13). This button allows you to assign property set definitions to the object style. When the Property Sets button is selected in an object style General tab, the Edit Schedule Data dialog box is displayed with the property set definitions assigned to that object style listed.

The object styles included with Architectural Desktop already have property set definitions assigned to them (see Figure 19.14). In many cases, these property set definitions include properties that are reporting on specific parts of the object. For example, the property set definition that is assigned to a window style includes properties that report on the inserted window's width and height.

Property set definitions assigned to a style contain properties that are common to objects of that type. For example, a property set definition assigned to an object style can contain a property for the type of window, width of door, or fire rating of a wall style.

One of the most significant differences between the default property set definitions for individual objects and the default property set definitions assigned to a style is the addition of automatic properties

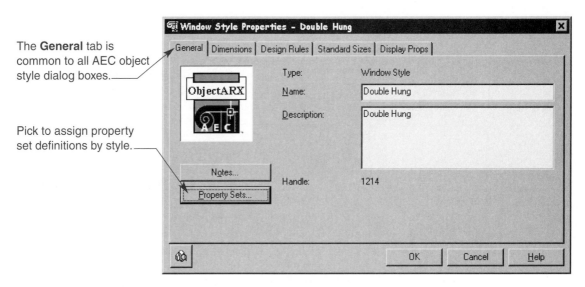

The **General** tab is common to all AEC object style dialog boxes.

Pick to assign property set definitions by style.

FIGURE 19.13   Each AEC object style dialog box includes a Property Sets button found under the General tab that allows you to assign property sets by style.

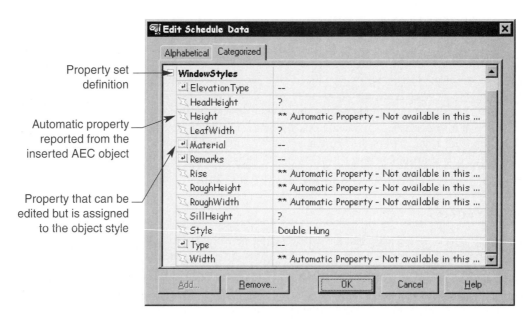

Property set definition

Automatic property reported from the inserted AEC object

Property that can be edited but is assigned to the object style

FIGURE 19.14   Property set definitions assigned to an AEC object style contain general properties for the type of object, such as size, material, and type.

(see Figure 19.14). *Automatic properties* are represented by a lightning bolt. These are properties that are automatically filled out in a schedule based on object information in the drawing. For example, the width and height of a door is an automatic value, as a single door style can be inserted as a 3′-0″ × 7′-0″ door or a 2′-4″ × 6′-8″ door. Although the sizes vary when an individual door is created in the drawing, the automatic property looks to the inserted door size and reports that in the schedule.

Automatic properties can be assigned to a property set definition that is attached to individual objects. In most default property set definitions the automatic properties are found in the property set definitions assigned to an object style. Custom property set definitions are discussed in Chapter 20.

Like property set definitions assigned to individual objects, the property set definitions assigned to a style include an Add button. If additional property set definitions are available for the style being edited the Add… button is available. If a property is edited for a style, such as the material property, that property is assigned to each object that uses that style.

Keep in mind that two separate property set definitions are assigned to objects: individual *object* property set definitions and *object style* property set definitions. If the individual property set definitions need to be edited, use the Edit Schedule Data command. If the style property set definition needs to be modified, use the object style dialog box.

Accessing Additional Property Set Definitions

Initially, when an Architectural Desktop template is used, a few property set definitions are available. As new objects are drawn or schedule tags are added, additional property set definitions are also added to the drawing. Architectural Desktop also includes additional property set definitions that can be added to the drawing.

Use one of the following options to add more property set definitions:

✓ Select the **Documentation** pull-down menu, pick the **Schedule Data** cascade menu, and select the **Property Set Definitions...** command.

✓ Right-click, and pick **Documentation, Schedule Data, Property Set Definitions...** from the cursor menu.

✓ Type **AECPROPERTYSETDEFINE** at the Command: prompt.

The Style Manager dialog box opened to the Property Set Definitions section, as shown in Figure 19.15, is displayed. Pick the **Open Drawing** button to open the PropertySetDefs drawing. In the Open Drawing dialog box, browse to the \\Autodesk Architectural Desktop 3\Content\Imperial\Schedules folder. Select the PropertySetDefs.dwg file, then press the Open button. This drawing contains additional property set definitions that can be added to the current drawing (see Figure 19.16).

A list of property set definitions is displayed below the Property Set Definitions section icon. Property set definitions can be dragged from the list in the left tree view to the current drawing folder in the tree-view list. Selecting the Property Set Definitions section icon lists all the property set definitions in the right preview pane. The definitions can also be dragged from the right pane into the current drawing folder in the left tree pane.

Selecting a definition from the left tree view provides a list of detailed information about the property set definition in the right pane (see Figure 19.16). The details of a property set indicate where the property set can be applied. For example, Figure 19.16 shows that the selected property set definition applies to Styles and Definitions, and the AEC object is the door object. In other words, this property set definition is applied to door styles. Other property set definitions can be applied to entities, which are individual objects.

The property set definition naming conventions also indicate where the property set definitions are applied. If the property set definition name is an object name only or the name of an AEC object followed by the word *Objects,* then the property set definition applies to individual objects. If the naming convention begins with the type of object followed by the word *Styles,* the property set definition can be applied only to object styles (see Figure 19.17).

FIGURE 19.15 The Style Manager dialog box is displayed when the Property Set Definitions command is entered.

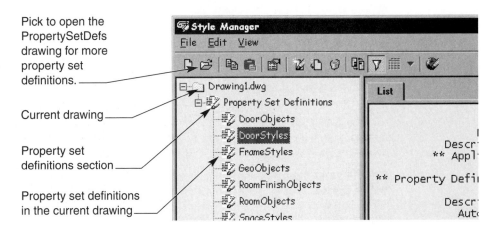

Pick to open the PropertySetDefs drawing for more property set definitions.

Current drawing

Property set definitions section

Property set definitions in the current drawing

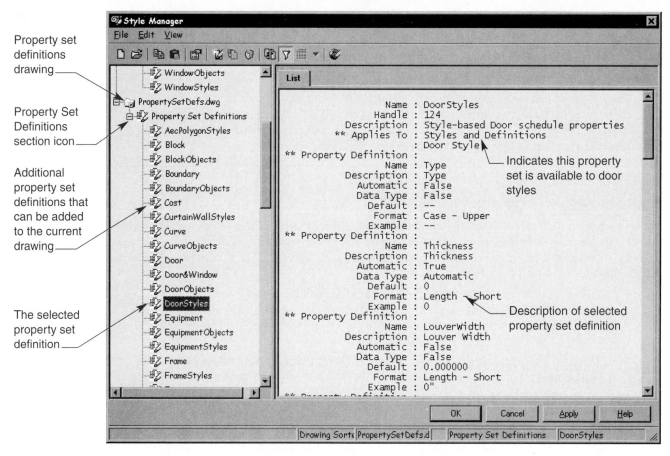

Property set definitions drawing

Property Set Definitions section icon

Additional property set definitions that can be added to the current drawing

The selected property set definition

Indicates this property set is available to door styles

Description of selected property set definition

FIGURE 19.16 The Style Manager can be used to open the PropertySetDefs.dwg, which contains additional property set definitions.

The selected property set definition applies to all individual objects.

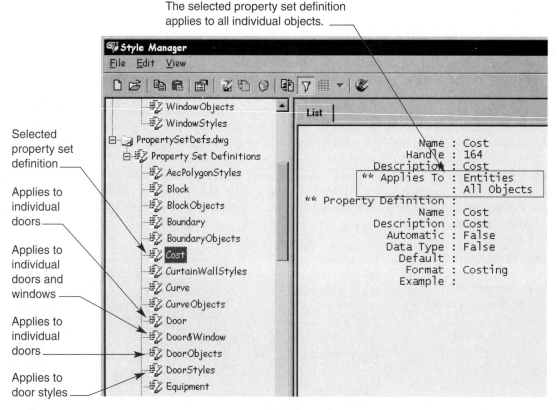

Selected property set definition

Applies to individual doors

Applies to individual doors and windows

Applies to individual doors

Applies to door styles

FIGURE 19.17 The property set definition naming convention indicates where the property set can be applied.

Exercise 19.3

1. Open EX19-2-Flr-Pln.dwg.
2. Access the Property Set Definitions command to access the Style Manager.
3. Use the Open Drawing button to open the PropertySetDefs.dwg.
4. Drag and drop the Cost property set definition into the current drawing folder within Style Manager.
5. Use the Edit Schedule Data command to add the cost property set definition to the appliances in the kitchen.
6. Save the drawing as EX19-3-Flr-Pln.dwg.

Browsing through Property Data

Property set definitions can be browsed through and properties modified quickly through the **Browse Property Data** dialog box, shown in Figure 19.18. This dialog box is used to browse through property data on a grand scale and edit any properties available, whether they are assigned to individual objects or to styles.

Use one of the following options to open the **Browse Property Data** dialog box:

✓ Select the **Documentation** pull-down menu, pick the **Schedule Data** cascade menu, then select **Browse Data....**

✓ Right-click, and pick **Documentation, Schedule Data, Browse Data...** from the cursor menu.

✓ Type **AECPROPERTYDATABROWSE** at the Command: prompt.

The dialog box is divided into two panes. The left pane lists the property set definitions available in the current drawing. Pressing the + sign next to a property set expands a tree list of data. If a property set that is attached to individual objects is expanded, a list of objects that have the property set attached is listed below the name of the property set. If a property set that is attached to an object style is expanded, the styles the property set is attached to are listed.

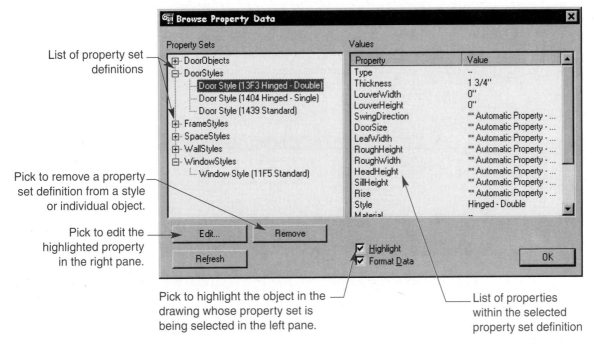

FIGURE 19.18 The Browse Property Data dialog box is used to browse through property values and edit them if necessary.

Selecting an individual object or a style from the expanded list in the left pane displays the list of properties in the property set in the right pane. If a property value needs to be modified, select the property from the right pane then pick the Edit... button. The Edit Schedule Data dialog box is displayed, where the property can then be modified.

Property set definitions also can be removed from individual objects or from styles by highlighting the property set in the left pane, then picking on the Remove button. The property set definition and any property values are removed from the object or from the style an object is using.

As property data are modified, they may not appear updated in the right pane. Use the **Refresh** button to refresh the data in the right pane. There are also two check boxes within this dialog box. If the **Highlight** check box is selected, picking on top of a property set in the left pane highlights any objects in the drawing with that property set definition attached. This is a good way to see quickly whether the objects in the drawing have the proper property set definitions attached. The **Format Data** check box displays property values as formatted values or as raw-data values. For example, the width of a window may be displayed in a formatted value of 3′-0″ or an unformatted value of 36.000. When you have finished browsing through the data, press the OK button to close the dialog box.

Creating a Schedule

Once schedule tags have been added and property values edited, a schedule can be created. The Architectural Desktop templates include a few preconfigured schedules such as a door schedule, a window schedule, and a room finish schedule. Additional schedules are also included for use or can be customized.

Use one of the following options to generate a schedule:

✓ Select the **Documentation** pull-down menu, pick the **Schedule Tables** cascade menu, then select **Add Schedule Table....**

✓ Pick the **Add Schedule Table** button in the Schedule - Imperial (or Metric) toolbar.

✓ Right-click, and pick **Documentation, Schedule Tables, Add Schedule Table...** from the cursor menu.

✓ Type **AECTABLEADD** at the Command: prompt.

The **Add Schedule Table** dialog box appears (see Figure 19.19). This dialog box is used to set the parameters for the type of schedule to be created. The following features are available in the dialog box:

■ **Schedule Table Style:** This drop-down list includes all the schedule styles available in the drawing. The templates include a few styles, and additional styles are available. Accessing the additional styles is covered later in this chapter. Select the schedule style that you want to generate.

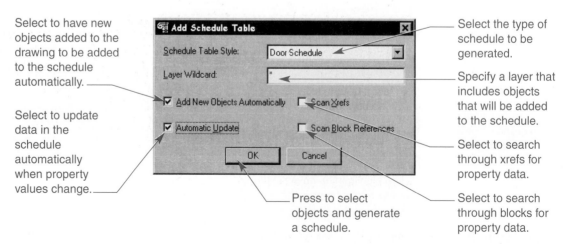

FIGURE 19.19   The Add Schedule Table dialog box is used to establish parameters for the new schedule.

- **Layer Wildcard:** This text box initially includes an asterisk [*] in the field, which indicates that all objects on any layer are looked at for information used by the schedule style. The asterisk represents any character in a layer name. If you want the schedule to look only through objects that are on specific layers, then enter the layer names separated by commas.

Field Notes

The schedule style selected is set up to look for specific types of objects. For example, the Door schedule looks only for doors in the drawing. If you are generating door schedules for each floor of a building from a series of reference drawings, you may want to use a layer filter that filters for doors on the specific layer on which a reference drawing was inserted.

- **Add New Objects Automatically** When selected, this check box adds new information to the schedule automatically as new objects are added to the drawing. For example, a window schedule may include 15 different windows. As new windows are added to the drawing, they are automatically added to the schedule, keeping the schedule up-to-date.

- **Automatic Update** This check box automatically updates the schedule as property values are edited. If the size or style of door is changed, the schedule is updated to reflect the latest changes.

- **Scan Xrefs** This check box allows you to scan through reference drawings for objects in order to generate the schedule. Schedules can be placed in the same drawing as the original objects, or a drawing containing the objects for the schedule can be referenced into a separate drawing, and a schedule generated from the xref can be created.

- **Scan Block References** When selected, this check box allows the schedule to scan through blocks for property values that are added into the schedule. When you have finished specifying the parameters for the schedule, press the OK button.

The Add Schedule Table dialog box closes and you are prompted to select objects. Select the objects to be used in the schedule. The schedule style looks for a specific type of object. For example, door schedules look for doors, and window schedules look for windows. If you accidentally select a wrong object, it is ignored. Additionally, the layer wildcard filters for the specific type of objects that are on a specified layer.

When you have finished selecting the objects to be included in the schedule, press the [Enter] key. An outline of the schedule is attached to the crosshairs. Pick a location for the upper left corner of the schedule. The schedule is inserted at that location. The next prompt ask you to specify the lower right corner of the schedule or to press Return. If a lower right corner of the schedule is selected, the schedule will not be scaled correctly in the drawing. Pressing the [Enter] key at this prompt scales the schedule according to the drawing scale specified in the Drawing Setup dialog box. The following prompt is used to place a schedule so that it is scaled correctly in the drawing:

Upper left corner of table: *(pick a location for the upper left corner of the schedule)*
Lower right corner *(or RETURN):* ↵

Architectural Desktop then inserts a schedule of information from the property values within the property set definitions that are attached to the objects in the drawing (see Figure 19.20).

FIGURE 19.20 A window schedule has been generated from the windows in a drawing. The property values within the windows property set definition are displayed in the schedule.

	SIZE				
MARK	WIDTH	HEIGHT	TYPE	MATERIAL	NOTES
1	5'-8"	1'-4"	TRANSOM PICTURE	WOOD	--
10	3'-0"	5'-0"	SINGLE-HUNG	WOOD	--
11	3'-0"	5'-0"	SINGLE-HUNG	WOOD	--
12	3'-0"	5'-0"	SINGLE-HUNG	WOOD	--
13	3'-0"	6'-6"	HALF ROUND PICTURE	WOOD	--
14	3'-0"	6'-6"	HALF ROUND PICTURE	WOOD	--
15	3'-0"	6'-6"	HALF ROUND PICTURE	WOOD	--
2	3'-0"	5'-0"	SINGLE-HUNG	WOOD	--
3	3'-0"	5'-0"	SINGLE-HUNG	WOOD	--
4	3'-0"	5'-0"	SINGLE-HUNG	WOOD	--
5	3'-0"	5'-0"	SINGLE-HUNG	WOOD	--
6	3'-0"	3'-6"	SINGLE-HUNG	WOOD	--
7	3'-0"	5'-0"	SINGLE-HUNG	WOOD	--
8	3'-0"	5'-0"	SINGLE-HUNG	WOOD	--
9	3'-0"	5'-0"	SINGLE-HUNG	WOOD	--

Title: WINDOW SCHEDULE

Exercise 19.4

1. Open EX19-3-Flr-Pln.dwg.
2. Add a door schedule to the drawing. Check the Add New Objects Automatically and the Automatic Update check boxes.
3. Insert the door schedule to the side of the drawing.
4. Add a window schedule to the drawing.
5. Insert the window schedule to the side of the drawing.
6. Save the drawing as EX19-4-Flr-Pln.dwg.

Updating Schedules

When you start to create a schedule, you have the option of manually updating the schedule as you make changes. If this option is not selected, any changes to the objects being reported on do not accurately reflect information in the schedule. The schedule is no longer current if an object is deleted, changes size, or its property values are changed.

To have the most up-to-date schedule in the drawing, use the **Update Schedule Table** command. Use one of the following options to access this command:

✓ Pick the **Documentation** pull-down menu, select the **Schedule Tables** cascade menu, then pick **Update Schedule Table...** from the list.

 ✓ Pick the **Update Schedule Table** button from the Schedule - Imperial (or Metric) toolbar.

✓ Right-click, and select **Documentation, Schedule Tables, Update Schedule Table...** from the cursor menu.

✓ Type **AECTABLEUPDATENOW** at the Command: prompt.

✓ Select the schedule, right-click, then pick **Update Table** from the shortcut menu.

FIGURE 19.21   In most situations, if a schedule is out of date, a diagonal line will run through the schedule, indicating that it needs to be updated.

WINDOW SCHEDULE					
MARK	SIZE		TYPE	MATERIAL	NOTES
	WIDTH	HEIGHT			
1	3'–0"	5'–0"	– –	– –	– –
2	8'–0"	7'–0"	– –	– –	– –
3	5'–0"	5'–0"	– –	– –	– –
4	8'–0"	7'–0"	– –	– –	– –
5	8'–0"	7'–0"	– –	– –	– –
6	8'–0"	7'–0"	– –	– –	– –
7	3'–0"	5'–0"	– –	– –	– –

As soon as this command is entered, the schedule is automatically updated. In most cases, if a schedule is not up-to-date, a line runs diagonally through it, indicating the schedule is not up-to-date (see Figure 19.21). If the schedule is supposed to always update automatically, after it is created, select the schedule, right-click, and pick Automatic Update from the shortcut menu to change the schedule from being nonautomatically updateable to being automatically updateable.

Field Notes

The only sure way to be certain the schedule contains up-to-date information is to close the drawing, then reopen it. Some changes made to drawings, such as through reference editing, are reflected only when the drawing is reopened.

Modifying the Properties of a Schedule Table

Each schedule remembers the properties that were set initially when the schedule was created. The schedule remembers the layer it is looking through for objects that are included in the schedule if it is automatically updateable, whether it is looking through xrefs, and its drawing scale.

You can modify these values at any time after the schedule is generated by using one of the following options:

✓ Select the schedule table, right-click, and select **Table Properties...** from the shortcut menu.

✓ Type **TABLEPROPS** at the Command: prompt.

The **Schedule Table Properties** dialog box, shown in Figure 19.22, is displayed. The following four tabs are included in this table:

- **General** This tab is used to provide a description for the schedule, to add notes to the schedule, and to assign property set definitions to the schedule.

- **Style** This tab displays the schedule table style currently being used by the selected schedule. If a different schedule is desired, it can be selected from the list.

- **Settings** This tab contains the settings for the selected schedule. This tab contains the following four options:

  - **Selection** This area defines how objects are selected to become a part of the schedule according to the settings specified at the time the schedule was created. These values can be changed by adding or removing the check mark from the appropriate check box.

Automatically updates the
schedule as objects are modified
or property values are changed ⎯⎯

Determines the scale of
the schedule table based
on the drawing scale⎯⎯

Adds new objects
automatically to the
schedule ⎯⎯

Looks through xrefs
for objects to add to
the schedule ⎯⎯

Looks through blocks
for objects to add to
the schedule ⎯⎯

Indicates the layers
being looked through for
objects that will be
added to the schedule⎯⎯

Schedule Table Properties

General | Style | Settings | Location

┌─ Selection ─────────────────────┐   ┌─ Update ──────┐
│ ☐ Add New Objects Automatically │   │ ☐ Automatic   │
│ ☐ Scan Xrefs   │   └───────────────┘
│ ☐ Scan Block References   │   ┌─ Scale ───────┐
│   │   │ Scale: 96.0000│
└──────────────────────────────────┘   └───────────────┘

┌─ Filter ──┐
│ Layer Wildcard: [*]   │
└──┘

[OK]   [Cancel]   [Help]

FIGURE 19.22   The Schedule Table Properties dialog box contains parameters set up for
the selected schedule.

- **Update**   This area controls whether the schedule updates automatically. Placing a
  check in the box causes the schedule to always update automatically.

- **Scale**   This text box includes the drawing scale–scale factor. Recall that the drawing
  scale is determined by the scale chosen in the Drawing Setup dialog box.
  Changing the value causes the schedule to change size.

- **Filter**   This text box includes the layer filter used to search for objects to be included
  in the schedule. Specify the layer names and separate them with commas to
  apply multiple layer filters.

- **Location**   This tab indicates the schedule location in the drawing.

Exporting a Schedule Table

Some offices prefer to create schedules with other software, such as Microsoft Excel. Information can be
generated from the property values attached to an object in Architectural Desktop then sent to an external
file, or to an existing schedule in the drawing can be exported to an external file.

To create an external file containing property values from an object use the **Export Schedule Table**
command. Use one of the following options to access this command:

✓ Select the **Documentation** pull-down menu, pick the **Schedule Tables** cascade menu, then
   select the **Export Schedule Table...** command.

✓ Pick the **Export Schedule Table** button in the Schedule - Imperial (or Metric) toolbar.

✓ Right-click, and pick **Documentation, Schedule Tables, Export Schedule Table...** from
   the cursor menu.

✓ Type **AECTABLEEXPORT** at the Command: prompt.

The Export Schedule Table dialog box, shown in Figure 19.23, is displayed. This dialog box has the
following two areas:

- **Output**   This area includes a drop-down list and a text box. The Save As Type drop-down list
  allows you to pick the type of external file to be created. Options include Microsoft

Pick to browse the hard
drive for a location for
the new file.

Select to use an
existing schedule
in the drawing.

Select to scan through
reference files.

Select to scan
through blocks.

Select the type of file to which
to export the property values.

Specify a name for
the external file.

If the **Use Existing Table** check
box is cleared, select the type of
schedule to use to obtain the
proper property values.

If the **Use Existing Table** check
box is cleared, enter layer filter to
search for property information.

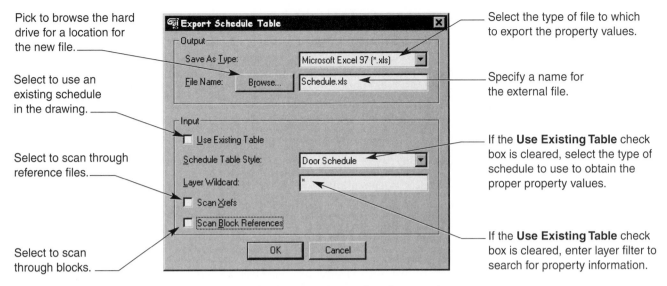

FIGURE 19.23   The Export Schedule Table dialog box contains settings for exporting
a schedule to an external file.

Excel 97 (*.xls), Text (Tab delimited)(*.txt), and CSV (Comma delimited)(*.csv).
An example of the results for each type of file are displayed in Figure 19.24.

The File Name text box is used to specify a name for the new file. Use the
Browse... button to browse the folders for a location for the new file.

- **Input**   This area includes several options for control. When the Use Existing Table check
box is selected, an existing schedule in the drawing must be selected to be
processed for exportation. Unchecking this check box allows you to select a
schedule table style for processing, a layer wildcard, and options to scan xrefs or
blocks.

When you have finished selecting the parameters for exporting the schedule information, the objects
are scanned, and an external file containing all the property values is created.

Field Notes

Information exported to an Excel file can be brought back into your drawing by using Object
Linking and Embedding (OLE). See the *Creating Compound Documents with OLE* section in
the AutoCAD User's Guide, in the Architectural Desktop Help.

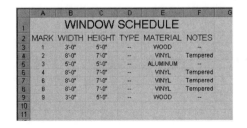

A	**B**	**C**
Excel file	Text file	CSV file

FIGURE 19.24   (A) A Microsoft Excel file generated from property values. (B) The same
information in a Text file. (C) The same information in a CSV file.

Accessing Additional Schedule Table Styles

When the Architectural Desktop templates are initially used, four schedule table styles are available. More table styles are included with Architectural Desktop and can be imported into the current drawing. The **Schedule Table Styles** command is used to import these tables. Use one of the following options to access this command:

✓ Pick the **Documentation** pull-down menu, select the **Schedule Tables** cascade menu, then pick **Schedule Table Styles....**

✓ Pick the **Schedule Table Styles** button in the Schedule - Imperial (or Metric) toolbar.

✓ Right-click, and select **Documentation, Schedule Tables, Schedule Table Styles...** from the cursor menu.

✓ Type **AECTABLESTYLE** at the Command: prompt.

The Style Manager, opened to the Schedule Table Styles section, is displayed as shown in Figure 19.25.

After the Style Manager has been opened, pick the Open Drawing button to open the drawing containing the additional schedule tables. Browse to the \\Autodesk Architectural Desktop 3\Content\Imperial\ Schedules folder. Select the Schedule Tables (Imperial).dwg from the folder and press the Open button to open the drawing. The drawing is opened in Style Manager (see Figure 19.26).

FIGURE 19.25   The Style Manager is used to import additional schedule table styles into the current drawing.

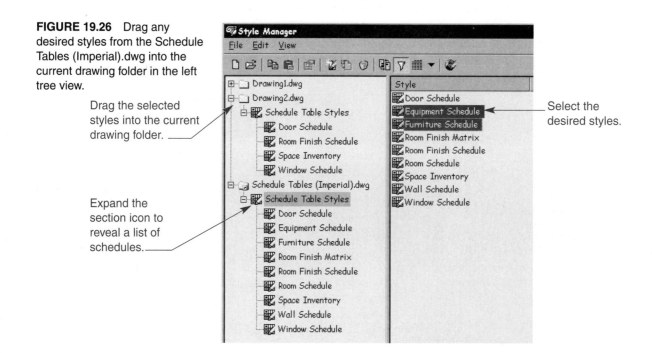

FIGURE 19.26   Drag any desired styles from the Schedule Tables (Imperial).dwg into the current drawing folder in the left tree view.

Expand the Schedule Tables section to list all the schedules available. Select any desired schedules, then drag and drop them onto the current drawing folder to import additional schedules into the drawing. Close the Style Manager dialog box, then add the new schedule to the drawing.

Field Notes

Custom schedules can also be dragged into the current drawing using the same process. The creation of custom schedules is covered in the Chapter 20.

Editing Schedule Information

The information being reported in the schedule table contains property values that have been assigned to the objects the schedule is querying. This information can be edited at any time using the methods described earlier in this chapter or through other Architectural Desktop commands. This session describes additional tools that can be used to modify the information found in the schedule.

Editing a Table Cell

Two types of property values are available within a property set definition: *automatic properties* and *user-defined properties*. Within any given schedule, the information present may be data that are being retrieved directly from the actual object, such as width, height, or style, or data that are provided by the user, such as material and remarks.

Information within a cell of the schedule that is completed by the user and is not automatic data can be edited from the schedule. The command to do this is the **Edit Table Cell** command. Use one of the following options to access this command:

✓ Pick the **Documentation** pull-down menu, select the **Schedule Tables** cascade menu, then pick the **Edit Table Cell** command.

✓ Right-click, and select **Documentation, Schedule Tables, Edit Table Cell** from the cursor menu.

✓ Type **AECTABLECELLEDIT** at the Command: prompt.

✓ Select the schedule, right-click, and pick **Edit Table Cell...** from the shortcut menu.

The following prompt is issued at the Command: line:

Select schedule table item *(or the border for all items)*:

From this prompt, select text within a cell of the schedule that is to be edited. The **Edit Schedule Property** dialog box, shown in Figure 19.27, is displayed. This dialog box displays the name of the property set to which the selected text belongs, the property name being modified, the Value text box, and the Formatted Value. Enter a new value in the Value text, and press the OK button to change the value in the table cell.

FIGURE 19.27 The Edit Schedule Property dialog box allows you to change the value of a single cell in the schedule table.

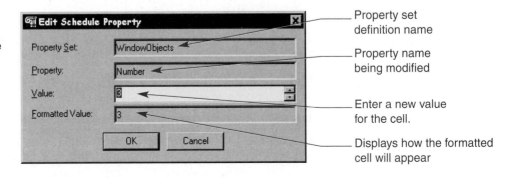

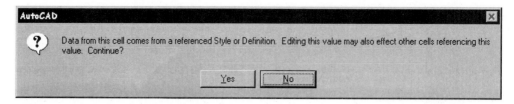

FIGURE 19.28   When a value is selected in a schedule a warning message is displayed indicating that changing the values in the selected column will globally change the values for other objects using the same property set.

If a column heading or border edge of a value is selected, the **Edit Schedule Data** dialog box is displayed, allowing you to modify the cell information. In addition to the Edit Schedule Property dialog box, some warning messages may be displayed. A warning is displayed if a cell value is selected that is a property assigned to the object's style (see Figure 19.28). Pressing <u>Y</u>es closes this dialog box and displays the Edit Schedule Property dialog box. Changing the property value globally changes the values for other objects of the same style.

If an automatic value is selected in the table, the entire line of information for that object is highlighted. The actual item in the drawing also is highlighted, allowing you to see the object in question so that it can be modified to change the automatic values. When you have finished editing the cells of the schedule press the [Enter] key to exit the command.

Adding Objects to and Removing Objects from a Schedule

When a new schedule is added to the drawing, one of the check box options is **Add New Objects Automatically**. If this option has been selected, new objects being reported on are automatically added to the schedule. As objects are removed from the drawing the schedule is also automatically updated, and the appropriate item is removed from the schedule.

If this option was not selected when the schedule was added, new objects can be added or removed from the schedule manually. Use one of the following options to add new objects in the drawing to an existing schedule:

✓ Select the **Doc<u>u</u>mentation** pull-down menu, pick the **Schedule <u>T</u>ables** cascade menu, then select **Add <u>T</u>able Selection.**

✓ Right-click, and select **Doc<u>u</u>mentation, Schedule <u>T</u>ables, Add <u>T</u>able Selection** from the cursor menu.

✓ Type **TABLESELECTIONADD** at the Command: prompt.

✓ Select the schedule, right-click, and select the **Selection** cascade menu, then **Add** from the shortcut menu.

Select the new objects that need to be added to the schedule. When you have finished, press the [Enter] key to end the command and add the new objects to the schedule.

Objects can also be removed from a table schedule—for example, when you want to separate information into a different table rather than keeping it all in one schedule. Use one of the following options to remove objects from a schedule:

✓ Select the **Doc<u>u</u>mentation** pull-down menu, pick the **Schedule <u>T</u>ables** cascade menu, then select **<u>R</u>emove Table Selection.**

✓ Right-click, and select **Doc<u>u</u>mentation, Schedule <u>T</u>ables, <u>R</u>emove Table Selection** from the cursor menu.

✓ Type **TABLESELECTIONREMOVE** at the Command: prompt.

✓ Select the schedule, right-click, and select the **Selection** cascade menu, then **Remove** from the shortcut menu.

You are prompted to select objects. Begin selecting objects in the drawing that will be removed from the schedule. When you have finished, press the [Enter] key to remove the items from the schedule.

Reselecting Objects for the Schedule

After a schedule table has been added to the drawing, the objects used in a schedule can be reselected for placement in the schedule. When objects are reselected, only those objects selected are reentered into the schedule. For example, suppose a door schedule has already been created in the drawing with doors numbered 1–10. If door numbers 1, 3, 5, 7, and 9 are reselected for placement in the schedule, then the other doors are removed from the schedule and only the selected doors remain in the schedule. Use one of the following options to access the **Reselect Table Selection** command:

✓ Pick the **Documentation** pull-down menu, select the **Schedule Tables** cascade menu, then pick the **Reselect Table Selection** command.

✓ Right-click, and select **Documentation, Schedule Tables, Reselect Table Selection** from the cursor menu.

✓ Type **TABLESELECTIONRESELECT** at the Command prompt.

✓ Select the schedule, right-click, and pick the **Selection** cascade menu, then **Select Again** from the shortcut menu.

Select the objects in the drawing that need to be reentered into the schedule table. Press the [Enter] key when you have finished to update the schedule.

Locating an Object in the Schedule

When schedules are created, the property values within the schedule may need to be updated or filled in over time. In some cases as objects are added to a drawing and the schedule is updated, you may see the schedule fill up with question marks, as shown in Figure 19.29. In these situations property values need to be modified within the objects, or property set definitions need to be added to objects that have question marks in the schedule cells. If the drawing includes many objects being reported on, it may be difficult to locate the specific object in the drawing that needs to be edited.

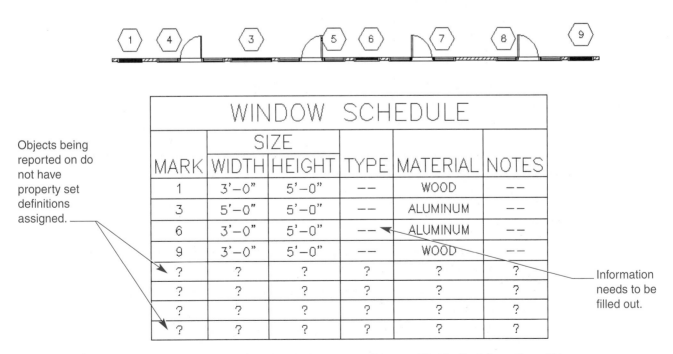

Objects being reported on do not have property set definitions assigned.

Information needs to be filled out.

WINDOW SCHEDULE					
MARK	SIZE		TYPE	MATERIAL	NOTES
	WIDTH	HEIGHT			
1	3'–0"	5'–0"	––	WOOD	––
3	5'–0"	5'–0"	––	ALUMINUM	––
6	3'–0"	5'–0"	––	ALUMINUM	––
9	3'–0"	5'–0"	––	WOOD	––
?	?	?	?	?	?
?	?	?	?	?	?
?	?	?	?	?	?
?	?	?	?	?	?

FIGURE 19.29 As objects are added to the drawing and property values are filled in, the information within the schedule may not reflect a desired schedule.

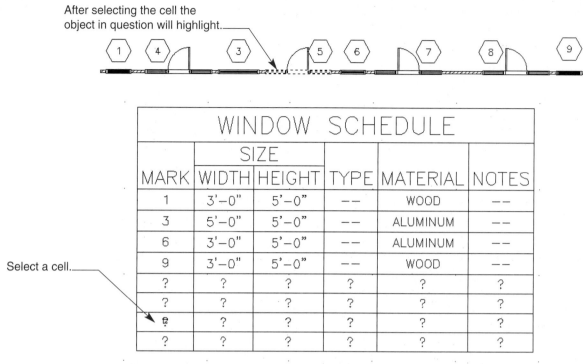

After selecting the cell the
object in question will highlight.

Select a cell.

FIGURE 19.30   Picking a cell using the Show Table Selection command highlights the desired
object in the drawing.

The **Show Table Selection** command can be used to locate the object in the drawing that is reporting
erroneous information in the schedule. This command allows you to select a cell in the schedule in order to
locate the object reporting the wrong value. When the cell is picked, the object providing the information is
highlighted (see Figure 19.30).

Use one of the following options to access this command:

✓ Select the **Documentation** pull-down menu, select the **Schedule Tables** cascade menu, then
pick the **Show Table Selection** command.

✓ Right-click, and select **Documentation, Schedule Tables, Show Table Selection** from the
cursor menu.

✓ Type **TABLESELECTIONSHOW** at the Command prompt.

✓ Select the schedule, right-click, and pick the **Selection** cascade menu, then **Show** from the
shortcut menu.

Pick the cell where the information needs to be modified. The object is highlighted in the drawing.
After identifying the object, use the Edit Schedule Data command to modify the property values or to add a
property set to the object.

✎ Field Notes

If a schedule reports a number of question marks, use the Edit Schedule Data dialog box to add
the appropriate property set definitions. You may also need to add a property set definition to
the object's style.

CHAPTER REVIEW

Use the CD-ROM to test your knowledge and skills.

Chapter Test

To check your understanding of the content provided in this chapter, access the Test file in the CH19 folder of the CD-ROM that accompanies this text.

Chapter Project

To practice the Architectural Desktop skills presented in this chapter, access the Project files in the CH19 folder of the CD-ROM that accompanies this text. The project files are in pdf format and include sample drawings and instructions for completing each project.

||||||||||||||||||||||||||||||||||||||||||||||||||||||||||||||||||||||||||||||

 Creating Custom Schedule Information

LEARNING GOALS

After completing this chapter, you will be able to:

◎ Create custom property set definitions.

◎ Attach property set definitions to styles and definitions.

◎ Add properties to property set definitions.

◎ Add manual and automatic data properties.

◎ Create a data format style.

◎ Construct custom schedules.

◎ Create custom schedule tags.

◎ Add custom tags to the DesignCenter.

The previous chapter described the commands available and the steps required to create a schedule of information. As indicated earlier, a number of default schedule tags, property set definitions, and schedules are included for use with your drawings. These default objects provide much of the information required by a set of construction documents; however, they may not conform to your standards or provide you with the exact information you need to accomplish your designs.

The advantage of using Architectural Desktop is that these default objects can be customized to look and provide information the way you desire. This chapter explains the procedures for creating custom property set definitions, schedules, and tags.

Custom Property Set Definitions

The importance of property set definitions and the information they contain was discussed in Chapter 19. Architectural Desktop includes many property set definitions that can be assigned to individual objects or object styles. These default property set definitions can be found in the PropertySetDefs.dwg located in the \\Autodesk Architectural Desktop 3\Content\Imperial\Schedules folder.

In addition to these property set definitions, custom property set definitions can be created to reflect any type of property value desired. The first step in creating a custom property set definition is to access the **Property Set Definitions** command using one of the following options:

✓ Select the **Documentation** pull-down menu, pick the **Schedule Data** cascade menu, and select the **Property Set Definitions** command.

✓ Right-click, and select **Documentation, Schedule Data, Property Set Definitions** from the cursor menu.

✓ Type **AECPROPERTYSETDEFINE** at the Command: prompt.

This entire chapter is Next Step Material.

737

Current drawing ⎯⎯⎯ Property set definition section ⎯⎯⎯ **New Style** button ⎯⎯⎯ Right-click the section icon to access the shortcut menu.

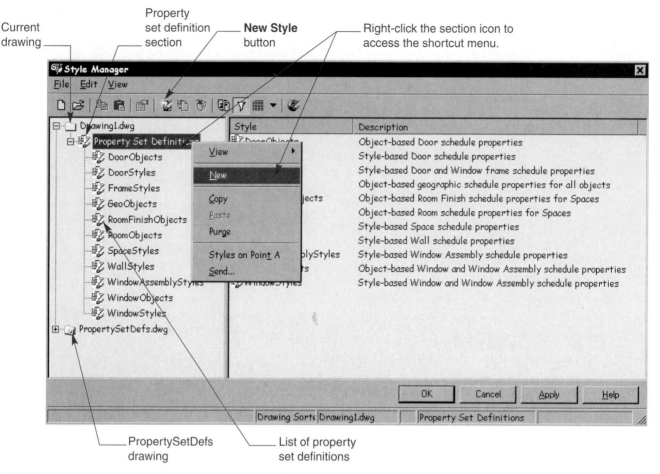

PropertySetDefs drawing ⎯⎯⎯ List of property set definitions

FIGURE 20.1 The Style Manager dialog box is displayed when the Property Set Definitions command is accessed.

The Style Manager dialog box opened to the Property Set Definitions section, shown in Figure 20.1, is displayed.

To create a custom property set definition, pick the **New Style** button within the Style Manager dialog box or select the Property Set Definitions section icon, then right-click and select New from the shortcut menu. A new property set definition is created and displayed in the right pane. Enter a new name for the property set definition. Double-click on top of the new style to begin editing the new property set definition. The **Property Set Definition Properties** dialog box is displayed.

This dialog box includes the following three tabs:

- **General** This tab is used to rename the property set definition, to provide a description for its use, or to apply design notes to it.

- **Applies To** This tab is used to determine how the property set definition is applied in the drawing (See Figure 20.2). The property set definition can be applied either to entities (individual objects) or to styles and definitions (by style). This tab is described in greater detail in the next section.

- **Definitions** This tab is used to create and modify available properties within the property set definition (See Figure 20.3). This tab is described in the following section.

Working with the Applies To Tab

After a new property set definition has been created, it must be applied to an object. There are two types of property set definitions: definitions that are applied or attached to individual objects or entities, and defini-

Choose how the property set definition
will be attached to objects

Checked items
indicate the
objects to which
the property set
definition can be
attached.

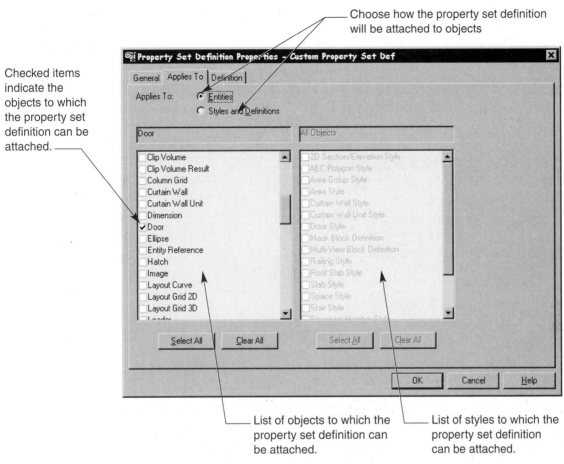

List of objects to which the
property set definition can
be attached.

List of styles to which the
property set definition
can be attached.

FIGURE 20.2   The Applies To tab is used to indicate the types of objects to which the property set
definition will be attached.

Definition tab

Custom
properties
added to the
new property
set definition

Configuration
area for defining
rules for each
property

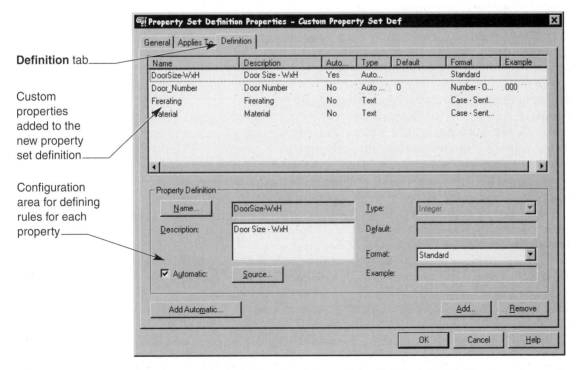

FIGURE 20.3   The Definition tab is used to create custom properties that are added to the
property set definition.

Determines the type of property
set definition being defined

List of AEC object styles to which
the property set can be applied.
This area correlates to the Style
and Definitions radio button.

Applies To tab

List of AEC
objects and
AutoCAD entities
to which the
definition can be
applied. This
area correlates
to the Entities
radio button.

Type of object to
which the property
set definition can
be attached.

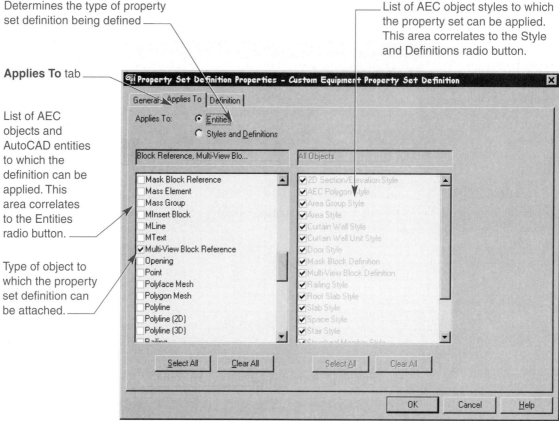

FIGURE 20.4　The Applies To button is used to select the objects to which the definition will be
applied, either individual objects or styles..

tions that are attached to an object style. The Applies To tab is used to select the type of object to which the property set definition is assigned.

At the top of this tab are two radio button options: **Entities** and **Styles and Definitions** (see Figure 20.4). The Entities radio button sets up the property set definition so it can be attached to individual AEC objects or AutoCAD entities. The Styles and Definitions radio button is used to set up the property set definition for use with AEC object styles.

Attaching Property Set Definitions to Objects and Entities

To make the property set definition available only to individual objects and entities select the Entities radio button. The list on the left is made available if the Entities radio button is selected. This list includes the objects and entities to which the property set definition can be applied. A check mark next to an object or entity name indicates that the property set definition can be attached only to that type of object. For example, in Figure 20.4, a checkmark beside the Multi-View Block Reference indicates that the property set definition can be attached only to multi-view blocks in the drawing.

A property set definition can be applied to any or all of the items in the list or to only one item in the list. The Select All button below the list on the left places a check in all the boxes in the list. The Clear All button below the list on the left, clears all of the checkmarks.

Once this type of property set definition has been defined, it can be attached to individual AEC objects or to AutoCAD entities through the **Edit Schedule Data** command.

Attaching Property Set Definitions to Styles and Definitions

To make the property set definition available only to AEC object styles select the Styles and Definitions radio button. When this option is selected, the list on the right side of the tab becomes available, and the left side is grayed out, as shown in Figure 20.5.

The list on the right side of the tab includes the object styles and definitions in Architectural Desktop to which the property set definition can be attached. A checkmark beside a style or definition name indi-

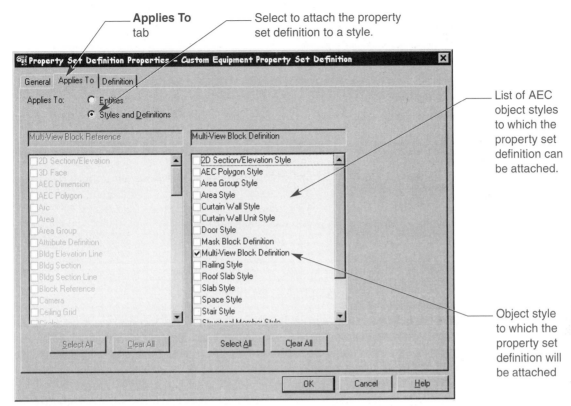

FIGURE 20.5 Selection of the Styles and Definitions radio button makes a list of AEC object styles available on the right side of the tab.

cates that the property set definition can be assigned to that style or definition. The Select All button below the list places a check in all the style check boxes. The Clear All button removes all the checkmarks beside the style names.

A property set definition can be attached to one or more styles by selecting the appropriate boxes from the list on the right. Once the property set definition has been defined, it can be attached to the style or definition by accessing the appropriate style/definition properties dialog box. In the General tab for the style, pick the Property Sets button to add the property set definition to the style (see Figure 20.6).

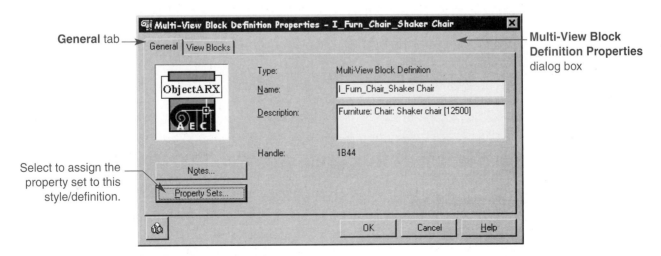

FIGURE 20.6 A property set definition is attached to the AEC style or definition by selecting the Property Sets... button.

Exercise 20.1

1. Open EX14-1-Fnd.dwg.
2. Create a new property set definition named Slab Objects, then double-click to edit it.
3. Enter the following description in the General tab: Use on individual slab objects.
4. In the Applies To tab, select the Entities radio button. Clear all the object and entity check boxes on the left side. Place a check next to the Slab object. Exit the dialog box by pressing OK.
5. Create another property set definition named Member Styles, then double-click to edit it.
6. Enter the following in the General tab: Use on Structural Member styles.
7. In the Applies To tab, select the Styles and Definitions radio button. Clear all the style and definition check boxes on the right side. Place a check next to the Structural Member Style. Exit the dialog box by pressing OK. Exit the Style Manager dialog box.
8. Save the drawing as EX20.1-Fnd.dwg.

Adding Properties to the Property Set Definition

After the type of property set definition to be created has been determined, properties can be assigned. As mentioned in Chapter 19, two types of properties are available: manual data and automatic data. The Definition tab within the Property Set Definition Properties dialog box is used to assign both manual data and automatic data (see Figure 20.7).

This tab is divided into two main areas: The top portion of the tab lists all properties that are assigned to the property set definition. The lower portion of the tab controls the configuration of an individual property.

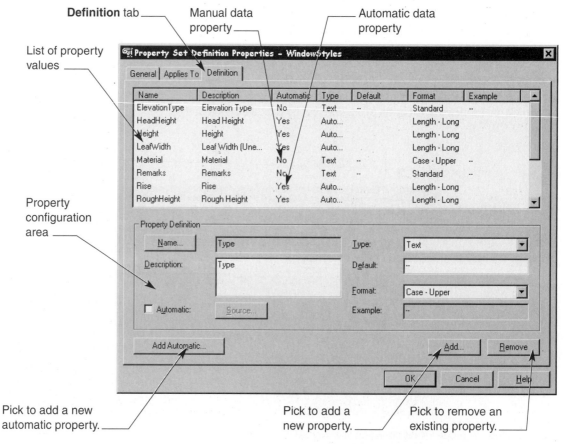

FIGURE 20.7　The Definitions tab is used to begin adding properties to the property set definition.

FIGURE 20.8 The New Property dialog box is used to create a new property name and allows configurations to be based on an existing property in the drawing.

Enter a name for the new property.

Choose from a list of properties currently in the drawing to use as a template.

Adding Manual Data Properties

Before a manual data property can be configured it needs to be added to the list area at the top of the tab. Use the Add... button in the lower right corner of the dialog box to add a new property to the property set definition. Pressing this button displays the **New Property** dialog box (see Figure 20.8). Enter a name for the new property. Below the Name: text box is the Start With: drop-down list that includes a list of all the properties currently defined in the drawing. Selecting one of these properties sets the parameters for the new property equal to the values for a previously defined property. Press the OK button when you have finished.

A new property is added to the list area of the dialog box (see Figure 20.9). In order to configure the new property, select the property from the list to highlight it. The defined configurations for the property are displayed in the Property Definition area at the bottom of the tab. The following are configuration settings that affect the manual data property:

- **Name:** Select this button to rename the highlighted property.

- **Description:** Enter a description for the property.

- **Automatic:** Check this box to change this property to an automatic data property.

- **Type:** This drop-down list establishes the type of information this property is supposed to be. The following options are included in this list:

  - **Auto Increment** This setting sequentially numbers or alphabetizes the value as new property set definitions are added to the drawing. This type is used to increment the door and window number properties.

New manual data property added to the property set definition

Press to rename the highlighted property.

Enter a description for the property.

Select to make the property an automatic property.

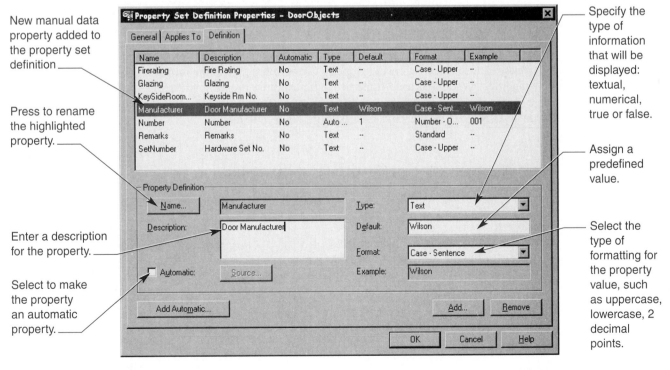

Specify the type of information that will be displayed: textual, numerical, true or false.

Assign a predefined value.

Select the type of formatting for the property value, such as uppercase, lowercase, 2 decimal points.

FIGURE 20.9 A new manual data property has been added to the property set definition and its settings configured.

- **Integer**   This type of formatting requires the input of a whole number, for example, −3, 12, or 0.

- **Real**   This formatting requires the input of a real number, for example, 1.25, 4.0, .75.

- **Text**   This type indicates that the property information is textual content, not numerical content.

- **True/False**   This type indicates that the property must be a true or false value.

- **Default:**   This text box allows you to enter a default value for the property. A value entered in this text box becomes the default value for this property the next time the property set definition is attached to an AEC object. The default value must match the property type selected above.

- **Format:**   This drop-down list includes different formatting styles available for the property values. For example, a Case-Upper style makes any information in the property all uppercase in the Edit Schedule Data dialog box and within the schedule table. Creating custom data format styles is discussed later in this chapter.

- **Example:**   This text box displays the Default property value in its formatted design.

When you have finished configuring the property you can create additional manual data properties by pressing the Add... button and formatting. If an Automatic property is desired, select Add Automatic.

Adding Automatic Data Properties

Automatic data properties can be added by selecting the Add Automatic button in the lower left corner of the tab. When this button is selected, the **Automatic Property Source** dialog box is displayed, as shown in Figure 20.10. This dialog box lists the available automatic properties for the type of object to which the property set definition is being applied. Figure 20.10 displays automatic properties for the door object, as the property set definition is being applied to door objects.

Scroll through the list until you find the type of automatic property that you would like to use in your property set definition. Place a check in the corresponding box. After selecting the type of property to be added to the property set definition, pick the OK button to add the property to the list (see Figure 20.11). The new property can now be configured.

Much of the configuration area is unavailable for an automatic data property; however, the description can be changed, a different automatic property can be selected, and the formatting for the information can all be modified. When you have finished configuring the property, press the OK button to accept the changes to the property set definition and return to the Style Manager. The property set definition is now ready to be assigned to the object(s) selected in the Applies To tab.

Object to which the property
set definition is applied ⎯

List of automatic properties
for the door object ⎯

Place a check in the box for
the property that will be
added to the properties list. ⎯

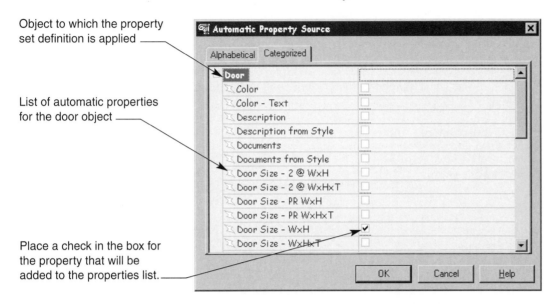

FIGURE 20.10 The Automatic Property Source dialog box displays the available properties for the type of object to which the property set definition is applied.

New automatic
property ⎯

Change the
default
description
if desired. ⎯

Press to change
the automatic
value being
reported on. ⎯

Select a type
of formatting.

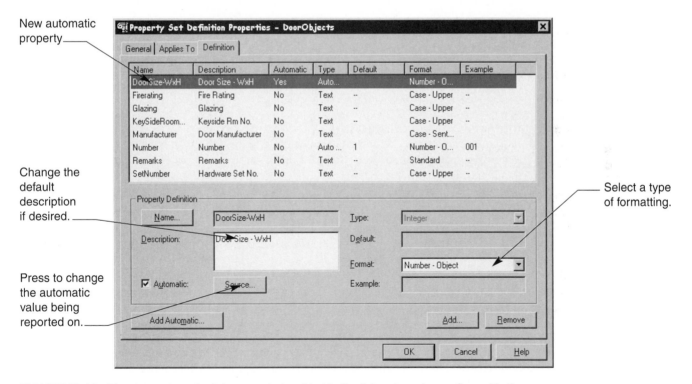

FIGURE 20.11 The new automatic data property is added to the list and can be configured in the Property Definition area.

Exercise 20.2

1. Open EX20.1-Fnd.dwg.
2. Edit the Slab Objects property set definition. Add the following automatic and manual data properties:

Name	Description	Automatic	Type	Default	Format	Example
Area-Gross	Area-Gross	Yes	Automatic		Area	
Description	Description	No	Text	4″ Conc. Slab	Case-Sentence	4″ Conc. Slab
Volume	Volume	Yes	Automatic		Standard	

3. Edit the Member Styles property set definition. Add the following automatic and manual data properties:

Name	Description	Automatic	Type	Default	Format	Example
2x4	2x4	No	True/False	True	Standard	True
2x6	2x6	No	True/False	True	Standard	True
2x8	2x8	No	True/False	True	Standard	True
2x10	2x10	No	True/False	True	Standard	True
2x12	2x12	No	True/False	True	Standard	True
4x4	4x4	No	True/False	True	Standard	True
4x8	4x8	No	True/False	True	Standard	True
Material	Type of Materials	No	Text	DF #2	Case-Sentence	DF #2
Length	Length	Yes	Automatic		Length-Nominal	
Style	Style	Yes	Automatic		Case-Upper	

4. Exit the Style Manager dialog box.
5. Assign the Member Styles property set definition to the 4x8 girder member style. Fill out the property set definition to be false for all the structural sizes except the 4x8, which should be true.
6. Assign the Member Styles property set definition to the Post and Pier member style. Fill out the property set definition to be false for all structural sizes except the 4x4, which should be true.
7. Save the drawing as EX20.2-Fnd.dwg.

Creating a Data Format Style

When configuring the properties in the property set definition, you have the option of choosing the type of formatting to apply to the data. The type of data formatting selected affects how information in the schedule tables, schedule tags, and the Edit Schedule Data dialog box appears. Architectural Desktop includes several different types of data formats and provides you with a tool to create your own types of formatting.

Use one of the following options to create a new data format style:

✓ Select the **Documentation** pull-down menu, pick the **Schedule Data** cascade menu, then select the **Data Format Styles...** command.

✓ Right-click, and select **Documentation, Schedule Data, Data Format Styles...** from the cursor menu.

✓ Type **AECPROPERTYFORMATDEFINE** at the Command: prompt.

The Style Manager dialog box is displayed opened to the Schedule Data Formats section, as shown in Figure 20.12.

Current
drawing

Schedule Data
Formats section

New Style button

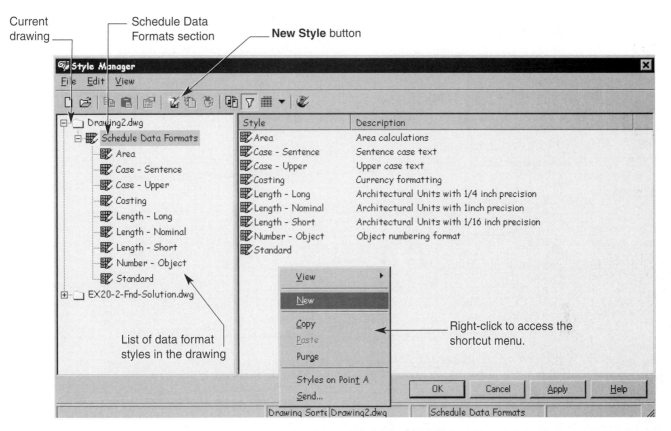

FIGURE 20.12   The Style Manager can be used to create and modify data format styles.

After the Style Manager has been opened, select the New Style button to create a new data format style. You can also right-click over the Schedule Data Formats icon or in the right window pane to display a shortcut menu from which you can access the New command. Enter a new name for the data format style. After the new data format style has been created, double-click the style, or right-click over the new style and select Edit... from the shortcut menu. The **Schedule Data Format Properties** dialog box, shown in Figure 20.13, is displayed.

This dialog box is organized into the following five areas:

- **General**   This area is used to control the general settings for data, and it includes the following three text boxes:

  - **Prefix:**   Enter a prefix for schedule data. For example, if you are creating a format style that will be used in a pricing property, you may use $ as a prefix before a dollar amount.

  - **Suffix:**   Enter a suffix for the schedule data. For example, if you are using the format style for an area property, you may use the suffix SF to represent square feet.

  - **Undefined:**   This text box is used as a default property value for any value that has not been defined in the property set definition. By default the symbol ? is used, causing the schedule to produce ? marks in the schedule if the property has not been filled out. You may want to change this symbol to a hyphen (-), as it looks better in a schedule.

- **Text**   This area includes options for controlling the appearance of text in a property. The following options are available in the Case drop-down list:

  - **As Is**   This option enters the data into the schedule and the property set definition as typed.

Enter a suffix for the data.

Enter a prefix for the data.

Specify types of units used by data.

Enter a precision for the data.

Specify a value for undefined data.

Specify how a fraction appears in Architectural or Fractional units.

Specify text formatting.

Specify the separator for decimal numbers.

Enter number of leading places in the data.

Select the suppression of 0's in the data.

Enter how true/false data will be reported.

Select a dimension style to configure the formatting settings.

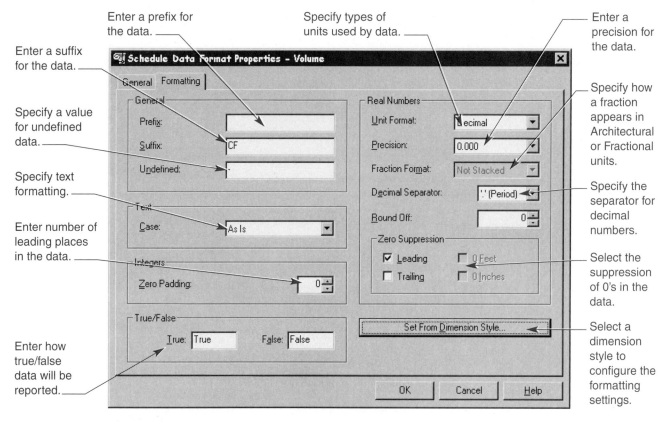

FIGURE 20.13 The Schedule Data Format Properties dialog box is used to configure the settings for a data format style.

■ **Sentence Case**   This option capitalizes the first letter of the first word in a string of text entered in the property value.

■ **Title Case**   This option capitalizes the first letter of each word in a string of text entered in the property value.

■ **Upper Case**   This option capitalizes all the letters of each word in a string of text entered in the property value.

■ **Lower Case**   This option displays the text as all lowercase letters in the text string entered in the property value.

■ **Integers**   This area is used to specify the maximum length of a number after it has been "padded" with leading zeros before the number. This area is used when identity numbers are created such as the numbers used for door and window tags.

■ **Zero Padding:**   Enter a number representing the total number of digits that will be used for the number. For example a padding value of 3 will cause the number 7 to appear as 007, or the value 21 as 021.

■ **True/False**   This area is used for property values that are using a true/false type value in the property. Enter the text that is to appear in the schedule for a true statement in the True text box, and the text that is to appear in a false statement in the False text box.

■ **Real Numbers**   This area is used to configure numerical property values. The following settings are available:

■ **Unit Format:**   Select the drop-down list to choose from several different types of units for formatting any numerical data in the property values. Options include Architectural, Decimal, Engineering, Fractional, and Scientific.

■ **Precision:** Enter a value for the precision level of the numerical formatting. This number will directly correspond to the types of units selected. For example, a precision of .25 is used with Decimal units, and a precision of 1/4″ is used with Architectural units.

■ **Fraction Format:** This area is available if the Architectural or Fractional units have been selected. This option allows you to control how fractional values are to appear.

■ **Decimal Separator:** This option controls how decimals are separated in a numerical property value.

■ **Round Off:** This area rounds off the numerical value specified to the nearest value entered in this text box. For example, a value of .25 rounds the numerical value in the property to the nearest 1/4″.

■ **Zero Suppression** This area controls the visibility of 0's before and after the numerical value. If a check is placed in the Leading (for decimal units) or 0 Feet (for Architectural units) box, zeros before the number are suppressed in the schedule. If a check is placed in the Trailing (for decimal units) or the 0 Inches (for Architectural units) box, zeros after the number are suppressed.

■ **Set From Dimension Style...** Selecting this button provides you with a list of dimension styles in the drawing that can be used to format the values in this dialog box.

When you have finished setting the values for the data format style, press the OK button to return to the Style Manager. After a new data format style has been created, it can be applied to the properties within a property set definition.

Exercise 20.3

1. Open EX20.2-Fnd.dwg.
2. Create a new data format style named Volume.
3. Change the suffix to read CF, and the Undefined text box to use a hyphen (-).
4. Use Decimal units and set the precision to two decimal places.
5. Press OK to return to the Style Manager.
6. Modify the Slab Objects property set definition. Pick the Volume property and change the format from Standard to the new Volume data format style.
7. Assign the Slab Objects property set definition to the slab in the garage and to the driveway and sidewalk slab.
8. Save the drawing as EX20.3-Fnd.dwg.

Creating Custom Schedules

The schedules that are included with Architectural Desktop provide a basic set of information that is often required within a schedule. Often, offices use their own standards regarding how a schedule must be organized and the types of information that must be included within the schedule. The schedule styles included can be modified to fit individual needs.

Custom schedule styles can also be created to report any type of information available in an AutoCAD or Architectural Desktop object. The only requirements for creating a custom schedule are the following:

1. The data must be available for the object. (These can be found by viewing the automatic properties of a particular object.)

2. The properties desired in the schedule must be included in a property set definition.

Once the appropriate property set definitions have been established, the custom schedule can be created, or an exiting schedule can be modified. Use one of the following options to create a schedule:

✓ Select the **Documentation** pull-down menu, pick the **Schedule Tables** cascade menu, then select the **Schedule Table Styles...** command.

✓ Pick the **Schedule Table Styles** button in the Schedule - Imperial (or Metric) toolbar.

✓ Right-click, and select **Documentation, Schedule Tables, Schedule Table Styles...** from the cursor menu.

✓ Type **AECTABLESTYLE** at the Command: prompt.

The Style Manager is displayed opened to the Schedule Table Styles section, as shown in Figure 20.14.

If you are going to edit a schedule style, first select the desired schedule table, then right-click, and pick Edit... from the shortcut menu, or pick the Edit Style button from the buttons at the top of Style Manager. If you are going to create a new style, select the New Style button, or right-click over the Schedule Table Styles icon or within the right window pane and select New from the shortcut menu (see Figure 20.14). After you have created a new style, enter a name for the style.

Once you have created the new style and entered the name, double-click the style to edit it. Alternatively, you can highlight the new style and select the Edit Style button, or right-click over the style and select Edit... from the shortcut list. The **Schedule Table Style Properties** dialog box is displayed. This dialog box includes the following tabs:

- **General**   This tab is used to rename the schedule table style and to provide a description for the schedule. Any additional notes or references can be added using the Notes... button in this tab.

- **Default Format:** This tab controls the formatting that applies to each cell within the schedule unless the cell is overridden. (see Figure 20.15). The following settings are available in this tab:

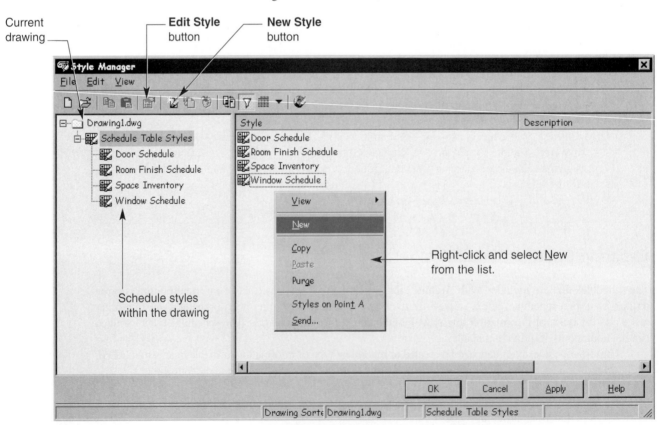

FIGURE 20.14   The Style Manager is used to create and edit schedule table styles.

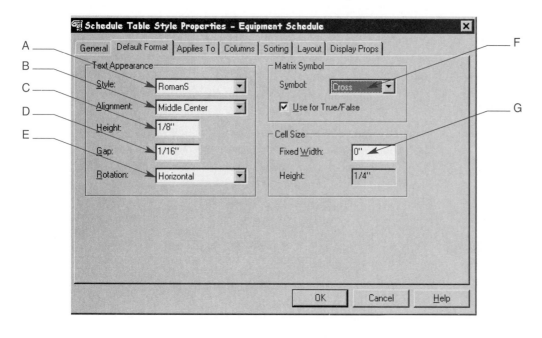

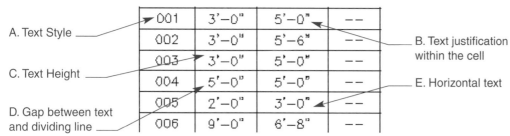

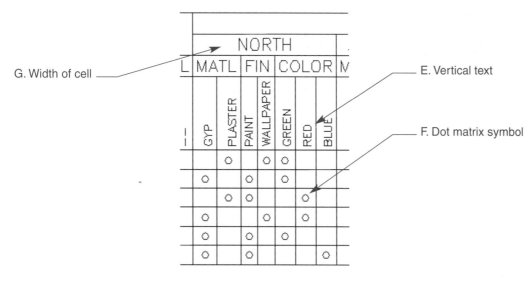

FIGURE 20.15 The Default Format tab is used to set up default formatting for the information within the cells of a schedule. These settings can be overridden in the Layout tab.

- **Style:** This drop-down list includes a list of text styles in the current drawing. Additional styles can be configured by using the **STYLE** command.

- **Alignment:** This drop-down list allows you to justify the text within a cell in the table. Options allow you to justify the text to the left or right, or to center the text within the cell.

- **Height:** This text box is used to specify the height of the text within the schedule table as it should appear when plotted. Like the documentation symbols, the schedule table is scaled by the drawing scale set up in the Drawing Setup dialog box.

- **Gap:** This text box allows you to specify the distance from the text to the cell border. This gap is applied all the way around the text.

- **Rotation:** This drop-down list allows you to set the text in the schedule to be horizontal or vertical.

- **Symbol:** This drop-down list is used to specify a symbol type when a schedule table is created that displays a matrix (see Figure 20.15). A matrix may be used when the value of a single property may have many different options. For example, the finish material property may be used in different rooms where different types of finishes are being applied.

- **Use for True/False:** This check box is used when a true/false property should be incorporated into a matrix schedule.

- **Fixed Width:** This text box allows you to set a fixed width for the cells in the schedule. If the value is set to 0″, then the cells of the schedule are determined automatically based on the length of text in the property set definitions.

- **Height:** The height in the Cell Size area is a value automatically determined by the combination of the text height and the gap values.

- **Applies To** Like property set definitions, the schedule table needs to be associated with an object(s) that will be reported on within the schedule table. Select the type of object(s) the schedule will be used for by placing a check mark beside the object name, as shown in Figure 20.16.

- **Columns** This tab is used to configure the actual information that will be reported on (see Figure 20.17). The table area displays the properties from the property set definitions that will be reported on. The properties can be organized such that the columns of information are arranged as desired. This tab includes several options for configuring the columns within the schedule.

  To begin adding columns of information, select the Add Column... button to display the **Add Column,** dialog box, where properties can be selected and added to the schedule (see Figure 20.18). Select a property from one of the property set definition lists and press the OK button to add the column to the schedule table. Continue pressing the Add Column... button to add columns to the schedule. Once you have added all the columns of information, you can move to the next tab. Adding and configuring the columns within the table is discussed in greater depth in the next section.

- **Sorting** This tab is used to sort the information within the schedule. Information can be sorted in ascending or descending order. This feature may be used when numerical or alphabetic information such as door numbers or equipment numbers is sorted. Use the

FIGURE 20.16 The Applies To tab is used to assign the schedule to the objects that will be reported on in the schedule table.

Applies To tab

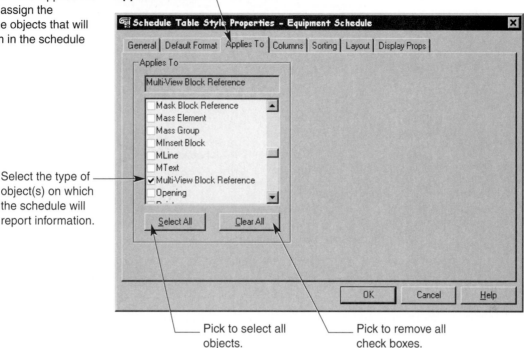

Select the type of object(s) on which the schedule will report information.

Pick to select all objects.

Pick to remove all check boxes.

FIGURE 20.17 The Columns tab is used to add and configure columns for the schedule table.

Columns tab

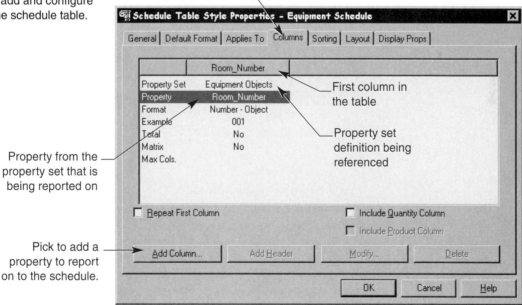

First column in the table

Property set definition being referenced

Property from the property set that is being reported on

Pick to add a property to report on to the schedule.

Add... button to add a column that will be sorted. Multiple columns can be sorted by picking the Add... button several times and selecting a new column each time (see Figure 20.19).

You can remove columns being sorted, and move them up or down the list by using the appropriate buttons at the top of the tab.

■ **Layout**

This tab is used to configure how the title, header titles, and text information appear in the schedule table (see Figure 20.20). The text box at top of this tab allows you to enter a title that appears at the top of the schedule. Below the title text box are three buttons labeled Override Cell Format. Each of these buttons allows you to override how the schedule text is formatted in the Default Format tab.

Property set definitions available for the type of object being reported on

Properties within each property set definition

Select one property to add as a column to the schedule.

Configure the properties for the column

Position in which the column will be placed within the schedule

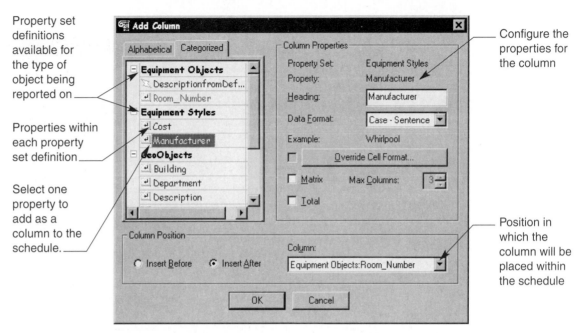

FIGURE 20.18   The Add Column dialog box is used to add properties to report on to the schedule.

FIGURE 20.19   The Sorting tab is used to sort information alphabetically or numerically within a column.

Sorting tab

Pick to select a column to sort.

Column/property to be sorted

Select the type of sorting for the selected column.

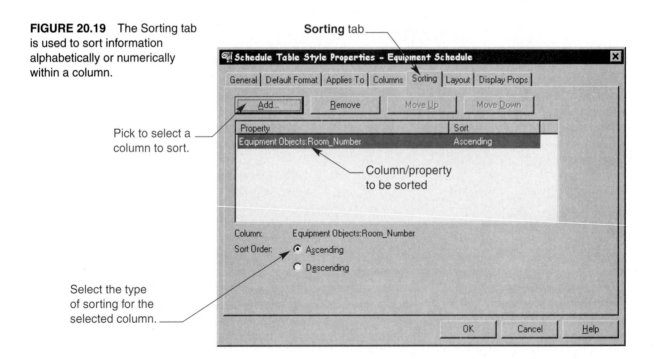

Picking one of the buttons displays the Cell Format Override dialog box. This dialog box contains the same information as found in the Default Format tab and allows you to format the title, column header titles, and the matrix Headers (information in the schedule).

■ **Display Props**

This tab is used to configure the display settings for the schedule object (see Figure 20.21). The General display representation is the only representation available from the drop-down list. Select the appropriate property source to edit and attach an override. Once the override has been set, pick the Edit Display Props... button to configure the display of components such as lines and text within the schedule.

Layout tab

Enter a name to be used for the title.

Pick to set text format overrides to the titles and information in the schedule.

The Cell Format Overridedialog box appears when overrides to the schedule table text are set.

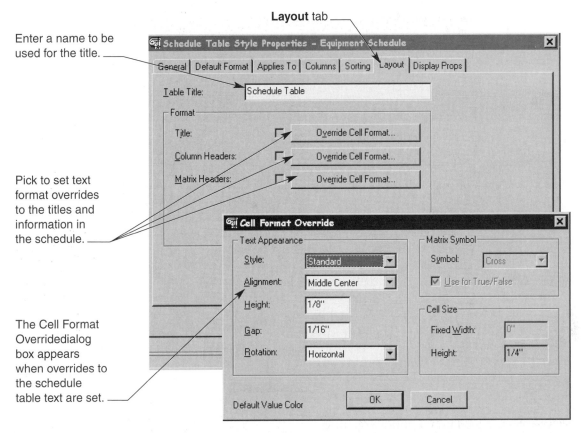

FIGURE 20.20   The Layout tab is used to override the settings found in the Default Format tab and to apply the overrides to the different parts of the table.

Display Props tab

The General display representation

Select the level of display control and attach an override.

Adjust the display properties of the components within the schedule table.

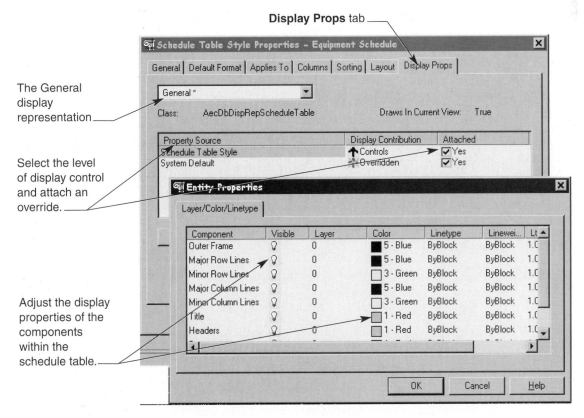

FIGURE 20.21   The Display Props tab is used to configure the display properties for the schedule table.

After you have configured the schedule table, press the OK button to return to the Style Manager. Within the Style Manager window, press the OK button to return to the drawing. The schedule table is now ready to be added to the drawing.

Exercise 20.4

1. Open EX20.3-Fnd.dwg.
2. Create a new schedule style named Slab Schedule.
3. Assign the schedule to the Slab Object in the Applies To tab.
4. In the Columns tab, select the <u>A</u>dd Column... button. In the Add Column dialog box, search for the Slab Object's property set definition. Pick the Description property and press the OK button to add the property as a column to the schedule table.
5. Press the <u>A</u>dd Column... button again and select the Area-Gross property from the Slab Object's property set definition. Place a checkmark in the Total check box. Press OK to add the new column to the schedule.
6. Press the <u>A</u>dd Column... button again and select the Volume property from the Slab Object's property set definition. Place a checkmark in the Total check box. Press OK to add the new column to the schedule.
7. In the Layout tab name the schedule Concrete Slab Schedule. Press OK until you return to the drawing.
8. Add the Slab Schedule to the drawing.
9. Save the drawing as EX20.4-Fnd.dwg.

Developing the Columns in the Schedule

One of the most important steps when creating or modifying a schedule is to determine what type of object or objects the schedule table will be reporting on. This task is accomplished in the Applies To tab in the **Schedule Table Style Properties** dialog box. Determining the types of objects that will be reported on makes the property set definitions and individual properties associated with those objects available in the Columns tab.

The Columns tab is used to add and configure the columns within the schedule to produce a table of information (see Figure 20.22). To begin adding columns to the schedule table, press the <u>A</u>dd Column... button located in the lower left corner of the tab. The <u>A</u>dd Column... dialog box, shown in Figure 20.23, is displayed.

FIGURE 20.22 The Columns tab is used to add columns of information to the schedule table.

Columns tab

Schedule Headers for the schedule table

Table of properties being reported on

Press to add a column to the drawing.

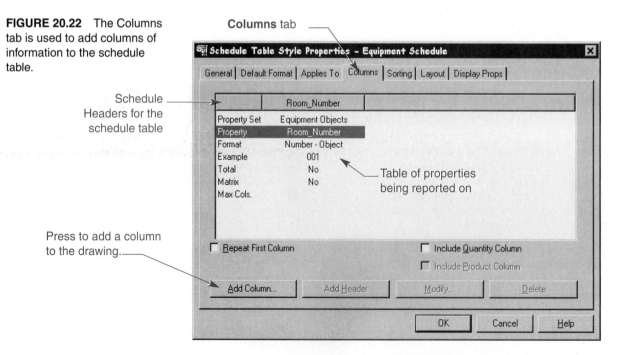

Header name ───

Formatting for information ───

Property set definitions available for the assigned object ───

Individual properties that can be reported on ───

User-defined properties ───

Automatic properties ───

Press to override the default cell format.

Enter the total number of cells for the matrix under the header.

Check to turn the property into a matrix cell.

Check to create a total from the values in the schedule.

Specify a location for the new header. ───

Choose the header in front of which or behind which the new column will be placed in the schedule. ───

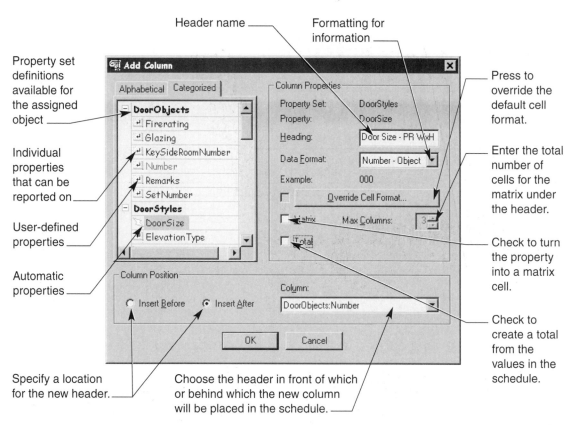

FIGURE 20.23 The Add Column dialog box is used to add a column to the schedule table.

Adding and Configuring a Column

The Add Column dialog box is used to add the property to be reported on to the schedule and to configure how the information will appear in the schedule (see Figure 20.23). The upper left corner includes a list of property set definitions that are assigned to the same individual objects and object styles to which the schedule table is assigned. Select a property from the list to add to the schedule table.

Once a property has been selected, its name appears in the Heading: text box on the right. The text entered here is the text that appears as the header title in the schedule. Use the Data Format: drop-down list to select a type of formatting to apply to the information in this column that appears in the schedule.

The Default Format tab was discussed earlier. This tab formats how all the data within the schedule will appear. The Override Cell Format... button is used to control how the information within the cells below the header will appear. Pressing this button displays the **Cell Format Override** dialog box. Make any adjustments to the text height, justification, and gap for the reported information.

Below the Override Cell Format... button is the Matrix: check box. Selecting this box makes the Max Columns: list available. Both these options are used to turn the column into a matrix type of schedule (see Figure 20.24).

When the column is turned into a matrix, any values entered for the property in the **Edit Schedule Data** dialog box are sorted into separate columns under the heading title. The Max Columns text box indicates the maximum number of values for the matrix column. For example, in Figure 20.24, the ceiling material property is being reported on. As the column is turned into a matrix the maximum number of values that can be reported on is three. If four objects have four different values for the ceiling material, only the first three will be displayed in the schedule.

Below the Matrix check box is the Total check box. This box is not available if the Matrix check box has been selected. Selecting this box provides a total at the bottom of the column. For example, suppose you are reporting on doors in the drawing, and you need to know the total number of doors by style. You can select the door style property to report on, then select the Total check box to report the total number of door styles.

The last type of configuration for the new column is its position within the schedule table. The Column Position section of the Add Column dialog box includes two options: Insert Before and Insert

Schedule Title —

Properties/Heading
title being reported on —

Matrix symbols —

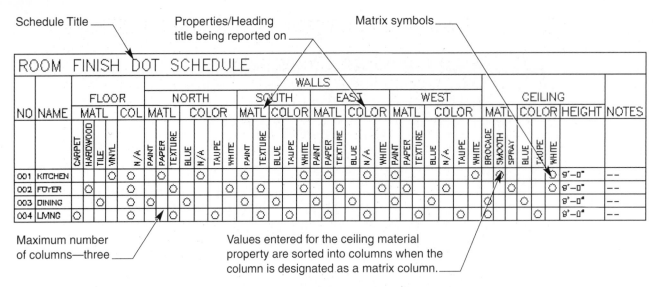

Maximum number
of columns—three —

Values entered for the ceiling material
property are sorted into columns when the
column is designated as a matrix column. —

FIGURE 20.24　Example of a matrix schedule.

FIGURE 20.25　Several
properties have been selected
to create columns in the schedule
table.

Columns tab in
the schedule —

Columns tab —

Pick on header
button to modify. —

Property set
definition being
referenced —

Property being
reported on —

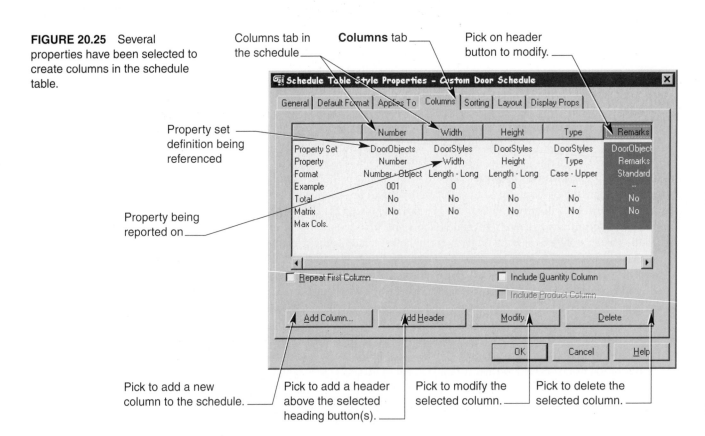

Pick to add a new
column to the schedule. —

Pick to add a header
above the selected
heading button(s). —

Pick to modify the
selected column. —

Pick to delete the
selected column. —

After. If this is the first column being added to the schedule, the column is placed at the beginning of the schedule. If this is an additional column, you have a choice of placing the column before or after a column currently in the schedule, as displayed in the Column drop-down list. Press the OK button when you have finished to add the column to the schedule. Figure 20.25 displays several columns added to a schedule.

Adding Additional Columns

After choosing the different properties to be used as columns in the schedule you can add additional built-in columns to the schedule table. The Columns tab includes the following three additional check boxes that add columns to the schedule table, as shown in Figure 20.26:

FIGURE 20.26 Three additional check boxes allow you to add predefined columns to the schedule.

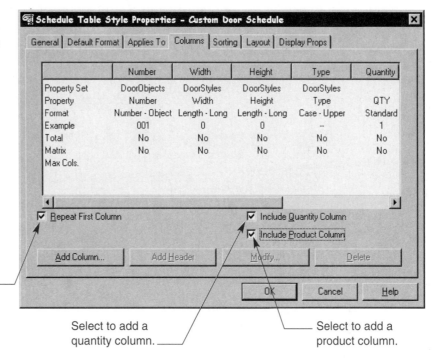

Select to add a copy of the first column to the end of the schedule.

Select to add a quantity column.

Select to add a product column.

- **Repeat First Column**   This check box adds an additional column to the end of the schedule that is a duplicate of the first column in the schedule. This option may be used when a schedule is a long table, and repeating the first column at the end of the schedule will make the schedule easier to read. The repeated column does not get added as a header in the Columns tab but is displayed at the end of the schedule in the drawing.

- **Include Quantity Column**   This column adds a column to the beginning of the schedule that reports the total quantity of a reproduced object within the schedule.

- **Include Product Column**   This column produces a value based on the total quantity of an object multiplied by another value. For example, if the total number of board feet used in structural members is desired, the Product column takes the quantity of each type of structural member and multiples it by the varying lengths of structural members. This column is always used in conjunction with the Quantity column.

     To configure the product column, or any part of the Product column after it has been added to the schedule, press the Modify... button. The **Edit Product Column** dialog box, shown in Figure 20.27, is displayed. Select a property value to be multiplied by the Quantity column from the Data Column drop-down list. The calculation between the quantity and the property value is made, yielding the product value in the schedule.

Finishing the Column Configurations

In addition to the Add Column... button, three more buttons are available in the configuration of the schedule columns. The Modify... button is used to modify any column in the schedule. First, select the column by picking on the header in the schedule or any text in the desired column, then pick the Modify... button. The Modify Column dialog box is displayed, where you can rename the header name, apply overrides to the column formatting and the text information, or change the column to a matrix or to a totaled column.

Enter a name for the
product header. ⎯⎯

2. Select product column. ⎯⎯

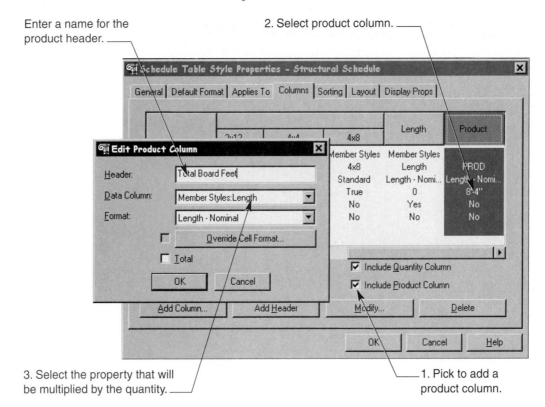

3. Select the property that will
be multiplied by the quantity. ⎯⎯/

⎯ 1. Pick to add a
product column.

FIGURE 20.27   The Edit Product Column dialog box is used to select the property that will be multiplied by the Quantity column.

The Delete button is used to delete a column from the schedule table. To remove a column from the schedule, first select the column, then pick the Delete button. You can delete multiple columns by holding down the [Ctrl] key and picking the header names in the table then picking the Delete button. The **Remove Columns/Headers** dialog box appears, allowing you to verify the columns to be deleted.

The Add Header button is used to add a header above one or more headers derived from the selected properties to be reported on. Select one column, then pick the Add Header button to add a new header over the selected header name (see Figure 20.28A). If multiple property headers are to be placed under a single header, select the desired columns by holding down the [Ctrl] key and selecting the headers, then pick the Add Headers button, as shown in Figure 20.28B. A new header is placed over the highlighted column(s) and gives you the opportunity to name the new header.

Columns can also be repositioned within the schedule by pressing and holding down the pick button over a header name and moving it to the desired location within the schedule table. When you are finished configuring the columns for the schedule, you can use the other tabs in the dialog box to make additional configurations in the schedule table.

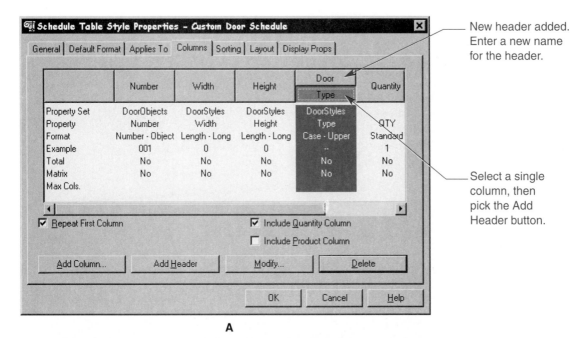

New header added.
Enter a new name
for the header.

Select a single
column, then
pick the Add
Header button.

A

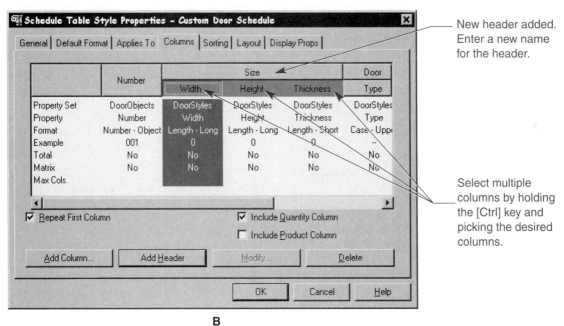

New header added.
Enter a new name
for the header.

Select multiple
columns by holding
the [Ctrl] key and
picking the desired
columns.

B

FIGURE 20.28 (A) Adding a header over a single column. (B) Adding a header over multiple columns.

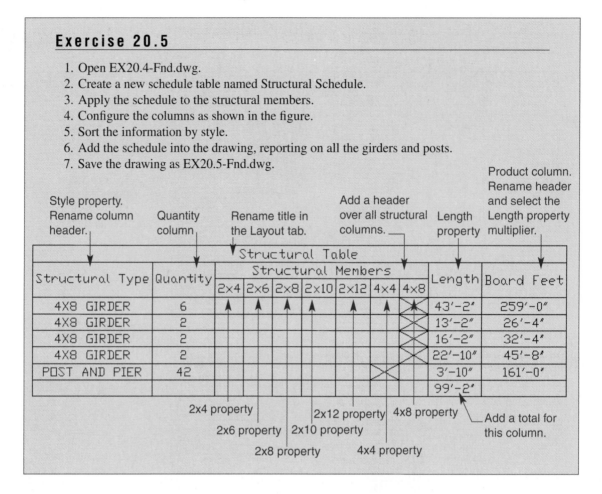

Exercise 20.5

1. Open EX20.4-Fnd.dwg.
2. Create a new schedule table named Structural Schedule.
3. Apply the schedule to the structural members.
4. Configure the columns as shown in the figure.
5. Sort the information by style.
6. Add the schedule into the drawing, reporting on all the girders and posts.
7. Save the drawing as EX20.5-Fnd.dwg.

Style property.
Rename column header.

Quantity column

Rename title in the Layout tab.

Add a header over all structural columns.

Length property

Product column. Rename header and select the Length property multiplier.

Structural Table

Structural Type	Quantity	Structural Members							Length	Board Feet
		2x4	2x6	2x8	2x10	2x12	4x4	4x8		
4X8 GIRDER	6	↑	↑	↑	↑	↑	↑	⊠	43'-2"	259'-0"
4X8 GIRDER	2							⊠	13'-2"	26'-4"
4X8 GIRDER	2							⊠	16'-2"	32'-4"
4X8 GIRDER	2							⊠	22'-10"	45'-8"
POST AND PIER	42						⊠		3'-10"	161'-0"
									99'-2"	

2x4 property
2x6 property   2x12 property   4x8 property
2x10 property
2x8 property   4x4 property

Add a total for this column.

Creating Custom Schedule Tags

Architectural Desktop includes a number of schedule tags that can be applied to different objects as documentation items and scheduling information. If the International Extensions have been installed, additional tags can be found in the MetricDACH folder.

Depending on the office standard or personal preference, these tags can be used as is. Occasionally, however, some offices prefer to use a different size or type of symbol than what is included with Architectural Desktop. In these situations the included tags can be modified to fit office standards. You can also create custom tags for your use. This section explains the different methods and rules for modifying or custom creating schedule tags in Architectural Desktop.

Modifying Existing Schedule Tags

Probably the easiest way to create custom tags is to modify an existing tag. To do this you can use the Windows Explorer program to make a copy of the desired tag. Enter **EXPLORER** at the Command: prompt to open the Microsoft® Windows Explorer (see Figure 20.29). Browse through the tree list on the left side of the window to your Architectural Desktop folder. By default, Architectural Desktop is installed to the \\Program Files\Autodesk Architectural Desktop 3 folder.

The imperial-scaled schedule tags can be found under the Imperial folder. The path to these blocks is \\Autodesk Architectural Desktop 3\Content\Imperial\Schedules folder (see Figure 20.29). The metric-scaled schedule tags are found in the \\Autodesk Architectural Desktop 3\Content\Metric\Schedules folder. If the International Extensions are installed, the path to these tags is found in the \\Autodesk Architectural Desktop 3\Content\Metric DACH\ Schedules folder. Within each of these schedule folders are additional folders that organize the different types of tags available.

In this discussion the door tag found in the Imperial folder will be copied and modified. The following processes are similar to those used to modify other tags in the Imperial, Metric, and Metric DACH folders.

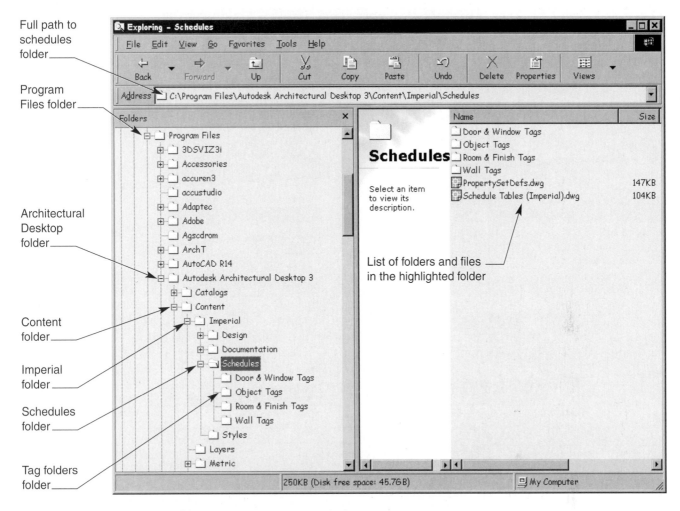

FIGURE 20.29   The Microsoft® Windows Explorer is a program that is used to manage files on your hard drive.

✎Field Notes

> Check with you instructor or manager before editing the default tags included with Architectural Desktop.

In order to create a new door tag from the existing door tag, you need to copy the existing Door Tag drawing and paste the copy into a folder somewhere under the Content root directory folder. To do this, browse to the Door & Window Tags folder. With the Door Tag.dwg and the Window Tag.dwg displayed in the right pane of Windows Explorer, highlight the Door Tag.dwg, right-click, and select Copy from the shortcut list (see Figure 20.30A) to copy the drawing file to the Clipboard program. You can then paste files in the Clipboard into another folder. It is easier to manage door tag drawings in the same folder, so the copied drawing is pasted into the same folder. Right-click in a blank area of the right pane and select Paste from the shortcut menu to paste the copied drawing into the same folder as the original drawing, and name the copied drawing Copy of Door Tag.dwg (see Figure 20.30B).

You may decide to rename the copied drawing to something other than Copy of Door Tag. To do this, select the copied drawing, right-click, and select Rename from the shortcut menu. Enter a new name for the file, making sure to not remove the .dwg file extension. If you accidentally remove the .dwg file extension, use the Rename option to add it to the end of the file. For the purposes of this discussion the tag Copy of Door Tag.dwg is renamed: Door Tag 2.dwg.

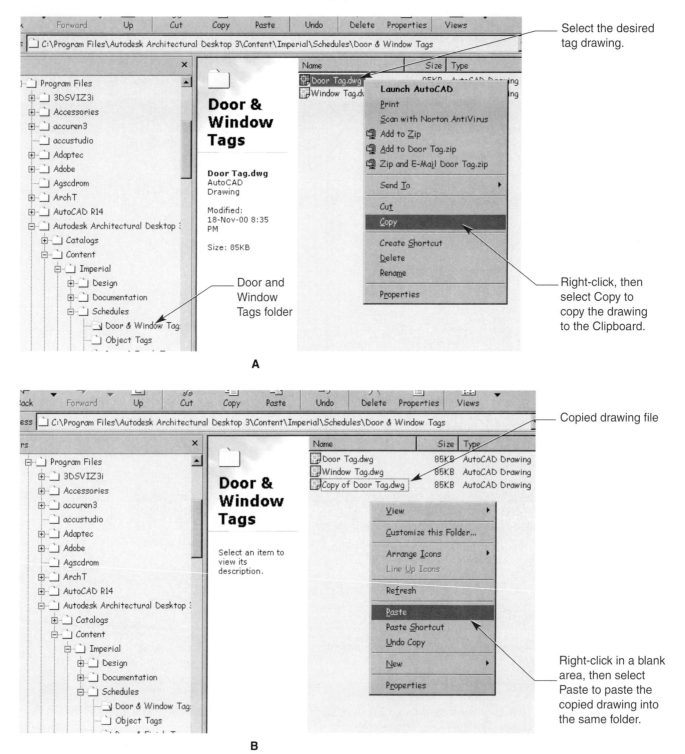

FIGURE 20.30 Copy, then Paste the tag drawing to be modified into the same folder.

Use caution when renaming files in Windows Explorer. If the file extension is removed, the file becomes invalid until the extension is added again.

Now that the tag has been copied, pasted, and renamed you no longer need Windows Explorer open. Close the window by selecting the File pull-down menu and selecting Close or by picking the [X] close button in the upper right corner of the window.

Exercise 20.6

1. Access the Microsoft(Windows Explorer program.
2. Browse to the \\Autodesk Architectural Desktop 3\Content\Imperial\Schedules\Door & Window Tags folder.
3. Copy the Door Tag.dwg, then paste it into the same folder.
4. Rename the copied drawing Door Tag 2.dwg.
5. Exit Windows Explorer.

Modifying the Copied Tag

Inside Architectural Desktop pick the Documentation pull-down menu, select the Schedule Tags cascade menu, then pick the Door & Window Tags... command. The DesignCenter window opened to the Door & Window Tags folder you were accessing in Windows Explorer is displayed (see Figure 20.31). The copied drawing tag should now be in this folder. Select the Door Tag 2 block icon, right-click, and select Open from the shortcut menu. The copied drawing is opened in the drawing screen.

FIGURE 20.31 Highlight the copied drawing block icon, right-click, and select Open to open the file in the drawing screen.

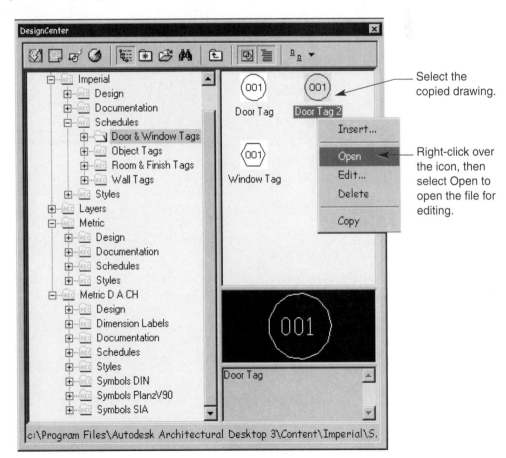

The geometry in the drawing is a block with text inside. This text is called an *attribute*. The information within the geometric symbol is variable based on information provided by the drafter or a property set definition. Before editing the block to a desired shape it is necessary to consider the current size of the block. The schedule tags are drawn eight times their desired size on paper. For example, if the text in the door tag block is to appear 1/8″ high in the finished plotted drawing, then it is created here as a 1″-high piece of text. This sizing deals with how tags are scaled appropriately when different drawing scales are chosen. The circle is drawn as a 3″-diameter circle, which reflects a 3/8″-diameter circle on the plotted drawing.

The geometry is grouped together as a block definition, so it cannot be edited with standard AutoCAD commands. You must use the **REFEDIT** command to edit the referenced block definition. The **REFEDIT** command was discussed in Chapter 15. Enter **REFEDIT** at the Command: prompt, then select the block reference in response to the prompt to select a reference. The **Reference Edit** dialog box is displayed (see Figure 20.32). This dialog box displays a preview of the block being modified. At the bottom of the dialog box select the Display attribute definitions for editing check box to enable the attributes in the drawing to be modified.

Press the OK button after selecting the check box. You are then prompted to select nested objects. Use a crossing window to select the circle and attribute text. This makes all the geometry used to create the block visible. Note that the attribute text within the circle is changed. This is the attribute tag name that is visible and not the attribute tag value. Also, a few additional attributes are visible below the circle. These were invisible attributes. The creation of attributes is explained later in this chapter (see Figure 20.33).

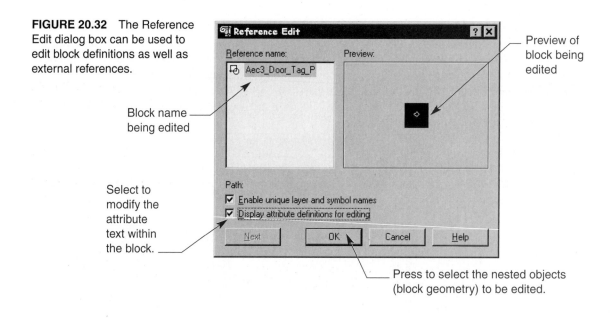

FIGURE 20.32   The Reference Edit dialog box can be used to edit block definitions as well as external references.

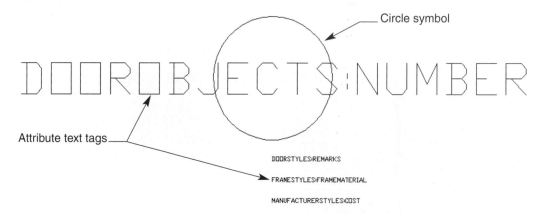

FIGURE 20.33   After the nested objects are selected, all the geometry for the block becomes visible.

After the nested objects are selected, the block geometry can be edited with standard AutoCAD commands. Note that the **Refedit** toolbar is visible in the drawing and can be used to save the changes made to the block.

It is desired to change the geometric symbol from a circle to a diamond. The **POLYGON** command can be used to draw a four-sided polygon. Use the following sequence to create a four-sided polygon around the circle, then rotate it 45°.

Command: **POLYGON** ⏎
Enter number of sides <4>: ⏎

Specify center of polygon or [Edge]: **CEN** ⏎
Of *(move mouse to edge of circle until the center osnap symbol is displayed, then pick)*
Enter an option [Inscribed in circle/Circumscribed about circle] <I>: **C** ⏎
Specify radius of circle: **QUAD**⏎
Of *(move mouse to edge of circle until a quadrant osnap symbol is displayed, then pick)*
Command: **ROTATE** ⏎
Current positive angle in UCS: ANGDIR=counterclockwise ANGBASE=0.00
Select objects: **L** ⏎
1 found
Select objects: ⏎
Specify base point: **CEN** ⏎
Of *(move mouse to edge of circle until the center osnap symbol is displayed, then pick)*
Specify rotation angle or [Reference]: **45** ⏎
Command:

A diamond is created around the outside (circumference) of the circle. Next, use the **ERASE** command to remove the circle and to make additional modifications if desired. Do not move the symbol, as it is centered on the insertion point for the block. When you have finished making the desired changes, type **REFCLOSE** at the Command: prompt, or select the **Save back changes to reference** button in the **Refedit** toolbar to save the modifications back to the block definition. You will receive a warning message that the changes will be saved to the reference. Press the OK button to save the changes. The block geometry is updated displaying the new geometric symbol.

A few additional things need to be modified before this drawing is ready to be used as a new schedule symbol. If this block is added to a drawing that already includes the default door tag block, the custom tag will not work properly. To avoid this situation, you must rename the block. Enter the **RENAME** command at the Command: prompt to display the **Rename** dialog box shown in Figure 20.34.

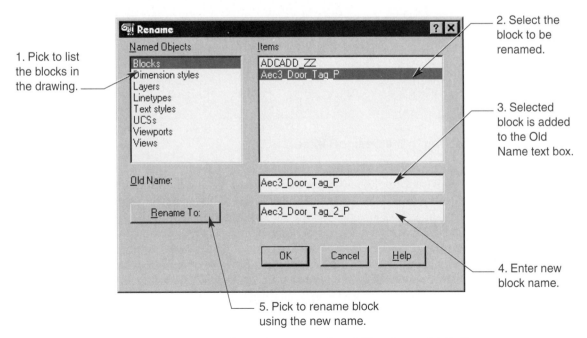

FIGURE 20.34　The Rename dialog box is used to rename AutoCAD styles and definitions.

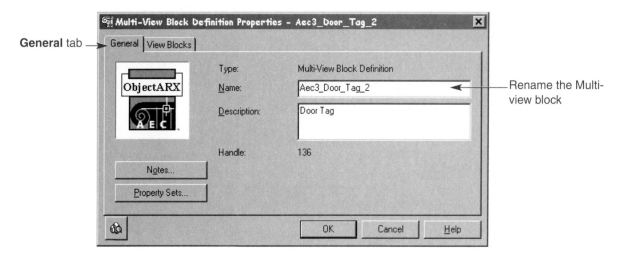

General tab

Type: Multi-View Block Definition

Name: Aec3_Door_Tag_2   ← Rename the Multi-view block

Description: Door Tag

Handle: 136

Notes...

Property Sets...

OK   Cancel   Help

FIGURE 20.35   Multi-view block is renamed in the Multi-view Block Definition Properties dialog box.

Select the Blocks text from the Named Objects list box to list the block names in the drawing. Pick the AEC3_Door_Tag_P block from the list to add the existing block name to the Old Name text box. Enter the new block name AEC3_Door_Tag_2_P in the Rename To: text box. Press the Rename To: button to rename the block. Press the OK button to exit the dialog box.

The final thing that you need to do before saving the new schedule tag is to rename the multi-view block definition or else it may have conflicts when it is inserted into a drawing that has the default door tag symbol. The creation of multi-view blocks was discussed in Chapter 7.

Select the Desktop pull-down menu, pick the Multi-View Blocks cascade menu, then select the Multi-View Block Definitions... command. The Style Manager dialog box is displayed. Double-click on the multi-view block definition to edit it. The Multi-View Block Definition Properties dialog box, shown in Figure 20.35, is displayed. In the General tab rename the multi-view block definition Aec3_Door_Tag_2. Press the OK button to close the dialog box. Press OK again in the Style Manager to return to the drawing screen. Finally, save and close the drawing.

The modifications to the copied tag are now complete.

Exercise 20.7

1. In Architectural Desktop, access the Door & Window Tags folder in the DesignCenter.
2. Select the Door Tag 2.dwg block icon, right-click, and select Open to open the drawing.
3. Modify the tag as outlined above using REFEDIT.
4. Rename the Block AEC3_Door_Tag_2_P.
5. Rename the multi-view block Aec3_Door_Tag_2.
6. Save and close the drawing.

Adding the Modified Tag to the DesignCenter

In order to use the tag in a new drawing, you must re-add the tag to the DesignCenter, by processing it through the Create AEC Content Wizard. To do this, access the Door and Window Tags folder in the DesignCenter. The modified tag block is displayed in the minipreview but is still displayed as the original circle block icon. Picking on top of the block reveals its modified shape in the larger preview pane (see Figure 20.36).

To process the modified tag, select the minipreview, right-click, and select the Edit... command from the shortcut menu. The Create AEC Content Wizard where the block can be processed is displayed (see Figure 20.37). The first page of the wizard displays the settings for the original door tag that was modified. Notice that the Custom Command option has been selected. Schedule tags are processed into the DesignCenter as commands. This allows the tag to be attached to an object and assigned a number or letter.

The first step is to add the block definition and the multi-view block definition to the Content File list area as shown in Figure 20.37. Select the AEC3_Door_Tag_2_P block and AEC3_Door_Tag_2 multi-view block from the Current Drawing list area and press the Add>>> button to add them to the other side (see Figure 20.37).

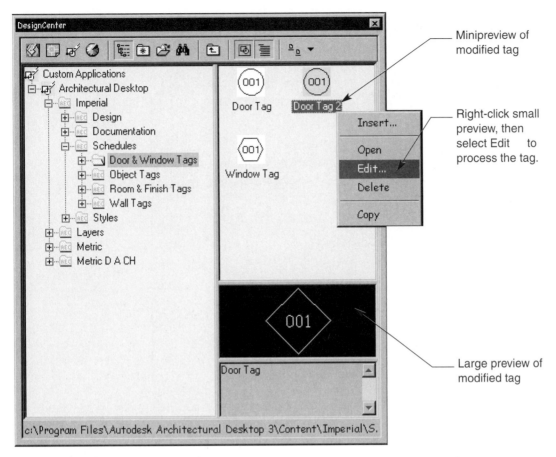

FIGURE 20.36 Right-click the minipreview to edit the copied door block and access the Create AEC Content Wizard.

The next thing to do is to ensure that the new multi-view block is inserted when this tag is attached to a door. At the bottom of the page is the Command String text box. This string contains the commands used to insert and attach the tag to the door object. Select the Expand... button to see the entire string (see Figure 20.38). In the Custom Command String dialog box, change the name of the old multi-view name (Aec3_Door_Tag) to the new modified multi-block name (Aec3_Door_Tag_2). This tells the command string to insert the new modified multi-view block into the drawing. Press the OK button to accept the string and return to the first page of the wizard. Press the Next> button at the bottom of the string to advance to the next page.

 Field Notes

Be careful when entering the multi-view block name. When the command is processed a space in the string is read by the command as an Enter and causes the command to be terminated.

This next page controls the properties of the inserted multi-view block. Values such as right-reading text, layer key, and insertion scaling options can be controlled here (see Figure 20.39). If any changes to how the tag will be inserted and appear in the drawing are desired, this page can be used to control these settings. The new schedule tag imitates the original door tag, so no changes are required in this page. Press the Next> button to advance to the final page.

The final page establishes how the minipreview of the tag appears in the DesignCenter. Select the Default Icon... button to display a minipreview of the actual modified tag. Pressing the New Icon... button allows you to select a .bmp image to use as a preview icon. Enter a description for the modified tag as shown in Figure 20.40. When you have finished, pick the Finish button to process the tag. Once the tag is processed it can be added to a drawing and attached to door objects.

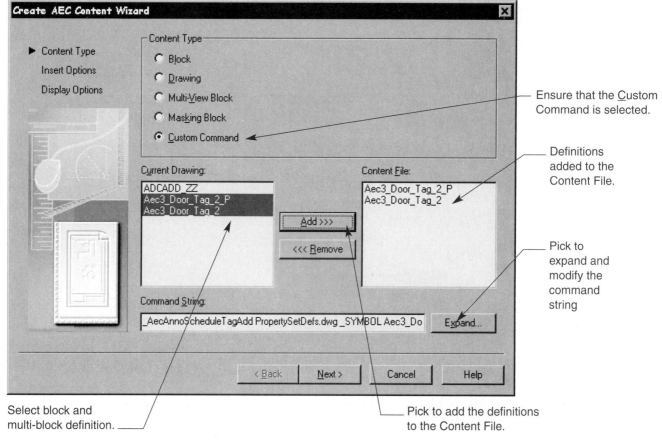

FIGURE 20.37 The first page of the wizard uses the criteria from the original tag that was modified.

Field Notes

- The block preview is displayed in the color of the drawing screen background. You may wish to change the background color to white before processing the block to make the preview a little easier to read in the DesignCenter.
- The previously discussed process is similar for all the tags included with Architectural Desktop. Follow the steps outlined to custom your own schedule tags.

Exercise 20.8

1. In the DesignCenter, select and right-click over the Door Tag 2 preview icon. Select the Edit... option to process the tag.
2. Add the Aec3_Door_Tag_2_P block and Aec3_Door_Tag_2 multi-view block definitions to the Content File.
3. Change the command string multi-view block name from Aec3_Door_Tag to Aec3_Door_Tag_2.
4. Select the Next> button to advance to the last page of the wizard.
5. Select the Default Icon... button to update the preview icon.
6. Enter a new description for the tag. Try adding the tag to some doors in a drawing.
7. Close the DesignCenter.

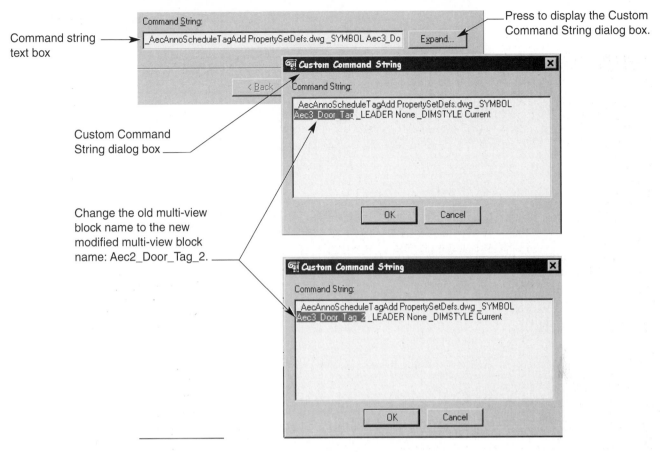

Command string text box → Command String:
_AecAnnoScheduleTagAdd PropertySetDefs.dwg _SYMBOL Aec3_Do | Expand...

Press to display the Custom Command String dialog box.

< Back

Custom Command String

Command String:

AecAnnoScheduleTagAdd PropertySetDefs.dwg _SYMBOL
Aec3_Door_Tag _LEADER None _DIMSTYLE Current

OK   Cancel

Custom Command String dialog box

Change the old multi-view block name to the new modified multi-view block name: Aec2_Door_Tag_2.

Custom Command String

Command String:

AecAnnoScheduleTagAdd PropertySetDefs.dwg _SYMBOL
Aec3_Door_Tag_2 _LEADER None _DIMSTYLE Current

OK   Cancel

FIGURE 20.38 In the command string, the name of the old multi-view block definition is changed to the new multi-view block definition name.

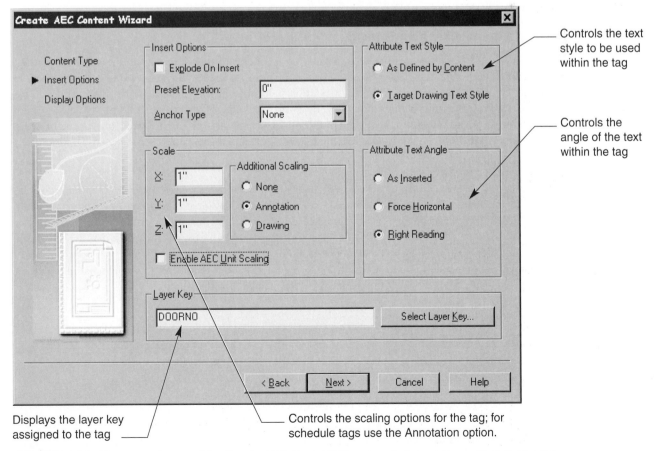

Create AEC Content Wizard

Content Type
▶ Insert Options
Display Options

Insert Options
☐ Explode On Insert
Preset Elevation: 0"
Anchor Type None

Attribute Text Style
○ As Defined by Content
● Target Drawing Text Style

Controls the text style to be used within the tag

Scale
X: 1"
Y: 1"
Z: 1"
☐ Enable AEC Unit Scaling

Additional Scaling
○ None
● Annotation
○ Drawing

Attribute Text Angle
○ As Inserted
○ Force Horizontal
● Right Reading

Controls the angle of the text within the tag

Layer Key
DOORNO   Select Layer Key...

< Back   Next >   Cancel   Help

Displays the layer key assigned to the tag

Controls the scaling options for the tag; for schedule tags use the Annotation option.

FIGURE 20.39 The second page of the Create AEC Content Wizard controls insertion options for the tag.

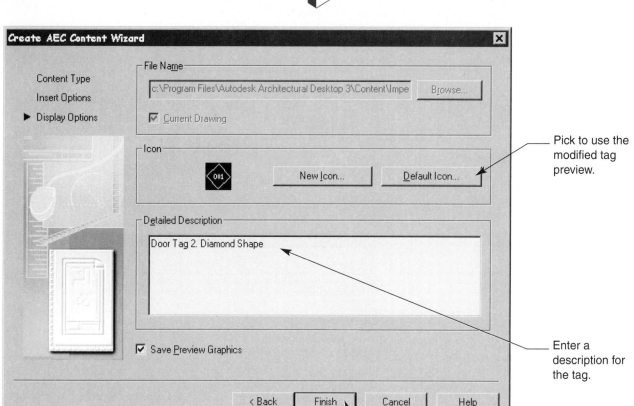

Pick to use the modified tag preview.

Enter a description for the tag.

Pick to process the tag.

FIGURE 20.40   The final page of the wizard establishes the block preview and the description for the tag.

Creating a Custom Tag from Scratch

The process used to create a customized schedule tag is similar to that described in the previous section; however, a few additional things need to be understood before the tag can be processed into a new schedule tag. When creating a custom schedule tag it is often helpful to begin a new drawing from scratch. This keeps the tag drawing from being cluttered with extra styles and layer names. To start a drawing from scratch enter the Today window and select the Create Drawings tab. From the **Select how to begin** drop-down list, pick Start from Scratch, and select the English (feet and inches) option or the Metric option.

Drawing the Tag Block Geometry
Schedule tags are "smart" blocks, because they understand how to be scaled in the drawing based on the chosen drawing scale and the type of units selected in the Drawing Setup dialog box. In order to replicate this process the new tag geometry must be created eight times larger than the size it will be plotted. For example, a 1/4″-diameter circle with 1/8″-high text in a plotted drawing needs to be drawn as a 2″-diameter circle with 1″-high text (1/4″ × 8 = 2″ and 1/8″ × 8 = 1″).

Chapter 7 discussed the creation of blocks. Remember that new geometry to be included into a block needs to be created on layer 0 using the ByBlock color, linetype, and lineweight.

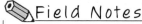 Field Notes

> For the purposes of this discussion, a custom window tag is created. Note that this process is similar for the creation of other custom schedule tags. Check with your instructor or manager before creating custom tags.

Begin by setting layer 0 current, and the color, linetype, and linweight to ByBlock. Create the desired geometric shape. The custom window tag will be a 3/8″ x 3/8″ square with a 1/8″-high window number inside on a plotted drawing. Remember that the original geometry needs to be eight times its actual plotted size. This makes the square geometry 3″ × 3″ in the drawing. Use the **RECTANG** or **PLINE** command to create the shape.

Exercise 20.9

1. Start a new drawing from scratch.
2. Set Layer 0 current and the color, linetype, and lineweight drop-down lists to ByBlock.
3. Draw a 3″ × 3″ square in the drawing screen.
4. Draw a line from one corner of the square diagonally to the opposite corner. This line will be used in the placement of the attribute text.
5. Save the drawing as Window Tag 2.dwg in the \\Autodesk Architectural Desktop 3\Content\Imperial\Schedules\Door & Window Tags folder. Keep the drawing open for the next exercise, or close the drawing.

Adding Attribute Text

An *attribute* is text incorporated into a block that allows you to specify what type of text is displayed when the block is added to the drawing. In Architectural Desktop the attribute text references a property set definition, which automatically fills out the attribute value in the tag. Examples can be found when adding a door or window tag to a drawing. The number is changed based on the value entered in the number property.

Use one of the following options to add an attribute into the drawing:

✓ Select the **Draw** pull-down menu, pick the **Block** cascade menu, then select **Define Attributes....**

✓ Type **ATTDEF** at the Command prompt.

The **Attribute Definition** dialog box, shown in Figure 20.41, is displayed. The main areas used in this dialog box when schedule tags are created are the Attribute, Insertion Point, and Text Options areas. The Mode area is usually used when attributes are created for standard block definitions. For more information concerning the attribute modes, look up the attributes section in the on-line help.

The next step is to determine the attribute tag name. Schedule tags rely on property set definition properties for the text to be displayed in the tag. For example, if the window number is to be displayed in the tag, the Number property within the WindowObjects property set definition is referenced. In order for the number to be displayed properly in the custom schedule tag, the attribute tag name needs to be WindowObjects:Number (see Figure 20.42).

FIGURE 20.41 The Attribute Definition dialog box is used to define attribute tags for blocks.

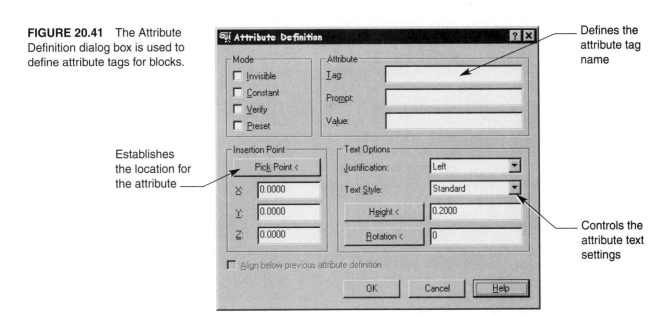

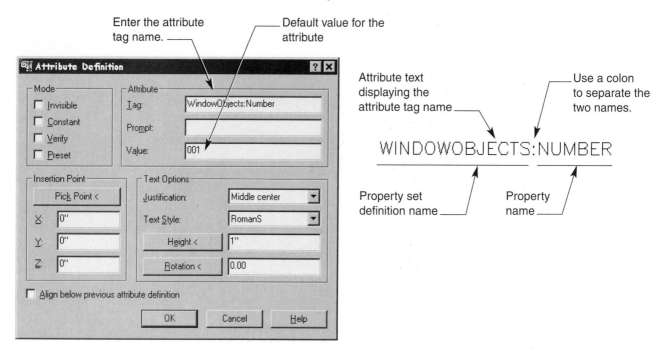

Enter the attribute tag name.

Default value for the attribute

Attribute text displaying the attribute tag name

Use a colon to separate the two names.

WINDOWOBJECTS:NUMBER

Property set definition name

Property name

FIGURE 20.42 When the schedule tag will reference a property value from a property set definition the attribute tag name must be entered in the *property set definition name:property* format.

No matter what property is displayed in the schedule tag, the attribute tag must always be named using the property set definition name followed by a colon, then the property name: *property set definition name:property.*

Field Notes

Before entering the attribute tag name you may want to determine the type of property set definition property that will be displayed in your custom tag. To do this, you can access the PropertySetDefs.dwg file and look through the different properties available for each property set definition.

After entering a name for the attribute tag, you need to enter a default value for the attribute. In the Value text box below the Prompt text box, enter a default value that will be displayed in the attribute text if the properties in the property set definition have not been filled out.

After the attribute tag name and a default value have been entered, the settings for the tag can be adjusted. The Justification drop-down list allows you to set the insertion point within the attribute. The default is Left, which inserts the attribute into the drawing from the lower left corner of the text. Some of the common justifications for attributes are **middle center, middle left,** and **middle right.** These place the attribute text in relation to the point selected for the insertion point (see Figure 20.43). Middle center justification is often used to center an attribute tag on a geometric shape. Middle left aligns the left edge of the attribute along a left edge, and middle right aligns the right edge of the text along a left edge.

Below the Justification drop-down list is the Height< button and text box. Enter the height for the attribute tag. Remember that the attribute text needs to be created eight times its plotted size. If the text is to be 1/8″, then the height needs to be set to 1″ (1/8″ × 8 = 1″).

Finally, the tag needs to be placed in the drawing in relationship to the geometric shape. Use the Pick Point< button in the Attribute Definition dialog box to return to the drawing to select a point. After the point has been selected, the Attribute Definition dialog box reappears. Press the OK button to add the attribute to the schedule tag. The attribute tag is displayed in relationship to the geometric shape (see Figure 20.44).

Additional attribute tags can be added in and around the symbol if desired. One suggestion for the custom window tag is to add the width and the height property tags below the symbol. To do this, add the WindowStyles:Width and WindowStyles:Height attributes tags to the drawing, separating the two with a standard text *X* (see Figure 20.45).

Insertion point Middle
center justification

WINDOWOBJECTS:NUMBER

Insertion point Middle
left justification ➜ WINDOWOBJECTS:NUMBE

WINDOWOBJECTS:NUMBER ← Insertion point Middle
right justification

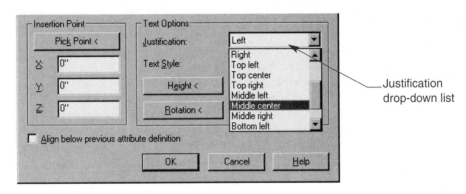

FIGURE 20.43 Some common justification points for attributes.

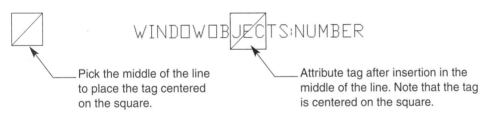

Pick the middle of the line
to place the tag centered
on the square.

WINDOWOBJECTS:NUMBER

Attribute tag after insertion in the
middle of the line. Note that the tag
is centered on the square.

FIGURE 20.44 Pick a location for the attribute text. When you have finished, the attribute tag will be
displayed in the drawing.

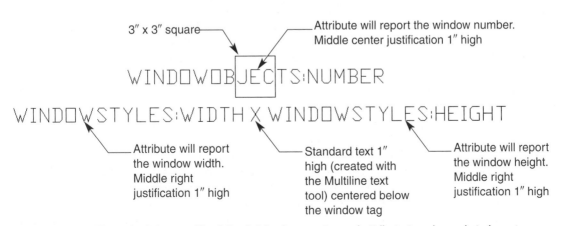

3″ x 3″ square

Attribute will report the window number.
Middle center justification 1″ high

WINDOWOBJECTS:NUMBER

WINDOWSTYLES:WIDTH X WINDOWSTYLES:HEIGHT

Attribute will report
the window width.
Middle right
justification 1″ high

Standard text 1″
high (created with
the Multiline text
tool) centered below
the window tag

Attribute will report
the window height.
Middle right
justification 1″ high

FIGURE 20.45 The schedule tag with all the finished geometry and attribute tags is ready to be
turned into a block.

Exercise 20.10

1. Continue from Exercise 20.9, or open the drawing by browsing to the Door & Window Tags folder and selecting the Window Tag 2.dwg.
2. Add a tag named WindowObjects:Number with a default value of 001, a justification of middle center and 1″ height at the midpoint of the dialog line. Erase the line when finished.
3. Add a tag named WindowStyles:Width with a default value of 3′-0″, using a middle right justification and a text height of 1″ below the square.
4. Add a tag named WindowStyles:Height with a default value of 7′-0″, using a middle left justification and a text height of 1″ below the square.
5. Use the Multiline Text tool to add an *X* below the square between the WindowStyles:Width and WindowStyles:Height attribute tags. Use the MOVE command as needed to separate the two attribute tags, leaving room for the *X* between them (see Figure 20.45).
6. Save the drawing, and keep it open for the next exercise.

Making the Block and Multi-View Block Definitions

After the tag geometry and the attribute tags have been drawn, all the geometry needs to be grouped together into a standard block definition, which is added to a multi-view block definition. Use the BLOCK command discussed in Chapter 7 to add the tag geometry and attribute tags together as shown in Figure 20.46.

Provide a name for the block that helps describe the block. For example, consider naming the block Window_Tag_2_P, where the *P* stands for plan view, because the window tag used as a model is placed in a plan view, and a window tag block is already included with Architectural Desktop. Do not use spaces between the block names, because this disrupts the command when the block is inserted into the drawing. Instead, use the underscore symbol (_) to separate the words.

Select a logical insertion point for the block. This block is used as a schedule tag, so the center of the tag geometry is probably the most desirable location. Use the **Insertion** object snap to snap to the insertion point of the attribute tag that is centered on the geometric shape.

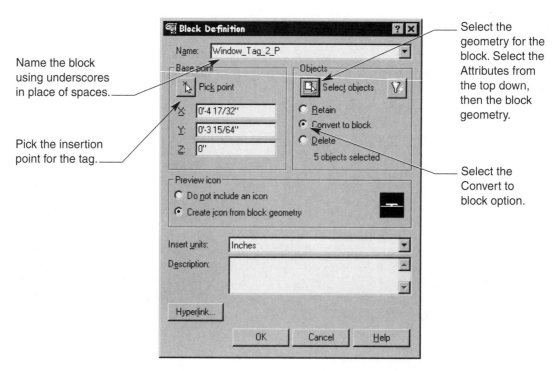

FIGURE 20.46   The BLOCK command is used to create a block from the drawn geometry.

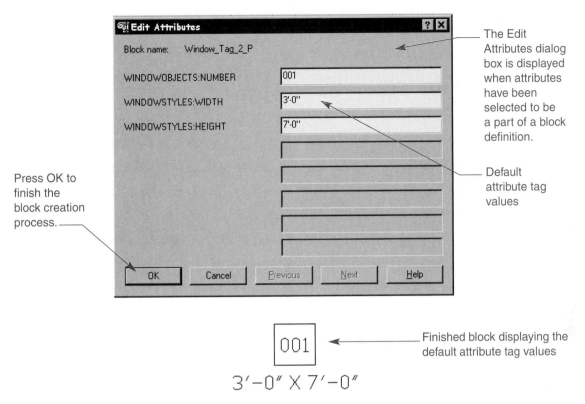

Press OK to finish the block creation process.

The Edit Attributes dialog box is displayed when attributes have been selected to be a part of a block definition.

Default attribute tag values

Finished block displaying the default attribute tag values

FIGURE 20.47 The Edit Attributes dialog box is displayed if attribute tags have been included in a block definition.

Next, select the items to be included in the tag. Begin by individually picking each attribute tag, starting from the top and working your way down from the left to right, to fill out attributes in order, then select the additional geometry such as the square and the *X* separating the two attributes. Select the **Convert to block** radio button to immediately change the geometry into a block within the drawing.

Press the OK button when you have finished to turn the geometry into a block definition. Because attributes are being included in the block definition, the **Edit Attributes** dialog box is displayed. This allows you to enter any value desired for each attribute tag to be displayed in the block. Press **OK** to finish the process of turning the geometry into a block. Notice that the geometry is now grouped together into a single block definition, and the attribute tags are now displaying the default values entered in the Edit Attributes dialog box (see Figure 20.47).

The final step before the block can be processed back into the DesignCenter is to add the block definition into a multi-view block definition. Creating multi-view block definitions was discussed in Chapter 7. To create the multi-view block, select the Desktop pull-down menu, pick the Multi-View Blocks cascade menu, then pick the Multi-View Block Definitions... command. The Style Manager dialog box is displayed. Create a new definition and give it the same name as the block definition name except leave off the "_P". Remember to use underscores in place of the spaces between words. In this case the multi-view block name should be Window_Tag_2.

Select the new multi-view block and edit the definition. In the View Blocks tab, select the General display representation (see Figure 20.48). Pick the Add... button, then select the block definition. In this case the Window_Tag_2_P block is selected. Assign this block to only the top viewing direction. The window tag is thus displayed only in the top viewing direction. Press the OK button to return to the Style Manager. Press OK again to return to the drawing. You are now ready to process the tag into the DesignCenter.

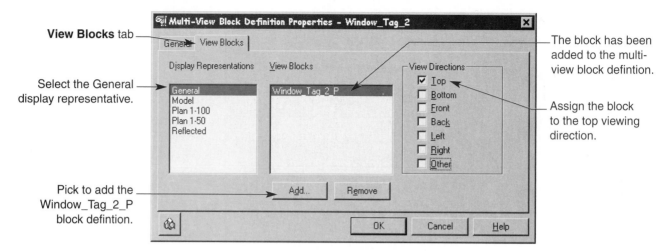

View Blocks tab

Select the General display representative.

Pick to add the Window_Tag_2_P block definition.

The block has been added to the multi-view block defintion.

Assign the block to the top viewing direction.

FIGURE 20.48 The newly defined block is added to the top viewing direction in the General display representation.

Exercise 20.11

1. Continue from Exercise 20.10, or open the drawing by browsing to the Door & Window Tags folder and selecting the Window Tag 2.dwg.
2. Turn the geometry into a block named Window_Tag_2_P. Specify the insertion point by using Insertion osnap and snap to insert the WindowObjects:Number attribute.
3. Select the geometry for the tag. Select the Convert to block option. Press OK when finished.
4. Create a new multi-view block definition named Window_Tag_2.
5. Edit the new multi-view block by assigning the Window_Tag_2_P block to the General display representation and having it display only in the top viewing direction.
6. Save and close the drawing.

Adding the Custom Tag to the DesignCenter

Now that the tag geometry has been used to create a block definition and a multi-view block definition, it needs to be processed into the DesignCenter. In order to process the block, you need to open the window tag drawing as a read-only drawing. To do this, use the **OPEN** command to display the Select File dialog box (see Figure 20.49). Browse to the folder to which the tag drawing was saved.

Select the drawing to highlight it. In the lower right corner of the dialog box next to the Open button is a drop-down arrow. Select this arrow and pick Open Read-Only to open the drawing as a read-only file.

Field Notes

Opening a drawing as a read-only drawing allows you to process the tag into the DesignCenter drawing. If the drawing is not opened as a read-only file, the Create AEC Content Wizard will not process the tag.

To process the tag, select the Desktop pull-down menu, and pick the Create AEC Content... command. The first page of the wizard is displayed. Select the Custom Command radio button to process the tag as a command. Add the Window_Tag_2_P block definition and the Window_Tag_2 multi-view block definition to the Content File area (see Figure 20.50).

In order for the tag to work, a command string needs to be added. A *command string* is a series of commands to insert the schedule tag and to attach the property set definitions to the tag. The commands are separated by spaces, which indicate to Architectural Desktop that the command is to be processed. This is

FIGURE 20.49 The Open dialog box is used to open the custom tag drawing as a read–only file.

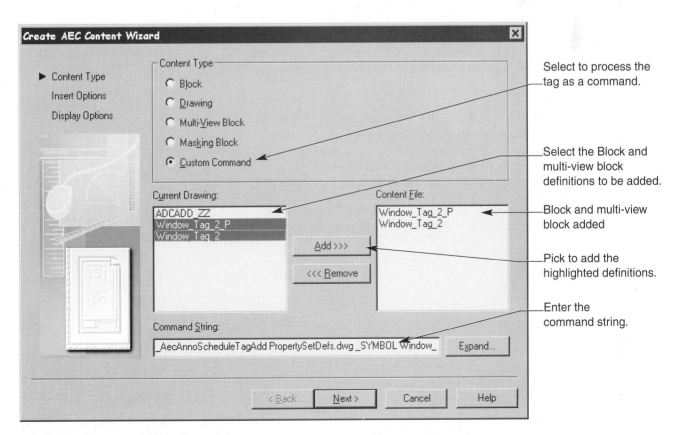

FIGURE 20.50 The Custom Content radio button in the first page of the wizard is used to process the tag as a command.

similar to entering a command at the Command: prompt then pressing the [Enter] key to process the command. The space represents the pressing of the [Enter] key. The following command string is used to create schedule tags:

> _AecAnnoScheduleTagAdd PropertySetDefs.dwg _SYMBOL Window_Tag_2 _LEADER None
>   _DIMSTYLE Current

The first part of the command string indicates the command being used to insert the schedule tag:

> _AecAnnoScheduleTagAdd

Preceded by a space, the next part of the command indicates the drawing housing the property set definitions referenced by the attribute tags:

> PropertySetDefs.dwg

Preceded by a space, the next command indicates that a symbol (multi-view block) will be inserted:

> _SYMBOL

Preceded by a space, the multi-view block name is entered as the name of the symbol to be inserted:

> Window_Tag_2

Preceded by a space the next command indicates that the Leader option in the AecAnnoScheduleTagAdd command is being accessed:

> _LEADER

Preceded by a space, the next part of the command string indicates that no leader line will be attached to the tag:

> None

Preceded by a space, the last option of the AecAnnoScheduleTagAdd command indicates the dimension style being referenced for the insertion of the tag:

> DIMSTYLE

Finally, the value for the DIMSTYLE option is entered. This value is preceded by a space, plus a space at the end of the command string ends the command:

> Current

If this command is typed at the Command: prompt, it appears as follows:

> Command: **AECANNOSCHEDULETAGADD** ↵
> Property set definitions drawing < PropertySetDefs.dwg>: **PROPERTYSETDEFS.DWG** ↵
> Select object to tag [Symbol/Leader/Dimstyle]: **SYMBOL** ↵
> Enter tag symbol name <Aec3_Window_Tag>: **WINDOW_TAG_2** ↵
> Select object to tag [Symbol/Leader/Dimstyle]: **LEADER** ↵
> Enter leader type [None/STraight/SPline] <None>: **NONE** ↵
> Select object to tag [Symbol/Leader/Dimstyle]: **DIMSTYLE** ↵
> Enter leader dimstyle CURRENT <Aec_Arch_I$7>: **CURRENT** ↵
> Select object to tag [Symbol/Leader/Dimstyle]: ↵
> Command:

Field Notes

Be cautious about entering the command string. Use spaces only to separate the command, command options, and option values, and use a space at the end of the command. Additional spaces or lack of spaces between the command options prevent the command from functioning. Also, note that the example command string is used to insert schedule tags. Change the multi-view block name portion of the command string as required.

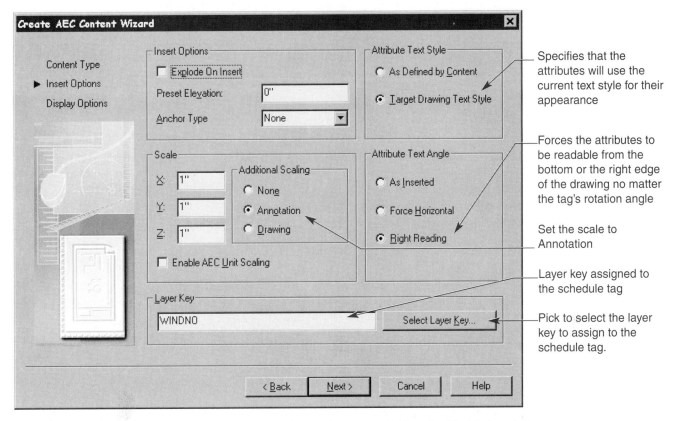

FIGURE 20.51 The second page of the wizard contains settings to control how the block is inserted and how the attribute text is to appear.

After you have configured all of the settings in the first page and entered the command string, press the Next> button to advance to the next page. This page controls the scale of the schedule tag, the layer key the tag is assigned to, and attribute options (see Figure 20.51).

In the Scale area, select the Annotation radio button to scale the schedule tag according to the drawing scale selected in the drawing. Under the Attribute Text Style area select the Target Drawing Text Style radio button to cause the attributes to use the current text style in the drawing for the tag fonts. In the Attribute Text angle area, select the Right Reading radio button to force the attribute text to be readable from the bottom or the right side of the plotted drawing. This avoids having upside-down text in the tag. Finally, pick the Select Layer Key... button to select a layer key to assign to the tag. In this case the WINDOW key is assigned, so the tag is associated with the A-Glaz_Iden layer. When you have finished press the Next> button to advance to the final page of the wizard.

The final page is used to assign a location and a name to the new content file (see Figure 20.52). Pick the Browse... button to browse to the folder where you wish to save the tag. Enter a name for the tag. The schedule tag was opened as a read-only file for the AEC Content command, so you can select the existing drawing name and location to save on top of the original. When you select the Save button you may get a message stating that the drawing will be overwritten. Press Yes to overwrite the file.

Pick the Default Icon... button to accept the default image for use as the preview in the DesignCenter. Enter a description as desired, then pick the Finish button to process the command. The block should now be processed and added to the DesignCenter for use. The last thing to do is to close the read-only schedule tag file. You will get a message asking if you want to save the file. Enter a No, because the command has already been processed into the DesignCenter. The new tag is now ready to be attached to the windows.

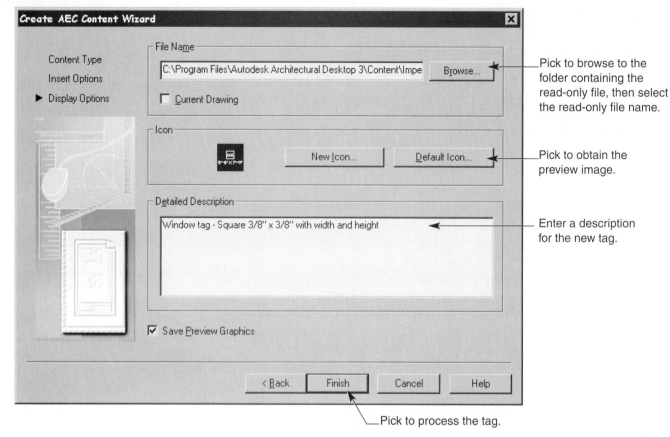

Pick to browse to the folder containing the read-only file, then select the read-only file name.

Pick to obtain the preview image.

Enter a description for the new tag.

Pick to process the tag.

FIGURE 20.52   The final page allows specification of the tag name and where the tag will reside, the preview image, and the description for the tag.

 Field Notes

Custom Schedule tags understand the objects they need to be attached to by the name of the attributes in the block definition. If a block includes the two attributes DoorObjects:Number and WindowStyles:Width, it understands that it can be used for both door and window objects. If the tag includes only DoorObjects or DoorStyles attributes, it is attached only to door objects.

Exercise 20.12

1. Open the Window Tag 2 drawing as a read-only file.
2. Use the Create AEC Content command to process the new tag into the DesignCenter for use as a new tag.
3. Use the preceding example for the tag settings.
4. Save the block into the \Autodesk Architectural Desktop3\Content\Imperial\Schedules\ Door & Window Tag folder. Specify the tag name as Window Tag 2.
5. Close the read-only drawing without saving when you have finished processing the tag.
6. Try the new tag on some window objects.

CHAPTER REVIEW

Use the CD-ROM to test your knowledge and skills.

Chapter Test

To check your understanding of the content provided in this chapter, access the Test file in the CH20 folder of the CD-ROM that accompanies this text.

Chapter Project

To practice the Architectural Desktop skills presented in this chapter, access the Project files in the CH20 folder of the CD-ROM that accompanies this text. The project files are in pdf format and include sample drawings and instructions for completing each project.

|||||||||||||||||||||||||||||||||||||||||||||||||||||||||||||||||||||||||||||||||||||||||||||||

CHAPTER 21

Layouts and Plotting

LEARNING GOALS

After completing this chapter, you will be able to:

◎ Answer questions related to model and layout space.

◎ Calculate scale factors.

◎ Set up a layout page.

◎ Establish viewports.

◎ Control the viewports, layers, linetypes, and color display.

◎ Add a title block to the layout.

◎ Add dimensions and annotations.

◎ Manage your layout.

◎ Use page setup.

◎ Set up plot styles.

◎ Preview your plot.

◎ Plot your drawings.

◎ Add a plot stamp.

◎ Plot to a file.

◎ Plot a DWF file.

The previous chapters covered many of the commands used to create a set of construction documents. All these commands and the geometry they establish have been used in the *model space* area of the drawing screen. Everything that is drawn in the model space area is always drawn at full scale. This helps in the development of a plan, because you can create the building model accurately without having to fuss with a scale.

When the drawing is ready to be printed or plotted, it is taken to the *layout space* of the drawing screen, scaled to fit on a full-size sheet of paper, and plotted. Layout space is also known as *paper space*, because the drawing is plotted on a sheet of paper. These terms are often used interchangeably. Once the layout space has been arranged to your satisfaction, the drawing can be sent to a printer or plotter and assembled into a hard copy of construction documents.

Model Space and Layout Space

Along the bottom edge of the drawing screen are a series of tabs known as *layout tabs* (see Figure 21.1). These tabs indicate the layout in which you are currently drawing. The Model tab indicates the model space area, where everything is drawn at 1:1, full scale. Selecting a different layout tab changes the drawing screen to layout space. Layout space is used to develop the appearance of a finished sheet in a set of construction documents.

Both the model space and the layout space area contain a user coordinate system (UCS) icon. This icon found in the lower left corner of the drawing screen indicates the current drawing space. (see Figure 21.2). The

Model space tab   Layout/paper
                                                                space tabs

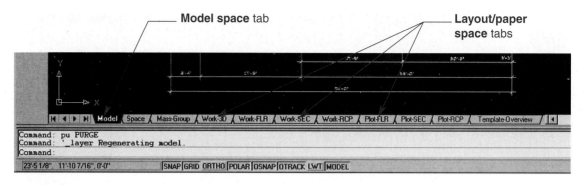

FIGURE 21.1   The tabs along the bottom edge of the drawing screen indicate the layouts available in the drawing.

FIGURE 21.2   UCS icons for model and layout space indicate the current drawing space.

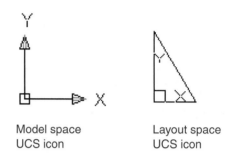

Model space   Layout space
UCS icon   UCS icon

model space UCS icon is represented as two arrows pointing in the positive X and Y directions. This icon is displayed when you are working in model space, in each tiled viewport in model space, or in floating viewports in layout space. The layout space UCS icon is represented as a right triangle with the letters X and Y indicating the positive axis directions. This icon is visible only when you are working on a layout sheet.

The model space UCS icon is available in both spaces, but the layout space UCS icon is available only in a layout. Similar to the tiled viewports in model space, *floating viewports* are found in layout space. These viewports allow you to arrange views of the drawing on the sheet of paper to be plotted. When you are working inside a floating viewport in a layout sheet, the model space icon is displayed in each floating viewport. When you are arranging the floating viewports for plotting in a layout sheet or filling out title bar information, the layout space icon is displayed (see Figure 21.3).

In addition to the UCS icons, which indicate the current space being worked in, the last button in the status bar indicates which space is active. If model space is the current space, the MODEL button is displayed. If layout/paper space is current the PAPER button is displayed. Pressing the button toggles between the two spaces in a layout tab. Layout tabs and floating viewports are discussed in greater detail later in this chapter.

The AEC Arch (Imperial), AEC Arch (Imperial - Int) and AEC Arch (Metric) templates include a number of tabs preconfigured to display different parts of the building model. Some of the tabs are designed to view floor plans, reflected ceiling plans, sections, and 3D views. Other tabs are set up for plotting and include a title block. The last tab named Template-Overview, displays a drawing representing each layout tab that describes the function of each tab (see Figure 21.4).

Understanding Layout Space

As previously mentioned, every component used to create the building model, details, floor plans, elevations, and sections should be drawn at full scale. The *only* exception to this should be annotation items such as text, dimension component sizes, and annotation blocks such as detail bubbles, section arrows, revision clouds, and leader lines. These items are drawn according to the drawing scale determined in the Drawing Setup dialog box.

Annotation objects are created by multiplying the desired finished paper size object by the drawing scale factor. For example, a detail bubble that is to appear as a 3/8″-diameter circle on a drawing that is scaled 1/4″=1′-0″ is actually drawn as an 18″-diameter circle in model space. The drawing scale factor is automatically determined for dimension size and annotation blocks when the drawing scale is set up in the Drawing Setup dialog box.

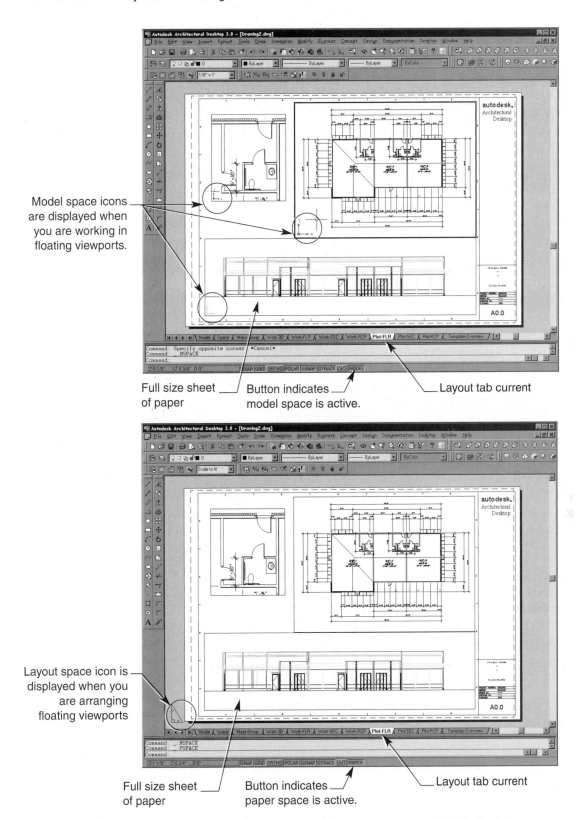

Model space icons are displayed when you are working in floating viewports.

Full size sheet of paper

Button indicates model space is active.

Layout tab current

Layout space icon is displayed when you are arranging floating viewports

Full size sheet of paper

Button indicates paper space is active.

Layout tab current

FIGURE 21.3 The model UCS is displayed in each floating viewport. The layout UCS is displayed in layout space.

Diagrams of each layout tab with
descriptions of the function of each

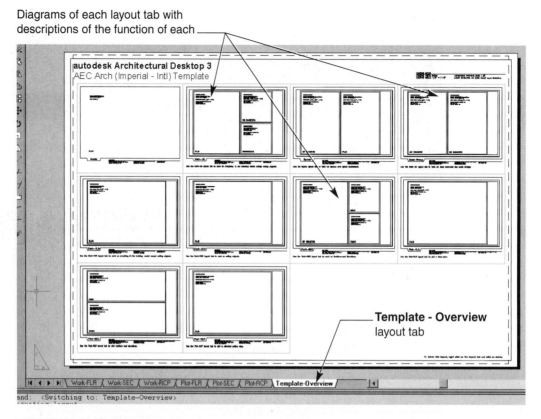

Template - Overview
layout tab

FIGURE 21.4   The Template Overview layout tab is helpful for understanding the function of each layout tab in a template.

Detail bubble: ¼″ = 1′-0″ drawing scale
                      25″ = 12″
                      12″ ÷ .25″ = scale factor: 48

                      3/8″ ϕ (circle on paper) X 48 (scale facto)r = 18″ ϕ (in model space)

If you are creating text, multiply the plotted text height by the scale factor to get the model space text height. For example, a piece of text that is to appear 1/8″ high on the finished drawing on a 1/8″=1′-0″ drawing should be drawn 12″ high in model space.

Text height   ⅛″ = 1′-0″ drawing scale
                    .125″ = 12″
                    12″ ÷ .125″ = scale factor: 96

                    ⅛″ (text height) X 96 (scale factor) = 12″ (high text in model space)

Layout space is where you set up a real-size sheet of paper in the drawing, draw a title block and notes, scale the model space drawing to fit on the paper, and arrange any views of the drawing on the sheet of paper. The geometry that is drawn directly in layout space is also drawn at full scale, like the building components in model space. For example, because the layout paper size is a full size sheet of paper, the text drawn directly on the layout sheet is drawn ⅛″ high and not the model space scaled height of 12″.

The only thing that is scaled in a layout is the view of the model within a floating viewport. These full-size views of the building are scaled down to fit on the layout sheet. The building model is not actually scaled in model space, but the viewing distance is stepped away from the building, making it smaller so that it will fit on a piece of paper (see Figure 21.5).

Although drawings can be plotted from model space, only one tiled viewport can be plotted at a time. The creation of tiled viewports was discussed in Chapter 4. The advantage to using a sheet of paper in layout space is that all viewports on a layout can be arranged and plotted at different scales at the same time. For example, a floor plan sheet may include a viewport in which the floor plan is displayed set to a scale of ¼″ = 1′-0″. Additionally, it may also include some details in other viewports that are set to a scale of ¾″ = 1′-0″ as shown in Figure 21.6.

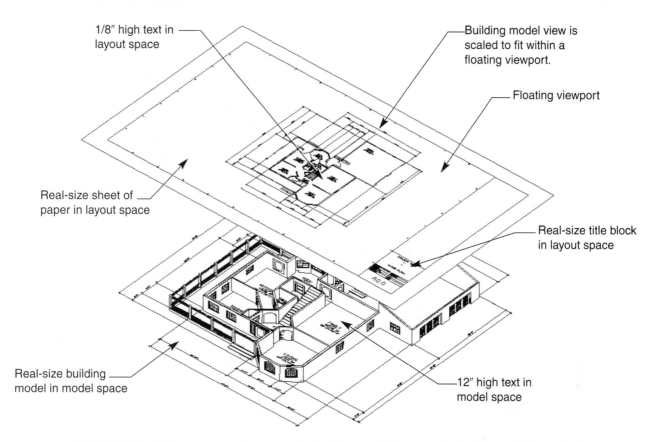

FIGURE 21.5   When a layout sheet view of a building model is set up the building model is scaled in order to fit onto a real-size sheet of paper.

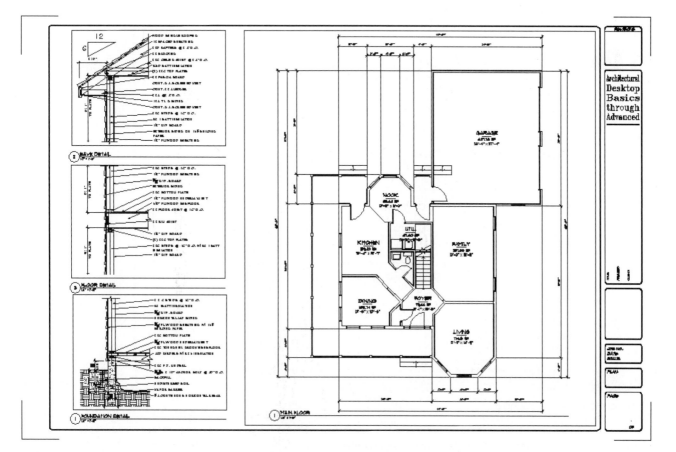

FIGURE 21.6   A layout sheet that is ready to be plotted may include different viewports at different scales that can be plotted on the same sheet of paper.

Basic Layout Setup

You may set a layout sheet up prior to drawing any building component in model space, or you may set it up after the drawing is completed. When you use the AEC Arch (Imperial) or AEC Arch (Imperial - Int) templates, three layout tabs specifically for plotting have been included. These templates are named Plot-FLR, Plot-SEC, and Plot-RCP.

To switch between model space and one of these tabs, pick on top of the desired tab at the bottom of the drawing screen. This switches the space you are currently working in to a different layout (see Figure 21.7). Layout space displays a white sheet of paper that represents a full-size sheet of paper. If a building model has been drawn in model space, then it is displayed in a floating viewport.

Layout space can be thought of as a full-size sheet of paper that is draped over the full-size building model. Cutting a hole in the paper allows you to see the building model through the paper. Moving the paper away from the model (zooming) allows the building to become smaller in the viewport, thus scaling the full-scale model smaller, and allows the model to fit onto a full-size sheet of paper.

The Plot-FLR, Plot-SEC, and Plot-RCP layout tabs have been preset to use a 36″ × 24″ sheet of paper and a plotting scale of ⅛″ = 1′-0″. Each of these tabs includes a generic title block, one or two floating viewports that display a different viewing direction, and display configurations assigned to each viewport. The Plot-FLR tab is set to the Plot display configuration, which displays the floor plan view of the drawing. The Plot-SEC tab includes two floating viewports. The top viewport displays a front view of the building model and uses the Plot display configuration; the bottom viewport displays a left view of the building model and also uses the Plot display configuration. The Plot-RCP tab displays a floor plan view but uses the Plot_Reflected display configuration, which displays the floor plan as a reflected ceiling plan.

The preconfigured plot layouts can be modified to use a different size of paper, to have a plotter assigned to them, and to assign plotted colors and lineweights to the building model geometry. These settings are controlled by the **Page Setup** command. Page Setup can be accessed by right-clicking on top of

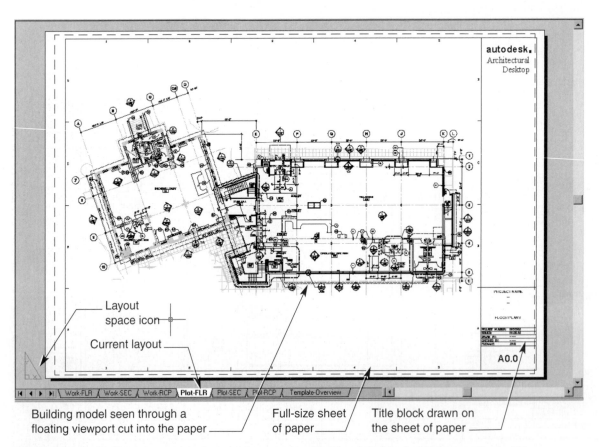

FIGURE 21.7 Picking a layout tab changes the space to display a full size sheet of paper. If a building model is already drawn in model space, it is displayed in a floating viewport.
Drawing courtesy SERA Architects Inc., Portland, OR.

the layout to be modified, then selecting Page Setup... from the shortcut menu. Page Setup is described in depth in the next section.

Additional layouts can be created and configured to plot on any size sheet of paper, with any scale, view, and arrangement desired. Use one of the following options to create a new layout sheet:

✓ Pick the **Insert** pull-down menu, select the **Layout** cascade menu, and then pick the **New Layout** command.

✓ Pick the **New Layout** button in the **Layouts** toolbar.

✓ Right-click on top of an existing layout tab, and select **New layout** from the shortcut menu.

✓ Type **LAYOUT** at the Command: prompt using the New option.

A new layout tab named Layout 1 is created, which is placed at the end of all the layouts. Once the new layout has been added or you have selected a layout to modify, you can adjust its setup.

Setting Up a Layout Page

The first time a new layout tab is selected, the Page Setup dialog box is displayed so that sheet settings can be assigned to the layout. Selecting an existing layout tab only makes the space active and does not display the Page Setup dialog box. Use one of the following options to access the Page Setup dialog box from an existing layout:

✓ Select the **File** pull-down menu, then pick the **Page Setup...** command.

✓ Pick the **Page Setup** button in the **Layouts** toolbar.

✓ Right-click over the current layout tab, and select **Page Setup...** from the shortcut menu.

✓ Type **PAGESETUP** at the Command: prompt.

The Page Setup dialog box is displayed for the current layout tab, as shown in Figure 21.8.

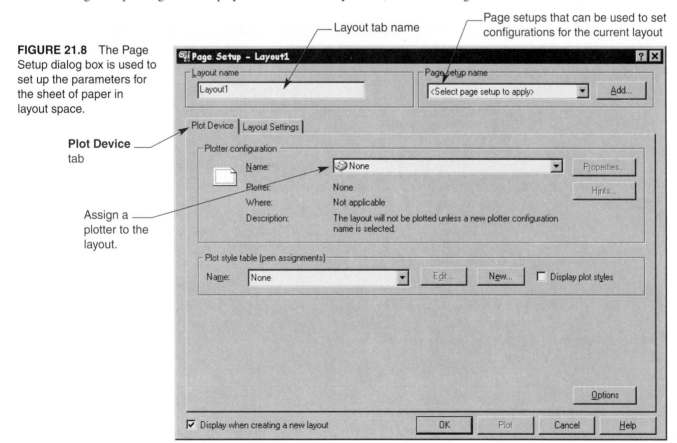

FIGURE 21.8 The Page Setup dialog box is used to set up the parameters for the sheet of paper in layout space.

Each layout tab includes its own configurations known as a ***page setup***. The page setup indicates the type of plotter assigned to the sheet, the size of the paper, and the scale of the sheet. These parameters allow the layout sheet to remember its settings at the time it is plotted. The Page Setup command is similar to other Windows programs page setup commands.

The Page Setup dialog box includes two tabs used to set up the layout sheet. First select the Plot Device tab (see Figure 21.9). In the Layout name area enter a name for the layout tab. This name appears on the layout tab at the bottom of the drawing screen. Names for layouts vary, ranging from a brief description of the sheet, such as Roof Plan, to the sheet name, such as A-FP01.

After entering a name for the layout, move to the Plotter configuration area. Select a plotter or printer from the Name: drop-down list. The printers and plotters listed are plotters and printers configured within the Windows operating system and some included with Architectural Desktop. The plotter or printer selected designates the paper sizes available when you assign a paper sheet size to the layout (see Figure 21.9).

After entering a name for the layout and selecting a plotter, you can begin configuring the layout settings. Select the Layout Settings tab to assign the paper size and scale to the layout sheet (see Figure 21.10). Select the Paper size: drop-down list to choose the paper size. The paper sizes available are based on the plotter specified in the Plot Device tab. If a paper size is not available, try choosing a different plotter.

The Plot scale area allows you to assign a scale for the layout being plotted. This setting does not affect the scale of the building model within the floating viewport but designates the scale of the entire layout sheet. If you are plotting out of model space, specifying a scale here will scale the model space view to the selected scale on the printed sheet of paper. Because you are setting up a layout sheet you want to be plotted to a real-size sheet of paper, so you should select the scale of 1:1.

When you have finished configuring the page setup, press the OK button to return to the layout. If this is an existing layout being modified, the paper size and any other settings will be updated. If this is a new layout being configured, the paper size selected is displayed, with a floating viewport cut into the paper and the building model displayed inside the viewport (see Figure 21.11). In addition to the paper size and floating viewport a dashed line representing the margin area for the paper is displayed. Any geometry within the margin area will be plotted. Anything outside the boundary is ignored and will not be plotted.

The final step in setting up the layout is to assign a scale and display configuration to the floating viewport(s). When a scale and a display configuration are assigned, the viewport to be scaled must be

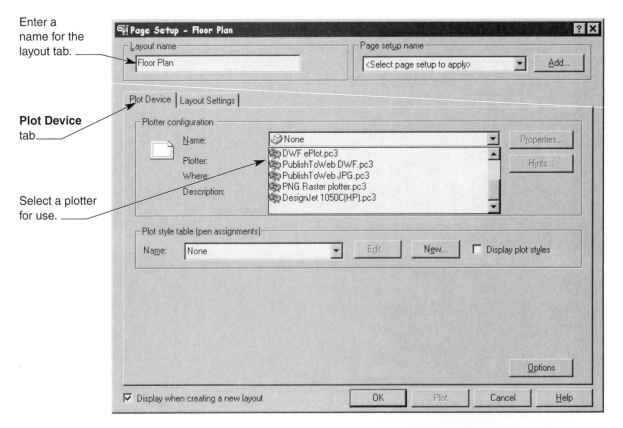

FIGURE 21.9   The plot device tab is used to enter a name for the layout and assign a plotter to the layout.

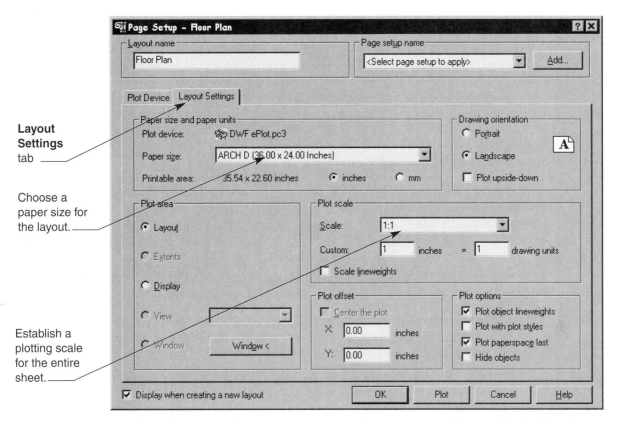

FIGURE 21.10　The Layout Settings tab is used to configure the paper size and sheet scale.

Layout
Settings
tab

Choose a
paper size for
the layout.

Establish a
plotting scale
for the entire
sheet.

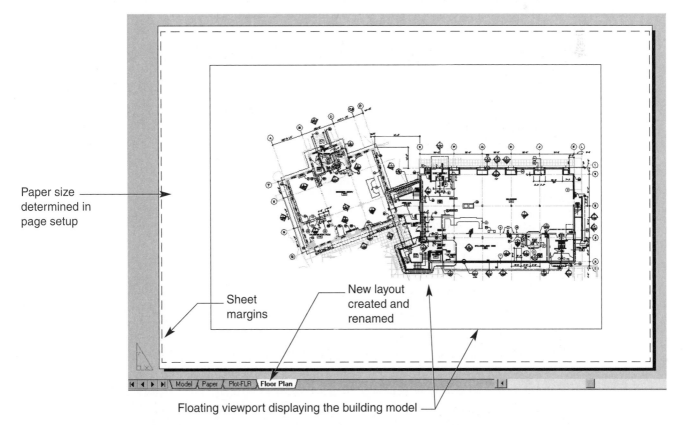

Paper size
determined in
page setup

Sheet
margins

New layout
created and
renamed

Floating viewport displaying the building model

FIGURE 21.11　The new layout is displayed with the full-size sheet of paper and a viewport cut into the paper displaying the building model behind.
Drawing courtesy SERA Architects Inc., Portland, OR.

active. Before scaling the viewport you need to understand how to activate the layout space corresponding to the floating viewport.

A few cues can help you determine whether the viewport is active or the layout sheet is active. An active viewport displays the outline of the viewport as a heavy line, and the last button in the Status bar displays the word MODEL. When the viewport is active, you are also constrained to moving the crosshairs only within the borders of the viewport. Moving the cursor outside the viewport causes the cursor to change from crosshairs to an arrow. In addition, an active viewport displays the model space UCS icon in the lower left corner of the viewport (see Figure 21.12A). An active viewport limits your work to the viewport itself and does not allow you to draw on the sheet of paper.

When a viewport is active the layout sheet space is inactive. A few cues also can help you determine whether the layout sheet space is active. If the layout space is active, the viewport edges are displayed as a thin border. The layout space icon is displayed in the lower left corner of the drawing screen. Crosshairs can be moved anywhere within the drawing area, and the last button in the status bar displays the word PAPER (see Figure 12-12B).

When a viewport is active you are working in the model space area through a layout space viewport. Any work that is done directly affects the model space objects and view. When the layout space is active, you can draw anywhere on the paper or in the layout space. Geometry that is created or modified does not affect geometry in model space.

You can toggle between the two areas by picking the MODEL/PAPER button in the status bar, entering **PSPACE** at the Command: prompt to enter the layout area or **MSPACE** at the Command: prompt to enter the viewport. Additionally, you can double-click the mouse within the viewport boundaries to make the viewport active, or double-click the mouse outside the viewport boundary to make the layout space active.

Setting the Scale

Once the viewport has been activated, the scale for the building model can be determined and a display configuration set. There are two methods of scaling the building within the viewport. The first requires that you know the scale factor of the drawing scale to be assigned to the viewport. The following formula can help you determine the scale factor for a drawing scale:

Drawing scale:	¼″ = 1′-0″	Drawing scale:	½″ = 1′-0″
	.25″ = 12″		.5″ = 12″
Scale factor:	12″ ÷ .25″ = 48	Scale factor:	12″ ÷ .5″ = 24

 Field Notes

> Appendix F provides tables of common architectural scales and their scale factors, text heights, and linetype scales for use in your drawings.

Once you have determined the scale factor you can apply it to the viewport. With the viewport active, perform a zoom extents of the building model to extend the geometry to the borders of the viewport. Next, use the PAN command or the ZOOM command with the Center option to center the part of the drawing that you want centered in the viewport.

After centering the geometry in the viewport you can apply the scale. Enter **ZOOM** at the Command: prompt, then enter 1/nxp where n is the scale factor to set the viewport scale. This setting zooms the view closer to or farther away from the building model in the viewport, thus scaling the view to the chosen scale. The formula 1/nxp is interpreted as "scale the model 1 nth the size of the model space units times paper space." The following prompt sequence scales the drawing in Figure 12.13 in the layout sheet viewport.

Command: **ZOOM** ⏎
Specify corner of window, enter a scale factor (nX or nXP), or
[All/Center/Dynamic/Extents/Previous/Scale/Window] <real time>: **1/96xp** ⏎
Command:

If the viewport cuts off part of the building model beneath, do not worry, this can be adjusted. Adjusting the viewport is discussed later in the chapter. The important point to understand is that the view-

Bold viewport
outline

Crosshairs are
constrained to
the interior of the
viewport

Model space
UCS icon

Model button
displayed

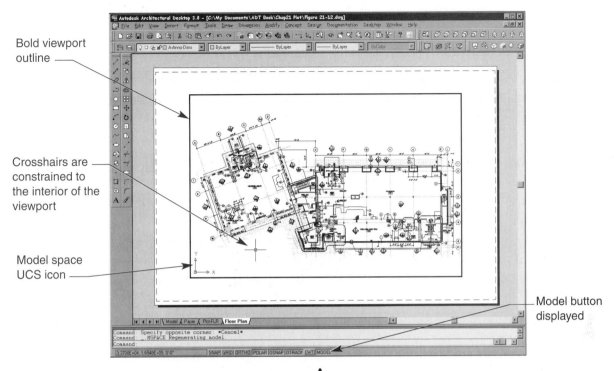

A
Viewport active

Thin viewport
outline

Crosshairs can
move anywhere in
the drawing screen.

Layout/Paper
space UCS
icon

Paper button
displayed

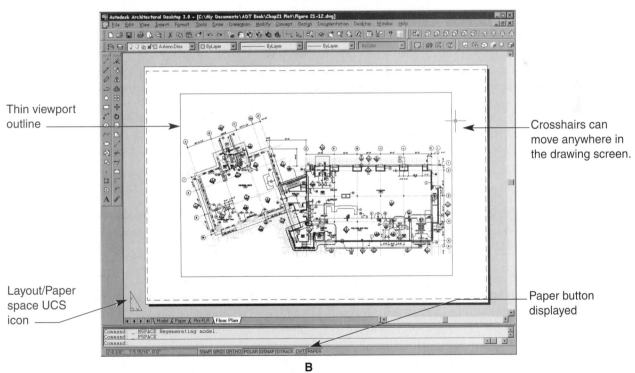

B
Layout active

FIGURE 21.12   Determining the current drawing space.
Drawing courtesy SERA Architects Inc., Portland, OR.

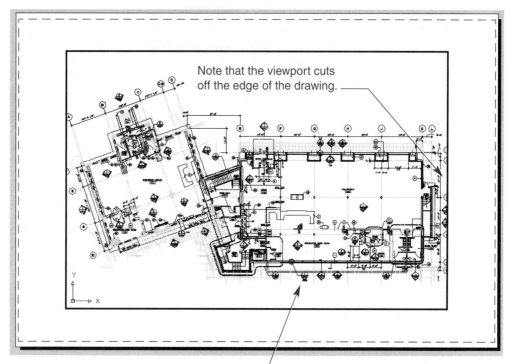

The view of the building model has been set
to a scale of 1/8″ = 1′-0″ (scale factor = 96).

FIGURE 21.13 Scaling the building model in the viewport.
Drawing courtesy SERA Architects Inc., Portland, OR.

port has been given a scale. Everything within the viewport (in model space) is now scaled in this viewport to the designated scale. It also is important to remember that because the ZOOM command was used to scale the drawing, any zooming performed within the viewport causes the building model to not be scaled any more, however, the view can be panned within the viewport without adjusting the scale.

 Field Notes

After a scale has been applied to a viewport **do not** use the ZOOM command unless a new scale will be applied. Zooming causes the viewport to not be scaled. Use the PAN command to slide the view of the model around to adjust it in the viewport.

The second method of scaling a viewport is to use the scale drop-down list in the **Viewports** toolbar. This toolbar can be activated by right-clicking over the top of an AutoCAD toolbar, then selecting Viewports from the shortcut list. The Viewports toolbar is displayed in your drawing screen. When a floating viewport is active the scale drop-down list can be selected and a new scale chosen for the contents within the viewport (see Figure 21.14).

If you accessed a layout from one of the Architectural Desktop Plot layout tabs, the scale drop-down list will be unavailable, because the viewport scale is locked. To unlock the viewport, double-click outside the viewport to return to layout space, select the viewport border, right-click and select the Display Locked cascade menu, then pick the No option from the shortcut menu. This will unlock the viewports scale. Double-click inside the viewport to make the viewport active again, then select a scale for the viewport.

When you have finished establishing a scale for the viewport you can assign a display configuration to the viewport. Display configurations were discussed in Chapter 5. Display configurations can be assigned to individual viewports within layout space or globally to all tiled viewports in model space. After assigning the display configuration double-click outside the viewport boundary to return to layout space, where you can adjust the viewport, add a title block and plot the layout sheet.

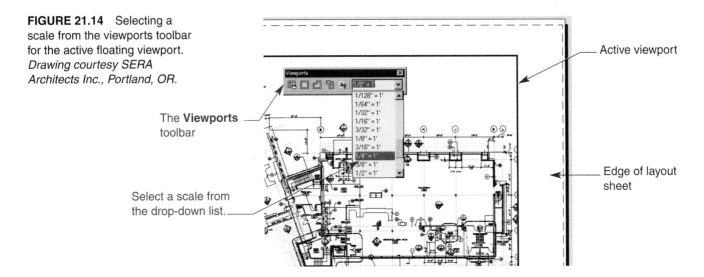

FIGURE 21.14 Selecting a scale from the viewports toolbar for the active floating viewport. *Drawing courtesy SERA Architects Inc., Portland, OR.*

The **Viewports** toolbar

Select a scale from the drop-down list.

Active viewport

Edge of layout sheet

Exercise 21.1

1. Start a drawing using the AEC Arch (Imperial) or the AEC Arch (Imperial - Int) template.
2. Create the following layers:
   - Xref-Floor
   - Xref-Fndn
   - Xref-Elev
   - Xref-Sect
3. Xref the following drawing files onto the appropriate layers. Use the Overlay option when xreffing. Arrange the floor plan and the foundation plan so they stack on top of each other correctly.
   - EX19-4-Flr-Pln.dwg
   - EX20-5-Fnd.dwg
   - EX17-12-Elev.dwg
   - EX-18-7-Sect.dwg
4. Create four separate layout tabs named Floor Plan, Foundation Plan, Elevations, and Sections.
5. Set each layout to use the DWF plotter. Assign the AIA(256) Scale 48.ctb pen table. Set the paper size to a 36″x24″ sheet of paper and a plot scale of 1:1.
6. Save the drawing as EX21.1-Plot.dwg.

Working with Viewports

As discussed earlier, viewports created in layout space are known as floating viewports. Floating viewports are similar to holes cut into a sheet of paper so that the model space building can be viewed under the paper. Floating viewports are separate objects that can be moved around and arranged or *floated* on the layout sheet.

Once a floating viewport has been created, display functions such as views, panning, camera views, display configurations, and zooming can be used to modify the view of the building model through the floating viewport. These functions allow you to control how the building model appears on the final printed sheet of paper. Editing functions such as moving, copying, erasing, and stretching can be used when working in the layout space area to arrange and modify the floating viewports and their borders.

One example of using editing commands on a floating viewport is the situation created in Figures 21.13 and 21.14. After the scale of the drawing has been specified in the viewport, the edge of the viewport cuts off part of the drawing. While in layout space use the grip stretch command to adjust the viewport border to be a little bit larger. Because the scale of the drawing was specified inside the viewport, adjusting the

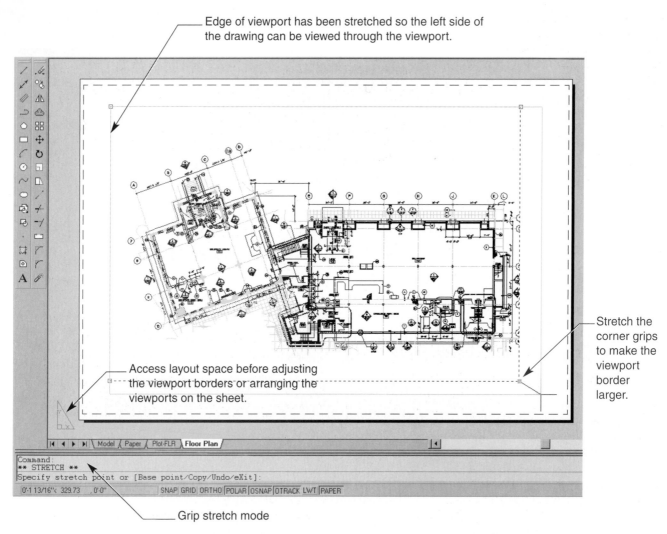

FIGURE 21.15   If the viewport border cuts off part of the model space drawing, grips are used to stretch the viewport border larger. The scale specified within the viewport is not changed. *Drawing courtesy SERA Architects Inc., Portland, OR.*

border maintains the scale while making the viewport border larger so the entire drawing can be viewed (see Figure 21.15).

Moving the viewport allows you to adjust where the viewport border is placed on the sheet. Moving a viewport maintains the current view and scale of the building within the viewport. Erasing a viewport will remove the "hole" from the paper so that new viewport configurations can be "cut" into the paper.

Creating Floating Viewports

The creation of multiple floating viewports is similar to the creation of multiple viewports in model space. By default the Plot tabs in the AEC Arch templates include one floating viewport along the edges of the title block border; any new layout sheets generate a single viewport. Individual floating viewports can be created with the **MVIEW** command, or multiple floating viewports can be created with the **Viewports** dialog box. The **Viewports** dialog box and the creation of tiled viewports in model space were discussed in Chapter 4.

Accessing the Viewports dialog box while in the Model tab creates multiple tiled viewports. Using the Viewports dialog box while in a Layout tab creates multiple floating viewports on the layout sheet. The only difference in the dialog box when in layout space is that the Apply To: drop-down list is replaced with a Viewport Spacing: text box (see Figure 21.16). This text box allows you to specify a spacing distance between the borders of the floating viewports.

After selecting the floating viewport configuration that you want to place on the layout sheet, press the OK button to return to the drawing screen. In order to create the floating viewport configuration, you

FIGURE 21.16 When layout space is active the Viewport Spacing text box is available for specifying a distance between floating viewports.

Spacing between floating viewports

Floating viewport preview

Enter a distance between floating viewports.

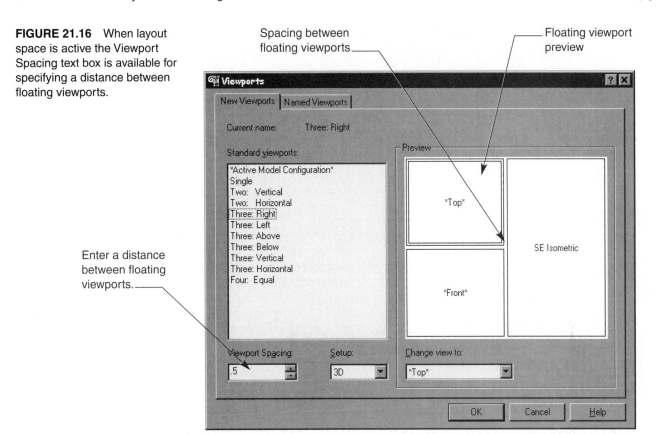

need to select two opposite diagonal corners of a box on the layout. (see Figure 21.17A). The imaginary box defines the outer limits for the floating viewports. Once the second point is selected, the floating viewport configuration is applied to the layout sheet. Architectural Desktop "cuts" the holes out of the paper, sets the selected views, and performs a zoom extents in each viewport (see Figure 21.17B). After the viewports are cut, access each viewport to set the viewport scale.

Field Notes

> When creating multiple viewports it is best to first erase the single default viewport. If this viewport is not erased, the multiple floating viewports may be cut on top of the existing viewport, causing viewports to lie over the top of one another.

Multiple floating viewports are beneficial in a set of construction documents, because several different views of the drawing can be laid out on the paper, with each viewport containing its own scale. This is particularly helpful when creating elevation, section, and detail sheets, where many views of different parts of the drawing are often located and may have different scales.

The floating viewport borders drawn on the layout sheet are created on the current layer, which causes them to be plotted. In many cases it is not desirable to have a border around each viewport on the finished plot. To avoid this situation, create a layer for the floating viewports that is set to be nonplottable. Set this layer current before drawing the viewports. The layer is a nonplottable layer, so it will not be plotted on the final sheet.

Another way to create floating viewports is to use the **MVIEW** command. This command allows you to create a single viewport on the layout sheet by picking two diagonally opposite corners of a rectangle. The rectangle is then turned into a floating viewport on the sheet. Layout space must be active, and you cannot be within a floating (active) viewport.

Use one of the following options to access the MVIEW command:

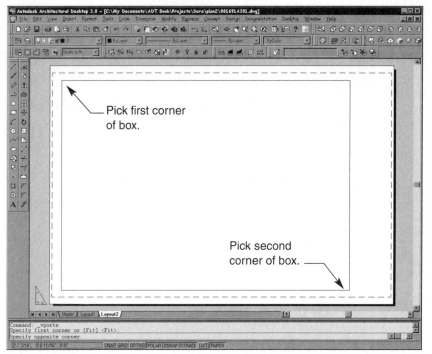

A

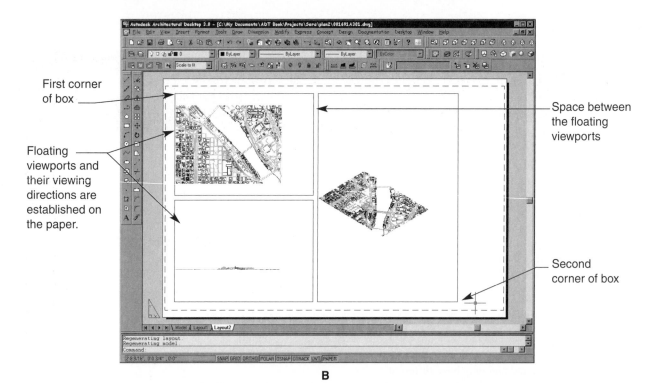

B

FIGURE 21.17 (A) Two opposite corners of an imaginary box are selected to define the outer edges of the floating viewports. (B) Once the second point has been selected, the viewport configuration is applied, and the viewports are cut into the paper.
Drawing courtesy SERA Architects Inc., Portland, OR.

✓ Select the **View** pull-down menu, pick the **Viewports** cascade menu, then select **1 Viewport** from the menu.

✓ Pick the **Single Viewport** button in the **Viewports** toolbar.

✓ Type **MVIEW** at the Command: prompt.

Once you have enetered the command, select two points in the drawing to establish the viewport. The following prompt sequence is used to establish the viewport in Figure 21.18.

Command: **MVIEW** ↵
Specify corner of viewport or
[ON/OFF/Fit/Hideplot/Lock/Object/Polygonal/Restore/2/3/4] <Fit>: *(pick first corner)*
Specify opposite corner: *(pick second corner)*
Regenerating model.
Command:

The **MVIEW** command includes the following options for creating and adjusting new or floating viewports:

■ **On and OFF:** These options turn the display of the contents on or off within the floating viewport. When the OFF option is selected, you are prompted to pick the viewports to be affected. The ON option is used to turn the model space geometry back on in the viewport. Turning a viewport off can enhance computer system performance, because Architectural Desktop is not trying to regenerate all the information in the viewport as you are working. If a viewport is turned off, turn on the viewport prior to plotting.

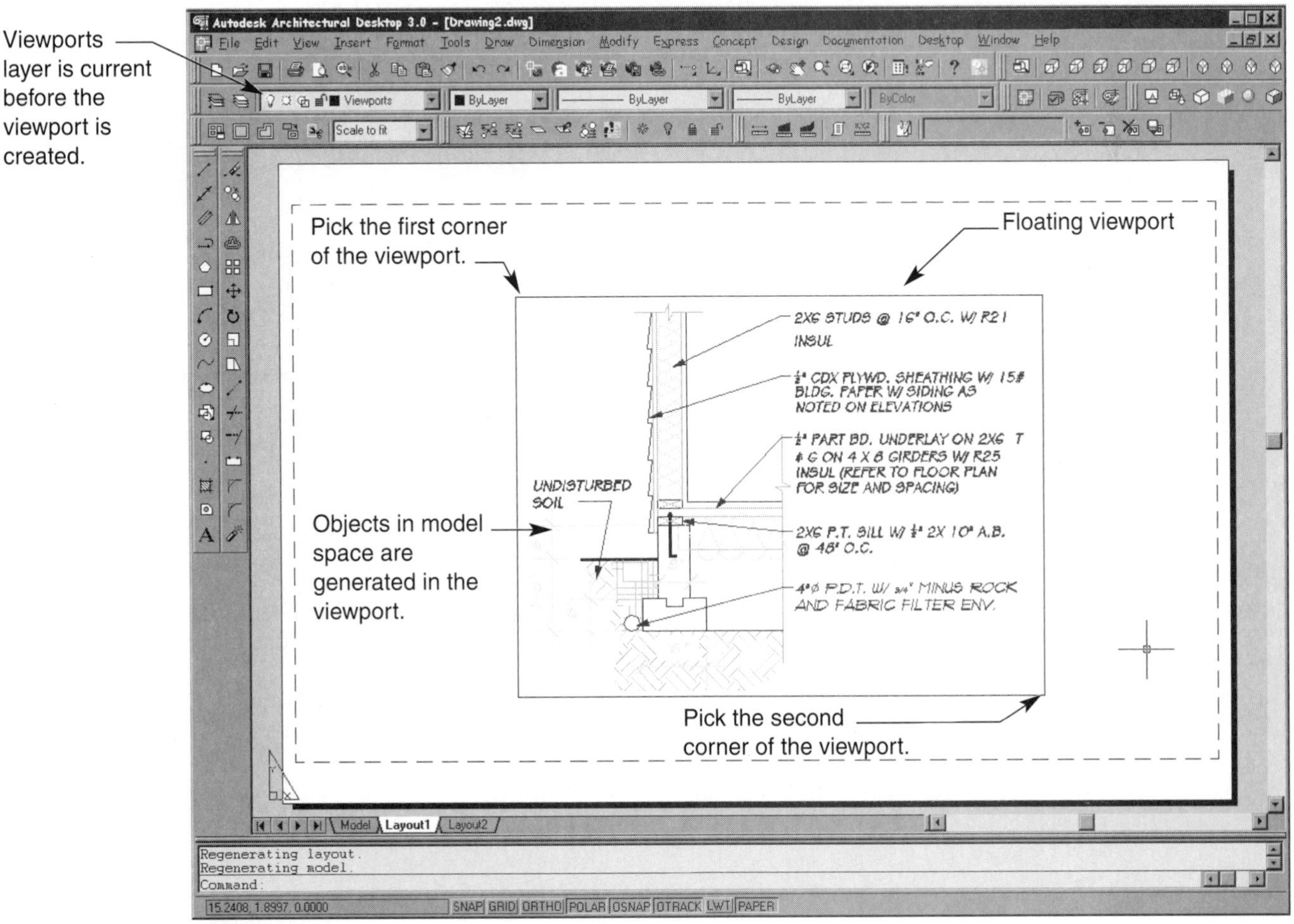

FIGURE 21.18 A single viewport is created on a blank sheet of paper in layout space.

- **Fit:**   This option creates a single viewport that fills the entire plottable area on a sheet. This is the area inside the margins on the sheet of paper.

- **Hideplot:**   This option hides lines behind 3D objects at the time of plotting in the floating viewport. Although you can manually perform a hide or shade within a viewport, the objects do not plot hidden or shaded. If the objects are to be plotted as a hidden view for views such as elevation views, isometric views, or perspective views, use the Hideplot option on the floating viewport. After selecting this option pick the viewports that are to be hidden.

- **Lock:**   This option locks the view of one or more floating viewports. This option is used after a viewport has been assigned a scale and has been arranged inside the viewport with the PAN command. This prevents accidental zooming, panning, or scale changing when you are working inside of the active floating viewport.

- **Object:**   This option is used to turn any closed object drawn on the layout sheet into a floatfloating viewport. Circles, ellipses, polygons, and closed polylines can all be converted into a floating viewport (see Figure 21.19). One of the following methods can also be used to accesss this option:

✓ Pick the **Convert Object to Viewport** button in the **Viewports** toolbar.

✓ Select the **View** pull-down menu, pick the **Viewports** cascade menu, then select **Object** from the menu.

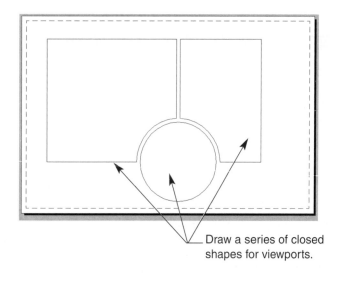

Draw a series of closed shapes for viewports.

A

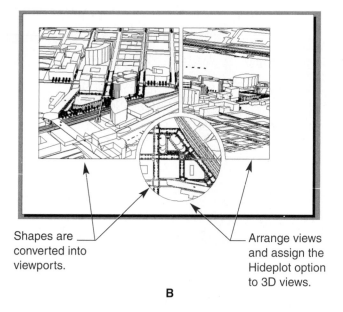

Shapes are converted into viewports.

Arrange views and assign the Hideplot option to 3D views.

B

FIGURE 21.19   (A) A series of closed objects to be used as viewports is drawn. The Convert Object to Viewport command or the Object option in the MVIEW command is used to convert the objects to viewports. (B) The finished viewports after conversion arrangement of the views, and assignment of the Hideplot option to the viewports.
Drawing courtesy SERA Architects Inc., Portland, OR.

- ■ **Polygonal:** This option is used to draw an irregular-shaped floating viewport by selecting points in a manner similar to drawing a polyline. The shape can be composed of linear or angular edges. This is similar to using a closed object for a viewport except the viewport is actually drawn on the paper. One of the following methods can also be used to access this option:

✓ Pick the **Polygonal Viewport** button in the **Viewports** toolbar.

✓ Select the **View** pull-down menu, pick the **Viewports** cascade menu and select the **Polygonal Viewport** command from the menu.

- ■ **Restore:** This option converts a named viewport configuration from the Viewports dialog box into individual floating viewports in the layout.

- ■ **2, 3, or 4:** These options allow you to create two, three, or four viewports similar to the method used from the Viewports dialog box. You need to pick two opposite corners of a box to create the multiple viewports.

In addition to the options found in the MVIEW command and Viewports toolbar, one additional tool is provided in the toolbar. The **Clip Existing Viewport** command can be used to clip away parts of a floating viewport by accessing one of the following options:

✓ Select the **Clip Existing Viewport** button from the **Viewports** toolbar.

✓ Type **VPCLIP** at the Command: prompt.

This command is similar to using the XCLIP command described in Chapter 15.

 Field Notes

Floating viewports can overlap one another. Each viewport contains its own view to model space, so overlapping viewports do not interfere with one another. If you create a viewport completely within another viewport, double-clicking in one of the viewports makes it active. If you cannot access the other viewport inside, use the [Ctrl]+[R] key combination to cycle through the active viewports on the sheet.

After you have drawn the viewports on the layout sheet, you can arrange them by using grip editing or the Move and Stretch commands. When working in layout space the only objects you can "touch" are the floating viewport edges; however, you can snap to building objects within the viewport using object snaps. If you are trying to line up views, such as elevation views, access the MOVE command, select the viewport edge to move, then snap to a known point on the building geometry within the viewport to use as a base point. Complete the moving operation by using object tracking or the perpendicular object snap referencing another object in a different viewport.

Exercise 21.2

1. Open EX21.1-Plot.dwg.
2. Erase the default-created viewports in the Floor Plan, Foundation Plan, Elevations, and Sections layout tabs.
3. Set the Viewports layer current and assign it to be a nonplotting layer.
4. Create viewports in the four layout sheets as shown in the figure. Leave some room on the right side for a title block that will be added later.
5. Adjust the views for each viewport through by zooming and panning. Assign the scales to the viewports as indicated in the diagram. Do not worry about the linework overlapping in the xrefs. We will address this issue in the next section.
6. Save the drawing as EX21.2-Plot.dwg.

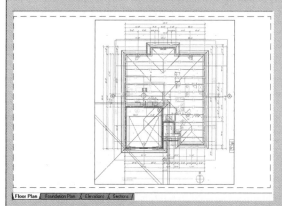

Floor Plan tab: Scale the viewport to 1/4″=1′-0″. Center the floor plan and foundation plan on the viewport.

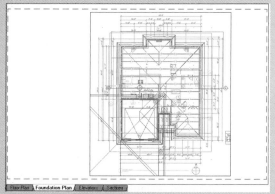

Foundation Plan tab: Scale the viewport to 1/4″=1′-0″. Center the foundation plan and floor plan on the viewport.

Elevations tab: Create four viewports viewing the top view (for the 2D front elevation), the left view, the right view, and the back view. Scale the viewports to 1/4″=1′-0″. Center the elevations on the viewports.

Sections tab: Create a viewport for the 2D section scaled to 1/4″=1′-0″. Create four viewports for the interior elevations scaled to 3/8″=1′-0″. Center the views on the viewports.

Controlling the Viewport Display

When a floating viewport is first created, the view into model space is zoomed to the extents of the geometry. This allows you to see the entire contents of model space through the viewport. When the viewport is scaled, the view is adjusted from the center of the viewport. If a part of the drawing is to be the focus of the viewport, use the PAN or ZOOM-Center to center the focus of the drawing on the viewport, then assign the scale. The part of the drawing you are focusing on remains centered on the viewport after the scale has been assigned.

After you have assigned a viewport scale, the ZOOM command should never be used in the viewport again, because zooming causes the viewport to not be scaled. If the view needs to be adjusted, use the pan command to slide the model around in the viewport. If parts of your geometry are still being cut off by the viewport border, access the layout space and adjust the viewport border using the grip stretch mode.

Once you have specified the correct viewing area, and assigned the scale, assign a display configuration to the viewport. With the viewport still active pick the **Set Current Display Configuration** button to assign a display configuration to the viewport (see Figure 21.20).

If the AEC Arch templates are used, some preconfigured plotting display configurations are available. Select the desired display configuration to be used in the viewport. Each floating viewport can have its own display configuration assigned. For example, if you are creating a layout sheet that will display a floor plan in one viewport and a reflected ceiling plan in another viewport, assign the appropriate display configurations to each viewport.

Once you have set the scale and display configurations for the viewport, you may decide to lock the viewport so that accidental zooming does not affect your scale. In addition to using the Lock option in the

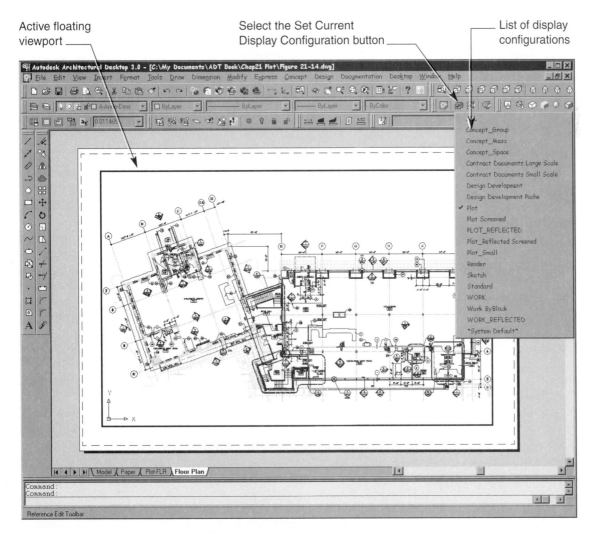

FIGURE 21.20   With the viewport still active, a display configuration is selected to be used in the viewport.

FIGURE 21.21   The Properties
window can be used to modify
the properties of viewports and
other selected objects in
Architectural Desktop.

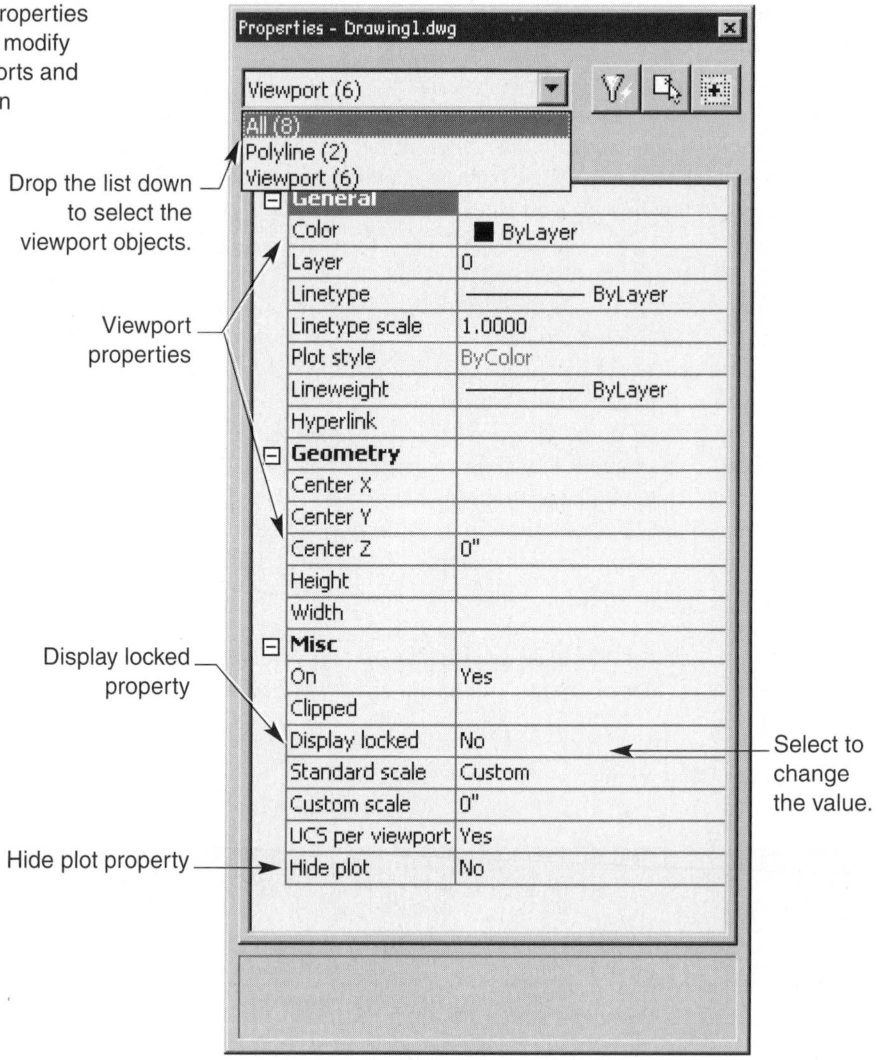

Drop the list down
to select the
viewport objects.

Viewport
properties

Display locked
property

Hide plot property

Select to
change
the value.

MVIEW command, you can use two other methods to lock a viewport. Both methods require that the layout space be active.

The first method is to select the viewport border to display the grip boxes, then right-click, and select the **Display Locked** cascade menu. The cascade menu includes **Yes** or **No** options. Selecting Yes locks the viewport, and No unlocks the viewport. This method allows you to select only one viewport at a time.

The second method allows you to lock or unlock multiple viewports at once. First, select the desired viewports to display the grip boxes, right-click, and select **Properties** from the bottom of the shortcut list. The Properties dialog box is displayed (see Figure 21.21). At the top of the dialog box is a drop-down list that contains all of the different types of objects selected for modifying with a number beside each indicating the total number of that type of object. Selecting the Viewport objects from the list provides you with a list of properties for the viewports. A property value labeled Display Locked is displayed at the bottom of the list. Select the box to the right of the label to choose yes or no to lock the viewports.

In addition to locking the viewports from the Properties window, you can turn the selected viewports on or off, assign viewports to be hidden by selecting the Hide plot property, and scale them. Note that if clipped viewports are selected, the selected object's drop-down list indicates "All" in the list. Drop the list down and select the viewport objects to make adjustments to the viewports.

Viewport Layer Display

Whenever a floating viewport is created, everything that is visible in model space is displayed in the floating viewport. Depending on how your drawings are organized and laid out, this can be a lot of geometry that is being sorted through in the viewport possibly causing linework to overlap.

Multiple reference files are
stacked on top of one another.

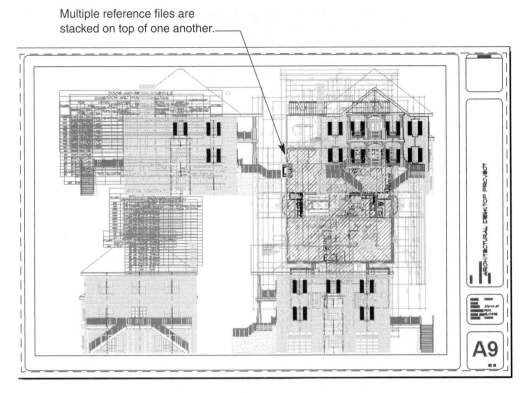

FIGURE 21.22   A multistory building with different information in each floor level in a viewport.

In a multistory building the floors are often drawn as separate levels, which are in turn referenced and stacked on top of one another to form the 3D building for the elevations (see Figure 21.22). When a floating viewport is generated, all the floors of the building and any additional information such as schedules, elevations, and sections may end up overlapping.

The viewport in Figure 21.22 is set up to display a single floor plan and not the other floors, elevations, and schedules. Other layout sheets also need to display only the appropriate floors, elevations, and schedules without overlapping drawings or floors. This pattern can be solved by freezing the layers that are not needed in this viewport, but that will also freeze the layers in the other viewports. An alternative is to use a tool known as **VPLAYER**.

VPLAYER stands for viewport layer and is used to freeze and thaw layers in a floating viewport. This command is different from the standard freeze/thaw commands because it allows you to control the display of objects on layers in a floating viewport. The standard freeze and thaw functions affect objects globally in all floating viewports and model space. Typically, a drawing being prepared for plotting has all the desired layers thawed, then each individual floating viewport freezes layers that are not required for that viewport.

In order to freeze a layer that is in a floating viewport, the viewport must be active. Layers inside floating viewports can be frozen by using the VPLAYER command, selecting the Freeze or thaw in the current viewport icon in the layer drop-down list on the Object Properties toolbar, or by selecting the Active VP Freeze icon in the Layer Manager dialog box (see Figure 21.23). The Active VP Freeze icons in the Layer Manager are not available unless you are working in a floating viewport.

Before freezing the layers, make active the floating viewport that will have layers frozen. Once the viewport is active, you can freeze or thaw the layers by picking on top of the freeze/thaw icon in the layer pull-down or in the Layer Manager. If you are using the VPLAYER command, you are prompted with the following options:

Command: **VPLAYER** ↵
Enter an option [?/Freeze/Thaw/Reset/Newfrz/Vpvisdflt]:

- **?**   This option lists the currently frozen layers in the active viewport.

- **Freeze:**   Enter the name of the layer(s) that are to be frozen. The following prompts are
               used when layer names are to be frozen:

FIGURE 21.23 (A) The Freeze or Thaw in the current viewport icon is used to quickly freeze a layer in the active floating viewport. (B) The Active VP Freeze icon in the Layer Manager dialog box is used to freeze a series of layers in the active viewport.

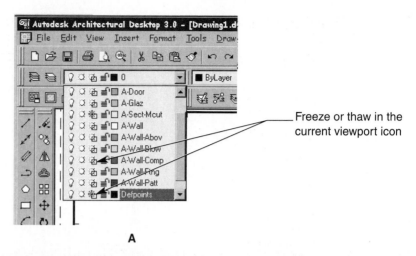

Freeze or thaw in the current viewport icon

A

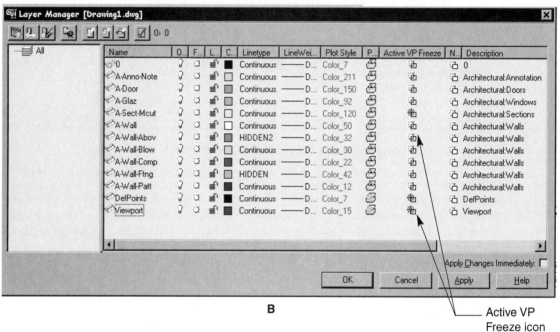

B

Active VP Freeze icon

Enter layer name(s) to freeze or <select objects>: (*enter the name of a layer or several layer names by separating them with commas*)
Enter an option [All/Select/Current] <Current>:

After entering the layer name(s) to be frozen you can apply the layer freezing to all the floating viewports in the current layout space, in a viewport by selecting it, or in the currently active viewport.

■ **Thaw**   Enter the name of the layer(s) that are to be thawed. The prompts for this option are similar to the prompts for the freeze option.

■ **Reset**   This option resets the layer visibility status in the active viewport to the layers that are frozen or thawed in the Layer Manager. Prompts similar to those for the freeze option are used for this option.

■ **Newfrz**   This option creates new layers that are frozen in all viewports.

■ **Vpvisdflt**   This option freezes or thaws layer names automatically in newly created viewports. This saves you the time of going back through each viewport to freeze the same layers.

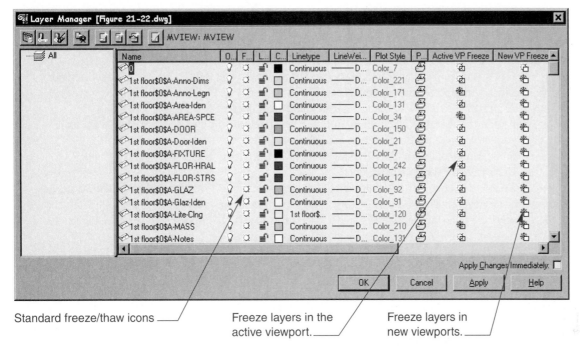

Standard freeze/thaw icons ———

Freeze layers in the active viewport. ———

Freeze layers in new viewports. ———

FIGURE 21.24 The Layer Manager dialog box can be used to freeze layers in the active viewport and in newly created viewports.

Layer visibility in floating viewports can also be controlled through the Layer Manager dialog box (see Figure 21.24). Before entering the Layer Manager, select the viewport in which you want to have layers frozen. The columns at the far right include the Active VP Freeze and the New VP Freeze.

Selecting an icon in the Active VP Freeze column will freeze or thaw the layer in the current viewport. Freezing is represented by a snowflake icon; thawing is represented by a shining sun. Selecting an icon in the New VP Freeze column will freeze or thaw the layer in new viewports not yet created.

Field Notes

> Do not freeze layer 0 in model space or in a viewport. Objects such as multi-view blocks are dependent on this layer. Freezing this layer affects objects that are on other layers.

When freezing layers in a viewport, you need to determine which layers are not desired in the current viewport. For example, if you are creating a viewport that displays 2D elevations, the layers that are probably frozen include the building model layers such as A-Door, A-Glaz, and A-Wall. If the viewport is an elevation view of the building model, you do not want to freeze these layers, but you may freeze the schedule layers.

Field Notes

> One of the benefits of placing reference files on their own layers when xref'ing them is that you can freeze the layer they are inserted on in the active viewport without having to freeze each of the individual xref layers.

Freezing layers in a viewport also helps system performance. Layers that are beyond the viewport boundaries are still calculated by Architectural Desktop. Freezing these layers saves time regenerating the viewport, because the layers are no longer being calculated. Figure 21.25 displays the layout sheet from Figure 21.22 after the xref layers that each floor was inserted on were frozen.

Exercise 21.3

1. Open EX21.2-Plot.dwg.
2. Freeze and thaw the appropriate layers in each viewport in the Floor Plan, Foundation Plan, Elevations, and Sections layout sheets.
3. Assign the three building model elevations to be hidden.
4. Assign the Plot display configuration, the building model viewports, and the Work display configuration to the 2D elevation and section viewports.
5. Freeze all the layers in the new viewports except layer 0.
6. Create a new layout named Notes.
7. Create viewports for the four schedules in the drawing. Freeze the appropriate layers in each viewport. *Note:* Do not freeze the xref layers in these viewports. This will cause the schedule to disappear. Instead, freeze each layer that is not a schedule layer, then determine the appropriate schedule layer for your viewports.
8. Save the drawing as EX21.3-Plot.dwg.

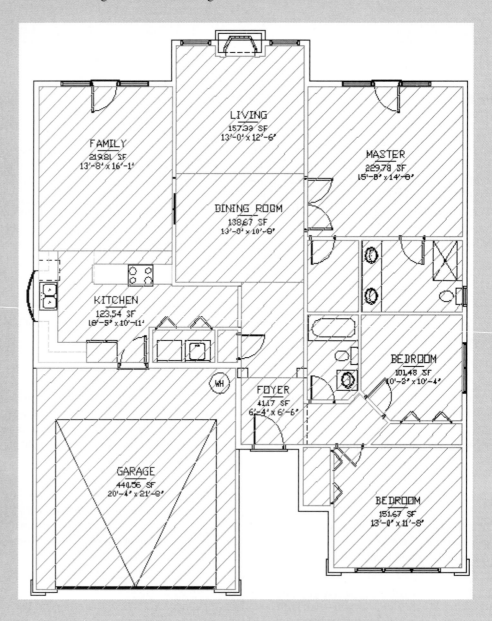

Xref layers
frozen in the
current
viewport

Elevation xref
layer is
thawed in the
viewport.

Active
viewport

Only geometry
within the
elevation xref
remains.

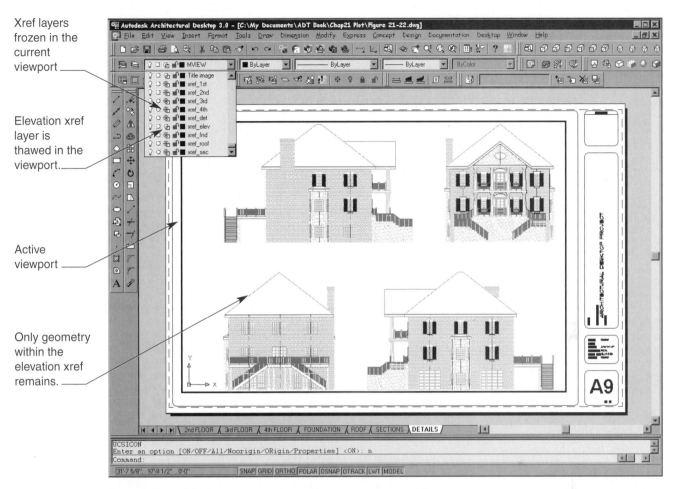

FIGURE 21.25   The layout sheet now displays only elevation layers after the xref floor layers were frozen.

Controlling Linetype Display

When a set of construction documents is produced, it is important to ensure that the drawings look consistent regardless of who or how many people have worked on them. However, when creating viewports that include different scales, the linetypes are scaled differently per the viewports' scale.

The **LTSCALE** system variable controls the scale of linetypes in a drawing. Typically, a LTSCALE value equals the drawing scale. For example, if you are working with a scale of 1/8″=1′-0″, the scale factor is 96. Thus the LTSCALE value should be 96. This scale is good in most situations especially when the drawing is being worked on in model space; however, when the drawing is taken to layout space and two viewports are assigned two different scales, the linetypes are also scaled differently (see Figure 21.26A).

It is often desirable on the finished drawing to have the dashes and dots of the linetypes appear the same length. The variable known as **PSLTSCALE** is used to do this. PSLTSCALE stands for *paper space linetype scale*. Setting this value to 1 and the LTSCALE value to 1 makes the linetypes appear the same size in the two floating viewports (see Figure 21.26B).

If you return to model space, the dashes and dots may not be displayed correctly, because they are so close together you cannot see the spaces between dashes without zooming in on them. Changing the **PSLTSCALE** variable to 0 turns off the linetype scaling in the viewports.

Adding a Title Block

The addition of a title block on the layout sheet provides information about the company. Additional information such as the project name, date created, revision numbers, name of drafter, and client information can all be included in the title block.

When geometry is created on the layout, the geometry is drawn at a 1:1 scale, similar to the geometry in model space. Use polylines, arcs, and lines to draw a title block directly on the layout sheet. When draw-

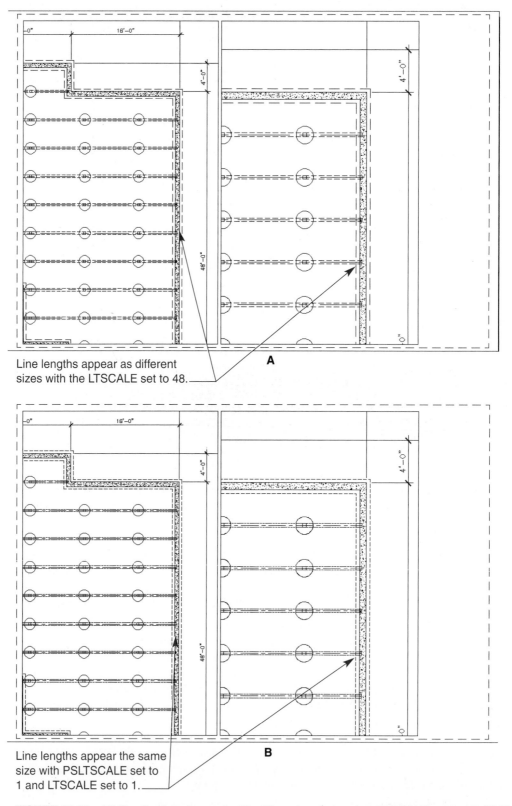

Line lengths appear as different
sizes with the LTSCALE set to 48.

A

Line lengths appear the same
size with PSLTSCALE set to
1 and LTSCALE set to 1.

B

FIGURE 21.26   (A) Two floating viewports with different scales and an LTSCALE set to 48. (B) The same
viewports with the PSLTSCALE set to 1 display the linetypes with the same length dashes.

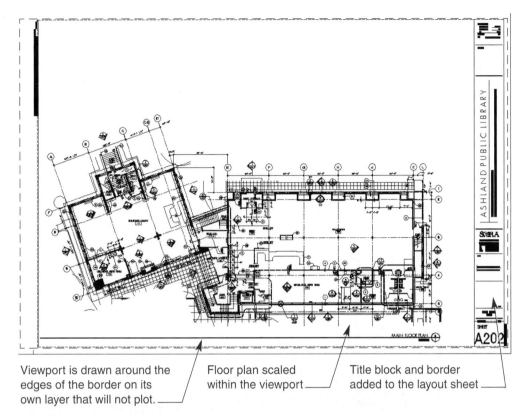

Viewport is drawn around the edges of the border on its own layer that will not plot.

Floor plan scaled within the viewport

Title block and border added to the layout sheet

FIGURE 21.27   The addition of a title block can provide general information about the project, the company that created the drawing, and any information pertaining to the drawing. *Drawing courtesy SERA Architects Inc., Portland, OR.*

ing a title block and associated information on the layout you may decide to establish layers for the geometry. For example, PS-Title Block, PS-Text, and PS-Viewports may be layer names established to indicate that these layers are used for geometry drawn in paper space/layout space.

Some companies prefer to draw a title block in a separate drawing and xref or insert the title block on the layout. If you do this, make sure that you have a layer where the xref is to be placed. This layer might be named Xref-Title. Remember to draw the title block inside the margin area. Geometry outside the margins does not print.

Figure 21.27 shows a drawing on a 36″ × 24″ sheet of paper, with a single scaled viewport and the title block added to the layout sheet.

Exercise 21.4

1. Open EX21.3-Plot.dwg.
2. Create the following layers: PS-Title and PS-Text.
3. Add a title block to the Floor Plan, Foundation Plan, Elevations, Sections, and Notes layouts.
4. Use the PS-Title layer for the linework. Use the PS-Text layer to add notations to the title block using the MTEXT command.
5. Save the drawing as EX21.4-Plot.dwg.

Adding Dimensions and Annotations

Scale factors and drawing scales have been discussed throughout this text. When the drawing scale is set up in the Drawing Setup dialog box, any annotation objects that are inserted are drawn real size multiplied by the scale factor. Setting a drawing scale also affects the current dimension style scale.

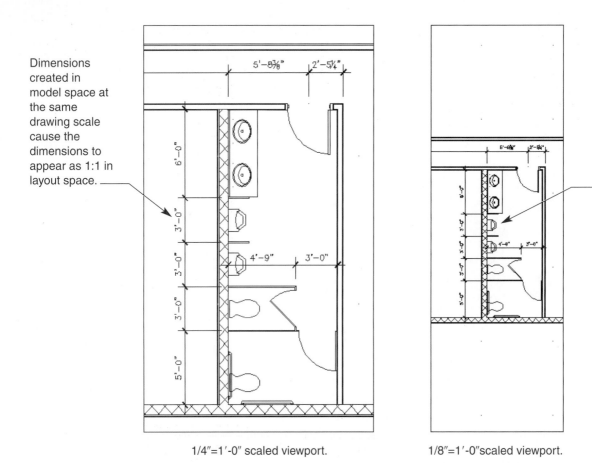

Dimensions created in model space at the same drawing scale cause the dimensions to appear as 1:1 in layout space.

Dimensions created in model space at 1/4″=1′-0″ scale appear as half-size dimensions in layout space.

1/4″=1′-0″ scaled viewport.

1/8″=1′-0″scaled viewport.

FIGURE 21.28   The dimensions in model space are drawn at a scale factor of 48 (1/4″=1′-0″). When taken into layout space the left viewport is set to a scale of 1/4″=1′-0″, creating dimensions that appear 1:1 on the layout. The right viewport is scaled to 1/8″=1′-0″ causing the same dimensions to appear half the size.

When a drawing scale is set, an override is applied to the current dimension style so that the dimension style creates dimensions scaled to the correct drawing scale size. The model space dimensions appear as a scaled-up dimension size. In layout space, like the linetype scale, the dimensions appear as two different scaled sizes in two viewports with different scales (see Figure 21.28).

This may become a problem when the same dimensions are used for differently scaled viewports. To overcome this problem, first create dimensions in model space using the most common scale in the construction document set, then for the viewports that are to be set to a different scale, create another set of dimensions using a different drawing scale.

This forces you to duplicate dimensions in model space, because you have one set of dimensions at one scale and another set of dimension at another scale. The only problem with this solution is that both sets of dimensions show up in all the viewports. If you attempt to freeze the dimension layer in a floating viewport, all the dimensions freeze. In order to work with the two sets of dimensions, create two layers: one for dimensions that are set at the main scale, and another for the other dimensions. This allows you to freeze the appropriate dimension layer in each viewport.

For example, assume that two scales are used in the drawing: ¼″=1′-0″ and ⅛″=1′-0″. Two sets of dimensions need to be created, one that is scaled 48 times and another that is scaled 96 times. Create the ¼″ scale dimensions on a layer named Dim48, and the ⅛″ scaled dimensions on a layer named Dim96. Then, in a ¼″ scaled floating viewport freeze the Dim96 layer. In the ⅛″ scaled floating viewport freeze the Dim48 layer. Figure 21.29 demonstrates how this method works.

Creating text in model space can produce the same results. Text created using the MTEXT command in model space needs to be the scaled-up size. For example, text that is to appear ⅛″ high on the layout needs to be ⅛″ multiplied by the scale factor in model space. In order for the text to appear correctly in the ¼″=1′-0″ viewport, the text in model space needs to be drawn 6″ high (⅛″ height × 48 = 6″ total text height). If the same notation is used for a ⅛″ scaled viewport, the note should be duplicated 12″ high (⅛″ height × 96 = 12″ total height).

FIGURE 21.29 (A) Model space includes duplicate dimensions at both scales on two separate layers. (B) The 1/4″=1′-0″ viewport has frozen the Dim96 layer. (C) The 1/8″=1′-0″ viewport has frozen the Dim48 layer.

Model space: Dimensions are duplicated at each scale and placed on their own layers.

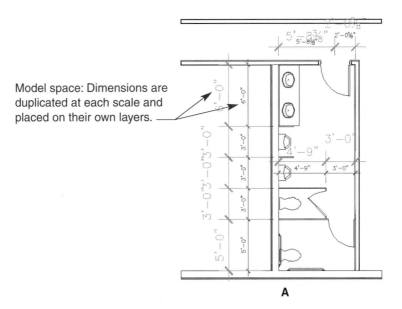

A

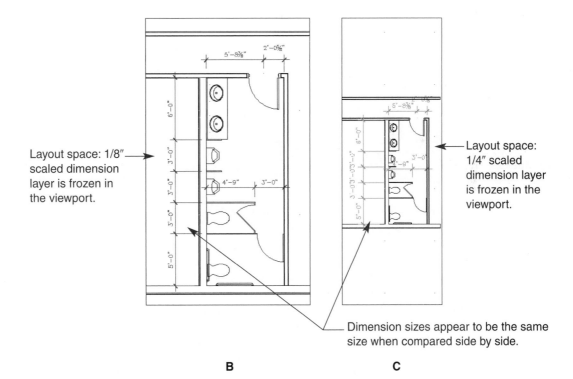

Layout space: 1/8″ scaled dimension layer is frozen in the viewport.

Layout space: 1/4″ scaled dimension layer is frozen in the viewport.

Dimension sizes appear to be the same size when compared side by side.

B   **C**

If both sets of text are created on their own layers, they can be frozen in the appropriate viewports, like the dimensions previously discussed. Use Appendix F to determine different scale factors and text heights for text placed in model space.

If text is to be placed directly on the layout, then it should be created full size. For example, if the text needs to appear ¼″ high in the final plot, then create the text ¼″ high on the layout. Figure 21.30 shows the final layout sheet with notes applied directly on the layout sheet.

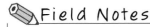Field Notes

If annotations from the DesignCenter are placed on the layout sheet, the drawing scale is ignored and the annotation objects are placed on the layout at full scale.

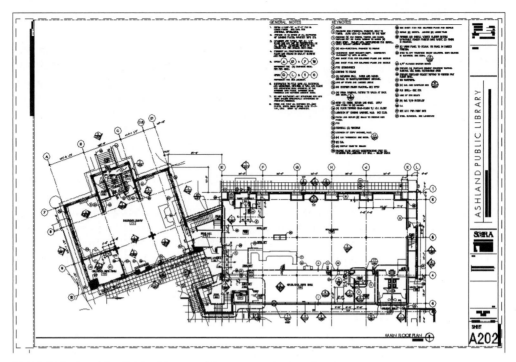

FIGURE 21.30   The finished layout with notes applied on the sheet.
Drawing courtesy SERA Architects Inc., Portland, OR.

Exercise 21.5

1. Open EX21.4-Plot.dwg.
2. Add a title mark under each floating viewport. Select the Documentation pull-down menu, the Documentation Content cascade menu, then Title Marks… from the list.
3. Name each title according to what is being viewed in the viewport.
4. Save the drawing as EX21.5-Plot.dwg.

Guidelines for Setting Up Layout Space

The following are recommended steps to follow when creating custom layout sheets:

1. Draw the building model in model space full scale, 1:1.

2. Save any views and perspectives in model space that you would like to use in a floating viewport.

3. Create a new Layout tab.

4. Enter page setup and name the tab, and select the plotter and paper size.

5. Draw a title block on the layout sheet.

6. Create and arrange floating viewports on the layout. Be sure to place the viewports on a non-plotting layer.

7. Restore the views saved in model space to the floating viewports as desired.

8. Apply the scale to each viewport and lock the viewports when you have finished setting up the view.

9. Freeze and thaw different layers in each viewport.

10. Add any annotations to the layout that are required.

Getting Ready to Plot

The final step in a building model project is to purchase a hard copy set of drawings so that the contractor, subcontractors, and building officials can refer to the drawings as the building is being constructed. The previous section discussed how to set up a layout sheet and get it ready for plotting. Before plotting you need to consider what type of final output is desired. For example, will the file be printed to a printer, a plot plotter, or a file that can be viewed over the Internet? This section discusses how to plot the layout sheet.

Managing Layouts

A drawing file can contain as many layout sheets as needed. In the previous exercises, you xrefed the entire set of drawing files into a new drawing template that is used strictly for plotting. Drawings can also be plotted directly from the individual drawing files themselves. With either method you need to manage your layout sheets through the **LAYOUT** command. This command can be accessed by entering **LO** or **LAYOUT** at the Command: prompt. Many of the options in this command can also be accessed by right-clicking on top of a layout tab and selecting an option from the shortcut menu. When this command is accessed the following options are available for use:

> Command: **LAYOUT** ↵
> Enter layout option [Copy/Delete/New/Template/Rename/SAveas/Set/?] <set>:

- **Copy**   This option is used to copy an existing layout and insert it into the drawing after the copied layout. This is a good option if one layout has been set up and the same options need to apply to the next layout.

- **Delete**   This option deletes a layout tab from the drawing. The Model tab cannot be deleted, and there must be at least one layout tab.

- **New**   This option creates a new layout tab based on default setup information.

- **Template**   This option creates a new layout from an existing layout within a template file. After selecting the template containing the desired layout you need to enter a layout name from the template that will be added to the current drawing.

- **Rename**   This option is used to rename a layout. Layout names cannot be duplicated and may contain a total of 255 characters. The first 32 characters in the name can be displayed in the tab.

- **SAveas**   This option allows you to save a layout to a new template file. When this command is accessed, the **Create Drawing File** dialog box is displayed, allowing you to create a new template file containing the selected layout.

- **Set**   This option sets a layout current in the drawing.

- **?**   This option lists all the layouts in the current drawing.

As previously indicated, many of these options can also be accessed by right-clicking over an existing layout tab (see Figure 21.31).

FIGURE 21.31 Right-clicking on top of a layout tab produces a shortcut list of layout options.

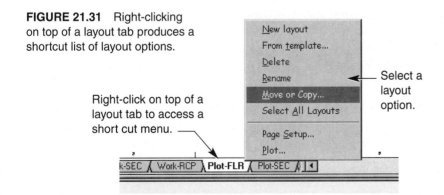

Right-click on top of a layout tab to access a short cut menu.

Select a layout option.

FIGURE 21.32 The
DesignCenter can be used to
import layout sheets into the
current drawing.

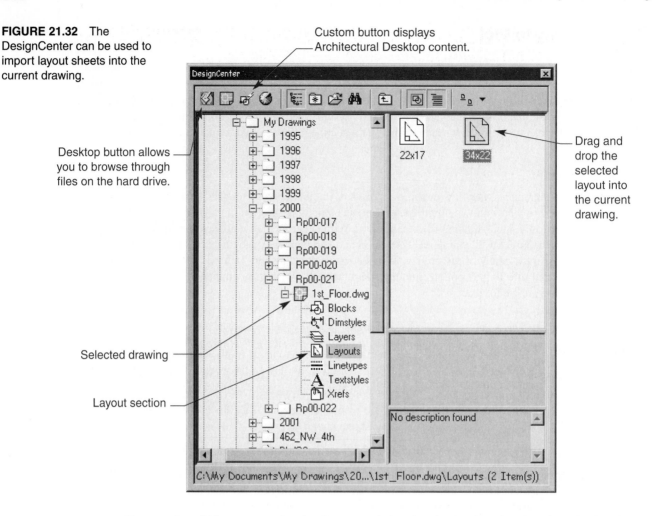

You can also add layouts to a drawing from an existing drawing by dragging them into the drawing
screen from the DesignCenter. First, access the DesignCenter, then select the Desktop button to browse for
an existing drawing that contains the desired layout (see Figure 21.32). Once you have found the drawing,
select the Layouts section to display the layouts in the drawing. Drag and drop the desired layouts into the
current drawing.

Exercise 21.6

1. Open EX21.5-Plot.dwg.
2. Create a new layout tab by copying the Floor Plan tab.
3. Rename the tab Roof Plan.
4. Within the viewport on the Roof Plan layout, freeze all the layers but the roof layers.
5. Adjust the title name by double-clicking on the title block symbol and renaming it Roof Plan.
6. In the Floor Plan layouts viewport freeze the roof layers.
7. Delete the following tabs: Space, Mass-Group, Work-3D, Work-FLR, Work-SEC, Work-
   RCP, Plot-FLR, Plot-SEC, Plot-RCP, and Template-Overview.
8. Save the drawing as EX21.6-Plot.dwg.

Using Page Setup

When you first access a new layout tab the **Page Setup** dialog box is displayed, allowing you to establish
the settings for the layout sheet. Each layout sheet can have its own unique setup. For example, you may
decide to create a floor plan layout on a 36″ x 24″ sheet of paper and a smaller version for records on a
17″ × 11″ sheet of paper.

Layout tab
name

Plot Device
and **Layout
Settings** tabs

Assign a
plotter to the
current
layout.

Assign a pen
style to the
layout.

Page Setup
name

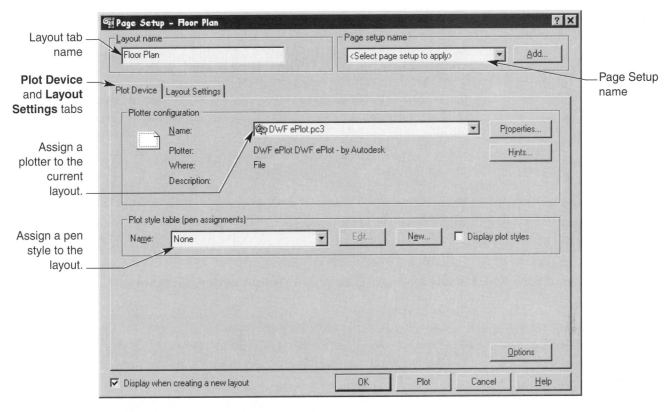

FIGURE 21.33   The Page Setup dialog box.

Once the Page Setup dialog box has been accessed, it no longer appears automatically. In order to adjust the settings for the layout later, you can make the layout active, then right-click on the layout tab and select Page Setup... from the shortcut menu. The Page Setup dialog box is displayed again so that you can modify any settings (see Figure 21.33).

At the top of the Page Setup dialog box are two settings. The **Layout name** text box is used to assign a name to the layout tab. The **Page setup name** drop-down list includes a list of page setups in the drawing from which you can select to assign settings to the current layout. Once a layout has been configured its settings can be saved as a page setup. Select the Add... button at the top of the dialog box to display the **User Defined Page Setups** dialog box (see Figure 21.34).

FIGURE 21.34   The User Defined Page Setups dialog box allows you to save your page setup configurations as a page setup for use in other layout sheets.

Enter a new name for a saved page setup.

List of previously saved page setups

Pick to save the current layout settings as a page setup.

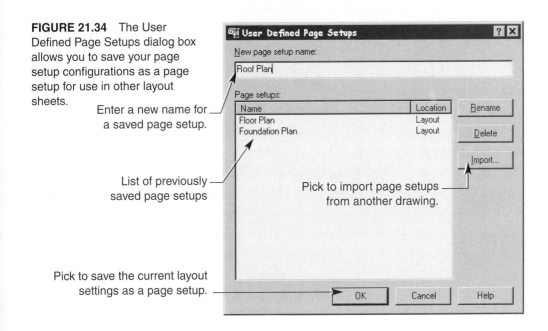

Pick to import page setups from another drawing.

To save the current layout settings as a page setup enter a new name in the <u>N</u>ew page setup name: text box. Press the OK button to save the page setup in the drawing. If you have created a page setup in another drawing that you would like to use, then select the Import... button to access a drawing that contains the desired page setup and import it into the current drawing. When you have finished saving a new page setup press the OK button to return to the Page Setup dialog box. You can then select the new page setup from the list, and any settings associated with the page setup will be automatically set in the current page setup.

Specifying the Plotter

The Page Setup dialog box includes the **Plot Device** and **Layout Settings** tabs. The Plot Device tab is used to configure the plotter and the pen styles that will be used to plot the layout.

In the Plotter configuration area, select a plotter from the Name drop-down list. The list of plotters in the list is a list of plotters configured within the Windows operating system. A few additional plotters are also configured by Architectural Desktop. One of these plotter configurations is a DWF plotter. The DWF plotter plots your drawing as an electronic file that can be viewed through a program known as *Volo View Express*. This is a free downloadable program from Autodesk that can be used by anyone who does not have Architectural Desktop and needs to view or redline your drawings.

The **P<u>r</u>operties...** button is used to display the **Plotter Configuration Editor** dialog box. Settings for the selected plotter can be adjusted and controlled here (see Figure 21.35).

Below the P<u>r</u>operties... button in the Page Setup dialog box is the **H<u>i</u>nts...** button. Selecting this button will provide you with information and tips regarding the selected plotter. When you have finished selecting the plotter you can assign the pen styles to the page setup.

 Field Notes

> Additional plotters or printers can be added to Architectural Desktop by selecting the **File** pull-down menu and picking the **Plotter <u>M</u>anager...** command. The **Plotters** window is displayed where the Add-a-Plotter Wizard icon can be selected to add a plotter configuration.

FIGURE 21.35 The Plotter Configuration Editor dialog box is used to modify the plotter configuration.

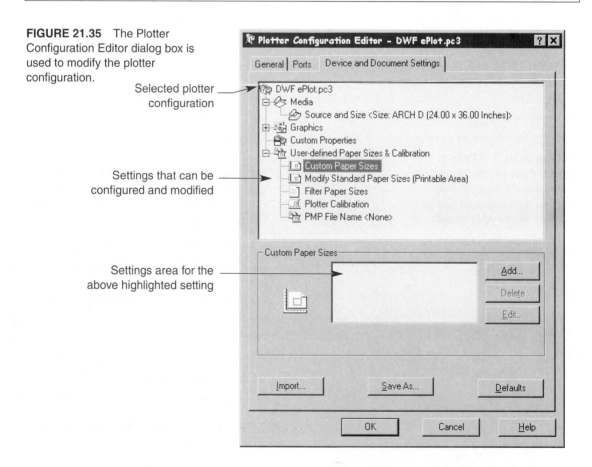

Pick to create a new
plot style table.

Pick to edit an existing
plot style table.

Select to display the plotting
effects to the layout view.

Select a plot style
table to assign to
the layout.

Pick to access
the Options
dialog box.

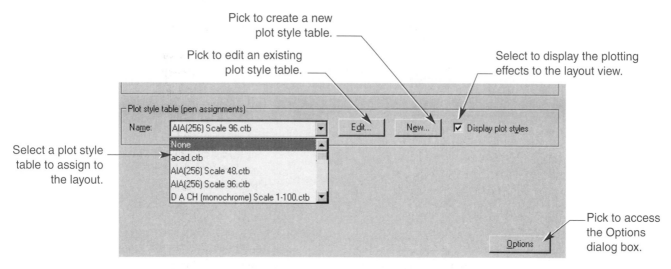

FIGURE 21.36   The Plot style tables (pen assignments) area of the Page Setup dialog
box is used to assign pens and lineweights for plotting.

Specifying Pen Settings

A *plot style,* also known as *pen setting*, is a type of setting that is applied to the layout and determines the colors and lineweights that will be plotted from the geometry on your layout sheet. A list of plot style tables is available under the Name: drop-down list in the Plot style table (pen assignments) area (see Figure 21.36).

The list includes the default pen style tables included with Architectural Desktop. Plot style tables interpret the colors seen in a layout and map the drawing colors to a color and lineweight on the finished plot. This list includes tables such as monochrome.ctb, which plots everything in the drawing with a black pen: Grayscale.ctb, which plots the linework in shades of gray: and acad.ctb, which plots the linework using the same colors specified in the drawing.

In addition to these standard plot style tables, two plot style tables are included for Architectural Desktop. The AIA(256) Scale 48.ctb and the AIA(256) Scale 96.ctb tables interpret the colors of Architectural Desktop objects and plot them using black and gray colors. The name of each suggests the scale of layout in which it should be assigned.

The **Edit...** button is used to modify the selected plot style table. Selecting this button displays the **Plot Style Table Editor** dialog box (see Figure 21.37). This dialog box is used to modify the settings of the colors and lineweights that will be used to plot the drawing. Configuring a plot style table is described later in this chapter.

The **New...** button is used to create a new plot style table. Picking this button displays a wizard that walks you through the steps of creating a new plot style table. The **Display plot styles** check box, when selected, displays the layout space geometry as it will appear in the finished plot. This is helpful in determining whether the pen settings will plot the layout space drawing correctly.

The **Options** button in the lower right corner of the dialog box is used to set up rules on how Architectural Desktop is to use plotters and plot style tables. For example, a default plotter and pen table can be selected in the Options dialog box. Selecting this button opens the Options dialog box, where these setting can be made.

Field Notes

The AutoCAD 2000 and 2002 engines provide two types of plot style tables: color tables(.ctb) and style tables(.stb). Architectural Desktop defaults to using the color tables. When plotting Architectural Desktop drawings you should use color tables to have more control over the final output of the drawing. This text deals strictly with the color tables. For more information regarding the style tables refer to your AutoCAD documentation or to AutoCAD Help.

FIGURE 21.37 The Plot Style Table Editor dialog box is used to adjust the settings of the plot style table.

Color being modified

Colors used by geometry in the drawing

Color that will be plotted

Set the percentage of screening to apply to the plotted color.

Assign a lineweight to the selected color.

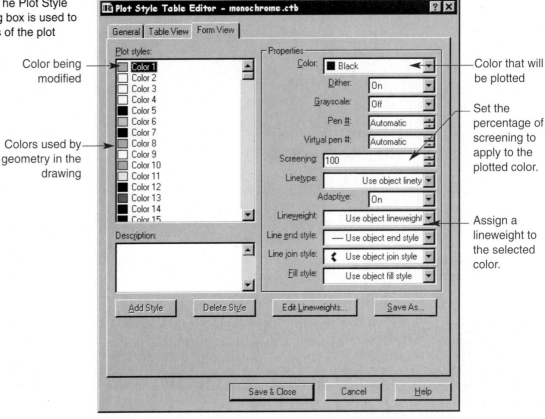

Specifying the Layout Settings

After you have made the settings for the Plot Device tab, you need to establish the settings for the layout. The Layout Settings tab is divided into the following six areas, as shown in Figure 21.38:

- **Paper size and paper units** — This area is used to specify the size of the sheet in the layout area. The sheet sizes available are based on the type of plotter selected in the Plot Device tab. Select a paper size from the Paper size: drop-down list.

- **Drawing Orientation** — This area controls the rotation of the layout on the sheet of paper. The options include Portrait and Landscape. The Portrait option orients the long side of the sheet vertically, and the Landscape option orients the long edge of the sheet horizontally. The Plot upside-down check box rotates the plot 180° on the finished plot.

- **Plot area** — This area specifies what area of the drawing will be plotted. There are five options in this area:

  - **Layout/Limits** — The Layout option is selected by default when the Page Setup dialog box is accessed from a layout tab. The Layout option plots everything that is displayed on the sheet of paper in layout space. This is the best option to use, as all of the sheet setup is done on the paper in layout space. The Limits option is displayed here if page setup is accessed from the Model tab. This option plots everything that is within the grid in model space.

  - **Extents** — This option plots all the geometry in the drawing. Much like with the Zoom Extents option, the area that will be plotted is all geometry within the edges of the farthest placed geometry.

Select the
sheet size.

Select the sheet
orientation.

Select the
plottable
area.

Assign a scale
to the layout.

Layout options

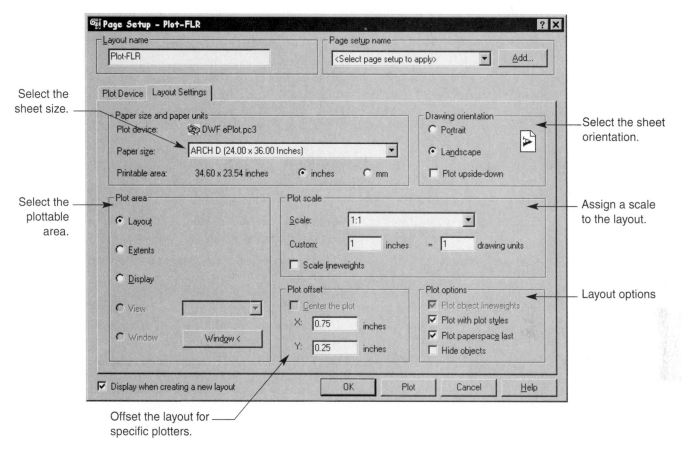

Offset the layout for
specific plotters.

FIGURE 21.38   The Layout Settings tab in the Page Setup dialog box is used to configure settings for the
layout sheet.

- **Display**   This option plots only the items that are currently being displayed in layout
  space. If you are zoomed into an area of the layout sheet, then the zoomed-in
  area will be plotted.

- **View**   This option plots user-defined views. A list of views in layout space is listed in the
  drop-down list beside this option if page setup has been accessed from a layout
  tab. This option is grayed out if there are no views specified in the current space.

- **Window**   This option is grayed out by default. Picking the Window< button allows you
  to specify a window area on the drawing. The points of the window specified
  designate an area to be plotted.

- **Plot scale**   This area is used to specify a scale for the area selected in the Plot area section.
  If the Layout area has been specified to be plotted, then select the 1:1 scale,
  because this will plot the layout sheet 1:1 on the plotter. This area does not plot
  the building model within a viewport to a scale, only the layout sheet itself to a
  scale. For example, if you have set up a 36″ × 24″ sheet of paper but select the
  1:2 scale, the layout sheet will be plotted half its actual size.

  You can also enter custom scales in the inches and drawing units text boxes.
  This area also includes the Scale lineweights check box. Under normal circum-
  stances lineweights plot using the size designated by the plot style table. If this
  box is selected, the lineweights are scaled relative to the scale selected in the
  Scale: drop-down list.

- **Plot offset**   This area is used to determine how far the drawing is offset from the lower left
  corner of the layout sheet. The origin used by a plotter is the lower left corner
  of the paper. To begin plotting away from the origin point, adjust the values in
  the X: and Y: text boxes to move the drawing away from the origin point.

■ **Plot options**   This area includes the following four options that affect how the drawing is plotted:

   ■ **Plot object lineweights**   This option is used when objects and layers have been assigned a lineweight. Selecting this option will plot these objects using the lineweights designated by the object or the layer instead of using the plot style tables. If an object or layer has not been assigned a lineweight, the plot style table will designate the lineweights to be plotted.

   ■ **Plot with plot styles**   This option plots the drawing using the plot styles applied to objects and layers defined in the plot style table. This setting is ignored if color tables are used for plotting.

   ■ **Plot paperspace last**   This option plots the objects on the layout sheet after the objects in the floating viewport have been plotted.

   ■ **Hide objects**   This option hides lines behind a surface when plotting. This option applies only to objects within the current space. This option *does not* plot a 3D view within a viewport as hidden. This option is best used when you are plotting from model space and the view in model space needs to be plotted hidden. If plotting views in layout space that need to be hidden, use the MVIEW Hideplot option in the viewports that are to be hidden.

Once you have finished setting up the page setup for the current layout, press the OK button to return to the drawing. If you create a page setup of this layout, you can apply the settings to another layout tab. It often advantageous to set up a few page setups so you do not have to set them up again later.

Exercise 21.7

1. Open EX21.6-Plot.dwg.
2. Access the Page Setup dialog box for the Floor Plan layout.
3. Make sure that the DWF plotter has been specified.
4. Assign the AIA(256) Scale 48.ctb plot style table. Turn on the Display plot styles check box.
5. In the Plot area select the Layout option.
6. Make sure the scale is set to1:1.
7. Set the orientation to Landscape.
8. Create a new page setup labeled 36x24.
9. Apply the page setup to the other layout tabs.
10. Save the drawing as EX21.7-Plot.dwg.

Working with Plot Styles

The previous section introduced the use of plot style tables. **Color tables** are used by default in Architectural Desktop and provide a method of mapping colors used in the drawing to pen colors and lineweights on the finished plot. As mentioned earlier, style tables are also a part of the AutoCAD engine running Architectural Desktop. Style tables are similar to color tables in that pens and lineweights are mapped to objects instead of colors. Thest tables provides little control over how AEC objects are plotted in Architectural Desktop. In contrast, the color tables give you the most control over plotting AEC objects.

Color Tables

Color table (.ctb) files assigned to a layout interpret the color used by objects in the drawing and plot the colors using settings specified in the table. The settings for the color table can be modified by selecting the Edit... button in the Page Setup dialog box.

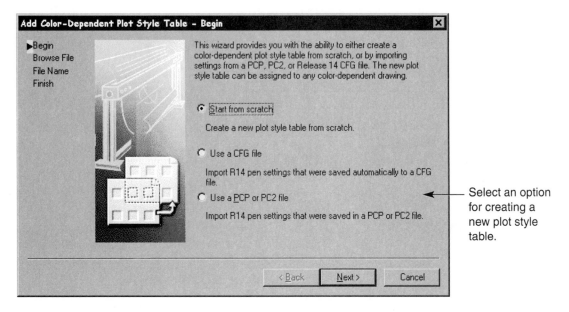

Select an option for creating a new plot style table.

FIGURE 21.39   The Add Plot Style Table wizard is used to create a new plot style table.

You can also create new color tables by selecting the N**e**w... button in the Page Setup dialog box, which displays the **Add Plot Style Table** wizard (see Figure 21.39). The first page of the wizard allows you to select one of the following methods for creating a new plot style table:

- **Start from scratch**   This option creates a new plot style table from default settings.

- **Use a CFG file**   This option is for previous users of AutoCAD. An existing acad.cfg file can be used to create the new plot style table.

- **Use a PCP or PC2 file**   This option also is for previous users of AutoCAD who have set up a .pcp file for their pen settings. This option allows a .pcp or .pc2 file to be imported as a template for the plot style table.

Choose the N**e**xt> button to advance to the next page of the wizard, where you can specify a name for the new plot style table. Enter a name in the F**i**le name text box (see Figure 21.40A). Press the N**e**xt> button to advance to the final page of the wizard (see Figure 21.40B). This page allows you to modify the settings for the plot style table and apply the table to the current layout sheet. Select the Plot Style Table E**d**itor... button to configure the plot style parameters.

The Plot Style Table Editor dialog box is displayed (see Figure 21.41). This dialog box includes three tabs. The **General** tab is used to provide a description for the table. The **Table View** and the **Form View** tabs contain the same setting parameters but are organized in two different ways (see Figure 21.41).

Each tab includes a list of colors. These are the colors within the drawing and the colors that the plot style table will map to a pen setting. For example, if the color red is used in the drawing, it can be mapped to print as a black line. To do this, select the color used by the drawing that you want to configure, then select the pen setting parameters. The following properties can be adjusted in either tab for each color inside the drawing:

- **Color**:   This setting is used as the color to be plotted. First, select the object color to be modified, then select a color that will be used for plotting. As in the preceding example, select the red color and assign it to print out as black. When the drawing is plotted, the color table finds all the red colors in the drawing and plots them out using a black color. The **Use object color** option is set by default and plots the drawing using the color specified in the drawing.

- **Dither:** or **Enable *Dithering*** is a form of simulating the depths of color by varying the **Dithering:**   spacing of a series of dots along the plotted line. Dithering can be turned on or off. Many offices disable dithering, because the varying dots pro-

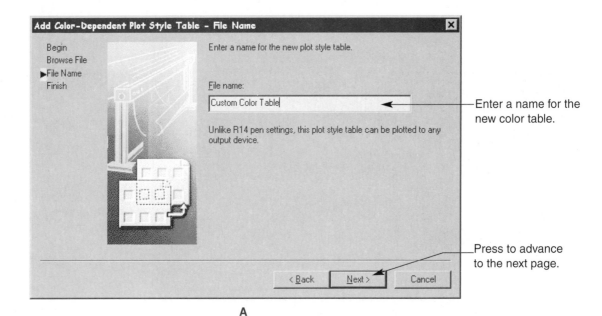

FIGURE 21.40   (A) Specify a name for the plot style. (B) The final page of the Plot Style Table wizard.

duce varied results. Perform test plots to determine if dithering should be turned on or off.

- **Grayscale:** or **Convert to grayscale**

  This option is also either turned on or off. If the grayscale is turned on, the linework in the drawing is plotted using shades of gray. As with the dithering option, perform test plots to determine whether or not to use grayscales.

- **Pen #:** or **Use assigned pen #**

  This setting is used only with plotters that use pens for plotting. A total of 32 pens can be assigned to the 255 AutoCAD colors. The pen number assigned corresponds to the pen number in the plotter. For example, pen number 3 may be assigned to the magenta color, which in turn plots with a .2 mm pen. Assigning this value to the Automatic pen causes AutoCAD to determine the pen number based on the plotter configuration.

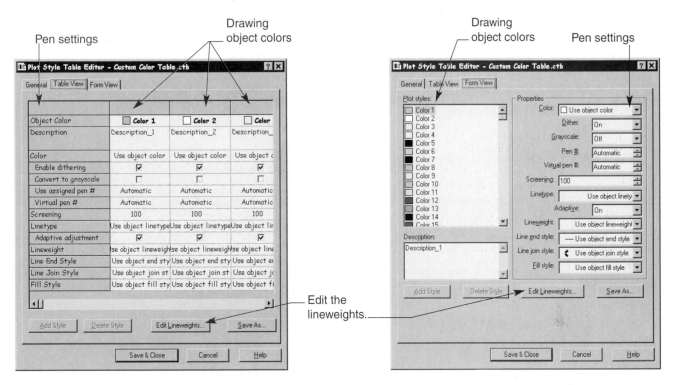

FIGURE 21.41 The Plot Style Table Editor dialog box is used to set up different parameters for the color table.

- **Virtual pen #:** or **Virtual pen #**

  This setting is used to simulate pen numbers for non-pen plotters such as ink-jet plotters. Virtual pens use pen numbers ranging from 1 to 255. Refer to the *AutoCAD Installation Guide* for information on configuring plotters to use virtual pens. If the setting is set to Automatic, AutoCAD determines the virtual pen to use from the AutoCAD Color Index (ACI).

- **Screening**

  This option controls the amount of ink that is applied on paper for each object's color. This setting fades or washes out colors that are plotted. A value of 100 plots the object color with the full intensity of color. A value of 50 plots the linework with half the intensity.

- **Linetype:**

  Specifying a value overrides the linetype assigned to the drawing. The *Use object linetype* option plots the object using the linetypes specified in the drawing.

- **Adaptive** or **Adaptive adjustment:**

  This setting adjusts the scale of a linetype to complete a repetitive pattern. If this setting is not selected, the linework may end before the pattern has been completed.

- **Lineweight:**

  This setting assigns a lineweight to an object color. Specifying a lineweight here overrides the lineweight specified in the drawing. Select the *Use object lineweight* option to have the layers and the objects designate the lineweights to use.

- **Line end style:**

  This setting is used to apply a line end style to the ends of all linework in the drawing.

- **Line join style:**

  This setting is used to configure the style that plots when linework is joined in the drawing.

- **Fill style:**

  This option is used to configure how filled objects appear on the final plot. A *filled object* may be a hatch pattern using the Solid Fill hatch or a TrueType font that includes a filled-in area. The default value **Use object fill style** plots fills using a filled color instead of a fill pattern.

Once you have configured the settings for each color (or the colors you are using in the drawing), press the Save & Close button to apply the configurations to the new plot style table. This will return you to the last page of the wizard. If the new color table is to be applied to the current layout, then select the appropriate option. Select the Finish button to create the plot style table. A permanent color table (.ctb) file is created for use in future drawings.

Plotting the Drawing

When you have finished setting up a plotter, assigning a plot style table, configuring the layout sheet size, and drawing the geometry on the layout sheet, you are ready to plot the drawing. Before the drawing can be plotted, the layout tab to be printed needs to be active. This tab can be the Model tab (if plotting from model space) or a layout tab.

Once the layout is active, the **PLOT** command can be used to plot the drawing using one of the following options:

✓ Select the **File** pull-down menu, then pick **Plot...** from the list.

✓ Select the **Plot (Ctrl+P)** button from the **Standard** toolbar.

✓ Right-click on top of the layout tab, and select **Plot...** from the shortcut menu.

✓ Type **PLOT** at the command prompt.

✓ Pick the [Ctrl]+[P] key combination.

The Plot dialog box, shown in Figure 21.42, is displayed.

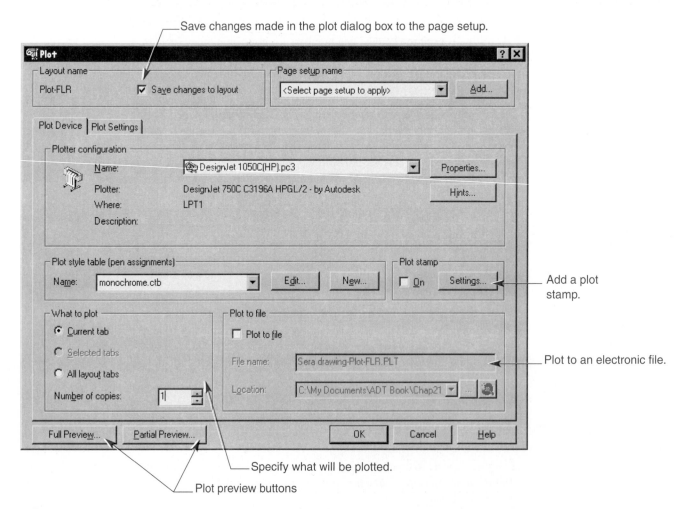

FIGURE 21.42   The Plot dialog box is used to verify the settings for the layout and plot the drawing.

This dialog box is similar to the Page Setup dialog box, with the following additional features: the Layout name area, the Plot stamp area, the What to plot area, and the Plot to file area.

You can make changes to the page setup at the time of plotting. For example, suppose you need to use a different plot style table instead of the one specified in page setup. Select the new plot style table from the list, and the drawing will be plotted using the new table. When this change has been made you can save the change back to the page setup by selecting the **Save changes to layout** check box at the top of the window.

Adding a Plot Stamp

The Plot stamp dialog box is used to add a plot stamp to the finished plot. A *plot stamp* can include information such as the drawing name, the date and time the drawing was plotted, and the plotter device plotted to. Selecting the On check box will add the plot stamp to the drawing. Picking the Settings... button allows you to control the parameters of the plot stamp through the **Plot Stamp** dialog box (see Figure 21.43).

In the **Plot stamp fields** area, select the type of items that you want to have included in the plot stamp. The **User defined fields** area allows you to define custom fields that you would like to add to the plot stamp. Use the Add/Edit button to add custom fields to the user-defined list. For example, the Drafter: and Ron custom fields have been added, then selected in the drop-down lists.

In order to add a plot stamp to the finished plot, you need to save the plot stamp as an external plot stamp parameter file. Select the Save As button in the **Plot stamp parameter file** area and specify a location and name for the file. Selecting the Advanced button allows you to configure additional parameters for the plot stamp in the **Advanced Options** dialog box (see Figure 21.44).

The following options are included in the Advanced options dialog box:

- **Location and offset**   This area is used to control where the plot stamp is located. This area includes a Location drop-down list that specifies a corner of the paper that will include the plot stamp. The Orientation drop-down list specifies whether the plot stamp will be plotted horizontally or vertically. The X offset and Y offset text boxes offset the plot stamp from the corner of the plottable area or the corner of the layout sheet.

- **Text properties**   Use the Font drop-down list to specify the font to use for the plot stamp. Specify a text height by entering a value in the Height text box, and specify whether the plot stamp will be printed as a single line or multiple lines in the Single line plot stamp check box.

- **Plot stamp units**   Select the type of units to use for the plot stamp text height. Values included inches, millimeters, and pixels.

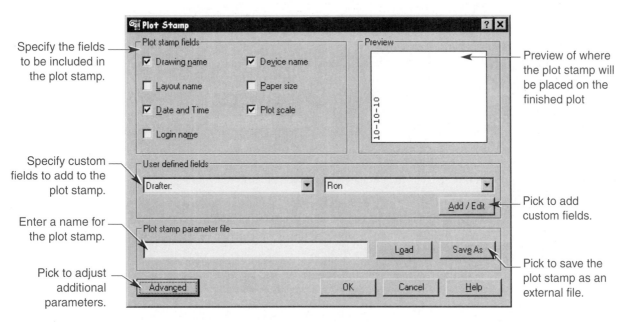

FIGURE 21.43   The Plot Stamp dialog box is used to configure the plot stamp to be added to the drawing.

Specify whether the plot stamp will be
printed horizontally or vertically.

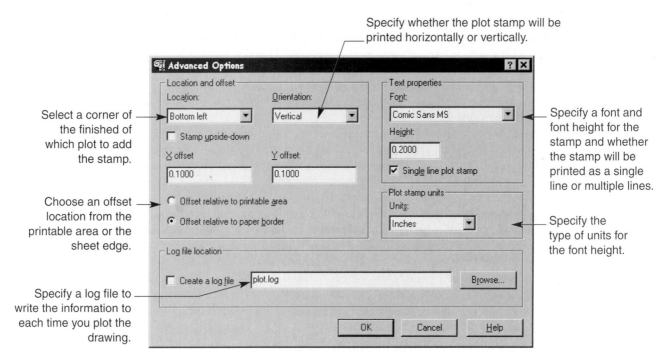

Select a corner of
the finished of
which plot to add
the stamp.

Choose an offset
location from the
printable area or the
sheet edge.

Specify a log file to
write the information to
each time you plot the
drawing.

Specify a font and
font height for the
stamp and whether
the stamp will be
printed as a single
line or multiple lines.

Specify the
type of units for
the font height.

FIGURE 21.44   The Advanced Options dialog box is used to set additional parameters for the plot stamp.

- **Log file location**   Select the Create a log file check box to keep a log each time the layout is plotted. Use the Browse... button to determine a path and a name for the log file.

Field Notes

A *log file* can be generated without generating a plot stamp. This file keeps track each time the layout is plotted and writes the information to a .log file that can be opened in a text editor such as the Windows Notepad program.

When you have finished setting the parameters for the plot stamp press the OK button to return to the Plot Stamp dialog box. If you will be generating a plot stamp, make sure the plot stamp parameters file has been specified, then press the OK button to return to the Plot dialog box.

Specifying What to Plot

The What to plot area in the Plot dialog box is used to determine what will be plotted. The following options are included in this area:

- **Current tab**:   This button plots the currently selected layout tab.

- **Selected tab**:   You can plot multiple tabs at once by holding down the [Ctrl] key and picking on top of the desired layouts before entering the Plot dialog box. Selecting this option plots the selected tabs using the parameters set up in the Plot dialog box.

- **All layout tabs**:   Picking this option plots all the layout tabs using the settings in the Plot dialog box.

- **Number of copies:**   This text box allows you to specify the total number of prints desired of one of the above settings.

Previewing the Plot

After you have specified the settings for the plotter you may want to preview how the drawing will be plotted before actually plotting the drawing. This allows you to double-check geometry and ensures that the drawing will be plotted correctly. If you discover a problem at the plot preview stage, you can cancel the plot and make modifications, then re-enter the plot command for plotting.

Two buttons are included for previewing the plot. The **Full Preview...** button provides you with a full preview of the entire drawing as it will be plotted (see Figure 21.45). The full preview window displays the drawing as it has been prepared. Notice that the viewport edges do not appear, because they have been assigned to a nonplotting layer. Pen settings and colors are also displayed in the full preview window. Right-clicking in the preview provides you with a shortcut menu of options that can be used to view the drawing and return to the plot dialog box. When you have finished previewing the plot, right-click and select Exit to return to the Plot dialog box, or Plot to plot the drawing.

The **Partial Preview** button is used to display a smaller preview of the drawing (see Figure 21.46). Selecting this button displays the **Partial Plot Preview** dialog box. This preview displays a minisheet of paper representing the selected layout sheet. The dashed line around the paper represents the allowable plotted area on the sheet.

The shaded blue area represents the extents of the geometry in the layout and where it is located on the sheet. This is known as the *effective plotting area*. Below the preview, information regarding the paper size, printable area, and the effective area is displayed. If there are any errors in the effective plot, they are displayed in the Warnings area at the bottom of the window. When you have finished previewing the drawing press the OK button to return to the Plot dialog box.

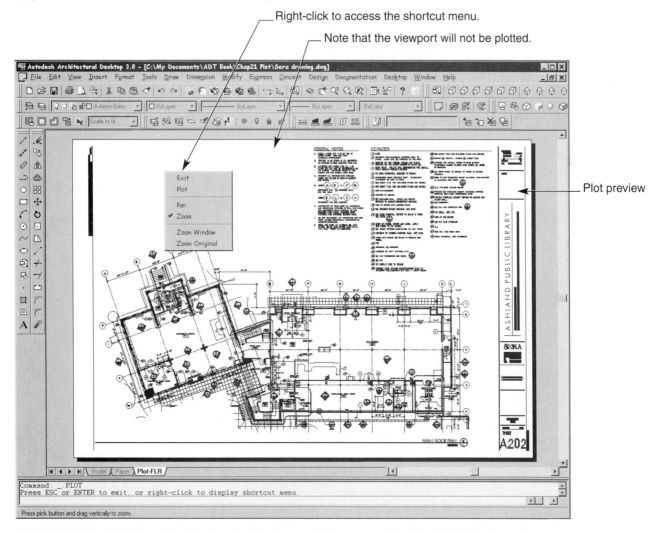

FIGURE 21.45 The full preview displays the drawing as it will appear on the final plot. Objects on layers that will not plot are not displayed.
Drawing courtesy SERA Architects Inc., Portland, OR.

FIGURE 21.46 The Partial Plot
Preview dialog box provides
information on where the
geometry in the layout is in
relation to the paper sheet.

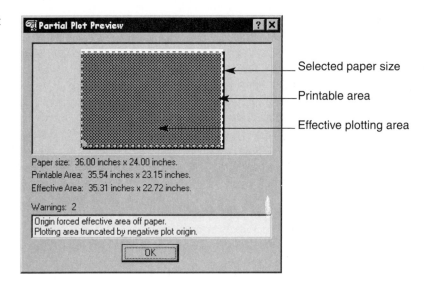

Selected paper size

Printable area

Effective plotting area

Paper size: 36.00 inches x 24.00 inches.
Printable Area: 35.54 inches x 23.15 inches.
Effective Area: 35.31 inches x 22.72 inches.

Warnings: 2

Origin forced effective area off paper.
Plotting area truncated by negative plot origin.

OK

Once you are ready to plot the drawing, select the OK button at the bottom of the Plot dialog box to
send the information to the selected plotter, and the plotter will begin to plot the drawing.

Plotting to a File

The **Plot to file** area is used to plot the layout to an electronic file. This area is grayed out by default if a
"real" plotter or printer has been selected. Selecting the Plot to file check box activates the File name: and
Location text boxes (see Figure 21.47).

The **File name** text box allows you to define a name for the electronic file. By default the name is
defined from the name of the drawing plus the layout name followed by the file extension .plt (.dwf if the
DWF plotter has been selected). A *.plt file* is a file that remembers all geometry displayed in the layout, the
plot style tables assigned, and the layout settings. This type of file can be plotted on a plot spooler.

A *plot spooler* is a type of disk drive with memory that allows you to copy a .plt file from a floppy
disk to it, and it in turn sends the information to the plotter for plotting. Many plotting services allow you to
send them a .plt file that they can plot out for you.

The **Location** text box specifies the folder where the .plt (or .dwf) will be saved. The ellipsis button
[...] can be used to browse through folders on local hard drives or on a network drive in which to save the
electronic file. Beside the ellipsis button is a button with a world icon. Selecting this button allows you to
browse the Internet for a location to save the file to.

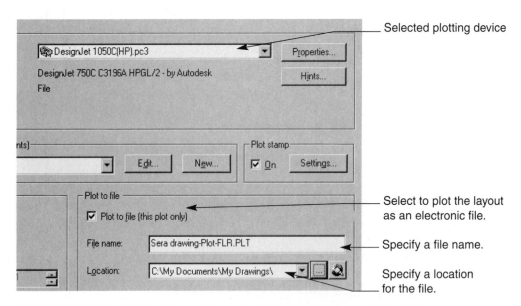

Selected plotting device

Select to plot the layout
as an electronic file.

Specify a file name.

Specify a location
for the file.

FIGURE 21.47 The Plot to file area is used to plot the layout to an electronic file.

Plotting a DWF File

Mentioned earlier in this chapter was a Drawing Web Format (DWF) file, which gives you the ability to plot the drawing to an electronic file that can be viewed over the Internet. Architectural Desktop provides you with a DWF plotter configuration. Selecting this plotter causes the Plot to file area of the drawing to become active. The File name area displays the name of the file with a .dwf extension. Specify a location for the file, and the layout will be plotted to the file.

The DWF file can be viewed in one of two ways. The first is through an Internet browser such as Microsoft Internet Explorer (see Figure 21.48). The only requirement to viewing this file in a browser is that a driver known as the **Whip!** driver is installed on your computer system. You can find this driver on the Internet by searching for the keyword *Whip!* Autodesk is dropping the support for the Whip! driver, but an alternative is available.

A program known as **Volo View Express** can be downloaded from the www.autodesk.com website. This program is a free download and allows you to view and redline DWF files. A full version of Volo View can also be purchased from Autodesk and provides you with additional redlining tools and the ability to plot a DWF file to a scale. Right-clicking on the DWF file in either of these programs provides you with a set of tools that you can use to turn layers on or off, display saved views of the drawing, and zoom and pan in the drawing.

The advantage to plotting a drawing to a DWF file is that the drawings can be shared with another person without their physically having AutoCAD or Architectural Desktop. Another advantage is that the drawings cannot be edited. This provides a good means of sending a drawing file to a consultant for review without their having the ability to edit the drawing. DWF files occupy very little memory in the actual electronic file and can therefore be sent through e-mail without much concern regarding file size.

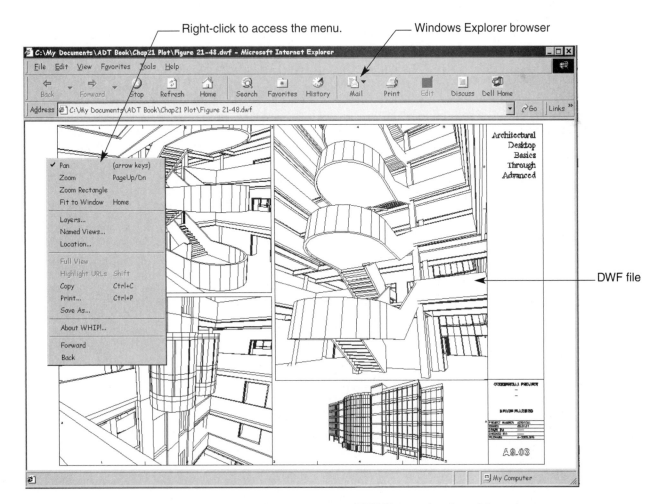

FIGURE 21.48 With the Whip! driver installed on your system, DWF files can be viewed through an Internet browser.

|||||||||||| CHAPTER REVIEW

 Use the CD-ROM to test your knowledge and skills.

Chapter Test

To check your understanding of the content provided in this chapter, access the Test file in the CH21 folder of the CD-ROM that accompanies this text.

Chapter Project

To practice the Architectural Desktop skills presented in this chapter, access the Project files in the CH21 folder of the CD-ROM that accompanies this text. The project files are in pdf format and include sample drawings and instructions for completing each project.

CHAPTER 22

||||||||||||||||||||||||||||||||||||||||||||||||||||||||||||||||||||||||||||||||||||||||||||

Conceptual Design

LEARNING GOALS

After completing this chapter, you will be able to:

◎ Build a mass model.

◎ Add custom shapes to the mass model.

◎ Modify mass elements.

◎ Reference an object.

◎ Group the mass objects.

◎ Add AEC objects to the mass group.

◎ Use the Model Explorer.

◎ Create floor plates.

◎ Export slices to another drawing.

◎ Create polylines from slices.

Until now you have been working with objects and geometry used in the design, design development, and documentation phases of the architectural work flow, usually for a building. In many cases, a new building begins with an idea. This idea may consist of sketches, preliminary drawings, and models. Architectural Desktop adds to these tools by providing the architect and designer with a series of options that can be used to create a conceptual three-dimensional model. The conceptual model can be used for design analysis, massing studies, sun studies, site studies, and approvals (see Figure 22.1). Once the conceptual model has been approved, it can be moved forward to the design and design development phases. From the conceptual model, walls and spaces can be generated to create the construction documents.

Although Architectural Desktop provides you with many tools for all phases of work, construction documents do not have to begin with a conceptual model; however, the conceptual model can be used as a starting point in the development of a building and carried all the way through to the finished plotted documents. This chapter explains the tools included to create a conceptual model and prepare it for the design phase.

Building a Mass Model

The first step in creating a conceptual design is beginning a new drawing. Architectural Desktop includes a template specifically designed for working with conceptual models named AEC (Imperial) Massing.dwt. Although any template can be used, this template includes some predefined profile shapes that can be used when designing, and a layout tab for working with the conceptual model.

When the conceptual model is created, 3D shapes are used to mass together a building idea. The 3D shapes included with Architectural Desktop have volume or mass and are known as *mass elements*. Mass elements come in an assorted number of 3D shapes and give you the ability to create custom shapes.

This entire chapter is Next Step Material.

835

Mass elements

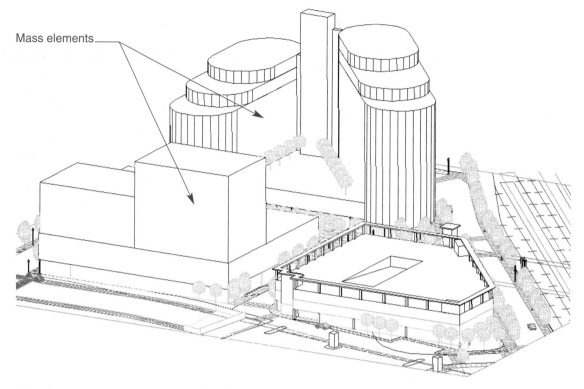

FIGURE 22.1　A hidden view of a conceptual model.
Drawing courtesy SERA Architects Inc., Portland, OR.

Using Mass Elements

Use one of the following options to begin adding mass elements:

- ✓ Select the **Concept** pull-down menu, pick the **Mass Elements** cascade menu, then select **Add Mass Element....**

- ✓ Right-click, and select **Concept, Mass Elements, Add Mass Element...** from the cursor menu.

- ✓ Type **AECMASSELEMENTADD** at the Command prompt.

The **Add Mass Element** modeless dialog box, shown in Figure 22.2, is displayed.
The following options are available in the Add Mass Element dialog box:

- ■ **Shape**　This drop-down list includes 12 predefined shapes that can be added to the drawing. The shapes include an Arch, Barrel Vault, Box, Doric Column, Cone Cylinder, Dome, Gable, Pyramid, Isosceles Triangle, Right Triangle, and Sphere.

  If you are not sure what the shape mass element will look like in the drawing, select the Floating viewer to preview the shape before adding it to the drawing. The last two shapes, Extrusion and Revolution, allow you to use a custom profile as a mass element shape.

- ■ **Profile:**　Initially this drop-down list is grayed out if a standard shape has been selected. If the Extrusion or the Revolution shape has been selected, this drop-down list becomes available. The items in the list are custom profile shapes defined in the drawing. Custom profile shapes were discussed in Chapter 7. The AEC (Imperial) Massing.dwt includes five basic profile shapes. Create a custom profile with or without additional rings to create custom mass elements for your drawings.

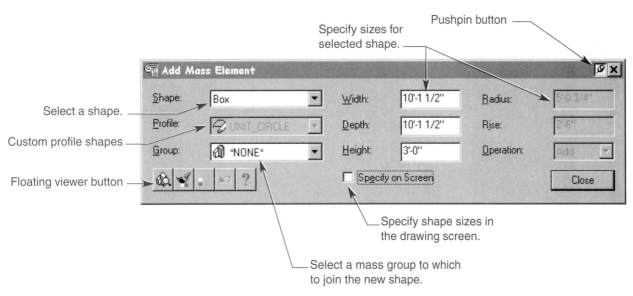

FIGURE 22.2 The Add Mass Element dialog box is used to add mass element shapes to a drawing to create a 3D design.

- **Group:** This drop-down list allows you to select a group to which to add the new mass element shape. Initially there are no mass groups available, so the drop-down list will read *NONE*. Creating and using mass groups is discussed later in this chapter.

- **Width:** This text box allows you to enter a width or the selected shape. This option is not available for all shapes.

- **Depth:** This text box allows you to specify a depth for the selected shape. Again, this value is not available for all the shapes.

- **Height:** This option is used to specify a height for the selected shape.

- **Radius:** This option is used to assign a radial size for curved shapes.

- **Rise:** This option is available only for the Gable shape. The rise is measured from the top of the peak down, as shown in Figure 22.3. Entering a value requires the height value to be greater than or equal to the rise height.

- **Operation:** This option becomes available when the mass element is assigned to a group and allows you to add the mass element to another mass element, subtract the mass element from another mass element, and create an intersection of two mass elements. These options are discussed later in this chapter when grouping of mass elements is discussed.

- **Specify on Screen:** This option allows you to define the sizes for the mass element as you add it to the drawing. If this option is selected, the size text boxes are grayed out. Unchecking this option requires you to add sizes in the text boxes.

Adding a Mass Element

After selecting the desired shape and setting any size parameters you can add the mass element to the drawing. If the pushpin button is pinned down on the dialog box, select the drawing once to make the drawing active, then pick a point for the insertion of the mass element. If the pushpin button is unpinned, move the cursor off the modeless dialog box to minimize it, then pick a point in the drawing to add the mass element.

If you are inserting a shape with its size values entered in the dialog box, a 2D representation of the selected shape is attached to the crosshairs. Picking a point in the drawing establishes the location for the

FIGURE 22.3 The Rise value is used to specify a rise height for the gable shape.

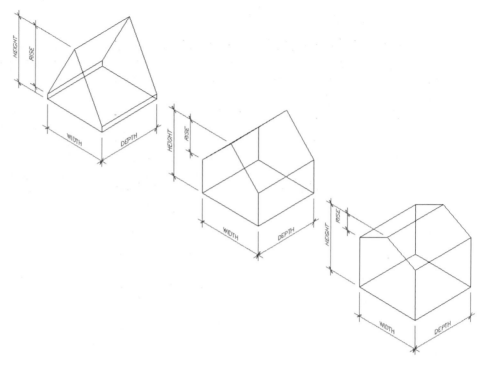

shape. After the insertion point has been located, you are prompted to enter a rotation angle for the shape. After the rotation angle has been entered, another shape is attached to the crosshairs for continued addition of mass elements (see Figure 22.4). Continue to add new mass elements, making changes in the Add Mass Elements dialog box as required. When you have finished adding mass elements pick the Close button on the dialog box, or press the [Enter] key to end the command.

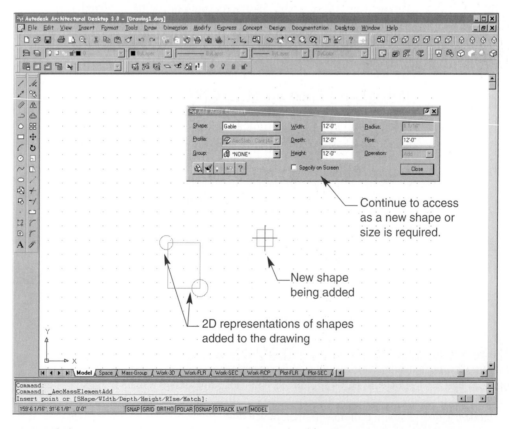

FIGURE 22.4 After adding a mass element you can continue to make changes to new shapes and to add them.

Field Notes

The AEC (Imperial) Massing.dwt is set up with a top view in model space. When you are designing a building it may be advantageous to split the drawing screen into a couple of viewports so that you can view the top view and an isometric view as you design.

If you select the Specify on Screen check box when adding shapes, you are prompted with some questions at the Command: line. Each shape can require a different set of techniques in order to be added. For example, the box requires you to select two opposite diagonal corners for the width and depth, a height, and a rotation angle. The cylinder requires a center point, radius, height, and rotation angle. When you are working in a top view, specifying the height value on screen can be difficult, because the shape is being extended perpendicular to your line of sight. It is often easier to add mass elements in an isometric view when specifying their sizes on screen (see Figure 22.5).

Field Notes

Note that when the mass element is added in a top view, the shape appears as a 2D shape with a magenta color. Switching to an isometric view causes the shape to appear as a gray color. Mass elements, like other AEC objects can be controlled through Entity Display.

As mass elements are added they are placed on the A-Mass layer. The shapes can overlap one another, be separated, and be placed inside one another. The Shade command can give you a general idea of the shape of your design.

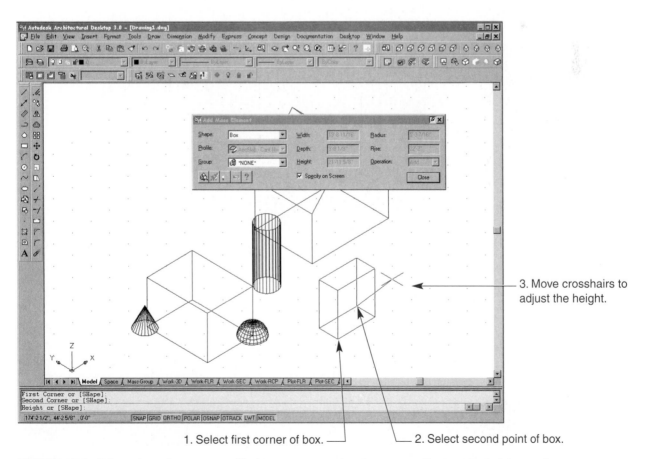

3. Move crosshairs to adjust the height.

1. Select first corner of box. ——   —— 2. Select second point of box.

FIGURE 22.5   When size values are specified on screen, moving the cursor will cause the height to adjust dynamically until a point is selected or a height value is entered in the Command prompt.

FIGURE 22.6 The Add Mass Element toolbar can be used to add mass element shapes into the drawing.

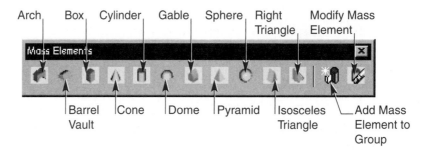

In addition to using the Concept pull-down menu to add mass elements, or entering the **AECMAS-SELEMENTADD** command, you can add mass elements by using the **Mass Elements** toolbar (see Figure 22.6). This toolbar includes buttons for most of the shapes and accesses the Add Mass Elements dialog box. The Add Mass Element to Group button and the Modify Mass Element are discussed later in this chapter.

Exercise 22.1

1. Start a new drawing using the AEC (Imperial) Massing.dwt.
2. In the Model tab divide the drawing screen into two vertical viewports. One viewport should reflect a top view, and the other an isometric view.
3. Add mass elements a–e in part A of the figure using the sizes specified. Arrange the mass elements as in part B.
4. Save the drawing as EX22-1-Mass.dwg.

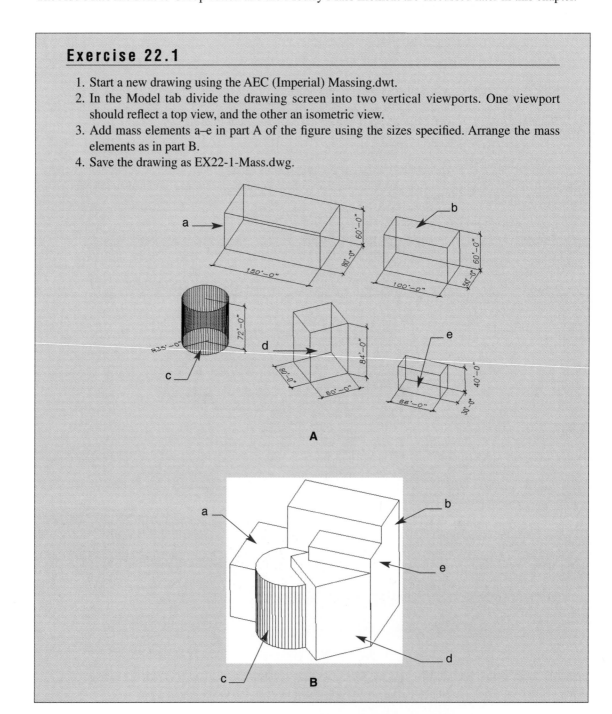

Adding a Custom Shape

You can also add custom mass elements to the drawing by using an AEC profile. First, draw the shape desired with a closed polyline, add any internal rings desired, then process the polylines into a profile definition (see Figure 22.7). Once the profile has been defined, it can be used in the Add Mass Element dialog box.

To use a profile as a mass element, access the Add Mass Element dialog box. In the Shape drop-down list, two types of shapes can be used to create the custom shape. Selecting either of these options causes the Profile drop-down list to become active. The custom profile is listed. Select the profile shape, then add the mass element to the drawing.

The Extrusion shape uses the profile and gives it a thickness (see Figure 22.8A). When an extrusion shape is created, the size of the actual profile is listed in the Width and Depth size text boxes. Enter a new height for the mass element, then add it to the drawing. The Revolution shape uses the selected profile shape and revolves it around an axis, as shown in Figure 22.8B. When the revolution shape is used, the height and the radius can be specified.

Field Notes

> An AutoCAD 3D shape created using solids can be converted into a mass element shape by using the MASSELEMENT command. Enter this command, then type C to convert solids into a mass element. Although a Convert option is not displayed at the Command: prompt, it can be used to create new mass element shapes. This is an undocumented command. The following prompt sequence demonstrates how this command works:

Command: **MASSELEMENT** ↵
Mass element [Add/Modify/Properties]: *(enter C to convert)* **C** ↵
Select objects to convert: *(select a 3D solid object)*
1 found

FIGURE 22.7 The shape desired is created, and any internal rings are added for voids. The polylines are then processed into an AEC profile.

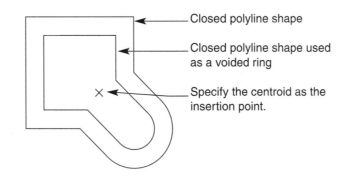

Closed polyline shape

Closed polyline shape used as a voided ring

Specify the centroid as the insertion point.

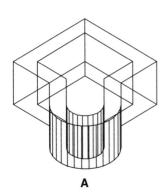

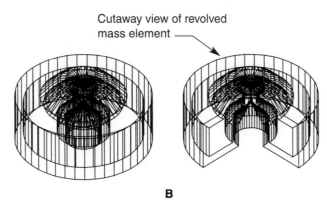

Cutaway view of revolved mass element

A   B

FIGURE 22.8 (A) The profile from Figure 22.7 used as an extrusion. (B) The same profile used as a revolution.

Select objects to convert: ↵
Name: *(name the new mass element)* **Custom_Shape** ↵
Erase layout geometry? [Yes/No] <Y>: **Y** ↵
Mass element [Add/Modify/Properties]: ↵
Command:

Exercise 22.2

1. Open EX22-1-Mass.dwg.
2. Create the polyline shape displayed in part A in the figure.
3. Process the polyline into a profile named Atrium.
4. Access the Add Mass Element dialog box.
5. Use the Extrusion shape to extrude the custom profile to a height of 24″-0″. Use the default width and depth values.
6. Use the Move command to move the atrium piece into place, as in part B.
7. Save the drawing as EX22-2-Mass.dwg.

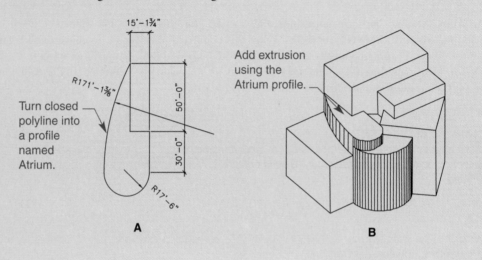

Modifying Mass Elements

Mass elements, like other AEC objects, can be modified in a number of ways. Common AutoCAD commands such as Move, Copy, Rotate, Mirror, Stretch, and Mirror can be used on mass elements. When using these commands it is helpful to use the object snaps to snap to specific points on the mass elements. For example, if you are moving a pyramid shape from the midpoint of one side to a corner of a box, use the Midpoint and Endpoint object snaps accordingly.

In addition to having the standard osnap points, mass elements include some specific node osnap locations, as shown in Figure 22.9. There is a node located at the bottom of the shape and one at the top centroid of the shape. Using the Node osnap on a mass element selects one of these points.

Mass elements can also be modified with grips. When a mass element is selected, the grip boxes are located at all the node locations on the mass element. Selecting a grip along the perimeter of a mass element and using the grip stretch mode stretches the width and depth of the mass element. Selecting the grip at the bottom centroid of the mass moves the mass element, and selecting the grip at the top centroid of the mass and using the stretch mode adjusts the height of the mass element (see Figure 22.10).

Modifying the Mass Element

Once a mass element has been placed in the drawing, its shape and size can be modified using the **Modify Mass Element** command. Use one of the following options to access this command:

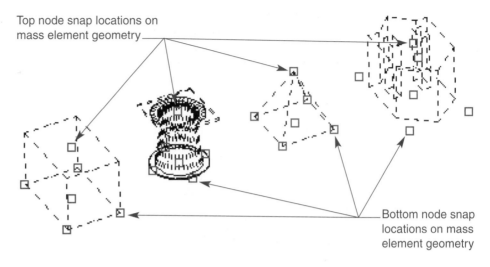

FIGURE 22.9 Mass elements include a node snap point at the bottom of the shape and one at the top centroid of the shape.

Top node snap locations on mass element geometry

Bottom node snap locations on mass element geometry

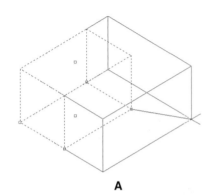

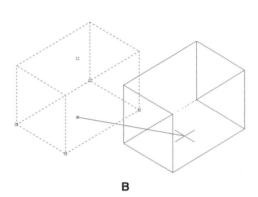

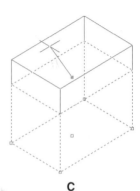

A   B   C

FIGURE 22.10   (A) Stretching the width and depth. (B) Moving the mass element. (C) Stretching the height.

✓ Pick the **Concept** pull-down menu, select the **Mass Elements** cascade menu, then pick the **Modify Mass Element...** command from the menu.

✓ Pick the **Modify Mass Element** button from the **Mass Elements** toolbar.

✓ Right-click, and select **Concept, Mass Elements, Modify Mass Element...** from the cursor menu.

✓ Type **AECMASSELEMENTMODIFY** at the Command: prompt.

✓ Select a mass element(s), right-click, then select **Element Modify...** in the shortcut menu.

The Modify Mass Element dialog box, shown in Figure 22.11, is displayed. This dialog box is similar to the Add Mass Element dialog box. Using the dialog box, change the shape and size of the selected mass element, or change the group to which the element belongs. Press the Apply button to verify that this is the change desired before pressing the OK button.

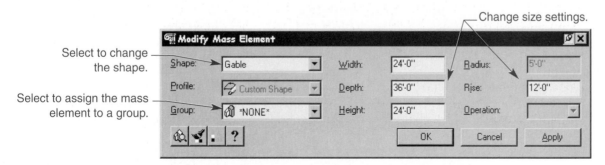

Change size settings.

Select to change the shape.

Select to assign the mass element to a group.

FIGURE 22.11   The Modify Mass Element dialog box is used to modify mass elements.

Adjusting the Mass Element Properties

Additional mass element properties can be adjusted using the **Mass Element Properties** dialog box. This dialog box can be accessed by first selecting the mass element, right-clicking, and picking **Element Properties...** from the shortcut menu, or by typing **MASSELEMENTPROPS** at the Command: prompt. This dialog box includes the following four tabs for adjusting the mass element being modified:

- **General**   This tab is used to enter any design notes and instructions for the mass element use by selecting the Notes... button. You can also assign a property set definition to the mass element by selecting the Property Sets... button. The description area is important when working with mass elements, as discussed later in this chapter. Enter a simple description that describes what this element will be used for, for example, Main Building, West Wing, Foyer, and Roof (see Figure 22.12).

- **Dimensions**   This tab is used to change the shape and the size of the selected mass element (see Figure 22.13). Selecting the Extrusion or the Revolution shape makes the Profile drop-down list available.

- **Mass Group**   This tab is used to assign the mass element to a group and to specify an operation of the element. When a mass element belongs to a group, it can

FIGURE 22.12   Notes, property set definitions, and a description for the mass element are entered in the General tab.

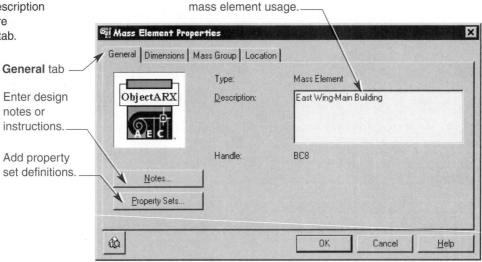

FIGURE 22.13   The Dimensions tab allows you to adjust the shape and the size of the selected mass element.

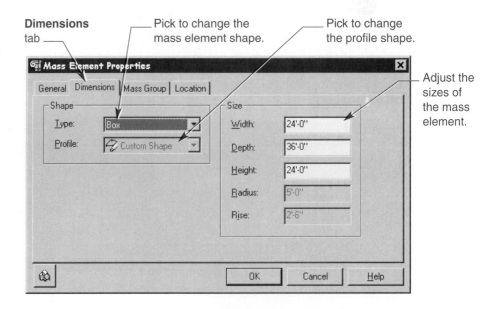

FIGURE 22.14   The Mass Group tab is used to assign the mass element to a mass group and to assign an operation.

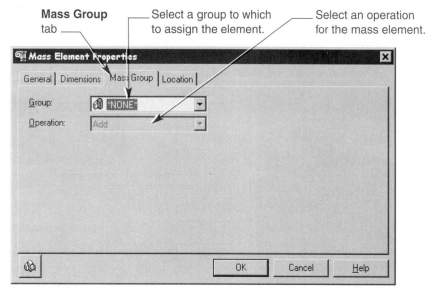

Mass Group tab — Select a group to which to assign the element. — Select an operation for the mass element.

FIGURE 22.15   The Location tab is used to specify X, Y, Z coordinates for the mass element.

Select to adjust the element location in "real-world" space. — **Location** tab — Enter a value to place the element higher in elevation.

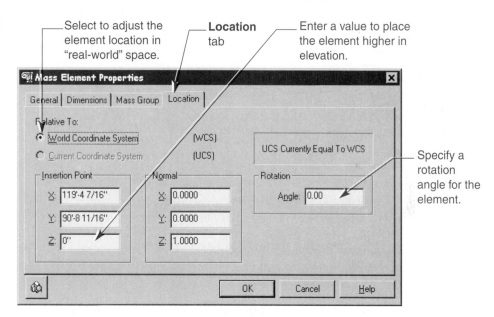

Specify a rotation angle for the element.

be an additive object that is joined to another object or it can be a subtractive element that is subtracted from another object. This topic is described later in this chapter (see Figure 22.14).

■ **Location**   This tab is used to specify the location for the mass element. Use the World Coordinate System radio button to specify where the element is in the drawing. Specifying a Z: value adjusts the elevation of the mass element. For example, if the mass element needs to be placed 60'-0" up in the air, then enter a value of 60'-0" in the Z: text box (see Figure 22.15).

Using Entity Display

Like other AEC objects, mass elements include their own display system. By default, mass elements appear as 2D shapes in a top view and as 3D shapes in an isometric, elevation, or perspective view. There are no styles for mass elements, so the display system cannot be controlled through a Style dialog box. Instead, the display of mass elements must be controlled through Entity Display.

To control the display of the mass elements, select one mass element, right-click, and pick Entity Display from the shortcut menu. The **Entity Display** dialog box, shown in Figure 22.16, is displayed. Select the Display Props tab to modify the display of the mass elements.

Select a display   **Display**   Select the level of
representation   **Props** tab   display control.
to adjust.

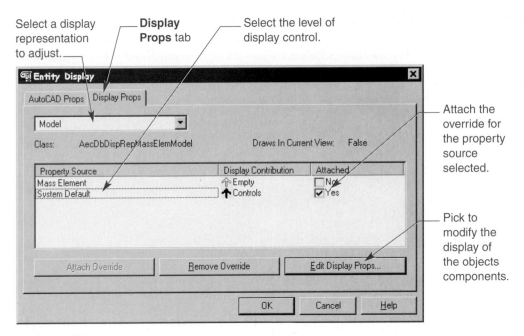

Attach the
override for
the property
source
selected.

Pick to
modify the
display of
the objects
components.

FIGURE 22.16   The Entity Display dialog box is used to modify the display of mass elements.

Field Notes

> The system default display properties for mass elements can also be controlled in the Drawing
> Setup dialog box by selecting the Display tab.

At the top of the dialog box is the display representation drop-down list. Select the display representation to be modified. The two most common representations to be modified are Model and Sketch. Next, select the level of display control in the Property Source column. Mass elements have only a System Default property source, which controls how all the mass elements will be displayed and the Mass Element property source, which controls how the single selected mass element is to be displayed.

Attach an override to the selected property source, then pick the Edit Display Props... button to modify the display. The **Entity Properties** dialog box is displayed (see Figure 22.17). Depending on the representation selected, two to three tabs are available for modification. The mass element includes the Entity, Bounding Box, and Hatch components.

Adjust the
display of the
mass element
components.

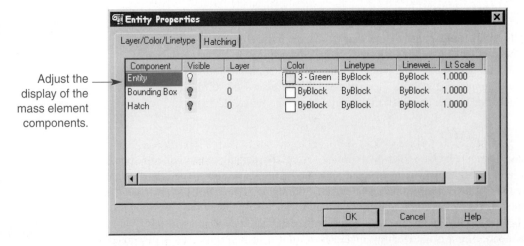

FIGURE 22.17   The Entity Properties dialog box is used to adjust the display of mass elements.

The Entity component controls how the mass element is to appear. You can change the color, linetype, and lineweight of the component. The Bounding Box component is a box that completely surrounds the extents of the mass element. By default this component is turned off, unless a 3D solid has been converted into a mass element. The Hatch component is available only in the sketch, reflected, and plan representations and turns on a hatch within the mass element.

Exercise 22.3

1. Open EX22-2-Mass.dwg.
2. Enter a name for each mass element in the description area of the Mass Element Properties dialog box.
3. Name each mass as indicated in the figure.
4. Change the model representation for the Atrium and the Glass Entrance mass elements to use a blue color.
5. Save the drawing as EX22-3-Mass.dwg

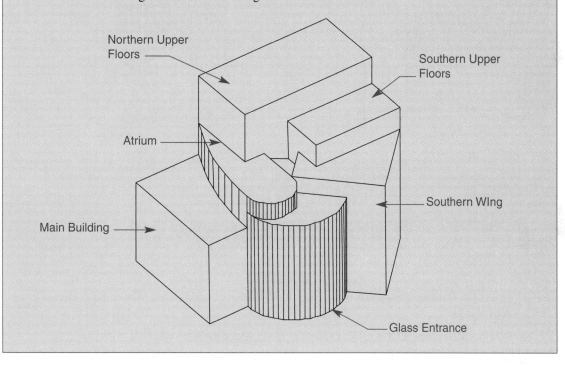

Using Mass Elements with Other Geometry

Mass elements are not limited to conceptual design. Like other AEC objects, mass elements can be integrated into the construction documents. For example, mass elements can be used as an interference condition within a wall, or as a custom tread and riser at the bottom of a stair (see Figure 22.18). Be creative when using mass elements; use them to develop a conceptual plan, or to enhance the design of a building.

Referencing an Object

As you are designing you may use several objects that are the same size and shape. If the design changes and multiple objects need to be adjusted, you have to modify each object to make the changes. Architectural Desktop provides a tool known as a ***reference object*** that can be used to make a copy of a "master" object. When changes are made to the master, the copies change accordingly. For example, suppose a number of columns will be used in the design. Instead of creating each column individually, create one of the columns, then reference it. Any changes made to the original will be reflected in the referenced copies.

Reference objects can be used on all AEC objects and polylines. Use one of the following options to reference an object:

Mass elements added as interference conditions

Mass elements added as custom riser and tread

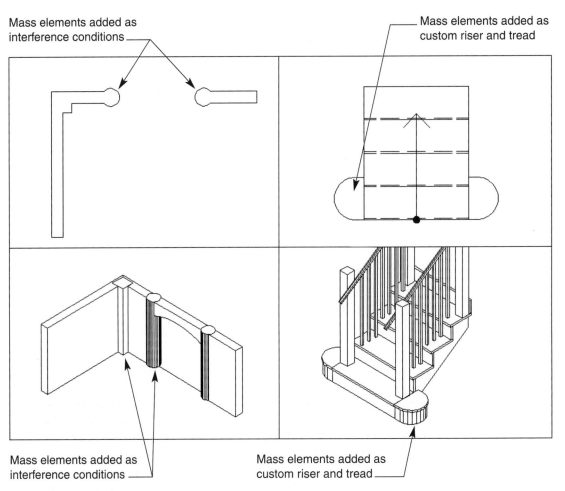

Mass elements added as interference conditions

Mass elements added as custom riser and tread

FIGURE 22.18 Mass elements can be used as an interference condition in a wall or as an enhancement to a stair.

✓ Select the **Desktop** pull-down menu, pick the **Utilities** cascade menu, then select **Reference AEC Objects....**

✓ Pick the **Reference AEC Objects** button in the **Utilities** toolbar.

✓ Right-click, and select **Desktop, Utilities, Reference AEC Objects...** from the cursor menu.

✓ Type **AECENTREF** at the Command: prompt.

When this command is entered, the following prompt is displayed: "Entity reference [ADd/ Properties/Insertion point/ATtach]:" These options are described next:

- **ADd** This option adds a reference to an existing object. When you enter AD at the prompt you are asked to select an object to reference, an insertion point for the copied reference, a location of the copy, and the rotation angle of the copy. The following prompt sequence is used to add a reference object in Figure 22.19:

Entity reference [ADd/Properties/Insertion point/ATtach]: **AD** ↵
Select an entity to reference: *(select an object to reference)*
Insertion point: *(select an insertion point on or away from the original object as the insertion point of the reference)*
Location: *(pick a location for the reference copy)*
Rotation angle <0.00>: *(enter the rotation angle for the reference copy)*
Entity reference [ADd/Properties/Insertion point/ATtach]: ↵
Command:

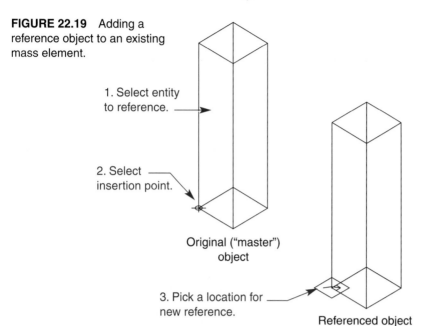

FIGURE 22.19 Adding a reference object to an existing mass element.

1. Select entity to reference.

2. Select insertion point.

Original ("master") object

3. Pick a location for new reference.

Referenced object

Once an object has been referenced, it can be copied to create several reference objects, which are all affected if the original object is modified.

- **Properties**   Typing P at the prompt opens the **Entity Reference Properties** dialog box, where you can enter property set definitions, notes, and descriptions. The Entity Reference Properties dialog box also includes a Location tab, so you can specify a location for the reference.

- **Insertion**   Selecting this option allows you to reestablish the insertion point for the reference. When specifying a new insertion point, you must select the point in relation to the original object being referenced (see Figure 22.20).

- **Attach**   This option allows you to select another object to use as a reference. First, select the reference object that will reference a different object, then select the object to be referenced.

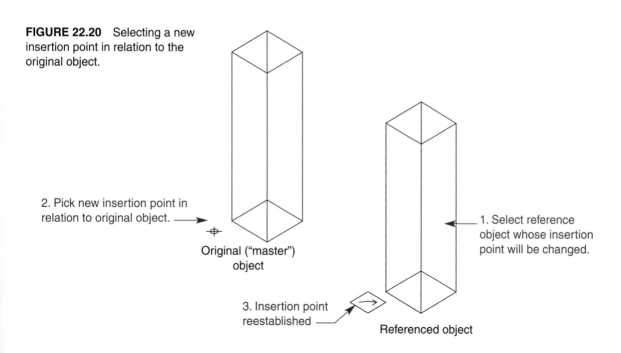

FIGURE 22.20   Selecting a new insertion point in relation to the original object.

2. Pick new insertion point in relation to original object.

Original ("master") object

1. Select reference object whose insertion point will be changed.

3. Insertion point reestablished

Referenced object

Exercise 22.4
_____

1. Start a new drawing using the AEC (Imperial) Massing.dwt.
2. Add a box mass element that is $24'' \times 24'' \times 8'\text{-}0''$.
3. Use the Reference AEC Objects command to reference the box.
4. Pick a location for the reference copy.
5. Copy the reference several times in the drawing.
6. Change the shape of the original mass element to a cylinder with a radius of $9''$.
7. Save the drawing as EX22-4.dwg.

Grouping the Mass Objects

When mass elements have been placed together to form a design, the result is called a ***massing study***. The elements are evaluated together as a group, which represents the building as a whole. During the development of the conceptual design, there may be areas that need to have holes or voids in the sides or top of the building, or when a Hide is performed on the building intersections between mass elements do not show up properly. In these situations, the mass elements can be grouped together to form one object.

The **Mass Group** command is used to group individual mass elements or other AEC objects together to form one object. Then, operations such as subtractions, additions, and intersections can be performed on this one object. Then, floors can be cut from the perimeter of the group of objects, and complex shapes can be formed in the development of the design.

Adding Mass Elements to a Mass Group

In order to group objects together a marker you need to place in the drawing to which the individual mass elements or AEC objects will be attached. Once the marker is placed in the drawing, the individual objects are attached to the marker to join all the individual objects together.

Use one of the following options to add the mass group marker:

✓ Select the **Concept** pull-down menu, access the **Mass Groups** cascade menu, then select **Add Mass Group.**

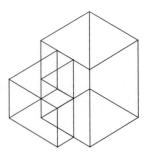 *(toolbar button icon)*

✓ Pick the **Add Mass Group** button on the **Mass Groups** toolbar.

✓ Right-click, and select **Concept, Mass Groups, Add Mass Group** from the cursor menu.

✓ Type **AECMASSGROUPADD** at the Command: prompt.

You are then prompted for a location and rotation angle for the marker. Place the marker near the outside of the objects that you want grouped together, and press the [Enter] key for the rotation angle. This adds the Mass Group Marker to the drawing, as shown in Figure 22.21.

FIGURE 22.21 Adding a Mass Group Marker so that objects can be attached to it to form a single object.

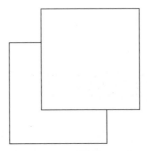

Mass Group Marker
(top view)

Mass Group Marker
(isometric view)

After the marker has been placed in the drawing, attach the objects using one of the following options:

✓ Access the **Concept** pull-down menu, select the **Mass Groups** cascade menu, then pick the **Attach Elements** command.

✓ Pick the **Attach Elements** button in the **Mass Groups** toolbar.

✓ Right-click, and select **Concept, Mass Groups, Attach Elements** from the cursor menu.

✓ Type **AECMASSGROUPATTACH** at the Command: prompt.

✓ Select the mass group marker, then right-click, and select **Attach Elements** from the shortcut menu.

You are then prompted to select a mass group. Select the mass group marker. The next prompt asks you to select elements to attach. Select any AEC object or 3D Solid object to be attached to the mass group marker. All the pieces are joined together, forming a single object.

At this point it may be difficult to work with the individual objects and the mass group in model space. Access the Mass-Group layout tab. This tab can be found in the AEC (Imperial) Mass.dwt, AEC Arch (Imperial).dwt, and the AEC Arch (Imperial - Int).dwt. This tab is divided into two viewports. The viewport on the left displays all the mass elements in model space. The right viewport displays the mass group object (see Figure 22.22).

Display configurations allow these two different views into the drawing. The left viewport is assigned the Concept_Mass display configuration, which displays only mass elements and a few other AEC objects. The right viewport is assigned to the Concept_Group display configuration, which displays only mass groups. Notice that in the right viewport the objects have been joined together, yet in the left viewport the individual objects remain.

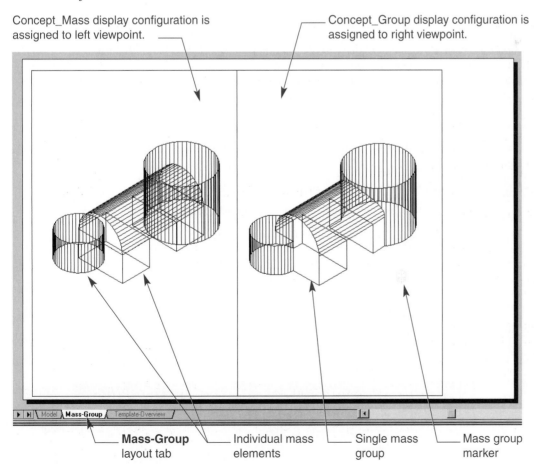

FIGURE 22.22 The Mass-Group layout tab displays the mass elements in the left viewport and the mass group in the right viewport.

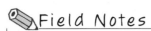

Once a mass group has been established, new mass elements can be added to the mass group. When a mass element is added, the Group drop-down list in the **Add Mass Element** dialog box includes the new mass group. The mass group is arbitrarily assigned a name, which may consist of a series of letters and numbers (see Figure 22.23). Select the group to which to attach the mass element, and the mass element is automatically added to the group. The Operation drop-down list is also available and allows you to set one of the following operators: add, subtract, or intersect. These affect how the new mass element will react with other elements within the group.

Another method of adding mass elements to a group is by using the **Add Mass Element to Group** command, which is accessed using one of the following options:

✓ Pick the **Concept** pull-down menu, select the **Mass Elements** cascade menu then pick **Add Mass Element to Group....**

✓ Pick the **Add Mass Element to Group** button in the **Mass Elements** toolbar.

✓ Right-click, and select **Concept, Mass Elements, Add Mass Element to Group...** from the cursor menu.

✓ Type **AECMASSGROUPADDELEM** at the Command: prompt.

You are prompted to select a group to which to add the new element. The Add Mass Element dialog box then appears with the group already selected in the Group drop-down list. Set the appropriate operation, then add the new mass element while working in the left viewport. If you add the mass element in the right viewport the element is displayed only in the left viewport.

If you accidentally added a mass element to a group, or if the design no longer requires a mass element, it can be detached from the group with the **Detach Elements** command. Access this command using one of the following options:

✓ Pick the **Concept** pull-down menu, select the **Mass Groups** cascade menu, then pick the **Detach Elements.**

✓ Pick the **Detach Element** button in the **Mass Groups** toolbar.

✓ Right-click, and select **Concept, Mass Groups, Detach Elements** from the cursor menu.

✓ Type **AECMASSGROUPDETACH** at the Command: prompt.

✓ Pick the mass group, right-click, and select **Detach Elements** from the shortcut menu.

You are then prompted to select a mass group. In the right viewport, select the mass group that will have an element removed. The next prompt asks you to select the elements to detach. In the left viewport, select any objects that you want to remove from the group.

FIGURE 22.23 After a mass group is added to the drawing, the name of the group appears in the Add Mass Element dialog box.

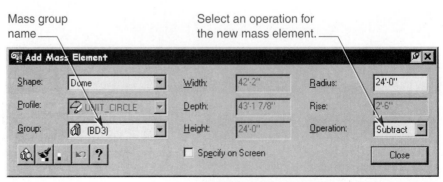

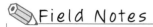

Field Notes

You can add as many mass groups into a drawing as required. You may want to do this when a design requires multiple mass groups that can be used independently.

Exercise 22.5

1. Open EX22-3-Mass.dwg.
2. Add a mass group marker to the drawing. Place the marker near the group of mass objects but not inside.
3. Attach all the elements to the mass group marker.
4. Select the Mass-Group layout tab.
5. Add a new mass element to the mass group. Select the box shape, setting the width to 8′-0″, depth to 18′-0″ and height to 12′-0″. Select the subtract operation and insert it at the bottom midpoint of the Main Building. See the figure for placement.
6. Add a box barrel vault above the box you just added to the same mass group. Set the width to 8′-0″ and the radius to 9′-0″. Set the operation to subtract. See the figure for placement.
7. Create a reference object of the box and the barrel vault. Set these to the side of the original. Copy the reference objects to the other side of the original.
8. Use the Attach Elements command to attach the references to the mass group.
9. Save the drawing as EX22-5-mass.dwg.

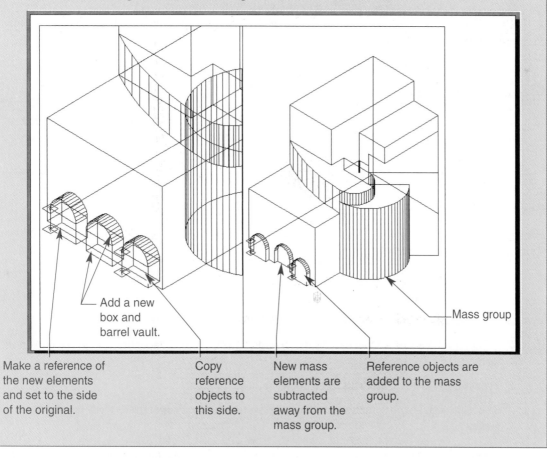

Add a new box and barrel vault.

Make a reference of the new elements and set to the side of the original.

Copy reference objects to this side.

New mass elements are subtracted away from the mass group.

Reference objects are added to the mass group.

Mass group

Adding AEC Objects to the Mass Group

In addition to mass elements, all AEC objects and AutoCAD 3D solids can be added to a mass group. This property is useful when an object such as a stair needs to be enhanced with additional modeling objects, or a complex 3D solid needs to be incorporated with the mass elements.

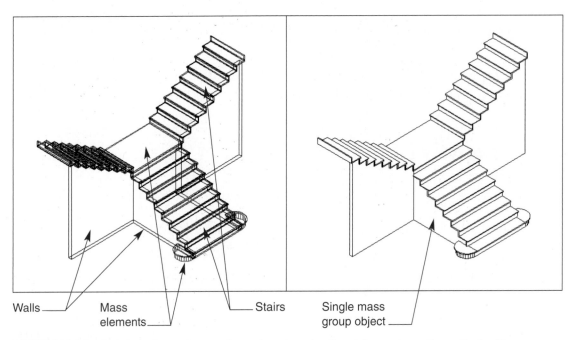

Walls — Mass elements — Stairs — Single mass group object —

FIGURE 22.24 A stair, walls, and mass elements enhancing the stair are drawn then attached to a mass group, forming a single mass group object.

Create a mass group marker, then use the Attach Elements command to attach the objects (see Figure 22.24). These objects can also be detached from the mass group as needed.

 Field Notes

> When attaching objects to a mass group you may want to create a custom display configuration to not display the entities making up the mass group or freeze the layer to which the entities are assigned.

Mass Group Properties

Mass groups like other objects can be modified through a Properties dialog box. Use one of the following options to access the mass group properties dialog box:

✓ Select the **Concept** pull-down menu, pick the **Mass Groups** cascade menu, then select **Mass Group Properties....**

✓ Pick the **Mass Group Properties** button in the **Mass Groups** toolbar.

✓ Right-click, and select **Concept, Mass Groups, Mass Group Properties...** from the cursor menu.

✓ Type **AECMASSGROUPPROPS** at the Command: prompt.

✓ Select the mass group, then right-click, and select **Group Properties...** from the shortcut menu.

This Mass Group Properties dialog box includes the standard General tab where notes, property set definitions, and a description for the group can be entered. As with mass elements, entering a description for the mass group can help identify the group later. If several mass groups are used in a design, entering a description displays the name of the group in the Group drop-down list in the Add Mass Element dialog box.

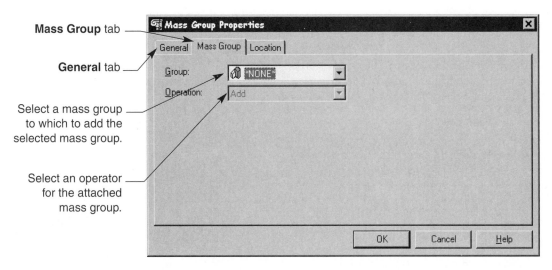

FIGURE 22.25 The Mass Group Properties dialog box allows you to control how the mass group behaves.

The Mass Group tab allows you to add the selected mass group into another mass group (see Figure 22.25). This feature allows you to combine mass groups to form complex shapes. The Operation drop-down list allows you to add an operator to the mass group when attaching it to another mass group. The Location tab lets you control the mass group marker location.

Entity Display

Because a mass group is an AEC object, it has the ability to be displayed differently depending on your viewing direction. Like mass elements, mass groups do not include styles, so the display of the mass group cannot be controlled through a Style dialog box. Instead, the mass group can be controlled by selecting the mass group, right-clicking, and selecting Entity Display... from the shortcut menu. Display can also be controlled through the Display tab in the Drawing Setup dialog box.

When the Entity Display dialog box is displayed, select the display representation that you want to modify (see Figure 22.26). The common display properties are Plan and Model. Next, select the level of display control from the Property Source column. Notice that there are only two options: System Default and Mass Group. System Default controls how all mass groups are to appear in the drawing, and Mass Group controls how the selected mass group is displayed.

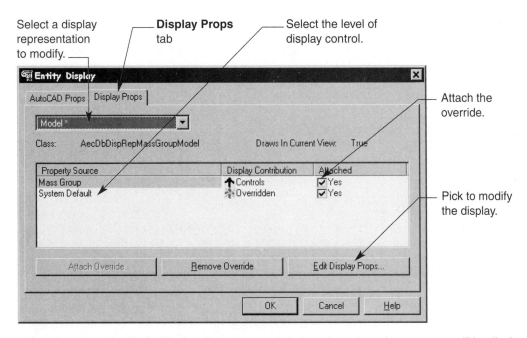

FIGURE 22.26 The Entity Display dialog box controls how the selected mass group will be displayed.

After determining the property source, attach the override and pick the Edit Display Props button to edit the display. When the display properties are adjusted, the mass group includes the entity and the marker. The entity controls the display of the grouped masses, and the marker controls the display of the mass group marker. You can decide to assign the mass group marker to a layer that does not plot. When you have finished adjusting the display properties for the mass group in the display representation that you want to control, press the OK button to return to the drawing screen.

Exercise 22.6

1. Open EX22-5-Mass.dwg.
2. Add a description to the mass group such as the name of the building. Use the Mass Group Properties dialog box to do this.
3. Change the entity display of the mass group so that the model representation is a gray color.
4. Save the drawing as EX22-6-Mass.dwg.

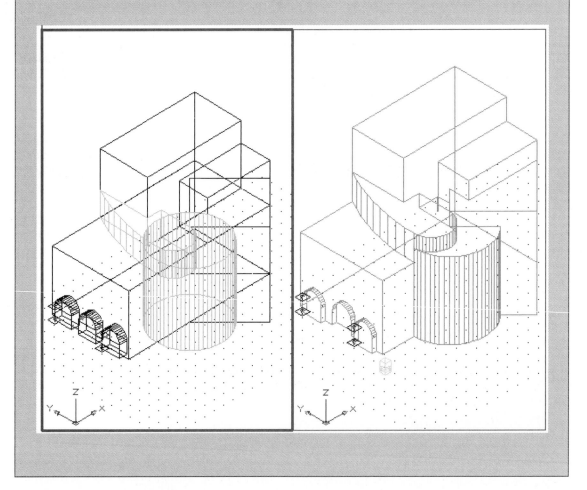

Using the Model Explorer

When an idea is being conceptualized, masses and mass groups are used to create the massing study. Along with these tools, another tool known as the **Model Explorer** can be used to manipulate the way the objects interact with one another. This command opens a separate window in Architectural Desktop where the elements within a mass group can be adjusted, new elements can be added, and operators on the elements can be controlled.

After adding a mass group to the drawing, you can use the Model Explorer for fine-tuning your drawing. Use one of the following options to enter the Model Explorer, shown in Figure 22.27:

Tree view toolbar controls the parameters of the elements in the mass group. —

Object viewer toolbar controls display in preview pane.

Open drawing in Architectural Desktop —

Mass element in drawing —

Description of mass group —

List of objects grouped together in the mass group —

Description of mass element —

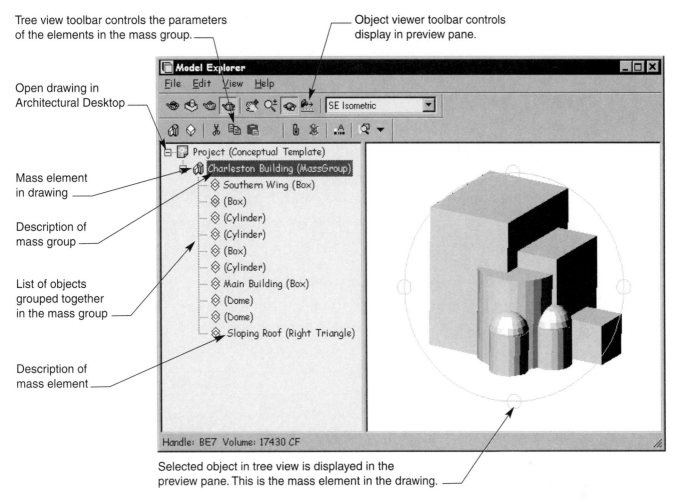

Selected object in tree view is displayed in the preview pane. This is the mass element in the drawing. —

FIGURE 22.27   The Model Explorer is a separate window that allows you to manipulate the elements within a mass group.

✓ Pick the **Concept** pull-down menu, then select **Show Model Explorer....**

✓ Right-click, and select **Concept, Show Model Explorer....**

✓ Type **AECMODELEXPLORER** at the Command: prompt.

The Model Explorer displays a tree along the left side that lists the open drawings in Architectural Desktop, the mass groups in each drawing, and the objects attached to the mass group. The right side of the window displays the selected element in the tree.

Model Explorer Menus

Initially two toolbars and a series of pull-down menus are displayed in the Model Explorer window. The first toolbar is the Object Viewer toolbar. This toolbar controls how the object selected in the tree view is displayed in the preview pane. The series of buttons in the toolbar allows you to control whether the preview is displayed as a wire frame, hidden, or shaded. Additional buttons control zooming and panning. The drop-down list at the end of the toolbar includes the predefined viewing directions (see Figure 22.28).

The Tree View toolbar allows you to modify the elements within the tree structure and to add new mass elements or mass groups. Additional tools can be used to make copies of mass elements within a group and paste them into a different mass group (see Figure 22.29).

The pull-down menus duplicate the commands found in the toolbars and allow you to control all the parameters of the elements within a mass group. Additionally, right-clicking over an element in the tree view list displays a shortcut menu specific to the object (see Figure 22.30A). Right-clicking in the preview pane provides a shortcut list that controls the display in the preview pane (see Figure 22.30B).

FIGURE 22.28 The buttons in
the Object Viewer toolbar control
the display of the preview pane.

 Wireframe: Displays the selected object without shading, and displays all the edges used to construct the shape.

 Hide: Hides the lines behind a surface.

 Shade: Shades the selected object using a flat shading effect.

 Render: Displays the selected object using realistic shading.

 Pan: Allows you to pan the current view in the preview pane around your model.

 Zoom Realtime: Allows you to zoom closer to and away from the selected object in the preview pane.

 Orbit: Allows you to orbit around the selected object.

 Adjust Distance: Adjusts the perspective distance in the preview pane.

 New Grouping: Creates a new mass group marker in the drawing. With the Model Explorer window still open, select a location in the drawing for the marker.

 New Element: Adds a new mass element to the drawing. With the Model Explorer window still open, select a location for the new mass element.

 Cut: Cuts the selected element from the tree view and places it on the Windows Clipboard for use in another mass group.

 Copy: Copies the selected element in the tree view list to the Windows Clipboard for use in another mass group.

 Paste: Pastes an item from the Windows Clipboard into the currently selected mass group in the tree view list.

 Delete Item: Removes the selected item in the tree view from the list, mass group, and drawing.

 Attach Items: Attaches existing mass elements and AEC objects to the selected mass group in the tree view list. With the Model Explorer window open select the object that needs to be added to the mass group.

 Detach Items: Detaches the selected element in the tree view list from its mass group.

 Properties: Accesses the Properties dialog box for the type of object selected in the tree view.

 Display Configuration: With its associated drop-down button changes the display configuration used in the preview pane.

FIGURE 22.29 The Tree View toolbar includes tools for modifying elements in a mass
group and for adding new mass elements or mass groups to the drawing.

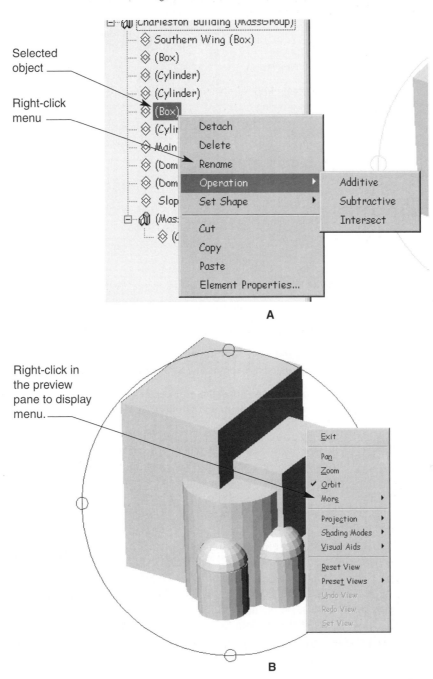

FIGURE 22.30 (A) Right-clicking on top of an element in the tree view list displays a menu specific to the selected object. (B) Right-clicking in the preview pane provides a menu list of items used to control the display.

Working with the Tree View

The tree view is the most important part of the Model Explorer window because this is the area where the elements are arranged and modified to create a conceptual design. The tree view list displays an icon representing each drawing open in Architectural Desktop (see Figure 22.31). Below each drawing icon is a mass group icon if the drawing includes a mass group. If the mass group was given a description in the Mass Group Properties dialog box, the name appears beside the icon.

Below each mass group icon is a list of the elements that are grouped within the mass group. As with the mass groups, if the element was given a description, it appears beside the icon. The elements in the tree view are placed in a hierarchical list. The element listed above another element is generated first in the drawing. This becomes important as operators are configured for the elements.

FIGURE 22.31   The tree view list indicates all the parameters of a mass group and its elements in a drawing.

Current drawing

Mass group in drawing

Elements grouped in the mass group

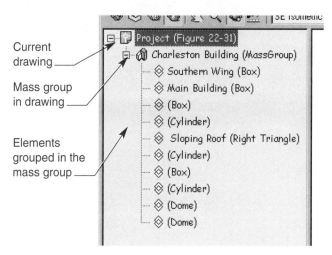

Beside each element is an icon representing its operational status. The icons represent additive, subtractive, and intersection. Each of these operators is described next:

- **Additive**   This icon indicates that the element is an additive object and is added to the object listed above it in the tree view list (see Figure 22.32A).

- **Subtractive**   This icon indicates that the element is a subtractive object and subtracts itself from the element listed above it in the tree view list (see Figure 22.32B).

- **Intersection** This icon indicates that the element is an intersection object. This type of operator creates a new object based on the geometric space shared by this element and the element above it in the tree view list (see Figure 22.32C).

Dragging one element in the list above another can reorganize the order in which elements appear and interact with one another. For example, suppose you have a subtracting element listed first in the tree view. This element is created before the additive elements; therefore, it will subtract itself out of the mass group before the additive elements create the group (see Figure 22.33A). To overcome this problem, drag an element in the list above the subtraction element (see Figure 22.33B). Thus, the additive element is created first, then the subtractive element subtracts itself from this additive element.

To drag an element in the list, press and hold the left mouse button on the desired element. Drag the mouse until it is above or on top of the element in the list that needs to be created after the element being dragged. This allows you to reorganize the hierarchical list to create a list of element operators that work with one another to form a new design.

Changing the Element Operator

The operator used on an element can be changed at any time during the process of designing, either from the **Modify Mass Element** dialog box or from within the Model Explorer. Changing the operator will affect how the mass group is created.

In the tree view, right-click on top of the element in the tree view list whose operator is to be changed (see Figure 22.34). The shortcut menu is displayed. Move to the Operation cascade menu and change the operator being used.

Additional tools, namely, the **Make Additive, Make Subtractive,** and **Make Intersection** commands, allow you to quickly change the operation of an element by picking on top of the element.

Use one of the following options to access the Make Additive command:

✓ Select the **Concept** pull-down menu, pick the **Mass Groups** cascade menu, then select **Make Additive.**

✓ Pick the **Make Element Additive** button in the **Mass Groups** toolbar.

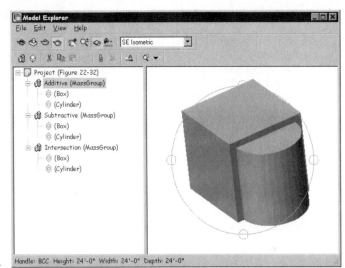

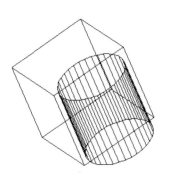

Additive joins the objects together.

A

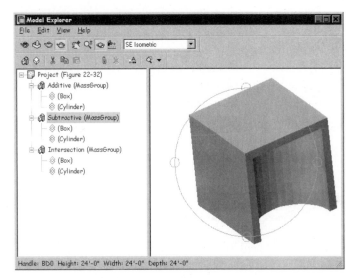

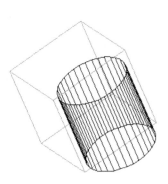

Subtractive removes an element from the previous element.

B

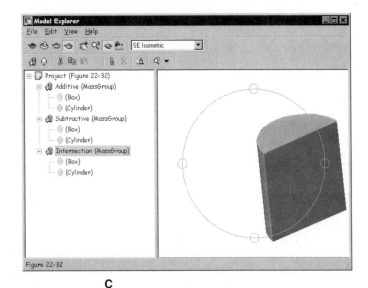

Intersection creates an object from the common space between two elements.

C

FIGURE 22.32 (A) The Additive operator joins two objects together. (B) The Subtractive operator subtracts one element from another. (C) The Intersection operator produces an object from the intersection of two elements.

FIGURE 22.33   Rearranging the elements in the tree view list to develop the design desired.

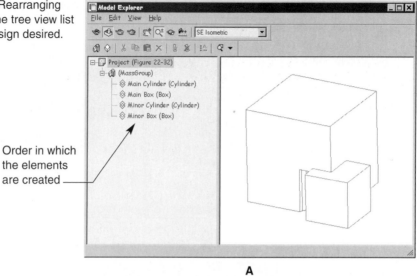

Order in which the elements are created ─

A

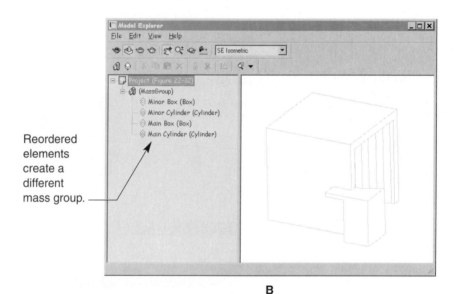

Reordered elements create a different mass group. ─

B

FIGURE 22.34   The shortcut menu in the tree view allows you to change the operation of an element.

Right-click over the element to be modified. ─

Select a new operator. ─

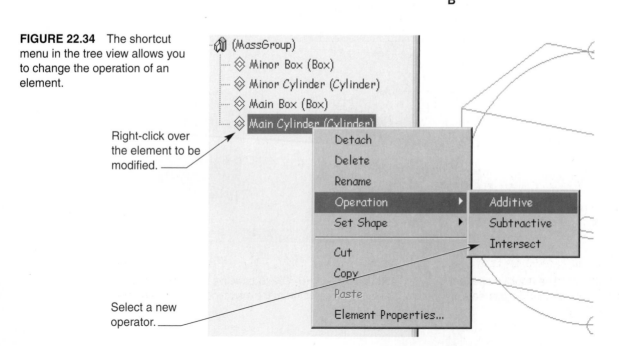

✓ Right-click, and select **Concept, Mass Groups, Make Additive** from the cursor menu.

✓ **Type AECMASSELEMENTOPADD** at the Command: prompt.

You are prompted to select massing elements. Select the elements that will be additive operators.
Use one of the following options to access the Make Subtractive command:

✓ Pick the **Concept** pull-down menu, select the **Mass Groups** cascade menu, then select **Make Subtractive.**

✓ Pick the **Make Element Subtractive** button in the **Mass Groups** toolbar.

✓ Right-click, and select **Concept, Mass Groups, Make Subtractive** from the cursor menu.

✓ Type **AECMASSELEMENTOPSUBTRACT** at the Command: prompt.

After entering the command select the elements that you want to be subtractive.
Use one of the following options to access the Make Intersection command:

✓ Select the **Concept** pull-down menu, pick the **Mass Groups** cascade menu, then select **Make Intersection.**

✓ Pick the **Make Element Intersection** button in the **Mass Groups** toolbar.

✓ Right-click, and select **Concept, Mass Groups, Make Intersection** from the cursor menu.

✓ Type **AECMASSELEMENTOPINTERSECT** at the Command: prompt.

Select the elements that will use the Intersect operator.

Exercise 22.7

1. Open EX22-6-Mass.dwg.
2. Access the Model Explorer.
3. Change the reference object's operator to Subtractive.
4. Place all the subtractive elements after the Main Building mass element.
5. Save the drawing as EX22-7-Mass.dwg.

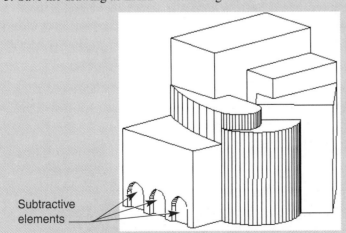

Subtractive elements

Once the mass model has been completed, it can be used in conjunction with other mass models referenced together for design studies and approvals, such as the model displayed in Figure 22.35. A conceptual design of a proposed building in Portland, Oregon, is placed with other mass models of the city to form a design study drawing.

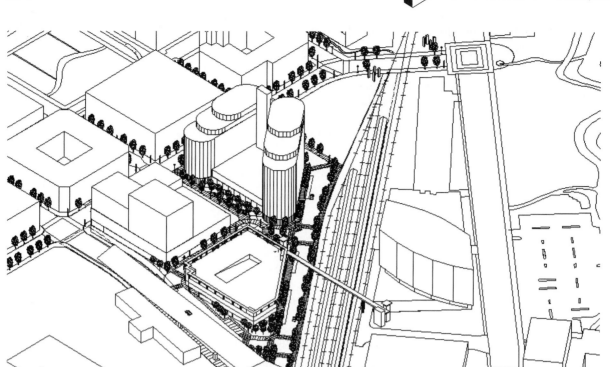

FIGURE 22.35   Groups of mass models are placed together to form a larger design study of a proposed building.
Drawing courtesy SERA Architects Inc., Portland, OR.

Creating Floor Plates

After the conceptual design has been completed, the massing drawing can be moved forward toward the design development phase. Before this can be done the mass model needs to be sliced into *floor plates* that represent each floor of the building. The slice floor plates routines are used to generate floor slices from the mass model. Before slicing the floors you need to determine how many floors to cut into the mass model and the distance between floors.

Generating Slices

Use one of the following options to add a slice:

- ✓ Select the **Concept** pull-down menu, pick the **Slice Floorplates** cascade menu, then select **Generate Slice.**

- ✓ Pick the **Generate Slice** button in the **Slice Floorplates** toolbar.

- ✓ Right-click, and select **Concept, Slice Floorplates, Generate Slice** from the cursor menu.

- ✓ Type **AECSLICECREATE** at the Command: prompt.

When a slice is added to the drawing, a slice marker is drawn representing each elevation of the floor, as shown in Figure 22.36. During the process of adding a slice marker, you are prompted for the first and second corners. These are the corners of the slice marker. You do not need to wrap the slice marker around the mass group. Instead, pick two corner points off to the side of the mass group. This marker is then used much like a mass group marker. The following prompt sequence adds a slice to the drawing and sets the distance in elevation between slices/floors for the drawing in Figure 22.37.

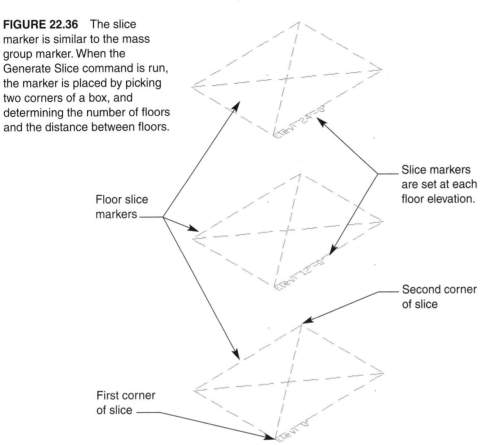

FIGURE 22.36 The slice marker is similar to the mass group marker. When the Generate Slice command is run, the marker is placed by picking two corners of a box, and determining the number of floors and the distance between floors.

Floor slice markers

Slice markers are set at each floor elevation.

Second corner of slice

First corner of slice

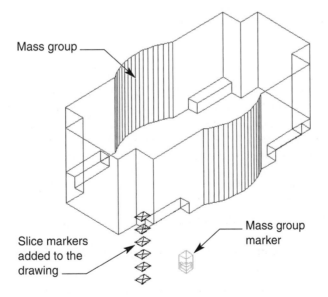

FIGURE 22.37 Slice markers have been added to the drawing with a spacing of 12'-0" between slices/floors.

Mass group

Mass group marker

Slice markers added to the drawing

Command: **AECSLICECREATE** ↵
Number of slices <1>: *(enter the total number of floors desired)* **6** ↵
First corner: *(select a point for the first corner of the marker)*
Second Corner or [Width]: *(select a point for the second corner of the marker, do not wrap the marker around the mass model)*
Rotation <0.00>: ↵
Starting height <0">: *(enter the first floor elevation)* ↵
Distance between slices <1'-0">: *(enter the distance between floors)* 12'
Command:

Field Notes

When placing the slice markers it is a good idea to place these away from the model so that they do not interfere with the model slice that is attached to the model. Placing them away from the model also makes it easier to select them when modifying them.

Attaching the Slices to the Mass Group

After adding slice markers to the drawing, you need to attach them to the mass group using one of the following options:

- ✓ Select the <u>C</u>oncept pull-down menu, pick the **Slice <u>F</u>loorplates** cascade menu, then select **<u>A</u>ttach Objects.**

- ✓ Pick the **Attach Objects** button in the **Slice Floorplates** toolbar.

- ✓ Right-click, and select **Concept, Slice Floorplates, Attach Objects** from the cursor menu.

- ✓ Type **AECSLICEATTACH** at the Command: prompt.

- ✓ Select the slice markers, right-click, and select **Attach Objects** from the shortcut menu.

You are then prompted to:

"Select elements to attach:"

Select the mass group objects. When an object is selected, the slice draws a perimeter around each object it finds. Attaching the slice to the mass group causes the perimeter of the building to be generated at each floor level. If you attach the slice to the mass elements, a perimeter is drawn around each mass element, defeating the purpose of creating the mass group. When you have finished selecting the mass groups to which the slice(s) will be attached, press the [Enter] key to end the command. The slices draw a perimeter outline around the mass group (see Figure 22.38).

FIGURE 22.38 The slice markers have been attached to the mass group, generating the perimeter floor levels.

Slices are generated at each elevation specified by the markers and wrap around the perimeter of the mass group.

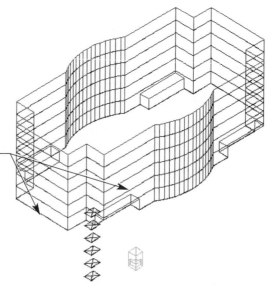

If a slice is not supposed to have cut through the mass group, it can be removed or detached from the massing with the **Detach Objects** command. Use one of the following options to access this command:

✓ Select the <u>C</u>oncept pull-down menu, pick the **Slice <u>F</u>loorplates** cascade menu, then select **Detach Objects.**

✓ Pick the **Detach Objects** button in the **Slice Floorplates** toolbar.

✓ Right-click, and select <u>C</u>oncept, Slice <u>F</u>loorplates, <u>D</u>etach Objects from the cursor menu.

✓ Type **AECSLICEDETACH** at the Command: prompt.

✓ Select the slice marker to be detached, right-click and select **Detach Objects** from the short-cut menu.

You are prompted to select the elements to detach. Select the mass groups or mass elements to which the slice is attached to remove the perimeter lines.

Adjusting the Slice Elevation

The elevations of individual floor plates can be adjusted after the slices have been placed. For example, suppose a series of slices is added to a mass group. The majority of the slices maintain a distance of 12'-0", but a few slices have a distance of 14'-0" between the floors. Use one of the following options to access the **Set Slice Elevation** command to adjust these slices after they have been added to the drawing and the mass group:

✓ Select the <u>C</u>oncept pull-down menu, pick the **Slice <u>F</u>loorplates** cascade menu, then select **Set Slice Elevation.**

✓ Pick the **Set Slice Elevation** button in the **Slice Floorplates** toolbar.

✓ Right-click, and select <u>C</u>oncept, Slice <u>F</u>loorplates, <u>S</u>et Slice Elevation from the cursor menu.

✓ Type **AECSLICEELEVATION** at the Command: prompt.

✓ Select the slice marker to be detached, right-click, and select **Set Elevation** from the short-cut menu.

You are prompted to select slices. Select a slice to adjust its current elevation. Note that if more than one slice object is selected, all the slice objects are placed at the same elevation. After selecting the slice, specify a new elevation. The slice is repositioned in elevation and the perimeter is adjusted as designated by the mass group.

At this point, design changes can still be made to the mass elements and the mass group. As long as the slices are still attached to the mass group, they are updated as the perimeter of the building changes.

Exercise 22.8

1. Open EX22-7-Mass.dwg.
2. Add a slice marker off to the side of the mass group.
3. Specify nine slices spaced 13'-0" apart.
4. Attach the slices to the mass group.
5. Adjust the second floor slice to be located at 16'-0".
6. Adjust the slices so that there is 13'-0" between them.
7. Save the drawing as EX22-8-Mass.dwg.

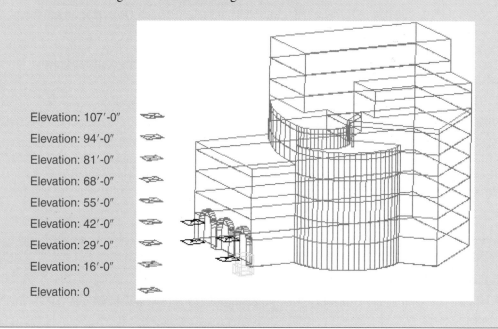

Elevation: 107'-0"
Elevation: 94'-0"
Elevation: 81'-0"
Elevation: 68'-0"
Elevation: 55'-0"
Elevation: 42'-0"
Elevation: 29'-0"
Elevation: 16'-0"
Elevation: 0

Exporting the Slices to Separate Drawings

After the floor slices have been created from a mass group, they can be used to generate space boundaries and spaces from which walls can be developed. Space boundaries and spaces are discussed in Chapter 23. The slices can also be converted into polylines, which in turn can be converted into walls. This topic is discussed in the next section.

The recommended method for developing a multistoried building is to create each floor in a separate file, then xref the floors together to create the building model, so the slices need to be exported into separate files. Use the **WBLOCK** command to create separate floor drawings for each slice. This command was discussed in Chapter 7.

In a model view, access the WBLOCK command. Enter a name for the wblock such as the floor number: 1st Floor Schematic, or 2nd Floor Schematic. When selecting the insertion point for the wblock, use the Insertion object snap and snap to the insertion point of the desired slice. Select the single slice object as the object to be exported. Continue to repeat this process for each slice of the building, which creates a separate file for each floor of the building.

When a slice is selected, the mass group and the mass elements are automatically exported, because the slice is dependent on them. Although each floor has been established as a wblock file with insertion points, the wblocks need to be inserted into a template where they can be used for drawing.

If one of the exported files is opened and used for drawing, the 0-Z axis elevation may be below the floor slice. To overcome this problem, start a drawing using one of the templates, then insert the desired floor slice at 0,0,0 using the **INSERT** command. Do this for each floor of the building, so that separate floors are created for the building model, with each floor sitting on the 0-Z axis. The walls and other objects can then be generated properly.

Exercise 22.9

1. Open EX22-8-Mass.dwg.
2. Wblock each slice out to a separate file.
3. Use the Insertion osnap to snap to the slice insertion point for the insertion point of the wblock.
4. Select one slice per file. Name the files EX22-9-First Floor Schematic, EX22-9-Second Floor Schematic, EX22-9-Third Floor Schematic, and so on.
5. Do this for all nine floors of the building.

Creating Polylines from Slices

Slices can be converted into polylines, which in turn can be converted into walls for the floor plans.
Use one of the following options to convert a slice into a polyline:

✓ Select the **Concept** pull-down menu, pick the **Slice Floorplates** cascade menu, then select **Convert to Polyline.**

✓ Pick the **Convert to Polyline** button in the **Slice Floorplates** toolbar.

✓ Right-click, and select **Concept, Slice Floorplates, Convert to Polyline** from the cursor menu.

✓ Type **AECSLICETOPLINE** at the Command: prompt.

✓ Select the slice marker to be detached, right-click ,and select **Convert to Polyline** from the shortcut menu.

You are then prompted to select slices. Select the slice to convert it into a polyline. The polyline is now ready to be converted into walls, slabs, and roofs.

Exercise 22.10

1. Start a drawing using the AEC Arch (Imperial).dwt or the AEC Arch (Imperial - Int).dwt.
2. Insert EX22-9 First Floor Schematic.dwg.
3. Set the insertion point values to X=0, Y=0, Z=0; scale to1, and the rotation to 0. Explode the drawing on insertion.
4. Freeze the A-Mass and A-Mass-Grps layers.
5. Convert the slice to a polyline.
6. Save the drawing as EX22-10-First Floor.
7. Repeat this process for the other eight floors, changing the file name as required.

✎ Field Notes

By inserting the exported slice files to separate drawing files, you are inserting the slice at the 0-Z axis elevation. This will help later as you develop the walls and you need to xref the floors together. When xref'ing the floors together, set the insertion point's Z axis at the appropriate floor level. This will stack the building correctly, lined up along the X and Y axes.

|||||||||||| CHAPTER REVIEW

 Use the CD-ROM to test your knowledge and skills.

Chapter Test

To check your understanding of the content provided in this chapter, access the Test file in the CH22 folder of the CD-ROM that accompanies this text.

Chapter Project

To practice the Architectural Desktop skills presented in this chapter, access the Project files in the CH22 folder of the CD-ROM that accompanies this text. The project files are in pdf format and include sample drawings and instructions for completing each project.

|||||||||||||||||||||||||||||||||||||||||||||||||||||||||||||||||||||||||||

CHAPTER 23

Spaces and Space Boundaries

LEARNING GOALS

After completing this chapter, you will be able to:

◎   Create a space style.

◎   Design and modify spaces.

◎   Apply interference.

◎   Generate space information.

◎   Create space schedules.

◎   Draw and modify ceiling grids.

◎   Create and edit column grids.

◎   Add column grid labels.

◎   Create and modify space boundaries.

◎   Generate walls.

Space planning is a design process in which the interior spaces of a building are established and arranged to create a functional design. The creation of spaces allows you to define boundaries around the spaces to help identify the shape of the building. Architectural Desktop provides the Spaces and Space Boundaries tools specifically for this design process.

A *space object* represents the interior of a room, which in turn is surrounded by space boundaries. *Space boundaries* are the outer edges of a space and are similar to walls, because they define an internal boundary. As the design process nears completion the space boundaries can be converted to walls so the construction documents can be developed.

Working with Spaces

A space object represents the internal space of a room or area. The object consists of a floor and a ceiling. These components are controlled by the display system and can be turned on or off as desired. When a space is added, a style is used. The style for a space designates the room name and the total square footage allotted to the style.

When spaces are added, they can be tagged using Room tags, which were introduced in Chapter 19 (see Figure 23.1). The space style name is then displayed in the tag name. The space style name is also used when room schedules and finish schedules are created.

Spaces can be added independently into the drawing for design purposes or can be dependent on a space boundary or walls for their size. If you have begun designing with walls and have defined the rooms with walls, you can generate space objects from walls so that the room name tags can be added. This section discusses the tools and options you have when working with spaces.

This entire chapter is Next Step Material.

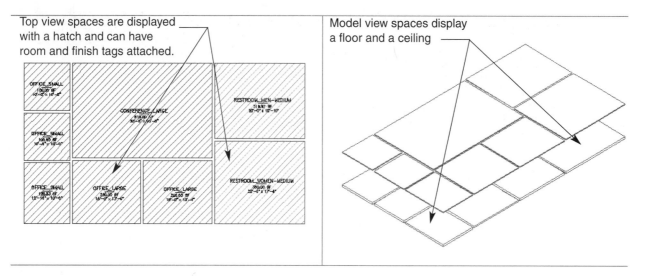

FIGURE 23.1 Room name tags are attached to space objects. The space objects consist of a floor and ceiling in a model view and a perimeter and area in the top view.

Creating a Space Style

Architectural Desktop includes a number of different space styles. The space styles included within the drawing templates are named for commercial purposes and are defined with common sizes. Use one of the following options to create a space style:

    ✓ Select the **Concept** pull-down menu, pick the **Spaces** cascade menu, then select **Space Styles...** from the menu.

    ✓ Pick the **Space Styles** button in the **Spaces** toolbar.

    ✓ Right-click, and select **Concept, Spaces, Space Styles...** from the cursor menu.

    ✓ Type **AECSPACESTYLE** at the Command: prompt.

The **Style Manager** dialog box is displayed, opened to the **Space Styles** section (see Figure 23.2). To create a new space style, pick the New Style button in the Style Manager window or right-click on the space style text in the tree view and select New. A new space style is created in the tree view list. Enter a new name for the style. The name entered is the name that will be displayed when a room tag is added to the space.

After the space style has been created and named it can be edited. Double-click the new style name, highlight the new style, and pick the Edit Style button, or right-click and select Edit... from the shortcut menu to edit the style. The Space Style Properties dialog box is displayed, and it includes the following options:

- **General**   This tab specifies the general properties of the space style (see Figure 23.3). The Name: text box displays the name of the style; rename the style as desired. Enter a description for the style in the Description: area. You can also attach external documents to the style such as finish material specifications by selecting the Notes... button. In order for the room tags to recognize the space style name, you need to attach a property set definition to the style, using the Property Sets... button. The property set definition that must be used for the room tags is the SpaceStyles property set. This property set definition must be loaded into the drawing before it can be assigned to a space style. Property set definitions were discussed in Chapters 19 and 20.

- **Dimensions**   This tab is used to specify the size parameters for the space style (see Figure 23.4). The Area: section allows you to specify a target area, a minimum area,

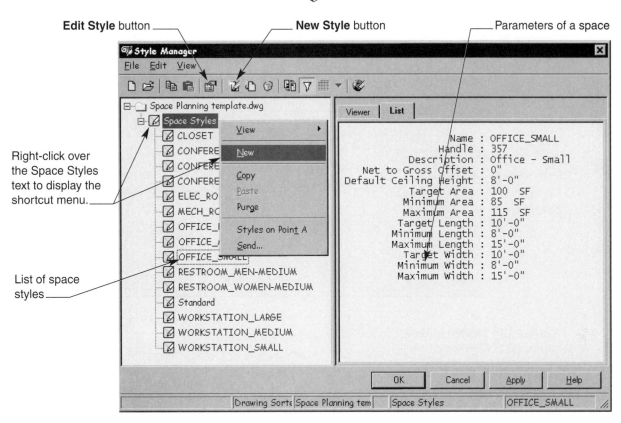

FIGURE 23.2 The Style Manager dialog box opened to the Space Styles section.

FIGURE 23.3 The General tab is used to rename the space style, to provide a description, and to attach the SpaceStyles property set definition.

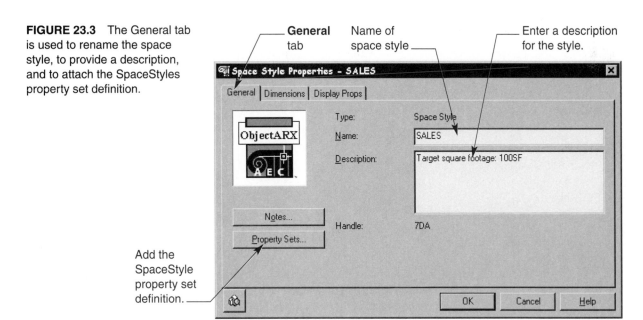

and a maximum area when adding spaces to the drawing. The Length and Width section allows you to specify target sizes, minimum sizes, and maximum sizes when adding new spaces to the drawing.

The Net to Gross Offset area is used when spaces adjacent to one another are added and the area between the spaces needs to be calculated. Setting a value greater than 0″ offsets the placement of a new space using this style when you are snapping to the node of an adjacent space. Figure 23.5 illustrates how a Net to Gross value of 6″ is used.

FIGURE 23.4 The Dimensions tab controls size parameters for the space style.

Dimensions tab   Set area sizes.

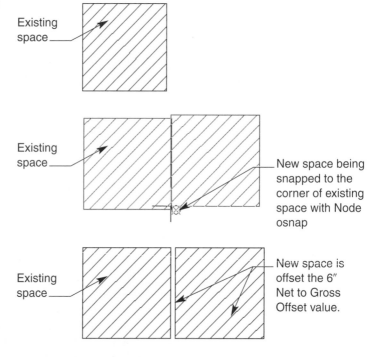

Set length and width sizes.

Set an offset value for placing spaces or leave at 0".

FIGURE 23.5 The Net to Gross Offset is set to 6″. When a new space is added and snapped to an adjacent space with Node osnap the new space will be offset 6″ away from the existing space.

Existing space

Existing space

New space being snapped to the corner of existing space with Node osnap

Existing space

New space is offset the 6″ Net to Gross Offset value.

- **Display Props** This tab controls the appearance of the space object in each display representation. The Model, Plan, and Reflected display representations are the most commonly adjusted display representations.

Once you have created the space style and set the parameters, press the OK button to return to the Style Manager. Additional Space styles are included in the Architectural Desktop Style folder. Use the Open Drawing button in the Style Manager to open one of the style drawings. These styles can be found by browsing to the \\Autodesk Architectural Desktop 3\Content\Imperial\Styles folder. There are four drawings full of space styles for your use. Open these in Style Manager and drag and drop the desired styles into your current drawing folder in the tree view.

Exercise 23.1

1. Open EX19-4-Flr-Pln.dwg.
2. Import the following styles from the Spaces - Residential (Imperial).dwg found in the Styles folder:
   - Bedroom
   - Bedroom_Master
   - Dining_Room
   - Entry_Foyer
   - Garage_2-Car
   - Kitchen
   - Living
3. Rename the Bedroom_Master, Master.
4. Rename the Entry_Foyer, Foyer.
5. Rename the Garage_2-Car, Garage.
6. Save the drawing as EX23-1-Flr-Pln.dwg.

Adding Spaces

There are several methods of adding spaces to a drawing. Spaces can be added to the drawing through the Add Space command, they can be converted from a polyline, or they can be generated from walls designating a closed boundary. Each of these methods has its own command and works a little differently from the others.

Using the Add Spaces Command

Once a space style has been defined, it can be added to the drawing using one of the following options:

✓ Select the **Concept** pull-down menu, pick the **Spaces** cascade menu, then select **Add Space...**

✓ Pick the **Add Space** button in the **Spaces** toolbar.

✓ Right-click, and select **Concept, Spaces, Add Space...** from the cursor menu.

✓ Type **AECSPACEADD** at the Command: prompt.

The **Add Space** dialog box, shown in Figure 23.6, is displayed.

In the Add Space dialog box, select the style of space to be added. When you select a style the square footage, length, and width values are inserted automatically based on the target sizes specified in the style. If you need to change the square footage value, enter a new value in the Area text box. The value that you enter must be within the parameters set in the style. If you enter a value greater than the maximum square footage area or smaller than the minimum square footage area, you will receive a message giving the range

FIGURE 23.6 The Add Space dialog box is used to add spaces to the drawing.

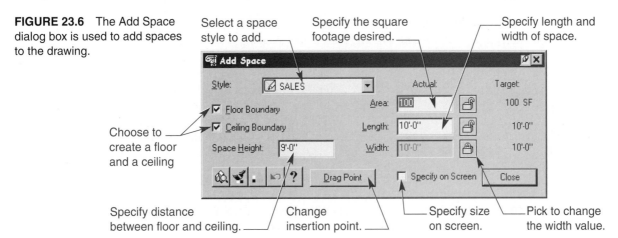

of appropriate square footage values. Press the OK button in the message box, and enter a value between the maximum and minimum square footages.

You can also adjust the length and width values by entering new values, but these, too, need to be between the maximum and minimum values set in the space style. The padlock buttons beside the area, length, and width values allow you to lock down a value so that the other values can be specified within the limits of the locked value. This feature allows you to make adjustments to the space being added without going outside the size parameters. You can unlock a value by picking the padlock button, which in turn locks one of the other values.

The Specify on Screen check box allows you to specify the size of the space as you are drawing it on the drawing screen. The space is still limited to the target values specified in the space style.

The Floor Boundary and Ceiling Boundary check boxes allow you to create a floor and a ceiling object that is displayed in a model view. The Space Height text box is used to specify the height between the floor and ceiling component. Initially when the space object is inserted into the drawing, the insertion point is located at the lower left corner of the space (see Figure 23.7A). Picking the Drag Point button changes the insertion point of the space to another corner. Continue to pick the Drag Point button until you reach the desired insertion point.

Once you have determined the parameters for the space, pick a point in the drawing screen to place the space. After specifying the location you need to specify a rotation angle for the space. A space is added with a diagonal hatch in the top view. Add as many spaces as required for the design, and change the styles as you design. When you have finished adding spaces to the drawing, press the Close button in the Add Space dialog box, or press the [Enter] key to exit the command.

Converting to Spaces

Spaces can also be converted from a closed polyline boundary. First, draw a polyline defining a boundary, which contains a space object. Once the polyline is drawn, use one of the following options to convert it:

✓ Select the **Concept** pull-down menu, pick the **Spaces** cascade menu, then select **Convert to Spaces....**

✓ Pick the **Convert to Spaces** button in the Spaces toolbar.

✓ Right-click, and select **Concept, Spaces, Convert to Spaces...** from the cursor menu.

✓ Type **AECSPACECONVERT** at the Command: prompt.

You are prompted to select a polyline. Select the polyline defining the boundary for the space. You are then prompted to erase the layout geometry. Remember that the polyline can later be converted to walls, slabs, and roofs if you leave it. After you answer this prompt the **Space Properties** dialog box is displayed, allowing you to set up any parameters for the space that is being created.

The Space Properties dialog box includes three tabs. The General tab is used to provide a description of the newly converted space object. The Style tab allows you to pick a space style from a list of the styles in the current drawing. The Dimensions tab is used to establish sizes for the new space (see Figure 23.8). It allows you to reestablish the length and width values of the converted space. Be careful when adjusting the length and width, because new values override the overall length and width established by the polyline.

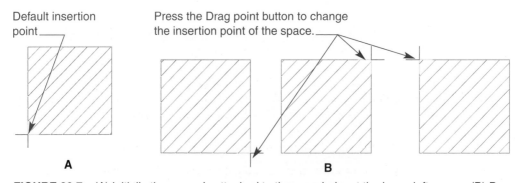

FIGURE 23.7 (A) Initially the space is attached to the crosshairs at the lower left corner. (B) Pressing the Drag Point button adjusts the insertion point for the space relative to the crosshairs.

Dimensions tab

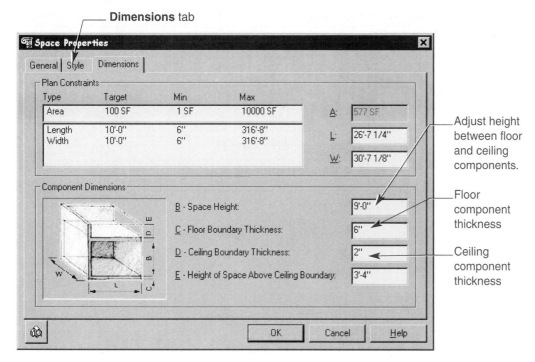

Adjust height between floor and ceiling components.

Floor component thickness

Ceiling component thickness

FIGURE 23.8 The Dimensions tab of the Space Properties dialog box is used to establish size parameters for the converted space.

You can also enter other values for the converted space such as the thickness of the floor, ceiling boundaries, and the distance between the floor and the ceiling. The Height of Space Above Ceiling Boundary text box establishes an imaginary space in the drawing that contains volume. This is the space above a finished ceiling measured up to the next floor. Entering a value here does not create any graphics in the drawing but calculates this information in the property set definition so it can be reported.

When you have finished specifying the values in the Space Properties dialog box, press the OK button to return to the drawing. The new space will be present inside the polyline.

Generating Spaces

Another method of creating spaces is by generating them from walls or linework. This method is similar to using the **BHATCH** command to hatch an internal area. In order to use this command, you must have a closed area. The area does not have to be a closed polyline but can be a room made up of walls or lines and arcs that form a closed boundary. Use one of the following options to enter the command:

✓ Pick the **Concept** pull-down menu, select the **Spaces** cascade menu, then pick **Generate Spaces....**

✓ Pick the **Generate Spaces** button in the Spaces toolbar.

✓ Right-click, and pick **Concept, Spaces, Generate Spaces...** from the cursor menu.

✓ Type **AECSPACEAUTOGENERATE** at the Command: prompt.

The Generate Spaces dialog box, shown in Figure 23.9, is displayed.

Pick a selection filter from the drop-down list. The selection filter that you select designates only which of the following options the generate space routine will look for as a boundary:

- **Walls only**   This option looks only to wall objects to use as a boundary for the space.

- **Walls, lines, arcs, polylines, and circles**   This option uses walls, lines, arcs, polylines and circles as boundaries for a space.

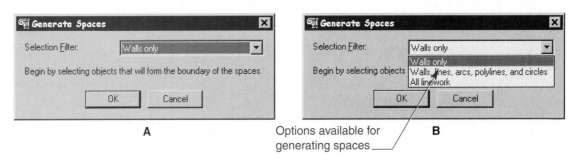

A   B

Options available for
generating spaces

FIGURE 23.9   The Generate Spaces dialog box allows you to determine what entities are evaluated in order to create the space.

- **All linework**   This option uses any type of object as a boundary, including all AEC objects, AutoCAD linework, and blocks.

Once you have selected the selection filter, press the OK button. You are returned to the drawing, where you are prompted to select objects. Select the objects in the drawing that you will use as a space boundary. When you have finished selecting the objects for the selection filter, the Generate Spaces dialog box is displayed (see Figure 23.10).

In the Generate Spaces dialog box, select the space style that will be generated. If you need to respecify the selection filter, pick the Modify Boundary Set button. Use the Update Space button on an existing space if it needs to be updated to fit inside a modified boundary.

If there currently is a room tag definition in the drawing, selecting the Tag Settings... button displays the **Tag Settings** dialog box (see Figure 23.11). This dialog box allows you to specify the room tag to be attached to the space, a property set definition to attach, and the properties within the property set definition

FIGURE 23.10   The Generate Spaces dialog box is used to select the type of space that will be generated from the boundaries selected.

Select a space style to generate.

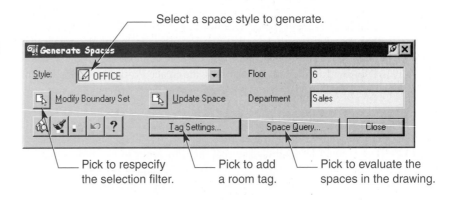

Pick to respecify
the selection filter.

Pick to add
a room tag.

Pick to evaluate the
spaces in the drawing.

FIGURE 23.11   The Tag Settings dialog box is used to specify a room tag and property set definition to attach to the space as it is generated.

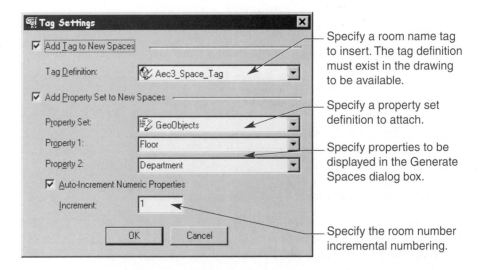

Specify a room name tag
to insert. The tag definition
must exist in the drawing
to be available.

Specify a property set
definition to attach.

Specify properties to be
displayed in the Generate
Spaces dialog box.

Specify the room number
incremental numbering.

to which you want to add information. Selecting two properties in the drop-down lists displays these properties in the Generate Spaces dialog box, where the property values can be edited.

The Space Query... button displays the **Space Information** dialog box (see Figure 23.12). This dialog box evaluates all the spaces in the drawing and provides you with information about the size and total square footage of each space or each type of space. This information can then be exported to a Microsoft Access database by selecting the Create MDB button.

When you have finished configuring the information for the space to be generated, pick a point inside the boundary to evaluate the selection filter, find the boundaries, and generate a space (see Figure 23.13). Continue to add spaces as needed. Press the Close button to exit the command.

FIGURE 23.12 The Space Information dialog box provides the total number of spaces and their square footage in the drawing.

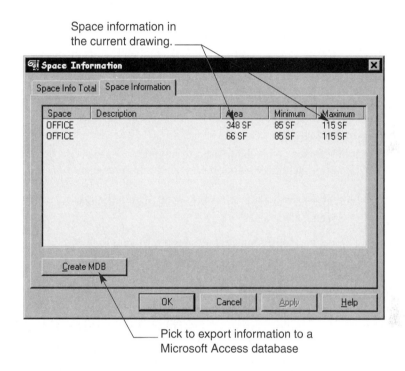

FIGURE 23.13 (A) Pick a point inside the boundary that will generate the space. (B) The space has been generated by using the walls as a selection filter.

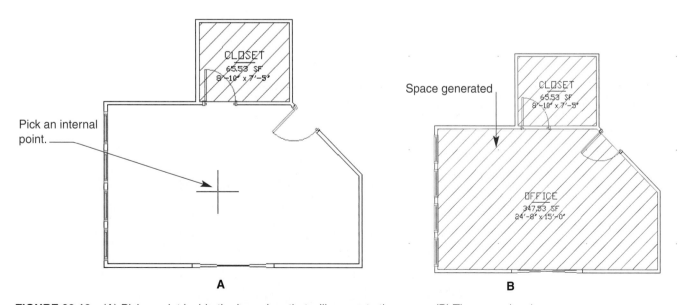

Exercise 23.2

1. Open EX23-1-Flr-Pln.dwg.
2. Generate spaces in the bedrooms, master bedroom, and garage using the appropriate space styles.
3. Use the Standard space style in the kitchen, living room, bathrooms, and closet areas of the drawing.
4. Add room name tags to the spaces.
5. Save the drawing as EX23-2-Flr-Pln.dwg.

Modifying Spaces

After spaces have been added to the drawing, the design may change, affecting the space plan, or the wrong space style may have been used when the space was initially inserted. There are a few commands that can be used to modify the space objects in order to stay current with the design needs. The tools used to modify spaces are discussed in this section.

Using Space Modify

The **Modify Space** command is used to modify the space style being used, and some of the size parameters. Use one of the following options to access this command:

✓ Select the **Concept** pull-down menu, pick the **Spaces** cascade menu, then select the **Modify Space...** command.

✓ Pick the **Modify Spaces** button in the **Spaces** toolbar.

✓ Right-click, and select **Concept, Spaces, Modify Space...** from the cursor menu.

✓ Type **AECSPACEMODIFY** at the Command: prompt.

✓ Select a space, right-click, and select **Space Modify...** from the shortcut menu.

The **Modify Space** dialog box, shown in Figure 23.14, is displayed. This dialog box allows you to change the space style being used, any size parameters, and whether a floor or ceiling is created. The length and width values refer to the space style for the maximum and minimum sizes. If an invalid size is entered, a warning dialog box is displayed, alerting you to use values within a range specified by the style.

Using Space Properties

Another tool for modifying a space is the **Space Properties** dialog box. Use one of the following options to access this dialog box:

✓ Select the space to be modified, right-click, and select **Space Properties...** from the shortcut menu.

✓ Type **SPACEPROPS** at the Command: prompt.

FIGURE 23.14 The Modify Space dialog box is used to modify a space object.

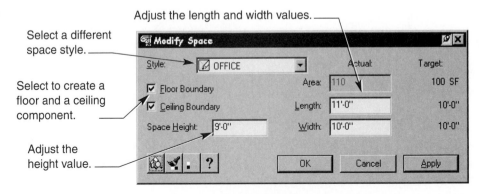

General   **Style**   **Dimensions**   **Location**
tab   tab   tab   tab

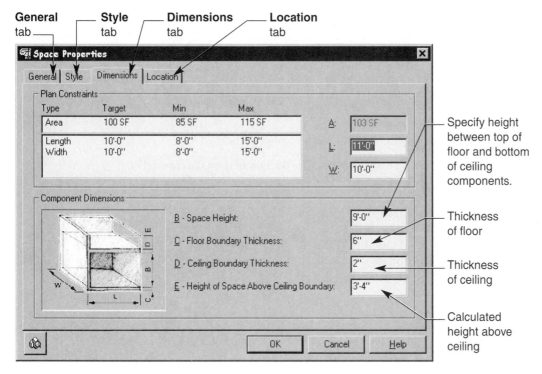

- Specify height between top of floor and bottom of ceiling components.
- Thickness of floor
- Thickness of ceiling
- Calculated height above ceiling

FIGURE 23.15 The Space Properties dialog box controls additional parameters not covered in the Modify Space dialog box.

The Space Properties dialog box includes four tabs used to adjust the space (see Figure 23.15). The **General** tab is used to assign property set definitions to the space, such as the RoomFinishObjects property set definition. The **Style** tab is used to select a different style for use. The **Dimensions** tab controls all the size parameters for the selected space. The **Location** tab is used to define where the object is in the World coordinate system.

Using Grips to Modify Spaces
Space objects, like other AEC objects, can be modified using standard AutoCAD commands. Spaces can be copied, moved, rotated, and stretched. In addition, grip editing can be used to modify a space. When a space is selected, a grip is located at each vertex location and midway between two vertex points.

Selecting a grip point and using the grip stretch mode allows you to relocate that point. If a grip between two vertex points is selected, the entire edge is stretched. If a space is stretched using grips, its size parameters from the style are ignored until you make a change in the Modify Space dialog box.

Joining and Dividing Spaces
In addition to the tools previously mentioned, two tools are avaliable to join spaces together and to divide a space into two parts. Two spaces can be joined together to form one space object and update the total being reported in a schedule. The spaces being joined do not have to be adjacent to each and can overlap each other.

Use one of the following options to join two spaces together:

- ✓ Select the **Concept** pull-down menu, access the **Spaces** cascade menu, then select **Join Spaces.**

- ✓ Pick the **Join Spaces** button in the **Spaces** toolbar.

- ✓ Right-click, and select **Concept, Spaces, Join Spaces** from the cursor menu.

- ✓ Type **AECSPACEJOIN** at the Command: prompt.

- ✓ Pick one space to be joined, right-click, and select **Join** in the shortcut menu.

You are then prompted to "Select neighboring space to join with:".

Select a space beside, on, or away from the selected space. The two space objects are joined together, and the total square footage is recalculated for the single object.

Spaces can also be divided, using the **Divide Spaces** command which can be accessed using one of the following options:

✓ Access the **C**oncept pull-down menu, select the **S**paces cascade menu, then pick **D**ivide **Spaces.**

 ✓ Pick the **Divide Spaces** button in the **Spaces** toolbar.

✓ Right-click, and select **C**oncept, **S**paces, **D**ivide **Spaces** from the cursor menu.

✓ Type **AECSPACEDIVIDE** at the Command: prompt.

✓ Pick one space to be divided, right-click, and select **Divide** in the shortcut menu.

You are then prompted to "Select a space:".

Select a space to be divided into two parts. The next prompt asks for a starting point then an ending point for the dividing line. The two points selected create the division line that separates the spaces.

Applying an Interference

An interference in a space is similar to an interference in a wall. You may apply an interference within a space where a structural column appears in the middle of the space area or where a stair may interfere with the space as it goes to the next floor. The interference object may be a polyline or an AEC object (see Figure 23.16).

Use one of the following options to access the Interference command:

✓ Select the **C**oncept pull-down menu, pick the **S**paces cascade menu, then select **In**terference **Condition.**

 ✓ Pick the **Interference Condition** button in the **Spaces** toolbar.

✓ Right-click, and pick **C**oncept, **S**paces, **In**terference **Condition** from the cursor menu.

✓ Type **AECSPACEINTERFERENCE** at the Command: prompt.

The following command sequence is used to create the interference in Figure 23.16:

Command: **AECSPACEINTERFERENCE** ↵
Space interference [Add/Remove]: **A** ↵
Select spaces: *(select the space object)*
1 found
Select spaces: ↵

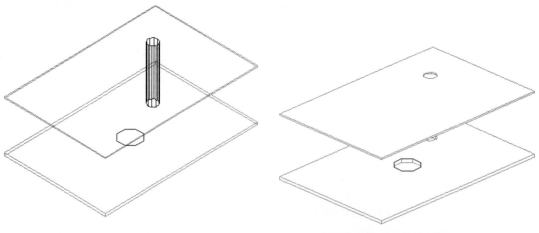

Original space, polyline,
and mass element objects

Holes cut into space object
with an inference condition

FIGURE 23.16 A polyline and a mass element are drawn on the space and used as an interference in the space.

Select plines or AEC Entities: *(select the polyline)*
1 found
Select plines or AEC Entities: *(select the mass element)*
1 found, 2 total
Select plines or AEC Entities: ↵
2 object(s) added to space 85A.
Space interference [Add/Remove]: ↵
Command:

Once the objects are added as an interference, holes are cut into the space removing that portion of area. The total square footage amount is then updated.

Controlling Entity Display

Like other AEC objects, space objects are controlled through the display system. The display properties for a space can be controlled at the system default level, which affects how all the space styles will appear; at the style level; or at the individual object level. Different components for the space can be controlled such as a hatch pattern and color in the top view versus the appearance of the space in a model view.

To control the display of a space object, select the object, right-click, and pick Entity Display from the shortcut menu. The Entity Properties dialog box, shown in Figure 23.17, is displayed. Select the appropriate display representation to control. Select the level of control from the Property Source column and attach an override. Pick the Edit Display Properties... button to edit the display.

The Plan and Reflected representations include the following components: Net Boundary, Gross Boundary, and Hatch. If you are creating a space plan for presentation and proposals, assign a solid hatch pattern to each space style with a different color (see Figure 23.18). The model representation includes a Floor and Ceiling component.

FIGURE 23.17 The Entity Properties dialog box is used to configure how the space object will appear in the drawing.

Display Props tab

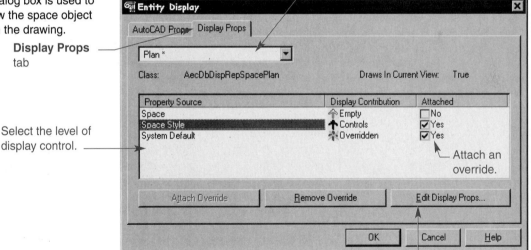

Select a display representation to modify.

Select the level of display control.

Attach an override.

Pick to edit the display of the spaces.

FIGURE 23.18 A space plan for a proposal using a solid hatch pattern with different colors.

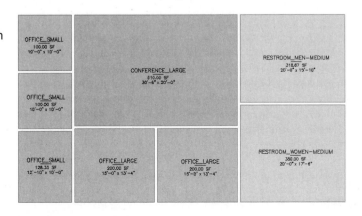

Exercise 23.3

1. Open EX23-2-Flr-Pln.dwg.
2. Divide the area between the kitchen and the family room. Modify the styles used for the kitchen and family styles as necessary. Note that you may need to adjust the space style sizes in order to modify the space in the drawing.
3. Divide the living room space into a living room, dining room, hall, and foyer. Modify the spaces styles as required.
4. Join the two hall spaces together.
5. Add room tags as required.
6. Save the drawing as EX23-3-Flr-Pln.dwg.

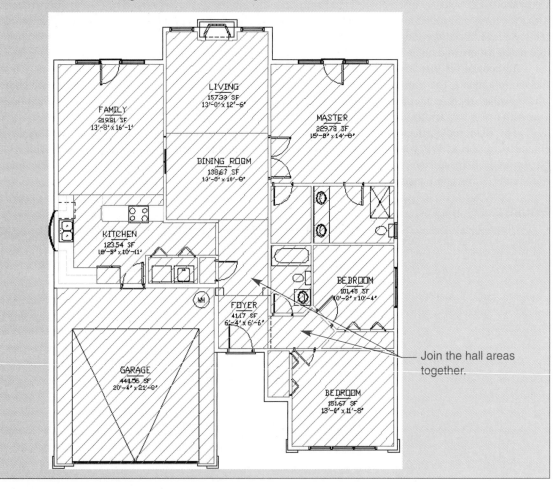

Join the hall areas together.

Generating Space Information

After the spaces have been added to the drawing and modified as needed, the area information can be generated. The **Space Inquiry** tool in Architecural Desktop evaluates all the spaces in the drawing and provides you with a detailed list of the space styles in the drawing. You can then export the list to a Microsoft Access database for further evaluation.

Use one of the following options to access this command:

✓ Select the **Concept** pull-down menu, pick the **Spaces** cascade menu, then select **Space Inquiry...** from the menu.

✓ Pick the **Space Inquiry** button from the **Spaces** toolbar.

✓ Right-click, then pick **Concept, Spaces, Space Inquiry...** from the cursor menu.

✓ Type **AECSPACEQUERY** at the Command: prompt.

FIGURE 23.19 The Space Information dialog box organizes the space information in the drawing.

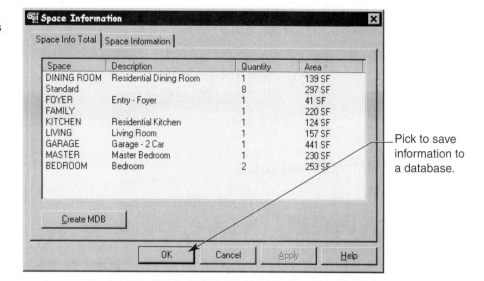

Pick to save information to a database.

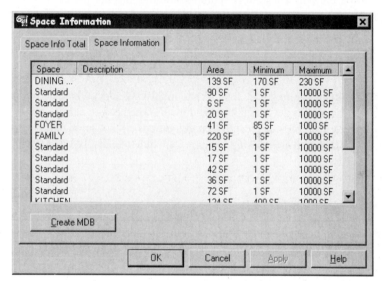

The **Space Information** dialog box is displayed. This dialog box includes the **Space Info Total** and **Space Information** tabs (see Figure 23.19). The Space Info Total tab provides the total square footage of all space styles used in the drawing. The Space Information tab lists each space style in the drawing. At the bottom of the dialog box is the Create MDB button. Pick this button to save the space information as an .mdb (Microsoft database) file.

Creating Space Schedules

Another method of retrieving space and area information from the drawing is to create a schedule. Generating schedules and creating custom schedules was discussed in Chapters 19 and 20. Architectural Desktop includes four schedules preconfigured for use. These can be found in the \\Autodesk Architectural Desktop\Content\Imperial\Schedules folder in the Schedule Tables (imperial).dwg.

These schedules include a Room Finish Matrix, Room Finish Schedule, Room Schedule, and a Space Inventory schedule. These schedules report on space objects and do not work without them. Also, the room and finish tags found in the Documentation menu require the use of spaces in the drawing.

Exercise 23.4

1. Open EX23-3-Flr-Pln.dwg.
2. Import the Room Schedule into the drawing.
3. Create the schedule reporting on all the spaces in the drawing.
4. Save the drawing as EX23-4-Flr-Pln.dwg.

Using Ceiling Grids

An important component of commercial drawings is the inclusion of ceiling grids in a reflected ceiling plan. The *ceiling grid* represents the acoustical panels on a suspended ceiling. The purpose of drawing these grids is to indicate to the contractor where the panels should be placed in a room. Architectural Desktop incorporates a utility for creating ceiling grids in the building model.

Before adding a ceiling grid to the drawing, set the Work_Reflected or Plot_Reflected display configuration current. These display configurations display the ceiling grids as they are typically drawn on a reflected ceiling plan. The ceiling grid can then be drawn.

 Field Notes

When a ceiling grid, light fixture multi-view blocks, and ceiling fan multi-view blocks are added, the objects get added into the drawing whether you can see them or not. If the Work_Reflected or the Plot_Reflected display configurations are not set in model space or in the floating viewport in which you are working, the objects will not be displayed in the drawing.

Adding a Ceiling Grid

Ceiling grids can be added to the drawing as independent objects or as objects that are dependent on a space or polyline for masking purposes. Use one of the following options to add a ceiling grid to the drawing:

✓ Select the **Design** pull-down menu, access the **Grids** cascade menu, then pick **Add Ceiling Grid....**

✓ Pick the **Add Ceiling Grid** button in the **Grids** toolbar.

✓ Right-click, and select **Design, Grids, Add Ceiling Grid...** from the cursor menu.

✓ Type **AECCEILINGGRIDADD** at the Command: prompt.

The **Add Ceiling Grid** dialog box, shown in Figure 23.20, is displayed. This dialog box includes the parameters for inserting a ceiling grid into the drawing. The overall size of the ceiling grid object is specified by entering a value in the X - Width and the Y - Depth text boxes. If you would rather specify the overall size of the ceiling grid object as it is drawn, then select the Specify on Screen check box.

The Divide By check boxes allow you to divide the lines in the ceiling grid along the width and depth into an equal number of divisions. The X - Baysize and Y - Baysize text boxes specify the actual spacing between grid divisions along the width and depth of the ceiling grid object. If the ceiling grid is added to the drawing at this point, it is an independent object that crosses through walls and other AEC objects (see Figure 23.21A). The Set Boundary< button allows you to use a closed polyline or a space object to mask the edges of the ceiling grid (see Figure 23.21B).

When a polyline or a space is used to mask a ceiling grid, the overall ceiling grid extends beyond the boundaries of the object being used as a mask. The ceiling grid can be repositioned and rotated to reflect the design needs. When a ceiling grid is masked by a closed polyline, it is placed at the same elevation as the polyline (see Figure 23.22A). When the ceiling grid is masked by a space object, the ceiling grid is automatically placed at the same elevation as the ceiling component (see Figure 23.22B).

FIGURE 23.20 The Add
Ceiling Grid dialog box is used to
add a ceiling grid to the drawing.

Specify overall size
of ceiling grid.

Specify equal number
of divisions along the
width and depth.

Specify actual
division size
of the panels.

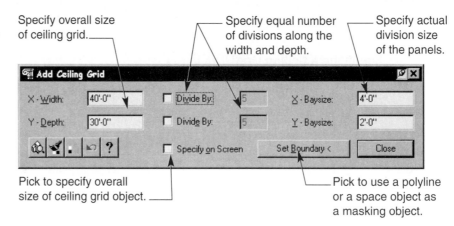

Pick to specify overall
size of ceiling grid object.

Pick to use a polyline
or a space object as
a masking object.

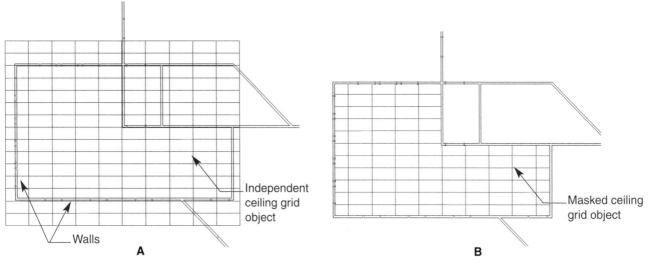

Independent
ceiling grid
object

Walls

A

Masked ceiling
grid object

B

FIGURE 23.21 (A) The ceiling grid is added to the drawing as an independent object. Note that it crosses
through walls. (B) A ceiling grid using a polyline to mask the edges. Note that the edges of the ceiling grid
stop at the edge of the walls.

FIGURE 23.22 (A) A ceiling
grid being masked by a closed
polyline. (B) A ceiling grid being
masked by a space object.

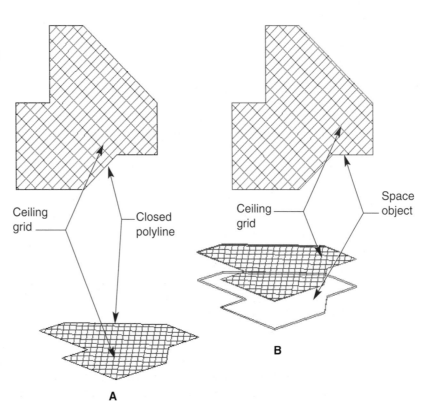

Ceiling
grid

Closed
polyline

Ceiling
grid

Space
object

A

B

Modifying a Ceiling Grid

As the design changes, a ceiling grid may also need to be changed. Suppose the size of the ceiling grid bays need to be modified. Use one of the following options to change the bay size:

- ✓ Select the **Design** pull-down menu, pick the **Grids** cascade menu, then select **Modify Ceiling Grid....**

- ✓ Pick the **Modify Ceiling Grid** button in the Grids toolbar.

- ✓ Right-click, and select **Design, Grids, Modify Ceiling Grid...** from the cursor menu.

- ✓ Type **AECCEILINGGRIDMODIFY** at the Command: prompt.

- ✓ Select the ceiling grid object, right-click, and pick **Ceiling Grid Modify...** from the shortcut menu.

The **Modify Ceiling Grid** dialog box, shown in Figure 23.23, is displayed. Make any changes to the bay size or the overall size, then press the OK button to return to the drawing screen. The ceiling grid will be updated to reflect your changes.

FIGURE 23.23   The Modify Ceiling Grid dialog box is used to modify a ceiling grid.

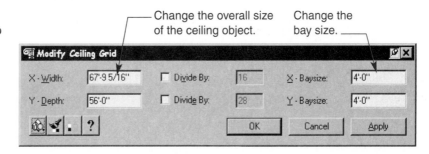

Modifying the Properties of a Ceiling Grid

The **Ceiling Grid Properties** dialog box can be used to further modify the divisions within the ceiling grid. This command can be accessed by first selecting a ceiling grid object, right-clicking, and selecting **Ceiling Grid Properties...**, or by typing **CEILINGGRIDPROPS** at the Command: prompt. The **Ceiling Grid Properties** dialog box is displayed. This dialog box includes the following five tabs:

- **General**   This tab is used to provide a description of the ceiling grid, and to add a property set definition and design notes to the object.

- **Dimensions**   This tab specifies the overall size of the ceiling grid, and the baysize spacing. If you want to specify the grid spacing manually, then uncheck the Automatic Spacing check box (see Figure 23.24). This allows you to modify the grid spacing through the X - Spacing and Y - Spacing tabs or by using grips on the drawing screen. Unchecking the Automatic Spacing check boxes adds grips to the ends of each grid line so that they can be stretched to a new size.

- **X - Spacing and Y - Spacing**   These tabs list the total number of grid lines created for the ceiling grid object. These tabs are unavailable if the Automatic Spacing check box has been selected in the Dimensions tab. If the Automatic Spacing check box has been cleared, then the spacing between grid lines can be modified. Pick on a value to adjust it (see Figure 23.25).

- **Location**   This tab is used to specify the ceiling grid location in relation to the World coordinate system.

Dimensions tab ─────

Overall size of ceiling grid

Uncheck to
specify the
grid spacing
manually. ─────

Offset the
starting and
ending grid
line a
distance from
the overall
grid size. ─────

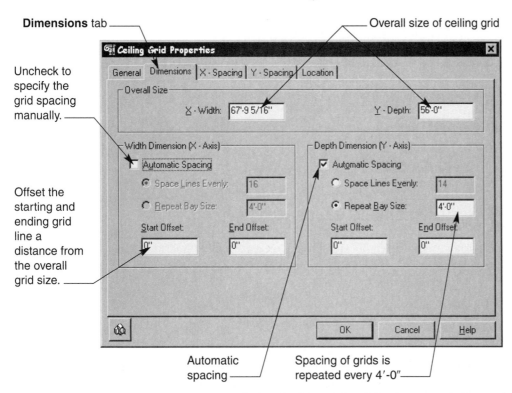

Automatic
spacing ─────

Spacing of grids is
repeated every 4'-0" ─────

FIGURE 23.24   The Dimensions tab in the Ceiling Grid Properties dialog box controls the spacing of
the grid lines in the ceiling grid.

FIGURE 23.25   The grid lines
can be adjusted manually by
specifying a new value.

X - Spacing
tab ─────

Y - Spacing
tab ─────

Pick on a value
to change it.

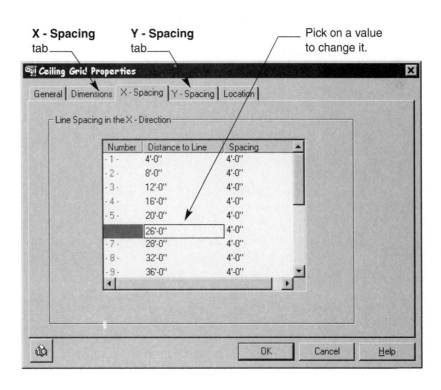

Clipping an Existing Ceiling Grid

In the earlier discussion of spaces it was mentioned that an interference condition can be applied to a space in order to cut a hole out of it. A similar process can be used to cut a hole out of a ceiling grid. First, a closed polyline or an AEC object needs to exist that will determine the hole. Once the object is in place on the drawing, access the **Clip Ceiling Grid** command using one of the following options:

✓ Pick the **Design** pull-down menu, select the **Grids** cascade menu, then pick the **Clip Ceiling Grid** command.

✓ Pick the **Clip Ceiling Grid** button in the **Grids** toolbar.

✓ Right-click, and select **Design, Grids, Clip Ceiling Grid.**

✓ Type **AECCEILINGGRIDCLIP** at the Command: prompt.

You are then prompted to select a ceiling grid. Select the ceiling grid to be clipped, enter **A** at the Command: prompt to add the hole, then select the polyline or AEC object to cut the hole in the ceiling grid. The following prompt sequence is used to add a hole in the ceiling grid in Figure 23.26.

```
Command: AECCEILINGGRIDCLIP ↵
Ceiling grid clip [Set boundary/Add hole/Remove hole]: A ↵
Select ceiling grids: (select the ceiling grid)
1 found
Select ceiling grids: ↵
Select a closed polyline or AEC entity  for hole: (select the polyline)
Ceiling grid clip [Set boundary/Add hole/Remove hole]: ↵
Command:
```

The options for this command are described next:

- **Set boundary**   This option allows you to set a closed polyline or a space as a masking object.

- **Add hole**   This option cuts a hole in the ceiling grid.

- **Remove hole**   This option removes a hole from the ceiling grid. The original object used to add the hole must be selected to be removed from the ceiling object.

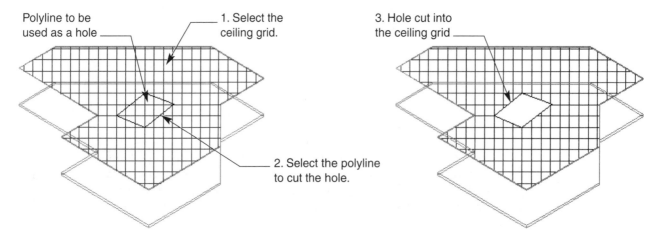

FIGURE 23.26   Adding a hole in a ceiling grid.

Exercise 23.5

1. Open EX23-4-Flr-Pln.dwg.
2. Change the display configuration to Work_Reflected, or pick and work in the Work-RCP layout tab.
3. Add a 2′-0″ × 2′-0″ ceiling grid to the kitchen, bathroom, and closet areas.
4. Use the space object as a masking object.
5. Turn off the space hatch in the Reflected display representation.
6. Save the drawing as EX23-5-Flr-Pln.dwg.

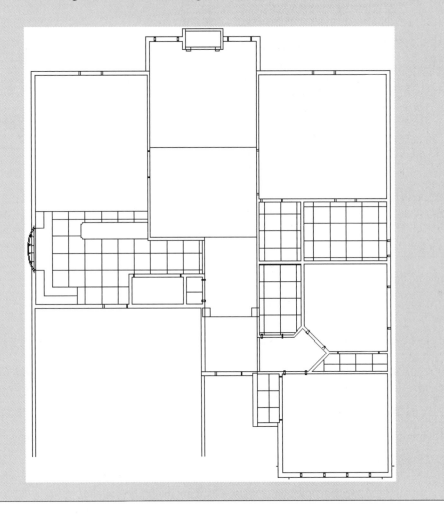

Using Column Grids

Another tool used by commercial drafting and design firms is the column or layout grid. These grids are used to space structural members across the building or to aid in designing across an area that is divided into specific bays or divisions. Often, the X- and Y- axis grid lines are labeled and dimensioned for quick and easy access by a person working on or referring to the plan. Figure 23.27 displays an example of a column grid in use on a floor plan.

Adding a Column Grid

Column grids can be added to the drawing at any stage of the design. They are usually added early in the design process so the drafter adding the structure, walls, and doors can lay out the objects in an organized system.

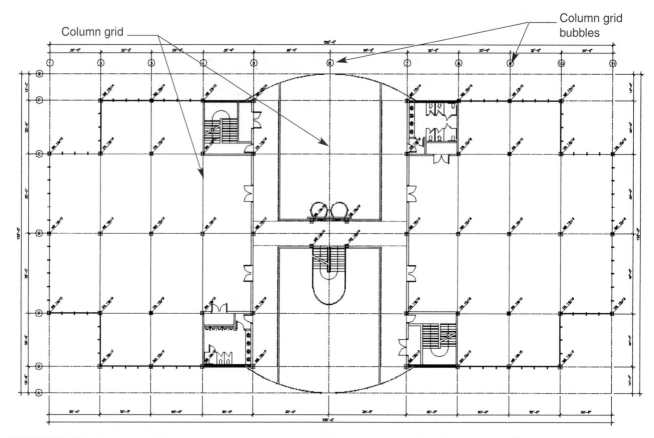

FIGURE 23.27   A column grid being used to space structural elements and design elements on a floor plan.

Use one of the following options to add a column grid to the drawing:

✓ Pick the **Design** pull-down menu, select the **Grids** cascade menu, then pick **Add Column Grid....**

✓ Pick the **Add Column Grid** button in the **Grids** toolbar.

✓ Right-click and select **Design, Grids, Add Column Grid...** from the cursor menu.

✓ Type **AECCOLUMNGRIDADD** at the Command: prompt.

The **Add Column Grid** dialog box is displayed. This dialog box is similar to the Add Ceiling Grid dialog box. At the top of the dialog box is a drop-down list that allows you to specify whether to create a rectangular column grid or a radial column grid.

Adding a Rectangular Grid

The rectangular column grid allows you to specify a total width and depth for the column grid by entering values in the X - Width and the Y - Depth text boxes. If you want to enter the total size of the column grid as you add it in the drawing, then select the Specify on Screen check box (see Figure 23.28).

Like ceiling grids, bays can be created by selecting the Divide By check boxes and entering division numbers in the associated text boxes. Actual bay sizes can be entered in the X - Baysize and Y - Baysize text boxes to place grid lines spaced the specified distance.

To add the column grid, pick a point in the drawing and specify a rotation angle. If you also want to add columns to the grid, then pick the Column< button before adding the grid to the drawing. The **Add Columns** dialog box, shown in Figure 23.29, is displayed. Select a structural style for the columns, then press the Close button to return to the Add Column Grid dialog box. Pick a point in the drawing and specify a rotation angle for the column grid. The new column grid is displayed with structural columns at each grid line intersection.

FIGURE 23.28 The Add Column Grid dialog box is similar to the Add Ceiling Grid dialog box.

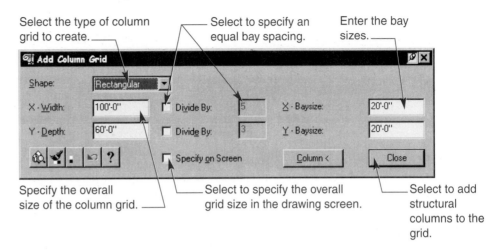

Select the type of column grid to create.

Select to specify an equal bay spacing.

Enter the bay sizes.

Specify the overall size of the column grid.

Select to specify the overall grid size in the drawing screen.

Select to add structural columns to the grid.

FIGURE 23.29 The Add Columns dialog box is used to pick a structural style to be used for the columns that will be added to the column grid.

Pick a structural style.

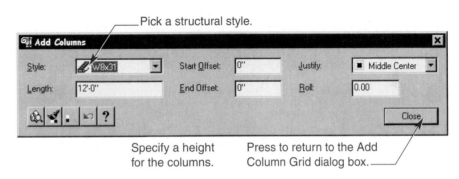

Specify a height for the columns.

Press to return to the Add Column Grid dialog box.

Adding a Radial Grid

When a radial grid is added to the drawing, settings in the Add Column Grid are adjusted to reflect a curve (see Figure 23.30). The R - Radius text box specifies the maximum radius for the grid. The A - Angle text box specifies the total angle of the grid. Enter a value for the Inside Radius. The X - Baysize controls the bay's offset from the inside radius. The B-Bayangle text box specifies the angles of the bays beginning with the first line of the angle grid.

As with the rectangular grid, if structural columns are to be added, pick the Column< button to select a structural column style. Pick the Close button to return to the Add Column Grid dialog box, then pick a point for the radial column grid.

Modifying a Column Grid

After a column grid has been added to the drawing, it can be modified to reflect new bay sizing or a different overall size. Use one of the following options to modify an existing column grid:

✓ Select the **Design** pull-down menu, pick the **Grids** cascade menu, then select **Modify Column Grid....**

✓ Pick the **Modify Column Grid** button in the **Grids** toolbar.

✓ Right-click, and pick **Design, Grids, Modify Column Grid...** from the cursor menu.

✓ Type **AECCOLUMNGRIDMODIFY** at the Command: prompt.

✓ Select the column grid, right-click, and pick **Column Grid Modify...** from the shortcut menu.

The **Modify Column Grid** dialog box, shown in Figure 23.31, is displayed. Make any required adjustments to the values in the dialog box. Press the Apply button to apply the changes to verify that they are what you want before closing the dialog box.

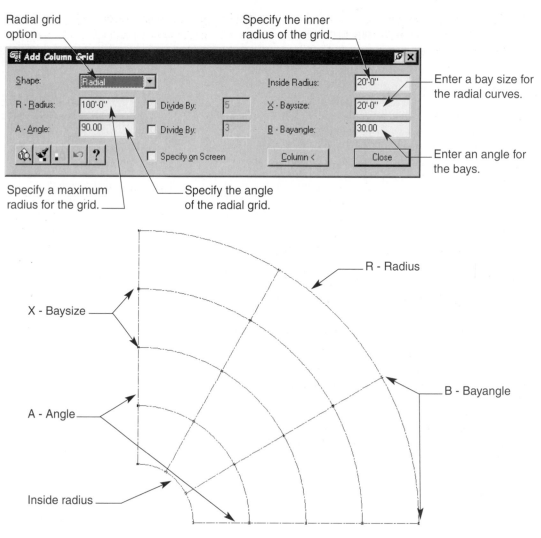

Radial grid option

Specify the inner radius of the grid.

Enter a bay size for the radial curves.

Specify a maximum radius for the grid.

Specify the angle of the radial grid.

Enter an angle for the bays.

R - Radius

X - Baysize

B - Bayangle

A - Angle

Inside radius

FIGURE 23.30   The Add Column Grid dialog box for a radial grid.

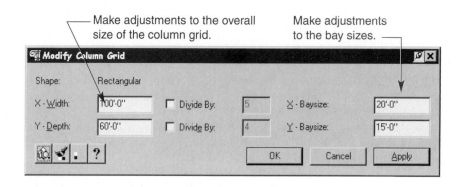

FIGURE 23.31   The Modify Column Grid dialog box is used to modify an existing column grid.

Make adjustments to the overall size of the column grid.

Make adjustments to the bay sizes.

Using Column Grid Properties

The column grid can also be modified through the **Column Grid Properties** dialog box. Use one of the following options to access this dialog box:

✓ Pick the column grid, right-click, and pick **Column Grid Properties...** from the shortcut menu.

✓ Type **COLUMNGRIDPROPS** at the Command: prompt.

This dialog box is exactly the same as the Ceiling Grid Properties dialog box. (Refer to Figures 23.24 and 23.25). The Dimensions tab controls the spacing of the bays. If the Automatic Spacing check boxes are cleared, the bay grid lines can be adjusted in the X - Spacing and Y - Spacing tabs.

Clipping a Column Grid

There may be times when you are laying out column grids that one column grid overlaps another (see Figure 23.32). To avoid confusion you may wish to clip one of the grids. In order to clip a grid, you must draw a polyline around the part of the column grid you wish to keep.

After drawing the polyline around the part of the column grid to be kept, use the Clip Column Grid command with one of the following options:

✓ Select the **Design** pull-down menu, pick the **Grids** cascade menu, then select **Clip Column Grid.**

✓ Pick the **Clip Column Grid** button in the **Grids** toolbar.

✓ Right-click, and select **Design, Grids, Clip Column Grid** from the cursor menu.

✓ Type **AECLAYOUTGRIDCLIP** at the Command: prompt.

The following prompt sequence is used to clip a column grid:

Command: **AECLAYOUTGRIDCLIP** ↵
Layout grid clip [Set boundary/Add hole/Remove hole]: **S** ↵
Select layout grids: *(select a column grid to clip)*
1 found
Select layout grids: ↵
Select a closed polyline for boundary: *(select the polyline to be used as a mask)*
Layout grid clip [Set boundary/Add hole/Remove hole]: ↵
Command:

Use the Set boundary option to select the polyline that will be masking away the unwanted portion of the column grid.

Adding Column Grid Labels and Dimensions

The final elements that need to be added to the column grid are labels and dimensions. The labels are bubbles with letters and numbers in them that refer to the section of a building. This makes it easier to find a part of the building when cross-referencing sections, elevations, and details.

FIGURE 23.32 Two column grids are crossing on top of each other.

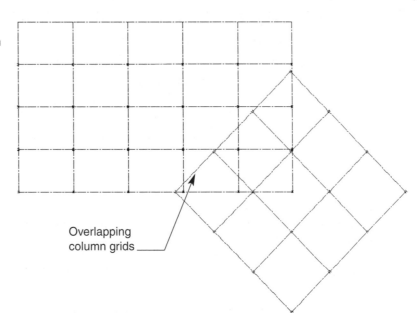

Overlapping
column grids

Use one of the following options to add column grid labels:

✓ Pick the **Design** pull-down menu, access the **Grids** cascade menu, then pick **Label Column Grid....**

✓ Pick the **Label Column Grid** button in the **Grids** toolbar.

✓ Right-click, and select **Design, Grids, Label Column Grid...** from the cursor menu.

✓ Type **AECCOLUMNGRIDLABEL** at the Command: prompt.

✓ Select a column grid, right-click, and pick **Labels...** from the shortcut menu.

The **Column Grid Labeling** dialog box, shown in Figure 23.33, is displayed. This dialog box includes two tabs: **X - Labeling** and **Y - Labeling**. The X - Labeling tab is used to label the grid lines along the width of the column grid, and the Y - Labeling tab is used to add labels along the depth of the column grid.

To add a label, pick the first (--) under the Number column. Enter a numerical or alphabetic value. Pick in the list box to accept this value. The rest of the grid liens are sequentially labeled. In the Automatic Labeling Rules area, specify whether the labeling should be in ascending or descending order. Select the Never Use Characters check box if there are characters you do not want to add to the labels.

In the Bubble Parameters area specify whether the labels are to be added to the top and bottom sides of the grid (if adjusting the X - Labeling tab) or the left and right sides (if adjusting the Y - labeling tab). Enter a value in the Extension text box to offset the label a distance away from the outer gridlines. Pick the Bubble...button to specify a detail bubble for use. Adjust the values for both the X - Labeling and the Y - Labeling tabs. When you have finished press the OK button to exit the dialog box and generate the labels.

Adding Dimensions to the Column Grid

Dimensions also can be added to the column grid using one of the following options:

✓ Pick the **Design** pull-down menu, select the **Grids** cascade menu, then pick **Dimension Column Grid.**

✓ Pick the **Dimension Column Grid** button in the **Grids** toolbar.

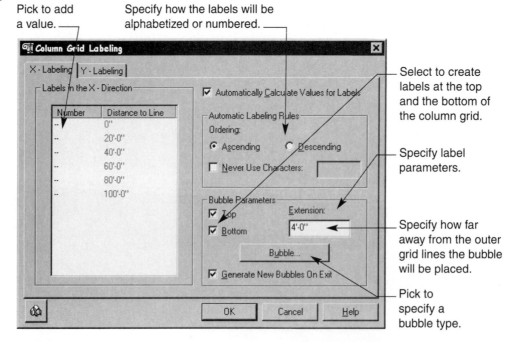

FIGURE 23.33   The Column Grid labeling dialog box is used to add labels at the ends of the column grid lines.

✓ Right-click, and select **Design, Grids, Dimension Column Grid** from the cursor menu.

✓ Type **AECCOLUMNGRIDDIM** at the Command: prompt.

✓ Select the column grid, right-click, and pick **Dimensions** from the shortcut menu.

You are then prompted to select a column grid. Select the column grid that is to be dimensioned. Next, enter the offset distance away from the outer grid lines where the dimension lines are to be placed. Dimensions are added around all the sides of the column grid.

Exercise 23.6

1. Open EX22-10-First Floor.dwg.
2. Add and adjust column grids around the slice object as shown in the figure. Add 16'-0"-long columns.
3. Clip the angled column grid away from the orthogonal column grid.
4. Add labels and dimensions.
5. Save the drawing as EX23-6-First Floor.dwg.

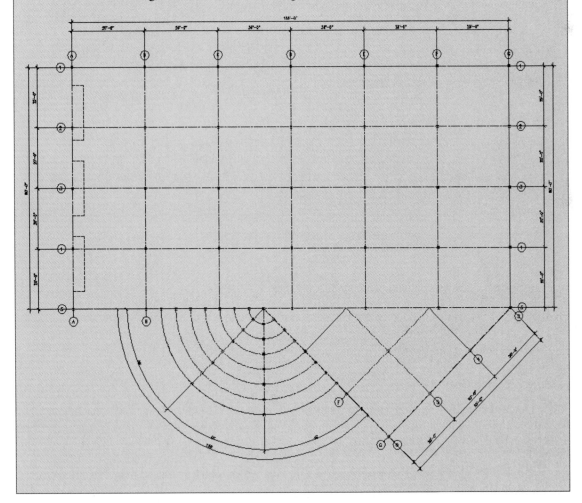

Working with Space Boundaries

Another method of space planning is by combining spaces and space boundaries. When designing using spaces you are typically designing from the inside of the building out. As you manipulate the space the exterior shape of the building is being developed.

The space objects in Architectural Desktop represent the area and volume space of a room within a building. Surrounding the spaces are areas that will be occupied by walls during the design development phase. During the space planning process, however, these areas can be occupied by space boundaries. As defined earlier, a space boundary is a type of object that can form a solid boundary between two spaces (similar to a wall), or a division between two adjacent spaces in the same room, similar to a division between two business departments occupying the same room (see Figure 23.34).

When solid form space boundaries are created, they appear very similar to wall objects. The difference is that all the space boundaries surrounding a space are considered one object instead of individual walls. This can be an advantage in designing, because a change to the width or height of a space boundary affects all the boundaries simultaneously. Another advantage is that the space is tied to the space boundary. If a change is made to the spaces size, the boundaries change proportionately. As a space changes size the walls need to be moved into place and adjusted to fit around the space.

An area separation space boundary is typically used when space is being arranged into separate usage areas within a building where the solid form boundaries have been defined.

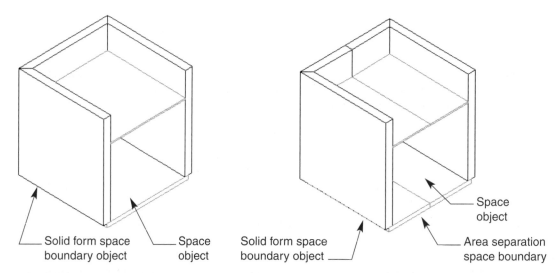

Solid form space boundary object

Space object

Solid form space boundary object

Space object

Area separation space boundary

FIGURE 23.34 A space boundary can be created as a solid boundary or as an area separation.

Adding a Space Boundary

Adding a space boundary is similar to drawing walls. The points you select define the boundary, and you have the option of creating the space contained within the boundary. Use one of the following options to add a space boundary into a drawing:

✓ Select the **Concept** pull-down menu, pick the **Space Boundaries** cascade menu, then select **Add Boundary....**

✓ Pick the **Add Boundary** button in the **Space Boundaries** toolbar.

✓ Right-click, and pick **Concept, Space Boundaries, Add Boundary...** from the cursor menu.

✓ Type **AECSPACEBOUNDARYADD** at the Command: prompt.

The Add Space Boundary dialog box, shown in Figure 23.35, is displayed. The dialog box includes the following options:

Select boundary type. — Enter height of boundary. —

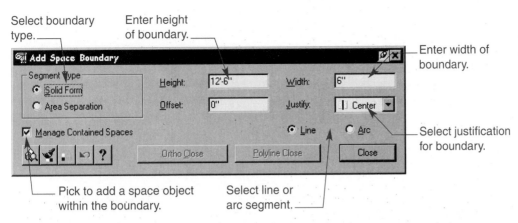

Enter width of boundary.

Select justification for boundary.

Pick to add a space object within the boundary.

Select line or arc segment. —

FIGURE 23.35   The Add Space Boundary dialog box is used to draw space boundaries.

- **Segment Type** This area contains the Solid Form and Area Separation options. The Solid Form creates boundaries that represent the walls of the building. The Area Separation creates a boundary without a width.

- **Manage Contained Spaces** This check box allows you to create a space object within the space boundaries. If this box is unchecked, a boundary is created with a voided space inside.

- **Height** This text box controls the height of the space boundary whether it is a solid-form or an area separation.

- **Offset** This option is similar to the Offset option in the Add Walls dialog box. The space boundary is offset a specified distance from the crosshairs as you pick points in the drawing.

- **Width** This text box is available when the solid form space boundary has been selected. It allows you to specify the width of the boundary.

- **Justify** This drop-down list is available when the solid form boundary is used. This allows you to specify a left, center, or right justification for the space boundary, similar to the justification for walls.

- **Line** This button draws the space boundary as a linear segment.

- **Arc** This button draws the space boundary as an arced segment.

- **Ortho Close** This button becomes available after three points (two boundary segments) have been drawn. Selecting this button adds two segments to the space boundary, closing it.

- **Polyline Close** This button also becomes available after three points have been selected in the drawing. Selecting this button closes the space boundary with one segment from the last point drawn to the first point drawn.

Once you have specified the parameters, begin picking points in the drawing to define the boundaries for a space. You can add as many enclosed spaces as needed to the drawing. As you pick points to establish room sizes you can switch between the Solid Form and Area Separation options to create space boundaries. When you have finished adding spaces on the drawing, pick the Close button or press the [Enter] key to end the command. Figure 23.36 displays several space boundaries used to create a space plan.

Solid form space
boundaries

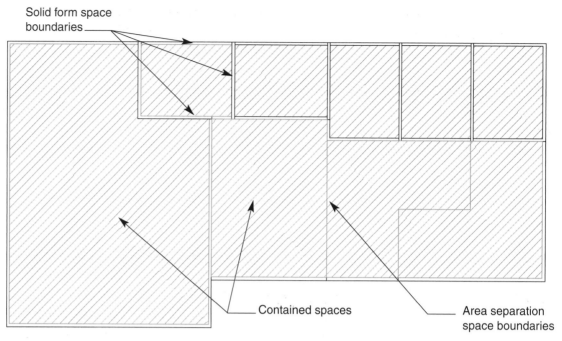

Contained spaces

Area separation
space boundaries

FIGURE 23.36 Space boundaries have been added to the drawing to create a space plan.

 Field Notes

Solid form space boundaries are similar to walls. They define the area the walls will occupy. The advantage of using solid form space boundaries by the space planner assembling a series of areas for a plan is that all the boundaries react globally to changes made to width and height for quick design changes. Later, the space boundaries can be converted into walls when the drafter works on the construction documents.

Converting Objects into Space Boundaries

In addition to being created on the fly, space boundaries can be created by converting polylines, spaces, and slices into space boundaries. The objects must exist in the drawing before they can be converted. Use one of the following options to convert an object into a space boundary:

✓ Select the **Concept** pull-down menu, pick the **Space Boundaries** cascade menu, then select **Convert to Boundaries....**

 ✓ Pick the **Convert to Boundaries** button in the **Space Boundaries** toolbar.

✓ Right-click, and select **Concept, Space Boundaries, Convert to Boundaries...** from the cursor menu.

✓ Type **AECSPACEBOUNDARYCONVERT** at the Command: prompt.

The following prompt sequence is issued at the Command: prompt, "Convert [Edges/SPace/SLice]:".
This prompt allows you to specify the type of objects to be selected for conversion. The following options are described next.

■ **Edges** This option allows you to convert lines, arcs, circles, or polylines into space boundaries. If the objects do not form a closed area, they are converted into a space boundary without a contained space. If the objects form a closed area, then they are converted into space boundaries with enclosed space objects.

- **SPace** This option converts individual spaces into space boundaries. The edges of the space are considered the inside edge of the room, and space boundaries are added around the edges of the space object.

- **Slice** This option converts a slice object from a mass model plan into space boundaries with contained spaces.

Enter the desired option, then select the objects to be converted. The **Space Boundary Properties** dialog box is displayed. This dialog box includes the following three tabs that control how the space boundaries and their contained spaces are created.

- **General** This tab allows you to add a description to the space boundary, as well as property set definitions and design notes.

- **Dimensions** This tab controls the dimensions of the space boundary (see Figure 23.37). Options include Segment Type (solid form or area separation), width of the boundary, and justification. Additionally, if a contained space is to be created, select the Manage Contained Spaces check box.

- **Design Rules** This tab is used to control the space boundary conditions at the ceiling and floors of the space object. Initially, the Automatically Determine from Spaces check boxes are selected for both conditions. Clearing these check marks allows you to modify the settings (see Figure 23.38).

The Boundary Conditions at Ceiling area controls the overall base height of the space object. Selecting the Ceiling Stops at Wall radio button allows you to set a distance above the finished ceiling to the top of the space boundary. The combined base height and upper extension determine the height of the space boundary. The Wall Stops at Ceiling radio button creates the space boundary only as high as the ceiling.

The Boundary Conditions at Floor area controls the placement and distances for the floor in relation to the space boundary. The Floor Stops at Wall radio button allows you to set a distance from the finished floor to the bottom of the space boundary. The combined base height, upper extension, and lower extension

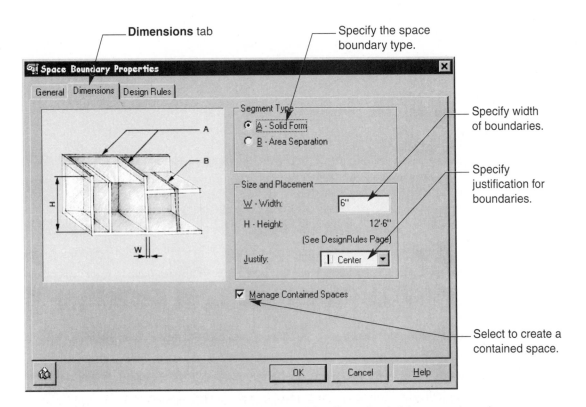

FIGURE 23.37   The Dimensions tab is used to control the dimensions of the space boundary.

Design Rules tab

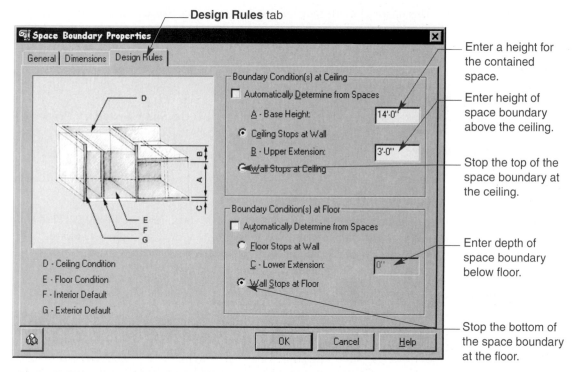

Enter a height for the contained space.

Enter height of space boundary above the ceiling.

Stop the top of the space boundary at the ceiling.

Enter depth of space boundary below floor.

Stop the bottom of the space boundary at the floor.

FIGURE 23.38 The Design Rules tab allows you to specify how the space boundary will interact with the space object.

create the total height of the space boundary. Selecting the Wall Stops at Floor places the bottom of the space boundary at the floor's elevation.

When you have finished determining the parameters for the space boundary and the contained space press the OK button to create the space boundary and contained space object.

Exercise 23.7

1. Open EX23-6-First Floor.dwg.
2. Convert the slice into a space boundary.
3. Set the width to 8″ and the Justification to Center.
4. Ensure that the Manage Contained Spaces check box is selected.
5. In the Design Rules tab, enter a base height of 12′-0″. Select the Ceiling Stops at Wall option and enter an upper extension of 4′-0″.
6. In the Boundary Conditions at Floor area, select the Wall Stops at Floor option.
7. Press OK to create the space boundary.
8. Save the drawing as EX23-7-First Floor.dwg.

Modifying Space Boundaries

As the space plan is developed you may find that some areas do not need a space boundary whereas others do need one, the space boundary height or width needs to be adjusted. The **Modify Boundary** command can be used to make the following changes. Use one of the following options to access this command:

- ✓ Select the **Concept** pull-down menu, pick the **Space Boundaries** cascade menu, then select **Modify Boundary....**

- ✓ Pick the **Modify Boundary** button in the **Space Boundaries** toolbar.

- ✓ Right-click, and pick **Concept, Space Boundaries, Modify Boundary...** from the cursor menu.

Specify the boundary height.

Specify the boundary width.

Select a new boundary type.

Select to link the contained space to the boundary.

Change the justification.

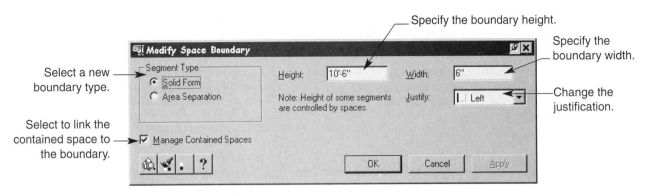

FIGURE 23.39   The Modify Space Boundary dialog box is used to make minor changes to a space boundary.

✓ Type **AECSPACEBOUNDARYMODIFY** at the Command: prompt.

✓ Select the space boundary to be modified, right-click, then pick **Boundary Modify...** from the short menu.

The **Modify Space Boundary** dialog box, shown in Figure 23.39, is displayed. This dialog box allows you to select a different boundary type, or boundary height and width and to adjust the justification of the selected space boundary. Additionally, the Manage Contained Spaces check box allows you to break the link between the contained space and the space boundary so that you can make size changes to the space boundary without updating the space object.

Adjusting Space Boundary Properties

In addition to using the Modify Space Boundary dialog box to adjust the parameters of the space boundary, you can use the **Space Boundary Properties** dialog box to control other aspects of the space boundary. Use one of the following options to access this command:

✓ Select the space boundary, right-click, and select **Boundary Properties...** from the shortcut menu.

✓ Type **SPACEBOUNDARYPROPS** at the Command: prompt.

The Space Boundary Properties dialog box is displayed. This is the same dialog box that is displayed when objects are converted into space boundaries. The only change is the addition of the Location tab for placement of the space boundary. Refer to Figures 23.37 and 23.38 for the values that can be changed.

Field Notes

Through the space planning process, windows, doors, and openings can be added to the space boundary for early design proposals. These items can get carried forward when the actual floor plans and elevations are created.

Working with Space Boundaries

There are a number of tools for manipulating the space boundaries, from merging boundaries together to splitting the spaces within a space boundary. Remember that as you work on a space boundary the entire boundary around a space is a single object that is turned into walls after the space plan is developed.

Attaching Spaces to a Boundary

There may be times as you are working with spaces and space boundaries separately that the two objects need to be joined together so that changes made to the space boundary control the space object. In these situations the space can be attached to a space boundary with the **Attach Spaces to Boundary** command. Use one of the following options to access this command:

✓ Pick the **Concept** pull-down menu, select the **Space Boundaries** cascade menu, then pick **Attach Spaces to Boundary.**

✓ Pick on the **Attach Spaces to Boundary** button in the **Space Boundaries** toolbar. Right-click, and pick **Concept, Space Boundaries, Attach Spaces to Boundary** from the cursor menu.

✓ Type **AECSPACEBOUNDARYMERGESPACE** at the Command: prompt.

✓ Select the space boundary, right-click, and select **Attach Spaces to Boundary** from the shortcut menu.

The following prompt sequence is used to attach a space to a space boundary:

Command: **AECSPACEBOUNDARYMERGESPACE** ↵
Select gaining space boundary: *(pick the space boundary that will gain the space)*
Select spaces to merge. *(pick the space that will be added)*
Select spaces:
1 found
Select spaces: ↵
Merged 1 space(s).
Command:

A space boundary is added around the space. The boundary around the space conforms to the rules applied to the entire space boundary.

Merging Boundaries

When several independent space boundaries are arranged, the boundaries may cross one another (see Figure 23.40A). As the location for these space boundaries becomes clear, you can merge the boundaries together to form a single object in which all the boundary size parameters can be controlled at the same time with the **Merge Boundaries** command. Use one of the following options to access this command:

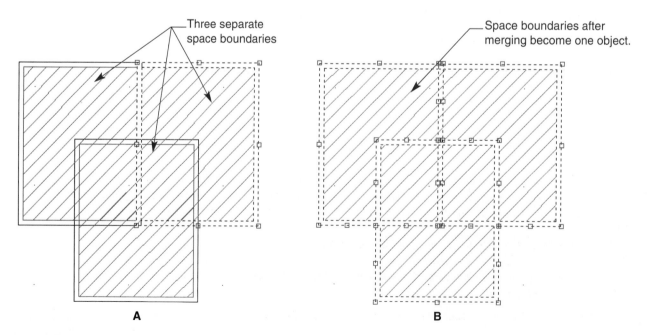

FIGURE 23.40 (A) Three separate space boundaries have been arranged on a plan. (B) The space boundaries have been merged into one object.

✓ Select the **Concept** pull-down menu, pick the **Space Boundaries** cascade menu, then access the **Merge Boundaries** command.

✓ Pick the **Merge Boundaries** button in the **Space Boundaries** toolbar.

✓ Right-click, and pick **Concept, Space Boundaries, Merge Boundaries** from the cursor menu.

✓ Type **AECSPACEBOUNDARYMERGE** at the Command: prompt.

✓ Select the space boundary, right-click, and select **Merge Boundaries** from the shortcut menu.

The following prompt sequence is used to merge the three space boundaries in Figure 23.40A to form one object:

Command: **AECSPACEBOUNDARYMERGE** ↵
Select gaining space boundary: *(select the space boundary that will gain another boundary)*
Select space boundary to merge: *(select the first space boundary that will be merged)*
Command: **AECSPACEBOUNDARYMERGE** ↵
Select gaining space boundary: *(select the space boundary that will gain another boundary)*
Select space boundary to merge: *(select the second space boundary that will be merged)*
Command:

Splitting Boundaries

Space boundaries that contain multiple contained spaces can be split into individual space boundaries. For example, suppose that during the process of designing space you merged all the space boundaries together only to decide later that one of the space boundaries surrounding a space object needed to be a separate object that could be manipulated. You would use the **Split Boundary** command to separate the space boundary around the space you wanted to modify.

Use one of the following options to access this command:

✓ Select the **Concept** pull-down menu, pick the **Space Boundaries** cascade menu, then select the **Split Boundary** command.

✓ Pick the **Split Boundary** button in the **Space Boundaries** toolbar.

✓ Right-click, and pick **Concept, Space Boundaries, Split Boundary** from the cursor menu.

✓ Type **AECSPACEBOUNDARYSPLIT** at the Command: prompt.

✓ Select the space boundary, right-click, and select **Split Boundary** from the shortcut menu.

You are then prompted to select a space. Select the space within the space boundary that you want to separate from the main space boundary (see Figure 23.41).

Adding Boundary Edges

As you define spaces and merge space boundaries, you may decide to add additional boundaries to the spaces within the main space boundary. The Add Space Boundary command does not merge the two boundaries together nor does it divide an internal space object; however, the **Add Boundary Edges** command can be used to add new boundaries to an existing space boundary and to divide the internal space it cuts through.

Use one of the following options to access this command:

✓ Pick the **Concept** pull-down menu, select the **Space Boundaries** cascade menu, then pick **Add Boundary Edges.**

✓ Pick the **Add Boundary Edges** button in the **Space Boundaries** toolbar.

✓ Right-click, and pick **Concept, Space Boundaries, Add Boundary Edges** from the cursor menu.

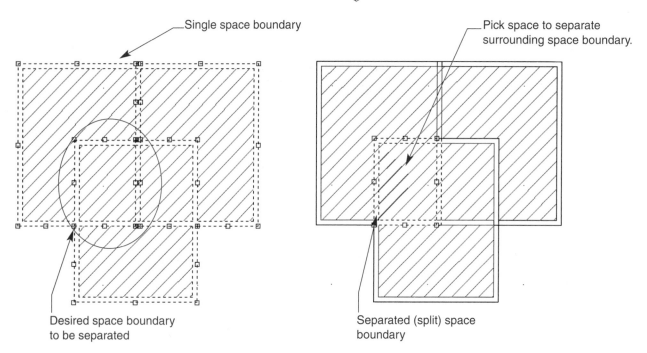

FIGURE 23.41   The Split Boundary command is used to separate a space boundary from a larger single space boundary.

✓ Type **AECSPACEBOUNDARYADDEDGES** at the Command: prompt.

✓ Select the space boundary, right-click, and select **Add Edges** from the shortcut menu.

You are then prompted to select a space boundary. Select the space boundary to which you would like to add new edges. The Add Space Boundary dialog box is displayed. This is the same dialog box used for the Add Space Boundary command shown in Figure 23.35. The difference is that the new edge becomes a part of the space boundary that you selected.

The new edge can be a solid form or an area separation boundary and can have a different height, width, and justification applied than the original space boundary. Select a starting point for the new edge and an ending point. If the new edge cuts through a space, the space is divided by the edge (see Figure 23.42).

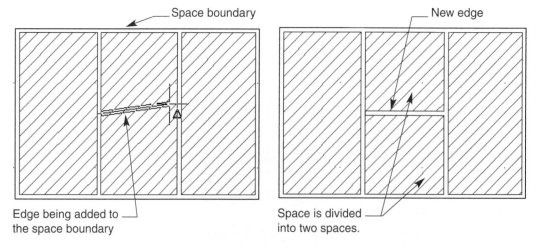

FIGURE 23.42   New edges can be added to an existing space boundary with their own size parameters. Adding a new edge divides the space into two pieces.

Field Notes

> Typically, the outer space boundary of a commercial building has a height that is measured from floor to floor. If you are adding interior edges that are not structural or bearing walls, you may want to create these walls the same height as the space object, as the space object is representing a suspended ceiling.

Editing Boundary Edges

You can edit individual space boundary edges to modify their segment type (solid form or area separation), height, width, and justification. For example, suppose a building uses 12″ wide boundaries throughout the construction, except in the office areas, where the boundaries are to be 6″ wide. You can modify the individual edges to change the overall space boundary width from 12″ to 6″ using one of the following options:

✓ Select the **Concept** pull-down menu, pick the **Space Boundaries** cascade menu, then select **Edit Boundary Edges.**

✓ Pick the **Edit Boundary Edges** button in the **Space Boundaries** toolbar.

✓ Right-click, and pick **Concept, Space Boundaries, Edit Boundary Edges** from the cursor menu.

✓ Type **AECSPACEBOUNDARYEDGE** at the Command: prompt.

✓ Select the space boundary, right-click, and select **Edit Edges...** from the shortcut menu.

You are prompted to select boundary edges. Select the edges of the space boundaries that you want to modify. When you have finished selecting edges to be modified, press the [Enter] key. The **Boundary Edge Properties** dialog box, shown in Figure 23.43, is displayed.

This dialog box is similar to the Space Boundary Properties dialog box. Make changes to the type of segment edge—either solid form or area separation, the width, and the justification in the Dimensions tab. Set any rules for how the space is to interact with this edge in the Design Rules tab.

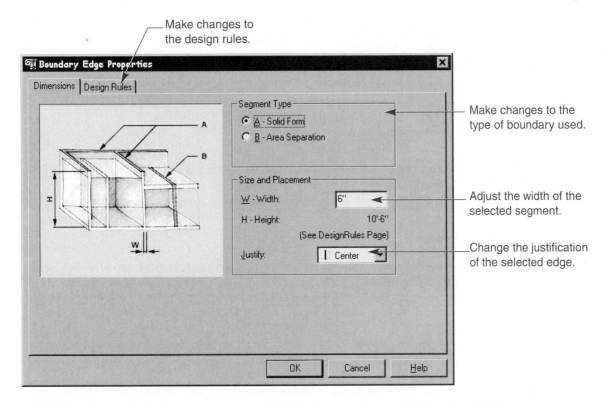

FIGURE 23.43   The Boundary Edge Properties dialog box is used to modify the selected edges.

Removing Boundary Edges

As you combine different space boundaries and adjust the space plan the design may call for the removal of some boundary edges to make larger open areas on the floor plan. Use the **Remove Boundary Edges** command to remove boundary edges by accessing one of the following options:

✓ Pick the **Concept** pull-down menu, access the **Space Boundaries** cascade menu, then select **Remove Boundary Edges.**

✓ Pick the **Remove Boundary Edges** button in the **Space Boundaries** toolbar.

✓ Right-click, and pick **Concept, Space Boundaries, Remove Boundary Edges** from the cursor menu.

✓ Type **AECSPACEBOUNDARYREMOVEEDGES** at the Command: prompt.

✓ Select the space boundary, right-click, and select **Remove Edges...** from the shortcut menu.

You are prompted to select the edges of a space boundary. Select any edge, then press the [Enter] key to remove the edge from the space boundary. If the edge is dividing a space into two parts, the spaces are joined together, forming one space object.

Attaching Objects to a Boundary Edge

During the design process you may decide to add design features such as columns or pilasters to the space boundary. These objects can then be anchored to the space object and can be moved freely within the constraints of the boundary itself. Any type of AEC object can be attached to the space boundary. Mass elements work very well for this type of process. Before you can use the **Anchor to Boundary** command, you must draw an object on the plan. Next, anchor it to the space boundary using one of the following options:

✓ Select the **Concept** pull-down menu, pick the **Space Boundaries** cascade menu, then select **Anchor to Boundary** command.

✓ Pick the **Anchor to Boundary** button in the **Space Boundaries** toolbar.

✓ Right-click, and select **Concept, Space Boundaries, Anchor to Boundary** from the cursor menu.

✓ Type **AECSPACEBOUNDARYANCHOR** at the Command: prompt.

The following prompt sequence is used to anchor an object to a space boundary:

Command: **AECSPACEBOUNDARYANCHOR** ↵
Space boundary anchor [Attach objects/Free objects]: **A**
Select object to be anchored: *(select the object to be anchored)*
Select a space boundary: *(select the space object)*
Command:

The object is anchored to the point closest to the crosshairs when you select the space boundary. This object is now anchored to the space boundary and cannot be moved outside the space boundary. If the object is to be released from the boundary, reenter the command and use the Free Objects option to release the object.

Generating Walls

Once the space plan has been finished and approvals received, the space plan can be moved forward to the design development phase. The boundaries can be converted into walls and the drawings finished. To convert the space boundaries into walls use the **Generate Walls** command. Use one of the following options to access this command:

✓ Pick the **Concept** pull-down menu, select the **Space Boundaries** cascade menu, then pick the **Generate Walls** command.

✓ Pick the **Generate Walls** button in the **Space Boundaries** toolbar.

✓ Right-click, and select **Concept, Space Boundaries, Generate Walls** from the cursor menu.

✓ Type **AECSPACEBOUNDARYGENERATEWALLS** at the Command: prompt.

✓ Select the space boundary, right-click, and select **Generate Walls** from the shortcut menu.

You are then prompted to select the space boundary. Pick the space boundary to be converted. You are then prompted to generate new openings. If you have windows, doors, or openings in the space boundary and want these items to be moved forward in the design, then answer Yes. Walls are created from the space boundary. Note that you may need to modify the wall style used and the wall cleanup radius to get the walls to clean up properly.

Exercise 23.8

1. Open EX23-7-First Floor.dwg.
2. Add some boundary edges along the grid lines making these bearing walls. These walls should be 12'-0" high with an upper extension of 4'-0".
3. Add some boundary edges for interior rooms and spaces. Make these boundaries 12'-0" without an upper extension.
4. Adjust some boundaries to create area separation boundaries.
5. Convert the space boundary into walls.
6. Save the drawing as EX23-8-First Floor.dwg.

CHAPTER REVIEW

Use the CD-ROM to test your knowledge and skills.

Chapter Test

To check your understanding of the content provided in this chapter, access the Test file in the CH23 folder of the CD-ROM that accompanies this text.

Chapter Project

To practice the Architectural Desktop skills presented in this chapter, access the Project files in the CH23 folder of the CD-ROM that accompanies this text. The project files are in pdf format and include sample drawings and instructions for completing each project.

CHAPTER 24

||||||||||||||||||||||||||||||||||||||||||||||||||||||||||||||||||||||||||||||||

Working with Areas

LEARNING GOALS

After completing this chapter, you will be able to:

◎ Add area objects to the drawing.

◎ Create an area style.

◎ Modify area objects.

◎ Create an area group style.

◎ Add an area group.

◎ Use calculations to modify the results of an area.

◎ Create an area group layout structure.

◎ Manage the area group structure.

◎ Create an area evaluation report.

During the course of developing a set of construction documents you may need to determine the total area being used in the building, the area being used by one department within a larger organization, or the area a building material will occupy and its total estimated cost. The Area feature in Architectural Desktop is used to define different types of areas in your drawing/building model, to group them together, and to provide calculations based on areas. This feature is part of the International Extensions that are installed by choice in Architectural Desktop 3.0 and automatically installed in Architectural Desktop 3.3 if a full installation is selected.

Area is a term that describes a two-dimensional space. In Architectural Desktop, an area is a 2D object that is displayed in a top view with a hatch pattern applied and contains perimeter information as well as the total area within its boundaries.

Field Notes

> An area object can be displayed when using the AEC Arch (Imperial or Metric).dwt. If the area object is not displayed in your drawing, create or modify a display configuration to display the Plan display representation for the area object. Creating and modifying display configurations was discussed in Chapter 5.

Once the design for the floor plan has been completed, areas can be added to the drawing for evaluation (see Figure 24.1). Calculation modifiers can be added to the area to determine building rent or the amount of hardwood needed on a floor, for example. Areas can also be grouped together into hierarchical structures to determine the total area of different departments in an office. Once the areas have been added to the drawing, area evaluation reports can be created and exported to a text file or to Microsoft Excel.

911

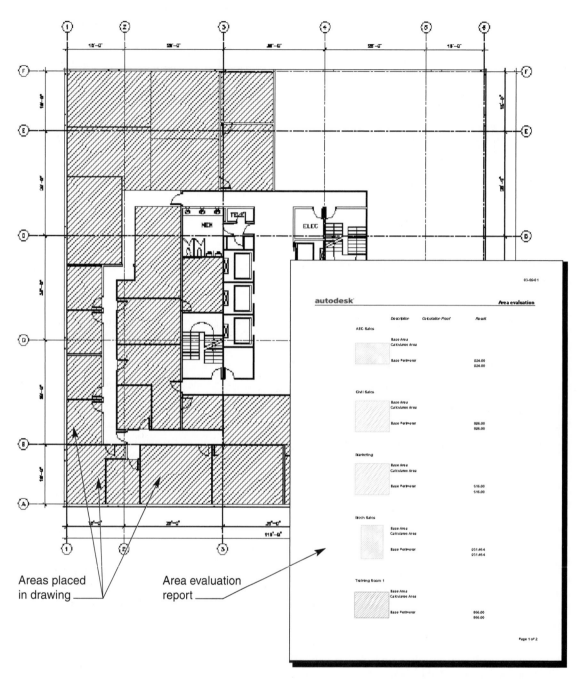

FIGURE 24.1   An example of area usage in an office building and a report created on the area usage.

This chapter discusses how to use the area tools to create accurate area calculations and reports for your drawings.

Adding Areas

When an area is created, points are selected in the drawing to define the edges or perimeter of the area. This edge is known as a ***ring*** and provides the boundaries for an area (see Figure 24.2A). Adding areas into the drawing is similar to drawing a polyline; the points are picked to establish the edges of the area. Complex areas consisting of positive and negative area values can also be added into the drawing. These are known as ***composed areas***. A composed area consists of two or more rings in which the area can be either a positive or a negative value (see Figure 24.2B).

Use one of the following methods to access the **Add Area** command:

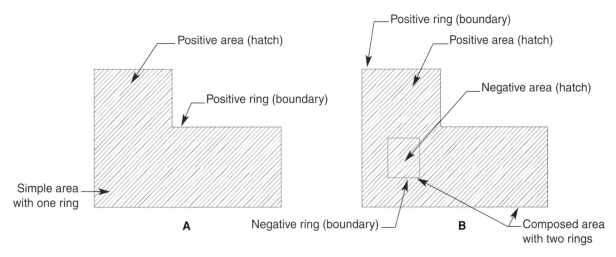

FIGURE 24.2 (A) An area ring with a hatch pattern representing the enclosed area. (B) A composed area consisting of two rings and associated hatches.

✓ Select the **Documentation** pull-down menu, pick the **Areas** cascade menu, then select the **Add Area...** command.

✓ Pick the **Add Area** button in the **Areas** toolbar.

✓ Right-click, and select **Documentation, Areas, Add Area...** from the cursor menu.

✓ Type **AECAREAADD** at the Command: prompt.

The **Add Area** dialog box, shown in Figure 24.3, is displayed. The Add Area dialog box includes the following settings:

- **Name:** Enter a name for the area if desired. This becomes the name of the area being reported on and displayed in an area tag.

- **Style:** Select an area style for use if there are styles available. A style controls the appearance of the area such as color and hatch pattern.

- **Tag:** Select an area tag to apply to the area after the area has been drawn. Area tags can be found in the Metric DACH\Schedules\Area Tags folder. To add a tag when creating an area, first drag and drop it into the drawing from the DesignCenter, then access the Add Area dialog box.

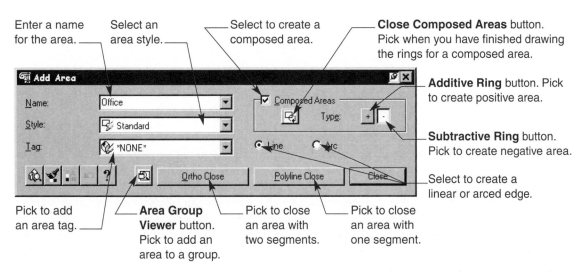

FIGURE 24.3 The Add Area dialog box is used to add simple and composed areas to a drawing.

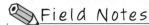Field Notes

The tags display the area in square meters. You can adjust this display by modifying the area data format style. Change the Unit Format to use Architectural Units in the drawing. You can also change the Suffix to use SQ FT instead of sqm. Data format styles were discussed in Chapter 20.

- **Area Group Viewer**   This button is used to attach the area to an area group. If there are no area groups in the drawing, this box will not contain a list of groups. An area can be attached to a group later if currently there are no groups available. Area groups will be discussed later in this chapter.

- **Composed Areas**   When this check box is selected, complex areas can be created in the drawing. When picking points for an area, you can create either a positive value or a negative value for the area by selecting the appropriate button in this section of the dialog box. When you have finished drawing the composed area, pick the Close Composed Areas button to create the previously drawn rings as one complex area with positive and negative areas.

- **Line and Arc**   These radio buttons create a linear ring edge or an arced ring edge.

- **Ortho Close**   This button becomes available after three or more points have been defined for the ring boundary. Press this button to add two edges to the area. This button is similar to the Ortho Close button for walls. The difference is that after pressing this button you can continue to add individual or composed areas to the drawing.

- **Polyline Close**   This button adds a single segment to the area and closes the area between the first and last points selected. Similar to the Ortho Close option, this button closes an area and allows you to draw additional areas or composed areas in the drawing.

Once you have established the area parameters, begin picking points in the drawing to establish ring edges. If the area you are creating will be a single ringed area, uncheck the Composed Areas check box, then select points in the drawing for the area. Pick the Polyline Close or the Ortho Close buttons to close the singled ringed area so that you can create another single ringed area. Press the Close button or the [Enter] key to end the command (see Figure 24.4).

Field Notes

Areas are created in the current XY plane. Areas can be obtained from vertical or sloped surfaces by changing the XY plane rotation. Use the UCS (User coordinate system) command to adjust the XY plane. For more information on how to adjust the UCS see the Defining a User Coordinate System in the AutoCAD 2000i or AutoCAD 2002 help section.

If the area you are creating will be a composed area consisting of several rings, then select the Composed Areas check box. Pick the appropriate positive area or negative area button, then begin picking points to establish the first ring. Use the Ortho Close or the Polyline Close buttons to close the first ring of the composed area. Again, pick the appropriate positive area or negative area button, select additional points to establish the second ring, and use Ortho Close or Polyline to close the second ring. Continue this process until all the composed rings have been created. When you have finished creating all the composed ring areas, pick the Close Composed Areas button to finish creating the composed area. A composed ring with multiple rings and positive and negative areas is created. The negative areas are subtracted from the positive area, leaving you with the total positive area (see Figure 24.5).

FIGURE 24.4 Creating single ringed areas by selecting points in the drawing.

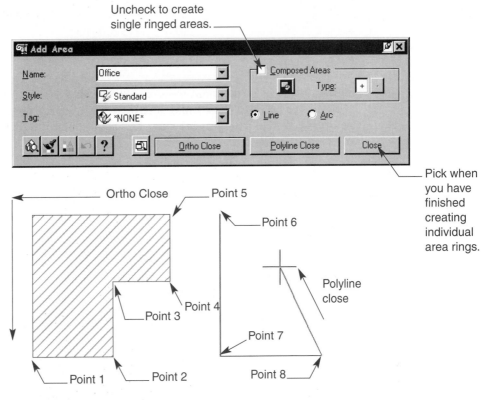

Uncheck to create single ringed areas.

Pick when you have finished creating individual area rings.

Ortho Close

Point 5

Point 6

Polyline close

Point 4

Point 3

Point 7

Point 1

Point 2

Point 8

FIGURE 24.5 Composed areas are created by selecting the positive and negative buttons, then picking points in the drawing.

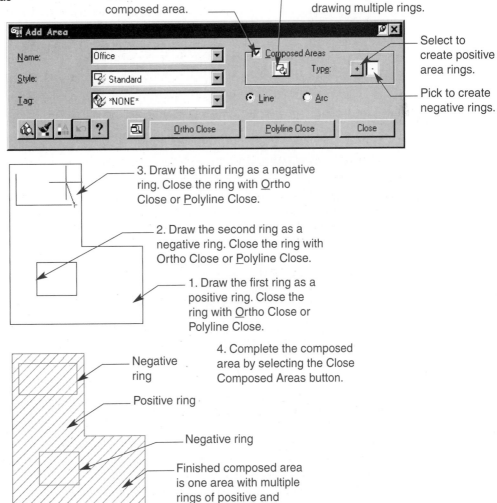

Select to create a composed area.

Pick when you have finished drawing multiple rings.

Select to create positive area rings.

Pick to create negative rings.

3. Draw the third ring as a negative ring. Close the ring with Ortho Close or Polyline Close.

2. Draw the second ring as a negative ring. Close the ring with Ortho Close or Polyline Close.

1. Draw the first ring as a positive ring. Close the ring with Ortho Close or Polyline Close.

4. Complete the composed area by selecting the Close Composed Areas button.

Negative ring

Positive ring

Negative ring

Finished composed area is one area with multiple rings of positive and negative values.

Field Notes

> Similar to creating single ringed areas, you can create many composed areas in one operation. Each time you press the Close Composed Areas button the rings you created are joined into that one composed area. If you begin picking new points in the drawing you will establish the next composed set of rings.

Exercise 24.1

1. Start a drawing using the AEC Arch (Imperial - Int).dwt.
2. Draw a 12'-0" × 10'-0" noncomposed area.
3. Create another area, this time selecting the Composed Areas check box.
4. Select the Additive Ring button and create a 12'-0" × 10'-0" ring. Create another ring inside the positive ring, this time selecting the Subtractive Ring button. Draw the subtractive ring 6'-0" × 4'-0".
5. Press the Close Composed Areas button.
6. Save the drawing as EX24-1.dwg.

Creating an Area Style

An *area style* defines how the area object is to appear in the drawing. The components of an area include an additive boundary and hatch and a subtractive boundary and hatch. Each of these components is displayed when a composed area is created. Use one of the following options to access the **Area Styles** command:

✓ Select the **Documentation** pull-down menu, pick the **Areas** cascade menu, then select the **Area Styles...** command.

✓ Pick the **Area Styles** button in the **Areas** toolbar.

✓ Right-click, and select **Documentation, Areas, Area Styles...** from the cursor menu.

✓ Type **AECAREASTYLE** at the Command: prompt.

The Style Manager dialog box opened to the Area Styles section is displayed (see Figure 24.6A). To create a new style, select the New Style button, or right-click on top of the Area Styles icon and select <u>N</u>ew. A new area style is created. Enter a name for the style. To edit the style, double-click on top of the new style, select the Edit Style button, or right-click on top of the style and select <u>E</u>dit... from the shortcut menu. The **Area Style** dialog box, shown in Figure 24.6B, is displayed.

Within the Area Style dialog box, select the General tab. Enter a description for the area style. You can attach notes and documents to the style by selecting the N<u>o</u>tes... button, and you can also attach property set definitions to the style. A property set definition is included with Architectural Desktop for area objects. This property set can be imported from the PropertySetDefsDACH.dwg found in the Autodesk Architectural Desktop 3\Content\Metric DACH\Schedules folder. This property set definition is assigned to individual area entities, so you may want to modify it to be attached to area styles.

Next, select the Display Props tab. At the top of the tab, select the display representation to be modified. Four display representations are available: Decomposed, Plan, Plan 1-100, and Plan 1-50. Check the display configurations in the drawing to see where each of these display representations is being used. Display representations and display configurations were discussed in Chapter 5.

Field Notes

> The Decomposed display representation is not assigned to a display configuration initially. This display representation displays the breakdown of an area and will be discussed later in this chapter.

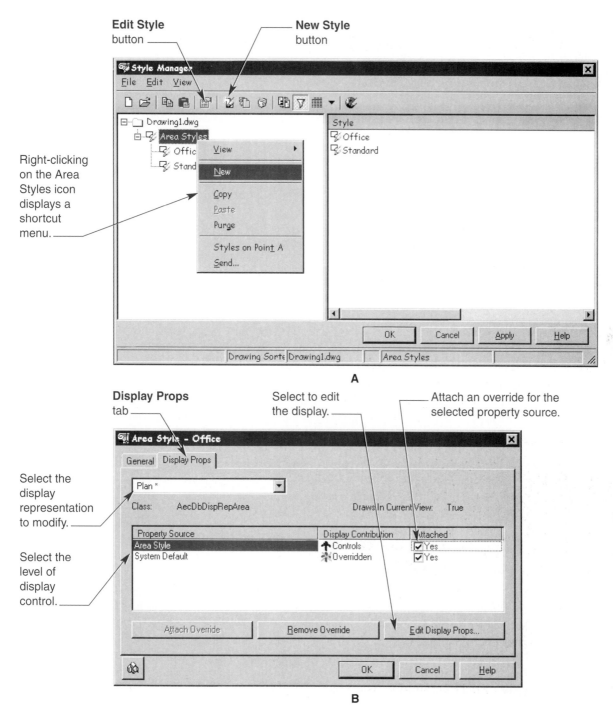

FIGURE 24.6 (A) The Style Manager dialog box is used to create a new area style. (B) When the style is edited, the Area Style dialog box is displayed.

After selecting the display representation, pick the level of display control from the Property Source column. Because this is a new style that is being defined you may decide to control the display at the Area Style level. Attach the override, then pick the Edit Display Props... button. The **Entity Properties** dialog box, shown in Figure 24.7, is displayed. Inside the dialog box, the colors used for the Additive Boundary and Hatch and the Subtractive Boundary and Hatch components can be assigned in the Layer/Color/ Linetype tab. Use the Hatching tab to configure the hatch patterns and properties for the hatch components. When you have finished press the OK button to return to the Style Manager to create additional area styles.

FIGURE 24.7 The Entity
Properties dialog box is used to
control the display of the area
object.

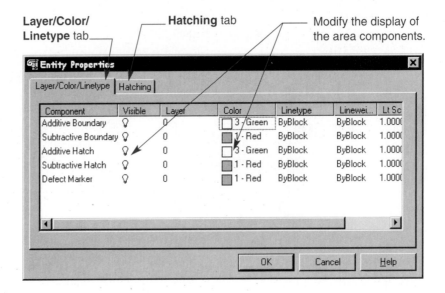

Layer/Color/
Linetype tab

Hatching tab

Modify the display of
the area components.

Exercise 24.2

1. Start a drawing using the AEC Arch (Imperial - Int).dwt.
2. Create an area style named Office. Assign the Additive Boundary and Hatch components a color. Change the Additive Hatch pattern rotation angle to 90.
3. Create an area style named Classroom. Assign the Additive Boundary and Hatch components a color. Change the Additive Hatch pattern rotation angle to 135.
4. Create an area style named Storage. Assign the Additive Boundary and Hatch components a color. Change the Additive Hatch pattern rotation angle to 0.
5. Create an area style named Walls. Assign the Additive Boundary and Hatch components a color. Change the Additive Hatch pattern rotation angle to 45.
6. Save the drawing as EX24-2.dwg.

Creating Areas from Objects

In addition to drawing areas into the drawing, you can generate areas from the exterior outline of an AEC object, from closed polylines or from the inside of a closed boundary, such as the walls forming a room. This is a convenient way of quickly creating areas within rooms to determine the total amount of usable area.

Use one of the following methods to generate an area from an object:

✓ Select the **Documentation** pull-down menu, pick the **Areas** cascade menu, then select the **Create Area from Object...** command.

✓ Pick the **Create Area from Object** button in the **Areas** toolbar.

✓ Right-click, and select **Documentation, Areas, Create Area from Object...** from the cursor menu.

✓ Enter **AECAREACREATEFROMOBJECT** at the Command: prompt.

The **Create Areas from Objects** dialog box, shown in Figure 24.8, is displayed. This dialog box is similar to the Add Area dialog box. Enter a name for the area to be created, select a style to generate, and pick a tag to attach if one is available in the drawing. If area groups exist in the drawing, then select the Area Group Viewer button to attach the area to a group. Area groups will be discussed later in this chapter.

Next, decide whether the area that will be created will be a composed area. If the area will be a composed area, then select the Composed Areas check box. Pick the Select Objects button to select an object(s) to generate areas. You are then prompted to enter a Cutplane height. Enter the height at which the area will be defined, basing the ring boundary on the edges of the object at the cutplane height (see Figure 24.9).

FIGURE 24.8 The Create Areas from Objects dialog box is used to generate an area from an object.

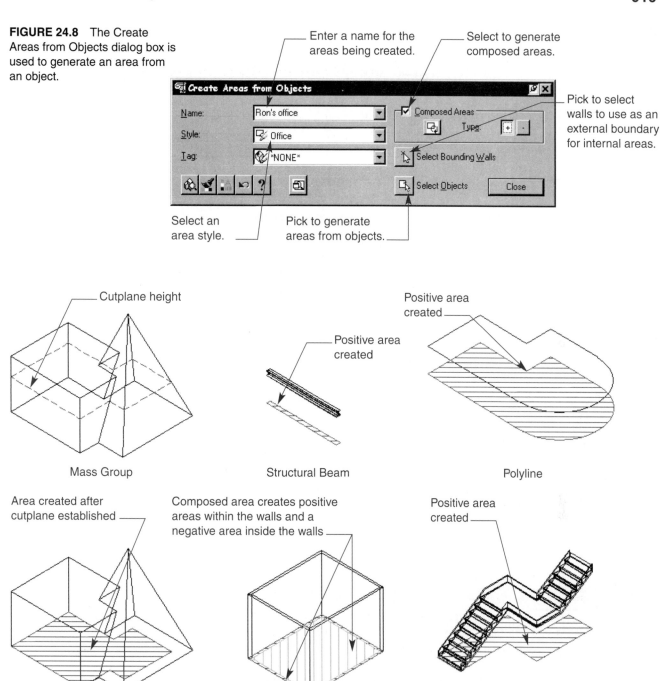

FIGURE 24.9 Several different types of objects used to generate areas.

When you select wall objects to create a noncomposed area, a positive area is drawn inside the walls so that you can calculate the total area the walls are occupying on a floor plan (see Figure 24.10A). If you select walls objects when creating a composed area, the positive area is created within the wall boundaries, and the negative area is calculated as the interior room area (see Figure 24.10B).

Another option for creating internal room areas from walls is to use the Select Bounding Walls button in the Create Areas from Objects dialog box (see Figure 24.8). When this button is selected, you must pick walls that form the boundaries for rooms. After selecting the walls, you are prompted to select internal points. Begin picking points inside a room to form an area (see Figure 24.11). If you create a noncomposed area, selecting inside a room will automatically create the area for that room.

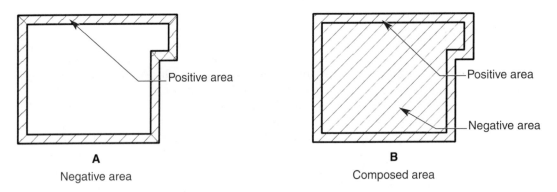

FIGURE 24.10   (A) Selecting walls for a noncomposed area generates areas within the walls. (B) Selecting walls for a composed area places the positive area within the walls and the negative area within the room.

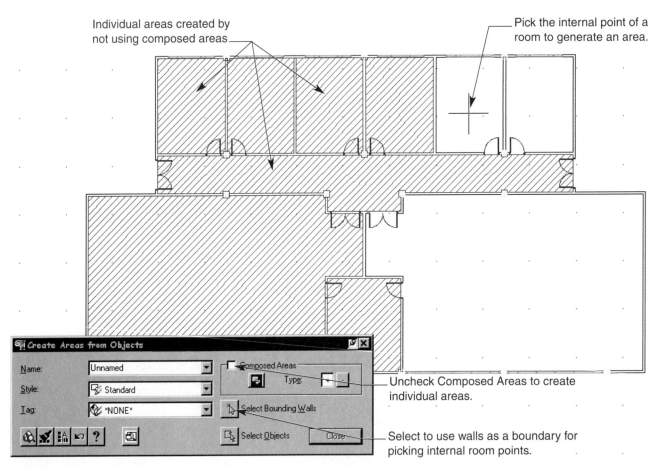

FIGURE 24.11   Creating noncomposed areas by picking the Select Bounding Walls button. Picking internal points of a room forms an area for the room.

If you choose a composed area, select the inside of the rooms you would like to have added together. Continue picking inside the rooms until you have finished selecting the rooms areas to be joined (composed). Press the [Enter] key in order to stop picking the interior of the rooms, then select the Close Composed Areas button to close the composed area, causing all the interior areas to be added together. When you have finished adding composed/noncomposed areas to the drawing, press the Close button to exit the command.

Field Notes

When you are creating composed areas, all objects or internal areas selected are joined together forming one composed area until you select the Close Composed Areas button. If you are creating room areas from the bounding walls, you probably will want to create noncomposed areas in order to obtain each individual room area rather than a total of all the room areas. Composed areas are beneficial when you are trying to obtain the total area of several objects. For example, if the total wall area is required, pick the Composed Areas check box, then the Select Objects button to select all your walls. When you have finished selecting the walls press [Enter] to stop picking objects, then pick the Close Composed Areas button to create the wall areas, which are grouped (composed) together.

Exercise 24.3

1. Open EX24-2.dwg.
2. Draw the building shown in the figure.
3. Uncheck the Composed Areas check box.
4. Pick the Select Bounding Walls button to pick internal points in each room. Create the room areas for each room using one of the styles created in Exercise 24.2. When you have finished, exit the dialog box.
5. Reenter the dialog box. Select the Walls style, then select the Composed Areas check box.
6. Pick the Select Objects button, then select each wall in the drawing. When you have finished, press the Close Composed Areas button to close the composed area. This will group all the wall areas together as one object.
7. Save the drawing as EX24-3.dwg.

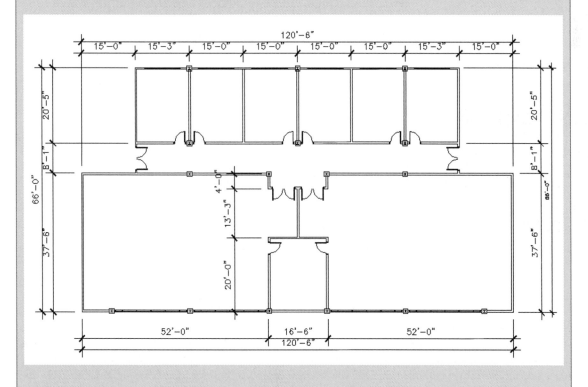

Modifying Areas

Once an area has been created, properties such as the area name, style, and area group to which it is attached can be modified through the Modify Area dialog box. Use one of the following methods to access the **Modify Area** command:

   ✓ Select the **Documentation** pull-down menu, pick the **Areas** cascade menu, then select the **Modify Area...** command.

   ✓ Pick the **Modify Area** button in the **Areas** toolbar.

   ✓ Right-click, and select **Documentation, Areas, Modify Area...** from the cursor menu.

   ✓ Enter **AECAREAMODIFY** at the Command: prompt.

   ✓ Select the area, then right-click and pick **Modify Area...** from the shortcut menu.

The **Modify Area** dialog box, shown in Figure 24.12, is displayed. In this dialog box, the name of the area or style can be changed, or the area group to which the area is assigned can be modified. Area groups will be discussed later in this chapter.

FIGURE 24.12   The Modify Area dialog box is used to modify basic parameters of the selected area.

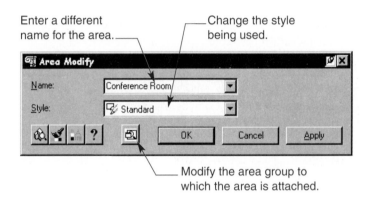

Editing Areas with Grips

Like other AEC objects and polylines, area objects can be modified with grip editing. The area objects include grip boxes at each vertex location and at each midpoint of a segment. Stretching the position of a grip box will increase or decrease the size of an area.

If a composed area is created with multiple rings, each ring includes its own grip points. Adjusting the grips of an internal ring will adjust the ring area plus the area's outer boundary (see Figure 24.13A). If the internal ring crosses over the outer ring, a faulty area is created (see Figure 24.13B). To fix this problem, use the grip boxes to move the internal ring back inside the outer ring.

Area Editing Tools

In addition to grip editing, a number of commands are available for modifying the area geometry. These functions include joining areas together, subtracting areas from other areas, creating new areas from the intersecting area of two area objects, adding or removing vertices, trimming areas, dividing areas, removing area rings, and reversing ring properties. These commands are described next.

Joining Areas
Areas inside a drawing can be joined together to form one area. You can join together areas that overlap each other, or areas that do not touch each other. When areas are to be joined, you are prompted to select areas to join. Select the first area that will be used to designate the new combined area name, style, and group to which the area is attached.

To join together areas in the drawing use one of the following options to access the Join command:

   ✓ Select the **Documentation** pull-down menu, pick the **Areas** cascade menu, then select the **Join** command.

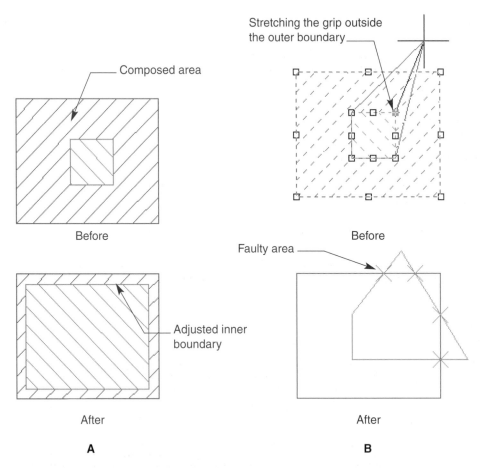

FIGURE 24.13 (A) Composed areas with internal rings can be adjusted to affect how area is calculated. (B) Moving a point on the internal ring outside the outer ring will create a faulty area.

✓ Pick the **Join** button in the **Areas** toolbar.

✓ Right-click, and select **Documentation, Areas, Join** from the cursor menu.

✓ Type **AECAREAOPJOIN** at the Command: prompt.

✓ Select the area, then right-click, select the **Operation** cascade menu, then pick the **Join** command from the shortcut menu.

When areas that overlap one another are joined, the ring or boundary is redefined to surround the two adjoining areas. The area name, style, and group of the first selected area become the properties set for the new single area (see Figure 24.14A).

If areas that are not touching each other are joined, their original boundaries are retained but become part of the total perimeter for the new single object. The new area is totaled from the areas selected to be joined (see Figure 24.14B).

Subtracting Areas

Areas inside a drawing can also be subtracted from one another. An area that overlaps another can be subtracted from the area it is overlapping, or an area completely inside another area can be subtracted from the area, forming a subtracted composed area (see Figure 24.15). Use one of the following options to access the Subtract command:

✓ Select the **Documentation** pull-down menu, pick the **Areas** cascade menu, then select the **Subtract** command.

✓ Pick the **Subtract** button in the **Areas** toolbar.

✓ Right-click, and select **Documentation, Areas, Subtract** from the cursor menu.

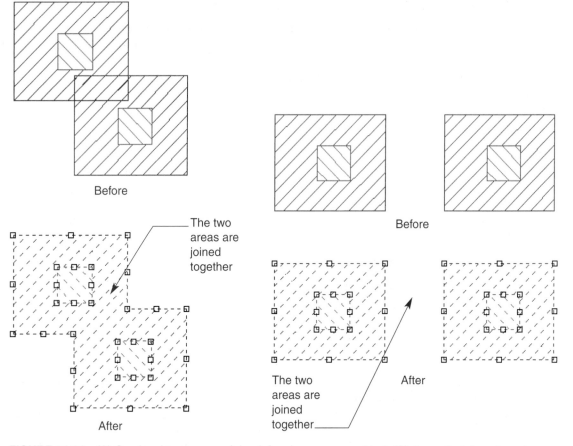

FIGURE 24.14   (A) Overlapping areas are joined, forming one area object. (B) Areas that do not overlap each other can be joined to form one area object.

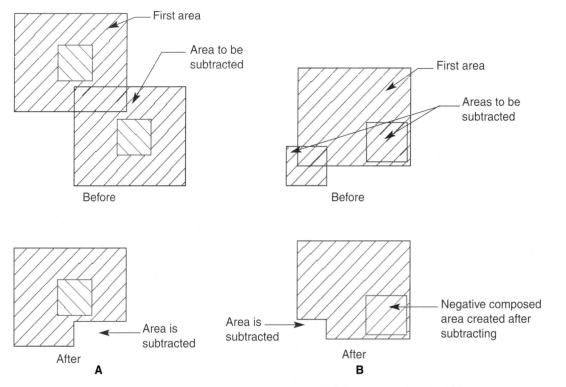

FIGURE 24.15   (A) An overlapping area is subtracted from the first area. (B) A ring on top of and inside another ring is subtracted, forming a subtractive composed area.

✓ Type **AECAREAOPSUBTRACT** at the Command: prompt.

✓ Select the area, then right-click, select the **Operation** cascade menu, then pick the **Subtract** command from the shortcut menu.

You are prompted to Select an area. The area you select becomes the area that will remain in the drawing. Next, select the areas that need to be subtracted from the first area. If you are subtracting individual areas from a composed area, the subtracted area becomes a negative area within the composed area.

Intersecting Areas

The Intersect command creates a new area from the common or shared area between two existing areas. Once the intersection has been accomplished, the two original areas are removed from the drawing, leaving behind an area formed from their shared space. Use one of the following options to access this command:

✓ Select the **Documentation** pull-down menu, pick the **Areas** cascade menu, then select the **Intersect** command.

✓ Pick the **Intersect** button in the **Areas** toolbar.

✓ Right-click, and select **Documentation, Areas, Intersect** from the cursor menu.

✓ Type **AECAREAOPINTERSECT** at the Command: prompt.

✓ Select the area, then right-click, select the **Operation** cascade menu, then pick the **Intersect** command from the shortcut menu.

You are prompted to select one of the areas. The area that you select will establish the area name, style, and group for the new area to be created. Next, select the area that will be used to create the intersected area. After the areas to be intersected are selected, the original areas are removed from the drawing and the new area created from the two objects intersection is created (see Figure 24.16).

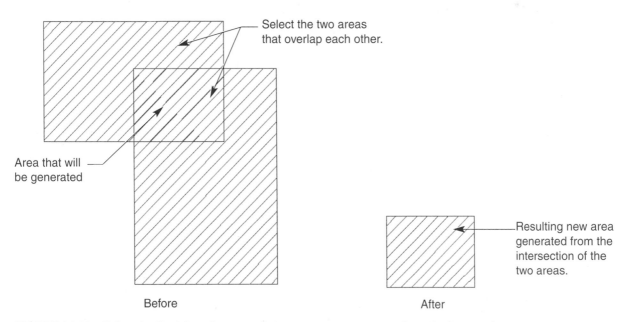

FIGURE 24.16   Select the first, then the second area to create a new area from the intersection.

Adding and Removing Vertices

Like slab objects, vertices on a slab can be added or removed for additional editing. Use one of the following options to access this command:

✓ Select the **Documentation** pull-down menu, pick the **Areas** cascade menu, then select the **Edit Vertices** command.

✓ Pick the **Edit Vertices** button in the **Areas** toolbar.

✓ Right-click, and select **Documentation, Areas, Edit Vertices** from the cursor menu.

✓ Type **AECAREAVERTEXMODIFY** at the Command: prompt.

✓ Select the area, then right-click, and select **Edit Vertices** from the shortcut menu.

To add vertices to an area, begin picking points along the boundary ring. If the ring is an internal ring, then select a point along the internal ring boundary to add vertices that can be grip edited. To remove a vertex from an area, press and hold the [Shift] key as you pick on top of an area vertex. A segment is formed between the vertices on each side of the vertex removed.

Trimming Areas

The Trim command for areas is used to trim away a portion of an area. Use one of the following options to access this command:

✓ Select the **Documentation** pull-down menu, pick the **Areas** cascade menu, then select the **Trim** command.

✓ Pick the **Trim** button in the **Areas** toolbar.

✓ Right-click, and select **Documentation, Areas, Trim** from the cursor menu.

✓ Type **AECAREATRIM** at the Command: prompt.

✓ Select the area, then right-click, select the **Operation** cascade menu, then pick the **Trim** command from the shortcut menu.

Select two points in the drawing to form the trimming line. After the line has been established, pick the area on the side of the trim line you wish to remove (see Figure 24.17).

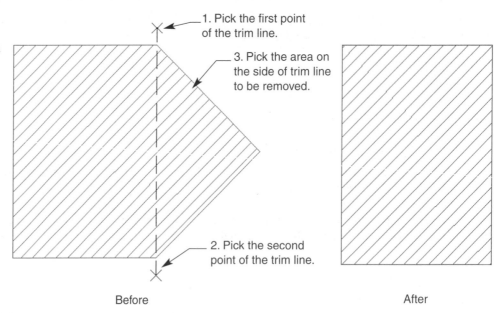

FIGURE 24.17 Establish a trimming line across an area, then select the side of the area that will be removed.

1. Pick the first point of the trim line.

3. Pick the area on the side of trim line to be removed.

2. Pick the second point of the trim line.

Before

After

Dividing Areas

Dividing an area is similar to trimming the area except that the line established divides the area into two pieces. Use one of the following options to access this command:

✓ Select the **Documentation** pull-down menu, pick the **Areas** cascade menu, then select the **Divide** command.

✓ Pick the **Divide** button in the **Areas** toolbar.

✓ Right-click, and select **Documentation, Areas, Divide** from the cursor menu.

✓ Type **AECAREADIVIDE** at the Command: prompt.

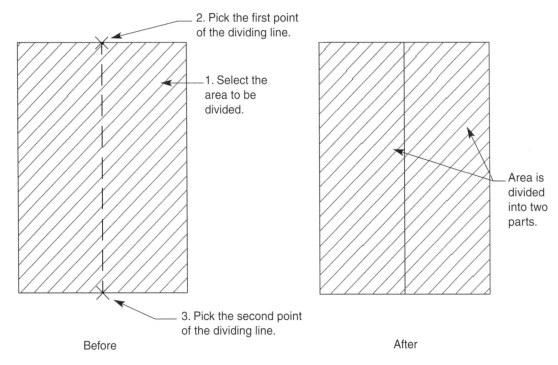

FIGURE 24.18   Using the Divide command to divide an area into two parts.

✓ Select the area, then right-click, select the **Operation** cascade menu, then pick the **Divide** command from the shortcut menu.

Pick the area to be divided. Next, pick the first point of the dividing line, then select the second point. The area is divided into two parts (see Figure 24.18).

Removing Area Rings

When composed areas are created, multiple rings are added to the area. These rings may consist of additive and subtractive areas. After a composed area has been created, you may later decide that a ring needs to be removed. Use one of the following options to access the **Remove Ring** command:

✓ Type **AREAREMOVERING** at the Command: prompt.

✓ Select the area then right-click. Pick the **Edit Rings** cascade menu then select the **Remove Ring** command from the shortcut menu.

You are prompted to pick on a ring to remove. Select the boundary or the hatch of the area that needs to be removed. The ring is removed from the boundary, and its space is filled in with the main area's additive or subtractive hatch.

Reversing Area Rings

It is also possible to change the additive or subtractive nature of a ring. Suppose that during the process of adding composed areas you forgot to change an internal ring from an additive state to a subtractive state. Instead of erasing the area then recreating it, you can reverse the internal area to be a subtractive ring. The **Reverse Ring** command works for reversing both additive or subtractive rings to the opposite state. Use one of the following options to access this command:

✓ Type **AREAREVERSERING** at the Command: prompt.

✓ Select the area, then right-click. Pick the **Edit Rings** cascade menu, then select the **Reverse Ring** command from the shortcut menu.

You are prompted to pick on top of the ring area whose state is to be reversed. The selected ring's state is immediately reversed.

Reversing all Rings in an Area

Similar to the Reverse Ring command, the **Reverse Profile** command reverses the state of an entire area object, transforming additive rings into subtractive rings, and subtractive rings into additive rings. Use one of the following options to access this command:

    ✓ Type **AREAREVERSEPROFILE** at the Command: prompt.

    ✓ Select the area, then right-click. Pick the **Edit Rings** cascade menu, then select the **Reverse Profile** command from the shortcut menu.

Converting an Area to a Polyline

If you are using areas as a space planning tool, you may later want to convert the area into walls, slabs, spaces, or roofs. In order to be converted into one of these types of objects a polyline must exist in the drawing. You can convert the area object into a polyline by using the **Create Polyline** command. Use one of the following options to access this command:

    ✓ Select the **Documentation** pull-down menu, pick the **Areas** cascade menu, then select the **Create Polyline** command.

    ✓ Pick the **Create Polyline** button in the Areas toolbar.

    ✓ Right-click, and select **Documentation, Areas, Create Polyline** from the cursor menu.

    ✓ Type **AECAREACREATEPLINE** at the Command: prompt.

You are promted to pick the area(s) that will generate a polyline. When an area is selected, every ring in the area generates a polyline.

Area Group Basics

Areas in a drawing represent an individual area or part of the building. Often these areas are smaller parts of a larger organization or group. Areas can be grouped together into a container known as an ***area group***. Area groups allow you to organize individual areas into a set of structured groups. For example, several individual office areas may be grouped together into a department group, which may be grouped together with other area groups to form a company group (see Figure 24.19).

Area groups are similar to the mass group marker or the slice markers described in Chapter 22. When the area group is added to the drawing, a marker representing the group is drawn (see Figure 24.19). The marker includes the area group name beside the marker for easy identification. Areas are then attached to the different groups, forming a hierarchical structure for the total area. Later these structured groups can be used to create an area evaluation of the floor plan. This section discusses the basics of adding an area group and attaching areas to the group. Later in this chapter other methods of organizing and adjusting the area group will be discussed.

Adding an Area Group

Area group markers can be added at any time during the development of an area plan by using the **Add Area Group** command. Use one of the following options to access this command:

    ✓ Select the **Documentation** pull-down menu, pick the **Area Groups** cascade menu, then select the **Add Area Group...** command.

    ✓ Pick the **Add Area Group** button in the **Area Groups** toolbar.

    ✓ Right-click, and select **Documentation, Area Groups, Add Area Group...** from the cursor menu.

    ✓ Type **AECAREAGROUPADD** at the Command: prompt.

The **Add Area Group** dialog box is displayed (see Figure 24.20). Enter a name for the area group and select a style from the Style: drop-down list. After setting the properties for the area group, pick a loca-

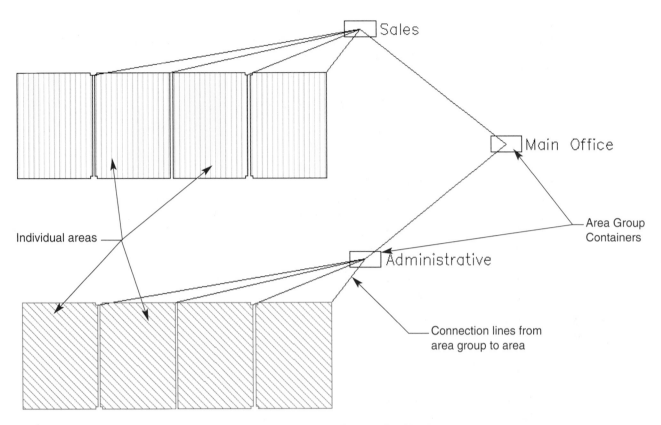

FIGURE 24.19   Areas can be added to area group containers (markers), which in turn can be added to other area groups.

FIGURE 24.20   The Add Area Group dialog box is used to add area group markers to the drawing.

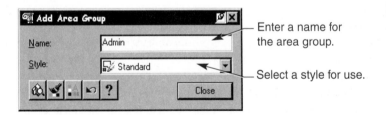

tion in the drawing to locate the area group marker. Continue to add as many markers as needed, accessing the Add Area Group dialog box when you need to change the name of the group or the style.

Attaching a Group to an Area

Once you have added an area group marker adding to the drawing, you can attach individual areas to the marker, which allows you to organize the areas into groups. One method of attaching an area to a group is through the Area Modify dialog box discussed previously. Select the area that you would like to attach to an area group, right-click, and select the Area Modify... command. Pick on the Area Group Viewer button to display the area group viewer (see Figure 24.21).

This dialog box displays all the area groups in the drawing with a check box beside each group name. Place a check mark in each box to which the area is to belong to attach the area to the area group. Press the Apply button to apply the changes to the areas, then press Close to exit of the dialog box.

When an area has been added to an area group, an outline is drawn around each area within the group. This allows you to select the area group marker and to see the areas that are attached to the group.

Another way of attaching an area to a group is to select the area to be attached, right-click, then select the **Attach to Group** command. You will then be prompted to select area groups. Select one or more group markers to which to attach the area. When you have finished press the [Enter] key to attach the areas to the selected groups.

FIGURE 24.21 The Area Group Viewer dialog box displays a list of area groups in the current drawing. Place a check in each group to which the area is to belong.

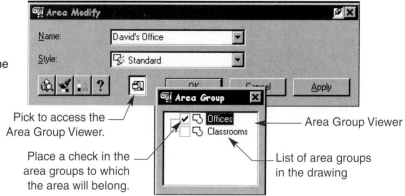

Pick to access the Area Group Viewer.

Place a check in the area groups to which the area will belong.

Area Group Viewer

List of area groups in the drawing

Field Notes

> When you use the Area Modify dialog box to attach an area to an area group, you can attach only one area at a time. If you want to attach several areas at once to an area group, select the areas, right-click, and pick the Attach to Group command from the shortcut menu. Pick the area group(s) to which the areas will be attached, then pick the [Enter] key to end the command.

Detaching a Group from an Area

In addition to attaching areas to area groups, you can detach areas from an area group or groups. Select the Area Group Viewer button in the Area Modify dialog box. If the area is attached to a group, there will be a check mark beside the group name. Clear the check mark to detach the area from the area group.

Another method of detaching the area from an area group is to select the area, right-click, then select the **Detach from Group** command. You will be prompted to select area groups. Pick on top of the area group markers under which you do not want the areas grouped to detach the area from the group.

Field Notes

> Depending on where you select the area, you may select the area or the area group. To overcome this problem and to pick the area object, use object cycling to cycle between the group and the area. You can also window around the area object, ensuring that you do not window around the area group marker, to select the object.

Exercise 24.4

1. Open EX24-3.dwg.
2. Add three area group markers. Name the groups using the following names: Classrooms, Offices, Hallways.
3. Attach the offices to the Offices area group.
4. Attach the two classrooms to the Classrooms area group.
5. Attach the Hallway to the Hallways area group.
6. Save the drawing as EX24-4.dwg.

Area Calculations

Calculations are often used in conjunction with areas. Often the purpose of obtaining an area is to process it through a formula in order to obtain a modified value. For example, suppose that when a site plan is created, 10% of the total area needs to be landscaped. A calculation modifier can be applied to the site plan area to determine the value of 10% of the area.

As calculation formulas are developed they can be applied to areas or to area groups. After the formulas are created and applied, an area evaluation report can be generated as part of the construction document set. Area evaluation reports are discussed later in this chapter.

Creating a Calculation Modifier Style

The calculation modifier formula can be applied to the total area, to the perimeter, or to both. The formula can be applied to an individual area or an area group. If the calculation modifier is applied to an area group, it affects all areas attached to that group. Use one of the following options to access the **Calculation Modifier Styles** command:

✓ Select the **Documentation** pull-down menu, pick the **Areas** cascade menu, then select the **Calculation Modifier Styles…** command.

✓ Pick the **Calculation Modifier Styles** button in the **Areas** toolbar.

✓ Right-click, and select **Documentation, Areas, Calculation Modifier Styles…** from the cursor menu.

✓ Type **AECAREACALCULATIONMODIFIERSTYLE** at the Command: prompt.

The **Style Manager** dialog box is displayed opened to the Area Calculation Modifier Styles section (see Figure 24.22). As with other AEC styles, pick the New Style button to create a new style. You can also right-click on the Area Calculation Modifier Styles section icon or in the right pane to display a shortcut menu where you can select New to create a new style. Enter a name for the calculation modifier style.

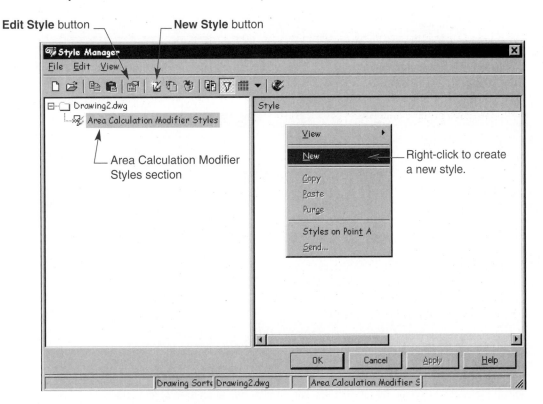

FIGURE 24.22   The Style Manager window is used to create an area calculation modifier style.

To edit the new style, highlight the style and select the Edit Style button, right-click on top of the new style name, or double-click on top of the style. The **Area Calculation Modifier Style Properties** dialog box is displayed. This dialog box includes the following tabs:

- **General**   This tab allows you to rename the style, provide a description, and attach notes.

- **Apply To**   This tab is used to specify the part of the area object to which the formula will apply (see Figure 24.23). Select the component to which you would like to attach the formula.

- **Definition**   This tab is used to enter a formula that modifies the component selected in the Apply To tab (see Figure 24.24). This tab has three main areas. The **Parameters** area allows you to set up variables that can be used in the formula. The **Formula** area is where the formula to be applied to the selected component is written. The **Test** area is where you can test the formula before applying it to an actual area object.

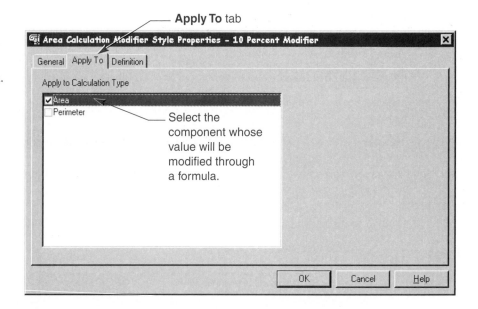

FIGURE 24.23 The Apply To tab is used to specify the component that will be used in the calculation modifier formula.

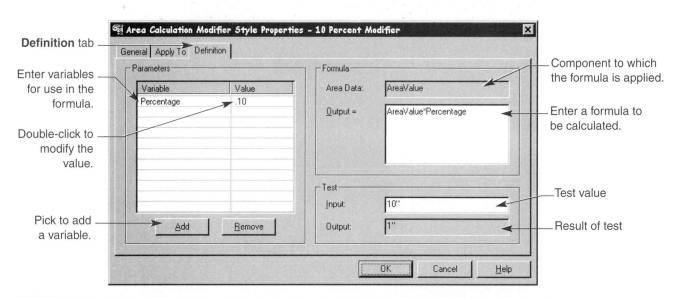

FIGURE 24.24 The Definition tab is used to write the formula that will be attached to an area object and will modify the value of the component to which it is applied.

Creating a Formula

After you have selected the component to which the modifier will be applied, you can create the formula. In the Definition tab you can create variables that will be applied to the formula. For example, if you are going to multiply the area by a percentage, you can create a Percentage variable and assign it a value (see Figure 24.24). Use the Add button to add a variable. Enter the name in the space provided. To assign the value to the variable, double-click in the Value space beside the variable name and enter a value for the variable.

After you have defined all your variables, you can write the formula. At the top of the Formula area the Area Data field lists the component variable that was selected in the Apply To tab. These variables *must* be a part of the formula. The following component variables can be listed in the Area Data field:

- **AreaValue** This variable is displayed if the calculation is being applied only to the area component.

- **PerimeterValue** This variable is displayed if the calculation is being applied only to the perimeter component.

- **Value** This variable is available if the calculation is being applied to both the area and the perimeter components.

Write the formula in the Output = area. If you are calculating the landscape area mentioned previously, then you may write the formula as AreaValue*Percentage. When applied to an area or area group object, the calculation modifier multiplies the total area by the percentage variable, producing the modified area value. You can test the formula by entering a value in the Input text box in the Test area. Figure 24.24 shows this formula being used and producing the correct result: 10% of a 10″ area is 1″. Although the input and output values are displayed using the current drawing units notation, the idea is correct, and the formula works.

If you decide to not use variables, you can write actual values in the Output = area. The preceding formula can be written as AreaValue* 10. The formula must use the Area Data variable in the formula, and the variable must be entered as it is displayed. For example, writing Area value*.10 makes the formula invalid.

You can write complex formulas, but the following characters are not allowed:

- Special characters such as ?, $, (), - cannot be used, but the underscore (_) can.

- Spaces are not allowed.

- Diacritical marks such as å, ö, ê, Ý cannot be used.

Some examples of calculation modifier styles follow:

Calculation Modifier Name	Description	Applies To	Variables	Value	Formula
Number of Bricks	Calculates total number of bricks with 5% spoilage. Area of one brick=.2083	Areas	OneBrick	.2083	AreaValue/OneBrick*.05+AreaValue
Landscape Area	Calculates 10% of an area	Areas	Percentage	.10	AreaValue*Percentage
Baseboard Cost	Calculates the cost of baseboards applied to the perimeter of an area	Perimeter	Cost	2.47	PerimeterValue*Cost
Concrete Volume and Cost	Calculates the volume of an 8″ high concrete area × $55 per cubic foot	Area	Cost Height	55 8	AreaValue*Height*Cost
Plaster Deductions	Calculates a 3% Plaster deduction along an area	Area	Plaster_Deduction	.97	AreaValue*Plaster_Deduction
Building Cost	Calculates the area × $165 per square foot	Area	Cost	165	AreaValue*Cost

Field Notes

> The value for one brick was established by multiplying the size of a 3-3/4″ × 8″ brick: 3.75 × 8 = 30 sq in. ÷ 144 (1 sq ft) = .20833. The area of a piece of 4″ × 4″ tile can be calculated the same way: 4 × 4 = 16 sq in. ÷ 144 = .111.

Field Notes

> When variables such as cost and number totals are calculated, the values are output as inches (or millimeters if using decimal units). The output 148′ 6-1/8″, for example, in a cost output would translate to $148.50. If the output was for the total number of tiles, you would round the value up to 149 tiles.

After the calculation modifier style has been created, it can be attached to an area, area group, or area group template. Attaching the calculation to area groups and area group templates will be discussed later in this chapter. Attaching the calculation modifier to an area is discussed in the next section.

Exercise 24.5

1. Open EX24-4.dwg.
2. Create an area calculation modifier style that estimates the total number of 8″ × 8″ floor tiles required to cover a floor.
3. Create an area calculation modifier style that multiplies the area by $1.47.
4. Save the drawing as EX24-5.dwg.

Using Area Properties

Modification of an area object was discussed earlier. The Area Modify dialog box was used to change the name, style, and group for a selected area. In addition to these properties, the attachment of area calculation modifier styles to the area can be controlled through the **Area Properties** dialog box. Use one of the following options to access the Area Properties dialog box:

✓ Select an area object, right-click, and select Area Properties… from the shortcut menu.

✓ Type **AreaProps** at the Command: prompt.

✓ Access the **Area Modify** dialog box, then select the Properties button.

This dialog box includes the following five tabs:

- **General**   This tab is used to rename the selected area, to provide a description for the area, and to attach notes and property set definitions to the area. A property set definition for area objects named AECArea can be found in the PropertySetDefsDACH.dwg in the \\Autodesk Architectural Desktop 3\ Content\Metric DACH\Schedules folder.

- **Style**   This tab displays the current area style assigned to the selected area. Choose a different style from the list to change the selected areas style.

- **Evaluation**   This tab is used to assign area calculation modifier styles to the area (see Figure 24.25). Use the Attach… button to attach a calculation modifier style

FIGURE 24.25 The Evaluation tab is used to add area calculation modifier styles to an area for processing.

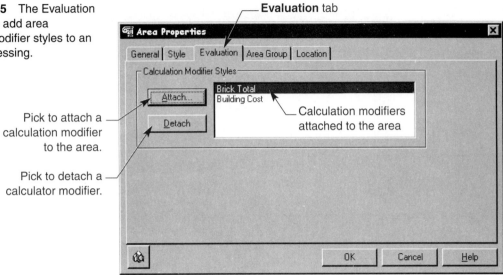

to the selected area. You can attach as many calculation modifiers to the area as needed. The order in which the modifiers are added will affect he outcome of the final area calculation.

When a calculation modifier is attached to an area, the modifier works through the formula determining the modified area result. If another modifier has been attached, the modified area result of the first calculation is used in the next formula. For example, if two modifiers are applied to an area, style A calculates the 10% landscape area of a 500′ x 800′ building site, and style B calculates the cost at $45.00 per sq ft for the landscaped area, the order in which the modifiers are applied affects the results of the calculation.

Style A attached before style B:

Style A: (AreaValue=500′×800′=400000) × .10 = 40000′-0″ (40,000 landscaped sq ft (modified area))

Style B: 40000 (modified area) × 45 = 1800000'-0" ($1,800,000.00)

Style B attached before Style A:

Style B: (AreaValue=500′×800′=400000) × 45 = 18000000′-0″ ($18,000,000 (modified area))

Style A: 18000000 (modified area) × .10 = 1800000′-0″ (1,800,000 landscaped sq ft)

- **Area Group** This tab is used to attach the selected area to an area group (see Figure 24.26). This dialog box is similar to the Area Group Viewer found in the Area Modify dialog box. Place a check next to the group to which the selected area will be attached. If more than one area is selected, this tab will not be available in the Area Properties dialog box.

- **Location** This tab is used to specify where the area object is located in the World coordinate system.

After you have assigned the calculation modifiers and groups to the areas, you can process an area evaluation report. This report breaks down the formulas as "proofs" into a text file or a spreadsheet. The creation of an evaluation report will be discussed later in this chapter.

FIGURE 24.26 The Area
Group tab is used to assign the
selected area to an area group.

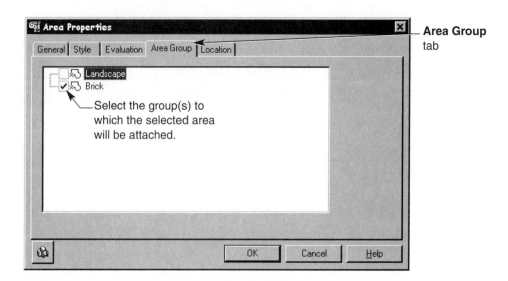

Area Group
tab

Select the group(s) to
which the selected area
will be attached.

Exercise 24.6

1. Open EX24-5.dwg.
2. Attach the calculation modifier that evaluates the total number of floor tiles.
3. Attach the calculation modifier that evaluates the cost of the tiles.
4. Save the drawing as EX24-6.dwg.

 ## Area Decomposition

In some building jurisdictions, an important part of an area evaluation is a visual breakdown of the areas
into subdivisions. This breakdown is also known as an ***area decomposition***. In some cases a diagram of the
decomposition is required for building approval (see Figure 24.27).

There are three types of decomposition available: Trapezoid, Triangle (Overlap), and Triangle. You
can adjust the display settings for each of these types by modifying the Decomposed display representation
for the area object. Although the Decomposed display representation can be modified, it will not be displayed in any of the standard display configurations in the AEC Arch (Imperial or metric) or the AEC Arch
(Imperial - Int) templates.

In order to view the decomposition of an area, you can turn on the display representation in the
Plot_Plan_Large and Plot_Plan_Small display sets. Another way of previewing the decomposition is to
import a display configuration named Proof of Areas, found in the AEC Arch (metric) DACH.dwt.
Importing and exporting display configurations was discussed in Chapter 5 of this text. Once the display
configuration has been imported, or the Decomposed display representation is turned on in one of your display sets, the decomposition can be displayed.

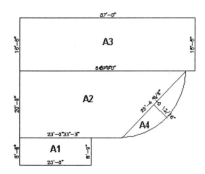

Trapezoid decomposition

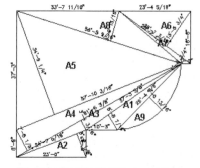

Triangle (overlap) decomposition

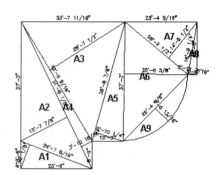

Triangle decomposition

FIGURE 24.27   Different area decomposition diagrams.

Editing the Decomposition Display

The display of the decomposed area is controlled through entity display. First, select the area object with the decomposition you intend to modify, right-click, and select Entity Display.... The **Entity Display** dialog box, shown in Figure 24.28, is displayed. Ensure that the Decomposed display representation is selected in the display representation drop-down list, select the level of display control in the Property Source column, attach an override, and pick the Edit Display Props...button. The Entity Properties dialog box is displayed. This dialog box contains the following tabs:

- **Layer/Color/Linetype** Use this tab to control the display of the different components within the decomposition.

- **Decomposition** This tab is used to specify a decomposition type to be displayed (see Figure 24.29). These include the Trapezoid, Triangle (overlap) and the Triangle types. The Explode Result area specifies the type of entities the decomposition will turn into if exploded.

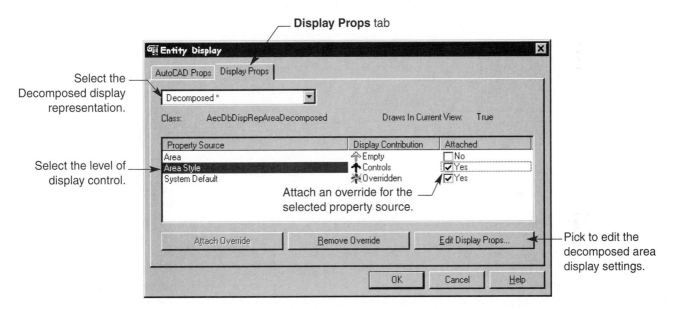

FIGURE 24.28 The Entity Display dialog box is used to modify the display of the decomposed area.

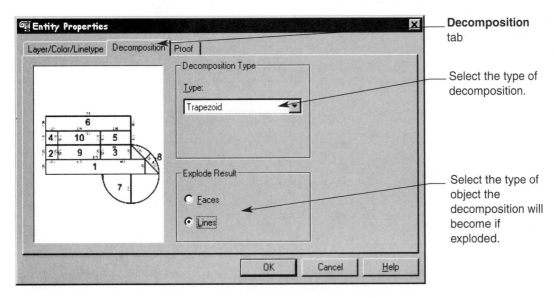

FIGURE 24.29 The Decomposition tab is used to control the type of decomposition to be displayed.

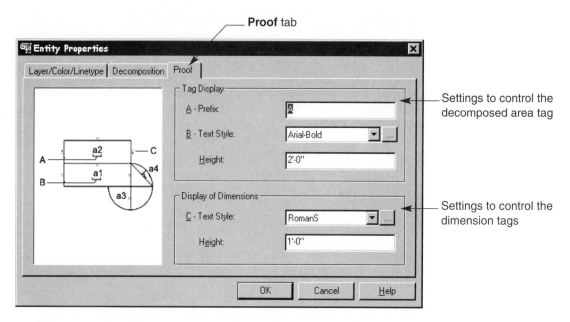

FIGURE 24.30   The Proof tab is used to control how the decomposed area tags and dimension tags will be displayed.

■ **Proof**

This tab is used to configure how the triangular or trapezoidal area tag and the dimension tags will be displayed (see Figure 24.30). When an area is broken up, it is decomposed into different subdivision areas. The subdivision is assigned a tag for easy identification of each decomposed subdivision area. The A - Prefix text box allows you to specify a prefix for the decomposed area tag name. You can choose a text style for the tag and enter a text height in the Tag Display area. The Display of Dimensions area can be used to control the dimension tags along each edge of a decomposed area. The text style and the height values can be modified.

Once these settings have been adjusted, the decomposed area will be displayed.

 Area Groups Advanced

As defined earlier, an area group is a container for areas and other area groups that are grouped together in a hierarchical or group structure (see Figure 24.31). Generally, an area represents the lowest level of the group structure, whereas area groups organize different levels for an entire building or business.

The following are a few basic steps in organizing a group structure:

1. Decide on all the groups and subgroups needed to complete the group structure.

2. Decide how the groups and subgroups are to relate to one another. Organize them logically.

3. Create the area group markers.

4. Create the group structure by attaching groups to subgroups.

Creating an Area Group Style

Before you begin adding area group markers to the drawing, you may wish to create area group styles. By creating styles you are able to configure how the total area of the group is to appear in the drawing. Initially, when an area group is attached to one or more areas, an outline is applied around each area. This outline or boundary can be assigned a color and linetype as well as have a hatch applied to its interior.

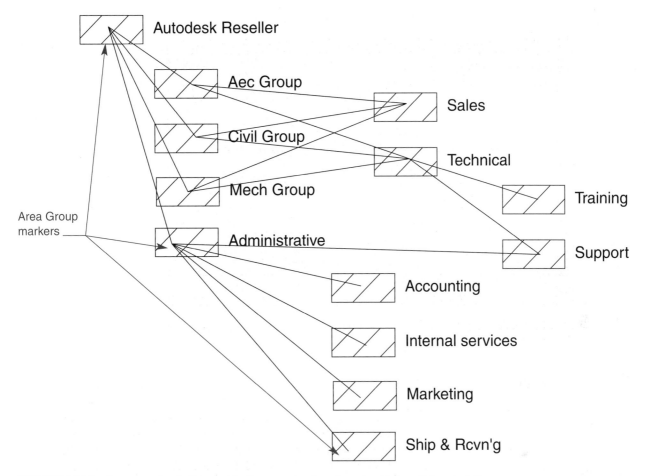

FIGURE 24.31   An example of a hierarchical group structure for a business within a building.

Use one of the following options to access the Area Group Styles command:

✓ Select the **Documentation** pull-down menu, pick the **Area Groups** cascade menu, then select the **Area Group Styles...** command.

✓ Pick the **Area Group Styles** button in the **Area Groups** toolbar.

✓ Right-click, and select **Documentation, Area Groups, Area Group Styles...** from the cursor menu.

✓ Type **AECAREAGROUPSTYLE** at the Command: prompt.

The Style Manager dialog box is displayed opened to the Area Group Styles section (see Figure 24.32). Create a new area group style by selecting the New Style button, or right-clicking on the Area Group Style icon or in the right window pane, and selecting the New command from the shortcut menu. Enter a name for the area group, then double-click on the new area group style, highlight the new style and pick the Edit Style button, or right-click on top of the new style and select the Edit... option from the shortcut menu.

The **Area Group Style Properties** dialog box, shown in Figure 24.33, is displayed. This dialog box includes two tabs: General and Display Props. The General tab is used to rename the style, to provide a description to the style, and to add notes and property set definitions to the style.

The Display Props tab is used to configure the display properties for the area group. The display representations drop-down list includes three separate display representations: Plan, Plan 1-100, and Plan 1-50. The plan representation is typically used in the Plot and Work display configurations. The Plan 1-100 display representation is typically used for the Small Scale display configuration, and the Plan 1-100 display representation is typically used in the Large Scale display configuration. Display configurations were discussed in Chapter 5.

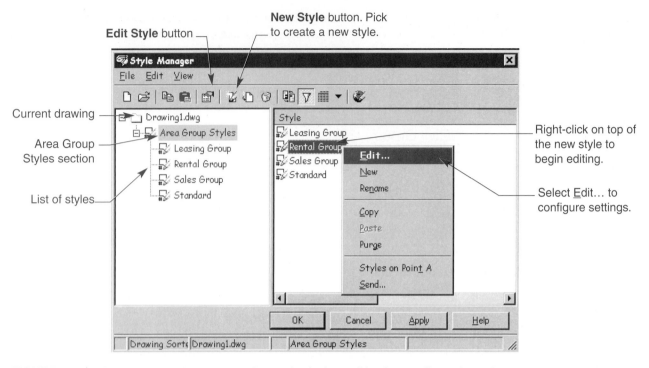

FIGURE 24.32   Creating a new area group style, naming it, then editing it to configure the style.

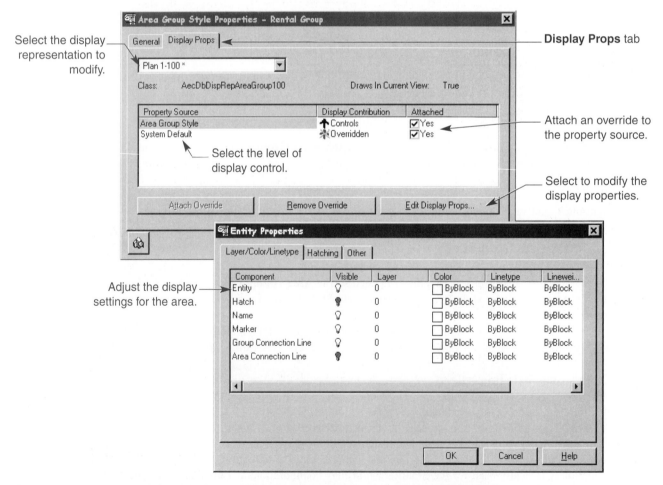

FIGURE 24.33   The Area Group Style Properties dialog box is used to configure the settings for an area group. The Entity Properties dialog box is used to adjust the display for the area.

After you have selected the display representation to be modified, pick the level of display control in the Property Source column. Because you are probably creating area group styles for each group marker in the drawing, select the Area Group Style property source. Attach an override, then pick the Edit Display Props... button. The Entity Properties dialog box, shown in Figure 24.33, is displayed.

The Entity Properties dialog box includes the following three tabs:

■ **Layer/Color/ Linetype**	This tab includes a list of components for the area group marker. Adjust the component's display, color, and linetype as desired. The following components control the display of different parts of the area group marker.
■ **Entity**	This component controls the display of the outline around the areas that are attached to the area group.
■ **Hatch**	This component turns the hatch inside the marker on or off. The hatch is also applied to the inside of the entity boundary.
■ **Name**	This component turns on the name assigned to the area group marker.
■ **Marker**	This tab turns the area group marker box on or off.
■ **Group Connection Line**	This component is a line that is drawn between area groups that are attached to one another.
■ **Area Connection Line**	By default this component is turned off. This component draws a line from the area group to the area(s) that are attached.
■ **Hatch**	If the hatch component has been turned on, this tab will control the display of the hatch pattern. Assigning a different hatch to each area group style helps identify the individual areas attached to each group.
■ **Other**	This tab controls additional properties for the area group marker (see Figure 24.34). The General area is used to set a size (width and height) for the rectangular marker. Selecting the Use Display Scale check box adjusts the size of the marker box relative to the current drawing scale. The text used for the area group name can also be assigned a text style and height. If the Draw All Areas check box is selected, the hatch pattern for the area group will be attached down through the area group structure to the area. Consequently, multiple hatch patterns can be created within one

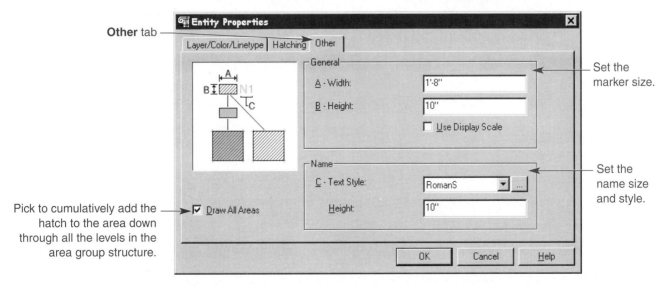

FIGURE 24.34 The Other tab is used to adjust additional properties for the area group marker.

area depending on how many levels of area groups the area is attached to. If this check box is cleared, only the first area group level hatch is applied to the areas that are attached.

When you have finished configuring the area group style, press the OK button to exit all the dialog boxes. A new area group using this style will now be available.

Attaching and Detaching Area Groups

After an area group style has been created, it can be attached to area objects and other area group markers. The group structure can be created by attaching groups to one another. Figure 24.31 shows area group markers attached to other group markers. An area group marker can be attached to any number of areas or other groups to form subgroups or be attached to other higher level groups.

One method of attaching an area group is to pick on the area group, then select the areas or groups that will form subgroups to the selected area group. Use one of the following options to access the **Attach** command:

✓ Select the **Documentation** pull-down menu, pick the **Area Groups** cascade menu, then select the **Attach** command.

✓ Pick the **Attach** button in the **Area Groups** toolbar.

✓ Right-click, and select **Documentation, Area Groups, Attach** from the cursor menu.

✓ Type **AECAREAGROUPATTACH** at the Command: prompt.

✓ Select the area group marker, right-click, then pick the **Attach Areas/Area Groups** command from the shortcut menu.

You are prompted to select first the area group that will be the higher level group, then to pick the subgroups and areas that are to be attached to the higher level group. The following prompt sequence is used to attach area groups and areas to the higher level group in Figure 24.35:

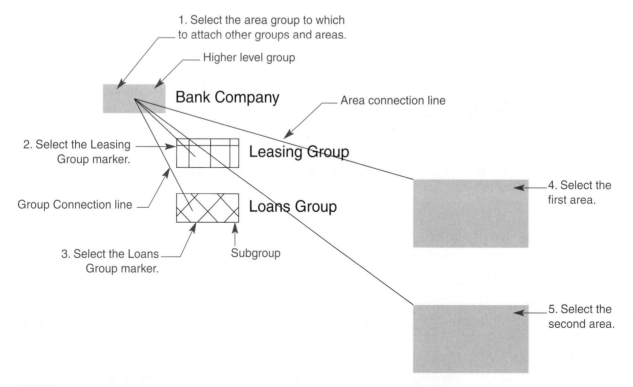

FIGURE 24.35   Attaching the Bank Company area group to other subgroups and areas.

Command: **AECAREAGROUPATTACH** ↵
Select Area Groups: *(select the Bank Company group)*
1 found
Select Area Groups: ↵
Select Areas and/or Area Groups to Attach: *(select the Leasing Company group)*
1 found
Select Areas and/or Area Groups to Attach: *(select the Loans Company group)*
1 found, 2 total
Select Areas and/or Area Groups to Attach: *(select first area)*
1 found, 3 total
Select Areas and/or Area Groups to Attach: *(select second area)*
1 found, 4 total
Select Areas and/or Area Groups to Attach: ↵
Attaching objects...
4 object(s) attached.
Command:

 Field Notes

> Depending on the display settings set up in the area group style, connector lines will be drawn
> between the higher level group and the groups and areas that are attached to it.

As you develop and arrange the area group structure you may decide that groups and areas that were
attached earlier no longer need to be attached to a particular group. In this case you can detach area groups
and area objects from a higher level group. Use one of the following options to access the **Detach** command:

✓ Select the **Documentation** pull-down menu, pick the **Area Groups** cascade menu, then
  select the **Detach** command.

✓ Pick the **Detach** button in the **Area Groups** toolbar.

✓ Right-click, and select **Documentation, Area Groups, Detach** from the cursor menu.

✓ Type **AECAREAGROUPDETACH** at the Command: prompt.

✓ Select the area group marker, right-click, then pick the **Detach Areas/Area Groups** command from the shortcut menu.

As with the Attach command, you are prompted to select first the higher level group, then to pick the
area groups and areas that will be detached. The following sequence is used to detach objects from a group.

Command: **AECAREAGROUPDETACH** ↵
Select Area Groups: *(select the higher level group marker)*
1 found
Select Area Groups: ↵
Select Areas and/or Area Groups to Detach: *(pick the area group or the areas to detach; you
    may need to use object cycling ([Ctrl]+ pick) to select an attached area)* <Cycle on> <Cycle
    off>
1 found
Select Areas and/or Area Groups to Detach: *(pick the area group or the areas to detach* <Cycle
    on>   <Cycle off>
1 found, 2 total
Select Areas and/or Area Groups to Detach: ↵
Detaching objects...
2 object(s) detached.
Command:

Editing Area Groups

After an area group has been added to the drawing, the area group name and style can be modified with the **Area
Group Modify** dialog box. Use one of the following techniques to access the **Modify Area Group** command:

FIGURE 24.36 The Area
Group Modify dialog box is used
to rename the selected area
group or to assign a different
area group style.

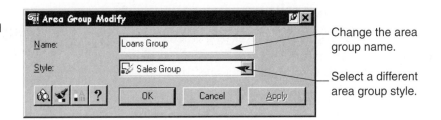

Change the area
group name.

Select a different
area group style.

✓ Select the **Documentation** pull-down menu, pick the **Area Groups** cascade menu, then
   select the **Modify Area Group...** command.

✓ Pick the **Modify Area Group** button in the **Area Groups** toolbar.

✓ Right-click, and select **Documentation, Area Groups, Modify Area Group...** from the
   cursor menu.

✓ Type **AECAREAGROUPMODIFY** at the Command: prompt.

✓ Select the area group marker, right-click, then pick the **Area Group Modify...** command
   from the shortcut menu.

The Area Group Modify dialog box, shown in Figure 24.36, is displayed. You can enter a new name
in the Name: text box, and you can change the area group style by selecting a new style from the Style:
drop-down list.

Modifying Additional Properties

Additional properties such as the attachment of calculation modifiers and the management of the group
structure can be edited through the Area Group Properties dialog box. To access this dialog box, select one
of the following options:

✓ Select the area group marker, right-click, and select **Area Group Properties...** from the
   shortcut menu.

✓ Type **AREAGROUPPROPS** at the Command: prompt.

✓ Enter the **Area Group Modify** dialog box and select the **Properties** button.

The Area Group Properties dialog box is displayed. This dialog box includes the following seven
tabs:

■ **General**   This tab is used to rename the selected area group, to enter a description for the
             group, and to add notes and property set definitions to the group. A property set
             definition named AecAreaGroup is included with Architectural Desktop and can
             be found in the \Autodesk Architectural Desktop3\Content\Metric DACH\
             Schedules folder in the PropertySetDefsDACH drawing.

■ **Style**   This tab can be used to change the style used by the selected area group.

■ **Content**   This tab is used to control how the area group will behave (see Figure 24.37).
             The Area Name Def: drop-down list allows you to associate the area group with
             an area name definition. The area name definition includes a list of areas names
             that can be assigned to an area object. Area name definitions will be discussed
             later in this chapter.

             The Contents area specifies the type of object that can be attached as a subgroup.
             This area includes an Area check box and a Group check box. The Access
             Rights area allows you to control the access rights for the selected group. If the
             Lock Group check box is selected, this group cannot be repositioned within the
             group structure or deleted. If the Lock Sub Group check box is selected, the sub-
             groups to the selected area group cannot be repositioned or deleted.

Content Tab

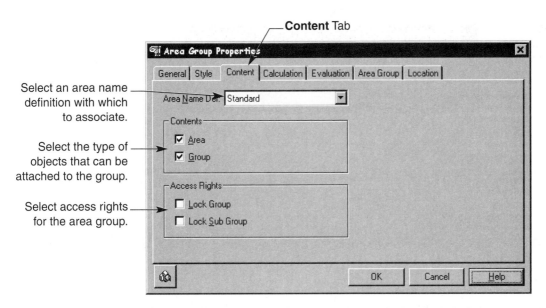

FIGURE 24.37   The Contents tab is used to control properties of the selected area group.

- **Calculation**   This tab is used to specify what parts of the total area attached to the area group can be processed by a calculation modifier (see Figure 24.38). The options include the Area and Perimeter check boxes.

- **Evaluation**   This tab allows you to assign a calculation modifier to the area group (see Figure 24.39). The totals for any areas or area groups attached to a higher level group are accumulated to obtain a total for the group. The Calculation modifier uses the total during its formula processing.

- **Area Group**   This tab is used to manage the selected group within the group hierarchical structure (see Figure 24.40). You can attach the area group being modified to other area groups within the structure to form subgroups by placing a check mark in the box beside an area group name. You can also detach area groups from other area groups by removing the check mark from a box.

Calculation tab

FIGURE 24.38   The Calculation tab specifies the components that can be used by a calculation modifier.

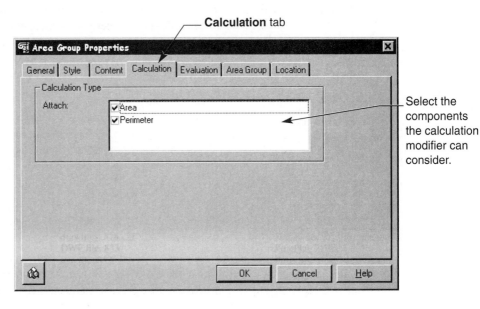

Select the components the calculation modifier can consider.

FIGURE 24.39 The Evaluation
tab is used to assign a
calculation modifier to the area
group that evaluates the total
area of the groups.

Pick to attach a
calculation modifier.

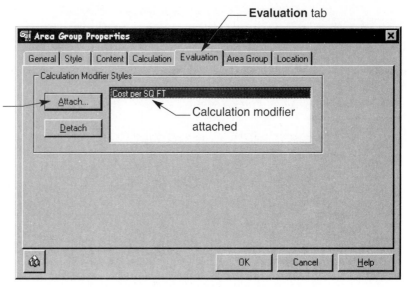

FIGURE 24.40 The Area
Group tab is used to organize
area groups within the group
hierarchical structure.

Select the area that will be
attached directly to the area
group being modified.

If a check box is applied
here, this subgroup will be
directly attached to the
group being edited.

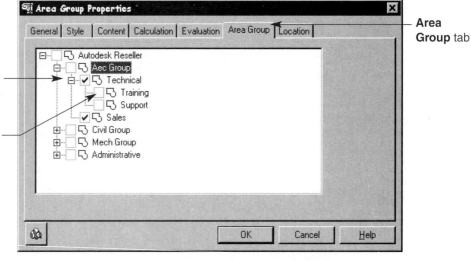

Field Notes

When organizing the group structure in the Area Group tab, you can attach the area group only
to other groups of the same level or to other subgroup levels. If the area group is to be a part of
a higher level group, then you need to edit the higher level group to include the subgroups.

Exercise 24.7

1. Open EX24-6.dwg.
2. Create four area group styles: Community College, Administrative, Instructional, and Common Ways.
3. Turn on the Hatch component for each style and assign a different hatch pattern and color to each style.
4. Add an area group marker named ADT University using the Community College style.
5. Modify the existing Offices area group marker to use the Administrative style. Attach it to the ADT University group.
6. Modify the existing Classrooms area group marker to use the Instructional style. Attach it to the ADT University group.
7. Modify the existing Hallways area group marker to use the Common Ways style. Attach it to the ADT University group.
8. Save the drawing as EX24-7.dwg.

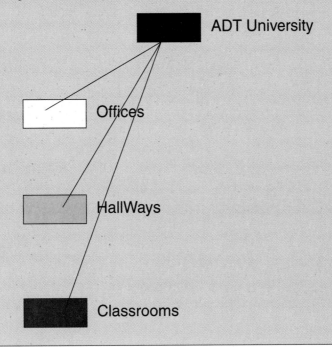

Creating a Group Layout

As area group markers are added and arranged in the drawing the hierarchical structure may become confusing (see Figure 24.41). The **Area Group Layout** command is used to sort the markers according to their logical/hierarchical positions. Use one of the following options to access this command:

✓ Select the **Documentation** pull-down menu, pick the **Area Groups** cascade menu, then select the **Area Group Layout...** command.

✓ Pick the **Area Group Layout** button in the **Area Groups** toolbar.

✓ Right-click, and select **Documentation, Area Groups, Area Group Layout...** from the cursor menu.

✓ Type **AECAREAGROUPLAYOUT** at the Command: prompt.

You are prompted to select area groups. Select the group that you would like to have organized. If you pick an area group that is a subgroup of a higher level group, then only the group markers of that subgroup will be organized. The **Area Group Layout** dialog box is displayed (see Figure 24.42).

FIGURE 24.41 (A) As area groups area added, attached, and organized the group structure may become difficult to review. (B) The same group structure after being organized using the Area Group Layout command.

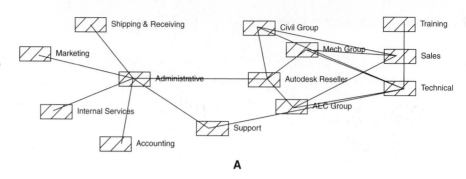

A

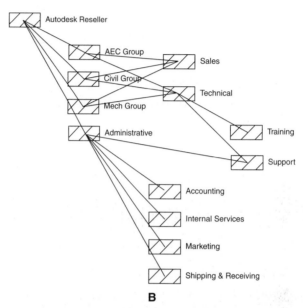

B

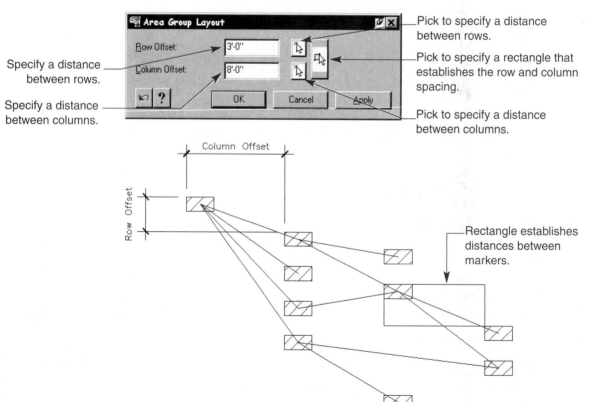

Pick to specify a distance between rows.

Pick to specify a rectangle that establishes the row and column spacing.

Pick to specify a distance between columns.

Specify a distance between rows.

Specify a distance between columns.

Rectangle establishes distances between markers.

FIGURE 24.42 The Area Group Layout dialog box allows you to specify the distances between the area group markers.

Use one of the methods in Figure 24.42 to establish the distances between group markers. The groups will be organized into alphabetical or numerical order in a hierarchical tree.

Creating a Polyline from a Group

The creation of a polyline from an area ring was described earlier in this chapter. Polylines can also be generated from all areas of a particular group. The polyline will be generated from the group boundary instead of each area boundary, which can be helpful if the polyline will be used to create walls, slabs, or roofs in the drawing. Use one of the following options to access the **Create Polyline** command:

✓ Select the **Documentation** pull-down menu, pick the **Area Groups** cascade menu, then select the **Create Polyline** command.

✓ Pick the **Create Polyline** button in the **Area Groups** toolbar.

✓ Right-click, and select **Documentation, Area Groups, Create Polyline** from the cursor menu.

✓ Type **AECAREAGROUPCREATEPLINE** at the Command: prompt.

You are prompted to select area groups. Select all the area groups from which you would like to generate polylines. You then have the option of creating the polylines from all the attached areas through the subgroups, from just the areas that are directly attached to the group, or from all the areas. The following prompt sequence is used to generate polylines:

```
Command: AECAREAGROUPCREATEPLINE ↵
Select area groups: (select an area group)
1 found
Select area groups: ↵
Create polylines from[All attached areas/Directly attached areas]<All>: ↵
1 polyline(s) created.
Command:
```

Area Group Templates

If you are creating area group structures that often use the same area group names, styles, and hierarchical organization, then you can create an area group template. The area group template allows you to establish common area group names, styles, and calculation modifiers that are attached to each level in the group structure, making it easier to add group structures to your drawings.

Creating Area Name Definitions

Before creating an area group template, you may want to establish all the area group styles and calculation modifiers that would typically get created and used in a group structure. After creating these styles you need to create the area name definition style.

The area name definition style is a list of possible names that can be applied to area objects. The list of names becomes available when an area is attached to an area group that is using the area definition style. The Area Group Properties dialog box was discussed earlier. Within this dialog box, the Content tab allows you to associate an area name definition style with a group, so that when an area is attached to the group the list of area names in the area name definition becomes available to choose from (see Figure 24.43).

Use one of the following options to access the **Area Name Definitions** command:

✓ Select the **Documentation** pull-down menu, pick the **Areas** cascade menu, then select the **Area Name Definitions...** command.

✓ Pick the **Area Name Definitions...** button in the **Areas** toolbar.

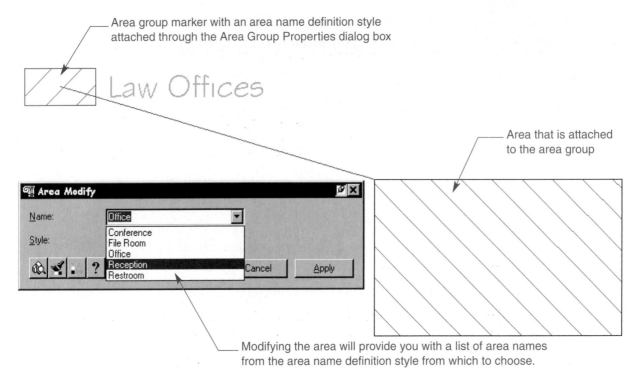

Area group marker with an area name definition style
attached through the Area Group Properties dialog box

Area that is attached
to the area group

Modifying the area will provide you with a list of area names
from the area name definition style from which to choose.

FIGURE 24.43   After an area group has been associated with an area name definition and has an area attached, editing the area will provide you with a list of area names from the area name definition style.

✓ Right-click, and select **Documentation, Areas, Area Name Definitions...** from the cursor menu.

✓ Type **AECAREANAMEDEF** at the Command: prompt.

The **Style Manager** dialog box is displayed, opened to the Area Name definitions section (see Figure 24.44). Pick the New Style button to create a new area name definition style. Enter a name for the style, highlight the new style, and select the Edit Style button, or right-click on top of the style, right-click, and select the Edit... option to edit the style.

Another way to edit the style is to double-click on top of the new style. The **Area Name Definition Properties** dialog box is displayed. This dialog box includes two tabs: **General** and **Content**. The General tab is used to rename the style, to provide a description for the style, and to add notes to the style.

FIGURE 24.44   The Style Manager dialog box opened to the Area Name Definitions section.

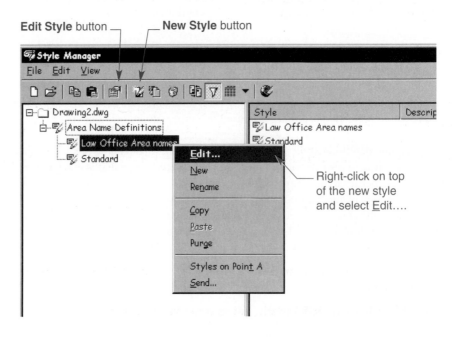

Content tab

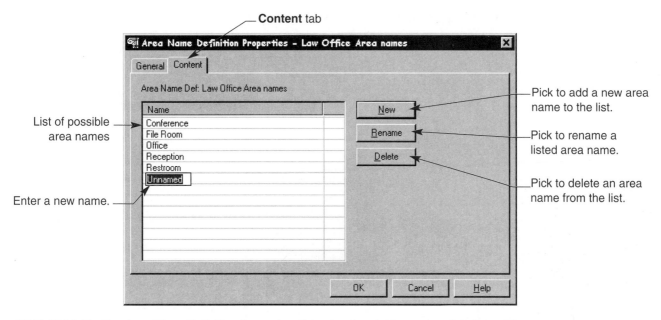

List of possible area names

Enter a new name.

Pick to add a new area name to the list.

Pick to rename a listed area name.

Pick to delete an area name from the list.

FIGURE 24.45   The Area Name Definition Properties dialog box is used to create a list of possible area names that can be chosen when editing an area that is attached to a group with an area name definition assigned.

The Content tab is used to add a list of area names that can be assigned to areas that are attached to area groups with the area name definition assigned (see Figure 24.45). Use the New button to add a possible area name to the list. Continue to pick the New button to add names. If you enter a name incorrectly or need to change a name, highlight the name in the list and pick the Rename button to rename it. You can also delete area names from the list by highlighting the desired name, then picking the Delete button.

Field Notes

Although an area name definition is often used in an area group template, it can also be assigned to individual groups, which then allows you to modify an area and choose the name for the area from the Name drop-down list (see Figure 24.43).

   Creating a Group Template

After you have created the area group styles, calculation modifiers, and area name definitions, you can assemble the area group template. The area group template can be used only to create a new group structure and does not maintain any ties to the group schedule. If the group structure is modified after it has been inserted, it will not affect the original template. Use one of the following options to access the **Area Group Templates** command:

✓ Select the **Documentation** pull-down menu, pick the **Area Groups** cascade menu, then select the **Area Group Templates...** command.

✓ Pick the **Area Group Templates** button in the **Area Groups** toolbar.

✓ Right-click, and select **Documentation, Area Groups, Area Group Templates...** from the cursor menu.

✓ Type **AECAREAGROUPTEMPLATE** at the Command: prompt.

FIGURE 24.46 Using the Style Manager to create a new area group template.

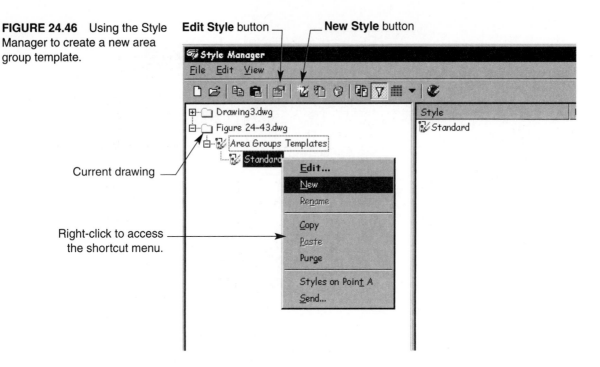

The Style Manager dialog box, opened to the Area Group Templates section, is displayed (see Figure 24.46). Create a new style using the New Style button, or right-click on the section icon and select <u>N</u>ew to create a new section. Enter a new name for the template. This name may reflect the type of group structure or the name of a company organization for easy identification of the style function.

After creating the new style, double-click on top of the style, highlight the style, and pick the Edit Style button, or right-click over the style and select the <u>E</u>dit... command to begin editing the template. The **Area Group Template Properties** dialog box is displayed.

This dialog box includes two tabs: General and Content. As in other style dialog boxes, the **General** tab is used to rename the style, to enter a description for the template, and to attach notes to the template. The **Content** tab is used to configure the layout of the group structure and to attach any properties to the different area group levels (see Figure 24.47).

The following are the different areas within the dialog box and the settings each controls:

- **Group Template**   This list box is used to create and organize the group structure. Initially, this list box includes one group level named "unnamed." Picking the unnamed group icon allows you to modify the properties assigned to the area group. Additional groups can be created by right-clicking over a group marker in the list box and selecting the New option. Enter a new name for the group level then pick it to begin editing.

    The group level you right-click on to create a new group level will determine the subgroups for the structure. As you create group sublevels, you can drag them from one area in the structure to another to form the structure. You can also hold down the [Ctrl] key and drag a group level marker to another part of the structure to copy it and its settings.

- **General**   This area includes the following four sections:

  - **Name**   This text box allows you to enter or rename a group level.

  - **Description**   This text box allows you to enter a description for your group structure.

  - **Style**   This drop-down list allows you to select an area group style from the styles available in the drawing to assign to the group level being edited.

  - **Area Name Def**   This drop-down list includes the name or names of area name definition styles in the drawing that can be attached to different groups. Recall

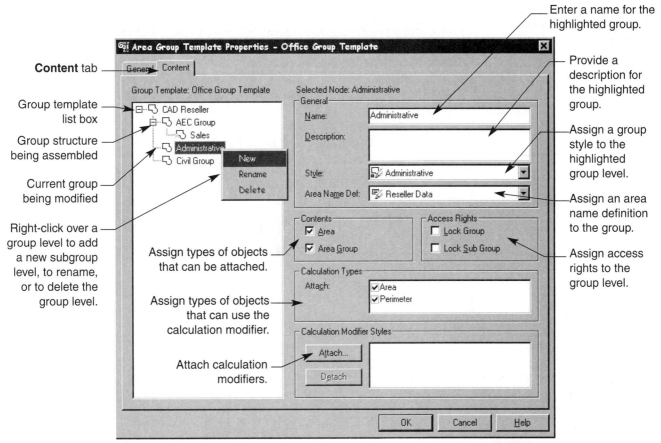

Enter a name for the highlighted group.

Provide a description for the highlighted group.

Assign a group style to the highlighted group level.

Assign an area name definition to the group.

Assign access rights to the group level.

Content tab

Group template list box

Group structure being assembled

Current group being modified

Right-click over a group level to add a new subgroup level, to rename, or to delete the group level.

Assign types of objects that can be attached.

Assign types of objects that can use the calculation modifier.

Attach calculation modifiers.

FIGURE 24.47   The Area Group Template Properties dialog box is used to configure a group structure template.

that all possible names for area objects will be listed if the area object is attached to the group and modified with the Area Modify dialog box.

- **Contents**   This area specifies the types of objects that can be attached to the group level being edited. The options include Area and Area Group.

- **Access Rights**   This area specifies any special access rights to the group level.

- **Calculation Types**   This area specifies the component of an area or area group to which the calculation modifier can be assigned.

- **Calculation Modifier Styles** This area allows you to attach a calculation modifier(s) to the area group being edited. Note that a small *x* is attached to the group marker, indicating that the calculation modifier has been assigned but it is not currently running the formula, as there are not any areas attached to the group markers.

Field Notes

Calculation modifiers accumulate through the structure. For example, if a 50% modifier is applied to a lower level group or area that is attached to a higher level area group with a 50% modifier, the end result will be a total that is 25% of the actual area instead of 50%.

When you have finished creating the structure and assigning the properties, you can add a new group structure to the drawing using this template. Figure 24.48 shows the finished group structure that can be used in this drawing or in other drawings if it is imported using the Style Manager dialog box.

FIGURE 24.48 The finished area group structure is ready to be created in the drawing.

Highlighted group marker indicates the group level being modified.

Assembled group structure

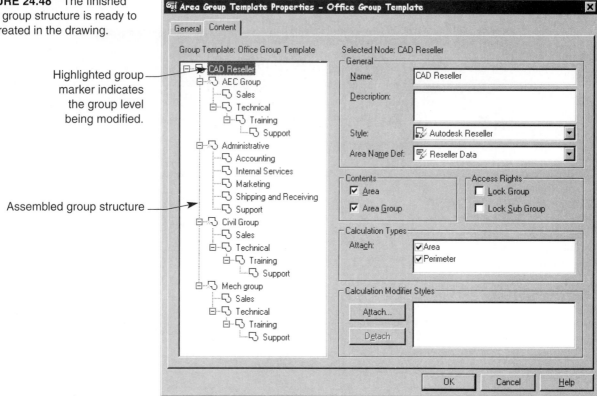

Creating an Area Group from a Template

Once the area group template has been created, it can be added to the drawing preassembled. When the group structure is added to the drawing, group connecting lines are drawn, and the group markers are ordered in the hierarchical order in which they appear in the Area Group Template Properties dialog box. Use one of the following options to access the **Create Area Groups from Template** command:

✓ Select the **Documentation** pull-down menu, pick the **Area Groups** cascade menu, then select the **Create Area Groups from Template...** command.

 ✓ Pick the **Create Area Groups from Template** button in the **Area Groups** toolbar.

✓ Right-click and select **Documentation, Area Groups, Create Area Groups from Template...** from the cursor menu.

✓ Type **AECAREAGROUPCREATEFROMTEMPLATE** at the Command: prompt.

The **Create Area Groups from Template** dialog box, shown in Figure 24.49, is displayed. Select the template to use, then specify the row and column offsets for the group markers. Next, pick a point in the drawing to place the highest level group marker. Figure 24.50 shows the final group structure inserted from a template.

After you have inserted the group structure into the drawing, you can attach areas to the group structure. Use the Area Group Layout command to readjust the group marker spacing if necessary.

FIGURE 24.49 The Create
Area Groups from Template
dialog box is used to specify the
template to be used and the row
and column offsets for the area
group markers.

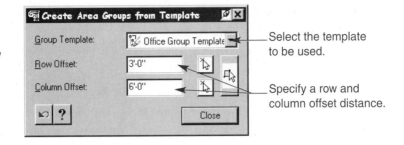

Select the template
to be used.

Specify a row and
column offset distance.

FIGURE 24.50 Group structure
inserted from a template.

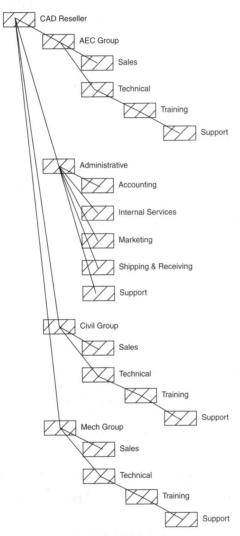

Finished group structure inserted
using a template

Exercise 24.8

1. Open EX24-7.dwg.
2. Use the Area Group Layout command to reorganize the existing group structure.
3. Create an area name definition style that reflects the names of the areas in the drawing.
4. Create an area group template that is organized like the existing group structure.
5. Assign any properties to the group levels as desired.
6. Save the drawing as EX24-8.dwg.

Area Evaluations

The final step when working with areas is to create an area evaluation report. As defined previously, an area evaluation is a report that calculates the area and area group information in your drawing and exports the information into a Microsoft Excel spreadsheet or word processing document.

Area evaluation reports are often required when the construction documents are being evaluated for a building permit. These reports can also be used for cost estimating, and for assigning jobs to subcontractors, and for management of the building by facilities personnel.

In order to create an area evaluation report, you need to include the areas and to configure the output options. Use one of the following options to access the **Area Evaluation** command.

✓ Select the **Documentation** pull-down menu, pick the **Areas** cascade menu, then select the **Area Evaluation...** command.

✓ Pick the **Area Evaluation** button in the **Areas** toolbar.

✓ Right-click, and select **Documentation, Areas, Area Evaluation...** from the cursor menu.

✓ Type **AECAREAEVALUATION** at the Command: prompt.

The **Area Evaluation** dialog box, shown in Figure 24.51, is displayed. This dialog box includes a tree view list along the left that lists the area group structures in all drawings that are currently open in Architectural Desktop.

When a group or area icon is selected in the tree view, information about the area is displayed in the upper right corner of the dialog box. If there are reference drawings attached to one of the open drawings, you can pick the Scan Xref Drawings button to list area group structures in the reference(s) within the tree view.

From the tree view list, place a check in the associated area group and area boxes that you would like to report on. Areas and area groups to be excluded from the report can be removed by picking on the check mark to remove it.

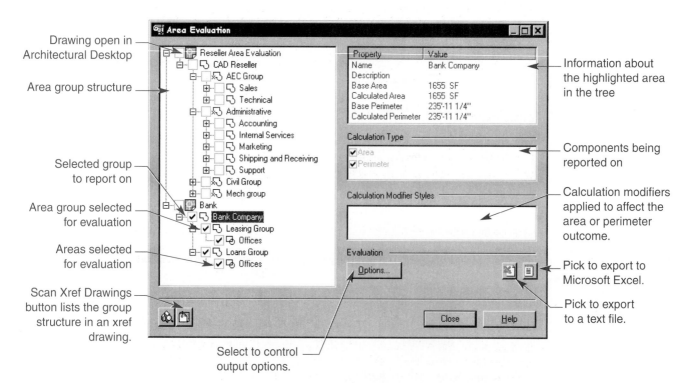

FIGURE 24.51   The Area Evaluation dialog box is used to display the total area of all the selected areas from the left tree view and to set up the parameters for the exported evaluation report.

When a check mark is placed in a box, it can be displayed in one of three colors: gray, black, or blue. The significance of each of these colors is as follows:

- **Black**   This color check mark indicates that the area or area group has been selected directly in the dialog box for evaluation. Picking on top of a black check mark will remove it from the evaluation.

- **Blue**   This color check mark is placed in subgroups and areas of an area group if a black check mark is placed in the area group above the subgroup. The blue check mark is automatically added because the area is part of a higher area group that was selected.

- **Gray**   This color check mark is displayed in higher level area groups if the area groups of a sub-level have been selected. Gray checked area groups are not included in the evaluation report unless they are picked directly and the gray check mark is turned into a black check mark.

The Calculation Type area in the middle right side of the dialog box indicates the components of an area that are being reported on. You cannot change this information. If you wanted to report on the area only and not on the perimeter, you would need to use the Area Group Properties dialog box to modify the reported components. The Calculation Modifier Styles area displays the calculation modifiers attached to an area group. Like the Calculation Type area, the information in this area can be modified only in the Area Group Properties dialog box.

Setting Up Output Parameters

After you have determined the areas and area groups that will be reported on, you need to configure the output configuration. Select the <u>O</u>ptions... button in the Area Evaluation dialog box. The **Evaluation Properties** dialog box is displayed. This dialog box includes the following four tabs used to configure the output settings:

- **Evaluation**   This tab includes a drop-down list for the area and area groups (see Figure 24.52). Depending on whether you choose the area or the area group from the drop-down list, a list of values is displayed that can be included in the evaluation report. Place a check in the associated box to include the value in the report. The following values are available:

Area Values

Value	Description
Name	Includes the name of the area
Description	Includes the description for the area from the Area Properties dialog box
Overview Image	Includes a graphical image of the area in the report
Base Area Label	Displays the prefix label *Base Area* in front of the base area value, e.g., Base Value 100
Base Area Result	Includes the total area without the calculation modifiers taken into account, e.g., Base Value 100
Calculated Area Label	Displays the prefix *Calculated Area* in front of the calculated area value, e.g., Calculated Area 80
Calculated Area Result	Includes the total area after processing through a calculation modifier, e.g., Calculated Area 80
Modifier Label	Displays the name of the calculation modifier for the area in front of the calculation modifier formula, e.g., Building Cost 80 * 189 = 151.2
Modifier Expression	Displays the formula for the calculation modifier being applied to the area, e.g., Building Cost 20 * 189 = 151.2
Modifier Result	Displays the resulting value of the area after processing through the calculation modifier, e.g., Building Cost 80 * 189 = 151.2
Proof Label	Displays the decomposed area label, e.g., A1: 151.2 * .97 = 146.664
Proof Expression	Displays the formula proof, e.g., A1: 151.2 * .97 = 146.664

continues on next page

Area Values continued

Proof Result	Displays the result of the proof, e.g., A1: 151.2 * .97 = 146.664
Decomposed Image	Includes an image of the decomposed area
Base Perimeter Label	Displays the prefix label *Base Perimeter* in front of the base perimeter value, e.g., Base Perimeter 56
Base Perimeter Result	Includes the total perimeter without the calculation modifier taken into account, e.g., Base Perimeter 56
Calculated Perimeter Label	Displays the prefix label *Calculated Perimeter* in front of the calculated perimeter value, e.g., Calculated Perimeter 44
Calculated Perimeter Result	Includes the calculated perimeter value after processing through the calculation modifier, e.g., Calculated Perimeter 44
Modifier Label	Displays the name of the calculation modifier for the perimeter in front of the calculation modifier formula, e.g., Plaster Deduction 44 * .97 = 42.68
Modifier Expression	Displays the formula for the calculation modifier being applied to the perimeter, e.g., Plaster Deduction 44 * .97 = 42.68
Modifier Result	Displays the resulting value of the perimeter after processing through the calculation modifier, e.g., Plaster Deduction 44 * .97 = 42.68

Area Group Values

Value	Description
Name	Includes the name of the area group
Description	Includes the description for the area from the Area Group Properties dialog box
Overview Image	Includes a graphical image of the area group in the report
Base Area Label	Displays the prefix label *Base Area* in front of the base area value, e.g., Base Value 1000
Base Area Result	Includes the total area without the calculation modifiers taken into account, e.g., Base Value 1000
Calculated Area Label	Displays the prefix *Calculated Area* in front of the calculated area value, e.g., Calculated Area 800
Calculated Area Result	Includes the total area after processing through a calculation modifier. Calculated Area 800
Modifier Label	Displays the name of the calculation modifier for the area in front of the calculation modifier formula, e.g., Building Cost 800 * 189 = 151200
Modifier Expression	Displays the formula for the calculation modifier being applied to the area, e.g., Building Cost 20 * 189 = 151200
Modifier Result	Displays the resulting value of the area after processing through the calculation modifier, e.g., Building Cost 80 * 189 = 151200
Base Perimeter Label	Displays the prefix label *Base Perimeter* in front of the base perimeter value, e.g., Base Perimeter 560
Base Perimeter Result	Includes the total perimeter without the calculation modifier taken into account, e.g., Base Perimeter 560
Calculated Perimeter Label	Displays the prefix label *Calculated Perimeter* in front of the calculated perimeter value, e.g., Calculated Perimeter 440
Calculated Perimeter Result	Includes the calculated perimeter value after processing through the calculation modifier, e.g., Calculated Perimeter 440
Modifier Label	Displays the name of the calculation modifier for the perimeter in front of the calculation modifier formula, e.g., Plaster Deduction 440 * .97 = 426.8
Modifier Expression	Displays the formula for the calculation modifier being applied to the perimeter, e.g., Plaster Deduction 44 * .97 = 426.8
Modifier Result	Displays the resulting value of the perimeter after processing through the calculation modifier, e.g., Plaster Deduction 44 * .97 = 426.8

Once you have selected the values to be included in the evaluation report, you can press the OK button to return to the Area Evaluation dialog box to export the data to the report, or continue configuring the output parameters.

FIGURE 24.52 The Evaluation
tab is used to select the values
from areas and area groups that
you would like included in an
evaluation report.

Select areas or area groups
to list the available values.

Evaluation tab

List of values for
areas that can
be included in
the report

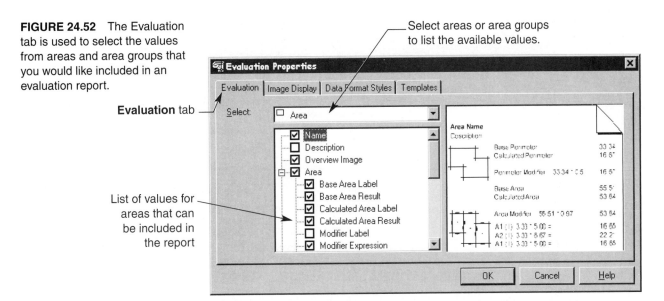

FIGURE 24.53 The Image Display tab is used to control how the graphical images of the areas will appear
in the final evaluation report.

Image Display tab

Select a drawing
from which to pull
images.

Select a display
configuration that will
be used when
extracting the image.

Select the color
depth for the
images.

Specify a final
size for the
image in pixels.

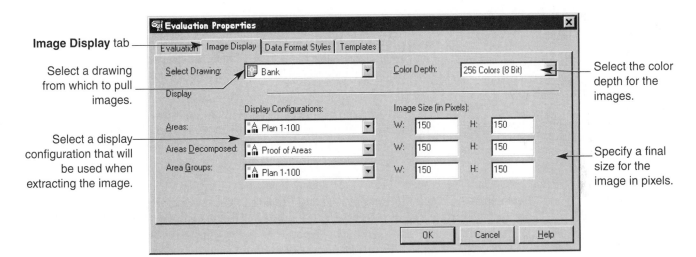

- **Image Display**   This tab is used to control the output display of the area, area group, and
decomposition graphical images in the report (see Figure 24.53). At the
top of the tab select the drawing from which the graphical images will be
extracted. In the Color Depth drop-down list choose the color depth for the
images. Selecting a high resolution will increase the clarity of the image in
the evaluation report but increase the file size.

  The three types of images that can be exported are listed at the bottom of
the dialog box: Areas, Areas Decomposed, and Area Groups. If one or all
three of these graphical image type values were selected in the Evaluation
tab they will be included in the final report. The Display Configurations
drop-down lists are assigned to each type of image. Select the display configuration that will be used to extract an image of area, decomposition,
and/or area group. Beside each display configuration drop-down list you
can specify the image size that will be displayed in the report in pixels.

- **Data Format
  Styles**   This tab allows you to assign a data format style to the area and perimeter
result values, to the area and perimeter expression values, and to the area
proof expression value (see Figure 24.54). This data format styles

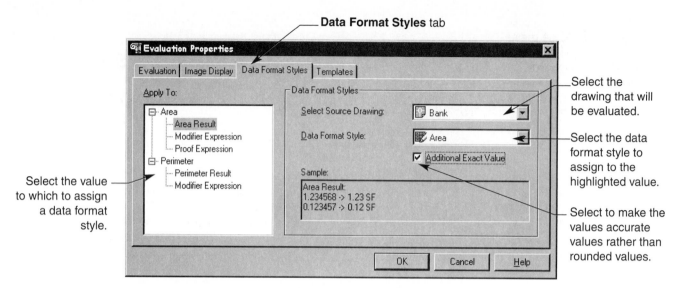

Data Format Styles tab

Select the value to which to assign a data format style.

Select the drawing that will be evaluated.

Select the data format style to assign to the highlighted value.

Select to make the values accurate values rather than rounded values.

FIGURE 24.54   The Data Format Styles tab allows you to assign formatting to the values being reported on for the evaluation report.

assigned designate how the final output for a value will appear in the report. Data format styles were discussed in Chapter 20 and allow you to configure prefixes and suffixes as well as textual formatting such as upper- or lowercase and unit data such as decimal, architectural, or engineering.

Select the values in the tree on the left, then select the drawing being evaluated and the type of formatting to apply to each value. In some cases a data format style may be set to round values to a whole number, to two decimal points, or to another value. If the report is to display the actual values of an area and not the rounded values, then select the Additional Exact Value check box.

- **Templates tab**   When the area evaluation is exported to a Microsoft Excel spreadsheet or to a word processing text file, a template must be used. The template organizes the values being reported on in the exported files. This tab specifies both an Excel template and a text template to which the report can be exported (see Figure 24.55). The text boxes list the path and template name that the report will be using for the final output. Additional templates can be used by selecting the ellipsis […] button to browse for a different template to use.

FIGURE 24.55   The Templates tab is used to specify a template that the values will be exported to and organized into.

Templates tab

Select an Excel Template to use.

Select a text document template to use.

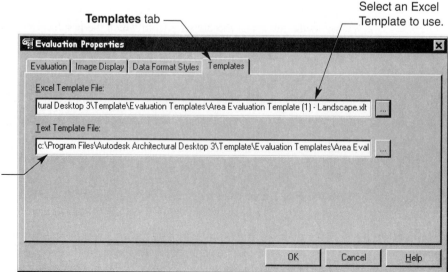

Field Notes

The Excel templates have an .xlt file extension. Additional Excel templates can be found in the \\Autodesk Architectural Desktop 3\Template\Evaluation Templates\ folder. The text templates have a .txt file extension and can be opened in any ASCII text editing software such as Microsoft Notepad or Word. There is only one included .txt template.

Creating the Area Evaluation

After selecting the areas and area groups to report on and configuring the output settings, you can create the evaluation report. To create a Microsoft Excel spreadsheet report, pick the Export Evaluation to Microsoft Excel button. The **Save Excel Evaluation file...** dialog box is displayed. Browse to a folder in which to save the file, enter a name for the evaluation spreadsheet, and select the Save button. You will be returned to the Area Evaluation dialog box, where Architectural Desktop processes the information and creates the evaluation report (see Figure 24.56).

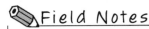Field Notes

If the spreadsheet is to be added to the construction document set drawings, use Object Linking and Embedding (OLE) to bring the Excel file into the drawing. For more information regarding OLE look up the *Embedding Objects in AutoCAD Drawings* section in the AutoCAD 2000i or AutoCAD 2002 help section.

If the evaluation report is to be exported as a text document, pick the Export Evaluation to Text Format. The **Save Evaluation Text file...** dialog box is then displayed. Select a folder into which to save the evalua-

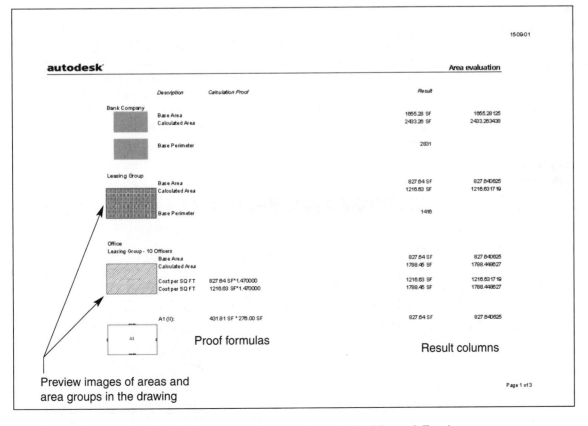

FIGURE 24.56 An example of an area evaluation report exported to Microsoft Excel.

tion, enter a name for the evaluation, then pick the Save button. This file can be opened with the Windows Notepad program, Word, or any other software that recognizes ASCII text files (see Figure 24.57).

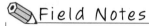Field Notes

If the text evaluation report is to be used in the construction documents, use the MTEXT command to import the .txt file as a piece of multiline text.

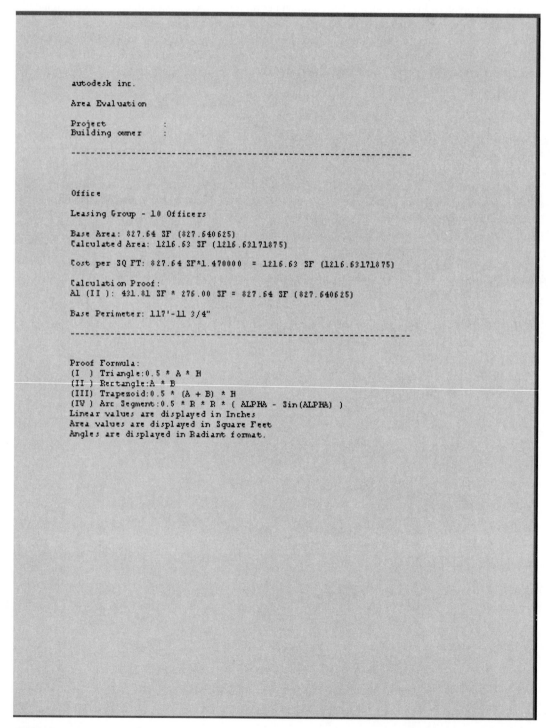

FIGURE 24.57   The evaluation report is exported as a .txt file and opened in the Windows Notepad program.

Exercise 24.9

1. Open EX24-8.dwg.
2. Use the Area Group Properties dialog box to assign the area name definition created in EX24-8 to each area group marker in the drawing.
3. Use the Modify Area command to provide a name for each area in the drawing from the area names listed in the <u>N</u>ame: drop-down list.
4. Create an area evaluation of the Classrooms, Offices, and Hallways area groups.
5. Choose the values on which you would like to report.
6. If you have access to Microsoft Excel, then export the report as an Excel spreadsheet. If not, export the report as a text document.
7. Save the drawing as EX24-9.dwg.

CHAPTER REVIEW

Use the CD-ROM to test your knowledge and skills.

Chapter Test

To check your understanding of the content provided in this chapter, access the Test file in the CH24 folder of the CD-ROM that accompanies this text.

Chapter Project

To practice the Architectural Desktop skills presented in this chapter, access the Project files in the CH24 folder of the CD-ROM that accompanies this text. The project files are in pdf format and include sample drawings and instructions for completing each project.

Architectural Desktop Toolbars

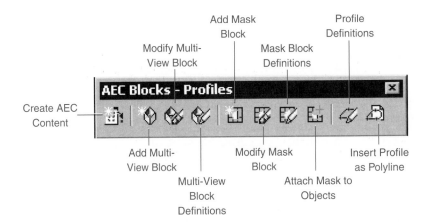

Create AEC Content — Modify Multi-View Block — Add Mask Block — Mask Block Definitions — Profile Definitions — Add Multi-View Block — Multi-View Block Definitions — Modify Mask Block — Attach Mask to Objects — Insert Profile as Polyline

AEC Blocks - Profiles

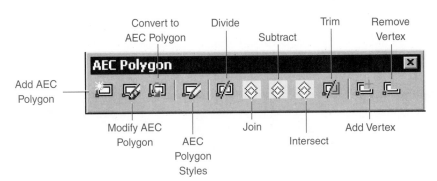

Add AEC Polygon — Convert to AEC Polygon — Divide — Subtract — Trim — Remove Vertex — Modify AEC Polygon — AEC Polygon Styles — Join — Intersect — Add Vertex

AEC Polygon

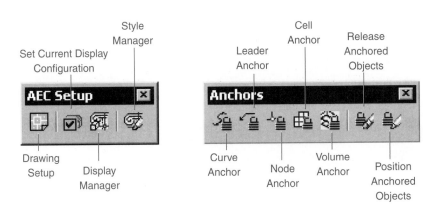

Set Current Display Configuration — Style Manager — Leader Anchor — Cell Anchor — Release Anchored Objects — Drawing Setup — Display Manager — Curve Anchor — Node Anchor — Volume Anchor — Position Anchored Objects

AEC Setup

Anchors

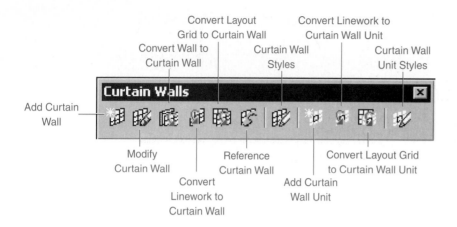

Curtain Walls

- Add Curtain Wall
- Modify Curtain Wall
- Convert Wall to Curtain Wall
- Convert Linework to Curtain Wall
- Convert Layout Grid to Curtain Wall
- Reference Curtain Wall
- Curtain Wall Styles
- Add Curtain Wall Unit
- Convert Linework to Curtain Wall Unit
- Convert Layout Grid to Curtain Wall Unit
- Curtain Wall Unit Styles

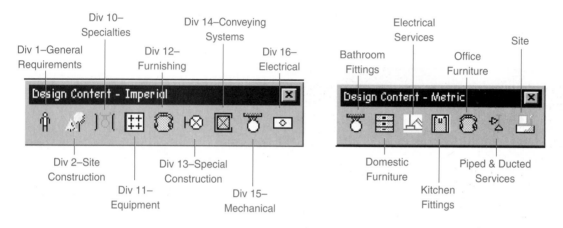

Design Content – Imperial

- Div 1–General Requirements
- Div 2–Site Construction
- Div 10–Specialties
- Div 11–Equipment
- Div 12–Furnishing
- Div 13–Special Construction
- Div 14–Conveying Systems
- Div 15–Mechanical
- Div 16–Electrical

Design Content – Metric

- Bathroom Fittings
- Domestic Furniture
- Electrical Services
- Kitchen Fittings
- Office Furniture
- Piped & Ducted Services
- Site

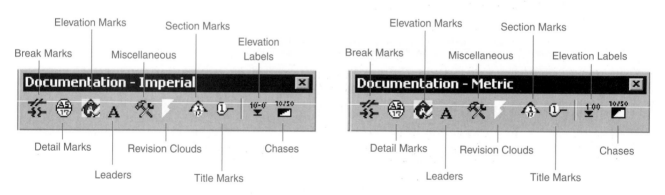

Documentation – Imperial

- Break Marks
- Detail Marks
- Elevation Marks
- Leaders
- Miscellaneous
- Revision Clouds
- Section Marks
- Title Marks
- Elevation Labels
- Chases

Documentation – Metric

- Break Marks
- Detail Marks
- Elevation Marks
- Leaders
- Miscellaneous
- Revision Clouds
- Section Marks
- Title Marks
- Elevation Labels
- Chases

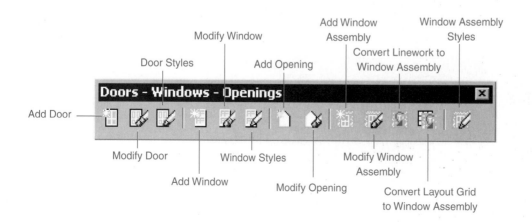

Doors – Windows – Openings

- Add Door
- Modify Door
- Door Styles
- Add Window
- Modify Window
- Window Styles
- Add Opening
- Modify Opening
- Add Window Assembly
- Convert Linework to Window Assembly
- Modify Window Assembly
- Convert Layout Grid to Window Assembly
- Window Assembly Styles

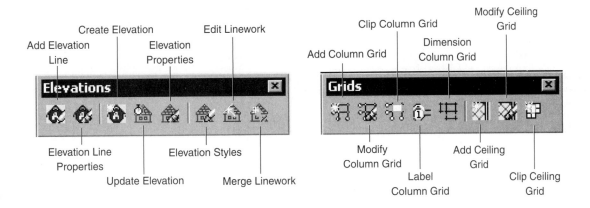

Add Elevation Line — Create Elevation — Elevation Properties — Edit Linework — **Elevations** — Elevation Line Properties — Update Elevation — Elevation Styles — Merge Linework

Add Column Grid — Clip Column Grid — Dimension Column Grid — Modify Ceiling Grid — **Grids** — Modify Column Grid — Label Column Grid — Add Ceiling Grid — Clip Ceiling Grid

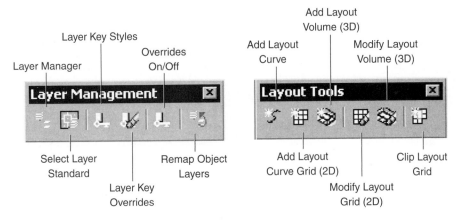

Layer Manager — Layer Key Styles — Overrides On/Off — **Layer Management** — Select Layer Standard — Layer Key Overrides — Remap Object Layers

Add Layout Curve — Add Layout Volume (3D) — Modify Layout Volume (3D) — **Layout Tools** — Add Layout Curve Grid (2D) — Modify Layout Grid (2D) — Clip Layout Grid

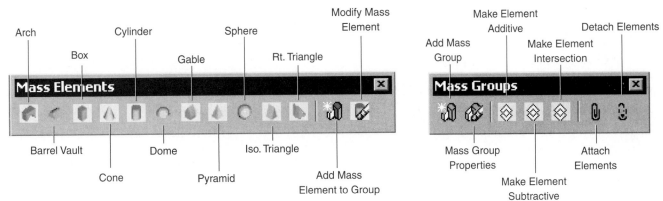

Arch — Box — Cylinder — Gable — Sphere — Rt. Triangle — Modify Mass Element — **Mass Elements** — Barrel Vault — Cone — Dome — Pyramid — Iso. Triangle — Add Mass Element to Group

Add Mass Group — Make Element Additive — Make Element Intersection — Detach Elements — **Mass Groups** — Mass Group Properties — Make Element Subtractive — Attach Elements

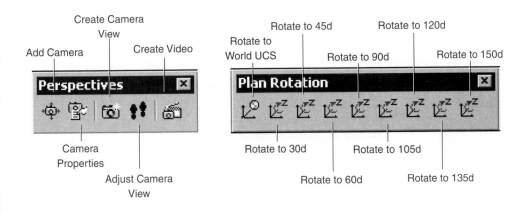

Add Camera — Create Camera View — Create Video — **Perspectives** — Camera Properties — Adjust Camera View

Rotate to World UCS — Rotate to 45d — Rotate to 90d — Rotate to 120d — Rotate to 150d — **Plan Rotation** — Rotate to 30d — Rotate to 60d — Rotate to 105d — Rotate to 135d

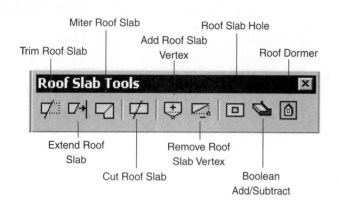

Trim Roof Slab · Miter Roof Slab · Add Roof Slab Vertex · Roof Slab Hole · Roof Dormer

Roof Slab Tools

Extend Roof Slab · Cut Roof Slab · Remove Roof Slab Vertex · Boolean Add/Subtract

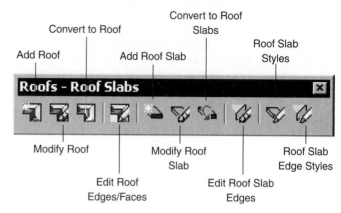

Add Roof · Convert to Roof · Add Roof Slab · Convert to Roof Slabs · Roof Slab Styles

Roofs – Roof Slabs

Modify Roof · Edit Roof Edges/Faces · Modify Roof Slab · Edit Roof Slab Edges · Roof Slab Edge Styles

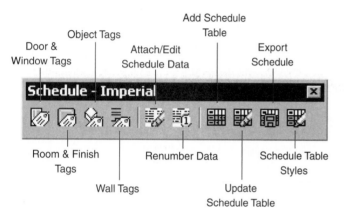

Door & Window Tags · Object Tags · Attach/Edit Schedule Data · Add Schedule Table · Export Schedule

Schedule - Imperial

Room & Finish Tags · Wall Tags · Renumber Data · Update Schedule Table · Schedule Table Styles

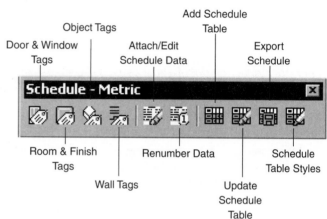

Door & Window Tags · Object Tags · Attach/Edit Schedule Data · Add Schedule Table · Export Schedule

Schedule - Metric

Room & Finish Tags · Wall Tags · Renumber Data · Update Schedule Table · Schedule Table Styles

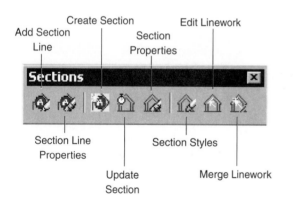

Add Section Line · Create Section · Section Properties · Edit Linework

Sections

Section Line Properties · Update Section · Section Styles · Merge Linework

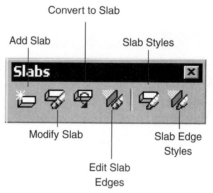

Add Slab · Convert to Slab · Slab Styles

Slabs

Modify Slab · Edit Slab Edges · Slab Edge Styles

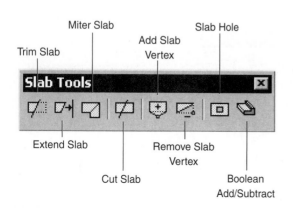

Trim Slab · Miter Slab · Add Slab Vertex · Slab Hole

Slab Tools

Extend Slab · Cut Slab · Remove Slab Vertex · Boolean Add/Subtract

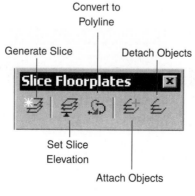

Generate Slice · Convert to Polyline · Detach Objects

Slice Floorplates

Set Slice Elevation · Attach Objects

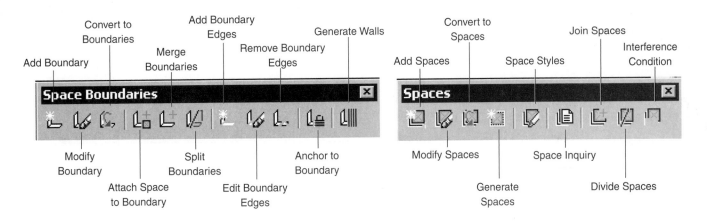

Space Boundaries toolbar: Add Boundary, Convert to Boundaries, Modify Boundary, Merge Boundaries, Attach Space to Boundary, Split Boundaries, Add Boundary Edges, Edit Boundary Edges, Remove Boundary Edges, Anchor to Boundary, Generate Walls

Spaces toolbar: Add Spaces, Convert to Spaces, Modify Spaces, Generate Spaces, Space Styles, Space Inquiry, Join Spaces, Divide Spaces, Interference Condition

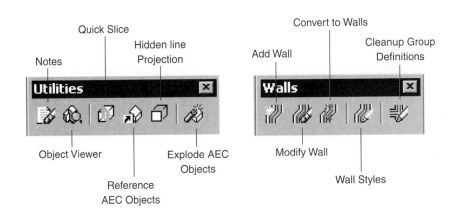

Stairs - Railings toolbar: Add Stair, Stair Styles, Modify Stair, Customize Edge, Add Railing, Convert to Railing, Modify Railing, Railing Styles, Anchor to Stair

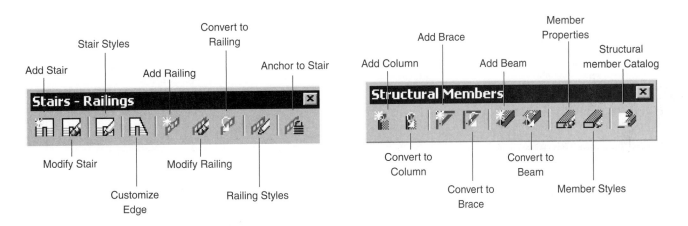

Structural Members toolbar: Add Column, Convert to Column, Add Brace, Convert to Brace, Add Beam, Convert to Beam, Member Properties, Member Styles, Structural member Catalog

Utilities toolbar: Notes, Quick Slice, Object Viewer, Reference AEC Objects, Hidden line Projection, Explode AEC Objects

Walls toolbar: Add Wall, Modify Wall, Convert to Walls, Wall Styles, Cleanup Group Definitions

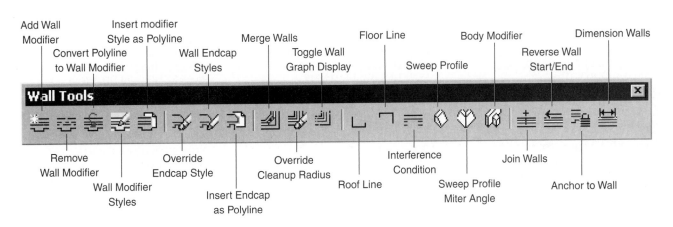

Wall Tools toolbar: Add Wall Modifier, Remove Wall Modifier, Convert Polyline to Wall Modifier, Wall Modifier Styles, Insert modifier Style as Polyline, Override Endcap Style, Wall Endcap Styles, Insert Endcap as Polyline, Merge Walls, Override Cleanup Radius, Toggle Wall Graph Display, Roof Line, Floor Line, Interference Condition, Sweep Profile, Sweep Profile Miter Angle, Body Modifier, Join Walls, Reverse Wall Start/End, Anchor to Wall, Dimension Walls

Index